Statistics for Business and Economics

FOURTH EDITION

Statistics for Business and Economics

FOURTH EDITION

David R. Anderson
UNIVERSITY OF CINCINNATI

Dennis J. Sweeney
UNIVERSITY OF CINCINNATI

Thomas A. Williams
ROCHESTER INSTITUTE OF TECHNOLOGY

WEST PUBLISHING COMPANY

Saint Paul   New York   Los Angeles   San Francisco

Copyediting: Linda Thompson
Composition: Science Press
Cover and Text Design: Kristen M. Weber
Cover Photo: Jean Claude Lozouet, The Image Bank

Minitab is a registered trademark of Minitab, Inc., 3081 Enterprise Drive, State College, PA 16801. (Telephone: 814/238-3280; telex: 881612; fax: 814/238-4383.)

___

Library of Congress Cataloging-in-Publication Data

Anderson, David Ray, 1941–
 Statistics for business and economics / David R. Anderson, Dennis
J. Sweeney, Thomas A. Williams.—4th ed.
  p. cm.
 Includes bibliographical references.
 ISBN 0-314-66500-5
 1. Commercial statistics. 2. Economics—Statistical methods.
3. Statistics. I. Sweeney, Dennis J. II. Williams, Thomas Arthur,
1944– . III. Title.
HF1017.A6 1990 519.5—dc20   89-38858
                           CIP

To Marcia, Cherri and Robbie

ABOUT THE AUTHORS

David R. Anderson is Professor of Quantitative Analysis in the College of Business Administration at the University of Cincinnati, where he has been teaching since 1968. Born in Grand Forks, North Dakota, he earned his B.S., M.S., and Ph.D. degrees from Purdue University. Professor Anderson has chaired the Department of Quantitative Analysis at the University of Cincinnati and has served as Acting Associate Dean of the College of Business Administration. In addition, he was coordinator of the College's first Executive Program.

At the University of Cincinnati, David Anderson has taught introductory statistics for business students as well as graduate level courses in regression analysis, multivariate analysis, and management science. He has also taught statistical courses at the Department of Labor in Washington, D.C. He has been honored with nominations and awards for excellence in teaching and excellence in service to student organizations.

Professor Anderson has co-authored six textbooks in the areas of statistics, management science, linear programming, and production/operations management. He is an active consultant in the field of sampling and statistical methods.

Dennis J. Sweeney is the C.B.A. Professor of Quantitative Analysis at the University of Cincinnati. Born in Des Moines, Iowa, he earned a B.S.B.A. degree from Drake University, graduating summa cum laude. He received his M.B.A. and D.B.A. degrees from Indiana University where he was an NDEA Fellow. Since receiving his doctorate in 1971, Professor Sweeney has spent all but two years at the University of Cincinnati. During 1978–79 he spent a year working in the management science group at Procter and Gamble; during 1981–82 he was a visiting professor at Duke University. For five years Professor Sweeney served as Chairman of the Quantitative Analysis Department and for four years served as Associate Dean at the University of Cincinnati.

Professor Sweeney has published over thirty articles in the general area of management science. The National Science Foundation, IBM, and Procter & Gamble have funded his research, which has been published in *Management Science, Operations Research, Mathematical Programming, Decision Science,* and other journals.

Together with David Anderson and Thomas Williams, he has published six textbooks.

Thomas A. Williams is Professor of Management Science at the Rochester State Institute of Technology, College of Business, in New York. Born in Elmira, New York, he earned his B.S. degree at Clarkson University. He did his graduate work at Rensselaer Polytechnic Institute, where he received his M.S. and Ph.D. degrees.

Before joining the College of Business at RIT, Professor Williams served for seven years as a faculty member in the College of Business at the University of Cincinnati, where he developed and served as the program coordinator for the undergraduate program in Information Systems. At RIT he was the first chairman of the Decision Sciences Department. He teaches courses in management science and statistics, as well as more advanced courses in regression and decision analysis.

Dr. Williams is the co-author of seven textbooks in the areas of management science, statistics, production and operations management, and mathematics. He has been a consultant for numerous Fortune 500 companies and has worked on projects ranging from the use of elementary data analysis, to the development of large-scale regression models for top-level decision makers.

CONTENTS

CHAPTER 5 ■■■ DISCRETE PROBABILITY DISTRIBUTIONS 155

CHAPTER 10 ■■ STATISTICAL INFERENCE ABOUT MEANS AND PROPORTIONS WITH TWO POPULATIONS 371

CHAPTER 15 ■■ REGRESSION ANALYSIS: MODEL BUILDING   595

PREFACE

The purpose of this book is to provide students, primarily in the fields of business administration and economics, with a sound conceptual introduction to the field of statistics and its many applications. The text is applications oriented and has been written with the needs of the nonmathematician in mind. The mathematical prerequisite is knowledge of algebra. The text material and problems have been developed with this prerequisite in mind.

APPLICATIONS AND METHODOLOGY INTEGRATED

Applications of data analysis and statistical methodology are an integral part of the organization and presentation of the material in the text. All statistical techniques are introduced in conjunction with problem scenarios where the techniques are helpful. The discussion and development of each technique is centered around an application setting with the statistical results providing insights to decisions and solutions to problems.

Although the book is applications oriented, we have taken care to provide a sound methodological development, and to utilize notation that is generally accepted for the topic being covered. Thus, students will find that the text provides good preparation for the study of more advanced statistical material. A bibliography that provides a guide to further study has been included as an appendix.

CHAPTER-ENDING APPLICATIONS

To further emphasize the applications of statistics, cases supplied by practitioners from business and government have been added at the end of each of the twenty chapters. Each application describes an actual organization and its current usage of the statistical methodology introduced in the chapter. Procter & Gamble, Polaroid, Monsanto, Xerox, Mead, Dow, and Colgate-Palmolive are a few of the companies that have contributed statistical applications included in this text. We feel these applications help motivate the student to learn the material and provide an appreciation for the wide range of statistical methods that are used in practice. The table at the end of the preface provides a list of the organizations and application descriptions presented in the text.

CHANGES IN THE FOURTH EDITION

We appreciate the acceptance and positive response to the earlier editions of this text. Accordingly, in making modifications for this new edition we have maintained the presentation style of previous editions. The more significant changes in this fourth edition are summarized below.

Examples and Problems Based on Real Data

Sources such as the *Wall Street Journal, Business Week, U.S.A. Today, Fortune, Forbes, Financial World* and *Barrons* have been used to provide real data and real case studies which demonstrate uses of statistics in business and economics. In some instances, data from the above sources enable the student to study methodology and solve problems based on real data. In other instances, the sources are used to provide data summaries and statistical information that are the bases for further questions and computations. The use of real data means that the student not only learns about the statistical methodology but also learns about an application and the types of studies encountered in practice.

Notes and Comments

At the end of many sections, we have provided "notes and comments" designed to give the student additional insights about the statistical methodology presented in the section. The various notes and comments include warnings and/or limitations about the methodology, recommendations for application, brief descriptions of additional technical considerations, and so on. It is hoped that this feature will expand the student's understanding of statistics and the student's ability to use the material.

Data Acquisition and Measurement

Chapter 1 has been expanded to include a discussion of how data are acquired and classified. The nominal, ordinal, interval and ratio scales of measurement are defined. In addition, the student learns how the scales of measurement lead to the classifications of qualitative and quantitative data. Chapter 1 also provides an introduction to the notions of a population, a sample, descriptive statistics and statistical inference.

Descriptive Statistics

The two chapters on descriptive statistics have been expanded to include the use of the standard deviation in the computation and interpretation of z-scores. Chebyshev's Theorem and the Empirical Rule have been added in Chapter 3 to further demonstrate the use of the standard deviation. Outliers and procedures for identifying outliers are now covered. In addition, the exploratory data analysis material on box plots and fences has been expanded. The presentation of frequency distributions has been simplified.

Probability and Probability Distributions

The hypergeometric probability distribution and examples of its use have been added to Chapter 5. The presentation of Bayes Theorem has been shortened and simplified.

Sampling and Sampling Methods

Chapter 7 includes a more thorough discussion of the important properties of point estimators. In addition, the section on other sampling methods such as stratified sampling and cluster sampling has been expanded.

Hypothesis Testing

Chapter 9 provides a revised treatment of hypothesis testing. More emphasis is placed on the proper formulation of hypotheses and reaching the proper conclusion when the hypothesis test leads to "Reject H_0" and "Do Not Reject H_0." The appropriate uses and limitations of hypothesis testing in research studies, in testing assumptions and in decision making are discussed. Several new examples and illustrations are presented. The use and interpretation of p-values is also covered.

Analysis of Variance

The presentation of analysis of variance in Chapter 12 has been simplified by reducing the emphasis on computational formulas. The detailed formulas have been removed from the text presentation and are available in the chapter appendix.

New Chapter on Model Building in Regression Analysis

The presentation of regression analysis has been expanded from two to three chapters. Simple linear regression and correlation is covered in Chapter 13. Multiple linear regression is covered in Chapter 14. A new Chapter 15 on regression analysis and model building has been added. The new chapter presents the general linear model and describes variable selection procedures such as stepwise regression and backward elimination. Additional uses of residual analysis to identify outliers, influential observations and autocorrelation are also covered.

New Chapter on Quality Control

A new Chapter 20 has been added to provide an introduction to the important role of statistics in the control of quality. Acceptance sampling and control charts are presented.

Microcomputer Software

Available to adopters of this edition is a revised and expanded version of *The Data Analyst*, an IBM-compatible microcomputer software package developed by the authors. *The Data Analyst* version 2.0 can be used to solve problems and analyze data using descriptive statistics (standard numerical measures, scatter diagrams, stem-and-leaf displays, and frequency distributions), confidence intervals for means and proportions of one and two populations, analysis of variance, simple and multiple linear regression, and more. Data sets from the text as well as additional data sets are available on a special data diskette accompanying *The Data Analyst*.

New Problems and Computer Exercises

Between 150 and 200 new problems have been added to this edition. As discussed earlier, many of the problems are based on real data and real case studies. Additional exercises with data sets which have been designed to be analyzed using a statistical computer package appear at the end of eleven chapters. *The Data Analyst* or commercially available packages such as MINITAB, SAS, SPSS or BMDP may be used to complete the computer assignments.

■ FLEXIBILITY

There is a reasonable amount of instructor flexibility in selecting material to satisfy specific course needs. As an illustration, a possible outline for a two-quarter sequence is as follows:

Possible Two-Quarter Course Outline

First Quarter	*Second Quarter*
Introduction (Chapter 1)	Hypothesis Testing (Chapter 9)
Descriptive Statistics (Chapters 2 and 3)	Two Population Cases (Chapter 10)
Introduction to Probability (Chapter 4)	Inferences About Population Variances (Chapter 11)
Probability Distributions (Chapters 5 and 6)	Analysis of Variance (Chapter 12)
Sampling and Sampling Distributions (Chapter 7)	Regression and Correlation (Chapters 13 and 14)
Interval Estimation (Chapter 8)	Tests of Goodness of Fit and Independence (Chapter 17)

Other possibilities exist for such a course, depending upon the time available and the background of the students. However, it is probably not possible to cover all the material in one semester or in two quarters unless some of the topics have been previously studied.

■ ANCILLARIES

Accompanying the text is a complete package of support materials. These include an Instructor's Manual, Study Guide, Test Bank, Demonstration Problems and Lecture Notes, and a Microcomputer Software Package.

The instructor's manual, prepared by the authors, includes learning objectives and completely worked solutions to all the problems. The study guide, developed by Mohammad Ahmadi (University of Tennessee at Chattanooga), provides an additional source of problems and explanations for students. The test bank, also prepared by Mohammad Ahmadi provides a series of multiple choice questions and problems that will aid in the preparation of exams. The set of demonstration problems and lecture notes, prepared by the authors, can be used as a complete set of transparency masters and/or lecture notes. Finally, *The Data Analyst* version 2.0, a software package developed by the authors for IBM-compatible microcomputers, is available to adopters.

We believe that the applications orientation of the text, combined with the package of support materials, provides an ideal basis for introducing students to statistics and statistical applications.

■ ACKNOWLEDGMENTS

We owe a debt to many of our colleagues and friends for their helpful comments and suggestions in the development of this manuscript. Among these are:

Robert Balough	Thomas McCullough
Harry Benham	Bette Midgarden
Michael Bernkopf	Glenn Milligan
John Bryant	Richard O'Connell
Peter Bryant	Al Palachek
John Cooke	Diane Petersen-Salameh
David W. Cravens	Ruby Ramirez
George Dery	Tom Ryan
Gopal Dorai	James R. Schwenke
Edward Fagerlund	William E. Stein
Nicholas Farnum	Willban Terpening
Jerome Geaun	Hiroki Tsurumi
Jamshid C. Hasseini	J. E. Willis
Ben Isselhardt	Donald Williams
Jeffrey Jarrett	Roy Williams
David Krueger	Greg Zimmerman

Our associates from business and industry who supplied the applications made a major contribution. They are recognized individually by a credit line on the first page of each application. In addition, we would like to acknowledge the cooperation of Minitab, Inc., for permitting the use of Minitab statistical software in this text. Finally, we are also indebted to our editor, Mary C. Schiller, and others at West Publishing Company for their editorial counsel and support during the preparation of this text.

David R. Anderson
Dennis J. Sweeney
Thomas A. Williams

An Overview of Chapter-ending Applications

	Chapter	Organization	Application
1	Introduction	Kings Island Inc.	Consumer Profile Sample Survey
2	Descriptive Statistics I. Tabular and Graphical Approaches	Colgate-Palmolive Company	Quality Assurance for Heavy Duty Detergents
3	Descriptive Statistics II. Measures of Location and Dispersion	Barnes Hospital	Time Spent in Hospice Program
4	Introduction to Probability	Morton International	Evaluation of Customer Service Testing Program
5	Discrete Probability Distributions	Xerox Corporation	Performance Test of an On-Line Computerized Publication System
6	Continuous Probability Distributions	Procter & Gamble Company	Manufacturing Strategy
7	Sampling and Sampling Distributions	Mead Corporation	Estimating the Value of Mead Forest Ownership
8	Interval Estimation	Dollar General	Sampling for Estimation of LIFO Inventory Costs
9	Hypothesis Testing	Harris Corporation	Testing for Defective Plating
10	Statistical Inference about Means and Proportions with Two Populations	Pennwalt Corporation	Evaluation of New Drugs
11	Inferences about Population Variances	U.S. General Accounting Office	Water Pollution Control
12	Analysis of Variance and Experimental Design	Burke Marketing	New Product Design
13	Simple Linear Regression and Correlation	Polaroid	Aging Study of Film
14	Multiple Regression	Champion International	Control of Pulp Bleaching Process
15	Model Building	Monsanto	Feed Development for Chickens
16	Time Series Analysis and Forecasting	The Cincinnati Gas & Electric Company	Forecasting Demand for Electricity
17	Tests of Goodness of Fit and Independence	United Way	Determining Community Perceptions of Charities
18	Nonparametric Methods	West Shell Realtors	Comparison of Real Estate Prices across Neighborhoods
19	Decision Analysis	Ohio Edison Company	Choice of Best Type of Particulate Control Equipment
20	Statistical Methods for Quality Control	Dow Chemical USA	Statistical Process Control

Statistics for Business and Economics

FOURTH EDITION

CHAPTER 1

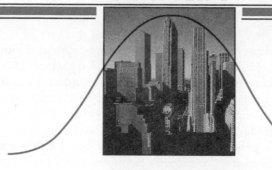

Data, Measurement,
and Statistics

F requently, we see news articles with statements such as the following:

■ Consumer prices rose 0.5% last month (*Business Week*, April 3, 1989).
■ The Dow Jones Industrial Average closed at 2291.97 (*Wall Street Journal*, April 10, 1989).
■ The unemployment rate was 5.0% in March (*Barrons*, April 10, 1989).
■ The median-priced house sells for $91,500 (*Money*, April, 1989).
■ Only 16% of men in their mid-60s remain in the labor force (*Financial World*, January 24, 1989).

The numerical facts in the preceding statements (.5%, 2291.97, 5.0%, $91,500, 16%) are called statistics. Thus, in everyday usage, the term *statistics* refers to numerical facts. However, the field, or subject, of statistics involves much more than simply the calculation and presentation of numerical facts.

The subject of statistics involves the study of how data are collected, how they are analyzed, how they are presented, and how they are interpreted. Particularly in business and economics, a major reason for collecting, analyzing, presenting, and interpreting data is to provide the information necessary to make good decisions. In this text we shall be oriented toward the use of statistics in a decision-making context.

Chapter 1 opens with some illustrations of the applications of statistics in business and economics. The remainder of the chapter is devoted to a discussion of how *data*, the raw material of statistics, are acquired, measured, and used. The two uses of data referred to as descriptive statistics and statistical inference are the subjects of Sections 1.5 and 1.6.

1.1 ▬ STATISTICAL APPLICATIONS IN BUSINESS AND ECONOMICS

The understanding and use of statistics are becoming more and more important in today's business world. Tremendous amounts of statistical information are available. The most successful managers and decision makers are the ones who can understand the information and make use of it effectively. In some cases, a statistical study will be undertaken to provide direct help with a current problem and pending decision. In other cases, statistics provide valuable information about the general business and economic environment within which decisions must be made. We provide examples that illustrate some uses of statistics in business and economics.

Accounting

Public accounting firms use statistical sampling procedures in performing audits for their clients. For instance, suppose an accounting firm wants to determine if the amount of accounts receivable shown on the client's balance sheet fairly represents the actual amount of accounts receivable. Usually, the number of individual accounts receivable is so large that it would be too time-consuming and too expensive to validate each one. In situations such as this, it is common practice for the audit staff to select and validate a sample of the individual accounts. After reviewing the sample results, the auditor will make a decision as to whether or not to accept the amount shown on the client's balance sheet. In this case, the sampling and statistical information is used directly in the decision-making process.

Finance

Financial planners and advisors use a variety of statistical information to guide investment decisions. These individuals study financial characteristics such as price-earnings ratios and dividend yields. By comparing the information for an individual stock with information on stock market averages, the financial manager will have help determining if an individual stock is over- or under-valued. For example, or April 10, 1989, *Barrons* reported that the average price-earnings ratio for the 30 stocks in the Dow Jones Industrial Average was 10.7. On the same day, Phillip Morris provided a price-earnings ratio of 12. The statistical information on price-earnings ratios showed that Phillip Morris had a higher price compared to its earnings than the average for the Dow Jones stocks. As a result, the financial manager might conclude Phillip Morris was slightly overpriced. This and other information about Phillip Morris can help the manager make buy and sell recommendations for the stock.

Marketing

Before marketing a new product nationally, companies conduct research studies designed to learn about market reaction and consumer acceptance of the product. Test markets and/or consumer panels may be identified. Statistical information on sales in the test markets and statistical summaries of consumer opinions will help the company decide whether or not to market the product nationally.

Production

Quality-control and lot-acceptance sampling are common applications of statistics in production. In recent years, U.S. manufacturers have been criticized for producing poor-quality products—especially when compared to Japanese manufacturers. Lot-acceptance sampling is a quality-control procedure used to ensure that purchased materials are of satisfactory quality. For instance, in a shipment of 10,000 components, a company might test only 100 in order to reach the decision of whether or not to accept the entire shipment. If a large number of components in the sample of 100 perform correctly, the entire shipment will be accepted. If this is not the case, the shipment will be rejected.

Economics

Economists are frequently asked to provide forecasts about the future of the economy or some part of it. In making such forecasts, a variety of statistical information is used. For instance, in forecasting inflation rates, economists use statistics on indicators such as the Producer Price Index, unemployment rate, and manufacturing capacity utilization. Often such statistics are used in computerized forecasting models, which predict inflation rates.

Applications of statistics, such as those just described, are an integral part of this text. Such examples provide an overview of the breadth of statistical applications. To supplement these examples, we have asked practitioners from the fields of business and economics to provide chapter-ending applications that serve to illustrate the material that has just been covered. We believe that these actual applications of statistics will provide an appreciation of the importance of statistics in several different types of decision-making situations. Table 1.1 provides an overview of these chapter-ending applications.

1.2 ■■■ DATA

Data are the facts and figures that are collected, analyzed, and summarized. All the data collected in a particular study is referred to as the *data set* for the study. Table 1.2 shows a data set for a group of shadow stocks with dividend reinvestment plans. The term *shadow* is used to indicate that the stocks are for small- to medium-sized firms that are not followed closely by the major brokerage houses. The data set in Table 1.2 was provided by the American Association of Individual Investors, *AAII Journal*, November 1988.

Elements, Variables, and Observations

The *elements* are the entities on which data are collected. For the data set in Table 1.2, each individual stock is an element. With 25 stocks, there are 25 elements in the data set.

TABLE 1.1   An Overview of Chapter-Ending Applications

	Chapter/Chapter Title	Organization	Application
1	Data, Measurement and Statistics	Kings Island Inc.	Consumer Profile Sample Survey
2	Descriptive Statistics I. Tabular and Graphical Approaches	Colgate Palmolive Company	Quality Assurance for Heavy Duty Detergents
3	Descriptive Statistics II. Measures of Location and Dispersion	Barnes Hospital	Time Spent in Hospice Program
4	Introduction to Probability	Morton International	Evaluation of Customer Service Testing Program
5	Discrete Probability Distributions	Xerox Corporation	Performance Test of an On-Line Computerized Publication System
6	Continuous Probability Distributions	Procter & Gamble Company	Manufacturing Strategy
7	Sampling and Sampling Distributions	Mead Corporation	Estimating the Value of Mead Forest Ownership
8	Interval Estimation	Dollar General Corporation	Sampling for Estimation of LIFO Inventory Costs
9	Hypothesis Testing	Harris Corporation	Testing for Defective Plating
10	Statistical Inference about Means and Proportions with Two Populations	Pennwalt Corporation	Evaluation of New Drugs
11	Inferences about Population Variances	U.S. General Accounting Office	Water Pollution Control
12	Analysis of Variance and Experimental Design	Burke Marketing	New Product Design
13	Simple Linear Regression and Correlation	Polaroid	Aging Study of Film
14	Multiple Regression	Champion International	Control of Pulp Bleaching Process
15	Model Building	Monsanto	Evaluating New Feed for Chickens
16	Time Series Analysis and Forecasting	The Cincinnati Gas & Electric Company	Forecasting Demand for Electricity
17	Tests of Goodness of Fit and Independence	United Way	Determining Community Perceptions of Charities
18	Nonparametric Methods	West Shell Realtors	Comparison of Real Estate Prices across Neighborhoods
19	Decision Theory	Ohio Edison Company	Choice of Best Type of Particulate Control Equipment
20	Statistical Methods for Quality Control	Dow Chemical Company	Quality Control for a Chemical Process

A *variable* is a characteristic of interest for the elements. Thus, the data set in Table 1.2 shows there are five variables, which are defined as follows:

■ *Exchange:* The stock exchange listing the stock—NYSE (New York Stock Exchange), AMEX (American Stock Exchange), and OTC (Over The Counter).

TABLE 1.2   A Data Set for 25 Shadow Stocks with Dividend Reinvestment Plans

Stock	Exchange	Ticker Symbol	Earnings per Share ($)	Dividend Yield (%)	Price-Earnings Ratio
Acme Electric	NYSE	ACE	0.17	5	40
American Filtrona	OTC	AFIL	1.73	4	13
Amer Recreation Ctrs	OTC	AMRC	0.24	2	33
Berkshire Gas	OTC	BGAS	1.35	8	12
Colonial Gas	OTC	CGES	1.76	9	11
Champion Products	AMEX	CH	2.65	1	13
Connecticut Energy	NYSE	CNE	2.32	8	9
Connecticut Wtr Sv	OTC	CTWS	1.91	7	11
Eastern Co.	AMEX	EML	1.74	4	8
Engraph Inc.	OTC	ENGH	0.81	1	18
Energy North Inc.	OTC	ENNI	1.41	7	10
Gorman-Rupp	AMEX	GRC	2.04	4	11
Marsh Supermarkets	OTC	MARS	1.45	2	13
Mobile Gas Service	OTC	MBLE	1.35	5	12
Medalist Indus	OTC	MDIN	1.43	4	11
Middlesex Water	OTC	MSEX	2.07	7	12
New Jersey Resources	NYSE	NJR	1.39	7	13
NECO Enterprises	AMEX	NPT	1.51	8	12
Peerless Tube	AMEX	PLS	0.39	6	18
Public Sv No. Car.	OTC	PSNC	1.65	7	9
United Cities Gas	OTC	UCIT	1.23	7	10
Upper Peninsula Pwr.	OTC	UPEN	3.73	9	7
UNTIL Corp.	AMEX	UTL	3.29	7	8
Valley Resources	AMEX	VR	1.15	3	20
Vulcan Corp.	AMEX	VUL	0.16	4	137

Source: *American Association of Individual Investors Journal,* November, 1988.

▓▓ *Ticker Symbol:* The abbreviation used to identify the stock on the exchange listing.

▓▓ *Earnings Per Share:* Net income after all expenses and taxes divided by the number of common shares outstanding.

▓▓ *Dividend Yield:* Most recent indicated annual dividend per share divided by share price.

▓▓ *Price-Earnings Ratio:* Market price per share divided by most recent 12 months' earnings per share.

Data are obtained by collecting measurements on each variable for every element in the study. The set of measurements collected for a particular element is called an *observation.* Referring to Table 1.2, we see that the first element (Acme Electric) has the observation NYSE, ACE, 0.17, 5, and 40. With 25 elements, there are 25 observations in the data set.

Most data that statisticians collect and work with are numeric. However, as the data set in Table 1.2 shows, data for some variables may be numeric and data for other variables may be nonnumeric. Specifically, the first two variables, Exchange and Ticker Symbol, have nonnumeric data. The nonnumeric symbols used are abbreviations for the stock exchange and stock identification of each element. The other three variables—earnings per share, dividend yield, and price-earnings ratio—have numeric data.

1.3 ███ SCALES OF MEASURMENT

The type of statistical analysis appropriate for the data on a particular variable depends upon the scale of measurement used for the variable. There are four scales of measurement: nominal, ordinal, interval and ratio. The scale of measurement determines the amount of information contained in the data and indicates the data summarization and statistical analyses that are most appropriate. We describe each of the four scales of measurement.

Nominal Scale

The scale of measurement for a variable is *nominal* when the observations for the variable are simply labels used to identify an attribute of each element. For example, referring again to the data set in Table 1.2, we see that the first variable, Exchange, is measured on a nominal scale. This is so because the observations of NYSE, AMEX, and OTC are labels used to identify the exchange that lists the stock. The second variable, Ticker Symbol, is also measured on a nominal scale with the abbreviation letters providing the identification label for the stock.

In cases where the scale of measurement is nominal, a numeric code as well as a nonnumeric symbol may be used. For example, the observations for the Exchange variable could use a numeric code with a value of 1 for the New York Stock Exchange, a value of 2 for the American Stock Exchange, and a value of 3 for Over the Counter. Such a code would facilitate recording and computer processing of the exchange data. However, it is important to remember that even though the data are numeric, the value of 1, 2, and 3 are simply labels used to identify the exchange. Thus, the scale of measurement is nominal.

Other examples of variables where data have a nominal scale are as follows:

███ Sex (male, female)
███ Marital status (single, married, widowed, divorced)
███ Religious affiliation (many possibilities)
███ Part-identification code (A13622, 12B63)
███ Employment status (employed, unemployed)
███ House number (5654, 2712, 624)
███ Occupation (many possibilities)

As the preceding examples show, the key feature of the nominal scale is that the observations are labels used to identify an attribute of the element.

Finally, it is important to note that arithmetic operations such as addition, subtraction, multiplication and division *do not* make sense for nominal data. Thus, even if the nominal data are numeric, computations such as summing and averaging the data are inappropriate.

Ordinal Scale

The scale of measurement for a variable is *ordinal* when

1. The data have the properties of nominal data, and
2. The data can be used to rank or order the observations for the variable.

Let us use the following example to illustrate the ordinal scale of measurement. Some restaurants place questionnaires on the tables to solicit customer opinions concerning the restaurant's performance in terms of food, service, atmosphere, and so on. A questionnaire used by the Lobster Pot Restaurant in Redington Shores, Florida, is shown in Figure 1.1. Note that the customers completing the questionnaire are asked to provide ratings for six different variables: food, drinks, service, waiter, captain, and hostess. The response categories are excellent, good, and poor for each variable. The observations for each variable possess characteristics of nominal data (each response rating is a label for excellent, good, or poor quality). In addition, the observations can be ranked, or ordered, with respect to quality. For example, consider the variable on food quality. After collecting the data, we can rank the observations in order of food quality by beginning with the observations of excellent, followed by the observations of good, and, finally, by the observations of poor. With only three response categories, we can expect to see many observations with tied rankings. Nonetheless, the observations can be ranked in terms of food quality.

Like data obtained from a nominal scale, data obtained from an ordinal scale may be either nonnumeric or numeric. For the Lobster Pot questionnaire, the

---

FIGURE 1.1 **Customer Opinion Questionnaire Used by the Lobster Pot Restaurant, Redington Shores, Florida (Used with Permission)**

	Excellent	Good	Poor	Comments
Food				
Drinks				
Service				
Waiter				
Captain				
Hostess				

Waiter's Name _____

Captain's Name _____

Other Comments _____

nonnumeric letters of E for excellent, G for good, and P for poor could be used to record the observations. Or, a numeric code with values of 1 for excellent, 2 for good, and 3 for poor could be used equally well. Finally, as with nominal data, it is important to remember that arithmetic operations *do not* make sense for ordinal data. Thus, even if the ordinal data are numeric, the computations of summing or averaging are inappropriate.

Interval Scale

The scale of measurement for a variable is *interval* when

1. The data have the properties of ordinal data, and
2. The interval between observations can be expressed in terms of a fixed unit of measure.

Temperature is a good example of a variable that uses an interval scale of measurement. The fixed unit of measure is a degree. An observation recorded at a particular point in time will be a numeric value specifying the amount, or quantity, of degrees. Such data possess the properties of ordinal data in that temperature observations of 35 degrees, 40 degrees, 85 degrees, and 90 degrees can be ranked, or ordered, from coldest to warmest. In addition, this interval scale has the property that the interval between observations can be expressed in terms of the fixed unit of measure (a degree). For example, the interval between 35 and 40 degrees is 5 degrees, the interval between 35 and 85 degrees is 50 degrees, and the interval between 85 and 90 degrees is 5 degrees. With nominal and ordinal data, such differences between observations are not meaningful.

The fixed unit of measure required by an interval scale means that the data *must always be numeric*. With interval data, the arithmetic operations of addition, subtraction, multiplication, and division are meaningful. As a result, data obtained using this scale lend themselves to more alternatives for statistical analysis than do data obtained from nominal or ordinal scales.

Ratio Scale

The scale of measurement for a variable is *ratio* when

1. The data have all the properties of interval data, and
2. The ratio of two observations is meaningful.

Variables such as distance, height, weight, and time use the ratio scale of measurement. A requirement of the ratio scale is that a zero value is inherently defined in the scale. Specifically, the zero value must indicate nothing exists for the variable at the zero point. Whenever we collect data on the cost of something, we use a ratio scale of measurement. Consider a variable indicating the cost of an automobile. The zero point is inherently defined in that a zero cost indicates the automobile is free (no cost). Then, comparing the $10,000 cost of one automobile with the $5,000 cost of a second automobile, the ratio property of the data shows that the first automobile is 10,000/5,000 = 2 times, or twice, the cost of the second automobile.

Since ratio data have all of the properties of interval data, ratio data are *always numeric* and enable meaningful arithmetic operations such as addition, subtraction, multiplication, and division. As with interval data, ratio data lend themselves to more alternatives for statistical analysis than do data obtained from nominal or ordinal scales.

Figure 1.2 provides a summary of the relationship between nonnumeric and numeric data and the four scales of measurement. We note that the nominal and ordinal scales can generate both nonnumeric and numeric data, but that interval and ratio scales generate only numeric data.

The amount of information in the data varies with the scale of measurement. Nominal data contain the least amount of information, followed by ordinal, interval, and then ratio data. Since arithmetic operations are meaningful only for interval and ratio data, it is important to know the measurement scale used in order to employ the most appropriate statistical procedures. There are many statistical procedures that can be used with interval and ratio data that are not meaningful with nominal and ordinal data.

Qualitative and Quantitative Data

Data can also be classified as being either qualitative or quantitative. *Qualitative data* provide labels or names for categories of like items. The categories for qualitative data may be identified by either nonnumeric descriptions or by numeric codes. Qualitative data are obtained from either a nominal or an ordinal scale of measurement. On the other hand, *quantitative data* indicate either "how much" or "how many" of something. Quantitative data are always numeric and are obtained from either an interval or a ratio scale of measurement.

■■ FIGURE 1.2   **The Relationship between Nonnumeric and Numeric Data and Scales of Measurement**

Scale of Measurement	Nonnumeric Data	Numeric Data
Nominal	Description indicates the category for the element	Numeric value indicates the category for the element
Ordinal	Description permits ranking or ordering of data	Numeric value permits ranking or ordering of data
Interval		Numeric values* are defined in fixed and equal units such that the interval between data values is meaningful
Ratio		Numeric values* have an inherently defined zero, and ratios of data values are meaningful

*Arithmetic operations are meaningful for these kinds of data.

In terms of the statistical methods used for summarizing data, qualitative data provided by nominal and ordinal scales employ similar methods, whereas quantitative data provided by interval and ratio scales employ similar methods. Tabular and graphical methods for summarizing qualitative data are presented in Section 2.1. Tabular and graphical methods for summarizing quantitative data are presented in Sections 2.2 and 2.3.

NOTES AND COMMENTS

1. An element is the entity on which measurements are obtained. It could be a company, a person, an automobile, and so on. An observation is the set of measurements obtained for each element, so the number of observations and the number of elements in a data set will always be the same. The number of measurements obtained on each element is the number of variables. Thus, the total number of data items in a data set is the number of elements times the number variables.

2. Data obtained from categorical responses are measured with either a nominal or ordinal scale. Such data are called qualitative data and are often nonnumeric. Assigning a numeric code to such data *does not* make it quantitative. Even when a numeric code is used, these data should not be subjected to arithmetic calculations.

3. For purposes of statistical analysis, the most important distinguishing characteristic of the types of data is that ordinary arithmetic operations are meaningful *only* with quantitative (interval and ratio-scaled) data. This is due to the fact that interval and ratio data share the property that the numeric values assigned are based on fixed and equal units of measurement.

1.4 ■ DATA ACQUISITION

We have introduced the concepts of elements, variables, and observations and how data collected in a particular study form a data set. In this section we describe the sources of data that are available and how the data can be acquired. The data needed for statistical analysis may be obtained through already-existing sources of data or through specially designed statistical studies that are undertaken to obtain new data.

Existing Sources of Data

In some cases, data needed for a particular application are already available within the firm or organization. For example, all companies maintain a variety of data on their employees and their business operations. Data on employee salaries, ages, years of experience, and so on can usually be obtained relatively easily from internal personnel records. Data on sales, advertising costs, distribution costs, inventory levels, production quantities, and so on are available from other internal information and record-keeping systems. Table 1.3 shows some of the data that are routinely available from internal information sources.

TABLE 1.3 Some Data Available from Internal Company Records

Company Record	Some of the Data Typically Available
Employee records	Name, address, social security number, salary, number of vacation days, number of sick days, bonus, and so on
Production records	Part or product number, quantity produced, direct labor cost, materials cost, and so on
Inventory records	Part or product number, number of units on hand, reorder level, economic order quantity, discount schedule, and so on
Sales records	Product number, sales volume, sales volume by region, sales volume by customer type, and so on
Customer credit	Customer name, address, phone number, credit limit, account receivable balance, and so on

In addition, substantial amounts of business and economic data are routinely collected and published in a variety of external sources. In some instances, a company may obtain access to a large data base through a leasing arrangement with an organization that specializes in maintaining a specific information system. Mead Corporation, Dun & Bradstreet, and Dow Jones & Company are some of the firms that provide data base services.

Governmental agencies provide another important source of existing data. For instance, the U.S. Department of Labor maintains considerable data on characteristics such as employment rates, wage rates, size of the labor force, and union memberships. Table 1.4 shows other selected governmental agencies and some of the data they provide. Data are also available from a variety of industry associations and special-interest organizations. The Travel Industry Association of America maintains travel-related information such as the number of tourists

TABLE 1.4 Selected Governmental Agencies and Some of the Data Available

Governmental Agency	Some of the Data Available
Bureau of the Census	Population data and its distribution; data on number of households and their distribution; data on household income and its distribution
Federal Reserve Board	Data on the money supply, installment credit, exchange rates, and discount rates
Office of Management and Budget	Data on revenue, expenditures, and debt of the federal government
Department of Commerce	Data on business activity—value of shipments by industry, level of profits by industry, and data on growing and declining industries

and travel expenditures by state. Such data would be of interest to firms and individuals participating in the travel industry. The Graduate Management Admission Council maintains data on MBA student characteristics and graduate management education programs. Most of the data from sources such as these are available to qualified users at a modest cost.

Statistical Studies

Sometimes, data are not readily available from existing internal or external sources. If the data are considered to be necessary, a statistical study will have to be conducted in order to obtain the data. Such statistical studies can be classified as being either *experimental* or *nonexperimental*.

In an experimental study, variables of interest are identified. Then one or more factors in the study are controlled so that data may be obtained about how the factors influence the variables. For example, a pharmaceutical firm might be interested in conducting an experiment designed to learn about how a new drug affects blood pressure. Blood pressure is the variable of interest in the study. The new drug is the factor that influences the blood pressure. To obtain data about the effect of the new drug, a sample of individuals will be selected. The dosage level of the new drug will be controlled, with different groups of individuals being given different dosage levels. Data on blood pressure will be collected for each group. Statistical analysis of the experimental data will help determine how the new drug affects blood pressure.

In nonexperimental, or observational, statistical studies, no attempt is made to control the influences of factors on the variable or variables of interest. A survey is perhaps the most common type of nonexperimental study. In a survey, research questions are identified. A questionnaire containing the research questions is designed and administered to a sample of individuals. In this way, data are obtained about the research variables but no attempt is made to control the factors that influence the variables.

Managers wishing to use data and statistical analyses as an aid to decision making must always be aware of the time and cost required to obtain the data. The use of existing data sources is desirable when data must be obtained in a relatively short period of time. If the data are not available from an existing source, advance planning will be necessary to provide the time to obtain the data from a satistical study. In all cases, it is desirable for the decision maker to consider the contribution of the data to the decision-making process. The cost of data acquisition and the subsequent statistical analysis should not be more than the savings generated by using the statistical information to make a better decision.

Possible Data-Acquisition Errors

Managers who use data should always be aware of the possibility of data errors. Using erroneous data and misleading statistical analyses would be worse than not using the data and statistical information at all. An error in data acquisition occurs whenever the data value obtained is not equal to the true or actual value that would have been obtained with a correct procedure. Such errors can occur in a number of ways. For example, an interviewer might make a recording error such as writing the age of a 24-year-old person as 42. In addition, the person answering an interview question might misinterpret the question and make an incorrect response.

Experienced data analysts take great care in collecting and recording data to ensure that errors are not made. Special procedures may be used to check for internal consistency of the data. Such procedures would indicate the analyst should review data for a respondent who is shown to be 22 years of age but reports 20 years of work experience. Data analysts also review data for unusually large and small values, called *outliers,* which are candidates for possible data errors. In Chapter 3, we present some of the methods statisticians use to identify outliers.

The point of this discussion is to alert the users of data to the possibility of errors occurring during data acquisition. Blindly using any data that happen to be available or using data that were acquired with little care can lead to poor and misleading information. However, taking steps to acquire accurate data can help provide reliable and valuable decision-making information.

1.5 ■■■ DESCRIPTIVE STATISTICS

Most of the statistical information in newspapers, magazines, reports and other publications comes from data that have been summarized and presented in a form that is easy for the reader to understand. These summaries of data, which may be tabular, graphical, or numerical, are referred to as *descriptive statistics.*

Refer again to the data set in Table 1.2. Twenty-five shadow stocks with dividend reinvestment plans are listed. Methods of descriptive statistics can be used to provide summaries of the information in this data set. For example, a tabular summary of the data for the exchange variable is shown in Table 1.5. A graphical summary of the same data is shown in Figure 1.3. The purpose of tabular and graphical summaries such as these is to make it easier to interpret the data. Referring to Table 1.5 and Figure 1.3, we can see easily that the majority of the stocks in the data set are listed on the over-the-counter exchange. On a percentage basis, 56% of the stocks are listed on the over-the-counter exchange. Also, 32% are listed on the American Stock Exchange and only 12% are listed on the New York Stock Exchange.

A graphical summary of the data on dividend yield of the 25 shadow stocks in Table 1.2 is provided by the histogram in Figure 1.4. From the histogram, it is easy to see that the dividend yields range from 1% to 9%, with the highest concentrations of dividend yields being 4% and 7%.

TABLE 1.5   **Frequencies and Percentages for the Exchange Listings of 25 Shadow Stocks with Dividend Reinvestment Plans**

Exchange	Frequency	Percent
New York Stock Exchange	3	12%
American Stock Exchange	8	32%
Over the Counter	14	56%
Total	25	100%

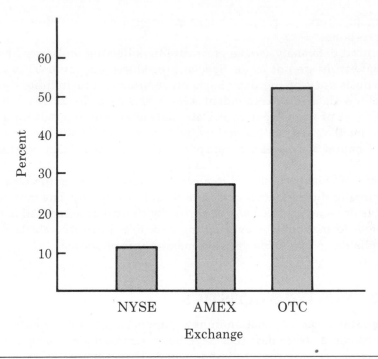

■ FIGURE 1.3   **Bar Graph of Exchange Listing for Shadow Stocks with Dividend Reinvestment Plans**

In addition to tabular and graphical displays, numerical descriptive statistics are often used to summarize data. The most common numerical descriptive statistic is the *average*, or *mean*. Using the data on dividend yield in Table 1.2, we can compute the average dividend yield by adding the yield data for all 25 stocks and dividing the sum by 25. Doing so tells us that the average dividend yield for

■ FIGURE 1.4   **Histogram of Dividend Yield for Shadow Stocks with Dividend Reinvestment Plans**

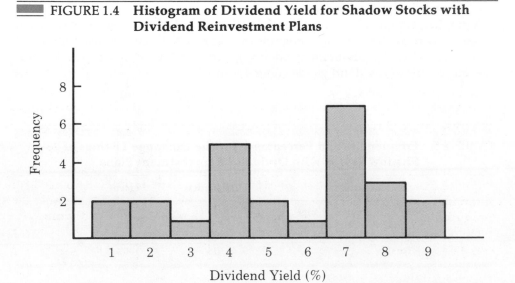

the stocks is 5.44%. This average is taken as a measure of the central value, or central location, of the data.

In recent years there has been a growing interest in statistical methods that can be used for developing and presenting descriptive statistics for data sets. Chapters 2 and 3 are devoted to the tabular, graphical, and numerical methods of descriptive statistics.

1.6 ▬▬ STATISTICAL INFERENCE

In many situations there exists a large group of elements (individuals, stocks, voters, households, products, customers and so on) about which data are sought. Because of time, cost, and other considerations, data are collected from only a small portion of the group. The larger group of elements in a particular study is called the *population,* and the smaller group is called the *sample.* Formally, we will use the following definitions:

Population

A *population* is the collection of all elements of interest in a particular study.

Sample

A *sample* is a subset of the population.

Much of the contribution of statistics is that statistical methods enable us to use data from a sample to make estimates and test claims about the characteristics of a population. This process is referred to as *statistical inference.*

As an example of statistical inference, let us consider the study conducted by Norris Electronics. Norris manufacturers a high-intensity light bulb that is used in a variety of electrical products. In an attempt to increase the useful life of the light bulbs, the product design department has developed a new light-bulb filament. In order to evaluate the advantages of the new filament, a sample of 200 new-filament bulbs was manufactured and tested. Data were collected on the number of hours each bulb operated before the filament burned out. The data collected are shown in Table 1.6.

The data available are based on a sample of 200 bulbs. The corresponding population is all bulbs that could be produced using the new filament design. Suppose that Norris is interested in using the data from the sample to make an inference about the lifetime for the population of all bulbs that could be produced. Adding the hours of useful life for the 200 bulbs and dividing the total by 200 provides the sample average, or sample mean, lifetime. Doing so for the data in Table 1.6 shows a sample average lifetime of 76 hours. Thus, we would use this sample result to estimate that the average lifetime for the bulbs in the population is 76 hours.

TABLE 1.6   Hours until Burnout for a Sample of 200 Bulbs for Norris Electronics

107	73	68	97	76	79	94	59	98	57
54	65	71	70	84	88	62	61	79	98
66	62	79	86	68	74	61	82	65	98
62	116	65	88	64	79	78	79	77	86
74	85	73	80	68	78	89	72	58	69
92	78	88	77	103	88	63	68	88	81
75	90	62	89	71	71	74	70	74	70
65	81	75	62	94	71	85	84	83	63
81	62	79	83	93	61	65	62	92	65
83	70	70	81	77	72	84	67	59	58
78	66	66	94	77	63	66	75	68	76
90	78	71	101	78	43	59	67	61	71
96	75	64	76	72	77	74	65	82	86
66	86	96	89	81	71	85	99	59	92
68	72	77	60	87	84	75	77	51	45
85	67	87	80	84	93	69	76	89	75
83	68	72	67	92	89	82	96	77	102
74	91	76	83	66	68	61	73	72	76
73	77	79	94	63	59	62	71	81	65
73	63	63	89	82	64	85	92	64	73

Whenever statisticians make an estimate, as in the Norris Electronics example, they usually provide a statement of the precision associated with the estimate. For the Norris example, the statistician might state that the estimate of the average lifetime for the population of new light bulbs is 76 hours with a precision of ±4 hours. Thus, an interval estimate of 72 hours to 80 hours would be provided for the average lifetime of the population of new bulbs to be produced. Using concepts from probability, the statistician can also state how confident he or she is that the interval of 72 hours to 80 hours contains the true average lifetime for the population. At this point, management should be ready to decide whether or not the results justify going ahead with the production of the new bulbs.

Figure 1.5 provides a graphical summary of the statistical inference process used in the Norris Electronics example. Chapters 7 through 18 describe a variety of statistical inference methods available to a decision maker.

1.7 ■■■ THE DATA ANALYST – A MICROCOMPUTER STATISTICAL SOFTWARE PACKAGE

Microcomputer software packages are making many statistical techniques easier to use. *The Data Analyst* is a statistical software package that has been developed to accompany this text. It is designed for solving problems in the text as well as small-scale problems that may be encountered in practice.

The Data Analyst consists of modules, or programs, that enable the user to perform statistical analysis in the following topical areas:

■■ Descriptive statistics
■■ Interval estimation for one population

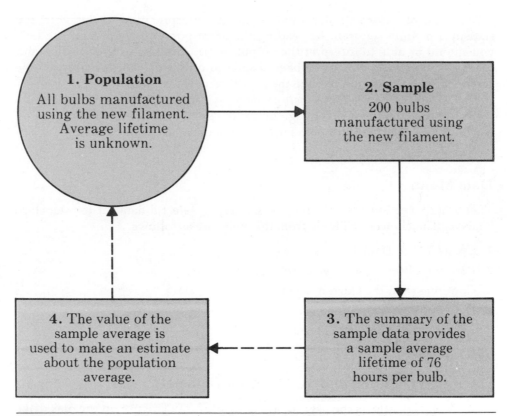

■■■ FIGURE 1.5   **The Process of Statistical Inference for the Norris Electronics Example**

■ Interval estimation for two populations
■ Analysis of variance
■ Regression and correlation

The use of *The Data Analyst* with this text is optional. The software package, itself, contains instructions for installing the system on your microcomputer; the remainder of this section provides an introductory description of the key features of the software package.

Top Level Menu

The Data Analyst is a menu-driven system; that is, users communicate with the system by selecting an option from a list provided. The user can direct the package to select a statistical routine, accept a data set, and display the statistical results. For example, the first menu that appears on the screen is the "Top Level Menu." The choices on this menu provide access to the corresponding statistical modules or programs that can be used to analyze a data set. As mentioned earlier the options for statistical analysis include descriptive statistics, interval estimation, analysis of variance, regression and correlation analysis. Once the desired option has been selected, the user can select the desired data set and the statistical results will be displayed on the screen.

The type of information provided and its interpretation varies with the statistical module selected. By reading the corresponding chapter in the text, you should be able to interpret the output information. After the solution and other output information have been displayed on the screen, the user will be provided with the option to have the results sent to a printer.

The advantage of *The Data Analyst's* menu system is that users do not have to learn a special command language to use the system; effort can be focused on learning how the computer package can aid in the analysis of data.

Data Menu

The "Data Selection Menu" enables the user to select a data set for statistical analysis. The choices available from the menu are as follows:

1. Create a New Data Set
2. Retrieve a Previously Saved Data Set
3. Continue with the Current Data Set
4. Delete a Previously Saved Data Set
5. Return to the Top Level Menu

Saving, Retrieving, and/or Deleting Data Sets

The Data Analyst allows the user to save data sets for future use on the disk drive that is specified by the user when the software package is initially started. When this option is selected, instructions for naming the data set will appear on the screen. The data set will then be saved automatically, using the name specified. An option is also provided to save the data as an ASCII file so it can be read by other software packages.

When reentering *The Data Analyst* at a later date, the user may select the retrieve option from the "Data Selection Menu" in order to recall a previously saved data set. In addition, an option is also provided to read ASCII data sets that have been created by other software packages. When the data set is no longer needed, the delete option can be used to erase the data set from the data disk.

Further Advice about Data Input

When using *The Data Analyst*, you may find the following data input suggestions helpful.

1. Do not enter commas (,) with your input data. For example, to enter the numerical value of 104,000, simply type the six digits 104000.
2. Do not enter the dollar sign ($) for profit or cost data. For example, a cost of $20.00 should be entered as 20.
3. Do not enter the percent sign (%) if percentage input is requested. For a percentage of 25%, simply enter 25. Do not enter 25% or 0.25.
4. If the computer did not interpret your input correctly (for example, you tried to input a comma), the message "Redo from start" may appear. This message

refers to the input question or prompt you are currently responding to. The message means to respond to the same question or prompt again.

5. For data values containing the digit zero, be sure to enter the numeric 0 rather than the letter O.

6. Occasionally data may be recorded with fractional values such as $\frac{1}{4}$, $\frac{2}{3}$, $\frac{5}{6}$, and so on. The data input for the computer must be in decimal form. The fraction of $\frac{1}{4}$ can be entered as .25. However, the fractions $\frac{2}{3}$ and $\frac{5}{6}$ have repeating decimal forms. In cases such as these we recommend the convention of rounding to five places. Thus, the corresponding decimal values of .66667 and .83333 should be entered.

7. Finally, we recommend that in general you attempt to scale extremely large input data so that smaller numbers may be input and operated on by the computer. For example, 2,500,000 may be scaled to 2.5 with the understanding that the data used in the problem reflect millions.

■ SUMMARY

In everyday usage, the term *statistics* is often taken to mean numerical facts. However, the field, or subject, of statistics requires a broader definition. In this sense, we define statistics as the science of collecting, analyzing, presenting, and interpreting data. Because nearly every college student majoring in business or economics is required to take a course in statistics, we began the chapter by describing typical statistical applications in business and economics.

Data were described as being the raw material of statistics. Thus, data are the facts and figures that are collected, analyzed, presented, and interpreted. A significant portion of the chapter was devoted to data, including how data are measured and how data are acquired.

Four scales of measurement are available for obtaining data on a particular variable: nominal, ordinal, interval, and ratio. The amount of information contained in the data depends upon the scale of measurement. Nominal data contains the least amount of information, followed by ordinal, interval, and ratio data. We showed that nominal and ordinal data are classified as qualitative data and may be recorded by either a nonnumeric description or a numeric code. Interval and ratio data are classified as quantitative data. Quantitative data are always numeric and indicate how much or how many for the variable of interest. Ordinary mathematical operations of addition, subtraction, multiplication, and division are meaningful only if the data are *quantitative*. As a result, many statistical procedures that are applicable for quantitative data are not appropriate for qualitative data.

In Sections 1.5 and 1.6 we introduced the topics of descriptive statistics and statistical inference. Descriptive statistics are the tabular, graphical, and numerical methods used to summarize data. Statistical inference is the process of using data obtained from a sample to make estimates or test claims about characteristics of a population. We concluded the chapter with an introduction to the statistical software package *The Data Analyst*.

■ GLOSSARY

Data   The facts and figures that are collected, analyzed, presented, and interpreted. Data may be numeric or nonnumeric.

Data set   All the data collected in a particular study.

Elements   The entities on which data are collected.

Variable   A characteristic of interest for the elements.

Observation   The set of measurements or data obtained for a single element.

Nominal scale   A scale of measurement that uses a label or category to define an attribute of an element. Nominal data may be recorded with a nonnumeric description or with a numeric code.

Ordinal scale   A scale of measurement that has the properties of a nominal scale and can be used to rank or order the observations. Ordinal data may be recorded with a nonnumeric description or with a numeric code.

Interval scale   A scale of measurement that has the properties of an ordinal scale and the interval between observations is expressed in terms of a fixed unit of measure. Interval data are always numeric.

Ratio scale   A scale of measurement that has the properties of an interval scale and the ratio of observations is meaningful. Ratio data are always numeric.

Qualitative data   Data obtained with a nominal or ordinal scale of measurement. Qualitative data may be recorded with a nonnumeric description or with a numeric code.

Quantitative data   Data obtained with an interval or ratio scale of measurement. Quantitative data are always numeric and indicate how much or how many for the variable of interest.

Descriptive statistics   Tabular, graphical, and numerical methods used to summarize data.

Population   The collection of all elements of interest in a particular study.

Sample   A subset of the population.

Statistical inference   The process of using data obtained from a sample to make estimates or test claims about the characteristics of a population.

■ EXERCISES

1. Discuss the difference between the concept of statistics as numerical facts and the concept of statistics as a discipline or field of study.

2. The data set in Table 1.7 (*Business Week*, May 1, 1989) provides a variety of information concerning executive compensation and profitability for seven firms in the aerospace industry.

TABLE 1.7   **Information Concerning Executive Compensation and Profitability for Seven Firms in the Aerospace Industry**

Company	CEO Salary ($000s)	Sales ($000,000s)	Return on Equity (%)	1986–88 Pay versus Corporation Profit Rating*
Boeing	846	16,962.0	11.4	3
General Dynamics	1041	9,551.0	19.7	3
Lockhead	1146	10,590.0	17.9	3
Martin Marietta	839	5,727.5	26.6	2
McDonnell Douglas	681	15,069.0	11.0	3
Parker Hannifin	765	2,397.3	12.4	3
United Technology	1148	18,000.1	13.7	4

*CEO's are assigned to five groupings according to pay versus corporate profitability. Those in group 1 are given a rating of 1, those in group 2 a rating of 2, and so on. A 1 is the best rating; a 5 is the worst.
Source: Business Week, May 1, 1989.

a. How many elements are in this data set?

b. How many variables are in this data set?

c. How many observations are in this data set?

3. Refer to the data set in Table 1.7.

a. Which of the variables are qualitative and which are quantitative?

b. What measurement scale is being used for each of the variables?

4. Refer to the data set in Table 1.7.

a. Compute the average salary for the chief executive officers.

b. The firms in Table 1.7 are the largest in the aerospace industry. Would you feel comfortable using the average salary computed in (a) as an estimate of the average salary for all chief executive officers in the aerospace industry?

5. Refer to the data set in Table 1.7.

a. What proportion of the firms received a rating of 3 on pay versus corporate profitability?

b. Why is it inappropriate to compute an average for the data on pay versus corporate profitability?

6. The data set in Table 1.8 (*Fortune*, April 24, 1989) provides a variety of information on a sample of 10 firms from the Fortune 500 Largest U.S. Industrial Corporations.

a. How many elements are in the data set?

b. What is the population from which this sample is drawn?

c. Compute the average sales for the sample.

d. Use the result of (c) to make an estimate of the average sales for the population.

7. Refer to the data set in Table 1.8.

a. How many variables are in the data set?

b. Which of the variables are qualitative and which are quantitative?

c. What type of measurement scale is being used for each of the variables?

8. A California state agency classifies worker occupations as professional, white collar, or blue collar.

a. The variable is worker occupation. Is it a qualitative or quantitative variable?

b. What type of measurement scale is being used for this variable?

TABLE 1.8 **A Sample of 10 of The Fortune 500 Largest U. S. Industrial Corporations**

Company	Sales ($000,000)	Rank	Profit ($000,000)	Assets ($000,000)	Industry Code
Raytheon	8,192.1	53	489.6	4,739.5	7
Texas Instruments	6,294.8	75	366.3	4,427.5	7
Clark Equipment	1,278.3	283	46.1	951.6	11
Aristech Chemical	1,065.2	321	188.2	676.1	5
Cons. Papers	896.8	355	149.9	934.6	9
Dexter	827.3	369	35.5	626.4	5
Cooper Tire	748.0	391	41.1	442.6	21
Shaklee	627.5	432	27.2	434.5	19
Grow Group	506.8	495	0.0	223.3	5
Chemed	500.6	500	24.1	322.7	5

Source: *Fortune*, April 24, 1989

9. A study of last-minute holiday shoppers conducted by Best Products Inc. (*USA Today*, December 22, 1988) asked individuals to indicate when they completed their holiday shopping. The response alternatives were as follows.

> Before Halloween
> Before Thanksgiving
> A few days before Christmas
> Christmas Eve

a. The variable of interest is time of completing holiday shopping. Is this a qualitative or quantitative variable?
b. What measurement scale is being used?
c. Would it make more sense to use averages or percentages as a summary of the data in this study? Explain.

10. State whether each of the following variables is qualitative or quantitative and indicate the measurement scale that is appropriate for each.

a. Age
b. Sex
c. Class rank
d. Make of automobile
e. Number of people favoring the death penalty

11. State whether each of the following variables is qualitative or quantitative and indicate the measurement scale being used.

a. Annual sales
b. Soft-drink size (small, medium, or large)
c. Employee classification (GS1 through GS18)
d. Earnings per share
e. Method of payment (cash, check, credit card)

12. The board of directors for a major corporation has asked its salary and compensation committee to make a recommendation concerning the appropriate salary for the corporations's chief executive officer.

a. As a member of the salary and compensation committee, what data would be helpful to you in making such a recommendation?
b. How would you recommend obtaining the necessary data?

13. A manager of a large corporation has recommended a $10,000 raise be given in order to keep a valued subordinate from moving to another company. What internal and external sources of data might be used to decide whether or not such a salary increase is appropriate?

14. The marketing group at your company has come up with a new diet soft drink that, it is claimed, will capture a large share of the young adult market.

a. What data would you want to see before deciding to invest substantial funds introducing the new product into the marketplace?
b. How would you expect the data mentioned in (a) to be obtained?

15. Define and give two examples of descriptive statistics.

16. Give two examples of statistical inference. That is, find two examples of cases where sample results have been used to make inferences about a population and its characteristics.

17. A sample of 10 grocery stores in Montgomery, Alabama, showed that the average price per pound for pork chops was $2.89.

a. What are the elements in the data set created by the sample?
b. How many variables are in the data set?

c. In the statistical inference process, we would like to estimate the average price per pound for pork chops for all grocery stores in Montgomery, Alabama. Suggest a value for such an estimate.

d. What is the population of interest in this study?

18. A study of 250 households in Cedar Bluff showed that a household produced an average of 4 pounds of garbage per day.

a. What is the sample in this study?

b. How many observations are there?

c. What is the variable of interest?

d. What is the population?

e. Discuss the role of statistical inference in the context of this example.

19. A *1988 Newsweek/Gallup* poll investigated whether adults preferred to stay at home or go out as their favorite way of spending time in the evening. The poll of 1500 adults concluded that the majority of adults (70%) indicated that "staying at home with family" was the favorite evening activity.

a. What is the population of interest in this study?

b. What is the variable being studied?

c. Is the variable being studied qualitative or quantitative?

d. What was the size of the sample used?

e. Where was a descriptive statistic used in this study?

f. Describe the process of statistical inference in this study.

20. In a recent study of causes of death in males 60 years of age and older, a sample of 120 men indicated that 48 had died due to some form of heart disease.

a. Develop a descriptive statistic that can be used as an estimate of the percentage of males 60 years of age or older who die from some form of heart disease.

b. Is the variable being studied qualitative or quantitative?

c. Discuss the role of statistical inference in this type of medical research.

21. The 25th Annual Report on Shoplifting in Supermarkets (*Commercial Service Systems, Inc., 1987*) used 391 supermarkets in Southern California to compile the following statistics on supermarket shoplifters caught in the act:

  ▬ Most items stolen were valued between $1 and $5
  ▬ Most shoplifters were male (56%)
  ▬ Most frequent age category of shoplifters was "under 30" (51%)

Answer the following questions assuming that the purpose of the study was to present statistical data on the national trend and impact of shoplifting in supermarkets.

a. Cite two descriptive statistics reported.

b. What warning would you issue if the results of the study were to be used to make a statistical inference about the national trend and impact of shoplifting in supermarkets?

22. Select a recent copy of the newspaper *USA Today*.

a. Note four examples of statistical information.

b. For each example, indicate the descriptive statistics used and discuss any statistical inferences made.

23. A 7-year medical research study (*Journal of the American Medical Association*, December 1984) reported that women whose mothers took the drug DES during pregnancy were *twice* as likely to develop tissue abnormalities that might lead to cancer as women whose mothers did not take the drug.

a. This study involved the comparison of two populations. What were the populations involved?

b. Do you suppose the data obtained here were the result of a survey or an experiment?

c. For the population of women whose mothers took the drug DES during pregnancy, a sample of 3980 women showed 63 developed tissue abnormalities that might lead to cancer. Provide a descriptive statistic estimating the number of women out of 1000 in this population who have tissue abnormalities.

d. For the population of women whose mothers did not take the drug DES during pregnancy, what is the estimate of the number of women out of 1000 who would be expected to have tissue abnormalities?

e. Medical studies of diseases and disease occurrence often use a relatively large sample (in this case, 3980). Why is this done?

24. A firm is interested in testing the advertising effectiveness of a new television commercial. As part of the test, the commercial is shown on a 6:30 P.M. local news program in Denver, Colorado. Two days later a market research firm conducts a telephone survey to obtain information on recall rates (percentage of viewers who recall seeing the commercial) and impressions of the commercial.

a. What is the population for this study?
b. What is the sample for this study?
c. Would you prefer a sample or a census of the entire population? Explain.

25. The Nielsen organization conducts weekly surveys of television viewing throughout the United States. The Nielsen statistical ratings indicate the size of the viewing audience for each major network television program. Rankings of the television programs and rankings of the network viewing audience market shares are published each week.

a. What is the Nielsen organization attempting to measure?
b. What is the population?
c. What would be a sample for this situation?
d. What kinds of decisions or actions are taken based on the Nielsen studies?

26. A sample of midterm grades for five students showed the following results: 72, 65, 82, 90, 76. Which of the following statements are correct and which should be challenged as being too generalized?

a. The average midterm grade for the sample of five students is 77.
b. The average midterm grade for all students who took the exam is 77.
c. An estimate of the average midterm grade for all students who took the exam is 77.
d. More than half of the students who take this exam will score between 70 and 90.
e. If five other students are included in the sample, their grades will be between 65 and 90.

APPLICATION

Kings Island, Inc.*

Kings Island, Ohio

Kings Island Family Entertainment Center is a 1600-acre year-round recreational, sports, and shopping complex located in southwestern Ohio. The Kings Island theme park, one of America's top parks, provides rides, entertainment, and other attractions, drawing nearly 3 million people annually. In addition to the theme park, the Kings Island Entertainment Center includes the National Football Foundation's College Football Hall of Fame; The Jack Nicklaus Sports Center, site of both professional golf and professional tennis tournaments; The Resort Inn by Kings Island; the Kings Island Campground; and the Factory Outlet Mall. In short, Kings Island Entertainment Center is unmatched in the Midwest for entertainment, sports, shopping and relaxation.

The Kings Island theme park contains six theme areas. They are as follows:

International Street—A colorful European boulevard of shops and restaurants with five European buildings—Italian, French, Spanish, Swiss, and German—encircling the Royal Fountain. A 330-foot tower, a one-third size replica of the Eiffel Tower in Paris, is the Kings Island landmark, located at the end of International Street.

Wild Animal Habitat—More than 350 wild animals from four continents roam freely on a 100-acre preserve. Air-conditioned monorail trains silently transport 2000 guests per hour on the 2-mile journey.

Oktoberfest—The unforgettable sights, sounds, and pageantry of Germany's famed Oktoberfest tradition are celebrated all season long in this area of the park. The centerpiece of Oktoberfest is a $2.5 million Festhaus featuring a hearty international menu and an original 30-minute musical production entitled World Cabaret.

Coney Island—A depiction of the famous amusement park that was located on the Ohio River for almost 100 years. It features the rides and attractions that made the 1920 amusement parks so popular. Roller coasters, Flying Carpets, Dodgems, the Scrambler, a double Ferris wheel, and numerous games and arcades are some of the attractions in this area of the park.

Rivertown—This scenic area depicts what life was like in the Ohio riverboat days, from thrilling action as seen from the Kings Island and Miami Valley Railroad to boat rides that shoot the rapids. The awesome Beast, however, is the main attraction. The biggest, "baddest" wooden roller coaster anywhere, it covers 7400 feet of track at speeds of 65 miles per hour; its two longest vertical drops are 135 and 141 feet.

Hanna-Barbera Land—This is a storybook kingdom that is brought to life for children and their families. The popular Enchanted Voyage boat ride and the Beastie roller coaster are in this area.

*The authors are indebted to Bill Mefford, Manager of Marketing Communications, and Tom Russell, Market Representative in Charge of Planning and Analysis, for providing this application.

Kings Island is committed to offering the best possible quality family entertainment and recreational activities for its visitors. In the future the park will continue to expand, offering more new rides and attractions.

■■■ KINGS ISLAND RESEARCH GROUP

Located on the Kings Island grounds but away from the crowds are the administrative offices of Kings Island. Here executives, managers, and staff plan the strategies and make the business decisions that continue to make Kings Island a successful business venture.

A particularly important part of the Kings Island organization is the research group, which operates with a staff of 11 individuals. This group is devoted to learning about the Kings Island consumer's behavior, attitudes, perceptions, and preferences and providing marketing inferences to Kings Island management. Such information guides operating policies and future plans for the park.

The major work done by the research group is statistical in nature. A population of park visitors can be easily identified; but the time, cost and inconvenience of using a census to collect data about the characteristics and preferences of this population is unacceptable. Yet to be effective and successful, Kings Island management must be knowledgeable about the needs and interests of its customers. Thus on a routine basis samples of park visitors are selected. The results from the samples provide the desired estimates and information about the population of park visitors.

The research group designs questionnaires, selects samples, conducts interviews, and analyzes data that provide the desired consumer attitudes and characteristics information. In total, the research group with its numerous special studies and samples will interview in the neighborhood of 30,000 visitors annually. One such sample, known as the consumer profile study, is described below.

■■■ THE CONSUMER PROFILE SAMPLE SURVEY

What types of individuals come to Kings Island, what their ages are, what their family size is, how far they drive, whether or not they have visited before, and so on comprise important information for Kings Island management. Such information helps determine advertising messages, identifies trends or changes in park attendees over time, and tells Kings Island how it is drawing from market areas such as Cincinnati, Dayton, Columbus,

Children enjoy a visit to the Yogi Bear Fountain in the Kings Island Storybook Kingdom of Hanna-Barbera Land

Lexington, Louisville, and Indianapolis as well as 12 other market areas throughout the Midwest.

Samples of park visitors are taken throughout the day. As the visitors enter the park, interviewers ask sampled individuals to answer a short questionnaire about themselves and why they came to the park today. Interest, cooperation, and participation by the park visitors is very good, and the number of visitor responses is high.

A variety of questions are asked in the 2- or 3-minute consumer-profile interview. Examples of data collected include home zip code, distance traveled to the park, type of admission ticket (group sales, special promotion, regular ticket, season pass, etc.), group size, number of previous visits to park, respondent's age, and so on. The interviewer also notes the day and time of the interview as well as the sex of the respondent.

Methods of descriptive statistics are used to summarize the sample results for Kings Island managers and the marketing department. Each piece of statistical data collected is of interest to someone in the organization. In particular, the home zip code provides an indicator of how each market area is doing in terms of sending visitors to the park. In short, the consumer profile sample is the way Kings Island learns about its visitors. A wide variety of plans, strategies, and decisions are based on the critical information provided by the sample results.

■■■ OTHER SAMPLES

In addition to the consumer profile sample, the research group also takes numerous special purpose samples to determine visitor attitudes toward a variety of park features, including food service, rides, shows, games, and so on. These attitudinal samples are usually taken as park visitors are leaving at the end of the day. Statistical summaries of these samples guide the park's Operations Department in identifying improvements and/or modifications that Kings Island management may want to consider.

To the visitor, Kings Island is a vast area of fun and excitement far removed from the busy world of business. However, behind the scenes an effective and efficient business organization operates Kings Island, much like its counterparts elsewhere in the business world. Consumer research through the use of samples and statistical methods plays a critical role in providing the data and information that keeps Kings Island one of the country's top theme parks and family entertainment centers.

CHAPTER 2

Descriptive Statistics I.
Tabular and Graphical Approaches

As indicated in Chapter 1, data may be classified as being either *qualitative* or *quantitative*. Qualitative data provides labels or names for categories of like items. The categories for qualitative data may be identified by nonnumeric descriptions or by numeric codes. Qualitative data are provided by either a nominal or ordinal scale of measurement. Quantitative data indicate 'how much' or 'how many'. These type of data are always numeric and are provided by either an interval or ratio scale of measurement.

The purpose of this chapter is to introduce several tabular and graphical procedures commonly used to summarize both qualitative and quantitative data. Tabular and graphical summaries of data can be found in annual reports, newspaper articles, research studies, and so on. Everyone is exposed to these types of presentations. Hence, it is important to understand how they are prepared and to know how they should be interpreted. We begin with tabular and graphical methods for summarizing qualitative data. Methods for summarizing quantitative data are presented in Section 2.2.

2.1 ▓ SUMMARIZING QUALITATIVE DATA

Frequency Distribution

We begin the discussion of how tabular and graphical methods can be used to summarize qualitative data with the definition of a *frequency distribution*.

Frequency Distribution

A *frequency distribution* is a tabular summary of a set of data showing the frequency (or number) of items in each of several nonoverlapping classes.

TABLE 2.1   **Data from a Sample of 50 New Car Purchases by Women**

Honda Accord	Ford Escort	Ford Taurus
Ford Taurus	Chevrolet Cavalier	Honda Accord
Honda Accord	Ford Escort	Ford Taurus
Honda Accord	Hyundai Excel	Hyundai Excel
Ford Escort	Hyundai Excel	Chevrolet Cavalier
Ford Taurus	Ford Escort	Ford Escort
Honda Accord	Chevrolet Cavalier	Chevrolet Cavalier
Ford Escort	Ford Escort	Chevrolet Cavalier
Honda Accord	Honda Accord	Hyundai Excel
Ford Taurus	Chevrolet Cavalier	Ford Escort
Honda Accord	Hyundai Excel	Hyundai Excel
Honda Accord	Ford Escort	Ford Escort
Ford Escort	Honda Accord	Hyundai Excel
Chevrolet Cavalier	Chevrolet Cavalier	Ford Taurus
Hyundai Excel	Ford Escort	Ford Escort
Chevrolet Cavalier	Ford Escort	Honda Accord
Ford Taurus	Ford Taurus	

The objective in developing a frequency distribution is to provide insights about the data that cannot be quickly obtained if we look only at the original data. To see how frequency distributions can be applied when dealing with qualitative data, let us consider the following illustration.

What models of automobiles are most often purchased by women? *USA Today* (December 13, 1988) reported that based on 1988 automobile sales data, Chevrolet Cavalier, Ford Escort, Ford Taurus, Honda Accord, and Hyundai Excel were the top choices among women making new-car purchase decisions. The *USA Today* article summarized the purchase choices of several thousand women. Assume that the data shown in Table 2.1 were collected from a sample of 50 women who made a recent purchase of one of the top five selling automobiles.

In order to develop a frequency distribution for the qualitative data in Table 2.1, we simply count the number of data items associated with each of the five purchase categories. This can be accomplished by preparing a tally sheet like the one shown in Table 2.2. A tally is recorded every time the category description appears in the data. Simply counting the tallies leads to the frequency distribution shown in Table 2.3.

TABLE 2.2   **Tally Sheet Used to Count the Number of Data Items in Each Purchase Category**

Automobile Purchased	Tally
Chevrolet Cavalier	⊬Ⱶ Ⅰ Ⅰ Ⅰ Ⅰ
Ford Escort	⊬Ⱶ ⊬Ⱶ Ⅰ Ⅰ Ⅰ Ⅰ
Ford Taurus	⊬Ⱶ Ⅰ Ⅰ Ⅰ
Honda Accord	⊬Ⱶ ⊬Ⱶ Ⅰ
Hyundai Excel	⊬Ⱶ Ⅰ Ⅰ Ⅰ

TABLE 2.3 **Frequency Distribution of New Car Purchases Based on a Sample of 50 Women**

Automobile Purchased	Frequency
Chevrolet Cavalier	9
Ford Escort	14
Ford Taurus	8
Honda Accord	11
Hyundai Excel	8
Total	50

The advantage of the frequency distribution is that it provides a better understanding of the preferences of the women in the sample than does the original data shown in Table 2.1. Using Table 2.3, we can see at a glance that 9 women purchased Chevrolet Cavalier, 14 purchased the Ford Escort, and so on. The Ford Escort and the Honda Accord were the first and second choices of women based upon the sample data. Ford models did very well, with Escort and Taurus accounting for $14 + 8 = 22$ of the 50 new car purchases. The information about the new-car purchases by women contained in Table 2.1 was much easier to grasp after the data had been systematically summarized in the frequency distribution of Table 2.3.

Relative Frequency Distribution

A frequency distribution shows the number (frequency) of data items in each of several nonoverlapping classes. However, often we are interested in knowing the fraction, or proportion, of the data items that fall within each class. The *relative frequency* of a class is simply the fraction, or proportion, of the total number of data items belonging to the class. For a data set with a total of n observations, or items, the relative frequency of each class is given by

Relative Frequency

$$\text{Relative Frequency of a Class} = \frac{\text{Frequency of the Class}}{n} \qquad (2.1)$$

A *relative frequency distribution* is a tabular summary of a set of data showing the relative frequency in each class. Using (2.1) we can compute a *relative frequency distribution* for the data presented in Table 2.1. For example, with the sample size $n = 50$, the relative frequency for the Chevrolet Cavalier is $9/50 = .18$. Computing the relative frequency for each automobile purchase category provides the relative frequency distribution shown in Table 2.4. The relative frequency of .28 for the Ford Escort shows this preferred automobile was

TABLE 2.4   **Relative Frequency Distribution of New-Car Purchases Based on a Sample of 50 Women**

Automobile Purchased	Relative Frequency
Chevrolet Cavalier	.18
Ford Escort	.28
Ford Taurus	.16
Honda Accord	.22
Hyundai Excel	.16
Total	1.00

selected by 28% of the women in the sample. Similarly, .22, or 22%, of the women selected the Honda Accord, .16, or 16%, selected the Hyundai Excel, and so on.

Bar Graphs and Pie Charts

A *bar graph* is a graphical device for depicting qualitative data that have been summarized in a frequency distribution or a relative frequency distribution. On the horizontal axis of the graph, we specify the labels that are used for each of

FIGURE 2.1   **Bar Graph of New Car Purchases Based on a Sample of 50 Women**

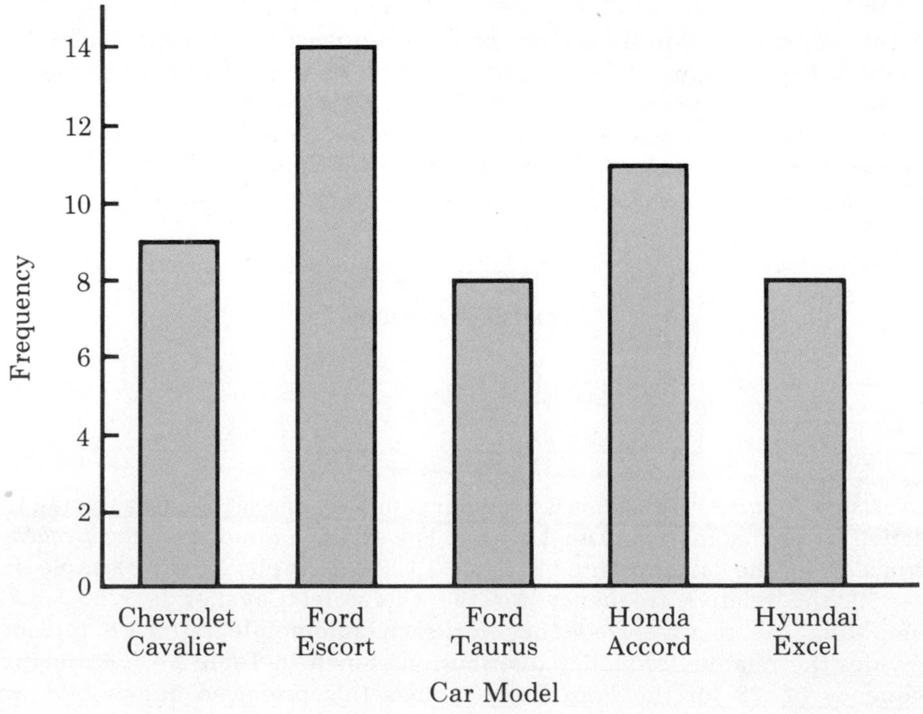

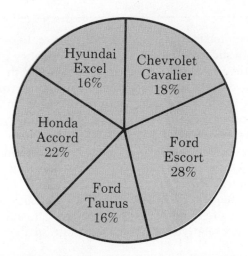

▬▬ FIGURE 2.2 **Pie Chart of New Car Purchases Based on a Sample of 50
Women**

the classes. Either a frequency scale or a relative frequency scale can be used for
the vertical axis of the graph. Then, using a bar of fixed width drawn above each
class label, we extend the height of the bar until we reach the frequency or
relative frequency of the class as indicated by the vertical axis. The bars are
separated to emphasize the fact that each class is a separate category. A bar graph
of the frequency distribution for the 50 new-car purchases by women is shown
in Figure 2.1. Note how the graphical presentation shows the Ford Escort and
the Honda Accord to be the two most preferred models.

The pie chart is a commonly used graphical device for presenting relative
frequency distributions for qualitative data. To draw a pie chart, first draw a
circle; then, use the relative frequencies to subdivide the circle into sectors, or
parts, that correspond to the relative frequency for each class. For example, since
there are 360 degrees in a circle and since the Ford Escort had a relative
frequency of .28, the sector of the pie chart labeled Ford Escort should consist of
.28 × 360 = 100.8 degrees. Similar calculations for the other classes yield the pie
chart for the sample of 50 new-car purchases by women shown in Figure 2.2. The
numerical values shown in each sector may be frequencies, relative frequencies,
or—as shown in Figure 2.2—percentages.

▬▬▬ NOTES AND COMMENTS ▬▬▬

1. Often the number of classes in a frequency distribution for qualitative
   data will be the same as the number of categories found in the data. This
   was the case for the automobile purchase data presented in this section.
   Since the data considered only five automobiles, or categories, a separate frequency distribution class was defined for each automobile. If the
   data set had included purchases of all models of automobiles, there

continued on next page

would have been too many models, or categories, to develop a separate class for each. In such a situation, the lower frequency categories may be grouped together to form an aggregate class. With the automobile purchase data, models such as Honda Civic, Chevrolet Corsica, and Toyota Camry could have been summarized in an aggregate class identified as "other models." As a general guideline, most statisticians recommend that from 5 to 20 classes be used in a frequency distribution. Whenever there are too many data categories to present a separate class for each, good judgment must be exercised in deciding which categories to group together.

2. The sum of the frequencies in any frequency distribution always equals the total number of elements in the data set. The sum of the relative frequencies in any relative frequency distribution always equals 1.00.

■ EXERCISES

1. What is the favorite beverage of Americans? *Psychology Today* (October, 1988) provided data on the consumption of beverage products including milk (M), fruit juice (F), soft drinks (S), beer (B), and bottled water (W). The following data show the results of a sample of 30 individuals who were asked to select their most frequently consumed beverage.

M	F	S	S	B	S	B	M	W	S
S	S	F	B	B	S	B	S	W	S
M	F	S	S	B	B	S	F	S	B

a. Comment on why these are qualitative data. Is the scale of measurement nominal or ordinal?

b. Provide a frequency distribution and a relative frequency distribution summary of the data.

c. Provide a bar graph and a pie chart summary of the data.

d. Based on this sample, what is America's favorite beverage?

2. Freshmen entering the college of business at Eastern University were asked to indicate their preferred major. The following data were obtained:

Major	Management	Accounting	Finance	Marketing
Number	55	51	28	82

Summarize the data by constructing

a. a relative frequency distribution
b. a bar graph
c. a pie chart

3. Employees at Electronics Associates are on a flextime system; under this system, the employees can begin their working day at 7:00, 7:30, 8:00, 8:30, or 9:00 A.M. The following data represent a sample of the starting times selected by the employees.

7:00	8:30	9:00	8:00	7:30	7:30	8:30	8:30	7:30	7:00
8:30	8:30	8:00	8:00	7:30	8:30	7:00	9:00	8:30	8:00

Summarize the data by constructing

   a. a frequency distribution
   b. a relative frequency distribution
   c. a bar graph
   d. a pie chart
   e. What do the summaries tell you about employee preferences concerning the flextime system?

4. What are the favorite movies of the year? The biggest box-office successes for 1988 were listed in *U.S. News and World Report*, December 26, 1988. Assume that sample data collected on movie preferences were summarized with the following letter codes:

        A   Coming to America
        B   Big
        D   Crocodile Dundee II
        R   Who Framed Roger Rabbit
        V   Good Morning, Vietnam
        O   Other motion picture preferred

The following sample data are available:

```
V   R   O   R   B   A   D   V   V   R   R   D   A   B   V
R   R   A   D   V   A   B   R   R   V   R   A   V   B   V
D   R   A   V   B   A   O   R   R   B   R   A   D   R   R
B   A   R   O   A   A   V   A   D   A   D   R   B   R   B
```

   a. Prepare a frequency distribution and a relative frequency distribution for the data set.
   b. Prepare a bar graph for the data set.
   c. Rank order the top five motion pictures for 1988. What motion picture appears to have been the most successful?

5. A national restaurant chain provides a card on each table that contains questions regarding the customers' opinions about the meal, the service, and so on. One question asks the customer to rate the quality of the service as poor, below average, average, above average, or outstanding. The following data represent the results obtained for one restaurant in the chain.

        Poor: 7
        Below average: 14
        Average: 33
        Above average: 67
        Outstanding: 19

Summarize the data by constructing
   a. a relative frequency distribution
   b. a bar graph
   c. a pie chart

6. Students in the College of Business Administration at the University of Cincinnati are asked to complete a course-evaluation questionnaire upon completion of their courses. There are a variety of questions that use a five-category response scale. One of the questions is as follows:

        Compared to other courses that you have taken, what is the overall quality of the course you are now completing?

     — —   — —   — —   — —   — —
     Poor   Fair   Good   Very Good   Excellent

A sample of 60 students completing a course in business statistics during the spring quarter of 1989 provided the following responses. To aid in computer processing of the questionnaire results, a numeric scale was used with 1 = Poor, 2 = Fair, 3 = Good, 4 = Very Good, and 5 = Excellent.

```
3   4   4   5   1   5   3   4   5   2   4   5
3   4   4   4   5   5   4   1   4   5   4   2
5   4   2   4   4   4   5   5   3   4   5   5
2   4   3   4   5   4   3   5   4   4   3   5
4   5   4   3   5   3   4   4   3   5   3   3
```

a. Comment on why these are qualitative data. Is the scale of measurement nominal or ordinal?

b. Provide a frequency distribution and a relative frequency distribution summary of the data.

c. Provide a bar graph and a pie chart summary of the data.

d. Based upon your summaries, comment on the overall evaluation of the course.

2.2 ▬ SUMMARIZING QUANTITATIVE DATA

Frequency Distribution

As defined in Section 2.1, a frequency distribution is a tabular summary of a set of data showing the frequency (or number) of items in each of several nonoverlapping classes. This definition holds for quantitative as well as for qualitative data. However, with quantitative data we have to be more careful in defining the nonoverlapping classes to be used in the frequency distribution.

For example, consider the quantitative data presented in Table 2.5. These data provide the time in days required to complete year-end audits for a sample of 20 clients of Sanderson and Clifford, a small public accounting firm. The three steps necessary to define the classes for a frequency distribution with quantitative data are as follows:

1. Determine the number of nonoverlapping classes.

2. Determine the width of each class.

3. Determine the class limits for each class.

Let us demonstrate these steps by developing a frequency distribution for the audit time data shown in Table 2.5.

TABLE 2.5 Year-End Audit Times (in Days)

12	14	19	18
15	15	18	17
20	27	22	23
22	21	33	28
14	18	16	13

Number of Classes. Classes are formed by specifying ranges of data values that will be used to group the elements in the data set. As a general guideline, we again recommend using between 5 and 20 classes. Data sets with a larger number of elements usually require a larger number of classes. Data sets with a smaller number of elements can often be summarized quite nicely with as few as 5 or 6 classes. The goal is to use enough classes to show the variation in the data, but not so many classes that there are only a few elements in many of the classes. Since the data set in Table 2.5 is relatively small ($n = 20$), we chose to develop a frequency distribution with 5 classes.

Width of the Classes. The second step in constructing a frequency distribution for quantitative data is to choose a width for the classes. As a general guideline, it is recommended that the width be the same for each class. Thus the choices of the number of classes and the width of the classes are not independent decisions. A larger number of classes means a smaller class width, and vice versa. In order to determine an approximate class width, we begin by identifying the largest data value and the smallest data value in the data set. Then, once the desired number of classes has been specified, the following expression can be used to determine the approximate class width.

$$\text{Approximate Class Width} = \frac{\text{Largest Data Value} - \text{Smallest Data Value}}{\text{Number of Classes}} \quad (2.2)$$

The class width given by (2.2) can be adjusted to a convenient width based on the preference of the person developing the frequency distribution. For example, a computed class width of 9.28 might be adjusted to a class width of 10 simply because 10 is a more convenient class width to use in constructing a frequency distribution.

For the data set involving the year-end audit times, the largest data value is 33 and the smallest data value is 12. Since we have decided to summarize the data set with 5 classes, using (2.2) provides an approximate class width of $(33 - 12)/5 = 4.2$. As a result, we decided to use a class width of 5 in the frequency distribution.

In practice, the number of classes and the appropriate class width are determined by trial and error. Once a possible number of classes is chosen, (2.2) is used to find the approximate class width. The process may be repeated for a different number of classes. Ultimately, the judgment of the analyst is used to determine the combination of the number of classes and class width that provides the best means for summarizing of the data.

Returning to the audit-time data in Table 2.5, we have decided to use 5 classes of width 5 days each to summarize the data. The next task is to specify the class limits for each of the 5 classes.

Class Limits. The *lower class limit* identifies the smallest possible data value assigned to the class. The *upper class limit* identifies the largest possible data value assigned to the class. Again, the judgment of the analyst is used, and a variety of acceptable class limits are possible.

For the data in Table 2.5, we defined the class limits for the 5 classes as follows: 10–14, 15–19, 20–24, 25–29, and 30–34. The smallest data value, 12, is included in the 10–14 class. The largest data value, 33, is included in the 30–34 class. Using the 10–14 class as an example, 10 is the lower class limit and 14 is the upper class limit. The difference between the lower class limits of adjacent classes provides the class width. Using the first two lower class limits of 10 and 15, we see that the class width is $15 - 10 = 5$.

The form of each lower class limit and each upper class limit depends upon the number of places to the right of the decimal point contained in the data. Since the audit times in Table 2.5 are integer, integer class limits of 10–14, 15–19, and so on are acceptable limits. If the audit times were recorded in tenths of days, such as 12.3, 14.4, 19.3, and so on, the class limits would also be stated in tenths. In this case, class limits of 10.0–14.9, 15.0–19.9, 20.0–24.9, and so on would have been appropriate. If the data were in hundredths, which is often the case with dollar and cents data, the class limits would also be stated in hundredths. In this case, class limits of 10.00–14.99, 15.00–19.99, 20.00–24.99, and so on would be appropriate. However the class limits are chosen, they should be defined in such a fashion that *each data value belongs to one and only one class.* For instance, class limits of 10–15, 15–20, and 20–25 are unacceptable, since data values of 15 and 20 belong to two different classes.

Once the number of classes, the class width, and the class limits have been determined, a frequency distribution for the data set can be obtained simply by counting the number of data items belonging to each of the classes. Table 2.6 shows a tally sheet for the audit time data from Table 2.5. Table 2.7 shows the frequency distribution for the same data. Using Table 2.7 we can easily see the following:

1. The most frequently occurring audit times are in the 15–19-day class. Eight of the 20 audit times belong to this class.
2. Only one audit required 30 or more days.

Other relevant observations are possible, depending upon the interests of the person viewing the frequency distribution. In this example, we see that the value of a frequency distribution is that it provides insights about the data that were not easily obtained by viewing the data in their original unorganized form.

TABLE 2.6　**Tally Sheet for the Audit-Time Data**

Tally Sheet

Audit Time	Tally
10–14	l l l l
15–19	ᴎᴛᴌ l l l
20–24	ᴎᴛᴌ
25–29	l l
30–34	l

TABLE 2.7	Frequency Distribution for the Audit-Time Data

Audit Time	Frequency
10–14	4
15–19	8
20–24	5
25–29	2
30–34	1
Total	20

Relative Frequency Distribution

We define the relative frequency distribution for quantitative data in the same manner that we did for qualitative data. First, recall that the relative frequency is simply the fraction or proportion of the total number of items belonging to a class. For a data set having n items,

$$\text{Relative Frequency of a Class} = \frac{\text{Frequency of the Class}}{n}$$

Based on the class frequencies shown in Table 2.7 and with $n = 20$, Table 2.8 shows the relative frequency distribution for the audit-time data. Note that .40, or 40%, of the audits required from 15 to 19 days. Only .05, or 5%, of the audits required 30 or more days. Again, additional interpretations and insights are possible using Table 2.8.

TABLE 2.8	Relative Frequency Distribution For The Audit-Time Data

Audit Time	Relative Frequency
10–14	.20
15–19	.40
20–24	.25
25–29	.10
30–34	.05
Total	1.00

Histogram

The most common graphical presentation of quantitative data is a *histogram.* This graphical summary may be prepared for data that have been previously summarized in either a frequency distribution or a relative frequency distribution. A histogram is constructed by placing the variable of interest on the horizontal axis and the frequency or relative frequency values on the vertical axis. The frequency or relative frequency of each class is shown by drawing a rectangle whose base is the class interval on the horizontal axis and whose height is the corresponding frequency or relative frequency.

A histogram for the audit-time data is shown in Figure 2.3. Note that the class with the greatest frequency is shown by the rectangle appearing above the class 15–19 days. The height of the rectangle shows that the frequency of this class is 8. A histogram for the relative frequency distribution of this data would look the same as the histogram in Figure 2.3 with the exception that the vertical axis would be labeled with relative frequency values.

As Figure 2.3 shows, the adjacent rectangles of a histogram touch one another. Unlike with a bar graph, there is no separation between the rectangles of adjacent classes. This is the usual convention for histograms. Since the class limits for the audit-time data were started as 10–14, 15–19, 20–24, 25–29, and 30–34, there appear to be one-unit intervals of 14 to 15, 19 to 20, 24 to 25 and 29 to 30 between the classes. These spaces are eliminated by drawing the vertical lines of the histogram halfway between the class limits. For example, the vertical lines for the 15–19 class are drawn above the values 14.5 and 19.5. Using this procedure for all classes, the vertical lines for the histogram in Figure 2.3 are drawn above the values 9.5, 14.5, 19.5, 24.5, 29.5, and 34.5. This minor adjustment to eliminate the spaces in a histogram helps show that, even though the data are rounded, all values between the lower limit of the first class and the upper limit of the last class are possible.

FIGURE 2.3 **Histogram for the Audit-Time Data**

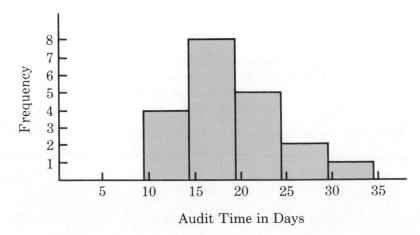

TABLE 2.9   **Frequency Distribution and Cumulative Frequency Distribution for the Audit-Time Data**

Audit Time (days)	Frequency	Cumulative Frequency
10–14	4	4
15–19	8	12
20–24	5	17
25–29	2	19
30–34	1	20

Cumulative Frequency and Cumulative Relative Frequency Distributions

A variation of the frequency distribution that provides another tabular summary of quantitative data is the *cumulative frequency distribution.* The cumulative frequency distribution uses the number of classes, class widths, and class limits that were developed for the frequency distribution. However, the numbers shown in the *cumulative* frequency distribution indicate the number of items *less than or equal to the upper class limit* of each class. In Table 2.9 we show the frequency distribution and the cumulative frequency distribution for the audit-time data.

To understand how the cumulative frequency distribution is constructed, consider the class interval 20–24. In order to determine the number of items with values less than or equal to the upper class limit, 24, we simply sum the frequencies for the classes with upper limits less than or equal to 24. Thus, using the frequency distribution in Table 2.9, the sum of the frequencies for classes 10–14, 15–19, and 20–24 indicate that there are $4 + 8 + 5 = 17$ items with values less than or equal to 24. Hence the cumulative frequency associated with class 20–24 shows 17 audits were complete in 24 days or less. Other observations for the cumulative frequency distribution show that 12 audits were completed in 19 days or less; 19 audits were completed in 29 days or less.

As a final point, we note that a *cumulative relative frequency distribution* shows the fraction, or proportion, of items with values less than or equal to the upper

TABLE 2.10   **Cumulative Relative Frequency Distribution for the Audit-Time Data**

Audit Time	Relative Frequency	Cumulative Relative Frequency
10–14	.20	.20
15–19	.40	.60
20–24	.25	.85
25–29	.10	.95
30–34	.05	1.00

limit for each class. The cumulative relative frequency distribution can be computed either by summing the relative frequencies in a relative frequency distribution or by dividing the cumulative frequencies (see Table 2.9) by the total number of items ($n = 20$). Either approach will provide the cumulative relative frequency distribution. The relative frequency distribution and the cumulative frequency distribution for the audit-time data are shown in Table 2.10. Note that the cumulative frequencies show that .85, or 85%, of the audits were completed in 24 days or less, .95, or 95%, of the audits were complete in 29 days or less, and so on.

NOTES AND COMMENTS

1. Given the class limits in a frequency distribution, we may occasionally want to know the *midpoints* of the classes. Each class midpoint is simply halfway between the lower and upper class limits. For example, with the class limits of 10–14, 15–19, 20–24, 25–29, and 30–34 in the audit-time example, the five class midpoints would be 12, 17, 22, 27 and 32, respectively.

2. An *open-end* class is one that has only a lower class limit or an upper class limit. For example, in the audit-time data of Table 2.5, suppose two of the audits had taken 58 and 65 days. Rather than continue with the classes of width 5 with class intervals 35–39, 40–44, 45–49, and so on, it would simplify the frequency distribution to show an open-end class of "35 or more." This class would have a frequency of 2. Most often the open-end class will appear at the upper end of the distribution. Sometimes an open-end class may appear at the lower end of the distribution, and occasionally it appears at both ends.

3. The last entry in a cumulative frequency distribution is always the total number of elements in the data set. The last entry in a cumulative relative frequency distribution is always 1.00.

■ EXERCISES

7. Earnings-per-share data for a sample of 20 companies from the *Fortune* 500 are as follows (*Fortune*, April 24, 1989):

Philip Morris	10.03	National Service	1.75
Monsanto	8.27	TOSCO	1.32
Amerada Hess	1.51	Hillenbrand	1.86
Armstrong World	3.51	Harvard Industries	0.80
Sequa	6.04	Chemed	2.60
Digital Equipment	9.90	AM International	−0.53
American Home Products	6.38	Fruit of the Loom	1.10
United Brands	1.47	Harley Davidson	3.00
Chicago Tribune	2.78	Carlisle	2.16
Himont	5.71	Texas Industries	1.09

Summarize the data by constructing

  a. a frequency distribution
  b. a relative frequency distribution
  c. a histogram
  d. Using these summaries, comment on what you learned about earnings per share from the sample data.

8. An industrial psychologist asked 25 managers to take a written examination designed to measure the manager's work satisfaction. The following data were obtained (higher scores indicate a higher level of satisfaction).

75	68	33	69	26
62	77	54	61	96
87	57	61	56	79
78	67	78	68	75
28	89	61	51	41

Summarize the data by constructing

  a. a frequency distribution
  b. a relative frequency distribution
  c. a histogram
  d. a cumulative frequency distribution

9. A doctor's office has studied the waiting times for patients who arrive at the office with a request for emergency service. The following data were collected over a 1-month period (the waiting times are in minutes):

$$2, 5, 10, 12, 4, 4, 5, 17, 11, 8, 9, 8, 12, 21, 6, 8, 7, 13, 18, 3.$$

Start with 0 and use a class width of 5.

  a. Show the frequency distribution.
  b. Show the relative frequency distribution.
  c. Show the cumulative frequency distribution.
  d. Show the cumulative relative frequency distribution.
  e. What proportion of patients needing emergency service have a waiting time of 9 minutes or less?

10. The personal computer has brought computer convenience and power into the home environment. But just how many hours a week are people actually using their personal computers at home? A study designed to determine the usage of personal computers at home (U.S. News and World Report, December 26, 1988) provided data in hours per week:

0.5	1.2	4.8	10.3	7.0	13.1	16.0	12.7	11.6	5.1
2.2	8.2	0.7	9.0	7.8	2.2	1.8	12.8	12.5	14.1
15.5	13.6	12.2	12.5	12.8	13.5	1.3	5.5	5.0	10.8
2.5	3.9	6.5	4.2	8.8	2.8	2.5	14.4	16.0	12.4
2.8	9.5	1.5	10.5	2.2	7.5	10.5	14.1	14.9	0.3

Summarize the data using

  a. a frequency distribution (use a class width of 3 hours)
  b. a relative frequency distribution
  c. a histogram
  d. Comment on what the data indicate about the usage of personal computers at home.

11. National Airlines accepts flight reservations by phone. Shown below are the call durations (in minutes) for a sample of 20 phone reservations. Construct the frequency and relative frequency distributions for the data. Also provide a histogram.

2.1	4.8	5.5	10.4
3.3	3.5	4.8	5.8
5.3	5.5	2.8	3.6
5.9	6.6	7.8	10.5
7.5	6.0	4.5	4.8

12. How much do executives of some of the largest corporations get paid? *Business Week* (May 1, 1989) reported executive compensation for 1988, including salary and bonus. The data reported in thousands of dollars for 25 chairpeople and chief executive officers are as follows:

Boeing	846	Delta Airlines	457
Whirlpool	563	Chrysler	1466
Bank of Boston	1200	Coca-Cola	2164
Sherwin-Williams	746	DuPont	1611
Bristol-Myers	824	Motorola	824
General Mills	1310	Marriott	1007
Sara Lee	1367	Honeywell	575
Eastman Kodak	1252	Exxon	1354
Apple Computer	2479	Scott Paper	1238
Bausch & Lomb	927	CBS	1253
K Mart	925	AT&T	1284
Goodyear	1279	Philip Morris	1660
Teledyne	860		

Summarize the data by constructing

a. a frequency distribution
b. a relative frequency distribution
c. a histogram
d. a cumulative frequency distribution
e. Using these summaries, comment on what you learned about executive salaries from the sample data.

13. The given data are the number of units produced by a production employee for the most recent 20 days:

160	170	181	156	176
148	198	179	162	150
162	156	179	178	151
157	154	179	148	156

Summarize the data by constructing

a. a frequency distribution
b. a relative frequency distribution
c. a cumulative frequency distribution
d. a cumulative relative frequency distribution

14. *Car and Driver* magazine (January 1989) selected the Honda Civic as one of the 10 best cars of the year. Data provided about the Civic included fuel-economy information stated in miles per gallon. Assume that the following miles-per-gallon data were obtained from a sample of actual mileage tests with the Civic.

30.2	29.0	27.5	28.3	29.2	32.1	33.8	25.2	34.3	30.6
30.5	28.3	26.0	28.5	29.4	30.3	30.8	29.2	25.9	26.4
27.7	33.9	30.4	29.4	29.4	30.2	28.8	27.5	30.8	30.0

Summarize the data by constructing

   a. a frequency distribution
   b. a relative frequency distribution
   c. a histogram
   d. *Car and Driver* magazine reported the fuel economy of the Honda Accord as being in the 22–27-miles-per-gallon range. Which of the two, the Accord or the Civic, appears to provide more miles per gallon?

2.3 ▬ EXPLORATORY DATA ANALYSIS

The techniques of *exploratory data analysis* focus on how simple arithmetic and easy-to-draw pictures can be used to summarize data quickly. In this section we study how one of these techniques—referred to as a *stem-and-leaf display*—can be used to rank order data and provide an idea of the shape of the underlying distribution of a set of quantative data.

One simple method of displaying data involves arranging the data in ascending or descending order. This process, referred to as rank ordering the data, provides some degree of organization. However, such an approach provides little insight concerning the shape of the distribution of data values. A stem-and-leaf display is a device that provides a display of both rank order and shape simultaneously.

To illustrate the use of a stem-and-leaf display in data analysis consider the data set presented in Table 2.11. These data show the results of an aptitude test (consisting of 150 questions) given to 50 individuals that were recently inter-

TABLE 2.11 Number of Questions Answered Correctly on an Aptitude Test

112	72	69	97	107
73	92	76	86	73
126	128	118	127	124
82	104	132	134	83
92	108	96	100	92
115	76	91	102	81
95	141	81	80	106
84	119	113	98	75
68	98	115	106	95
100	85	94	106	119

viewed for a position at Haskens Manufacturing. The data represent the number of questions answered correctly.

To develop a stem-and-leaf display for the data in Table 2.11, we proceed as follows. The first digits of each data item are arranged to the left of a vertical line. To the right of the vertical line we record the last digit for each item as we pass through the scores in the order they were recorded. The last digit for each item is placed on the horizontal line corresponding to its first digit:

```
 6 | 9  8
 7 | 2  3  6  3  6  5
 8 | 6  2  3  1  1  0  4  5
 9 | 7  2  2  6  2  1  5  8  8  5  4
10 | 7  4  8  0  2  6  6  0  6
11 | 2  8  5  9  3  5  9
12 | 6  8  7  4
13 | 2  4
14 | 1
```

Given this organization of the data, it is a simple matter to rank order the second digits on each horizontal line. Doing so leads to the following stem-and-leaf display of the data:

```
 6 | 8  9
 7 | 2  3  3  5  6  6
 8 | 0  1  1  2  3  4  5  6
 9 | 1  2  2  2  4  5  5  6  7  8  8
10 | 0  0  2  4  6  6  6  7  8
11 | 2  3  5  5  8  9  9
12 | 4  6  7  8
13 | 2  4
14 | 1
```

Each line in this display is referred to as a *stem*, and each piece of information on a stem is a *leaf*. For example, consider the first line

```
6 | 8   9
```

The meaning attached to this line is that there are two items in the data set whose first digit is six: 68 and 69. Similarly, the second line,

```
7 | 2  3  3  5  6  6
```

specifies that there are six items whose first digit is seven: 72, 73, 73, 75, 76, and 76. Thus we see that the data values in this stem-and-leaf display are separated into two parts. The label for each stem is the one- or two-digit first part of the number (that is, 6, 7, 8, 9, 10, 11, 12, 13, or 14)—to which we refer as the starting point—and the leaf is the single-digit second part (that is, 0, 1, 2, . . . , 8, 9). The vertical line simply serves to separate the two parts of each number listed.

To focus on the shape indicated by the stem-and-leaf display, let us use a rectangle to depict the "length" of each stem. Doing so we obtain the following:

```
 6 │  8  9
 7 │  2  3  3  5  6  6
 8 │  0  1  1  2  3  4  5  6
 9 │  1  2  2  2  4  5  5  6  7  8  8
10 │  0  0  2  4  6  6  6  7  8
11 │  2  3  5  5  8  9  9
12 │  4  6  7  8
13 │  2  4
14 │  1
```

Rotating this page counterclockwise onto its side provides a picture of the data very similar to that provided by a histogram of the data with classes of 60–69, 70–79, 80–89, and so on.

Although the stem-and-leaf display may appear at first glance to offer little more information about the data set than that provided by a histogram, there are two primary advantages:

1. The stem-and-leaf display is easier to construct; in fact, it is easier to construct such a display than to describe how to do it.
2. Within an interval, the stem-and-leaf display provides more information than the histogram, since the stem-and-leaf shows the actual data values.

Just as there is no right number of classes in a frequency distribution or histogram, there is no right number of rows or stems in a stem-and-leaf display. If we believe that our original stem-and-leaf display has condensed the data too much, it is a simple matter to stretch the display by using two or more stems for each starting point. For example, to use two lines for each starting point, we place all data values ending in 0, 1, 2, 3, or 4 on one line and all values ending in 5, 6, 7, 8 and 9 on a second line. To illustrate this approach, consider the following stretched stem-and-leaf display:

```
 6 |
 6 | 8  9
 7 | 2  3  3
 7 | 5  6  6
 8 | 0  1  1  2  3  4
 8 | 5  6
 9 | 1  2  2  2  4
 9 | 5  5  6  7  8  8
10 | 0  0  2  4
10 | 6  6  6  7  8
11 | 2  3
11 | 5  5  8  9  9
12 | 4
12 | 6  7  8
13 | 2  4
13 |
14 | 1
14 |
```

Note that data 72, 73, and 73 have leaves in the 0–4 range and are shown with the first stem value of 7. The data 75, 76, and 76 have leaves in the 5–9 range and are shown with the second stem value of 7. This stretched stem-and-leaf display is similar to a frequency distribution with intervals of 60–64, 65–69, 70–74, 75–79, and so on. A portion of the stem-and-leaf output provided by the Minitab computer software package is shown in Figure 2.4.

Although our illustration of a stem-and-leaf display shows data with up to three digits, stem-and-leaf displays for data with more digits are possible. For

---

■■■ FIGURE 2.4 **A Portion of the Stem-and-Leaf Output Provided by the Minitab Computer Software Package**

```
 6  89
 7  223
 7  566
 8  011234
 8  56
 9  12224
 9  556788
10  0024
10  66678
11  23
11  55899
12  4
12  678
13  24
13
14  1
```

example, consider the following data, which show the number of hamburgers sold by a fast-food restaurant for each of 15 weeks.

1852	1644	1766	1888	1912
2044	1812	1790	1679	2008
1565	1852	1967	1954	1733

Following is a stem-and-leaf display of these data:

15	6
16	47
17	369
18	1558
19	156
20	04

Note that only the first three digits of each number are used in the display. The wording "Multiple Stem and Leaf by 10" can be included at the bottom of the display to indicate the original data had four digits.

■■■ EXERCISES

15. A psychologist developed a new test of adult intelligence. The test was administered to 20 individuals, and the following data were obtained.

114	99	131	124	117
102	106	127	119	115
98	104	144	151	132
106	125	122	118	118

Construct a stem-and-leaf display for these data.

16. Develop a stem-and-leaf display for the earnings per share data from a sample of *Fortune 500* companies (*Fortune*, April 24, 1989). See data in Exercise 7.

17. In a study of job satisfaction, a series of tests were administered to 50 subjects. The following data were obtained, where high scores represent considerable dissatisfaction.

87	76	67	58	92	59	41	50	90	75
80	81	70	73	69	61	88	46	85	97
50	47	81	87	75	60	65	92	77	71
70	74	53	43	61	89	84	83	70	46
84	76	78	64	69	76	78	67	74	64

Construct a stem-and-leaf display for these data.

18. The net profit margins for oil companies were reported in the *Forbes* 41st Annual Report on American Industry (*Forbes*, January 9, 1989). Data on net profit margin are as follows:

Exxon	6.8	Phillips Petroleum	4.8
AMOCO	9.8	Quaker State	1.6
Du Pont	6.6	Ashland	2.2
Chevron	7.1	Union Pacific	8.9
Mobil	4.1	Kerr McGee	3.4
Occidental	1.8	Crown Central	6.5
Getty	2.1	Pacific Resources	1.9
Union Texas	9.2	American Petrofina	4.6
Atlantic Richfield	8.9	Coastal Corp	1.5

a. Develop a stem-and-leaf display for these data.

b. Use the results of the stem-and-leaf display to develop a frequency distribution and relative frequency distribution for the data.

19. Develop a stem-and-leaf display for the miles-per-gallon data provided in Exercise 14.

■■■ SUMMARY

A set of data, even if modest in size, is often difficult to interpret directly in the form in which it is gathered. Tabular and graphical procedures provide means of organizing and summarizing the data so that patterns are revealed and the data are more easily interpreted. Frequency distributions, relative frequency distributions, bar graphs, and pie charts were presented as tabular and graphical procedures for summarizing qualitative data. Frequency distributions, relative frequency distributions, histograms, cumulative frequency distributions, and cumulative relative frequency distributions were presented as ways of summarizing quantitative data. The chapter concluded with a brief introduction to exploratory data analysis. A stem-and-leaf display was presented as an exploratory data analysis technique that can be used to summarize quantitative data. Figure 2.5 provides a summary of the tabular and graphical methods presented in this chapter.

■■■ GLOSSARY

Qualitative data   Data that provide labels or names for categories of like items. These data are provided by either a nominal or ordinal scale of measurement.

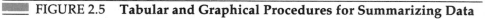

■■■ FIGURE 2.5   Tabular and Graphical Procedures for Summarizing Data

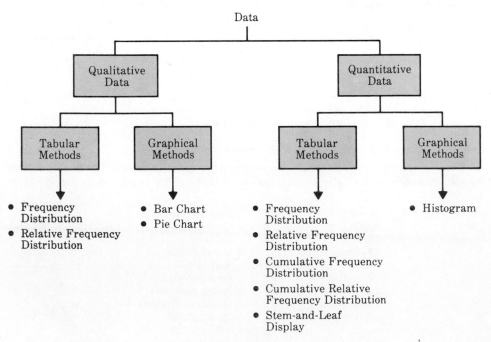

Quantitative data Data that indicate how much or how many. These data are provided by either an interval or ratio scale of measurement.

Frequency distribution A tabular summary of a set of data showing the frequency (or number) of items in each of several nonoverlapping classes.

Relative frequency distribution A tabular summary of a set of data showing the relative frequency—that is, the fraction or proportion—of the total number of items in each of several nonoverlapping classes.

Bar graph A graphical device for depicting the information presented in a frequency distribution or relative frequency distribution of qualitative data.

Pie chart A pictorial device for presenting qualitative data summaries based upon subdividing a circle into sectors that correspond to the relative frequency for each class.

Histogram A graphical presentation of a frequency distribution or relative frequency distribution of quantitative data constructed by placing the class intervals on the horizontal axis and the frequencies or relative frequencies on the vertical axis.

Cumulative frequency distribution A tabular summary of a set of quantitative data showing the number of items having values less than or equal to the upper class limit of each class.

Cumulative relative frequency distribution A tabular summary of a set of quantitative data showing the fraction or proportion of the items having values less than or equal to the upper class limit of each class.

Class midpoint The point in each class that is halfway between the lower and upper class limits.

Exploratory data analysis The use of simple arithmetic and easy-to-draw pictures to present data more effectively.

Stem-and-leaf display An exploratory data analysis technique that simultaneously rank orders quantitative data and provides insight into the shape of the underlying distribution.

▬ KEY FORMULAS

Relative Frequency

$$\frac{\text{Frequency of the Class}}{n} \tag{2.1}$$

Approximate Class Width

$$\frac{\text{Largest Data Value—Smallest Data Value}}{\text{Number of Classes}} \tag{2.2}$$

▬ SUPPLEMENTARY EXERCISES

20. Each of the *Fortune* 500 companies is classified as belonging to one of several industries (*Fortune*, April 24, 1989). Shown next is a sample of 20 companies with their corresponding industry classification.

Coca-Cola (Beverage)	McDonnell Douglas (Aerospace)
Union Carbide (Chemicals)	Morton Thiokol (Chemicals)
General Electric (Electronics)	Quaker Oats (Food)
Motorola (Electronics)	Pepsico (Beverage)
Beatrice (Food)	Maytag (Electronics)
Kellogg (Food)	Pillsbury (Food)
Dow Chemical (Chemicals)	Lockheed (Aerospace)
Campbell Soup (Food)	RJR Nabisco (Food)
Square D (Electronics)	Westinghouse (Electronics)
Ralston Purina (Food)	TRW (Electronics)

a. Provide a frequency distribution showing the number of companies in each industry.

b. Provide a relative frequency distribution.

c. Provide a bar graph for this data.

21. Frequent airline travelers were asked to indicate the airline they believed offered the best overall service. The four choices were American Air(A), East Coast Air(E), Suncoast(S), and Great Western(W). The following data were obtained.

```
E   A   E   S   W   W   E   S   W   E
W   E   E   A   S   S   W   E   A   W
W   S   E   E   A   E   E   S   W   A
S   E   A   W   A   A   W   E   S   W
```

Summarize the data by constructing

a. a frequency distribution

b. a relative frequency distribution

c. a bar graph

d. a pie chart

e. Which airline appears to offer the best service?

22. Voters participating in a recent election exit poll in Michigan were asked to state their political party affiliation. Coding the data a 1 for Democrat, 2 for Republican, and 3 for Independent, the data collected are as follows:

```
1   2   2   1   3   1   2   2   2   1   2   3   2   3   2   1   1   2   1   2
2   1   1   1   2   1   2   3   1   1   2   1   3   1   1   2   1   2   3   2
```

a. Show a frequency distribution and a relative frequency distribution for the data.

b. Show a bar graph for the data.

c. Comment on what the data suggest about the strengths of the political parties in this voting area.

23. A psychological test was designed to measure a subject's ability to anticipate the next item in a series while viewing or hearing its immediate predecessor. The number of items correctly anticipated was recorded.

```
21   13    8   10   17
15    6   23    3   19
12   17    8   12   14
```

Use appropriate tabular and graphical methods to summarize the data.

24. The following data represent quarterly sales volumes for 40 selected corporations.

```
17,864,000   15,065,000   42,200,000   13,523,000
49,747,000   20,510,000    5,520,000    7,985,000
 3,624,000   11,556,000    1,855,000    9,023,000
 3,804,000    5,933,000   23,900,000    6,145,000
 9,232,000    2,979,000    1,059,000   42,789,000
 5,143,000   33,380,000   20,779,000    6,145,000
 2,141,000   17,768,000   18,017,000   42,800,000
 5,090,000   41,626,000   12,003,000    6,840,000
 3,669,000   37,738,000   40,765,000   21,946,000
13,614,000   39,914,000    7,846,000   25,837,000
```

a. Construct a frequency distribution to summarize these data. Use a class width of $5,000,000.
b. Develop a relative frequency distribution for the data.
c. Construct a cumulative frequency distribution for the data.
d. Construct a cumulative relative frequency distribution for the data.
e. Construct a histogram as a graphical representation of the data.

25. The given data show home mortgage loan amounts (in dollars) handled by a particular loan officer in a savings and loan company. Use a frequency distribution, relative frequency distribution, and histogram to help summarize these data.

20,000	38,500	33,000	27,500	34,000
12,500	25,999	43,200	37,500	36,200
25,200	30,900	23,800	28,400	13,000
31,000	33,500	25,400	33,500	29,200
39,000	38,100	30,500	45,500	30,500
52,000	40,500	51,600	42,500	44,800

26. Given are the closing prices of 40 common stocks. (*Investor's Daily*, April 25, 1989).

$29^5/_8$	34	$43^1/_4$	$8^3/_4$	$37^7/_8$
$8^5/_8$	$7^5/_8$	$30^3/_8$	$35^1/_4$	$19^3/_8$
$9^1/_4$	$16^1/_2$	38	$53^3/_8$	$16^5/_8$
$1^1/_4$	$48^3/_8$	18	$9^3/_8$	$9^1/_4$
10	37	18	8	$28^1/_2$
$24^1/_4$	$21^5/_8$	$18^1/_2$	$33^5/_8$	$31^1/_8$
$32^1/_4$	$29^5/_8$	$79^3/_8$	$11^3/_8$	$38^7/_8$
$11^1/_2$	52	14	9	$33^1/_2$

a. Construct frequency and relative frequency distributions for these data.
b. Construct cumulative frequency and cumulative relative frequency distributions for these data.
c. Construct a histogram for the data.
d. Using your summaries, make some comments and observations about the price of common stock.

27. The grade point averages for 30 students majoring in economics are given.

2.21	3.01	2.68	2.68	2.74
2.60	1.76	2.77	2.46	2.49
2.89	2.19	3.11	2.93	2.38
2.76	2.93	2.55	2.10	2.41
3.53	3.22	2.34	3.30	2.59
2.18	2.87	2.71	2.80	2.63

a. Construct a relative frequency distribution for the data.
b. Construct a cumulative relative frequency distribution for the data.
c. Construct a histogram for the data.

28. The annual cost of room and board at a sample of 50 accredited United States Colleges and Universities is given (*The 1988 Information Please Almanac*).

Univ. of Alabama-Birmingham	2,898	Kansas State University	2,286
Alcorn State University	1,650	Lynchburg College	3,400
Amherst College	3,600	Marshall University	2,932
Appalachian State University	1,650	McNeese State University	1,650
Arizona State University	2,500	University of Miami	4,080
Baylor University	3,336	Minnesota Bible College	3,800
Bennington College	3,140	Morgan State University	3,220
Boise State University	2,415	Mount Holyoke College	3,475
Brigham Young University	2,500	University of Nebraska	2,170
Univ. of Calif.—Los Angeles	2,950	State U. of New York—Buffalo	3,080
Case Western Reserve Univ.	4,000	University of North Carolina	3,055
Catholic University of America	3,500	Ohio Wesleyan University	3,661
Central Michigan University	2,616	Oregon State University	2,445
Clemson University	2,150	Pepperdine University	4,535
University of Colorado	2,962	University of Pittsburgh	2,930
Drake University	3,130	Providence College	3,700
Eastern Kentucky University	2,150	St. Mary of the Woods College	2,805
Emory University	3,612	University of South Carolina	2,500
Fairleigh Dickinson University	3,955	University of South Florida	2,520
University of Florida	3,240	Syracuse University	4,430
Furman University	3,328	University of Texas	3,200
Georgia Institute of Technology	3,390	Washington and Lee University	3,308
Grand Valley State College	2,750	University of Washington	2,590
University of Illinois	3,370	William Penn College	2,180
Indiana State University	2,312	University of Wisconsin	2,900

Develop the following summaries:

a. a frequency distribution
b. a relative frequency distribution
c. a histogram
d. Use your summaries to make some comments about the annual room-and-board costs associated with attending college.

29. Hospital records show the following number of days of hospitalization for 20 patients.

5	7	7	15
21	15	22	10
10	6	8	18
14	5	7	8
3	8	4	10

a. Construct frequency and relative frequency distributions for the data.
b. Construct a cumulative relative frequency distribution for the data.
c. Construct a histogram.

30. The conclusion from a 41-state poll conducted by the Joint Council on Economic Education (*Time*, January 9, 1989) is that students do not learn enough economics. The findings were based on test results from eleventh- and twelfth-grade students who took a 46-question, multiple-choice test on basic economic concepts such as profit and the law of supply and demand. The following sample data represent a portion of the data on the number of questions answered correctly.

12	31	24	22	8	10	14	16	18	25
16	15	21	30	22	24	19	13	16	15
12	17	20	26	33	14	9	12	18	24
18	19	22	16	17	23	28	18	14	19

Summarize the following data using

  a. a stem-and-leaf display
  b. a frequency distribution
  c. relative frequency distribution
  d. a cumulative frequency distribution
  e. Based on these data, do you agree with the claim that students are not learning enough economics? Explain.

31. The final examination scores in a section of statistics resulted in the following data.

56	77	84	82	42
61	44	95	98	84
93	62	96	78	88
58	62	79	85	89
89	97	53	76	75

Provide a stem-and-leaf display for these data.

32. The daily high and low temperatures for 24 cities are given (*USA Today*, April 11, 1989).

City	High	Low	City	High	Low
Tampa	78	59	Birmingham	59	31
Kansas City	54	36	Minneapolis	45	23
Boise	65	34	Portland	71	42
Los Angeles	72	61	Memphis	57	37
Philadelphia	47	27	Buffalo	41	29
Milwaukee	47	32	Cincinnati	49	27
Chicago	49	35	Charlotte	54	30
Albany	42	24	Boston	46	31
Houston	62	48	Tulsa	58	41
Salt Lake City	56	38	Washington, D.C.	51	33
Miami	84	66	Las Vegas	87	62
Cheyenne	52	26	Detroit	45	33

  a. Prepare a stem-and-leaf display for the high temperatures.
  b. Prepare a stem-and-leaf display for the low temperatures.
  c. Compare the stem-and-leaf display from (a) and (b) and make some comments about the differences between daily high and low temperatures.
  d. Use the stem-and-leaf display from (b) to determine the number of cities having a low temperature below freezing (32°F).
  e. Provide frequency distributions for both the high-temperature and low-temperature data.

████ ██

████ COMPUTER EXERCISE

Consolidated Foods, Inc. has opened several new grocery stores at a variety of locations over the past 2 years. One of the special services at these new stores is that customers may pay for their purchases using Visa or MasterCard credit cards as well as using either cash or an approved check. In order to better understand how customers are using the new payment feature, a sample of 100 customers was selected over a 1-week period. Data collected for each customer included how much was spent during the shopping trip and the method of payment. The data collected are shown in the accompanying table.

Amount Spent ($)	Method of Payment	Amount Spent ($)	Method of Payment
84.12	Check	86.34	Check
34.66	Credit card	20.23	Credit card
37.27	Credit card	108.70	Check
38.82	Credit card	45.36	Credit card
46.50	Credit card	83.31	Check
99.67	Check	64.45	Credit card
70.18	Check	54.33	Credit card
99.21	Check	16.78	Cash
138.42	Check	115.96	Check
93.68	Check	95.83	Check
120.89	Check	19.76	Cash
10.14	Cash	35.37	Cash
74.51	Check	111.98	Check
17.91	Check	103.95	Check
49.59	Check	90.40	Credit card
4.74	Cash	6.68	Cash
48.14	Cash	32.09	Credit card
65.67	Credit card	79.70	Credit card
89.66	Check	96.08	Credit card
96.40	Check	20.60	Cash
54.16	Credit card	78.81	Check
79.55	Check	123.62	Check
67.95	Check	125.01	Check
30.69	Cash	41.58	Credit card
151.89	Check	36.73	Credit card
130.41	Check	52.07	Credit card
98.80	Check	19.78	Cash
23.59	Cash	66.44	Check
104.67	Check	5.08	Cash
90.04	Check	50.15	Credit card
77.62	Check	114.42	Check
36.01	Cash	97.26	Credit card
88.17	Check	22.75	Cash
66.76	Credit card	53.63	Credit card
23.50	Cash	132.31	Check
127.34	Check	105.54	Check
26.02	Cash	66.09	Check
79.77	Check	62.24	Check
29.35	Check	97.93	Check
71.31	Credit card	10.57	Cash
43.57	Credit card	51.21	Credit card
76.18	Credit card	90.17	Check
59.38	Credit card	24.08	Credit card
72.99	Credit card	42.72	Cash
19.24	Cash	97.72	Check
80.20	Check	112.67	Check
55.79	Cash	14.30	Cash
134.27	Check	28.76	Credit card
64.68	Credit card	81.85	Check
75.54	Check	56.84	Credit card

■ QUESTIONS

1. Develop tabular and graphical summaries of the method-of-payment data.

2. Develop tabular and graphical summaries of the amount-spent data.

3. Develop tabular summaries of the amount spent for the following three situations:

    a. Method of payment is by check.
    b. Method of payment is by cash.
    c. Method of payment is by credit card.

4. Discuss the results of your summaries. Does there appear to be any relationship between the amount spent and the method of payment?

APPLICATION

The Colgate-Palmolive Company*

New York, New York

The Colgate-Palmolive Company dates back to 1806, or 3 years after the Louisiana Purchase. It started as a small shop in New York City that made and sold candles and soap. The business grew and prospered. Colgate and Company merged with the Palmolive Company in the late 1920s to form the present-day corporation, which now has annual sales in excess of $5 billion.

Colgate products can be found around the globe, with international operations in over 55 countries. While best known for its traditional product line of soaps, detergents, and toothpastes, subsidiary operations include the Kendall Company, Riviana Foods, Etonic, Bike Athletic Company, and others.

Colgate operates three large factories in the United States, with locations in Jersey City, New Jersey; Jeffersonville, Indiana; and Kansas City, Kansas. The Jersey City and Jeffersonville plants are area landmarks, with giant clocks that face the downtown metropolitan areas of nearby New York City and Louisville, Kentucky, respectively.

Most of Colgate's products are in the low-price, high-volume consumer market, which is known for its intense competition. Since the possibility of increasing a product's price is limited by the competition, Colgate's profitability is determined largely by the efficiency of its management techniques and production operations.

The use of statistics touches most areas of the business. Market research, forecasting, and quality control, among other areas, frequently use statistical procedures. In the following example we look at a quality assurance application at the point of manufacture.

▬ QUALITY ASSURANCE FOR HEAVY-DUTY DETERGENTS

The slogan on Colgate's crest reads "Quality Products since 1806," and every department is touched by the demands of this central business principle. The Quality Assurance and Improvement Department within Colgate devotes full time toward achieving this goal and includes a wide variety of special services. At each manufacturing plant, a group of chemists, inspectors, engineers, and managers are involved with product quality levels. The data that are collected must be communicated to others throughout the organization. The format of the data summary is often vital to achieving the desired results.

A variety of statistical techniques are used. Relative frequency distributions and graphical techniques such as histograms are some of the most useful tools for communicating data and ideas. As an example of the use of these statistical techniques, consider

*The authors are indebted to Mr. William R. Fowle, Manager of Quality Assurance, Colgate-Palmolive Company for providing this application.

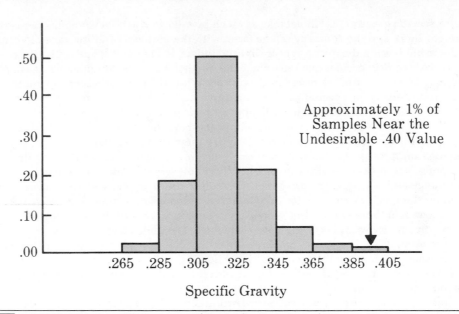

Approximately 1% of
Samples Near the
Undesirable .40 Value

Specific Gravity

▰▰▰ FIGURE 2A.1.   **Histogram of Specific Gravity of Heavy Duty Detergent**

the production of the familiar heavy duty detergent used for home laundries. Television advertisements claiming the relative merits of the competing brands are widely seen throughout the country.

The regular-size carton of the detergent has a stated weight of 20 ounces. In the manufacturing process great pains are taken to ensure that the label specifications of 20 ounces of detergent per carton are maintained. However, meeting the 20-ounce-per-carton weight is not the only aspect of quality assurance addressed in the manufacture of this product. Of particular concern, from the point of view of quality, is the density of the detergent powder that is placed in the carton. Even with rigid quality control standards in the powder production process, at times the powder varies in its weight per unit volume. For example, if the weight of the powder is on the heavy side (a

Statistical summaries of product weights aid a Colgate-Palmolive management team during a meeting on quality assurance

high specific gravity), it will not take as much powder to reach the 20-ounce-per-carton weight limit, and the company can be faced with the problem of filling cartons with 20 ounces but having the carton appear slightly underfilled when it is opened by the user.

To reduce this problem and maintain the quality standards, the powder is sampled periodically prior to being placed in the cartons. When the powder reaches an unacceptably high density or specific gravity, corrective action is taken to reduce the specific gravity of the powder before the filling operation is permitted to resume.

Repeated samples provide more and more data about the specific gravity of powder. At some point, various parties in the company are interested in knowing how the powder production process is doing in terms of meeting density guidelines. Tabular and graphical summaries provide convenient ways to present the data to production, quality assurance and management personnel. Table 2A.1 shows a relative frequency distribution for the specific gravity of 150 samples taken over a 1-week period. Figure 2A.1 shows a histogram of these sample data. Note that the specific gravity of the powder varies, with a specific gravity of .32 occurring most frequently. The undesirably high specific gravity occurs around .40. Thus the summaries show that the operation is meeting its quality guidelines, with practically all the data showing values less than .40. Production management personnel would be pleased with the quality aspect of the powder product as indicated by these statistical summaries.

In cases where the relative frequency distribution and/or histogram summaries do not support the above conclusion, managers and quality assurance personnel begin to closely monitor the powder production process. Engineers may be consulted on ways of reducing the specific gravity to a more satisfactory level. After making any change in the process, data are collected and summarized in similar tabular and graphical forms to determine how the modifications are affecting the quality of the product. The engineers' work continues until the statistical summaries show that the high quality level is once again obtained. The use of tabular and graphical methods of descriptive statistical summaries is essential in communicating data to the engineers, inspectors, and managers whose job it is to assure quality products at Colgate-Palmolive Company.

■■ A SUMMARY NOTE

While there are many other examples of the use of statistics that could be shared, many of the ideas presented above would be the same. Statistical methods such as the tabular and

TABLE 2A.1 Relative Frequency Distribution Showing the Specific Gravity of Heavy-Duty Detergent (Based on 150 Sample Results)

Specific Gravity	Relative Frequency
.27–.28	.02
.29–.30	.18
.31–.32	.50
.33–.34	.21
.35–.36	.06
.37–.38	.02
.39–.40	.01

graphical summaries presented above provide a basis for communicating data that help guide the decision-making process. It deserves mention here that people who are selected for promotion within our company are often good communicators, and those working with data are often good at presenting statistical results to management. It may seem simple when others do it effectively, but it is not always easy to pull together data and present them in a form that is understandable and easily grasped. This is a talent that improves with practice.

CHAPTER 3

Descriptive Statistics II. Measures of Location and Dispersion

I n Chapter 2 we discussed tabular and graphical methods used to summarize data. These procedures are effective in written reports and as visual aids when making presentations to a group of people. In this chapter, we present several numerical measures of location and dispersion that provide additional alternatives for summarizing data.

In Chapter 3, we consider data sets consisting of a single variable. Whenever the data for a single variable, such as age, salary, or the like, has been obtained from a sample of n elements, the data set will contain n items, or data values. The numerical measures of location and dispersion will be computed using all n data values. If there is more than one variable, the numerical measures presented here must be computed separately for each variable.

Several numerical measures of location and dispersion are introduced. The most important measures will be the mean, the variance, and the standard deviation. If the measures are computed for data from a sample, they are called *sample statistics*. If the measures are computed for data from a population, they are called *population parameters*.

3.1 ▦ MEASURES OF LOCATION

Mean

Perhaps the most important numerical measure of location is the *mean*, or average value, for a variable. The mean provides a good measure of central location. It is obtained by adding all the data values and dividing by the number of items. If the data are from a sample, the mean is denoted by $\bar{x}$; if the data are from a population the mean is denoted by the Greek letter μ.

In specifying statistical formulas, it is customary to denote the value of the first data item by x_1, the value of the second data item by x_2, and so on. (In general, the ith data value is denoted by x_i.) Using this notation, the formula for the sample mean is as follows:

Sample Mean

$$\bar{x} = \frac{\Sigma x_i}{n} \qquad (3.1)$$

---

where n = number of items in the sample. In this formula, the numerator is the sum of all n data values. That is,

$$\Sigma x_i = x_1 + x_2 + \cdots + x_n$$

The Greek letter Σ is the summation sign.

In order to illustrate the computation of a sample mean, let us consider the following class-size data for a sample of 5 college classes:

$$46 \quad 54 \quad 42 \quad 46 \quad 32$$

Using the notation x_1, x_2, x_3, x_4, x_5 to represent the number of students in each of the five classes, we have:

$$x_1 = 46$$
$$x_2 = 54$$
$$x_3 = 42$$
$$x_4 = 46$$
$$x_5 = 32$$

Thus to compute the sample mean, we can write

$$\bar{x} = \frac{\Sigma x_i}{n} = \frac{x_1 + x_2 + x_3 + x_4 + x_5}{5} = \frac{46 + 54 + 42 + 46 + 32}{5} = 44$$

Thus for the five classes sampled, the mean class size is 44 students.

Another illustration of the computation of a sample mean is given in the following situation. Suppose that a college placement office sent a questionnaire to a sample of business school graduates requesting information on starting salaries. Table 3.1 shows the data that have been collected. The mean monthly starting salary for the sample of 12 business college graduates is computed as shown:

$$\bar{x} = \frac{\Sigma x_i}{n} = \frac{x_1 + x_2 + \cdots + x_{12}}{12}$$

$$= \frac{1650 + 1750 + \cdots + 1680}{12}$$

$$= \frac{20{,}880}{12} = 1740$$

TABLE 3.1 **Monthly Starting Salaries for a Sample of 12 Business School Graduates**

Graduate	Monthly Salary	Graduate	Monthly Salary
1	1650	7	1690
2	1750	8	1930
3	1850	9	1740
4	1680	10	2125
5	1555	11	1720
6	1510	12	1680

Equation (3.1) shows how the mean is computed for a sample with n items. The formula for computing the mean of a population is the same, but we use different notation to indicate that we are dealing with the entire population. The number of elements in the population is denoted by N, and, as we mentioned previously, the symbol for the population mean is μ.

Population Mean

$$\mu = \frac{\Sigma x_i}{N}$$

(3.2)

Occasionally, a variable will have one or more unusually small and/or unusually large data values that significantly influence the value of the mean. With this influence, the mean may provide a poor description of the central location of the data. In order to remove the effect of the unusually small and/or large data values, we can eliminate, or trim, a percentage of small and large data values from the data set. The mean of the remaining data is called the *trimmed mean.* The intent is for the trimmed mean to be a better indicator of the central location of the data. For example, a 5% trimmed mean removes the smallest 5% of the data values *and* the largest 5% of the data values. The 5% trimmed mean is then computed as the mean of the middle 90% of the data. In general, an α percent trimmed mean is obtained by trimming α percent of the items from each end of the data and computing the mean for the remaining items.

Median

The *median* is another measure of central location for a variable. The median is the value falling in the middle when the data items are arranged in ascending order (rank ordered from smallest to largest). If there is an odd number of items, the median is the middle item. If there is an even number of data items, there is no single middle value. In this case, we follow the convention of defining the median to be the average of the middle two values. For convenience this definition is restated as follows.

> **Median**
>
> If there is an odd number of items, the median is the value of the middle item when all items are arranged in ascending order.
>
> If there is an even number of items , the median is the average value of the two middle items when all items are arranged in ascending order.

Let us apply this definition to compute the median class size for the sample of five college classes. Arranging the five data values in ascending order provides the following rank-ordered list.

$$32 \quad 42 \quad 46 \quad 46 \quad 54$$

Since $n = 5$ is odd, the median is the middle item in the rank-ordered list. Thus the median is 46. Even though there are two values of 46, each value is treated as a separate item when we arrange the data in ascending order and determine the median.

Suppose we also compute the median starting salary for the business college graduates shown in Table 3.1. Arranging the 12 items in ascending order provides the following:

1510   1555   1650   1680   1680   1690   1720   1740   1750   1850   1930   2125

Middle Two Values

Since $n = 12$ is even, we have identified the middle two items. The median is the average of these two values:

$$\text{Median} = \frac{1690 + 1720}{2} = 1705$$

Although the mean is the more commonly used measure of central location, there are some situations in which the median is preferred. As we stated previously, the mean is influenced by extremely small or large values. For instance, suppose that one of the graduates had earned a starting salary of $10,000 per month (maybe the individual's family owns the company). If we change the highest monthly starting salary in Table 3.1 from $2125 to $10,000 and recompute the mean, the sample mean changes from $\bar{x} = 1740$ to $\bar{x} = 2396$. The median, however, is unchanged, since 1690 and 1720 are still the middle two items. With the extremely high starting salary included, the median provides a better measure of central location than the mean. We can generalize to say that whenever there are extreme values, the median is often the preferred measure of central location.

Mode

A third measure of location is the *mode*. The mode is defined as follows.

Mode

The mode is the data value that occurs with greatest frequency.

To illustrate the computation of the mode, consider the sample of five class sizes. The only value that occurs more than once is 46. Since this value, occurring with a frequency of 2, has the greatest frequency, it is the mode. As another illustration, consider the sample of starting salaries for the business school graduates. The only monthly starting salary that occurs more than once is 1680. Since this value has the greatest frequency, it is the mode.

Situations can arise for which the greatest frequency occurs at two or more different values. In these instances more than one mode existed. If the data have exactly two modes, we say that the data are *bimodal*. If data have more than two modes, we say that the data are *multimodal*. In multimodal cases the mode is almost never reported, since listing three or more modes would not do a very good job of describing a location for the data. In the extreme case, every data value can be different, and an argument can be made that every value is a mode. We take the position that in this situation the mode is not an appropriate measure of location, and thus no value would be reported.

The mode is an important measure of location for qualitative data. For example, the qualitative data set in Table 2.1 resulted in the following frequency distribution for automobile purchases by women.

Automobile Purchase	Frequency
Chevrolet Cavalier	9
Ford Escort	14
Ford Taurus	8
Honda Accord	11
Hyundai Excel	8
Total	50

The mode, or most frequently purchased automobile, is the Ford Escort. For this type of data it obviously makes no sense to speak of the mean or median. But the mode does provide a good indicator of what we are interested in, the most frequently purchased automobile.

Percentiles

A *percentile* is a measure that locates values in the data that are not necessarily central locations. A percentile provides information regarding how the data items are spread over the interval from the smallest value to the largest value. With data that do not have numerous repeated values, the pth percentile divides the data into two parts. Approximately p percent of the items have values less than the pth percentile; approximately $(100 - p)$ percent of the items have values greater than the pth percentile. The pth percentile is formally defined as follows:

Percentile

The pth percentile is a value such that *at least p* percent of the items take on this value or less and *at least* $(100 - p)$ percent of the items take on this value or more.

Admission test scores for colleges and universities are frequently reported in terms of percentiles. For instance, suppose an applicant obtains a raw score of 54 on the verbal portion of an admissions test. It may not be readily apparent how this student performed relative to other students taking the same test. However, if the raw score of 54 corresponds to the 70th percentile, then we know that approximately 70% of the students had scores less than this individual and approximately 30% of the students had scores greater than this individual.

The following procedure can be used to compute the pth percentile.

Calculating the pth Percentile

Step 1. Arrange the data values in ascending order (rank order from smallest value to largest value).

Step 2. Compute an index i as follows:

$$i = \left(\frac{p}{100}\right)n$$

where p is the percentile of interest and n is the number of items.

Step 3. (a) If i *is not an integer, round up.* The next integer value *greater* than i denotes the position of the pth percentile.

(b) If i *is an integer,* the pth percentile is the average of the data values in positions i and $i + 1$.

As an illustration of this procedure, let us determine the 85th percentile for the starting salary data shown in Table 3.1.

Step 1. Arrange the 12 data values in ascending order:

1510 1555 1650 1680 1680 1690 1720 1740 1750 1850 1930 2125

Step 2.

$$i = \left(\frac{p}{100}\right)n = \left(\frac{85}{100}\right)12 = 10.2$$

Step 3. Since i is not an integer, *round up.* The position of the 85th percentile is the next integer greater than 10.2, the 11th position.

Returning to the data, we see that the 85th percentile corresponds to the 11th data value, or 1930.

As another illustration of the above procedure, let us consider the calculation of the 50th percentile. Applying step 2, we obtain

$$i = \left(\frac{50}{100}\right) 12 = 6$$

Since i is an integer, step 3b states that the 50th percentile is the average of the 6th and 7th data values; thus the 50th percentile is $(1690 + 1720)/2 = \$1705$. Note that the *50th percentile is also the median.*

Quartiles and Hinges

It is often desired to divide data into four parts, with each part containing approximately one-fourth, or 25%, of the items. Figure 3.1 shows a data set divided into four parts. The division points are referred to as the *quartiles* and are defined as follows:

Q_1 = first quartile, or 25th percentile

Q_2 = second quartile, or 50th percentile (also the median)

Q_3 = third quartile, or 75th percentile

The monthly starting salary data are again arranged in ascending order, as follows. Q_2, the median, has already been identified as 1705.

1510   1555   1650   1680   1680   1690   1720   1740   1750   1850   1930   2125

The compuations of Q_1 and Q_3 require the use of the rule for finding the 25th and 75th percentiles. These calculations are as follows:

For Q_1,

$$i = \left(\frac{p}{100}\right) n = \left(\frac{25}{100}\right) 12 = 3$$

Since i is an integer, step 3(b) indicates that the first quartile, or 25th percentile,

---

▬▬▬ FIGURE 3.1   **Location of the Quartiles**

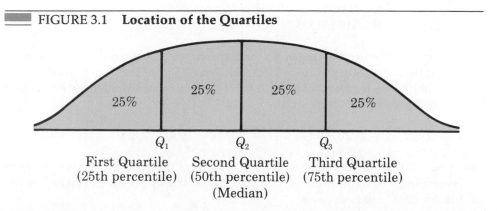

First Quartile   Second Quartile   Third Quartile
(25th percentile)   (50th percentile)   (75th percentile)
(Median)

is the average of the 3rd and 4th data values; thus, $Q_1 = (1650 + 1680)/2 = 1665$.

For Q_3,

$$i = \left(\frac{p}{100}\right) n = \left(\frac{75}{100}\right) 12 = 9$$

Again, since i is an integer, step 3(b) indicates that the third quartile, or 75th percentile is the average of the 9th and 10th data values; thus, $Q_3 = (1750 + 1850)/2 = 1800$.

As shown next, the quartiles have divided the 12 data values into four parts, with each part consisting of 25% of the items.

1510 1555 1650 | 1680 1680 1690 | 1720 1740 1750 | 1850 1930 2125

$Q_1 = 1665$ $Q_2 = 1705$ $Q_3 = 1800$

(Median)

We have defined the quartiles as the 25th, 50th, and 75th percentiles. Thus, we have computed the quartiles in the same fashion as for other percentiles. However, there are some variations in the conventions used to compute quartiles; the actual value computed may vary slightly depending on the convention used. Nevertheless, the objective of all procedures for computing quartiles is to divide data into roughly four equal parts.

Another approach used to divide data into four equal parts has been recently developed by proponents of exploratory data analysis. A *lower hinge* (lower 25%) and an *upper hinge* (upper 25%) are computed. To find these hinges, the items are first arranged in ascending order. Then the data are divided into two equal parts: data in positions less than or equal to the median position and data in positions greater than or equal to the median position. The median for the data in positions *less than or equal to the median position* is the lower hinge. The median for the data in positions *greater than or equal to the median position* is the upper hinge.

Referring to the sample of 12 monthly starting salaries for business school graduates, the median has been found to be 1705. From the listing of the data in ascending order, the position of the median is 6.5, or halfway way between the 6th and 7th data values. The data in positions less than or equal to 6.5 are the data in positions 1 through 6:

1510 1555 1650 1680 1680 1690

Following the rule for determining a median, we see that the median of these six values is $(1650 + 1680)/2 = 1665$. Thus, the *lower hinge* of the data is 1665.

Again with the median position at 6.5, the data in positions greater than or equal to 6.5 are the data in positions 7 to 12:

1720 1740 1750 1850 1930 2125

The median of these six values is $(1750 + 1850)/2 = 1800$. Thus, the *upper hinge* of the data is 1800. The lower hinge, the median and the upper hinge can be used to divide the data into four parts.

For the salary data, the lower hinge is equal to the first quartile and the upper hinge is equal to the third quartile. However, this should not be expected to hold for all data. In some cases, the hinges and the quartiles may take on slightly different values because they are based on slightly different computational procedures.

NOTES AND COMMENTS

In computing the hinges for data with an odd number of items, the median position is included in the computation of the lower hinge *and* in the computation of the upper hinge. For example, with 9 elements, the median position is 5. The median of the data in positions 1 to 5, the data value in position 3, is the lower hinge, and the median of the data in positions 5 to 9, the data value in position 7, is the upper hinge. The median position 5 is used in both computations.

■ EXERCISES

1. Manufacturers of Japanese automobiles established export quotas for automobiles to be shipped to the United States. While the quotas pleased the Detroit automobile manufacturers, the quotas meant Japanese cars would be in short supply and more expensive for the U.S. consumer. *The Wall Street Journal* (January 11, 1989) listed the Japanese export quotas for each of the 12 months of 1988. The data shown below are in terms of thousands of automobiles.

145   135   100   220   170   145   190   155   210   200   205   180

    a. What are the mean, median and mode for the automobile quota data?
    b. Compute and interpret the first and third quartiles for this data set.

2. *American Demographics* (December 1988) reported that 25 million Americans get up each morning and go to work in their offices at home. The growing use of personal computers is suggested as one of the reasons more people can operate at-home business-es. The article presented data on the ages of individuals who work at home. Assume the following is a sample of age data for these individuals.

22   58   24   50   29   52   57   31   30   41
44   40   46   29   31   37   32   44   49   29

    a. Compute the mean and mode.
    b. Compute a 5% and a 10% trimmed mean.
    c. The median age of the population of all adults is 40.5 years. Use the median age of the preceding data to comment on whether the at-home workers tend to be younger or older than the population of all adults.
    d. Compute the first and third quartiles.
    e. Compute and interpret the 32nd percentile.

3. A quality control inspector found the following number of defective parts on 16 different days:

11   14   18   14   21   17   13   21   25   19   17   13   28   13   17   18

Compute the mean, median, mode, and 90th percentile for these data.

4. A bowler has the following scores for six games:

182   168   184   190   170   174

Using these data as a sample, compute the following descriptive statistics.

a. mean
b. median
c. mode
d. 75th percentile

5. Monthly sales data for car telephone units for the RC Radio Corporation are: ·

80   115   82   102   94   90   88   91   89   95   105   108

Compute the mean, median, and mode for monthly sales.

6. The *Los Angeles Times* (January 6, 1989) reported the air quality index for various areas of Southern California. Index ratings of 0–50 are considered good, 51–100 are considered moderate, 101–200 unhealthy, 201–275 very unhealthy, and over 275, hazardous. Recent air quality indexes for Pomona were 28, 42, 58, 48, 45, 55, 60, 49, and 50.

a. Compute the mean, median and mode for the data. Could the Pomona air quality index be considered good?
b. Compute the 25th percentile and 75th percentile for the Pomona air quality data.
c. Compute the lower and upper hinges. Compare your result with the answers to part b.

7. The following data show the number of automobiles arriving at a toll booth during 20 intervals, each of 10 minutes' duration. Compute the mean, median, mode, first quartile, and third quartile for the data.

26   26   58   24
22   22   15   33
19   27   21   18
16   20   34   24
27   30   31   33

8. In automobile mileage and gasoline-consumption testing, 13 automobiles were road tested for 300 miles in both city and country driving conditions. The following data were recorded for miles-per-gallon performance:

City: 16.2, 16.7, 15.9, 14.4, 13.2, 15.3, 16.8, 16.0, 16.1, 15.3, 15.2, 15.3, 16.2
Country: 19.4, 20.6, 18.3, 18.6, 19.2, 17.4, 17.2, 18.6, 19.0, 21.1, 19.4, 18.5, 18.7

Use the mean, median, and mode to make a statement about the difference in performance for city and country driving.

9. A sample of 15 college seniors showed the following credit hours taken during the final term of the senior year:

15   21   18   16   18   21   19   15   14   18   17   20   18   15   16

a. What are the mean, median, and mode for credit hours taken? Compute and interpret.
b. Compute the first and third quartiles.
c. Compute the lower and upper hinges. Compare your answers to part b.
d. Compute and interpret the 70th percentile.

10. The Nielsen organization provides data on television viewing in the United States. A study (*USA Today*, December 13, 1988) reported that the mean number of hours of television viewing per week was increasing. Suppose that the following data provide the hours of television viewing per week for a sample of 16 college students.

14   9   12   4   20   26   17   15   18   15   10   6   16   15   8   5

    a. Compute the mean, median, and mode
    b. Compute the 10th and 80th percentiles.
    c. Compute the quartiles and hinges.

3.2 ▦ MEASURES OF DISPERSION

Whenever data are collected, whether for a sample or a population, it is desirable to consider the dispersion, or variability, in the data values. For example, assume that you are a purchasing agent for a large manufacturing firm and that you regularly place orders with two different suppliers. Both suppliers indicate that approximately 10 working days are required to fill your orders. After several months of operation you find that the number of days required to fill orders is indeed averaging around 10 days for both of the suppliers. The histograms summarizing the number of working days required to fill orders from each supplier are shown in Figure 3.2. Although the mean number of days required to fill orders is roughly 10 for both suppliers, do both suppliers possess the same degree of reliability in terms of making delivery on schedule? Note the dispersion, or variability, in the histograms. Which supplier would you prefer?

For most firms, receiving materials and supplies on schedule is an important part of the purchasing agent's responsibility. At times the early 7- or 8-day deliveries shown for J. C. Clark Distributors might indicate that the purchasing agent is doing a good job. However, a few of the slow 13 to 15-day deliveries could be disastrous in terms of keeping a work force busy and production on schedule. This example illustrates a situation where the dispersion, or variability, in the delivery times may be an overriding consideration in selecting a supplier. For most purchasing agents, the lower dispersion shown for Dawson Supply, Inc. would make Dawson the more consistent and preferred supplier.

We turn now to a discussion of some commonly used numerical measures of the dispersion, or variability, in data.

Range

Perhaps the simplest measure of dispersion for data is the *range:*

Range

The range is the difference between the largest and smallest data values.

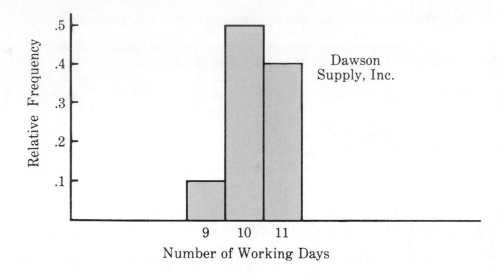

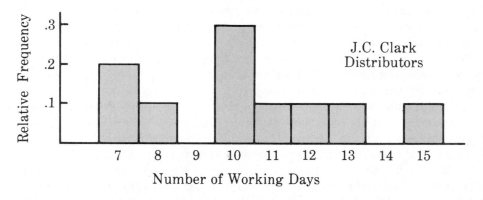

FIGURE 3.2   **Historical Data Showing the Number of Days Required to Fill Orders**

Let us refer to the data on monthly starting salaries for business school graduates in Table 3.1. The largest starting salary is $2125, and the smallest is $1510. The range is 2125 − 1510 = 615.

Although the range is the easiest of the measures of dispersion to compute, it is not widely used. The reason is that the range is based on only two of the items and thus is influenced too much by extreme data values. Suppose, as we did in the previous section, that one of the graduates obtained a starting salary of $10,000. In this case the range would be 10,000 − 1510 = 8490. This large value for the range would not be very descriptive of the variability in the data, since 11 of the 12 starting salaries are closely grouped between 1510 and 1930.

Interquartile Range

A measure of disperson that overcomes the dependency upon extreme data values is the *interquartile range* (IQR). This measure of dispersion is simply the difference between the third quartile, Q_3, and the first quartile, Q_1. That is,

---

Interquartile Range

$$\text{IQR} = Q_3 - Q_1 \qquad (3.3)$$

For the data on starting monthly starting salaries, we found the quartiles were $Q_3 = 1800$ and $Q_1 = 1665$. Thus the interquartile range is $1800 - 1665 = 135$.

Variance

The *variance* is a measure of dispersion that utilizes all the data values. The variance is based on the difference between each data value and the mean. The difference between each data value x_i and the mean ($\bar{x}$ for a sample, μ for a population) is called a *deviation about the mean*. For a sample, a deviation is writtin $(x_i - \bar{x})$; for a population, it is written $(x_i - \mu)$. In the computation of the variance, the deviations about the mean are *squared*.

If the data set involved is a population, the average of the squared deviations is called the *population variance*. The population variance is denoted by the Greek symbol σ^2. Given a population of N items and using μ to denote the population mean, the definition of the population variance is as follows:

Population Variance

$$\sigma^2 = \frac{\Sigma(x_i - \mu)^2}{N} \qquad (3.4)$$

In most statistical applications, the data set being analyzed is from a sample. When we compute a measure of variability for a sample, we are often interested in using the sample statistic obtained as an estimate of the population parameter σ^2. At this point it might seem that the average of the squared deviations about the sample mean would provide a good estimate of the population variance. However, statisticians have found that the average squared deviation for the sample has the undesirable feature of providing a biased estimate of the population variance σ^2; specifically, it tends to underestimate the population variance.

Although it is beyond the scope of this text, it can be shown that if the sum of the squared deviations about the sample mean is divided by $n - 1$, and not n, then the resulting value provides an unbiased estimate of the population variance. For this reason, the *sample variance* denoted by s^2, is defined as follows.

Sample Variance

$$s^2 = \frac{\Sigma(x_i - \bar{x})^2}{n - 1} \qquad (3.5)$$

TABLE 3.2 **Computation of Deviations and Squared Deviations About the Mean for the Class Size Data**

Number of Students in Class (x_i)	Mean Class Size $\bar{x}$	Deviation About the Mean ($x_i - \bar{x}$)	Squared Deviation About the Mean ($x_i - \bar{x}$)^2
46	44	2	4
54	44	10	100
42	44	−2	4
46	44	2	4
32	44	−12	144
		0	256
		$\Sigma(x_i - \bar{x})$	$\Sigma(x_i - \bar{x})^2$

To illustrate the computation of the variance for a sample, we will use the data on class size for the sample of five college classes. A summary of the data, including the computation of the deviations about the mean and the squared deviations about the mean, is shown in Table 3.2. The sum of squared deviations about the mean is $\Sigma(x_i - \bar{x})^2 = 256$. Hence, with $n - 1 = 4$, the sample variance is

$$s^2 = \frac{\Sigma(x_i - \bar{x})^2}{n - 1} = \frac{256}{4} = 64$$

Note that in Table 3.2, we show both the sum of the deviations and the sum of the squared deviations about the mean. For any data, whether a population or a

TABLE 3.3 **Computation of the Sample Variance for the Starting Salary Data**

Monthly Salary (x_i)	Sample Mean ($\bar{x}$)	Deviation About the Mean ($x_i - \bar{x}$)	Squared Deviation About the Mean ($x_i - \bar{x}$)^2
1650	1740	−90	8,100
1750	1740	10	100
1850	1740	110	12,100
1680	1740	−60	3,600
1555	1740	−185	34,225
1510	1740	−230	52,900
1690	1740	−50	2,500
1930	1740	190	36,100
1740	1740	0	0
2125	1740	385	148,225
1720	1740	−20	400
1680	1740	−60	3,600
		0	301,850
		$\Sigma(x_i - \bar{x})$	$\Sigma(x_i - \bar{x})^2$

Using (3.5),

$$s^2 = \frac{\Sigma(x_i - \bar{x})^2}{n - 1} = \frac{301,850}{11} = 27,440.91$$

sample, the sum of the deviations about the mean will *always equal zero*. Thus, as shown in Table 3.2, $\Sigma (x_i - \bar{x}) = 0$. This is true because the positive deviations and negative deviations always cancel each other, causing the sum of the deviations about the mean to equal zero.

As another illustration of computing a sample variance, consider the starting salaries for the 12 business school graduates provided in Table 3.1. The computation of the sample variance is shown in Table 3.3. Recall we previously computed the sample mean to be $\bar{x} = 1740$.

Standard Deviation

The *standard deviation* is defined to be the positive square root of the variance. Following the notation we adopted for a sample variance and a population variance, we use s to denote the sample standard deviation and σ to denote the population standard deviation. The standard deviation is derived from the variance in the following manner:

Standard Deviation

$$\text{Sample Standard Deviation} = s = \sqrt{s^2} \qquad (3.6)$$

$$\text{Population Standard Deviation} = \sigma = \sqrt{\sigma^2} \qquad (3.7)$$

Recall that the sample variance for the sample of class sizes in five college classes is $s^2 = 64$. Thus the sample standard deviation is $s = \sqrt{64} = 8$. For the data set consisting of starting salaries, the sample standard deviation is $s = \sqrt{27,440.91} = 165.65$.

What is gained, then, by converting the variance to its corresponding standard deviation? Note that the units, $(x_i - \bar{x})^2$, being summed in the variance calculation are squared. For example, the sample variance for the monthly starting salaries of the business graduates was $s^2 = 27,440.91$ (units of dollars squared). The fact that the units for variance are squared is the biggest reason why it is difficult to obtain an intuitive appreciation for the numerical value of the variance. Since the standard deviation is simply the square root of the variance, dollars squared in the variance is converted to dollars in the standard deviation. The standard deviation of business graduates starting salaries is 165.65 dollars. In other words, the standard deviation is measured in the same units as the original data. For this reason the standard deviation often is more easily compared to the mean and other statistics which are measured in the units of the original data.

Coefficient of Variation

In some situations we may be interested in a relative measure of the dispersion in the data. For example, a standard deviation of 1 inch would be considered very large for a batch of motor-mount bolts used in automobiles. However, a standard deviation of 1 inch would be considered small for the length of telephone poles. The difficulty is that when the means for data sets differ greatly we do not get an accurate picture of the relative dispersion in the two data sets by comparing the two standard deviations. A measure of dispersion that overcomes these difficulties is the *coefficient of variation*. The formula for the coefficient of variation is as follows:

Coefficient of Variation

$$\frac{\text{Standard Deviation}}{\text{Mean}} \times 100 \qquad (3.8)$$

For sample data the coefficient of variation is $(s/\bar{x}) \times 100$, and for a population it is $(\sigma/\mu) \times 100$. For the sample of business graduates the coefficient of variation is $(165.65/1740) \times 100 = 9.5$. In words, we could say that the standard deviation of the sample is 9.5% of the value of the sample mean.

The coefficient of variation may be helpful in comparing the relative dispersion in several variables that have different means and different standard deviations. However, we caution that this measure should be used only with data involving all, or nearly all, positive values. We can see why by referring to (3.8). A mean equal to or near zero could be obtained when both positive and negative values are present in the data. A mean near zero could cause the coefficient of variation to be very large even with a small standard deviation.

■■ NOTES AND COMMENTS ■■

1. Rounding the values of the sample mean $\bar{x}$ and the values of the squared deviations $(x_i - \bar{x})^2$ may introduce rounding errors in the values of the variance and standard deviation. We recommend carrying at least six significant digits during intermediate calculations. The resulting variance or standard deviation may then be rounded to fewer digits.

2. An alternate formula is available for the computation of the variance. This alternate formula for the sample variance is written as follows:

$$s^2 = \frac{\Sigma x_i^2 - n\bar{x}^2}{n - 1}$$

where $\Sigma x_i^2 = x_1^2 + x_2^2 + \cdots + x_n^2$. Using this formula eases the computational burden slightly and helps reduce rounding errors. Problem 13 requires using this alternate formula to compute the sample variance.

■■ EXERCISES

11. The *Washington Post* (January 7, 1989) reported on overcrowding in the Virginia prison system. Use the following data as the population of capacities of the five Virginia state prisons.

233   164   587   52   175

Compute the range, variance, and standard deviation for this population.

12. Exercise 6 summarized data from the *Los Angeles Times,* which reported the air quality index for various areas of Southern California. A sample of air quality index values for Pomona provided the following data: 28, 42, 58, 48, 45, 55, 60, 49 and 50.

   a. Compute the range and interquartile range.
   b. Compute the sample variance and sample standard deviation.
   c. A sample of air quality index readings for Anaheim provided a sample mean of 48.5, a sample variance of 136, and a sample standard deviation of 11.66. What comparisons can you make between the air quality in Pomona and Anaheim based on these descriptive statistics?

13. The Davis Manufacturing Company has just completed five weeks of operation using a new process that is supposed to increase productivity. The number of parts produced each week is

410   420   390   400   380

Compute the sample variance and sample standard deviation using the definition of sample variance (3.5) as well as the alternate formula provided in the Notes and Comments.

14. Assume that the data used to construct the histograms of the number of days required to fill orders for Dawson Supply, Inc. and J. C. Clark Distributors (see Figure 3.2) are as follows:

   Dawson Supply Days for Delivery: 11, 10, 9, 10, 11, 11, 10, 11, 10, 10
   Clark Distributors Days for Delivery: 8, 10, 13, 7, 10, 11, 10, 7, 15, 12

Use the range and standard deviation to support the earlier observation that Dawson Supply provides the more consistent and reliable delivery times.

15. In Exercise 4 a bowler's scores for six games were as follows:

182   168   184   190   170   174

Using these data as a sample, compute the following descriptive statistics:

   a. range
   b. variance
   c. standard deviation
   d. coefficient of variation

16. Given next are the yearly household incomes (in dollars) for a sample of ten families in Grimes, Iowa.

   10,648   17,418
   16,517   13,555
   14,821   19,226
   152,936   11,800
   18,527   12,222

   a. Compute the range as a measure of variability.
   b. Compute the interquartile range as a measure of variability.
   c. Compute the standard deviation as a measure of variability.
   d. Which of the above measures do you feel is the best measure of variability in the data? Why?

17. A production department uses a sampling procedure to test the quality of newly produced items. The department employs the following decision rule at an inspection station: If a sample of 14 items has a variance of more than .01, the production line must

be shut down for repairs. Suppose that the following data have just been collected:

3.43 3.45 3.43
3.48 3.52 3.50
3.39 3.48 3.41
3.38 3.49 3.45
3.51 3.50

Should the production line be shut down? Why or why not?

18. The following times were recorded by the quarter-mile and mile runners of a university track team (times are in minutes):

Quarter-mile times: .92, .98, 1.04, .90, .99
Mile times: 4.52, 4.35, 4.60, 4.70, 4.50

After viewing this sample of running times, one of the coaches commented that the quarter-milers turned in the more consistent times. Use the standard deviation and the coefficient of variation to summarize the variability in the data. Does the use of the coefficient of variation measure indicate that the coach's statement should be qualified?

3.3 ■■ SOME USES OF THE MEAN AND THE STANDARD DEVIATION

We have described several measures of location and dispersion for data. The mean is the most widely used measure of location, while the standard deviation and variance are the most widely used measures of dispersion. Using only the mean and the standard deviation, we can learn much about a data set.

z-scores

Using the mean and standard deviation, we can determine the relative location of any data value. Suppose we have a sample of n items, with the values denoted by $x_1, x_2, \ldots, x_n$. In addition, assume that the sample mean, $\bar{x}$, and the sample standard deviation, s, have been computed. Associated with each data value, x_i, is another value called its z score. Equation (3.9) shows how the z-score is computed for data value x_i.

---

z-score

$$z_i = \frac{x_i - \bar{x}}{s} \tag{3.9}$$

where z_i = the z-score for item i

$\bar{x}$ = the sample mean

s = the sample standard deviation

The z-score is often called the *standardized value* for the item. The standardized value z_i can be interpreted as the *number of standard deviations x_i is from the mean $\bar{x}$*.

TABLE 3.4 *z*-Scores for the Class Size Data

Number of Students in Course (x_i)	Deviation About the Mean ($x_i - \bar{x}$)	z-Score $\left(\dfrac{x_i - \bar{x}}{s}\right)$
46	2	2/8 = .25
54	10	10/8 = 1.25
42	−2	−2/8 = −.25
46	2	2/8 = .25
32	−12	−12/8 = −1.50

For example, $z_1 = 1.2$ would indicate x_1 is 1.2 standard deviations above, or larger than, the sample mean. Similarly, $z_2 = -0.5$ would indicate x_2 is .5, or 1/2, standard deviation below, or less than, the sample mean. As can be seen from equation (3.9), z-scores greater than zero occur for items with values greater than the sample mean and z-scores less than zero occur for items with values less than the mean. A z-score of zero indicates that the value of the item is equal to the mean.

The z-score for any item can be interpreted as a measure of the relative location of the item in a data set. Indeed, items in two different data sets with the same z-score can be said to have the same relative location in terms of being the same number of standard deviations from the mean.

The z-scores for the class size data are shown in Table 3.4. Recall that the sample mean $\bar{x} = 44$ and sample standard deviation $s = 8$ have been computed previously. The z-score of −1.50 for the fifth item shows it is farthest from the mean, falling 1.50 standard deviations below the mean.

Chebyshev's Theorem

Chebyshev's theorem permits us to make statements about the percentage of items that must be within a specified number of standard deviations from the mean. The statement of Chebyshev's theorem is as follows:

Chebyshev's Theorem

At least $(1 - 1/k^2)$ of the items in any data set must be within k standard deviations of the mean, where k is any value greater than 1.

Some of the implications of this theorem, using $k = 2$, 3, and 4 standard deviations, are as follows:

■ At least .75, or 75%, of the items must be within $k = 2$ standard deviations of the mean.

■ At least .89, or 89%, of the items must be within $k = 3$ standard deviations of the mean.

■ At least .94, or 94%, of the items must be within $k = 4$ standard deviations of the mean.

For an example using Chebyshev's theorem, assume that the midterm test scores for 100 students in a college business statistics course had a mean of 70 and a standard deviation of 5. How many students had test scores between 60 and 80? How many students had test scores between 50 and 90?

For the test scores between 60 and 80, we note that the value of 60 is two standard deviations below the mean and the value of 80 is two standard deviations above the mean. Using Chebyshev's theorem we see that at least 75% of the items must have values within two standard deviations of the mean. Thus, at least 75 of the students must have scored between 60 and 80.

For the range 50 to 90, we see that 50 is four standard deviations below the mean and 90 is four standard deviations above the mean. Thus, at least 94% of the students must have scored between 50 and 90.

The Empirical Rule

One of the advantages of Chebyshev's theorem is that it applies to any data set regardless of the shape of the distribution of the data. In practical applications, however, it has been found that many data sets have a mound-shaped, or bell-shaped, distribution like the one shown in Figure 3.3. When it is believed that the data approximates this distribution, the *empirical rule* can be used to determine the percentage of items that must be within a specified number of standard deviations of the mean.*

Empirical Rule

For data having a bell-shaped distribution,

- Approximately 68% of the items will be within one standard deviation of the mean.
- Approximately 95% of the items will be within two standard deviations of the mean.
- Almost all the items will be within three standard deviations of the mean.

For example, liquid detergent cartons are filled automatically on a production line. Filling weights frequently have a bell-shaped distribution. If the mean filling weight is 16 ounces and the standard deviation is .25 ounces, we can use the empirical rule to conclude the following:

■ Approximately 68% of the filled items will have weights between 15.75 and 16.25 ounces (that is, within one standard deviation of the mean).
■ Approximately 95% of the filled items will have weights between 15.50 and 16.50 ounces (that is, within two standard deviations of the mean).
■ Almost all filled items will have weights between 15.25 and 16.75 ounces (that is, within 3 standard deviations of the mean).

*The empirical rule is based on the normal probability distribution, which is presented in detail in Chapter 6.

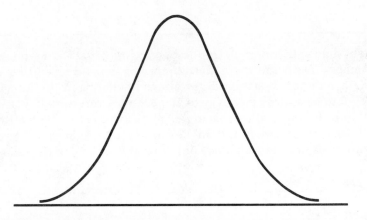

■ FIGURE 3.3   **A Mound-Shaped, or Bell-Shaped, Distribution**

Detecting Outliers

Sometimes a set of data will have one or more items with unusually large or unusually small values. Extreme values such as these are called *outliers*. Experienced statisticians take steps to identify outliers and then review each one carefully. An outlier may be an item for which the value has been incorrectly recorded. If so, the value can be corrected before proceeding with further analysis. An outlier may also be an item that was incorrectly included in the data set; if so it can be removed. Finally, an outlier may just be an unusual item that has been correctly recorded and does belong in the data set. In such cases the item should remain.

Standardized values (z-scores) can be used to help identify outliers. Recall that the empirical rule allows us to conclude that for data with a bell-shaped distribution, almost all the items will be within three standard deviations of the mean. Thus, when using z-scores to identify outliers, we recommend treating any item with a z-score less than −3 or greater than 3 as an outlier. Such items can then be reviewed for accuracy and to determine whether or not they belong in the data set.

Refer to the z-scores for the class size data shown in Table 3.4. The z-score of −1.50 shows the fifth item is farthest from the mean. However, this standardized value is well within the −3 to +3 guideline for outliers. Thus the z-scores indicate outliers with unusually small or large data values are not present in the class size data.

■■■ NOTES AND COMMENTS ■■■

1. Before analyzing a data set, statisticians usually make a variety of checks to ensure the validity of data. In a large study it is not uncommon for errors to be made in recording data values or in inputting the values at a computer terminal. Identifying outliers is one tool used to check the validity of data.

Continued on next page

> **2.** Chebyshev's theorem is applicable for any data set and makes a statement about the minimum number of items that will be within a certain number of standard deviations of the mean. If the data set is known to be approximately bell-shaped, more can be said. For instance, the empirical rule allows us to say that *approximately* 95% of the items will be within two standard deviations of the mean; Chebyshev's theorem allows us to conclude only that at least 75% of the items will be in this interval.

■■■ EXERCISES

19. Use the data on household incomes for 10 families in Grimes, Iowa (Exercise 16). What is the z-score for the income of $152,936? Interpret this standardized value and use the empirical rule to comment on whether or not this income value should be considered an outlier.

20. A sample of 10 NCAA men's college basketball scores (*USA Today*, January 5, 1989) provided the following winning teams and the number of points scored:

Rutgers 87	Tulsa 70
Niagara 79	Texas-El Paso 82
Mississippi 80	Stanford 83
Western Kentucky 64	Iowa 93
Purdue 75	Montana 62

a. Compute the sample mean and sample standard deviation for this data.

b. In another game, Penn State beat Massachusetts by a score of 110 to 79. Use the z-score to determine if the Penn State score should be considered an outlier. Explain.

c. Assume that the distribution of the points scored by winning teams has a mound-shaped distribution. Estimate the percentage of all NCAA basketball games in which the winning team will score 87 or more points. Estimate the percentage of all NCAA basketball games in which the winning team will score 58 or less points.

21. A 1989 salary survey (*Working Woman*, January 1989) listed the average salary of elementary and secondary school teachers as $28,085. Assume that the standard deviation of salaries is $4500.

a. Janice Herbranson was identified as a teacher in a one-room schoolhouse in McLeod, North Dakota. Her salary was reported to be $8100 per year. What is the z-score associated with $8100? Comment on whether or not this salary figure is an outlier.

b. Compute the z-score for each of the following salaries: $33,500, $25,200, $28,985 and $39,000. Should any of these be reviewed as possible outliers?

22. Use the salary data in Exercise 21 and Chebyshev's theorem to find the percentage of elementary school teachers that must have salaries in the following ranges:

a. $19,085 to $37,085

b. $14,585 to $41,585

c. Repeat parts a and b if it can be assumed that the distribution of teacher salaries is approximately bell-shaped.

23. The relationship between IQ scores and birth rates was cited in the *Atlantic Monthly*,

May 1989. IQ scores have a bell-shaped distribution with a mean of 100 and a standard deviation of 15.

    a. What percentage of the population should have an IQ score between 85 and 115?
    b. What percentage of the population should have an IQ score between 70 and 130?
    c. What percentage of the population should have an IQ score of more than 130?
    d. A person with an IQ score of more than 145 is considered a genius. Does the empirical rule support this statement? Explain.

3.4 ■■■ EXPLORATORY DATA ANALYSIS

In Chapter 2 we introduced exploratory data analysis. Recall that the focus of exploratory data analysis is on using simple arithmetic and easy-to-draw pictures to summarize data. In this section we continue our introduction of exploratory data analysis by considering 5-number summaries and box plots.

Five-Number Summary

In a five-number summary, the following 5 numbers are used to summarize the data:

1. Smallest value

2. First quartile (Q_1)

3. Median

4. Third quartile (Q_3)

5. Largest value

    The monthly starting salaries for a sample of 12 business school graduates were shown in Table 3.1. These data are as follows:

1650	1690
1750	1930
1850	1740
1680	2125
1555	1720
1510	1680

The median of 1705 and the quartiles $Q_1 = 1665$ and $Q_3 = 1800$ were computed in Section 3.1. Reviewing the preceding data shows a smallest value of 1510 and a largest value of 2125. Thus the five-number summary for the salary data is

    1510   1665   1705   1800   2125

Approximately one-fourth, or 25%, of the data values are between adjacent numbers in a five-number summary.

Box Plot

The *box plot* is a relatively recent development in the area of graphical summaries of data. Key to the development of a box plot is the computation of

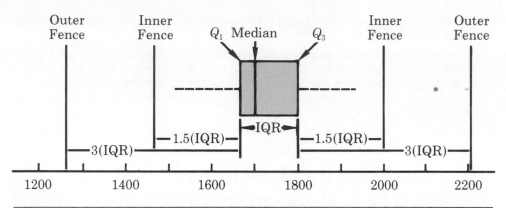

■■■ FIGURE 3.4   **Box Plot of the Monthly Starting Salaries of Business School Graduates With Lines Showing the Inner and Outer Fences**

the median and the quartiles, Q_1 and Q_3. The interquartile range, IQR = $Q_3 - Q_1$, is also used. Figure 3.4 shows the box plot for the monthly starting salary data. The steps used to construct the box plot are as follows:

1. A box is drawn with the ends of the box located at the 1st and 3rd quartiles. For the salary data, $Q_1 = 1665$ and $Q_3 = 1800$. This box contains the middle 50% of the data.

2. A vertical line is drawn in the box at the location of the median (1705 for the salary data). Thus the median line divides the data into two equal parts.

3. Using the interquartile range, IQR = $Q_3 - Q_1$, fences are located. The *inner fences* are located 1.5(IQR) below Q_1 and 1.5(IQR) above Q_3. The *outer fences* are located 3(IQR) below Q_1 and 3(IQR) above Q_3. For the salary data, IQR = $Q_3 - Q_1 = 1800 - 1665 = 135$. Thus, the inner fences are $1665 - 1.5(135) = 1462.5$ and $1800 + 1.5(135) = 2002.5$. The outer fences are $1665 - 3(135) = 1260$ and $1800 + 3(135) = 2205$. The fences are important aids in identifying outliers. Data falling between the inner and outer fences are considered *mild outliers*. Data falling outside the outer fences are considered *extreme outliers*.

■■■ FIGURE 3.5   **Box Plot of the Monthly Starting Salaries of Business School Graduates**

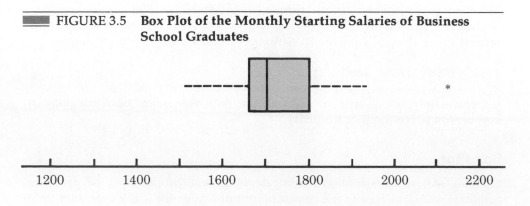

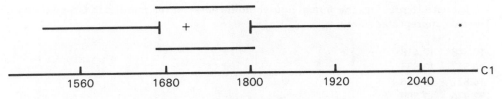

■ FIGURE 3.6   **Box Plot of the Monthly Starting Salary Data Provided by Minitab**

4. The dashed lines in Figure 3.4 are called *whiskers*. The whiskers are drawn from the ends of the box to the smallest and largest data values *inside the inner fences*. Thus the whiskers end at salary data values of 1510 and 1930.

5. Finally, the location of mild outliers are shown with the symbol ∗ and extreme outliers are shown with the symbol ○. In Figure 3.4 we see there is one mild outlier—the data value 2125. There are no extreme outliers in the salary data.

In Figure 3.4 we have included lines showing the location of the fences. These lines were drawn to show how fences are computed and where they are located for the salary data. Although the fences are always computed, generally they are not drawn on the box plots. Figure 3.5 shows the usual appearance of a box plot for the salary data.

Computer packages are capable of providing box plots of data. Figure 3.6 shows a Minitab printout of the box plot for the monthly salary data of the 12 business school graduates. Minitab uses the character ∗ to identify the location of the mild outliers.

NOTES AND COMMENTS

1. When using fences to identify outliers, we may or may not select the same items as when using z-scores less than −3 and greater than 3 to identify outliers. However, the objective of both approaches is simply to identify items that should be reviewed to ensure the validity of the data. Thus, outliers identified using either procedure should be reviewed.

2. An advantage of the exploratory data analysis procedures is that they are easy to use; few numerical calculations are necessary. We simply need to put the items in ascending order and identify the median and quartiles Q_1 and Q_3 in order to obtain the five-number summary. The fences and the box plot can then easily be determined. It is not necessary to compute the mean and the standard deviation for the data.

■ EXERCISES

24. Data on Japanese automobile manufacturer quotas were presented in Exercise 1. Provide a five-number summary and box plot for these data.

25. Exercise 16 provides the annual household incomes for ten families in Grimes, Iowa. The data are repeated here:

10,648	17,418
16,517	13,555
14,821	19,226
152,936	11,800
18,527	12,222

a. Provide a five-number summary for the data.
b. Compute the location of the fences. Does the income of $152,936 appear as a mild or extreme outlier?
c. Show the box plot for this data.

26. *Forbes* (January 9, 1989) published its 41st annual report on American industry. Data on the percent growth in sales for the past 12 months were presented for 26 companies in the paper industry. These data are as follows:

16.1	49.9	23.7	15.6	1.9
10.8	20.4	12.2	22.4	4.9
13.4	6.8	15.8	12.1	19.3
10.0	46.1	27.0	7.0	12.5
6.1	15.6	6.3	16.7	10.1
55.9				

a. Provide a five-number summary.
b. Compute the inner and outer fences.
c. Do there appear to be outliers? How would this information be helpful to a financial analyst?
d. Show a box plot.

27. Exercise 2 provides a sample of ages for 20 individuals who work at home (*American Demographics*, December 1988). Show a five-number summary and a box plot for these data.

28. Annual sales in millions of dollars for 17 companies in the chemical industry are as follows (*Forbes*, January 9, 1989):

484	2,731	598	2,472
3,261	1,220	4,514	32,249
15,980	8,030	3,122	8,258
1,061	2,188	2,366	1,049
636			

a. Provide a five-number summary of these data.
b. Compute the inner and outer fences.
c. Do there appear to be outliers? What does outlier information tell you in this case? *Note:* Companies associated with some of the preceding data are as follows: Valspar (484), Dow Chemical (15,980), and Du Pont (32,249).
d. Show a box plot.

29. The Highway Loss Data Institute "Injury and Collision Loss Experience" (September 1988) rates car models based on the number of insurance claims filed after accidents.

Index ratings near 100 are considered average. Lower ratings are better, and the car model is considered safer. Shown are ratings for 20 midsize cars and 20 small cars.

Midsize cars:	81	91	93	127	68	81	60	51	58	75
	100	103	119	82	128	76	68	81	91	82
Small cars:	73	100	127	100	124	103	119	108	109	113
	108	118	103	120	102	122	96	133	80	140

Summarize the data for the midsize and small cars separately.

a. Provide a five-number summary for midsize cars and for small cars.
b. Show the box plots.
c. Make a statement about what your summaries indicate about the safety of midsize cars compared to small cars.

3.5 ▓▓▓ COMPUTING MEASURES OF LOCATION AND DISPERSION FROM GROUPED DATA

In most cases, measures of central location and dispersion are computed using the individual data values. However, sometimes we are presented with data in a grouped or frequency distribution form. This section describes how approximations of the mean, variance, and standard deviation can be obtained directly from a frequency distribution.

Mean

Recall that in order to compute the sample mean using the individual data values, we simply sum all the values and divide by n, the sample size. If the data are available only in frequency distribution form, we will have to approximate the sum of the data values.

To do this, we treat the midpoint of each class as if it were the mean of the items in the class. Let M_i denote the midpoint for class i and f_i denote the frequency of the class. Then an approximation to the sum of the items in class i is given by $f_i M_i$. Summing these values over all classes, we obtain $\Sigma f_i M_i$, which approximates the sum of all the data values.

Once this approximation of the sum of all the data values is obtained, an approximation of the mean is computed by dividing this sum by the total number of data items. The following formula can be used to compute the sample mean from grouped data.

Sample Mean from Grouped Data

$$\bar{x} = \frac{\Sigma f_i M_i}{n} \qquad (3.10)$$

We do not expect the calculations based on the ungrouped data and grouped data to provide exactly the same numerical result. Thus $\bar{x}$ calculated using (3.10) is an approximation of $\bar{x}$ calculated when all data values are available. The difference between the two values of $\bar{x}$ is known as *grouping error*.

TABLE 3.5 Frequency Distribution of Audit Times

Audit Time (Days)	Frequency
10–14	4
15–19	8
20–24	5
25–29	2
30–34	1
Total	20

In Section 2.2 we provided a frequency distribution of the time in days required to complete year-end audits for the public accounting firm of Sanderson and Clifford. The frequency distribution of audit times based on a sample of 20 clients was developed. This frequency distribution is shown again in Table 3.5.

What is the sample mean audit time based on the grouped data shown in Table 3.5? Recall that the class midpoints, M_i, are located halfway between the class limits. Thus, the first class of 10–14 has a midpoint located at $(10 + 14)/2 = 12$. Since this class has a frequency $f_i = 4$, the value $f_i M_i = 4 \times 12 = 48$. The five class midpoints and the computations necessary to determine the sample mean using (3.10) are shown in Table 3.6. As can be seen, the sample mean audit time is 19 days.

Variance

The approach to compute the variance from grouped data is to use a slightly altered version of the formula for the variance provided in Equation (3.5). In (3.5), the squared deviations of the data values about the sample mean $\bar{x}$ were

TABLE 3.6 Computation of the Sample Mean Audit Time from Grouped Data

Audit Time (Days)	Frequency f_i	Class Midpoint M_i	$f_i M_i$
10–14	4	12	48
15–19	8	17	136
20–24	5	22	110
25–29	2	27	54
30–34	1	32	32
	20		380
			$\Sigma f_i M_i$

$$\text{Sample mean } \bar{x} = \frac{\Sigma f_i M_i}{n} = \frac{380}{20} = 19 \text{ days}$$

TABLE 3.7　Computation of the Sample Variance of Audit Times from Grouped Data (Sample Mean $\bar{x} = 19$)

Audit Time	Frequency f_i	Class Midpoint M_i	Deviation $(M_i - \bar{x})$	Squared Deviation $(M_i - \bar{x})^2$	$f_i(M_i - \bar{x})^2$
10–14	4	12	−7	49	196
15–19	8	17	−2	4	32
20–24	5	22	3	9	45
25–29	2	27	8	64	128
30–34	1	32	13	169	169
	20				570
					$\Sigma f_i(M_i - \bar{x})^2$

$$\text{Sample variance } s^2 = \frac{\Sigma f_i(M_i - \bar{x})^2}{n - 1} = \frac{570}{19} = 30$$

written $(x_i - \bar{x})^2$. However, with grouped data, the individual data values, x_i, are not known. In this case, we treat the class midpoint, M_i, as being a representative value for the x_i values in the corresponding class. Thus the squared deviations about the sample mean, $(x_i - \bar{x})^2$, are replaced by $(M_i - \bar{x})^2$. Then, just as we did with the sample mean calculations for grouped data, we weight each value by the frequency of the class, f_i, and sum for all classes. The sum of the squared deviations about the mean for all the data is approximated by $\Sigma f_i(M_i - \bar{x})^2$. Using this approach, the formula for the sample variance from grouped data is

---

Sample Variance from Grouped Data

$$s^2 = \frac{\Sigma f_i(M_i - \bar{x})^2}{n - 1} \qquad (3.11)$$

The calculation of the sample variance for audit times based on the grouped data from Table 3.5 is shown in Table 3.7.

Standard Deviation

The standard deviation computed from grouped data is simply the square root of the variance computed from grouped data. For the audit-time data, the sample standard deviation computed from grouped data is $s = \sqrt{30} = 5.48$.

Population Summaries from Grouped Data

Before closing this section on computing measures of location and dispersion from grouped data, we note that the formulas in this section were presented only for data sets constituting a sample. Population summary measures are computed in a similar manner. The grouped data formulas for a population mean and variance are as follows.

Population Mean from Grouped Data

$$\mu = \frac{\Sigma f_i M_i}{N}$$

(3.12)

Population Variance from Grouped Data

$$\sigma^2 = \frac{\Sigma f_i (M_i - \mu)^2}{N}$$

(3.13)

■■■ EXERCISES

30. The *Journal of Personal Selling and Sales Management* (August 1988) reported a study investigating the use of a persuasion technique in selling. A sample of 56 participants was used in the study. Assume that the frequency distribution below shows the number of sales presentations made per week by the 56 participants.

Number of Presentations	Frequency
10–12	· 5
13–15	9
16–18	22
19–21	12
22–24	8
Total	56

a. What is the mean number of sales presentations made per week for participants in this study?

b. What are the variance and the standard deviation?

31. The following frequency distribution for the first examination in operations management was posted on the department bulletin board.

Examination Grade	Frequency
40–49	3
50–59	5
60–69	11
70–79	22
80–89	15
90–99	6
Total	62

Treating these data as a sample, compute the mean, variance, and standard deviation.

32. *Psychology Today* (November 1988) reported on the characteristics of individuals caught in the act of shoplifting in a supermarket. Data from 9832 cases showed 56% of the shoplifters were male and 44% were female. The value of most items stolen was in the $1 to $5 range. The following frequency distribution shows the age of the shoplifters.

Age of Shoplifter	Frequency
6–11	364
12–17	1249
18–29	3392
30–59	3962
60–80	865
Total	9832

a. What is the sample mean age of supermarket shoplifters?

b. What are the variance and standard deviation of the ages?

33. A service station has recorded the following frequency distribution for the number of gallons of gasoline sold per car in a sample of 680 cars.

Gasoline (gallons)	Frequency
0–4	74
5–9	192
10–14	280
15–19	105
20–24	23
25–29	6
Total	680

Compute the mean, variance, and standard deviation for these grouped data. If the service station expects to service about 120 cars on a given day, what is an estimate of the total number of gallons of gasoline that will be sold?

34. Scores obtained by a sample of patients on a depression level test are summarized in the following frequency distribution:

Depression Level Score	Frequency
25–34	3
35–44	1
45–54	2
55–64	6
65–74	4
75–84	6
85–94	2
95–104	1
Total	25

Using the above grouped data, compute the following.

a. mean

b. variance

c. standard deviation

3.6 ▬ THE ROLE OF THE COMPUTER IN DESCRIPTIVE STATISTICS

In this section we describe the role of computers and computer software packages in descriptive statistics by showing how a statistical package can be used to generate descriptive statistics for a data set. In future chapters we provide further illustrations showing how statistical computing systems can support the analysis and interpretation of data.

In the 1960s there were relatively few computerized statistical packages for analyzing data. Since that time, however, the situation has changed dramatically. Today the user has a choice of packages such as SAS (Statistical Analysis System), SPSS (Statistical Package for the Social Sciences), BMDP (UCLA Biomedical statistical package), and Minitab, to name just a few. In this text we use the Minitab system to illustrate the application of statistical computing systems.

Minitab is a general-purpose statistical computing system that can be used on a variety of mainframe and personal computers. It has been designed for users who have had little or no previous computer experience. Although it is very easy to use, it offers a great deal of power for data summarization and statistical analysis.

In the illustrations of Minitab, we describe its use in what is referred to as *interactive mode*. In this mode the user enters data and commands from a computer terminal or personal computer keyboard; Minitab carries out each user command as soon as it is given. In the figures containing the steps of a Minitab session, we show the computer responses from the Minitab system in black and the input by the user in color.

Minitab consists of a worksheet of rows and columns in which data are stored. The columns are denoted with labels c1, c2, and so on, unless the user elects to name the columns with more descriptive labels. Each column corresponds to a variable. The rows of the worksheet correspond to the elements of the data set. That is, a separate row is used for each element. The Minitab system consists of about 150 commands, which can be used to analyze the data stored in the worksheet. To provide an illustration of how Minitab works, we use the data originally presented in Table 2.12. Recall that these data show the number of questions answered correctly by each of 50 individuals who took an aptitude test. For convenience, these data are shown in Table 3.8.

Refer to Figure 3.7. We assume that the user has loaded the Minitab system into the computer. The MTB > symbol appears on the terminal screen, which indicates that the Minitab system is waiting for a command from the user. The user then inputs read c1, indicating that the system is to take the data from the lines that follow and store the data in column 1 of the worksheet. Note that the

TABLE 3.8　**Data Showing Aptitude Test Scores**

112	72	69	97	107
73	92	76	86	73
126	128	118	127	124
82	104	132	134	83
92	108	96	100	92
115	76	91	102	81
95	141	81	80	106
84	119	113	98	75
68	98	115	106	95
100	85	94	106	119

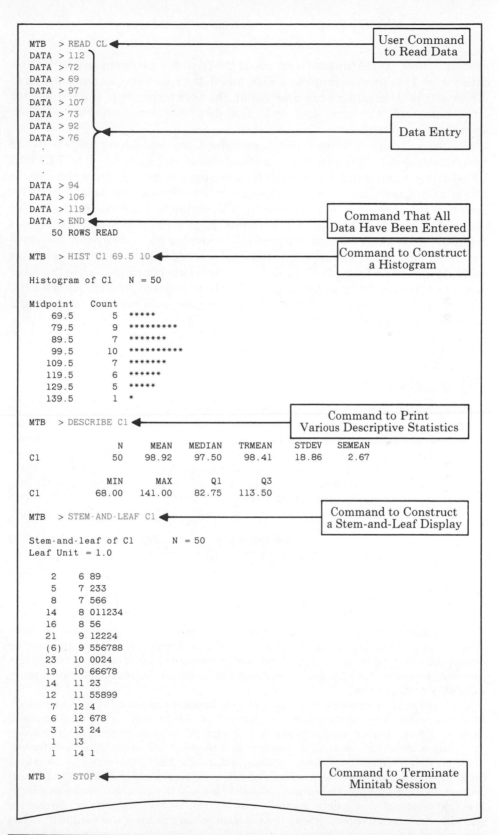

```
MTB  > READ Cl                          ┌──────────────────────┐
DATA > 112 ┐                            │   User Command       │
DATA > 72  │                            │   to Read Data       │
DATA > 69  │                            └──────────────────────┘
DATA > 97  │
DATA > 107 │                            ┌──────────────────────┐
DATA > 73  │                            │                      │
DATA > 92  │                            │     Data Entry       │
DATA > 76  │                            └──────────────────────┘
      .    │
      .    │
      .    │
DATA > 94  │
DATA > 106 │
DATA > 119 ┘                            ┌──────────────────────┐
DATA > END                              │ Command That All     │
     50 ROWS READ                       │ Data Have Been Entered│
                                        └──────────────────────┘
MTB  > HIST Cl 69.5 10                   ┌──────────────────────┐
                                        │ Command to Construct │
                                        │    a Histogram       │
Histogram of Cl   N = 50                └──────────────────────┘


Midpoint    Count
   69.5        5    *****
   79.5        9    *********
   89.5        7    *******
   99.5       10    **********
  109.5        7    *******
  119.5        6    ******
  129.5        5    *****
  139.5        1    *

MTB  > DESCRIBE Cl                       ┌──────────────────────┐
                                        │  Command to Print    │
                                        │Various Descriptive Statistics│
              N      MEAN   MEDIAN   TRMEAN   STDEV   SEMEAN
Cl           50     98.92    97.50    98.41   18.86    2.67


             MIN     MAX      Q1       Q3
Cl          68.00  141.00   82.75    113.50
                                        ┌──────────────────────┐
MTB  > STEM-AND-LEAF Cl                  │ Command to Construct │
                                        │a Stem-and-Leaf Display│
Stem-and-leaf of Cl       N = 50        └──────────────────────┘
Leaf Unit = 1.0

    2      6 89
    5      7 233
    8      7 566
   14      8 011234
   16      8 56
   21      9 12224
   (6)     9 556788
   23     10 0024
   19     10 66678
   14     11 23
   12     11 55899
    7     12 4
    6     12 678
    3     13 24
    1     13
    1     14 1
                                        ┌──────────────────────┐
MTB  >  STOP                             │ Command to Terminate │
                                        │  Minitab Session     │
                                        └──────────────────────┘
```

FIGURE 3.7   Computer Analysis of Apptitude Test Scores Using Minitab

next line shows that Minitab's response is DATA>; the user then enters the first data value. This process continues with one data value being entered per line. When all the data values have been input, the user responds to the request for more data with the command end. The data now resides in the Minitab worksheet in column 1.

Now that the data is in the Minitab worksheet, the user can continue with the analysis of the data. The next user command shown in Figure 3.7 is HIST C1 69.5 10, which is a command for Minitab to construct a histogram from the data in column 1; the midpoint of the first class interval is 69.5 and the class width is 10. Note that the output is produced directly after giving this command. In this way the analyst can quickly look at the output, form some initial judgments, and continue. Data analysis done interactively provides quick computer response and is an important reason why statistical packages such as Minitab are so valuable to an analyst. The session with Minitab continues with the user command DESCRIBE C1 which produces a variety of descriptive statistics for the data. Note that this session with Minitab concludes with the user requesting a stem-and-leaf display of the data.

■■■ SUMMARY

In this chapter we introduced several statistical measures that can be used to describe the location and dispersion of data. Unlike the tabular and graphical procedures for summarizing data, the measures introduced in this chapter summarize the data in terms of numerical values. When the numerical values obtained are for a sample, they are called sample statistics. When the numerical values obtained are for a population, they are called population parameters. Some of the notation used for sample statistics and population parameters are summarized below.

	Sample Statistic	Population Parameter
Mean	$\bar{x}$	μ
Variance	s^2	σ^2
Standard deviation	s	σ

As measures of central location, we defined the mean, median, and mode for both sample and population data. Then the concept of a percentile was used to describe the location of other values in the data. Next, we presented the range, interquartile range, variance, standard deviation, and coefficient of variation as statistical measures of variability or dispersion.

A discussion of two exploratory data analysis techniques that can be used to summarize data more effectively was included in Section 3.4. Specifically, we showed how to develop a five-number summary and a box plot in order to provide simultaneous information about the location, dispersion, and shape of the underlying distribution. Then, we described how the mean, variance, and standard deviation could be computed for grouped data. However, we recommend using the measures based on the individual data values unless the grouped format is the only manner in which the data are available. The chapter concluded with a discussion of the role of the computer in descriptive statistics. An interactive session with the software package Minitab was used to illustrate how statistical computing systems can support the analysis and interpretation of data.

■■ GLOSSARY

Population parameter A numerical value used as a summary measure for a population of data (e.g., the population mean, μ, the population variance, σ^2, and the population standard deviation, σ).

Sample statistic A numerical value used as a summary measure for a sample (e.g., the sample mean, $\bar{x}$, the sample variance, s^2, and the sample standard deviation, s).

Mean A measure of central location for a data set. It is computed by summing all the data values and dividing by the number of items.

Trimmed mean The mean of the data remaining after α percent of the smallest and α percent of the largest items have been removed. The purpose of a trimmed mean is to provide a measure of central location that has eliminated the effect of extremely large and extremely small data values.

Median A measure of central location. It is the value which splits the data into two equal groups—one with values greater than or equal to the median, and one with values less than or equal to the median.

Mode A measure of location, defined as the most frequently occurring data value.

Percentile A value such that at least p percent of the items are less than or equal to this value and at least $(100 - p)$ percent of the items are greater than or equal to this value. The 50th percentile is the median.

Quartiles The 25th, 50th, and 75th percentiles of the data referred to as the first quartile, the second quartile (median), and third quartile, respectively. The quartiles can be used to divide the data into four parts, with each part containing approximately 25% of the data.

Hinges The value of the lower hinge is approximately the first quartile, or 25th percentile. The value of the upper hinge is approximately the third quartile, or 75th percentile. The values of the hinges and quartiles may differ slightly due to differing computational conventions.

Range A measure of dispersion, defined to be the difference between the largest and smallest data values.

Interquartile range A measure of dispersion, defined to be the difference between the third and first quartiles.

Variance A measure of dispersion for a data set, found by summing the squared deviations of the data values about the mean and then dividing the total by N if the data is from a population or by $n - 1$ if the data is from a sample.

Standard deviation A measure of dispersion for a data set, found by taking the positive square root of the variance.

Coefficient of variation A measure of relative dispersion for a data set, found by dividing the standard deviation by the mean and multiplying by 100.

z-score For each data item, a value found by dividing the deviation about the mean $(x_i - \bar{x})$ by the standard deviation s. A z-score is referred to as a standardized value and denotes the number of standard deviations a data value x_i is from the mean.

Chebyshev's theorem A theorem applying to any data set that can be used to make statements about the percentage of items that must be within a specified number of standard deviations of the mean.

Empirical rule A rule that states the percentages of items that are within one, two, and three standard deviations from the mean for mound-shaped, or bell-shaped, distributions.

Outlier An unusually small or unusually large data value.

Box plot A graphical summary of data. A box, drawn from the first to the third quartiles, shows the location of the middle 50% of the data. Dashed lines, called whiskers, extending from the ends of the box, show the location of data greater than the third quartile and data less than the first quartile. The locations of any outliers are also noted.

Fences Values used to identify outliers. Inner fences are located 1.5(IQR) below the first quartile and 1.5(IQR) above the third quartile. Outer fences are located 3(IQR) below the first quartile and 3(IQR) above the third quartile. Data falling between the inner and outer fences are considered mild outliers. Data falling outside the outer fences are considered extreme outliers.

5-number summary An exploratory data analysis technique that uses the following 5 numbers to summarize the data set: smallest value, first quartile, median, third quartile and largest value.

Grouped data Data available in class intervals as summarized by a frequency distribution. Individual values of the original data are not recorded.

■ KEY FORMULAS

Sample Mean

$$\bar{x} = \frac{\Sigma x_i}{n} \tag{3.1}$$

Population Mean

$$\mu = \frac{\Sigma x_i}{N} \tag{3.2}$$

Interquartile Range

$$\text{IQR} = Q_3 - Q_1 \tag{3.3}$$

Population Variance

$$\sigma^2 = \frac{\Sigma (x_i - \mu)^2}{N} \tag{3.4}$$

Sample Variance

$$s^2 = \frac{\Sigma (x_i - \bar{x})^2}{n - 1} \tag{3.5}$$

Standard Deviation

$$\text{Sample Standard Deviation} = s = \sqrt{s^2} \tag{3.6}$$

$$\text{Population Standard Deviation} = \sigma = \sqrt{\sigma^2} \tag{3.7}$$

Coefficient of Variation

$$\left(\frac{\text{Standard Deviation}}{\text{Mean}} \right) \times 100 \tag{3.8}$$

z-score

$$z_i = \frac{x_i - \bar{x}}{s} \tag{3.9}$$

Sample Mean from Grouped Data

$$\bar{x} = \frac{\Sigma f_i M_i}{n} \tag{3.10}$$

Sample Variance from Grouped Data

$$s^2 = \frac{\Sigma f_i(M_i - \bar{x})^2}{n - 1} \qquad (3.11)$$

Population Mean from Grouped Data

$$\mu = \frac{\Sigma f_i M_i}{N} \qquad (3.12)$$

Population Variance from Grouped Data

$$\sigma^2 = \frac{\Sigma f_i(M_i - \mu)^2}{N} \qquad (3.13)$$

■■ SUPPLEMENTARY EXERCISES

35. A sample of six recent home mortgage loans showed the following interest rates.

12.5   13.2   11.2   13.0   12.0   12.5

Compute the following descriptive statistics for the data set

a. mean
b. median
c. mode
d. 25th percentile
e. range
f. interquartile range
g. variance
h. standard deviation
i. coefficient of variation

36. *Time* (January 9, 1989) published an article on the academic ability of college athletes. The article noted that some of the most successful athletic programs (citing the University of Notre Dame and Duke University) have athletes with very good college board scores. Assume that the following sample data are typical of college board scores for Notre Dame football players:

1100   970   1000   1250   880   790   1300   1050   900   950   1120

a. Compute the mean, median and mode.
b. Compute the range and interquartile range.
c. Compute the variance and standard deviation.
d. Using z-scores, state whether or not there are any outliers in this data set.

37. The following data show home mortgage loan amounts handled by a particular loan officer in a savings and loan association:

20,000	38,500	33,000	27,500	34,000
12,500	25,900	43,200	37,500	36,200
25,200	30,900	23,800	28,400	13,000
31,000	33,500	25,400	33,500	20,200
39,000	38,100	30,500	45,500	30,500
52,000	40,500	51,600	42,500	44,800

a. Find the mean, median, and mode.
b. Find the first and third quartiles.

38. *Newsweek* (January 9, 1989) reported statistics on the number of visits a couple makes to a therapist while undergoing marriage counseling. With data based on this article, assume that the following show the number of visits for a sample of 9 couples:

12   8   3   13   18   20   10   9   18

Compute the following descriptive statistics for this data.
   a. mean
   b. median
   c. mode
   d. 40th percentile
   e. range
   f. variance
   g. standard deviation

39. A sample of ten stocks on the New York Stock Exchange (May 22, 1989) shows the following price-earnings ratios:

9   4   6   7   3   11   4   6   4   7

Using these data, compute the mean, median, mode, range, variance, and standard deviation.

40. A sample of recent oil drilling locations shows oil found at the following depths (feet):

1500   1200   1600   1700   1500   2000

Compute the mean, median, mode, range, variance, and standard deviation for the drilling depth data.

41. Soft-drink purchases at the Wright Field concession stands show the following 1-day totals:

Drink	Units Purchased
Cola	4553
Diet cola	2125
Uncola	1850
Orange soda	1288
Root beer	1572

What is the mode for these sample data?

42. The Hawaii Visitors' Bureau reports visitors coming to Hawaii from the U.S. mainland spend an average of $102 per day (*St. Petersburg Times*, December 11, 1988). Assume the standard deviation is $27 and the distribution of expenditures is approximately bell-shaped. Answer the following:

   a. What percentage of visitors will have an average daily expenditure of between $75 and $129 per day?
   b. What percentage of visitors will have an average daily expenditure of between $48 and $156 per day?
   c. Should an average daily expenditure of $225 per day be considered an outlier?

43. The cost of new homes in selected cities throughout the United States was reported in *U.S. News and World Report* (September 1988). Assume the cost of homes in Chicago, Illinois, has a mean of $100,000 and a standard deviation of $40,000.

   a. Should a home selling for $200,000 be considered an outlier? Explain.
   b. Use Chebyshev's theorem to determine the percentage of homes selling between $40,000 and $160,000.

44. Public transportation and an automobile are two methods an employee has of getting to work each day. Samples of times recorded for each method are shown. Times are in minutes.

> Public Transportation: 28, 29, 32, 37, 33, 25, 29, 32, 41, 34
> Automobile: 29, 31, 33, 32, 34, 30, 31, 32, 35, 33

a. Compute the sample mean time to get to work for each method.
b. Compute the sample standard deviation for each method.
c. Based on your results from parts a and b, which method of transportation should be preferred? Explain.
d. Develop a box plot for each method. Does a comparison of the box plots support your conclusion in part c?

45. Final examination scores for 25 statistics students are as follows:

56	77	84	82	42
61	44	95	98	84
93	62	96	78	88
58	62	79	85	89
89	97	53	76	75

a. Provide a five-number summary.
b. Provide a box plot.

46. The results of a 150-question social-awareness test that was given to a group of 20 first-year college students show the number of questions answered correctly by each of the students. Data are as follows:

121	114	94	136	144	126	98	103	118	127
135	97	119	117	122	138	142	141	102	105

a. Provide a five-number summary.
b. Provide a box plot.

47. The following data shows the total yardage accumulated over the football season for 20 receivers.

744	652	576	1112	971	451	1023	852	809	596
941	975	400	711	1174	1278	820	511	907	1251

a. Provide a five-number summary.
b. Provide a box plot.
c. Identify any outliers.

48. A frequency distribution for the duration of 20 long-distance telephone calls (rounded to the nearest minute) is shown below:

Call Duration	Frequency
4–7	4
8–11	5
12–15	7
16–19	2
20–23	1
24–27	1
Total	20

Compute the mean, variance, and standard deviation for the above data.

49. Dinner check amounts at La Maison French Restaurant have the following frequency distribution:

Dinner Check (Dollars)	Frequency
25–34	2
35–44	6
45–54	4
55–64	4
65–74	2
75–84	2
Total	20

Compute the mean, variance, and standard deviation for the given data.

50. Automobiles traveling on the New York State Thruway are checked for speed by a state police radar system. A frequency distribution of speeds is shown.

Speed (Miles per Hour)	Frequency
45–49	10
50–54	40
55–59	150
60–64	175
65–69	75
70–74	15
75–79	10
Total	475

a. What is the mean speed of the automobiles traveling on the New York State Thruway?

b. Compute the variance and the standard deviation.

■■■ COMPUTER EXERCISE

A national association of nurses has sponsored a study to determine the job satisfaction of nurses employed in hospitals. As part of the study, 50 nurses were asked to indicate their degree of satisfaction in their work, in their pay, and in their opportunities for promotion. Each of the three aspects of satisfaction were measured on a scale of 0 to 100, with larger values indicating higher degrees of satisfaction. Data were also collected on the type of hospital where each nurse was employed. The hospital types considered in the study were investor-owned hospitals, Veterans Administration (VA) hospitals, and university hospitals. The following data were collected:

Satisfaction Scores			Type of Hospital	Satisfaction Scores			Type of Hospital
Work	Pay	Promotion		Work	Pay	Promotion	
71	49	58	VA	72	76	37	VA
84	53	63	University	71	25	74	Investor-owned
84	74	37	VA	69	47	16	University
87	66	49	University	90	56	23	University
72	59	79	University	84	28	62	Investor-owned
72	37	86	VA	86	37	59	VA
72	57	40	Investor-owned	70	38	54	Investor-owned
63	48	78	VA	86	72	72	VA

Satisfaction Scores			Type of	Satisfaction Scores			Type of
Work	Pay	Promotion	Hospital	Work	Pay	Promotion	Hospital
84	60	29	VA	87	51	57	Investor-owned
90	62	66	Investor-owned	77	90	51	University
73	56	55	VA	71	36	55	University
94	60	52	VA	75	53	92	University
84	42	66	Investor-owned	74	59	82	Investor-owned
85	56	64	Investor-owned	76	51	54	University
88	55	52	University	95	66	52	VA
74	70	51	University	89	66	62	Investor-owned
71	45	68	Investor-owned	85	57	67	Investor-owned
88	49	42	Investor-owned	65	42	68	VA
90	27	67	VA	82	37	54	VA
85	89	46	University	82	60	56	VA
79	59	41	University	89	80	64	University
72	60	45	Investor-owned	74	47	63	Investor-owned
88	36	47	Investor-owned	82	49	91	Investor-owned
77	60	75	Investor-owned	90	76	70	VA
64	43	61	Investor-owned	78	52	72	VA

███ QUESTIONS

1. Develop descriptive measures of location for each of the three job-satisfaction variables. What aspect of the job is the most satisfying for the nurses? What appears to be the most critical issue, or issue of lowest satisfaction for nurses? Explain.

2. Develop descriptive measures of dispersion for each of the three measures of job satisfaction. Which measure of satisfaction appears to have the greatest difference of opinion among the nurses?

3. Develop summary information for the type of hospital variable.

4. Show how exploratory data analysis can help summarize these data.

APPLICATION

Barnes Hospital*

St. Louis, Missouri

Since its establishment in 1914, Barnes Hospital, at Washington University Medical Center, has been the leading provider of health care for the people of St. Louis and neighboring areas. The 1200 bed acute care hospital is nationally recognized as one of the best in the United States. In January of 1986, Barnes Hospital opened its Hospice Program, adding another speciality to the continuum of medical and supportive services offered to patients and families.

Barnes Hospital's Hospice Program focuses upon improving the quality of life for terminally ill patients and their families. The Hospice helps patients and families continue to live together at home, in private and familiar surroundings, near loved ones and friends. Hospice care represents hope. Not the hope of cure, but rather the hope of caring and comfort, the hope of relief from physical, emotional, and spiritual pain.

The word hospice is derived from the Latin *hospes* or *hospitum*, which means a place of rest. Hospices date back to A.D. 475 in Rome, where inns were established to care for sick and weary travelers. In the 15th and 16th centuries hospices were operated throughout Europe, giving shelter and refreshments to travelers, and offering nursing care to the sick, wounded, and dying. St. Joseph's, the first modern day hospice devoted exclusively to the care of the terminally ill, opened in the early 1900s in London.

The American Hospice movement began in 1974, when the Connecticut Hospice in New Haven was established. By the end of 1979, there were more than 500 hospices either in operation or development. Two years later the number doubled, making hospice one of the fastest-growing specialties in the health care system. In 1986 there are an estimated 1800 to 2000 hospice programs in the United States alone, delivering care to over 100,000 patients and families each year.

The Hospice Team at Barnes is composed of a medical director, a coordinator, RN supervisor, home and inpatient RN's, home health aids, social workers, chaplains, dietitians, trained volunteers, a secretary, and professionals from other ancillary services, as needed. The Team meets at least weekly to discuss each current and prospective patient and family, and to review and update the Interdisciplinary Care Plan for each. Through the coordinated efforts of Hospice Team members, patients and families are given the guidance and support necessary to cope with the strains created by serious illness, separation, and death.

▰ STATISTICAL INFORMATION

In the coordination and administration of the Hospice Program at Barnes Hospital, monthly reports and quarterly summaries help team members review the ongoing

*The authors are indebted to Ms. Paula H. Gianino, Hospice Coordinator, Barnes Hospital for providing this application.

TABLE 3A.1 Length of Stay
in Hospice

Length of Stay (Days)	Frequency
0–24	39
25–49	11
50–74	8
75–99	3
100–124	1
125–149	2
150–174	2
175–199	1
	67

provision of medical, nursing, and support services. This information, likewise, is utilized as a basis for implementing policy changes and for future planning.

As an illustration of the type of statistical information used by the Barnes' Hospice, Table 3A.1 shows a frequency distribution providing the length of stay in the Hospice program for a sample of 67 patients.

While the frequency distribution is useful, the following descriptive statistics (computed using the original data) provide valuable information about the length of stay.

Mean:	35.7 days
Median:	17 days
Mode:	1 day

Note that the mean (average) length of stay in the program is a little over one month. However, the median shows that half of the patients are in the program for just over half a month. The mode (1 day) is consistent with the belief that the terminally ill hospice patients have a relatively short stay in the program.

Statistical summaries also include the number of admissions to the Hospice Program, the number of patient days spent at home versus in the inpatient unit, the number of discharges from the inpatient unit, the number of patient deaths at home and in the

Barnes Hospital, Washington University Medical Center, St. Louis, Missouri

inpatient unit, etc. In addition, information such as the above is analyzed according to patient age and insurance; i.e., private insurance versus Medicare populations. These statistics provide working information for analyzing the flow of services provided and aid in assisting team members with planning for staffing increases in specific service areas.

In the past five years, as the hospice movement has grown and become more specialized, national, state, local and private standards of care have been established to insure greater quality of care. Organizations such as the Joint Commission on the Accreditation of Hospitals (JCAH), have begun surveying hospice programs, insisting that programs, such as Barnes, provide a standard set of interdisciplinary services with appropriate and accurate documentation of such. In addition, Hospice Care has been added as a permanent part of the national Medicare system placing even greater demands upon hospice programs to improve the quality of services provided while containing costs. In all, hospice programs across the country are being challenged to continue to offer humane and personalized services while meeting the demands toward increased standardization and specialization. Timely and accurate statistics gathering and analysis have become more than useful tools. Today, they are an integral part of program development, implementation, and future planning.

CHAPTER 4

Introduction to Probability

Throughout our lives we are faced with decision-making situations that involve uncertainty. Perhaps you will be asked for an analysis of one of the following situations:

1. What is the "chance" that sales will decrease if the price of the product is increased?
2. What is the "likelihood" that the new assembly method will increase productivity?
3. How "likely" is it that the project will be completed on time?
4. What are the "odds" that the new investment will be profitable?

The subject matter most useful in effectively dealing with such uncertainties is contained under the heading of probability. In everyday terminology, *probability* can be thought of as a numerical measure of the chance or likelihood that a particular event will occur. For example, if we consider the event "rain tomorrow," we understand that when the television weather report indicates "a near-zero probability of rain" there is almost no chance of rain. However, if a 90% probability of rain is reported, we know that it is very likely or almost certain that rain will occur. A 50% probability indicates that rain is just as likely to occur as not.

Probability values are always assigned on a scale from 0 to 1. A probability near 0 indicates that the event is very unlikely to occur; a probability near 1 indicates that the event is almost certain to occur. Other probabilities between 0 and 1 represent varying degrees of likelihood that the event will occur. Figure 4.1 depicts this view of probability.

Probability is important in decision making because it provides a mechanism for measuring, expressing, and analyzing the uncertainties associated with future events. In this chapter we introduce the fundamental concepts of probability and begin to illustrate their use as decision-making tools. In

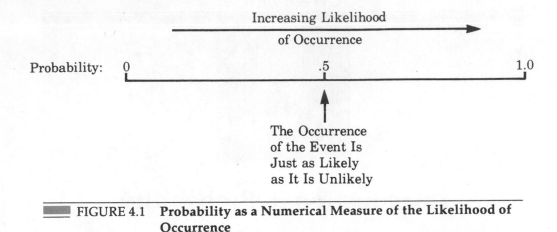

■■■■ FIGURE 4.1 **Probability as a Numerical Measure of the Likelihood of Occurrence**

subsequent chapters we will extend these basic notions of probability and demonstrate the important role that probability plays in statistical inference.

4.1 ■■■ EXPERIMENTS AND SAMPLE SPACE

Using the terminology of probability, we define an *experiment* to be any process which generates well defined outcomes. By this we mean that on any single repetition of the experiment *one and only one* of the possible experimental outcomes will occur. Several examples of experiments and their associated outcomes are as follows:

Experiment	Experimental Outcomes
Toss a coin	Head, tail
Select a part for inspection	Defective, nondefective
Conduct a sales call	Purchase, no purchase
Roll a die	1, 2, 3, 4, 5, 6
Play a football game	Win, lose, tie

The first step in analyzing a particular experiment is to carefully define the experimental outcomes. When we have defined *all* possible experimental outcomes, we have identified the *sample space* for the experiment. That is, the sample space is defined as the set of all possible experimental outcomes. Any one particular experimental outcome is referred to as a *sample point* and is an element of the sample space.

Let us consider the experiment of tossing a coin, as mentioned above. The experimental outcomes are defined by the upward face of the coin—a head or a tail. If we let S denote the sample space, we can use the following notation to describe the sample space and sample points for the coin-tossing experiment:

$$S = \{\text{Head, Tail}\}$$

Using this notation, the experiment of selecting a part for inspection would have a sample space with sample points as follows:

$$S = \{\text{Defective, Nondefective}\}$$

Finally, suppose that we consider the experiment of rolling a die, where the experimental outcomes are defined as the number of dots appearing on the upward face of the die. In this experiment, the numerical values 1, 2, 3, 4, 5, and 6 represent the possible experimental outcomes or sample points. Thus the sample space is denoted by

$$S = \{1, 2, 3, 4, 5, 6\}$$

Let us extend our discussion of experiments, sample points, and sample spaces to a slightly more involved illustration. Consider an experiment of tossing two coins, with the experimental outcomes defined in terms of the pattern of heads and tails appearing on the upward faces of the two coins. How many experimental outcomes (sample points) are possible for this experiment?

We can view the experiment of tossing two coins as a two-step experiment: Step 1 corresponds to tossing the first coin and Step 2 corresponds to tossing the second coin. A graphical device that is helpful in visualizing a multiple-step experiment and enumerating sample points is a *tree diagram*. Figure 4.2 provides a tree diagram for the two-coin-tossing experiment. Each of the points on the right-hand end of the tree corresponds to a sample point or experimental outcome. Thus we see that there are 4 experimental outcomes.

In Figure 4.2 the notation (H, H) is used to denote the sample point corresponding to a head on the first coin and a head on the second coin. Similarly, (H, T) denotes the sample point with a head on the first coin and a tail on the second coin, and so on. Thus, using this notation, we can describe the sample space S for the two-coin-tossing experiment as follows:

$$S = \{(H, H), (H, T), (T, H), (T, T)\}$$

A rule that is helpful in determining the number of sample points for an experiment consisting of multiple steps is stated next.

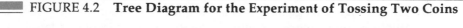

FIGURE 4.2 Tree Diagram for the Experiment of Tossing Two Coins

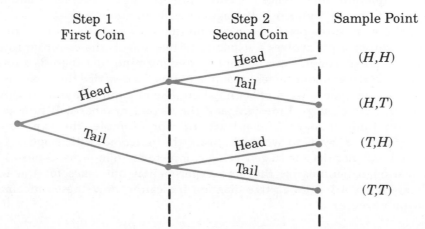

A Counting Rule for Multiple-Step Experiments

If an experiment can be described as a sequence of k steps in which there are n_1 possible outcomes on the first step, n_2 possible outcomes on the second step, and so on, then the total number of experimental outcomes is given by $(n_1)(n_2) \cdots (n_k)$. That is, the number of outcomes for the overall experiment is found by multiplying the number of outcomes on each step.

Referring again to the experiment of tossing two coins, we see that the two coins can be viewed as two steps with $n_1 = 2$, and $n_2 = 2$; thus, the counting rule states that there are $(n_1)(n_2) = (2)(2) = 4$ sample points, or experimental outcomes.

Let us now see how the concepts introduced thus far can be used in the analysis of the capacity expansion problem faced by the Kentucky Power and Light Company. We will begin by showing how the company's situation can be viewed as an experiment. Then we will attempt to define the appropriate sample points and the sample space for the experiment.

The Kentucky Power and Light Problem

The Kentucky Power and Light Company (KP&L) currently is starting work on a project designed to increase the generating capacity of one of its plants in Northern Kentucky. The project is divided into two sequential stages: stage 1 (design) and stage 2 (construction). While each stage will be scheduled and controlled as closely as possible, management cannot predict beforehand the exact elapsed time for each stage of the project. An analysis of similar construction projects over the past 3 years has shown completion times for the design stage of 2, 3, or 4 months and completion times for the construction stage of 6, 7, or 8 months. Thus management has decided to use these figures as the estimated completion times for the current project. In addition, because of the critical need for additional power, management has set a goal of 10 months for the total project completion time. Hence the entire project will be completed late if the total elapsed time to complete both stages exceeds 10 months.

Since there are three possible completion times for each stage of the project, and since the project involves a sequence of two stages, the counting rule for multiple-step experiments can be applied to determine that there is a total of $(3)(3) = 9$ experimental outcomes or sample points. To describe the experimental outcomes we will use a two-number notation. The first number will provide the completion time of stage 1 (design), and the second number will provide the completion time of stage 2 (construction). For example, the experimental outcome denoted by (2,6) would indicate the outcome with stage 1 being completed in 2 months and stage 2 being completed in 6 months. Table 4.1 uses this notation to summarize the nine experimental outcomes for the KP&L problem. Figure 4.3 shows a tree diagram indicating how these outcomes or sample points occur.

TABLE 4.1   Listing of Experimental Outcomes or Sample Points for the KP&L Problem

Completion Time (months)			
Stage 1 (Design)	*Stage 2 (Construction)*	**Notation for Experimental Outcome**	**Total Project Completion Time (months)**
2	6	(2, 6)	8
2	7	(2, 7)	9
2	8	(2, 8)	10
3	6	(3, 6)	9
3	7	(3, 7)	10
3	8	(3, 8)	11
4	6	(4, 6)	10
4	7	(4, 7)	11
4	8	(4, 8)	12

■■■ FIGURE 4.3   Tree Diagram for the KP&L Project

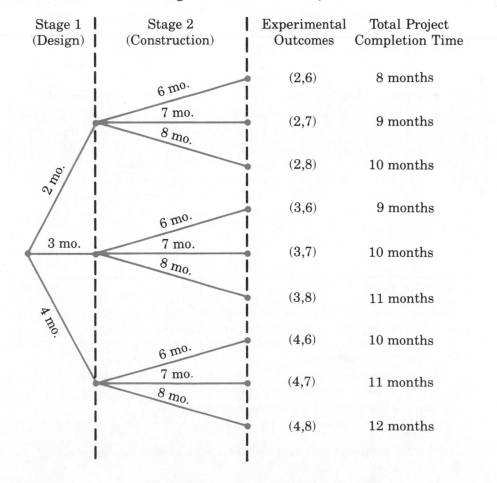

The process of identifying the sample space for the KP&L project helps the project manager visualize the various outcomes and completion times that are possible. However, probability information will be essential in order for the manager to understand the likelihood of the various outcomes. From the information in Figure 4.3, we know that the project will be completed in 8 to 12 months. However, the project manager will undoubtedly be interested in questions such as the following:

1. What is the probability that the project can be completed in 10 months or less?

2. What is the most likely or most probable project completion time?

3. What is the probability the project will take a year (12 months) to complete?

Recall that management's goal is to have the project completed within 10 months. Thus high probabilities for early and/or on-time completions will give the manager confidence that his or her goal will be met. However, high probabilities for late completions may lead to revised planning for the project, including corrective actions such as scheduling overtime, adding to the workforce, and so on. In any case, the probability information will be critical in helping the project manager understand the uncertainties associated with the project.

In order to provide the desired probability information for the KP&L problem, we need to consider how probability values can be assigned to the various experimental outcomes. This is the topic of the next section.

■■■ NOTES AND COMMENTS ■■■

In statistics, the notion of an experiment is somewhat different than in the laboratory sciences. In the laboratory sciences, it is assumed that each time an experiment is repeated in exactly the same way, the same outcome will occur. For the type of experiment we study in statistics, the outcome is determined by chance. Even though the experiment might be repeated in exactly the same way, a different outcome may occur. Because of this difference, the experiments we study in statistics are sometimes called random experiments.

■■■ EXERCISES

1. In a quality control process an inspector selects a completed part for inspection. The inspector then determines whether the part has a major defect, a minor defect, or no defect. Consider the selection and classification of the part as an experiment. List the sample points for the experiment.

2. Consider the experiment consisting of conducting three sales calls. On each of the calls there will be either a purchase or no purchase.

   a. Construct a tree diagram for this three-step experiment.

   b. Identify each sample point and the sample space. How many sample points are there?

   c. How many sample points would there be if the experiment consisted of four sales calls?

3. An airline has offered a special vacation package to Hawaii. The length of stay is either for 3 days or 7 days, and the type of accommodations can be either economy, regular, or deluxe. Consider the experiment of observing the choices made by the next person signing up for the package.

    a. How many experimental outcomes are possible?

    b. Develop a tree diagram for this experiment

4. A major resort in Florida is concerned about weather conditions in both the Northeast as well as in Florida. In characterizing the temperature in both areas, the following three categories are used: below average, average, or above average. A combination of below-average temperatures in the Northeast with above-average temperatures in Florida means an increased volume of business for the resort. For the experiment of observing the weather conditions on a particular day, answer the following questions.

    a. How many experimental outcomes are possible?

    b. Develop a tree diagram for this experiment

5. As part of a special new car loan, a savings and loan association offers customers two down-payment options (10% or 20%), as well as two length-of-loan options (36 months or 48 months). Consider the choices made by the next loan applicant as an experiment.

    a. How many experimental outcomes are possible?

    b. Develop a tree diagram for this experiment.

6. A department store has advertised a special sale for a particular television model at two of its regional warehouses. There are four sets in inventory at warehouse 1 and three sets in inventory at warehouse 2. How many experimental outcomes are possible if the experiment consists of two steps: observing the number of sets sold at warehouse 1 and the number of sets sold at warehouse 2? Be sure to note that 0 sales is also a possibility. Develop a tree diagram for this experiment.

7. In the city of Milford applications for zoning changes go through a two-step process: a review by the planning commission and a final decision by the city council. At step 1 the planning commission will review the zoning change request and make a positive or negative recommendation concerning the change. At step 2 the city council will review the planning commission's recommendation and then vote to approve or to disapprove the zoning change. In some instances the city council vote has agreed with the planning commission's recommendation. However, in other instances the council vote has been the opposite of the planning commission's recommendation. An application for a zoning change has just been submitted by the developer of an apartment complex. Consider the application process as an experiment.

    a. How many sample points are there for this experiment? List the sample points.

    b. Construct a tree diagram for the experiment.

8. An investor has two stocks: stock A and stock B. Each stock may increase in value, decrease in value, or remain unchanged. Consider the experiment as the investment in the two stocks.

    a. How many experimental outcomes are possible?

    b. Show a tree diagram for the experiment.

    c. How many of the experimental outcomes result in an increase in value for at least one of the two stocks?

    d. How many of the experimental outcomes result in an increase in value for both of the stocks?

9. Consider the experiment of rolling a pair of dice. Each die has six possible results (the number of dots on its face).

    a. How many sample points are possible for this experiment?

    b. Show a tree diagram for the experiment.

    c. How many experimental outcomes provide a sum of 7 for the dots on the dice?

10. Many states design their automobile license plates such that space is available for up to six letters or numbers.

   a. If a state decides to use only numerical values for the license plates, how many different license plate numbers are possible? Assume that 000000 is an acceptable license plate number, although it will be used only for display purposes at the license bureau. (*Hint:* Use the counting rule.)

   b. If the state decides to use two letters followed by four numbers, how many different license plate numbers are possible? Assume that the letters I and O will not be used because of their similarity to numbers 1 and 0.

   c. Would larger states, such as New York and California, tend to use more or fewer letters in license plates? Explain.

4.2 ■■■ ASSIGNING PROBABILITIES TO EXPERIMENTAL OUTCOMES

We have an understanding of the concept of an experiment and of the sample space as the set of all experimental outcomes. Let us now see how probabilities for the experimental outcomes (sample points) can be determined. Recall the discussion at the beginning of this chapter which stated that the probability of an experimental outcome is a numerical measure of the likelihood that the experimental outcome will occur. In the assigning of probabilities to the experimental outcomes there are various acceptable approaches; however, regardless of the approach taken the following two basic requirements must be satisfied:

1. The probability values assigned to each experimental outcome (sample point) must be between 0 and 1. That is, if we let E_i indicate experimental outcome i and $P(E_i)$ indicate the probability of this experimental outcome, we must have

$$0 \leq P(E_i) \leq 1 \qquad \text{for all } i \tag{4.1}$$

2. The sum of *all* of the experimental outcome probabilities must be 1. For example, if a sample space has k experimental outcomes, we must have

$$P(E_1) + P(E_2) + \cdots + P(E_k) = \Sigma P(E_i) = 1 \tag{4.2}$$

Any method of assigning probability values to the experimental outcomes which satisfies these two requirements and results in reasonable numerical measures of the likelihood of the outcomes is acceptable. In practice, one of the following three methods can be used:

1. Classical method
2. Relative frequency method
3. Subjective method

Classical Method

To illustrate the classical method of assigning probabilities, let us again consider the experiment of flipping a coin. On any one flip, we will observe one of two experimental outcomes: head or tail. It would seem reasonable to assume that the two possible outcomes are equally likely. Therefore, since one of the two equally likely outcomes is a head, we logically should conclude that the probability of observing a head is $\frac{1}{2}$, or .50. Similarly, the probability of observing a tail is also .50. When the assumption of equally likely outcomes is used as a basis for assigning probabilities, the approach is referred to as the *classical method*. If an experiment has n possible outcomes, the classical method would assign a probability of $1/n$ to each experimental outcome.

As another illustration of the classical method, consider again the experiment of rolling a die. In Section 4.1 we described the sample space and sample points for this experiment with the following notation:

$$S = \{1, 2, 3, 4, 5, 6\}$$

It would seem reasonable to conclude that the six experimental outcomes are equally likely, and hence each outcome is assigned a probability of $\frac{1}{6}$. Thus, if $P(1)$ denotes the probability that one dot appears on the upward face of the die, then $P(1) = \frac{1}{6}$. Similarly, $P(2) = \frac{1}{6}$, $P(3) = \frac{1}{6}$, $P(4) = \frac{1}{6}$, $P(5) = \frac{1}{6}$, and $P(6) = \frac{1}{6}$. Note that this probability assignment satisfies the two basic requirements for assigning probabilities. In fact, requirements (4.1) and (4.2) must be satisfied when the classical method is used, since each of the n sample points is assigned a probability of $1/n$.

The classical method was developed originally in the analysis of gambling problems, where the assumption of equally likely outcomes often is reasonable. In many business problems, however, this assumption is not valid. Hence alternative methods of assigning probabilities are required.

Relative Frequency Method

As an illustration of the relative frequency method, consider a firm that is preparing to market a new product. In order to estimate the probability that a customer will purchase the product, a test market evaluation has been set up wherein salespeople will call on potential customers. For each sales call conducted, there are two possible outcomes: the customer purchases the product or the customer does not purchase the product. Since there is no reason to assume that the two experimental outcomes are equally likely, the classical method of assigning probabilities is inappropriate.

Suppose that in the test market evaluation of the product, 400 potential customers were contacted; 100 actually purchased the product, but 300 did not. In effect, then, we have repeated the experiment of contacting a customer 400 times and have found that the product was purchased 100 times. Thus we might decide to use the relative frequency of the number of customers that purchased the product as an estimate of the probability of a customer making a purchase. Hence we could assign a probability of $100/400 = .25$ to the experimental outcome of purchasing the product. Similarly, $300/400 = .75$ could be assigned to the experimental outcome of not purchasing the product. This approach to assigning probabilities is referred to as the *relative frequency method*.

Subjective Method

The classical and relative frequency methods cannot be applied to all situations where probability assessments are desired. For example, there are situations where the experimental outcomes are not equally likely and where relative frequency data are unavailable. For example, consider the next football game that the Pittsburgh Steelers will play. What is the probability that the Steelers will win? The experimental outcomes of a win, a loss, or a tie are not necessarily equally likely. Also, since the teams involved have not played several times previously this year, there are no relative frequency data available that are relevant to this upcoming game. Thus if we want an estimate of the probability of the Steelers winning, we must use a subjective opinion of its value.

With the subjective method of assigning probabilities to the experimental outcomes, we may use any data available as well as our experience and intuition. However, after we consider all available information, a probability value that expresses the *degree of belief* that the experimental outcome will occur must be specified. This method of assigning probability is referred to as the *subjective method*. Since subjective probability expresses a person's "degree of belief," it is personal. Different people can be expected to assign different probabilities to the same event. Nonetheless, care must be taken when using the subjective method to ensure that requirements (4.1) and (4.2) are satisfied. That is, regardless of a person's "degree of belief," the probability value assigned to each experimental outcome must be between 0 and 1 and the sum of all the experimental outcome probabilities must equal 1.

Even in situations where either the classical or relative frequency approach can be applied, management may want to provide subjective probability estimates. In such cases, the best probability estimates often are obtained by combining the estimates from the classical or relative frequency approaches with the subjective probability estimates.

Probabilities for the KP&L Problem

To perform further analysis for the KP&L problem, we must develop probabilities for each of the nine experimental outcomes listed in Table 4.1. Based on

TABLE 4.2   Completion Results for 40 KP&L Projects

Completion Time (months)		Sample Point	Number of Past Projects Having These Completion Times
Stage 1	*Stage 2*		
2	6	(2, 6)	6
2	7	(2, 7)	6
2	8	(2, 8)	2
3	6	(3, 6)	4
3	7	(3, 7)	8
3	8	(3, 8)	2
4	6	(4, 6)	2
4	7	(4, 7)	4
4	8	(4, 8)	6
		Total	40

TABLE 4.3 Probability Assignments for the KP&L Problem Based on The Relative Frequency Method

Sample Point	Project Completion Time	Probability of Sample Point
(2, 6)	8 months	$P(2, 6) = 6/40 = .15$
(2, 7)	9 months	$P(2, 7) = 6/40 = .15$
(2, 8)	10 months	$P(2, 8) = 2/40 = .05$
(3, 6)	9 months	$P(3, 6) = 4/40 = .10$
(3, 7)	10 months	$P(3, 7) = 8/40 = .20$
(3, 8)	11 months	$P(3, 8) = 2/40 = .05$
(4, 6)	10 months	$P(4, 6) = 2/40 = .05$
(4, 7)	11 months	$P(4, 7) = 4/40 = .10$
(4, 8)	12 months	$P(4, 8) = 6/40 = \underline{.15}$
		Total 1.00

experience and judgment, management concluded that the experimental outcomes were not equally likely. Hence the classical method of assigning probabilities could not be used. Management then decided to conduct a study of the completion times for similar projects undertaken by KP&L over the past three years. The results of a study of 40 similar projects are summarized in Table 4.2.

After reviewing the results of the study, management decided to employ the relative frequency method of assigning probabilities. Management could still have provided subjective probability estimates, but it was felt that the current project was quite similar to the 40 previous projects. Thus the relative frequency method was judged best.

In using the data in Table 4.2 to compute probabilities we note that outcome (2,6)—stage 1 completed in 2 months and stage 2 completed in 6 months—occurred 6 times in the 40 projects. Thus we can use the relative frequency method to assign a probability of $6/40 = .15$ to this outcome. Similarly, outcome (2,7) also occurred in 6 of the 40 projects, providing a $6/40 = .15$ probability. Continuing in this manner, we obtain the probability assignments for the sample points of the KP&L project shown in Table 4.3. Note that $P(2,6)$ represents the probability of the sample point (2,6), $P(2,7)$ represents the probability of the sample point (2,7), and so on.

■ EXERCISES

11. Consider the experiment of selecting a card from a deck of 52 cards.

a. How many sample points are possible?
b. Which method (classical, relative frequency, or subjective) would you recommend for assigning probabilities to the sample points?
c. What is the probability assignment for each card?
d. Show that your probability assignments satisfy the two basic requirements for assigning probabilities.

12. In a survey of new matriculants to MBA programs ("School Selection by Students," *GMAC Occasional Papers*, March 1988), the following data were obtained on the marital status of the students.

Marital status	Frequency
Never married	1106
Married	826
Other (Separated, widowed, divorced)	106
Total	2038

Consider the experiment of observing the marital status of a new MBA student.

a. What method would you recommend for assigning probabilities to the experimental outcomes?

b. Show your probability assignments.

13. A small-appliance store in Madeira has collected data on refrigerator sales for the last 50 weeks. The data are as follows:

Number of Refrigerators Sold	Number of Weeks
0	6
1	12
2	15
3	10
4	5
5	2
	50

Suppose that we are interested in the experiment of observing the number of refrigerators sold in 1 week of store operations.

a. How many experimental outcomes are there?

b. Which approach would you recommend for assigning probabilities to the experimental outcomes?

c. Assign probabilities and verify that your assignments satisfy the two basic requirements.

14. Strom Construction has made a bid on two contracts. The owner has identified the possible outcomes and subjectively assigned probabilities as follows:

Experimental Outcome	Obtain Contract 1	Obtain Contract 2	Probability
1	Yes	Yes	.15
2	Yes	No	.15
3	No	Yes	.30
4	No	No	.25

a. Are these valid probability assignments? Why or why not?

b. What would have to be done to make the probability assignments valid?

15. An investor forecasts that the probabilities that a certain stock will either go down, remain the same, or go up are .20, .60, and .30, respectively. Does this seem reasonable? Explain.

16. Faced with the question of determining the probability of obtaining either 0 heads, 1 head, or 2 heads when flipping a coin twice, an individual argued that since it seems reasonable to treat the outcomes as equally likely, the probability of each event is $1/3$. Do you agree? Explain.

17. Planes flying from New York City to Chicago are listed as either arriving early, on time, or late. Discuss how you could develop estimates of the probabilities for each of these events.

18. A company that manufactures toothpaste is studying five different package designs. Assuming that one design is just as likely to be selected by a consumer as any other design, what probability would you assign to each of the package designs being selected? In an actual experiment, 100 consumers were asked to pick the design they preferred. The following data were obtained.

Design	1	2	3	4	5
Total	5	15	30	40	10

Do the data appear to confirm the belief that one design is just as likely to be selected as another? Explain.

4.3 ■■■ EVENTS AND THEIR PROBABILITIES

Until now we have used the term *event* much as it would be used in everyday language. However, at this point we introduce the formal definition of an event as it relates to probability. This definition is as follows:

Event

An *event* is a collection of sample points.

(like a subset of sample points)

For an example, let us return to the KP&L problem and assume that the project manager is interested in the event that the entire project can be completed in 10 months or less. Referring to Table 4.3, we see that six sample points (2, 6), (2, 7), (2, 8), (3, 6), (3, 7), and (4, 6) provide a project completion time of 10 months or less. Let C denote the event that the project is completed in 10 months or less; we write

$$C = \{(2, 6), (2, 7), (2, 8), (3, 6), (3, 7), (4, 6)\}$$

Event C is said to occur if *any one* of the six sample points shown above appears as the experimental outcome.

Other events that might be of interest to KP&L management include the following:

L = the event that the project is completed in *less* than 10 months

M = the event that the project is completed in *more* than 10 months

Using the information in Table 4.3 we see that these events consist of the following sample points:

$$L = \{(2, 6), (2, 7), (3, 6)\}$$
$$M = \{(3, 8), (4, 7), (4, 8)\}$$

A variety of additional events can be defined for the KP&L problem, but in each case the event must be identified as a collection of sample points for the experiment.

Given the probabilities of the sample points shown in Table 4.3, we can use the following definition to compute the probability of any event that KP&L management might want to consider:

Probability of an Event

The probability of any event is equal to the sum of the probabilities of the sample points in the event.

Using this definition, we calculate the probability of a particular event by adding the probabilities of the experimental outcomes that make up the event. We can now compute the probability that the project will take 10 months or less to complete. Since this event is given by $C = \{(2, 6), (2, 7), (2, 8), (3, 6), (3, 7), (4, 6)\}$, the probability of event C is shown below (note that P is used to denote the probability of the corresponding event or sample point):

$$P(C) = P(2, 6) + P(2, 7) + P(2, 8) + P(3, 6) + P(3, 7) + P(4, 6)$$

Refer to the sample point probabilities in Table 4.3; we have

$$P(C) = .15 + .15 + .05 + .10 + .20 + .05 = .70$$

Similarly, since the event that the project is completed in less than 10 months is given by $L = \{(2, 6), (2, 7), (3, 6)\}$, the probability of this event is given by

$$P(L) = P(2, 6) + P(2, 7) + P(3, 6)$$
$$= .15 + .15 + .10 = .40$$

Finally, for the project to be completed in more than 10 months, we have $M = \{(3, 8), (4, 7), (4, 8)\}$ and thus

$$P(M) = P(3, 8) + P(4, 7) + P(4, 8)$$
$$= .05 + .10 + .15 = .30$$

Using the above probability results, we can now tell KP&L management that there is a .70 probability that the project will be completed in 10 months or less, a .40 probability that the project will be completed in less than 10 months, and a .30 probability the project will be completed in more than 10 months. This procedure of computing event probabilities can be repeated for any event of interest to the KP&L management.

Any time that we can identify all the sample points of an experiment and assign the corresponding sample point probabilities, we can compute the probability of an event of interest to a decision maker. However, in many experiments the number of sample points is large and the identification of the sample points, as well as determining their associated probabilities, becomes extremely cumbersome if not impossible. In the remaining sections of this chapter we present some basic probability relationships that can often be used to compute the probability of an event without requiring knowledge of sample point probabilities.

NOTES AND COMMENTS

1. The sample space, S, is an event. Since it contains all the experimental outcomes, it has a probability of 1; that is, $P(S) = 1$.
2. When the classical method is used to assign probabilities, the assumption is that the experimental outcomes are equally likely. In such cases, the probability of an event can be computed by counting the number of experimental outcomes in the event and dividing the result by the total number of experimental outcomes.

▬ EXERCISES

Use the Kentucky Power and Light Company sample point and sample point probabilities in Table 4.3 to answer the following:

a. The design stage (stage 1) will run over budget if it takes 4 months to complete. List the sample points in the event the design stage is over budget.
b. What is the probability that the design stage is over budget?
c. The construction stage (stage 2) will run over budget if it takes 8 months to complete. List the sample points in the event the construction stage is over budget.
d. What is the probability that the construction stage is over budget?
e. What is the probability that both stages are over budget?

20. In Exercise 11 we considered the experiment of selecting a card from a deck of 52 cards. Each card corresponded to a sample point with a 1/52 probability.

a. List the sample points in the event an ace is selected.
b. List the sample points in the event a club is selected.
c. List the sample points in the event a face card (jack, queen, or king) is selected.
d. Find the probabilities associated with each of the events in parts a, b, and c.

21. Suppose that a manager of a large apartment complex provides the following subjective probability estimate about the number of vacancies that will exist next month:

Vacancies	Probability
0	.05
1	.15
2	.35
3	.25
4	.10
5	.10

List the sample points in each of the following events and provide the probability of the event:

a. No vacancies.

b. At least four vacancies.

c. Two or fewer vacancies.

22. Consider the experiment of rolling a pair of dice. Suppose that we are interested in the sum of the face values showing on the dice.

a. How many sample points are possible? (Hint: use the counting rule for multiple-step experiments.)

b. List the sample points.

c. What is the probability of obtaining a value of 7?

d. What is the probability of obtaining a value of 9 or greater?

e. Since there are six possible even values (2, 4, 6, 8, 10, and 12) and only five possible odd values (3, 5, 7, 9, and 11), the dice should show even values more often than odd values. Do you agree with this statement? Explain.

f. What method did you use to assign the probabilities requested above?

23. The manager of a furniture store sells from zero to four china hutches each week. Based on past experience, the following probabilities are assigned to sales of zero, one, two, three, or four hutches:

$$P(0) = .08$$
$$P(1) = .18$$
$$P(2) = .32$$
$$P(3) = .30$$
$$P(4) = \underline{.12}$$
$$1.00$$

a. Are these valid probability assignments? Why or why not?

b. Let A be the event that two or fewer are sold in one week. Find $P(A)$.

c. Let B be the event that four or more are sold in one week. Find $P(B)$.

24. A sample of 100 customers of Montana Gas and Electric resulted in the following frequency distribution of monthly charges.

Amount $	Number
0–49	13
50–99	22
100–149	34
150–199	26
200–249	5

a. Let A be the event that monthly charges are $150 or more. Find $P(A)$.

b. Let B be the event that monthly charges are less than $150. Find $P(B)$.

25. A survey of 50 students at Tarpon Springs College regarding the number of extracurricular activities resulted in the following data.

Number of activities	0	1	2	3	4	5
Frequency	8	20	12	6	3	1

a. Let A be the event that a student participates in at least 1 activity. Find $P(A)$.

b. Let B be the event that a student participates in 3 or more activities. Find $P(B)$.

c. What is the probability a student participates in exactly 2 activities?

4.4 ▄▄▄ SOME BASIC RELATIONSHIPS OF PROBABILITY

Complement of an Event

Given an event A, the *complement* of A is defined to be the event consisting of all sample points that are *not* in A. The complement of A is denoted by A^c. Figure 4.4 provides a diagram, known as a *Venn diagram,* which illustrates the concept of a complement. The rectangular area represents the sample space for the experiment and as such contains all possible sample points. The circle represents event A and contains only the sample points that belong to A. The shaded region of the diagram contains all sample points not in event A, which is by definition the complement of A.

In any probability application, either event A or its complement A^c must occur. Therefore, we have

$$P(A) + P(A^c) = 1 \qquad (4.3)$$

Solving for $P(A)$, we obtain the following result:

Computing Probability Using the Complement

$$P(A) = 1 - P(A^c) \qquad (4.4)$$

Equation (4.4) shows that the probability of an event A can be easily computed if the probability of its complement, $P(A^c)$, is known.

As an example, consider the case of a sales manager who, after reviewing sales reports, states that 80% of new customer contacts result in no sale. By letting A denote the event of a sale and A^c denote the event of no sale, the manager is stating that $P(A^c) = .80$. Using (4.4), we see that

$$P(A) = 1 - P(A^c) = 1 - .80 = .20$$

This shows that there is a .20 probability that a sale will be made on a new customer contact.

In another example, a purchasing agent states that there is a .90 probability that a supplier will send a shipment that is free of defective parts. Using the

---

▄▄▄ FIGURE 4.4   **Complement of Event A**

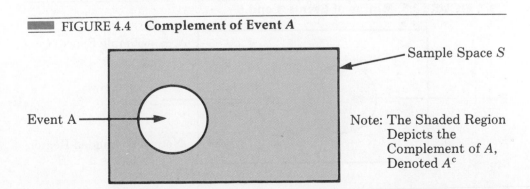

Sample Space S

Event A

Note: The Shaded Region Depicts the Complement of A, Denoted A^c

complement, we can conclude that there is a $1 - .90 = .10$ probability that the shipment will contain defective parts.

Addition Law

The addition law is a helpful probability relationship when we have two events and are interested in knowing the probability that at least one of the events occurs. That is, with events A and B we are interested in knowing the probability that event A or event B or both occur.

Before we present the addition law, we need to discuss two concepts concerning the combination of events: the *union* of events and the *intersection* of events.

Given two events A and B, the union of A and B is defined as follows:

Union of Two Events

The *union* of A and B is the event containing *all* sample points belonging to *A or B or both*. The union is denoted by $A \cup B$.

The Venn diagram shown in Figure 4.5 depicts the union of events A and B. Note that the shaded region contains all the sample points in event A as well as all the sample points in event B. The fact that the circles overlap indicates that there are some sample points contained in both A and B.

The definition of the intersection of two events A and B is as follows:

Intersection of Two Events

Given two events A and B, the *intersection* of A and B is the event containing the sample points belonging to *both A and B*. The intersection is denoted by $A \cap B$.

The Venn diagram depicting the intersection of the two events is shown in Figure 4.6. The area where the two circles overlap is the intersection; it contains the sample points that are in both A and B.

■ FIGURE 4.5　**Union of Events A and B**

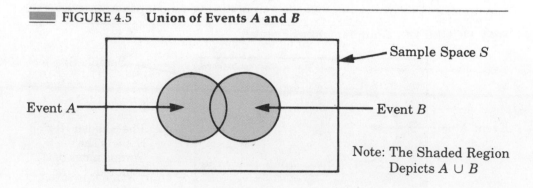

Event A　　　　　　　　　Event B

Sample Space S

Note: The Shaded Region Depicts $A \cup B$

Let us now continue with a discussion of the addition law. The addition law provides a way to compute the probability of event A or B or both occurring. In other words, the addition law is used to compute the probability of the union of two events, $A \cup B$. The addition law is written as follows:

Addition Law

$$P(A \cup B) = P(A) + P(B) - P(A \cap B) \qquad (4.5)$$

To obtain an intuitive understanding of the addition law, note that the first two terms in the addition law, $P(A) + P(B)$, account for all the sample points in $A \cup B$. However, since the sample points in the intersection $A \cap B$ are in both A and B, when we compute $P(A) + P(B)$ we are in effect counting each of the sample points in $A \cap B$ twice. We correct for this by subtracting $P(A \cap B)$.

In order to present an application of the addition law, let us consider the case of a small assembly plant with 50 employees. Each worker is expected to complete work assignments on time and in such a way that the assembled product will pass a final inspection. On occasion, some of the workers fail to meet the performance standards by completing work late and/or assembling defective products. At the end of a performance evaluation period, the production manager found that 5 of the 50 workers had completed work late, 6 of the 50 workers had assembled defective products, and 2 of the 50 workers had both completed work late *and* assembled defective products. Let

$$L = \text{the event that the work is completed late}$$

$$D = \text{the event that the assembled product is defective}$$

The above relative frequency information leads to the following probabilities:

$$P(L) = \frac{5}{50} = .10$$

$$P(D) = \frac{6}{50} = .12$$

$$P(L \cap D) = \frac{2}{50} = .04$$

■■■ FIGURE 4.6 **Intersection of Events A and B**

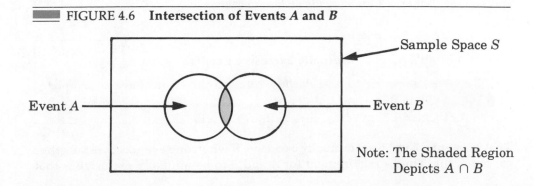

Note: The Shaded Region Depicts $A \cap B$

After reviewing the performance data, the production manager decided to assign a poor performance rating to any employee whose work was either late or defective; thus the event of interest is $L \cup D$. What is the probability that the production manager assigned an employee a poor performance rating?

Note that the probability question is about the union of two events. Specifically, we want to know $P(L \cup D)$. Here is where the addition law can be helpful. Using (4.5), we have

$$P(L \cup D) = P(L) + P(D) - P(L \cap D)$$

Knowing values for the three probabilities on the right-hand side of this expression, we can write

$$P(L \cup D) = .10 + .12 - .04 = .18$$

This tells us that there is a .18 probability that an employee will receive a poor performance rating.

As another example of the addition law, consider a recent study conducted by the personnel manager of a major computer software company. It was found that 30% of the employees that left the firm within 2 years did so primarily because they were dissatisfied with their salary, 20% left because they were dissatisfied with their work assignments, and 12% of the former employees said that *both* their salary and dissatisfaction with their work assignments were the primary reasons for leaving. What is the probability that an employee that leaves within 2 years does so because of being dissatisfied with salary, dissatisfied with the work assignment, or both?

Let

S = the event that the employee leaves due to salary

W = the event that the employee leaves due to work assignment

We have $P(S) = .30$, $P(W) = .20$, and $P(S \cap W) = .12$. Using (4.5), the addition law, we have

$$P(S \cup W) = P(S) + P(W) - P(S \cap W) = .30 + .20 - .12 = .38$$

This shows that there is a .38 probability that an employee leaves because of salary or work assignment reasons.

Before we conclude our discussion of the addition law, let us consider a special case that arises for *mutually exclusive events:*

Mutually Exclusive Events

Two events are said to be *mutually exclusive* if the events have no sample points in common.

That is, events A and B are mutually exclusive if when one event occurs the other cannot occur. Thus a requirement for A and B to be mutually exclusive is that their intersection must contain no sample points. The Venn diagram depicting two mutually exclusive events *A* and *B* is shown in Figure 4.7. In this case $P(A \cap B) = 0$; hence the addition law can be written as follows:

Addition Law for Mutually Exclusive Events

$$P(A \cup B) = P(A) + P(B)$$

▬ EXERCISES

26. A pharmaceutical company conducted a study to evaluate the effect of an allergy relief medicine; 250 patients with symptoms that included itchy eyes and a skin rash were given the new drug. The results of the study are as follows: 90 of the patients treated experienced eye relief, 135 had their skin rash clear up, and 45 experienced both relief from itchy eyes and the skin rash. What is the probability that a patient who takes the drug will experience relief for at least one of the two symptoms?

27. In a study of 100 students that had been awarded university scholarships, it was found that 40 had part-time jobs, 25 had made the dean's list the previous semester, and 15 had both a part-time job and had made the dean's list. What was the probability that a student had a part-time job or was on the dean's list?

28. A survey of the subscribers to *Fortune* magazine showed that 46% have mutual funds, 63% have money market funds, and 74% have mutual funds and/or money market funds (*Fortune Subscriber Portrait*, 1988). What is the probability a subscriber will have investments in both money market and mutual funds? What is the probability a subscriber will not have investments in either type of fund?

29. The survey of subscribers to *Fortune* magazine referred to in Exercise 28 (*Fortune Subscriber Portrait*, 1988) showed 54% rented a car in the past 12 months for business reasons, 51% rented a car for personal reasons, and 72% rented a car for either business or personal reasons.

   a. What is the probability a subscriber will have rented a car in the past 12 months for business reasons and for personal reasons?
   b. What is the probability a subscriber did not rent a car in the past 12 months?
   c. What is the probability a subscriber rented a car for business reasons only during the past 12 months?

▬ FIGURE 4.7 **Mutually Exclusive Events**

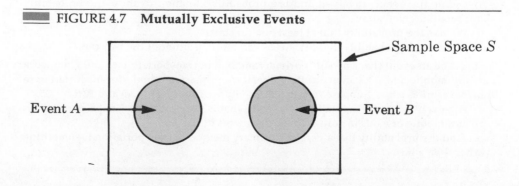

30. Suppose that we have a sample space $S = \{E_1, E_2, E_3, E_4, E_5, E_6, E_7\}$, where $E_1, E_2, \ldots,$ E_7 denote the sample points. The following probability assignments apply:

$$P(E_1) = .05$$
$$P(E_2) = .20$$
$$P(E_3) = .20$$
$$P(E_4) = .25$$
$$P(E_5) = .15$$
$$P(E_6) = .10$$
$$P(E_7) = \underline{.05}$$
$$1.00$$

Let

$$A = \{E_1, E_4, E_6\}$$
$$B = \{E_2, E_4, E_7\}$$
$$C = \{E_2, E_3, E_5, E_7\}$$

a. Find $P(A)$, $P(B)$, and $P(C)$.
b. Find $(A \cup B)$ and $P(A \cup B)$.
c. Find $(A \cap B)$ and $P(A \cap B)$.
d. Are events A and C mutually exclusive?
e. Find B^c and $P(B^c)$.

31. Let

A = the event that a person runs 5 miles or more per week

B = the event that a person dies of heart disease

C = the event that a person dies of cancer

Further, suppose that $P(A) = .01$, $P(B) = .25$, and $P(C) = .20$.

a. Are events A and B mutually exclusive? Can you find $P(A \cap B)$?
b. Are events B and C mutually exclusive? Find the probability that a person dies of heart disease or cancer.
c. Find the probability that a person dies from causes other than cancer.

32. During winter in Cincinnati, Mr. Krebs experiences difficulty in starting his two cars. The probability that the first car starts is .80, and the probability that the second car starts is .40. There is a probability of .30 that both cars start.

a. Define the events involved and use probability notation to show the probability information given above.
b. What is the probability that at least one car starts?
c. What is the probability that Mr. Krebs cannot start either of the two cars?

33. Let A be an event that a person's primary method of transportation to and from work is an automobile and B be an event that a person's primary method of transportation to and from work is a bus. Suppose that in a large city we find $P(A) = .45$ and $P(B) = .35$.

a. Are events A and B mutually exclusive? What is the probability that a person uses an automobile or a bus in going to and from work?
b. Find the probability that a person's primary method of transportation is something other than a bus.

4.5 ▩ CONDITIONAL PROBABILITY

Often, the probability of an event is influenced by whether or not a related event has occurred. Suppose that we have an event A with probability $P(A)$. If we obtain new information or learn that another, possibly related event, denoted by B, has occurred, we will want to take advantage of this information in calculating a new probability for event A. This new probability of event A is written $P(A \mid B)$. The "$\mid$" is used to denote the fact that we are considering the probability of event A *given* the condition that event B has occurred. Thus the notation $P(A \mid B)$ is read "the probability of A given B."

As an illustration of the application of *conditional probability*, consider the situation of the promotional status of male and female officers of a major metropolitan police force in the eastern United States. The police force consists of 1200 officers, 960 men and 240 women. Over the past 2 years, 324 officers on the police force have been awarded promotions. The specific breakdown of promotions for male and female officers is shown in Table 4.4.

After reviewing the promotional record, a committee of female officers raised a discrimination case on the basis that 288 male officers had received promotions, but only 36 female officers had received promotions. The police administration has argued that the relatively low number of promotions for female officers is due not to discrimination but to the fact that there are relatively few female officers on the police force. Let us show how conditional probability could be used to analyze the discrimination charge.

Let

$$M = \text{event an officer is a man}$$
$$W = \text{event an officer is a woman}$$
$$A = \text{event an officer is promoted}$$
$$A^c = \text{event an officer is not promoted}$$

Dividing the data values in Table 4.4 by the total of 1200 officers permits us to summarize the available information in the following probability values:

$$P(M \cap A) = 288/1200 = .24 = \text{probability that a randomly selected officer is a man } and \text{ is promoted}$$

TABLE 4.4   **Promotional Status of Police Officers Over the Past 2 Years**

	Men	Women	Totals
Promoted	288	36	324
Not Promoted	672	204	876
Totals	960	240	1200

$P(M \cap A^c) = 672/1200 = .56 =$ probability that a randomly selected officer is a man *and* is not promoted

$P(W \cap A) = 36/1200 = .03 =$ probability that a randomly selected officer is a woman *and* is promoted

$P(W \cap A^c) = 204/1200 = .17 =$ probability that a randomly selected officer is a woman *and* is not promoted

Since each of these values gives the probability of the intersection of two events, the probabilities are given the name of *joint probabilities*. Table 4.5, which provides a summary of the probability information for the police officer promotion situation, is referred to as a *joint probability table*.

The values in the margins of the joint probability table provide the probabilities of each event separately. That is, $P(M) = .80$, $P(W) = .20$, $P(A) = .27$, and $P(A^c) = .73$. These probabilities are referred to as *marginal probabilities* because of their location in the margins of the joint probability table. Thus we see that 80% of the force is male, 20% of the force is female, 27% of all officers received promotions, and 73% were not promoted.

Let us begin the conditional probability calculations by computing the probability that an officer is promoted given that the officer is a man. In conditional probability notation we are attempting to determine $P(A \mid M)$. In order to calculate $P(A \mid M)$, we first realize that this notation simply means that we are considering the probability of the event A (promotion) given that the condition designated as event M (the officer is a man) is known to exist. Thus $P(A \mid M)$ tells us that we are now concerned only with the promotional status of the 960 male officers. Hence since 288 of the 960 male officers received promotions, the probability of being promoted given that the officer is a man is $288/960 = .30$. In other words, given that an officer is a man there has been a 30% chance of receiving a promotion over the past 2 years.

The above procedure was easy to apply in our illustration because the data values in Table 4.4 show the number of officers in each category. We now want to demonstrate how conditional probabilites such as $P(A \mid M)$ can be computed directly from probability information rather than the frequency data of Table 4.4.

TABLE 4.5　Joint Probability Table for Promotions

Joint Probabilities Appear in the Body of the Table	Men (*M*)	Women (*W*)	Totals
Promoted (*A*)	.24	.03	.27
Not Promoted (*A^c*)	.56	.17	.73
Totals	.80	.20	1.00

Marginal Probabilities Appear in the Margins of the Table

We have shown that $P(A \mid M) = 288/960 = .30$. Let us now divide both the numerator and denominator of this fraction by 1200, the total number of officers in the study. Thus,

$$P(A \mid M) = \frac{288}{960} = \frac{288/1200}{960/1200} = \frac{.24}{.80} = .30$$

Hence we also see that the conditional probability $P(A \mid M)$ can be computed as $.24/.80$. Refer to the joint probability table shown in Table 4.5. Note in particular that .24 is the joint probability of A and M; that is, $P(A \cap M) = .24$. Also note that .80 is the marginal probability that a randomly selected officer is a man; that is, $P(M) = .80$. Thus the conditional probability $P(A \mid M)$ can be computed as the ratio of the joint probability $P(A \cap M)$ to the marginal probability $P(M)$. That is,

$$P(A \mid M) = \frac{P(A \cap M)}{P(M)} = \frac{.24}{.80} = .30$$

The fact that conditional probabilities can be computed as the ratio of a joint probability to a marginal probability provides the following general formula for conditional probability calculations for two events A and B.

Conditional Probability

$$P(A \mid B) = \frac{P(A \cap B)}{P(B)} \tag{4.6}$$

or

$$P(B \mid A) = \frac{P(A \cap B)}{P(A)} \tag{4.7}$$

Let us return to the issue of discrimination against the female officers. The probabilities in Table 4.5 show that the probability of promotion of an officer is $P(A) = .27$ (regardless of whether that officer is male or female). However, the critical issue in the discrimination case involves the two conditional probabilities $P(A \mid M)$ and $P(A \mid W)$. That is, what is the probability of a promotion *given* that the officer is a man, and what is the probability of a promotion *given* that the officer is a woman? If these two probabilities are equal, there is no basis for a discrimination argument, since the chances of a promotion are the same for male and female officers. However, if the two conditional probabilities differ, there will be support for the position that male and female officers are treated differently when it comes to promotions.

We have already determined that $P(A \mid M) = .30$. Let us now use the probability values in Table 4.5 and the basic relationship of conditional probability (4.6) to compute the probability that a randomly selected officer is promoted given that the officer is a woman; that is, $P(A \mid W)$. Using (4.6), we obtain

$$P(A \mid W) = \frac{P(A \cap W)}{P(W)} = \frac{.03}{.20} = .15$$

What conclusions do you draw? The probability of a promotion given that the officer is a man is .30, twice the .15 probability of a promotion given that the officer is a woman. While the use of conditional probability does not in itself prove that discrimination exists in this case, the conditional probability values offer support for the argument presented by the female officers.

Independent Events

In the above illustration we saw that $P(A) = .27$, $P(A \mid M) = .30$, and $P(A \mid W) = .15$. This shows that the probability of a promotion (event A) is affected or influenced by whether the officer is male or female. In particular, since $P(A \mid M) \neq P(A)$, we would say that events A and M are *dependent* events. That is, the probability of event A (promotion) is altered or affected by knowing whether or not M (the officer is a man) occurs. Similarly, with $P(A \mid W) \neq P(A)$, we would say that events A and W are *dependent* events. On the other hand, if the probability of event A were not changed by the existence of event M—that is, $P(A \mid M) = P(A)$—we would say that events A and M are *independent* events. This leads us to the following definition of the independence of two events:

Independent Events

Two events A and B are independent if

$$P(A \mid B) = P(A) \tag{4.8}$$

or

$$P(B \mid A) = P(B) \tag{4.9}$$

otherwise, the events are dependent.

Multiplication Law

While the addition law of probability is used to compute the probability of a union of two events, we can now show how the multiplication law can be used to find the probability of an intersection of two events. The multiplication law is based upon the definition of conditional probability. Using (4.6) and (4.7) and solving for $P(A \cap B)$, we obtain the *multiplication law*:

Multiplication Law

$$P(A \cap B) = P(B)P(A|B) \qquad (4.10)$$

or

$$P(A \cap B) = P(A)P(B|A) \qquad (4.11)$$

To illustrate the use of the multiplication law, consider a newspaper circulation department where it is known that 84% of the newspaper's customers subscribe to the daily edition of the paper. If we let D denote the event that a customer subscribes to the daily edition, $P(D) = .84$. In addition, it is known that the probability that a customer who already holds a daily subscription also subscribes to the Sunday edition (event S) is .75; that is, $P(S|D) = .75$. What is the probability that a customer subscribes to both the Sunday and daily editions of the newspaper? Using the multiplication law, we compute the desired $P(S \cap D)$ as follows:

$$P(S \cap D) = P(D)P(S|D) = .84 \,(.75) = .63$$

This tells us that 63% of the newspaper's customers take both the Sunday and daily editions.

Before concluding this section, let us consider the special case of the multiplication law when the events involved are independent. Recall that earlier in this section we defined independent events to exist whenever $P(A|B) = P(A)$ or $P(B|A) = P(B)$. Hence using (4.10) and (4.11) for the special case of independent events, the multiplication law becomes

---

Multiplication Law — Independent Events

$$P(A \cap B) = P(A)P(B) \qquad (4.12)$$

Thus to compute the probability of the intersection of two independent events we simply multiply the corresponding probabilities. Note that the multiplication law for independent events provides another way to determine if A and B are independent. That is, if $P(A \cap B) = P(A)P(B)$, then A and B are independent; if $P(A \cap B) \neq P(A)P(B)$, then A and B are dependent.

As an application of the multiplication law for independent events, consider the situation of a service station manager who knows from past experience that 80 percent of the customers use a credit card when they purchase gasoline. What is the probability that the next two customers purchasing gasoline will each use a credit card? If we let

> A = the event that the first customer uses a credit card

> B = the event that the second customer uses a credit card

then the event of interest is $A \cap B$. Given no other information, it seems reasonable to assume that A and B are independent events. Thus

$$P(A \cap B) = P(A)P(B) = (.80)(.80) = .64$$

■■ NOTES AND COMMENTS ■■

Do not confuse the notion of mutually exclusive events with that of independent events. Two events with nonzero probabilities cannot be both mutually exclusive and independent. If one mutually exclusive event is known to occur, the probability of the other occurring is reduced to zero. Thus they cannot be independent.

■■ EXERCISES

34. A Daytona Beach nightclub has the following data on the age and marital status of 140 customers:

		Marital Status	
		Single	Married
Age	Under 30	77	14
	30 or Over	28	21

a. Develop a joint probability table using this data.
b. Use the marginal probabilities to comment on the age of customers attending the club.
c. Use the marginal probabilities to comment on the marital status of customers attending the club.
d. What is the probability of finding a customer who is single and under the age of 30?
e. If a customer is under 30, what is the probability that he or she is single?
f. Is marital status independent of age? Explain, using probabilities.

35. In a survey of MBA students, the following data were obtained on "Students' first reason for application to the school in which they matriculated" ("School Selection by Students," *GMAC Occasional Papers*, Stolzenberg and Giarrusso, March, 1988).

		Reason for Application			
		School Quality	School Cost or Convenience	Other	Totals
Enrollment Status	Full time	421	393	76	890
	Part time	400	593	46	1039
	Totals	821	986	122	1929

a. Develop a joint probability table using this data.

b. Use the marginal probabilities of school quality, cost/convenience, and other to comment on the most important reason for choosing a school.

c. If a student goes full time, what is the probability school quality will be the first reason for choosing a school?

d. If a student goes part time, what is the probability school quality will be the first reason for choosing a school?

e. Let A be the event that a student is full-time and let B be the event that the student lists school quality as the first reason for applying. Are events A and B independent? Justify your answer.

36. A survey of automobile ownership was conducted for 200 families in Houston. The results of the study showing ownership of automobiles of United States and foreign manufacture are summarized as follows:

| | | Do You Own a U.S. Car? | | Totals |
		Yes	No	
Do you own a foreign car?	Yes	30	10	40
	No	150	10	160
Totals		180	20	200

a. Show the joint probability table for the above data.

b. Use the marginal probabilities to compare U.S. and foreign car ownership.

c. What is the probability that a family will own both a U.S. car and a foreign car?

d. What is the probability that a family owns a car, U.S. or foreign?

e. If a family owns a U.S. car, what is the probability that it also owns a foreign car?

f. If a family owns a foreign car, what is the probability that it also owns a U.S. car?

g. Are U.S. and foreign car ownership independent events? Explain.

37. The probability that Ms. Smith will get an offer on the first job she applies for is .5, and the probability that she will get an offer on the second job she applies for is .6. She thinks that the probability that she will get an offer on both jobs is .15.

a. Define the events involved, and use probability notation to state the probability information given above.

b. What is the probability that Ms. Smith gets an offer on the second job given that she receives an offer for the first job?

c. What is the probability that Ms. Smith gets an offer on at least one of the jobs she applies for?

d. What is the probability that Ms. Smith does not get an offer on either of the two jobs she applies for?

e. Are the job offers independent? Explain.

38. Shown are data from a sample of 80 families in a midwestern city. The data shows the record of college attendance by fathers and their oldest sons.

| | | Son | |
		Attended College	Did Not Attend College
Father	Attended College	18	7
	Did Not Attend College	22	33

a. Show the joint probability table.

b. Use the marginal probabilities to comment on the comparison between fathers and sons in terms of attending college.

c. What is the probability that a son attends college given that his father attended college?

d. What is the probability that a son attends college given that his father did not attend college?

e. Is attending college by the son independent of whether or not his father attended college? Explain, using probability values.

39. The Texas Oil Company provides a limited partnership arrangement whereby small investors can pool resources in order to invest in large scale oil exploration programs. In the exploratory drilling phase, locations for new wells are selected based on the geologic structure of the proposed drilling sites. Experience shows that there is a .40 probability of a type A structure present at the site given a productive well. It is also known that 50% of all wells are drilled in locations with type A structure. Finally, 30% of all wells drilled are productive.

a. What is the probability of a well being drilled in a type A structure *and* being productive?

b. If the drilling process begins in a location with a type A structure, what is the probability of having a productive well at the location?

c. Is finding a productive well independent of the type A geologic structure? Explain.

40. Assume that we have two events, A and B, that are mutually exclusive. Assume further that it is known that $P(A) = .30$ and $P(B) = .40$.

a. What is $P(A \cap B)$?

b. What is $P(A|B)$?

c. A student in statistics argues that the concepts of mutually exclusive events and independent events are really the same and that if events are mutually exclusive they must be independent. Do you agree with this statement? Use the probability information in this problem to justify your answer.

d. What general conclusion would you make about mutually exclusive and independent events given the results of this problem?

41. A purchasing agent has placed a rush order for a particular raw material with two different suppliers, A and B. If neither order arrives in 4 days the production process must be shut down until at least one of the orders arrives. The probability that supplier A can deliver the material in 4 days is .55. The probability that supplier B can deliver the material in 4 days is .35.

a. What is the probability that both suppliers deliver the material in 4 days? Since two separate suppliers are involved, we are willing to assume independence.

b. What is the probability that at least one supplier delivers the material in 4 days?

c. What is the probability the production process is shut down in 4 days because of a shortage in raw material (that is, both orders are late)?

4.6 ■■■ BAYES' THEOREM

In the discussion of conditional probability we indicated that revising probabilities when new information is obtained is an important phase of probability analysis. Often, we begin our analysis with initial or *prior* probability estimates for specific events of interest. Then, from sources such as a sample, a special report, a product test, and so on we obtain some additional information about

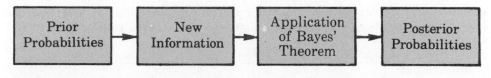

■ FIGURE 4.8 **Probability Revision Using Bayes' Theorem**

the events. Given this new information, we update the prior probability values by calculating revised probabilities, referred to as *posterior probabilities. Bayes' theorem* provides a means for making these probability calculations. The steps in this probability revision process are shown in Figure 4.8.

As an application of Bayes' theorem, consider a manufacturing firm that receives shipments of parts from two different suppliers. Let A_1 denote the event that a part is from supplier 1 and A_2 denote the event that a part is from supplier 2. Currently, 65% of the parts purchased by the company are from supplier 1, while the remaining 35% are from supplier 2. Thus if a part is selected at random, we would assign the prior probabilities $P(A_1) = .65$ and $P(A_2) = .35$.

The quality of the purchased parts varies with the source of supply. Based upon historical data, the quality ratings of the two suppliers are as shown in Table 4.6. Thus if we let G denote the event that a part is good and B denote the event that a part is bad, the information in Table 4.6 provides the following conditional probability values:

$$P(G|A_1) = .98 \qquad P(B|A_1) = .02$$
$$P(G|A_2) = .95 \qquad P(B|A_2) = .05$$

With the prior probabilities $P(A_1) = .65$ and $P(A_2) = .35$ and the above conditional probability information, Bayes' theorem can be used to revise the prior probabilities in light of new information. Suppose, for instance, that the parts from the two suppliers are used in the firm's manufacturing process and that a machine breaks down because it attempts to process a bad part. Given the information that the part causing the problem is bad, what is the probability that the part is from each of the suppliers? That is, letting B denote the bad part, what are the posterior probabilities $P(A_1|B)$ and $P(A_2|B)$?

Bayes' theorem is applicable when the events for which we want to compute posterior probabilities are mutually exclusive and their union is the entire

TABLE 4.6 **Conditional Probabilities based on Historical Quality Levels of Two Suppliers**

	Percentage Good Parts	Percentage Bad Parts
Supplier 1	98%	2%
Supplier 2	95%	5%

sample space.* To compute the posterior probabilities for events A_1 and A_2 when event B represents sample information obtained, we use the following:

Bayes' Theorem (Two-Event Case)

$$P(A_1|B) = \frac{P(A_1)P(B|A_1)}{P(A_1)P(B|A_1) + P(A_2)P(B|A_2)} \qquad (4.13)$$

$$P(A_2|B) = \frac{P(A_2)P(B|A_2)}{P(A_1)P(B|A_1) + P(A_2)P(B|A_2)} \qquad (4.14)$$

Using (4.13) and the probability values provided in our example, we have

$$P(A_1|B) = \frac{P(A_1)P(B|A_1)}{P(A_1)P(B|A_1) + P(A_2)P(B|A_2)}$$

$$= \frac{(.65)(.02)}{(.65)(.02) + (.35)(.05)} = \frac{.0130}{.0130 + .0175}$$

$$= \frac{.0130}{.0305} = .426$$

In addition, using (4.14), we find $P(A_2|B)$ as follows:

$$P(A_2|B) = \frac{(.35)(.05)}{(.65)(.02) + (.35)(.05)}$$

$$= \frac{.0175}{.0130 + .0175} = \frac{.0175}{.0305} = .574$$

Note that in this application we initially started with a probability of .65 that a part selected at random was from supplier 1. However, given information that the part is bad, the probability that the part is from supplier 1 drops to .426. In fact, if the part is bad, there is a better than 50–50 chance that the part came from supplier 2; that is, $P(A_2|B) = .574$.

Bayes' theorem can be extended to the case where there are n mutually exclusive events $A_1, A_2, \ldots, A_n$ whose union is the entire sample space. In such a case Bayes' theorem for the computation of any posterior probability $P(A_i|B)$ appears as follows:

Bayes' Theorem

$$P(A_i|B) = \frac{P(A_i)P(B|A_i)}{P(A_1)P(B|A_1) + P(A_2)P(B|A_2) + \cdots + P(A_n)P(B|A_n)} \qquad (4.15)$$

---

*If the union of events is the entire sample space, the events are often called *collectively exhaustive*.

TABLE 4.7 **Summary of Bayes' Theorem Calculations for the Two-Supplier Problem**

(1) Events A_i	(2) Prior Probabilities $P(A_i)$	(3) Conditional Probabilities $P(B\|A_i)$	(4) Joint Probabilities $P(A_i \cap B)$	(5) Posterior Probabilities $P(A_i\|B)$
A_1	.65	.02	.0130	.0130/.0305 = .426
A_2	.35	.05	.0175	.0175/.0305 = .574
	1.00		$P(B) = .0305$	1.000

With prior probabilities $P(A_1), P(A_2), \ldots, P(A_n)$ and the appropriate conditional probabilities $P(B|A_1), P(B|A_2), \ldots, P(B|A_n)$, Eq. (4.17) can be used to compute the posterior probability of an event $A_1, A_2, \ldots, A_n$.

The Tabular Approach ✓ *Calculating Bayes' theorem*

A tabular approach is helpful in conducting the Bayes' theorem calculations. Such an approach is shown in Table 4.7 for the parts supplier problem. The computations shown there are conducted as follows:

Step 1. Prepare the following three columns:
Column 1—The mutually exclusive events for which posterior probabilities are desired.
Column 2—The prior probabilities for the events.
Column 3—The conditional probabilities of the new information *given* each event.

Step 2. In column 4 compute the joint probabilities for each event and the new information B by using the multiplication law. These joint probabilities are found by multiplying the prior probabilities in column 2 by the corresponding conditional probabilities in column 3—that is, $P(A_i \cap B) = P(A_i)P(B|A_i)$.

Step 3. Sum the joint probabilities in column 4. The sum is the probability of the new information, $P(B)$. Thus we see that in the above example there is a .0130 probability of a bad part and supplier 1 and there is a .0175 probability of a bad part and supplier 2. Since these are the only two ways in which a bad part can be obtained, the sum .0130 + .0175 shows that there is an overall probability of .0305 of finding a bad part from the combined shipments of both suppliers.

Step 4. In column 5 compute the posterior probabilities using the basic relationship of conditional probability

$$P(A_i|B) = \frac{P(A_i \cap B)}{P(B)}$$

Note that the joint probabilities $P(A_i \cap B)$ are found in column 4, whereas the probability $P(B)$ appears as the sum of column 4.

NOTES AND COMMENTS

1. Bayes' Theorem is used extensively in decision analysis (Chapter 19). The prior probabilities are often subjective estimates provided by a decision maker. Sample information is obtained and posterior probabilities are computed for use in developing a decision strategy.
2. An event and its complement are mutually exclusive, and their union is the entire sample space. Thus, Bayes' theorem is always applicable for computing posterior probabilities of an event and its complement.

■ EXERCISES

42. The prior probabilities for events A_1, A_2, and A_3 are $P(A_1) = .20$, $P(A_2) = .50$, and $P(A_3) = .30$. The conditional probabilities of event B given A_1, A_2, and A_3 are $P(B|A_1) = .50$, $P(B|A_2) = .40$, and $P(B|A_3) = .30$.

a. Compute $P(B \cap A_1)$, $P(B \cap A_2)$, and $P(B \cap A_3)$.
b. Apply Bayes' theorem, equation (4.15), to compute the posterior probability $P(A_2|B)$.
c. Use the tabular approach to applying Bayes' theorem to compute $P(A_1|B)$, $P(A_2|B)$, and $P(A_3|B)$.

43. A consulting firm has submitted a bid for a large research project. The firm's management initially felt there was a 50–50 chance of getting the bid. However, the agency to which the bid was submitted has subsequently requested additional information on the bid. Past experience indicates that on 75% of the successful bids and 40% of the unsuccessful bids the agency requested additional information.

a. What is your prior probability the bid will be successful (i.e., prior to receiving the request for additional information)?
b. What is the conditional probability of a request for additional information given that the bid will ultimately be successful?
c. Compute a posterior probability that the bid will be successful given that a request for additional information has been received.

44. A local bank is reviewing its credit card policy with a view toward recalling some of its credit cards. In the past approximately 5% of cardholders have defaulted, and the bank has been unable to collect the outstanding balance. Thus management has established a prior probability of .05 that any particular cardholder will default. The bank has further found that the probability of missing one or more monthly payments for those customers who do not default is .20. Of course the probability of missing one or more payments for those who default is 1.

a. Given that a customer has missed a monthly payment, compute the posterior probability that the customer will default.
b. The bank would like to recall its card if the probability that a customer will default is greater than .20. Should the bank recall its card if the customer misses a monthly payment? Why or why not?

45. In a major eastern city, 60% of the automobile drivers are 30 years of age or older, and 40% of the drivers are under 30 years of age. Of all drivers 30 years of age or older, 4% will have a traffic violation in a 12-month period. Of all drivers under 30 years of age, 10% will have a traffic violation in a 12-month period. Assume that a driver has just been charged with a traffic violation; what is the probability that the driver is under 30 years of age?

46. A certain college football team plays 55% of its games at home and 45% of its games away. Given that the team has a home game, there is a .80 probability that it will win. Given that the team has an away game, there is a .65 probability that it will win. If the team wins on a particular Saturday, what is the probability that the game was played at home?

■■ SUMMARY

In this chapter we have introduced basic probability concepts and illustrated how probability analysis can be used to provide helpful decision-making information. We described how probability can be interpreted as a numerical measure of the likelihood that an event will occur. In addition, we saw that the probability of an event could be computed either by summing the probabilities of the experimental outcomes (sample points) comprising the event or by using the relationships established by addition, conditional probability, and multiplication laws of probability. For cases where additional information is available, we showed how Bayes' theorem could be used to obtain revised or posterior probabilities.

■■ GLOSSARY

Probability A numerical measure of the likelihood that an event will occur.

Experiment Any process which generates well defined outcomes.

Sample space The set of all possible sample points (experimental outcomes).

Sample points The individual outcomes of an experiment.

Tree diagram A graphical device helpful in defining sample points of an experiment involving multiple steps.

Basic requirements of probability Two requirements which restrict the manner in which probability assignments can be made:
   a.   For each experimental outcome E_i we must have $0 \le P(E_i) \le 1$.
   b.   If there are k experimental outcomes, then $\Sigma P(E_i) = 1$.

Classical method A method of assigning probabilities which assumes that the experimental outcomes are equally likely.

Relative frequency method A method of assigning probabilities based upon experimentation or historical data.

Subjective method A method of assigning probabilities based upon judgment.

Event A collection of sample points.

Complement of event A The event containing all sample points that are not in A.

Venn diagram A graphical device for representing symbolically the sample space and operations involving events.

Union of events A and B The event containing all sample points that are in A, in B, or in both. The union is denoted A $\cup$ B.

Intersection of A and B The event containing all sample points that are in both A and B. The intersection is denoted A $\cap$ B.

Addition law A probability law used to compute the probability of a union, $P(A \cup B)$. It is $P(A \cup B) = P(A) + P(B) - P(A \cap B)$. For mutually exclusive events, since $P(A \cap B) = 0$, it reduces to $P(A \cup B) = P(A) + P(B)$.

Mutually exclusive events Events that have no sample points in common; that is, $A \cap B$ is empty and $P(A \cap B) = 0$.

Conditional probability The probability of an event given that another event has occurred. The conditional probability of A given B is $P(A|B) = P(A \cap B)/P(B)$.

Independent events Two events A and B where $P(A|B) = P(A)$ or $P(B|A) = P(B)$; that is, the events have no influence on each other.

Multiplication law   A probability law used to compute the probability of an intersection, $P(A \cap B)$. It is $P(A \cap B) = P(A)P(B|A)$ or $P(A \cap B) = P(B)P(A|B)$. For independent events it reduces to $P(A \cap B) = P(A)P(B)$.

Prior probabilities   Initial estimates of the probabilities of events.

Posterior probabilities   Revised probabilities of events based on additional information.

Bayes' theorem   A method used to compute posterior probabilities.

▆▆ KEY FORMULAS

Computing Probability Using the Complement

$$P(A) = 1 - P(A^c) \tag{4.4}$$

Addition Law

$$P(A \cup B) = P(A) + P(B) - P(A \cap B) \tag{4.5}$$

Conditional Probability

$$P(A|B) = \frac{P(A \cap B)}{P(B)} \tag{4.6}$$

$$P(B|A) = \frac{P(A \cap B)}{P(A)} \tag{4.7}$$

Multiplication Law

$$P(A \cap B) = P(B)P(A|B) \tag{4.10}$$

$$P(A \cap B) = P(A)P(B|A) \tag{4.11}$$

Multiplication Law for Independent Events

$$P(A \cap B) = P(A)P(B) \tag{4.12}$$

Bayes' Theorem

$$P(A_i|B) = \frac{P(A_i)P(B|A_i)}{P(A_1)P(B|A_1) + P(A_2)P(B|A_2) + \cdots + P(A_n)P(B|A_n)} \tag{4.15}$$

▆▆ SUPPLEMENTARY EXERCISES

47. A research scientist is experimenting with a new drug that contains varying amounts of two chemicals. If the percentage of the first chemical used is either 1%, 2%, or 3% and the percentage of the second chemical used is either 1%, 2%, 3%, 4%, or 5%, how many different outcomes with varying combinations of the chemicals are possible?

48. The Food and Drug Administration (FDA) places new drug applications in one of three categories:

A = Potential breakthrough

B = Improvement over existing product

C = Me-too drug

During 1985, the FDA approved 3 A's, 15 B's, and 12 C's (*Financial World*, January 24, 1989).

   a. Consider the experiment of observing the category to which a new FDA approved drug is assigned. How many experimental outcomes are there?

   b. Using the data for 1985, assign probabilities to the experimental outcomes.

49. A financial manager has just made two new investments—one in the oil industry, and one in municipal bonds. After a 1-year period, each of the investments will be classified as either successful or unsuccessful. Consider the making of the two investments as an experiment.

   a. How many sample points exist for this experiment?

   b. Show a tree diagram and list the sample points.

   c. Let O = the event that the oil investment is successful and M = the event that the municipal bond investment is successful. List the sample points in O and in M.

   d. List the sample points in the union of the events $(O \cup M)$.

   e. List the sample points in the intersection of the events $(O \cap M)$.

   f. Are events O and M mutually exclusive? Explain.

50. A survey of 2125 subscribers to *Fortune* magazine indicated the following with respect to the number of subscribers owning various credit cards (*Fortune Subscriber Portrait*, 1988).

Type of Card	Number Holding
American Express	1360
Diners Club	234
Telephone Credit Card	1530
Gasoline Credit Card	1424
MasterCard	1466
VISA	1679
Discover	425

   a. If 2083 subscribers indicated they have at least one credit card, what is the probability a *Fortune* subscriber does not have any credit cards?

   b. What is the probability a subscriber holds an American Express card?

   c. Suppose it were known that 55% have both a MasterCard and a VISA card. What is the probability an individual holds one or the other or both?

   d. After studying the data, an analyst concluded that *at least* 956, or about 45%, of the subscribers must hold both an American Express and VISA card. Does this make sense? Why or why not?

51. A survey of new matriculants to MBA programs was conducted by the Graduate Management Admissions Council during 1985 ("School Selection by Students," *GMAC Occasional Papers*, March 1988). The following data show the number of schools to which students applied.

Number of Schools	Number of Students
1	1230
2	304
3	184
4	118
5	78
6	51
7	25
8	13
9	20
10	8
11	9
12	6
Total	2046

a. Use these data to assign probabilities to the number of schools to which an MBA student applies.

b. What is the probability a student will apply to only one school?

c. What is the probability a student will apply to three or more schools?

d. What is the probability a student will apply to more than six schools?

52. A telephone survey was used to determine viewer response to a new television show. The following data were obtained.

Rating	Frequency
Poor	4
Below average	8
Average	11
Above average	14
Excellent	13

a. What is the probability that a randomly selected viewer rates the new show as average or better?

b. What is the probability that a randomly selected viewer rates the new show below average or worse?

53. A bank has observed that credit-card account balances have been growing over the past year. A sample of 200 customer accounts resulted in the following data.

Amount Owed $	Frequency
0–99	62
100–199	46
200–299	24
300–399	30
400–499	26
500 and over	12

a. Let A be the event that a customer's balance is less than $200. Find $P(A)$.

b. Let B be the event that a customer's balance is $300 or more. Find $P(B)$.

54. Consider an experiment where eight experimental outcomes exist. We will denote the experimental outcomes as $E_1, E_2, \ldots, E_8$. Suppose that the following events are identified:

$$A = \{E_1, E_2, E_3\}$$

$$B = \{E_2, E_4\}$$

$$C = \{E_1, E_7, E_8\}$$

$$D = \{E_5, E_6, E_7, E_8\}$$

Determine the sample points making up the following events:

a. $A \cup B$

b. $C \cup D$

c. $A \cap B$

d. $C \cap D$

e. $B \cap C$

f. A^c

g. D^c

h. $A \cup D^c$

i. $A \cap D^c$.

j. Are A and B mutually exclusive?

k. Are B and C mutually exclusive?

55. Referring to Exercise 54 and assuming that the classical method is an appropriate way of establishing probabilities, find the following probabilities:

a. $P(A), P(B), P(C)$, and $P(D)$

b. $P(A \cap B)$

c. $P(A \cup B)$

d. $P(A|B)$

e. $P(B|A)$

f. $P(B \cap C)$

g. $P(B|C)$

h. Are B and C independent events?

56. Additional data from the GMAC MBA new-matriculants survey (see Exercise 51) shows the following:

		Applied to More than One School	
		Yes	No
Age Group	23 and under	207	201
	24–26	299	379
	27–30	185	268
	31–35	66	193
	36 and over	51	169

a. Prepare a joint probability table for the experiment consisting of observing the age and number of schools to which a randomly selected MBA student applies.

b. What is the probability an applicant will be 23 or under?

c. What is the probability an applicant will be older than 26?

d. What is the probability an applicant applies to more than one school?

57. Refer again to the data from the GMAC new-matriculants survey in Exercise 56.

a. Given that a person applied to more than one school, what is the probability the person is 24 to 26 years old?

b. Given that a person is in the 36-and-over age group, what is the probability the person applied to more than one school?

c. What is the probability a person is 24–26 years old *or* applied to more than one school?

d. Suppose a person is known to have applied to only one school. What is the probability the person is 31–35 years old?

e. Is the number of schools applied to independent of age? Explain.

58. Suppose that $P(A) = .30, P(B) = .25$, and $P(A \cap B) = .20$.

a. Find $P(A \cup B), P(A|B)$, and $P(B|A)$.

b. Are events A and B independent? Why or why not?

59. Suppose that $P(A) = .40, P(A|B) = .60$, and $P(B|A) = .30$.

a. Find $P(A \cap B)$ and $P(B)$.

b. Are events A and B independent? Why or why not?

60. Suppose that $P(A) = .60, P(B) = .30$, and events A and B are mutually exclusive.

a. Find $P(A \cup B)$ and $P(A \cap B)$.

b. Are events A and B independent?

c. Can you make a general statement about whether or not mutually exclusive events can be independent?

61. A market survey of 800 people found the following facts about the ability to recall a television commercial for a particular product and the actual purchase of the product:

	Could Recall Television Commercial	Could Not Recall Television Commercial	Totals
Purchased the Product	160	80	240
Had Not Purchased the Product	240	320	560
Totals	400	400	800

Let T be the event of the person recalling the television commercial and B the event of buying or purchasing the product.

a. Find $P(T)$, $P(B)$, and $P(T \cap B)$.

b. Are T and B mutually exclusive events? Use probability values to explain.

c. What is the probability that a person who could recall seeing the television commercial has actually purchased the product?

d. Are T and B independent events? Use probability values to explain.

e. Comment on the value of the commercial in terms of its relationship to purchasing the product.

62. A research study investigating the relationship between smoking and heart disease in a sample of 1000 men over 50 years of age provided the following data:

	Smoker	Nonsmoker	Totals
Record of Heart Disease	100	80	180
No Record of Heart Disease	200	620	820
Totals	300	700	1000

a. Show a joint probability table that summarizes the results of this study.

b. What is the probability a man over 50 years of age is a smoker and has a record of heart disease?

c. Compute and interpret the marginal probabilities.

d. Given that a man over 50 years of age is a smoker, what is the probability that he has heart disease?

e. Given that a man over 50 years of age is a nonsmoker, what is the probability that he has heart disease?

f. Does the research show that heart disease and smoking are independent events? Use probability to justify your answer.

g. What conclusion would you draw about the relationship between smoking and heart disease?

63. A large consumer goods company has been running a television advertisement for one of its soap products. A survey was conducted. On the basis of this survey probabilities were assigned to the following events:

B = individual purchased the product

S = individual recalls seeing the advertisement

$B \cap S$ = individual purchased the product and recalls seeing the advertisement

The probabilities assigned were $P(B) = .20$, $P(S) = .40$, and $P(B \cap S) = .12$. The following problems relate to this situation:

a. What is the probability of an individual's purchasing the product given that the individual saw the advertisement? Does seeing the advertisement increase the probability the individual will purchase the product? As a decision maker, would you recommend continuing the advertisement (assuming that the cost is reasonable)?

b. Assume that those individuals who do not purchase the company's soap product buy from its competitors. What would be your estimate of the company's market share? Would you expect that continuing the advertisement will increase the company's market share? Why or why not?

c. The company has also tested another advertisement and assigned it values of $P(S) = .30$ and $P(B \cap S) = .10$. What is $P(B|S)$ for this other advertisement? Which advertisement seems to have had the bigger effect on customer purchases?

64. A large company has done a careful analysis of a price promotion that it is currently testing. Some 20% of the people in a large sample of individuals in the test market both were aware of the promotion and made a purchase. It was further found that 80% were aware of the promotion and that prior to the promotion 25% of all people in the sample were purchasers of the product.

a. What is the probability that a person will make a purchase given that he or she is aware of the price promotion?

b. Are the events "made a purchase" and "aware of the price promotion" independent? Why or why not?

c. On the basis of these results, would you recommend that the company introduce this promotion on a national scale? Why or why not?

65. Cooper Realty is a small real estate company located in Albany, New York, and specializing primarily in residential listings. They have recently become interested in the possibility of determining the likelihood of one of their listings being sold within a certain number of days. An analysis of company sales of 800 homes for the previous years produced the accompanying data:

Initial Asking Price	Days Listed Until Sold			Totals
	Under 30	31–90	Over 90	
Under $50,000	50	40	10	100
$50,000–99,999	20	150	80	250
$100,000–150,000	20	280	100	400
Over $150,000	10	30	10	50
Totals	100	500	200	800

a. If A is defined as the event that a home is listed for over 90 days before being sold, estimate the probability of A.

b. If B is defined as the event that the initial asking price is under $50,000, estimate the probability of B.

c. What is the probability of $A \cap B$?

d. Assuming that a contract has just been signed to list a home that has an initial asking price of less than $50,000, what is the probability the home will take Cooper Realty more than 90 days to sell?

e. Are events A and B independent?

66. In the evaluation of a sales training program, a firm found that of 50 salespersons making a bonus last year, 20 had attended a special sales training program. The firm has

200 salespersons. Let B = the event that a salesperson makes a bonus and S = the event a salesperson attends the sales training program.

   a. Find $P(B)$, $P(S|B)$, and $P(S \cap B)$.

   b. Assume that 40% of the salespersons have attended the training program. What is the probability that a salesperson makes a bonus given that the salesperson attended the sales training program, $P(B|S)$?

   c. If the firm evaluates the training program in terms of the effect it has on the probability of a salesperson's making a bonus, what is your evaluation of the training program? Comment on whether B and S are dependent or independent events.

67. A company has studied the number of lost-time accidents occurring at its Brownsville, Texas plant. Historical records show that 6% of the employees had lost-time accidents last year. Management believes that a special safety program will reduce the accidents to 5% during the current year. In addition, it is estimated that 15% of those employees having had lost-time accidents last year will have a lost-time accident during the current year.

   a. What percentage of the employees will have lost-time accidents in both years?

   b. What percentage of the employees will have at least one lost-time accident over the 2 year period?

68. In a study of television viewing habits among married couples, a researcher found that for a popular Saturday night program 25% of the husbands viewed the program regularly and 30% of the wives viewed the program regularly. The study found that for couples where the husband watches the program regularly 80% of the wives also watch regularly.

   a. What is the probability that both the husband and wife watch the program regularly?

   b. What is the probability that at least one—husband or wife—watches the program regularly?

   c. What percentage of married couples do not have at least one regular viewer of the program?

69. A statistics professor has noted from past experience that students who do the homework for the course have a .90 probability of passing the course. On the other hand, students who do not do the homework for the course have a .25 probability of passing the course. The professor estimates that 75% of the students in the course do the homework. Given a student who passes the course, what is the probability that she or he completed the homework?

70. A salesperson for Business Communication Systems, Inc. sells automatic envelope-addressing equipment to medium- and small-size businesses. The probability of making a sale to a new customer is .10. During the initial contact with a customer, sometimes the salesperson will be asked to call back later. Of the 30 most recent sales, 12 were made to customers who initially told the salesperson to call back later. Of 100 customers who did not make a purchase, 17 had initially asked the salesperson to call back later. If a customer asks the salesperson to call back later, should the salesperson do so? What is the probability of making a sale to a customer who has asked the salesperson to call back later?

71. Migliori Industries, Inc. manufactures a gas-saving device for use on natural gas forced-air residential furnaces. The company is currently trying to determine the probability that sales of this product will exceed 25,000 units during next year's winter sales period. The company believes that sales of the product depend to a large extent on the winter conditions. Management's best estimate is that the probability that sales will exceed 25,000 units if the winter is severe is .8. This probability drops to .5 if the winter

conditions are moderate. If the weather forcast is .7 for a severe winter and .3 for moderate conditions, what is Migliori's best estimate that sales will exceed 25,000 units?

72. The Dallas IRS auditing staff is concerned with identifying potential fraudulent tax returns. From past experience they believe that the probability of finding a fraudulent return given that the return contains deductions for contributions exceeding the IRS standard is .20. Given that the deductions for contributions do not exceed the IRS standard, the probability of a fraudulent return decreases to .02. If 8% of all returns exceed the IRS standard for deductions due to contributions, what is the best estimate of the percentage of fraudulent returns?

73. An oil company has purchased an option on land in Alaska. Preliminary geologic studies have assigned the following prior probabilities:

$$P(\text{high quality oil}) = .50$$
$$P(\text{medium quality oil}) = .20$$
$$P(\text{no oil}) = .30$$

a. What is the probability of finding oil?
b. After 200 feet of drilling on the first well, a soil test is taken. The probabilities of finding the particular type of soil identified by the test are as follows:

$$P(\text{soil} \mid \text{high quality oil}) = .20$$
$$P(\text{soil} \mid \text{medium quality oil}) = .80$$
$$P(\text{soil} \mid \text{no oil}) = .20$$

How should the firm interpret the soil test? What are the revised probabilities, and what is the new probability of finding oil?

74. In the setup of a manufacturing process, a machine is either correctly or incorrectly adjusted. The probability of a correct adjustment is .90. When correctly adjusted, the machine operates with a 5% defective rate. However, if it is incorrectly adjusted, a 75% defective rate occurs.

a. After the machine starts a production run, what is the probability that a defect is observed when one part is tested?
b. Suppose that the one part selected by an inspector is found to be defective. What is the probability that the machine is incorrectly adjusted? What action would you recommend?
c. Before your recommendation in part b above was followed, a second part is tested and found to be good. Using your revised probabilities from part b as the most recent prior probabilities, compute the revised probability of an incorrect adjustment given that the second part is good. What action would you recommend now?

75. The Wayne Manufacturing Company purchases a certain part from three suppliers A, B, and C. Supplier A supplies 60% of the parts, B 30%, and C 10%. The quality of parts is known to vary among suppliers, with A, B, and C parts having .25%, 1%, and 2% defective rates, respectively. The parts are used in one of the company's major products.

a. What is the probability that the company's major product is assembled with a defective part?
b. When a defective part is found, which supplier is the likely source?

76. A Bayesian approach can be used to revise probabilities that a prospect field will produce oil (*Oil & Gas Journal*, January 11, 1988). In one case, geological assessment

indicates a 25% chance the field will produce oil. Further, there is an 80% chance that a particular well will strike oil given that oil is present on the prospect field.

a. Suppose that one well is drilled on the field and it comes up dry. What is the probability the prospect field will produce oil?

b. If two wells come up dry, what is the probability the field will produce oil?

c. The oil company would like to keep looking as long as the chances of finding oil are greater than 1%. How many dry wells must be drilled before the field will be abandoned?

Morton International*

Chicago, Illinois

| MORTON |

Morton Norwich combined with Thiokol Corporation in 1982 to form the company Morton Thiokol, Inc., now Morton International. From a salt business first started in Chicago in 1848, Jay Morton named his firm the Morton Salt Company in 1910. In 1914 the Morton Girl and slogan "When it rains, it pours" established Morton Salt as a recognized name to the consumer. In the following years, Morton added specialty chemicals businesses and Texize, a manufacturer of consumer household products, to its rapidly growing salt business.

Thiokol Corporation began in 1928 as a manufacturer of specialty polysulfide polymers that found widespread use as sealants. Other specialty chemical businesses that serve plastic, electronic, and other industries were added later. As an outgrowth of the expanding chemical business, the company became involved in the manufacture of solid rocket propellants and today builds a variety of rocket motors used in the United States space programs.

The combination of Morton and Thiokol has resulted in a company with strong businesses in salt and household products, rocket motors, and specialty chemicals. In particular, the specialty chemicals group now consists of several decentralized manufacturing operations serving their particular market segment.

PROBABILITY ANALYSIS GUIDES MANAGEMENT DECISION MAKING

Managers at the various divisions throughout Morton make decisions in light of uncertain outcomes. Often the managers' "feel" for the chances associated with the outcomes and/or subjective estimates of probabilities suggest a specific decision or course of action. Issues such as how much inventory to maintain, forecasts of demand, and estimates about order quantities from specific customers are instances where managers use probability considerations as part of the decision making process. In the following example we describe how probability considerations aided a customer service decision at Carstab Corporation, a subsidiary of Morton.

Carstab Corporation provides a variety of specialty chemical products for its customers. Because of the diversity of customer applications, customers differ substantially in terms of the unique specifications they require for the product. One approach to servicing customers would be for Carstab to wait until a customer order is received and then make the product to the exact customer specifications. A disadvantage of this approach is that the customer's order would have to wait until Carstab could schedule production for the special order. In some instances this would require an undesirable waiting period for the customer. In addition, this approach also has the disadvantage of requiring Carstab to make relatively short production runs for specific customer orders thus losing the economic advantage of larger production runs. In order to avoid these disadvantages,

*The authors are indebted to Michael Haskell of Morton Thiokol's International for providing this application.

Carstab makes relatively large production runs for many of its products and holds goods in inventory awaiting customers' orders. In doing this the company realizes that not all of its inventory will meet the unique specification requirements for all customers.

In one instance, a customer made small but repeated orders for an expensive catalyst product used in its chemical processing. Because of the nature of its operation, the customer placed unique specifications on the product. Some, but not all of the lots produced by Carstab would meet the customer's exact specifications.

The customer agreed to test each lot as it was received to determine whether or not the catalyst would perform the desired function. Carstab agreed to ship lots to the customer with the understanding the customer would perform the test and return the lots that did not pass the customer's specification test. The problem encountered was that only 60% of the lots sent to the customer would pass the customer's test. This meant that although the product was still good and usable to other customers, approximately 40% of the shipments sent to this particular customer were being returned.

Carstab explored the possibility of duplicating the customer's test and only shipping lots that passed the test. However, the test was unique to this one customer, and it was infeasible to purchase the expensive testing equipment needed to perform the customer's test.

Therefore, in order to improve the customer service Carstab chemists designed a new test, one that was believed to indicate whether or not the lot would eventually pass the customer's test. The question was: Would the Carstab test increase the probability that a lot shipped to the customer would pass the customer's test. The probability information sought was: What is the probability that a lot will pass the customer's test given it has passed the company's test?

A sample of lots was tested under both the customer's procedure and the company's proposed procedure. Results were that 55% of the lots passed the company's test and 50% of the lots passed both the customer's and the company's test. In probability notation, we have

A = the event the lot passes the customer's test

B = the event the lot passes the company's test

The Morton Salt Girl and the famous slogan.

where

$$P(B) = .55 \quad \text{and} \quad P(A \cap B) = .50$$

The probability information sought was the conditional probability $P(A|B)$ which was given by

$$P(A|B) = \frac{P(A \cap B)}{P(B)} = \frac{.50}{.55} = .909$$

Prior to the company's test the probability a lot would pass the customer's test was .60. However, the new results showed that given a lot passed the company's test it had a .909 probability of passing the customer's test. This was good supporting evidence for the use of the test prior to shipment. Based on this probability analysis, the preshipment testing procedure was implemented at the company. Immediate results showed an improved level of customer service. A few lots were still being returned; however, the percentage was greatly reduced. The customer was more satisfied and return shipping costs were reduced.

As seen in this example, probability did not make the decision for the manager. However, some basic probability considerations provided important decision-making information and were a significant factor in the decision to implement the new testing procedure which resulted in improved service to the customer.

CHAPTER 5

Discrete Probability Distributions

I n this chapter we continue the study of probability by introducing the concepts of random variables and probability distributions. The focus of this chapter is on discrete probability distributions. Three special discrete probability distributions, the binomial, Poisson, and hypergeometric are studied.

5.1 ▨ RANDOM VARIABLES

In Chapter 4 we defined the concept of an experiment and its associated experimental outcomes. A random variable provides a means of assigning numerical values to experimental outcomes. The definition of a random variable is as follows.

Random Variable

A *random variable* is a numerical description of the outcome of an experiment.

For any experiment a random variable can be defined such that each possible experimental outcome generates one and only one numerical value for the random variable. The particular numerical value that the random variable takes on depends upon the outcome of the experiment. That is, the value of the random variable is not known until the experimental outcome is observed.

Suppose we consider the experiment of selling automobiles for one day at a particular dealership. In this case if we let x = number of cars sold, then x is a

TABLE 5.1 **Examples of Random Variables**

Experiment	Random Variable (x)	Possible Values for the Random Variable
Make 100 sales calls	Total number of sales	0, 1, 2, . . . , 100
Inspect a shipment of 70 radios	Number of defective radios	0, 1, 2, . . . , 70
Work 1 year on a project to build a new library	Percentage of project completed after 6 months	$0 \leq x \leq 100$
Operate a restaurant	Number of customers entering in one day	0, 1, 2, . . .
Observe cars passing a checkpoint	Number of cars passing the checkpoint	0, 1, 2, . . .

random variable whose possible values are 0, 1, 2, ...* For another example, consider the fact that to receive state certification as a medical lab technician, candidates must pass a series of three examinations. If we let x = number of examinations passed, then x is a random variable that may assume the values 0, 1, 2, and 3. Some additional examples of experiments and associated random variables are given in Table 5.1.

Although many experiments such as those listed in Table 5.1 have experimental outcomes that are numerical values, others do not. For example, the outcome for the experiment of tossing a coin one time is either a head or a tail, neither of which has a natural numerical value. However, we still may want to express the outcome numerically. Thus, we need a rule that can be used to assign a numerical value to each of the experimental outcomes. One possibility is to let the random variable $x = 1$ if the experimental outcome is a head and $x = 0$ if the experimental outcome is a tail. While the numerical values for the random variable x are arbitrary, they are acceptable in terms of the definition of a random variable—namely, x is a random variable because it provides a numerical description of the outcome of the experiment.

A random variable can be classified as either *discrete* or *continuous* depending upon the numerical values it can assume. A random variable that may assume either a finite number of values or an infinite sequence (e.g., 1, 2, 3, . . .) of values is referred to as a *discrete random variable*. The number of units sold, the number of defects observed, and the number of customers that enter a bank during one day of operation are examples of discrete random variables. The first two and last two random variables listed in Table 5.1 are discrete random variables. Random variables such as weight, time, and temperature, which may take on all values in a certain interval or collection of intervals, are referred to as *continuous random variables*. For instance, the third random variable in Table 5.1 (percentage of project completed after 6 months) is a continuous random variable because it may take on any value in the interval from 0 to 100 (e.g., 56.33 or 64.227). Continuous random variables will be discussed further in Chapter 6.

---

*More advanced texts make a distinction between the random variable, denoted by X, and the values it can take on, denoted by x. For our purposes, such a distinction is unnecessary.

NOTES AND COMMENTS

One way to determine whether a random variable is discrete or continuous is to think of the values of the random variable as points on a line segment. If the entire line segment between any two of these points also represents values the random variable may assume, the random variable is continuous.

■■■ EXERCISES

1. Three students have interviews scheduled for summer employment at the Brookwood Institute. In each case the result of the interview will either be that a position is offered or not offered. Experimental outcomes are defined in terms of the results of the three interviews.

   a. List the experimental outcomes.
   b. Define a random variable that represents the number of offers made. Is this a discrete or continuous random variable?
   c. Show the value of the random variable for each of the experimental outcomes.

2. Subscribers of the Turner Broadcasting System may select one or more of the following services: Cable News Network, Superstation WTBS, Headline News, and Turner Network Television (*Business Week*, April 17, 1989). Suppose an experiment is designed to take a sample of five subscribers. The random variable x is defined as the number in the sample who receive the Cable News Network service. What values may the random variable take on?

3. In order to perform a certain type of blood analysis, lab technicians have to perform two procedures. The first procedure requires either 1 or 2 separate steps, and the second procedure requires either 1, 2, or 3 steps.

   a. List the experimental outcomes associated with performing an analysis.
   b. If the random variable of interest is the total number of steps required to do the complete analysis, show what value the random variable will assume for each of the experimental outcomes.

4. Consider the experiment of tossing a coin twice.

   a. List the experimental outcomes.
   b. Define a random variable that represents the number of heads occurring on the two tosses.
   c. Show what value the random variable would assume for each of the experimental outcomes.

5. Listed is a series of experiments and associated random variables. In each case identify the values that the random variable can take on and state whether the random variable is discrete or continuous.

Experiment	Random Variable (x)
a. Take a 20-question examination	Number of questions answered correctly
b. Observe cars arriving at a tollbooth for 1 hour	Number of cars arriving at tollbooth
c. Audit 50 tax returns	Number of returns containing errors
d. Observe an employee's work	Number of nonproductive hours in an 8-hour work day
e. Weigh a shipment of goods	Number of pounds

5.2 ■■■ DISCRETE PROBABILITY DISTRIBUTIONS

The *probability distribution* for a random variable describes how the probabilities are distributed over the values of the random variable. For a discrete random variable x, the probability distribution is defined by a *probability function*, denoted by $f(x)$. The probability function provides the probability for each value of the random variable. As an illustration of a discrete random variable and its probability distribution, we consider a study of 300 households in a village on the coast of Maine. As part of this study, data were collected showing the number of children in each household. The following results were obtained: 54 of the households had no children, 117 had 1 child, 72 had 2 children, 42 had 3 children, 12 had 4 children, and 3 had 5 children.

Suppose we consider the experiment of randomly selecting one of these households to participate in a follow-up study. If we let x denote the number of children in the household selected, possible values of x are 0, 1, 2, 3, 4, and 5. Thus $f(0)$ provides the probability that a randomly selected household has no children, $f(1)$ provides the probability that a randomly selected household has 1 child, and so on. Since 54 of the 300 households have no children, we assign the value $54/300 = .18$ to $f(0)$. Similarly, since 117 of the 300 households have 1 child, we assign the value $117/300 = .39$ to $f(1)$. Continuing in this fashion for the other values of random variable x, we obtain the discrete probability distribution shown in Table 5.2.

In the development of the probability function for a discrete random variable, the following two conditions must be satisfied:

Required Conditions for a Discrete Probability Function

$$f(x) \geq 0 \tag{5.1}$$

$$\Sigma f(x) = 1 \tag{5.2}$$

TABLE 5.2 **Probability Distribution for the Number of Children per Household**

x	$f(x)$
0	.18
1	.39
2	.24
3	.14
4	.04
5	.01
Total	1.00

Table 5.2 shows that the probabilities for the random variable x satisfy condition (5.1), since $f(x)$ is greater than or equal to 0 for all values of x. In addition, since the probabilities sum to 1 and (5.2) is satisfied, the probability function is a valid discrete probability function.

We can also present probability distributions graphically. In Figure 5.1 the values of the random variable x are shown on the horizontal axis and the probability that x assumes these values is shown on the vertical axis.

For some discrete random variables, the probability distribution can be given as a formula that yields $f(x)$ for every possible value of x. Consider the random variable x and its probability distribution, as shown by the following table.

x	$f(x)$
1	1/10
2	2/10
3	3/10
4	4/10

This probability distribution can also be given by the formula

$$f(x) = {}^{x}\!/_4 \qquad \text{for } x = 1, 2, 3, 4$$

■■■ FIGURE 5.1 **Graphical Representation of the Probability Distribution for Number of Children per Household**

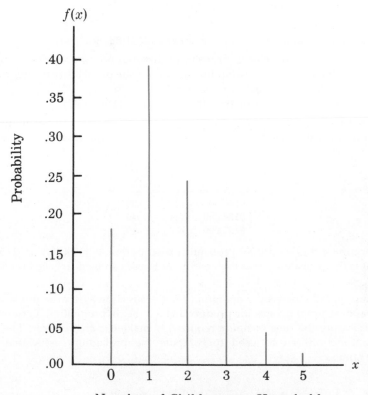

Number of Children per Household

The more widely used discrete probability distributions are usually specified by formulas. Three important cases are the binomial, Poisson and hypergeometric probability distributions, which are discussed later in the chapter.

EXERCISES

6. The following data were collected by counting the number of operating rooms in use at Tampa General Hospital over a 20-day period: On 3 of the days only 1 operating room was used, on 5 of the days 2 were used, on 8 of the days 3 were used, and on 4 days all 4 of the hospital's operating rooms were used.

   a. Use the relative frequency approach to construct a probability distribution for the number of operating rooms in use on any given day.
   b. Draw a graph of the probability distribution.
   c. Show that your probability distribution satisfies the required conditions for a valid discrete probability distribution.

7. According to *Dataquest*, the personal computer market during 1986 was as follows:

Company	Company Identification Code	Market Share (Percentage)
IBM	1	29.8
Apple	2	9.2
COMPAQ	3	6.6
Zenith	4	3.5
Tandy	5	6.6
Commodore	6	1.9
Other	7	42.4

Let x be a random variable based on the company identification code.

   a. Use these data to develop a probability distribution. Specify the values for the random variable and the corresponding values for the probability function, $f(x)$.
   b. Draw a graph of the probability distribution.
   c. Show that the probability distribution satisfies (5.1) and (5.2).

8. QA Properties is considering making an offer to purchase an apartment building. Management has subjectively assessed a probability distribution for x, the purchase price:

x	f(x)
$148,000	.20
$150,000	.40
$152,000	.40

   a. Determine if this is a proper probability distribution. (Check (5.1) and (5.2).)
   b. What is the probability that the apartment house can be purchased for $150,000 or less?

9. The cleaning and changeover operation for a production system requires from 1 to 4 hours, depending upon the specific product that will begin production. Let x be a random variable indicating the time in hours required to make the changeover. The following probability function can be used to compute the probability associated with any changeover time x:

$$f(x) = \frac{x}{10} \quad \text{for } x = 1, 2, 3, \text{ or } 4$$

a. Show that the probability function meets the required conditions of (5.1) and (5.2).
b. What is the probability that the changeover will take 2 hours?
c. What is the probability that the changeover will take more than 2 hours?
d. Graph the probability distribution for the changeover times.

10. The director of admissions at Lakeville Community College has subjectively assessed a probability distribution for x, the number of entering students.

x	$f(x)$
1000	.15
1100	.20
1200	.30
1300	.25
1400	.10

a. Is this a valid probability distribution?
b. What is the probability there will be 1200 or fewer entering students?

11. A psychologist has determined that the number of hours required to obtain the trust of a new patient is either 1, 2, or 3. Let x be a random variable indicating the time in hours required to gain the patient's trust. The following probability function has been proposed.

$$f(x) = \frac{x}{6} \quad \text{for } x = 1, 2, \text{ or } 3$$

a. Is this a valid probability function? Explain.
b. What is the probability that it takes exactly 2 hours to gain the patient's trust?
c. What is the probability that it takes at least 2 hours to gain the patient's trust?

12. Shown below is a partial probability distribution for the MRA Company's projected profits for the first year of operation (the negative value denotes a loss).

Profit in 1000s	
x	$f(x)$
−100	.10
0	.20
50	.30
100	.25
150	.10
200	

a. What is the value of $f(200)$? What is your interpretation of this value?
b. What is the probability that MRA will be profitable?
c. What is the probability that MRA will make at least $100,000?

5.3 ▓▓▓ EXPECTED VALUE AND VARIANCE

Expected Value

The *expected value*, or mean, of a random variable is a measure of the central location for the random variable. The mathematical expression for the expected value of a discrete random variable x is as follows.

Expected Value of a Discrete Random Variable

$$E(x) = \mu = \Sigma \, xf(x) \qquad\qquad (5.3)$$

Both the notations $E(x)$ and μ can be used to denote the expected value of a random variable.

Equation (5.3) shows that in order to compute the expected value of a discrete random variable, we must multiply each value of the random variable by the corresponding probability $f(x)$ and then add the resulting products. Recall the problem involving a study of 300 households in a village on the coast of Maine. Table 5.3 shows the calculation of the expected value of the number of children in a household. We see that the expected value or mean is 1.50 children per household.

Variance

While the expected value provides the mean value for the random variable, we often need a measure of dispersion, or variability, for the random variable. Just as we used variance in Chapter 3 to summarize the dispersion in a data set, we now use the variance measure to summarize the variability in the values of a random variable. The mathematical expression for the variance of a discrete random variable is as follows.

Variance of a Discrete Random Variable

$$\text{Var}(x) = \sigma^2 = \Sigma(x - \mu)^2 f(x) \qquad\qquad (5.4)$$

As (5.4) shows, an essential part of the variance formula is the deviation, $x - \mu$, which measures how far a particular value of the random variable is from the

---

TABLE 5.3 Expected Value of Children per Household

x	$f(x)$	$xf(x)$
0	.18	$0(.18) =$ .00
1	.39	$1(.39) =$ .39
2	.24	$2(.24) =$ .48
3	.14	$3(.14) =$ .42
4	.04	$4(.04) =$ .16
5	.01	$5(.01) =$ .05
		1.50

$$E(x) = \mu = \Sigma \, xf(x)$$

expected value or mean, μ. In computing the variance of a random variable, the deviations are squared and then weighted by the corresponding value of the probability function. The sum of these weighted squared deviations for all values of the random variable is referred to as the *variance*. Both the notations var(x) and σ^2 are used to denote the variance of a random variable.

The calculation of the variance for the probability distribution of the number of children per household is summarized in Table 5.4. We see that the variance for the number of children per household is 1.25. The *standard deviation, σ,* is defined as the positive square root of the variance. Thus the standard deviation of the number of children per household is

$$\sigma = \sqrt{1.25} = 1.118$$

The standard deviation is measured in the same units as the random variable ($\sigma = 1.118$ children per household); for this reason σ is often preferred in describing the variability of a random variable. The variance (σ^2) is measured in squared units and is thus more difficult to interpret.

Expected Value of the Sum of Random Variables

Occasionally it is desired to compute the expected value of a sum of random variables. For instance, DiCarlo Motors has dealerships in Saratoga and Albany. Let x = the number of cars sold per day at the Saratoga dealership and y = the number of cars sold per day at the Albany dealership. Suppose DiCarlo is interested in the expected total daily sales for both dealerships. To compute this expected value we need to calculate the expected value of the sum of the random variables x and y (i.e., $E(x + y)$). Equation (5.5) shows that the expected value of the sum of two random variables is given by the sum of expected values.

Expected Value of the Sum of Two Random Variables

$$E(x + y) = E(x) + E(y) \tag{5.5}$$

TABLE 5.4   **Calculation of Variance for Number of Children per Household**

x	$x - \mu$	$(x - \mu)^2$	$f(x)$	$(x - \mu)^2 f(x)$
0	$0 - 1.50 = -1.50$	2.25	.18	$2.25(.18) = .4050$
1	$1 - 1.50 = -.50$	.25	.39	$.25(.39) = .0975$
2	$2 - 1.50 = .50$	.25	.24	$.25(.24) = .0600$
3	$3 - 1.50 = 1.50$	2.25	.14	$2.25(.14) = .3150$
4	$4 - 1.50 = 2.50$	6.25	.04	$6.25(.04) = .2500$
5	$5 - 1.50 = 3.50$	12.25	.01	$12.25(.01) = \underline{.1225}$

$$1.2500$$

$$\sigma^2 = \Sigma(x - \mu)^2 f(x)$$

Suppose that the expected number of daily sales at the Saratoga dealership is 2.50 automobiles and the expected number of daily sales at the Albany dealership is 2.85 automobiles. Using (5.6), the expected total daily sales for the dealerships is

$$E(x + y) = E(x) + E(y)$$

$$= 2.50 + 2.85 = 5.35$$

Equation (5.5) can be extended to provide a method for computing the expected value of the sum of any number of random variables. The result is that the expected value of the sum of any number of random variables is equal to the sum of their individual expected values. For example, if DiCarlo Motors had a third dealership and if we let z represent the daily sales at the third dealership, then the expected value of the total daily sales for all three dealerships would be $E(x + y + z) = E(x) + E(y) + E(z)$.

Variance of the Sum of Independent Random Variables

In the previous chapter we said that two events were independent if the occurrence of one of the events did not affect the probability of the other event occurring. Similarly, two random variables are said to be independent if the value that one takes on does not affect the probabilities associated with the value that the other can take on. Equation (5.6) provides the variance for the sum of two independent random variables:

Variance of the Sum of Two Independent Random Variables

$$Var(x + y) = Var(x) + Var(y) \qquad (5.6)$$

As an illustration of (5.6) let us compute the variance of DiCarlo's total sales at the Saratoga and Albany dealerships. Suppose that the variance of daily sales at Saratoga is 1.55 and at Albany is 2.13. If DiCarlo feels that the sales at the dealerships are independent, the variance of total sales is given by

$$Var(total sales) = Var(Saratoga sales) + Var(Albany sales)$$

$$= 1.55 + 2.13 = 3.68$$

Note that the standard deviation of total sales is given by $\sigma = \sqrt{3.68} = 1.92$; it is not the sum of the individual standard deviations.

Equation (5.6) can be generalized to provide a method for computing the variance of the sum of any number of independent random variables. The result is that the variance of the sum is equal to the sum of the variances. For example, if DiCarlo Motors had a third dealership and we let z be the daily sales at the third dealership, then the variance of the total sales is $Var(x + y + z) = Var(x) + Var(y) + Var(z)$.

Of course, in order to apply this result to DiCarlo Motors, we must be convinced that daily sales of all three dealerships are independent.

▦ EXERCISES

13. A volunteer ambulance service handles from 0 to 5 service calls on any given day. The following probability distribution for the number of service calls is assumed:

Number of Service Calls	Probability
0	.10
1	.15
2	.30
3	.20
4	.15
5	.10

a. What is the expected number of service calls?

b. What is the variance in the number of service calls? What is the standard deviation?

14. Individual investors allocated their funds among four different investment categories in 1988 (*Money*, January 1989). The annual return for the categories and the proportion of investments in each category are as follows:

Investment Category	Annual Return	Proportion of Total Investments
Stocks	26.0%	.31
Bonds	8.6%	.23
CDs and Money Funds	7.7%	.45
Real Estate and Gold	−2.9%	.01

Let x be a random variable indicating the annual return percentage for a $1 investment. Assume the probabilities of each investment category, $f(x)$, are provided by the proportion of total investments data.

a. What is the expected annual return percentage on a $1 investment?

b. What are the variance and standard deviation?

c. Suppose the annual return on stocks drops to 5%. Recompute the expected return and the standard deviation of the return on a $1 investment.

15. A roulette wheel at a Las Vegas casino has 18 red numbers, 18 black numbers, and 2 green numbers. Assume that a $5 bet is placed on the black numbers. If a black number comes up, the player wins $5; otherwise the player loses $5.

a. Let x be a random variable indicating the player's net winnings on one bet. Thus, $x = 5$ if the player wins and $x = -5$ if the player loses. Show the probability distribution of x.

b. What is the expected amount won on a bet? What is your interpretation of this value?

c. What is the variance in the amount won on a bet? What is the standard deviation?

d. If a player places 100 bets of $5 each, what are the expected winnings? Comment on why casinos like a high volume of betting.

16. The probability distribution for collision insurance payments made by the Newton Automobile Insurance Company is as follows.

Payment ($)	Probability
0	.90
200	.04
500	.03
1000	.01
2000	.01
3000	.01

a. Use the expected collision payment to determine the collision insurance premium that would allow the company to break even on the collision portion of the policy.

b. The insurance company charges an annual rate of $130 for the collision coverage. What is the expected value of the collision policy for a policyholder? (*Hint:* It is the expected payments from the company minus the cost of coverage.) Why does the policyholder purchase a collision policy with this expected value?

17. The number of dots observed on the upward face of a die has the following probability function.

$$f(x) = \frac{1}{6} \qquad \text{for } x = 1, 2, 3, 4, 5, 6$$

a. Show that this probability function possesses the properties necessary for probability distributions.

b. Draw a graph of the probability distribution.

c. What is the expected value? What is the interpretation of this value?

d. What are the variance and the standard deviation for the number of dots?

18. The demand for a product of Carolina Industries varies greatly from month to month. Based on the past 2 years of data, the following probability distribution shows the company's monthly demand.

Unit demand	Probability
300	.20
400	.30
500	.35
600	.15

a. If the company places monthly orders based on the expected value of the monthly demand, what should Carolina's monthly order quantity be for this product?

b. Assume that each unit demanded generates $70 in revenue and that each unit ordered costs $50. How much will the company gain or lose in a month if it places an order based on your answer to part (a) and the actual demand for the item is 300 units?

19. What are the variance and the standard deviation for the number of units demanded in Exercise 18?

20. The J.R. Ryland Computer Company is considering a plant expansion that will enable the company to begin production of a new computer product. The company's president must determine whether to make the expansion a medium- or large-scale project. An uncertainty involves the demand for the new product, which for planning purposes may be low demand, medium demand, or high demand. The probability estimates for the demands are .20, .50 and .30, respectively. Letting *x* indicate the annual profit in $1000s,

the firm's planners have developed profit forecasts for the medium- and large-scale expansion projects.

		Medium-Scale Expansion Profits		Large-Scale Expansion Profits	
		x	$f(x)$	y	$f(y)$
	Low	50	.20	0	.20
Demand	Medium	150	.50	100	.50
	High	200	.30	300	.30

a. Compute the expected value for the profit associated with the two expansion alternatives. Which decision is preferred for the objective of maximizing the expected profit?

b. Compute the variance for the profit associated with the two expansion alternatives. Which decision is preferred for the objective of minimizing the risk or uncertainty?

21. A brokerage firm has offices in Houston and Dallas. The daily commissions at the Houston offices show $E(x)$ = $2000 and a standard deviation of $300, whereas daily commissions at Dallas show $E(y)$ = $1500 and a standard deviation of $400.

a. What are the total expected daily commissions for the two offices?

b. Assuming that daily commissions at the two offices are independent, determine the standard deviation of total sales for the two offices.

22. The expected production times and variances for a three-stage assembly are as follows:

	Expected Time	Variance
Stage A	5.4	1.00
Stage B	3.2	0.64
Stage C	8.4	1.69

a. What is the expected assembly time for a part passing through all three stages?

b. Assuming the assembly stages operate independently, what is the variance and standard deviation of the total assembly time.

5.4 ▬ THE BINOMIAL PROBABILITY DISTRIBUTION

The *binomial probability distribution* is a discrete probability distribution that has many applications. It is associated with a multiple-step experiment that we call the binomial experiment.

Properties of a Binomial Experiment

For a probability experiment to be classified as a *binomial experiment*, it must have the following four properties.

1. The experiment consists of a sequence of n identical trials.

2. Two outcomes are possible on each trial. We refer to one outcome as a *success* and the other as a *failure*.

3. The probability of a success on one trial, denoted by p, does not change from trial to trial. Consequently, the probability of failure on one trial, denoted by $1 - p$, does not change from trial to trial.

4. The trials are independent.

If properties 2, 3, and 4 are present we say the trials are generated by a Bernoulli process. If, in addition, property 1 is present, we say we have a *binomial experiment*. Figure 5.2 depicts one possible sequence of outcomes of a binomial experiment involving 8 trials.

In a binomial experiment, our interest is in the *number of successes occurring in the n trials*. If we let x denote the number of successes occurring in the n trials, we see that x can assume the values of 0, 1, 2, 3, . . . , n. Since the number of values is finite, x is a *discrete* random variable. The probability distribution associated with this random variable is called the *binomial probability distribution*. For an example, consider the experiment of tossing a coin 5 times and on each toss observing whether the coin lands with a head or a tail on its upward face. Suppose we are interested in counting the number of heads appearing during the 5 tosses. Does this experiment have the properties of a binomial experiment? What is the random variable of interest? Note that:

1. The experiment consists of five identical trials, where each trial involves the tossing of one coin.

2. There are two outcomes possible for each trial. The possible outcomes are a head and a tail. We can designate head a success and tail a failure.

3. The probability of a head and the probability of a tail are the same for each trial, with $p = .5$ and $1 - p = .5$.

4. The trials or tosses are independent, since the outcome on any one trial is not affected by what happens on other trials or tosses.

Thus the properties of a binomial experiment are satisfied. The random variable of interest is x = the number of heads appearing in the five trials. In this case, x can take on the values of 0, 1, 2, 3, 4, or 5.

As another example consider an insurance salesperson who pays a visit to 10 randomly selected families. The outcome associated with each visit is classified as a success if the family purchases an insurance policy and a failure if the family does not. From past experience, the salesperson knows the probability that a

---

■ FIGURE 5.2 **Diagram of an 8-Trial Binomial Experiment**

Property 1: The experiment consists of $n = 8$ identical trials.

Property 2: Each trial results in either success (S) or failure (F).

Trials ⟶ 1 2 3 4 5 6 7 8

Outcomes ⟶ S F F S S F S S

randomly selected family purchases an insurance policy is .10. Checking the properties of a binomial experiment, we observe the following:

1. The experiment consists of 10 identical trials, where each trial involves contacting one family.

2. There are two outcomes possible on each trial: The family purchases a policy (success) or the family does not purchase a policy (failure).

3. The probabilities of a purchase and a nonpurchase are assumed to be the same for each family, with $p = .10$ and $1 - p = .90$.

4. The trials are independent since the families are randomly selected.

Since the four assumptions are satisfied, this is a binomial experiment. The random variable of interest is the number of sales obtained in contacting the 10 families. In this case, x can assume the values of 0, 1, 2, 3, 4, 5, 6, 7, 8, 9, and 10.

Property 3 of the binomial experiment is called the *stationarity assumption* and is sometimes confused with Property 4, independence of trials. To see how they differ, consider again the case of the salesperson calling on families to sell insurance policies. If, as the day wore on, the salesperson got tired and lost enthusiasm, then the probability of success (selling a policy) might drop to .05, for example, by the tenth call. In such a case Property 3 (stationarity) would not be satisfied, and we would not have a binomial experiment. This would be true even if Property 4 held—that is, the purchase decisions of each family were made independently.

In applications involving binomial experiments, a special mathematical formula, called the *binomial probability function*, can be used to compute the probability of x successes in the n trials. Using probability concepts introduced in Chapter 4, we will show how the formula can be developed in the context of an illustrative problem.

The Nastke Clothing Store Problem

Let us consider the purchase decisions of the next three customers who enter the Nastke Clothing Store. Based on past experience the store manager estimates the probability that any one customer will make a purchase is .30. What is the probability that two of the next three customers will make a purchase?

Using a tree diagram (Figure 5.3) we can see that the experiment of observing the three customers each making a purchase decision has 8 possible outcomes. Using S to denote success (a purchase) and F to denote failure (no purchase) we are interested in experimental outcomes involving two successes in the three trials (purchase decisions). Next let us verify that the experiment involving the sequence of three purchase decisions can be viewed as a binomial experiment. Checking the four requirements for a binomial experiment, we note the following:

1. The experiment can be described as a sequence of three identical trials, one trial for each of the three customers that will enter the store.

2. Two outcomes—the customer makes a purchase (success) or the customer does not make a purchase (failure)—are possible for each trial or customer.

3. The probability the customer makes a purchase (.30) or does not make a purchase (.70) is assumed to be the same for all customers.

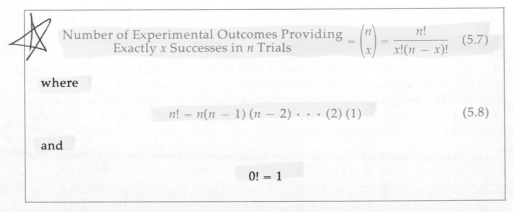

First Customer	Second Customer	Third Customer	Outcomes	Value of x
		S	(S, S, S)	3
	S	F	(S, S, F)	2
S	F	S	(S, F, S)	2
		F	(S, F, F)	1
	S	S	(F, S, S)	2
F		F	(F, S, F)	1
	F	S	(F, F, S)	1
		F	(F, F, F)	0

S = Purchase
F = No Purchase
x = Number of Customers Making a Purchase

■ FIGURE 5.3   **Tree Diagram for the Nastke Clothing Store Problem**

4. The purchase decision of each customer is independent of the decisions of the other customers.

Thus the properties of a binomial experiment are present.

The number of experimental outcomes resulting in exactly x successes in n trials can be computed from the following formula.*

$$\text{Number of Experimental Outcomes Providing Exactly } x \text{ Successes in } n \text{ Trials} = \binom{n}{x} = \frac{n!}{x!(n-x)!} \quad (5.7)$$

where

$$n! = n(n-1)(n-2) \cdots (2)(1) \quad (5.8)$$

and

$$0! = 1$$

---

*This formula is commonly used to determine the number of combinations of n objects selected x at a time. For the binomial experiment, this combinatorial formula provides the number of experimental outcomes (sequences of n trials) resulting in x successes.

Now let us return to the Nastke experiment involving three customers' purchase decisions. Equation (5.7) can be used to determine the number of experimental outcomes involving two purchases; that is, the number of ways of obtaining $x = 2$ successes in the $n = 3$ trials. From (5.7) we have

$$\binom{n}{x} = \binom{3}{2} = \frac{3!}{2!(3-2)!} = \frac{(3)(2)(1)}{(2)(1)(1)} = \frac{6}{2} = 3$$

Formula (5.7) shows that three of the outcomes yield two successes. From Figure 5.3 we see these three outcomes are denoted by SSF, SFS, and FSS.

Using (5.7) to determine how many experimental outcomes have three successes (purchases) in the three trials, we obtain:

$$\binom{n}{x} = \binom{3}{3} = \frac{3!}{3!(3-3)!} = \frac{3!}{3!0!} = \frac{(3)(2)(1)}{(3)(2)(1)(1)} = \frac{6}{6} = 1$$

From Figure 5.3 we see that the one experimental outcome with three successes is identified by SSS.

We know that (5.7) can be used to determine the number of experimental outcomes that result in x successes. But, if we are to determine the probability of x successes in n trials, we must also know the probability associated with each of these experimental outcomes. Since the trials of a binomial experiment are independent, we can simply multiply the probabilities associated with each trial outcome to find the probability of a particular sequence of outcomes.

The probability of purchases by the first two customers and no purchase by the third customer is given by

$$pp(1 - p)$$

With a .30 probability of a purchase on any one trial, the probability of a purchase on the first two trials and no purchase on the third is given by

$$(.30)(.30)(.70) = (.30)^2(.70) = .063$$

There are two other sequences of outcomes resulting in two successes and one failure. The probabilities for all three sequences involving two successes are shown.

Trial Outcomes

1st Customer	2nd Customer	3rd Customer	Success-Failure Notation	Probability of Experimental Outcome
Purchase	Purchase	No purchase	SSF	$pp(1-p) = p^2(1-p)$ $= (.30)^2(.70) = .063$
Purchase	No purchase	Purchase	SFS	$p(1-p)p = p^2(1-p)$ $= (.30)^2(.70) = .063$
No purchase	Purchase	Purchase	FSS	$(1-p)pp = p^2(1-p)$ $= (.30)^2(.70) = .063$

Observe that all three outcomes with two successes have exactly that same probability. This observation holds in general. In any binomial experiment, each sequence of trial outcomes yielding x successes in n trials has the *same probability* of occurrence. The probability of each sequence of trials yielding x successes in n trials is as follows:

Probability of a Particular
Sequence of Trial Outcomes $= p^x(1 - p)^{(n-x)}$ (5.9)
with x Successes in n Trials

For the Nastke Clothing Store this formula shows that any outcome with two successes has a probability of $p^2(1 - p)^{(3-2)} = p^2(1 - p)^1 = (.30)^2(.70)^1 = .063$, as shown.

Since (5.7) shows the number of outcomes in a binomial experiment with x successes and since (5.9) gives the probability for each sequence involving x successes, we combine (5.7) and (5.9) to obtain the following *binomial probability function.*

Binomial Probability Function

$$f(x) = \binom{n}{x} p^x(1 - p)^{(n-x)}$$ (5.10)

where

$f(x) =$ the probability of x successes in n trials

$n =$ the number of trials

$\binom{n}{x} = \dfrac{n!}{x!(n - x)!}$

$p =$ the probability of a success on any one trial

$(1 - p) =$ the probability of a failure on any one trial

In the Nastke Clothing Store example, let us compute the probability of each of the following: no customer makes a purchase, exactly one customer makes a purchase, exactly two customers make a purchase, and all three customers make a purchase. The calculations are summarized in Table 5.5. This table provides a tabular presentation for the probability distribution of number of customer purchases. A graph of this probability distribution is shown in Figure 5.4.

The binomial probability function can be applied to *any* binomial experiment. If we are satisfied that a situation has the properties of a binomial experiment and if we know the values of $n, p,$ and $(1 - p)$, (5.10) can be used to compute the probability of x successes in the n trials.

If we consider variations of the Nastke experiment, such as 10 customers rather than 3 entering the store, the binomial probability function given by (5.10) is still applicable. For example, the probability of making exactly 4 sales to

TABLE 5.5 **Probability Distribution for the Number of Customers Making a Purchase**

x	f(x)
0	$\dfrac{3!}{0!3!}(.30)^0(.70)^3 = .343$
1	$\dfrac{3!}{1!2!}(.30)^1(.70)^2 = .441$
2	$\dfrac{3!}{2!1!}(.30)^2(.70)^1 = .189$
3	$\dfrac{3!}{3!0!}(.30)^3(.70)^0 = \underline{.027}$
	1.000

10 potential customers entering the store is

$$f(4) = \frac{10!}{4!6!}(.30)^4(.70)^6 = .2001$$

This is a binomial experiment with $n = 10$, $x = 4$, and $p = .30$.

Using Tables of Binomial Probabilities

Tables have been developed that give the probability of x successes in n trials for a binomial experiment. These tables are generally easy to use and quicker than Equation (5.10), especially when the number of trials involved is large. A table of binomial probabilities is provided as Table 5 of Appendix B. A portion of this table is given in Table 5.6. In order to use this table it is necessary to specify the values of n, p, and x for the binomial experiment of interest. In the example at the top of Table 5.6, we see that the probability of $x = 3$ successes in a binomial experiment with $n = 10$ and $p = .40$ is .2150. You might want to use (5.10) to verify that this is the answer you would obtain using the binomial probability function directly.

Also check the use of this table by using it to verify the probability of four successes in ten trials for the Nastke Clothing Store problem. Note that the value of $f(4) = .2001$ can be read directly from the table of binomial probabilities, making it unnecessary to perform the calculations required by (5.10).

While the tables of binomial probabilities are relatively easy to use, it is impossible to have tables that show all possible values of n and p that might be encountered in a binomial experiment. However, with today's calculators, it is not too difficult to calculate the desired probability using (5.10), especially if the

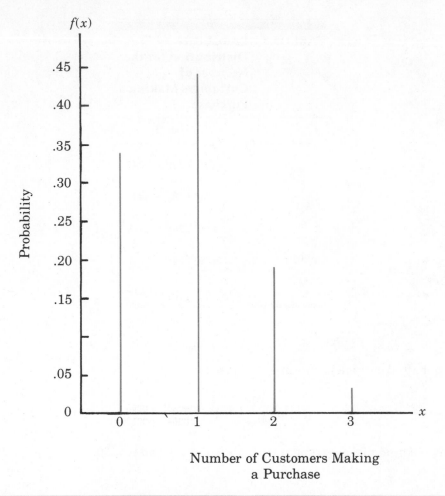

■■■ FIGURE 5.4   **Graphical Representation of the Probability Distribution for the Nastke Clothing Store Problem**

number of trials is not too large. In the exercises, you should practice using (5.10) to compute the binomial probabilities unless the problem specifically requests that you use the binomial probability table.

The Expected Value and Variance for the Binomial Probability Distribution

In Section 5.3 we provided formulas for computing the expected value and variance of a discrete random variable. In the special case where the random variable has a binomial probability distribution with a known number of trials (n) and a known probability of success (p), the general formulas for the expected value and variance can be simplified; the results are as follows.

> ### Expected Value and Variance for the Binomial Probability Distribution
>
> $$E(x) = \mu = np \qquad\qquad (5.11)$$
>
> $$\text{Var}(x) = \sigma^2 = np(1 - p) \qquad\qquad (5.12)$$

TABLE 5.6   Selected Values From the Binomial Probability Table
EXAMPLE: $n = 10, x = 3, p = .40; f(3) = .2150$

p

n	x	.05	.10	.15	.20	.25	.30	.35	.40	.45	.50
9	0	.6302	.3874	.2316	.1342	.0751	.0404	.0207	.0101	.0046	.0020
	1	.2985	.3874	.3679	.3020	.2253	.1556	.1004	.0605	.0339	.0176
	2	.0629	.1722	.2597	.3020	.3003	.2668	.2162	.1612	.1110	.0703
	3	.0077	.0446	.1069	.1762	.2336	.2668	.2716	.2508	.2119	.1641
	4	.0006	.0074	.0283	.0661	.1168	.1715	.2194	.2508	.2600	.2461
	5	.0000	.0008	.0050	.0165	.0389	.0735	.1181	.1672	.2128	.2461
	6	.0000	.0001	.0006	.0028	.0087	.0210	.0424	.0743	.1160	.1641
	7	.0000	.0000	.0000	.0003	.0012	.0039	.0098	.0212	.0407	.0703
	8	.0000	.0000	.0000	.0000	.0001	.0004	.0013	.0035	.0083	.0176
	9	.0000	.0000	.0000	.0000	.0000	.0000	.0001	.0003	.0008	.0020
10	0	.5987	.3487	.1969	.1074	.0563	.0282	.0135	.0060	.0025	.0010
	1	.3151	.3874	.3474	.2684	.1877	.1211	.0725	.0403	.0207	.0098
	2	.0746	.1937	.2759	.3020	.2816	.2335	.1757	.1209	.0763	.0439
	3	.0105	.0574	.1298	.2013	.2503	.2668	.2522	**.2150**	.1665	.1172
	4	.0010	.0112	.0401	.0881	.1460	.2001	.2377	.2508	.2384	.2051
	5	.0001	.0015	.0085	.0264	.0584	.1029	.1536	.2007	.2340	.2461
	6	.0000	.0001	.0012	.0055	.0162	.0368	.0689	.1115	.1596	.2051
	7	.0000	.0000	.0001	.0008	.0031	.0090	.0212	.0425	.0746	.1172
	8	.0000	.0000	.0000	.0001	.0004	.0014	.0043	.0106	.0229	.0439
	9	.0000	.0000	.0000	.0000	.0000	.0001	.0005	.0016	.0042	.0098
	10	.0000	.0000	.0000	.0000	.0000	.0000	.0000	.0001	.0003	.0010
11	0	.5688	.3138	.1673	.0859	.0422	.0198	.0088	.0036	.0014	.0005
	1	.3293	.3835	.3248	.2362	.1549	.0932	.0518	.0266	.0125	.0054
	2	.0867	.2131	.2866	.2953	.2581	.1998	.1395	.0887	.0531	.0269
	3	.0137	.0710	.1517	.2215	.2581	.2568	.2254	.1774	.1259	.0806
	4	.0014	.0158	.0536	.1107	.1721	.2201	.2428	.2365	.2060	.1611
	5	.0001	.0025	.0132	.0388	.0803	.1321	.1830	.2207	.2360	.2256
	6	.0000	.0003	.0023	.0097	.0268	.0566	.0985	.1471	.1931	.2256
	7	.0000	.0000	.0003	.0017	.0064	.0173	.0379	.0701	.1128	.1611
	8	.0000	.0000	.0000	.0002	.0011	.0037	.0102	.0234	.0462	.0806
	9	.0000	.0000	.0000	.0000	.0001	.0005	.0018	.0052	.0126	.0269
	10	.0000	.0000	.0000	.0000	.0000	.0000	.0002	.0007	.0021	.0054
	11	.0000	.0000	.0000	.0000	.0000	.0000	.0000	.0000	.0002	.0005

For the Nastke Clothing Store problem with 3 customers we may utilize (5.11) to compute the expected number of customers making a purchase.

$$E(x) = np = 3(.30) = .9$$

Suppose that for the next month Nastke's Clothing Store forecasts 1000 customers will enter the store. What is the expected number of customers who will make a purchase? The answer is $\mu = np = (1000)(.3) = 300$. Thus in order to increase the expected number of sales Nastke's must induce more customers to enter the store and/or somehow increase the probability that any individual customer will make a purchase after entering.

For the Nastke Clothing Store problem with three customers, we see that the variance and standard deviation for the number of customers making a purchase are

$$\sigma^2 = np(1 - p) = 3(.3)(.7) = .63$$
$$\sigma = \sqrt{.63} = .79$$

For the next 1000 customers entering the store, the variance and standard deviation for the number of customers making a purchase are

$$\sigma^2 = np(1 - p) = 1000(.3)(.7) = 210$$
$$\sigma = \sqrt{210} = 14.49$$

NOTES AND COMMENTS

The binomial tables show values of p only up to and including $p = .50$. Thus, it would appear they cannot be used when the probability of success exceeds $p = .50$. However, the tables can be used by noting that the probability of $n - x$ failures is also the probability of x successes. So when the probability of success is greater than $p = .50$, we just compute the probability of $n - x$ failures instead. The probability of failure, $1 - p$, will then be less than .50.

■■■ EXERCISES

23. The New York State Bar Examination is the basis for admitting law school graduates into the law profession. Historically, 30% of the individuals taking the examination pass on their first attempt. Suppose a group of 15 individuals will be taking the examination for the first time and that we are interested in the number of individuals in this group who will pass the exam. Describe the conditions necessary for this situation to be a binomial experiment.

24. When a new machine is functioning properly, only 3% of the items produced are defective. Assume that we will randomly select two parts produced on the machine and that we are interested in the number of defective parts found.

a. Describe the conditions under which this situation would be a binomial experiment.

b. Draw a tree diagram similar to Figure 5.3 showing this as a two-trial experiment.

c. How many experimental outcomes result in exactly one defect being found?

d. Compute the probabilities associated with finding no defects, exactly 1 defect, and 2 defects.

25. Ten percent of American truck drivers are women (*Statistical Abstract of the United States*, 1987). Suppose ten truck drivers are randomly selected to be interviewed about quality of work conditions.

a. Is the selection of the ten drivers a binomial experiment? Explain.

b. What is the probability three of the drivers will be women?

c. What is the probability none will be women?

d. What is the probability at least one will be a woman?

26. A baseball player with a batting average of .300 comes to bat four times in a game. What is the probability the player obtains exactly 1 hit? Exactly 2 hits? No hits?

27. National Oil Company conducts exploratory oil drilling operations in the southwestern United States. In order to fund the operation, investors form partnerships, which provide the financial support necessary to drill a fixed number of oil wells. Each well drilled is classified as a producer well or a dry well. Past experience shows that this type of exploratory operation provides producer wells for 15% of all wells drilled. A newly formed partnership has provided the financial support for drilling at 12 exploratory locations.

a. What is the probability that all 12 wells will be producer wells?

b. What is the probability that all 12 wells will be dry wells?

c. What is the probability that exactly 1 well will be a producer well?

d. In order to make the partnership venture profitable, at least 3 of the exploratory wells must be producer wells. What is the probability that the venture will be profitable?

28. Military radar and missile detection systems are designed to warn a country against enemy attacks. A reliability question deals with the ability of the detection system to identify an attack and issue the warning. Assume that a particular detection system has a **.90** probability of detecting a missile attack. Answer the following questions using the binomial probability distribution.

a. What is the probability that a single detection system will detect an attack?

b. If two detection systems are installed in the same area and operate independently, what is the probability that at least one of the systems will detect the attack?

c. If three systems are installed, what is the probability that at least one of the systems will detect the attack?

d. Would you recommend that multiple detection systems be operated? Explain.

29. Assume that the binomial distribution applies for the case of a college basketball player shooting free throws. Late in a basketball game, a team will sometimes foul intentionally in the hope that the player shooting the free throw will miss and the team committing the foul will get the ball. Assume that the best player on the opposing team has a .82 probability of making a free throw and that the worst player has a .56 probability of making a free throw.

a. What are the probabilities that the best player makes 0, 1, and 2 points if fouled and given two free throws?

b. What are the probabilities that the worst player makes 0, 1, and 2 points if fouled and given two free throws?

c. Does it make sense for a coach to have a preset plan about which player to intentionally foul late in a basketball game? Explain.

30. A firm estimates the probability of having employee disciplinary problems on any day to be .10.

a. What is the probability that the company experiences 5 days without a disciplinary problem?

b. What is the probability of exactly 2 days with disciplinary problems in a 10-day period?

c. What is the probability of at least 2 days with disciplinary problems in a 20-day period?

31. At a particular university it has been found that 20% of the students withdraw without completing the introductory statistics course. Assume that 20 students have registered for the course this quarter.

a. What is the probability that two or fewer will withdraw?

b. What is the probability that exactly four will withdraw?

c. What is the probability that more than three will withdraw?

d. What is the expected number of withdrawals?

32. For the special case of a binomial random variable we stated that the variance measure could be computed from the formula $\sigma^2 = np(1 - p)$. For the Nastke Clothing Store problem data in Table 5.5, we found $\sigma^2 = np(1 - p) = .63$. Use the general definition of variance for a discrete random variable, Equation (5.4), and the data in Table 5.4 to verify that the variance is in fact .63.

33. Suppose a salesperson makes a sale on 20% of customer contacts. A normal work week will enable the salesperson to contact 25 customers. What is the expected number of sales for the week? What is the variance for the number of sales for the week? What is the standard deviation for the number of sales for the week?

34. Eighty-five percent of the next-day express mailings handled by the U.S. Postal Service are actually received by the addressee 1 day after the mailing. What is the expected value and variance for the number of 1-day deliveries in a group of 250 express mailings?

35. Betting on the color red in the game of roulette has a $^{18}/_{38}$ chance of winning. What is the expected value and variance for the number of wins in a series of 100 bets on red?

5.5 ▬ THE POISSON DISTRIBUTION

In this section we consider a discrete random variable that is often useful when dealing with the number of occurrences of an event over a specified interval of time or space. For example, the random variable of interest might be the number of arrivals at a car wash in 1 hour, the number of repairs needed in 10 miles of highway, or the number of leaks in 100 miles of pipeline. If the two assumptions are satisfied, the number of occurrences is a random variable described by the *Poisson probability function*. The two assumptions are

1. The probability of an occurrence of the event is the same for any two intervals of equal length.

2. The occurrence or nonoccurrence of the event in any interval is independent of the occurrence or nonoccurrence in any other interval.

The Poisson probability function is given by equation (5.13).

Poisson Probability Function

$$f(x) = \frac{\mu^x e^{-\mu}}{x!} \tag{5.13}$$

where

$f(x) =$ the probability of x occurrences in an interval

$\mu =$ expected value or average number of occurrences in an interval

$e = 2.71828$

Before we consider a specific example to see how the Poisson distribution can be applied, note that there is no upper limit on x, the number of occurrences. It is a discrete random variable that may assume an infinite sequence of values ($x = 0, 1, 2, \ldots$). The Poisson random variable has no specific upper limit.

A Poisson Example Involving Time Intervals

Suppose that we are interested in the number of arrivals at the drive-in teller window of a bank during a 15-minute period on weekday mornings. If we can assume that the probability of a car arriving is the same for any two time periods of equal length and that the arrival or nonarrival of a car in any time period is independent of the arrival or nonarrival in any other time period, the Poisson probability function is applicable. Suppose that these assumptions are satisfied and that an analysis of historical data shows that the average number of cars arriving in a 15-minute period of time is 10; then the following probability function applies:

$$f(x) = \frac{10^x e^{-10}}{x!}$$

The random variable here is $x =$ number of cars arriving in any 15-minute period.

If management wanted to know the probability of exactly 5 arrivals in 15 minutes, we would set $x = 5$ and thus obtain*

$$\begin{array}{c} \text{Probability of Exactly} \\ \text{Five Arrivals in 15 minutes} \end{array} = f(5) = \frac{10^5 e^{-10}}{5!} = .0378$$

Although the above probability was determined by evaluating the probability function with $\mu = 10$ and $x = 5$, it is often easier to refer to tables for the Poisson probability distribution. These tables provide probabilities for specific values of

---

*Values of $e^{-\mu}$ can be found in Table 6 of Appendix B.

x and μ. We have included such a table as Table 7 of Appendix B. For convenience we have reproduced a portion of this table as Table 5.7. Note that in order to use the table of Poisson probabilities we need know only the values of x and μ. Thus from Table 5.7 we see that the probability of five arrivals in a 15-minute period is found by locating the value in the row of the table corresponding to $x = 5$ and the column of the table corresponding to $\mu = 10$. Hence we obtain $f(5) = .0378$.

Although our illustration has involved computing the probability of 5 arrivals in a 15-minute period, other time periods may be used. Suppose we wanted to compute the probability of 1 arrival in a 3-minute period. Since 10 is the expected number of arrivals in a 15-minute period, we see that $^{10}/_{15} = ^{2}/_{3}$ is the expected number of arrivals in a 1-minute period and that $(2/3)(3 \text{ minutes}) = 2$ is the expected number of arrivals in a 3-minute period. Thus, the probability of x arrivals in a 3-minute time period with $\mu = 2$ is given by the following Poisson probability function:

$$f(x) = \frac{2^x e^{-2}}{x!}$$

TABLE 5.7 **Selected Values from the Poisson Probability Tables**
Example: $\mu = 10$, $x = 5$; $f(5) = .0378$

x	9.1	9.2	9.3	9.4	9.5	9.6	9.7	9.8	9.9	10
0	.0001	.0001	.0001	.0001	.0001	.0001	.0001	.0001	.0001	.0000
1	.0010	.0009	.0009	.0008	.0007	.0007	.0006	.0005	.0005	.0005
2	.0046	.0043	.0040	.0037	.0034	.0031	.0029	.0027	.0025	.0023
3	.0140	.0131	.0123	.0115	.0107	.0100	.0093	.0087	.0081	.0076
4	.0319	.0302	.0285	.0269	.0254	.0240	.0226	.0213	.0201	.0189
5	.0581	.0555	.0530	.0506	.0483	.0460	.0439	.0418	.0398	**.0378**
6	.0881	.0851	.0822	.0793	.0764	.0736	.0709	.0682	.0656	.0631
7	.1145	.1118	.1091	.1064	.1037	.1010	.0982	.0955	.0928	.0901
8	.1302	.1286	.1269	.1251	.1232	.1212	.1191	.1170	.1148	.1126
9	.1317	.1315	.1311	.1306	.1300	.1293	.1284	.1274	.1263	.1251
10	.1198	.1210	.1219	.1228	.1235	.1241	.1245	.1249	.1250	.1251
11	.0991	.1012	.1031	.1049	.1067	.1083	.1098	.1112	.1125	.1137
12	.0752	.0776	.0799	.0822	.0844	.0866	.0888	.0908	.0928	.0948
13	.0526	.0549	.0572	.0594	.0617	.0640	.0662	.0685	.0707	.0729
14	.0342	.0361	.0380	.0399	.0419	.0439	.0459	.0479	.0500	.0521
15	.0208	.0221	.0235	.0250	.0265	.0281	.0297	.0313	.0330	.0347
16	.0118	.0127	.0137	.0147	.0157	.0168	.0180	.0192	.0204	.0217
17	.0063	.0069	.0075	.0081	.0088	.0095	.0103	.0111	.0119	.0128
18	.0032	.0035	.0039	.0042	.0046	.0051	.0055	.0060	.0065	.0071
19	.0015	.0017	.0019	.0021	.0023	.0026	.0028	.0031	.0034	.0037
20	.0007	.0008	.0009	.0010	.0011	.0012	.0014	.0015	.0017	.0019
21	.0003	.0003	.0004	.0004	.0005	.0006	.0006	.0007	.0008	.0009
22	.0001	.0001	.0002	.0002	.0002	.0002	.0003	.0003	.0004	.0004
23	.0000	.0001	.0001	.0001	.0001	.0001	.0001	.0001	.0002	.0002
24	.0000	.0000	.0000	.0000	.0000	.0000	.0000	.0001	.0001	.0001

To find the probability of one arrival in a 3-minute period, we can either use Table 7 in Appendix B or compute it directly.

$$\text{Probability of Exactly} \atop \text{1 Arrival in 3 Minutes} = f(1) = \frac{2^1 e^{-2}}{1!} = .2707$$

A Poisson Example Involving Length or Distance Intervals

Let us illustrate the variety of applications where the Poisson probability distribution is useful. Suppose that we are concerned with the occurrence of major defects in a section of highway 1 month after resurfacing. We will assume that the probability of a defect in this section of highway is the same for any two intervals of equal length and that the occurrence or nonoccurrence of a defect in any one interval is independent of the occurrence or nonoccurrence in any other interval. Thus the Poisson probability distribution can be applied.

Suppose we learn that major defects 1 month after resurfacing occur at the average rate of two per mile. Let us find the probability that there will be no major defects in a particular 3 mile section of the highway. Since we are interested in an interval with a length of 3 miles, $\mu = (2 \text{ defects}/\text{mile})(3 \text{ miles}) = 6$ represents the expected number of major defects over the 3-mile section of highway. Thus by using (5.13) or Table 7 in Appendix B we see that the probability of no major defects is 0.0025. Thus it is very unlikely that there will be no major defects in the 3 mile section. In fact, there is a $1 - .0025 = .9975$ probability of at least one major defect in the highway section.

■■■ EXERCISES

36. A certain restaurant has a reputation for good food. Restaurant management boasts that on a Saturday night, groups of customers arrive at the rate of 15 groups every half-hour.

  a. What is the probability that 5 minutes will pass with no groups of customers arriving?
  b. What is the probability that 8 groups of customers will arrive in 10 minutes?
  c. What is the probability that more than 5 groups will arrive in a 10-minute period of time?

37. During rush hours accidents occur in a particular metropolitan area at the rate of 2 per hour. The morning rush period lasts for 1 hour 30 minutes and the evening rush period lasts for 2 hours.

  a. On a particular day what is the probability that there will be no accidents during the morning rush period?
  b. What is the probability of 2 accidents during the evening rush period?
  c. What is the probability of 4 or more accidents during the morning rush period?
  d. On a particular day what is the probability there will be no accidents during both the morning and evening rush periods?

38. Airline passengers arrive randomly and independently at the passenger-screening

facility at a major international airport. The mean arrival rate is 10 passengers per minute.

   a. What is the probability of no arrivals in a 1-minute period?
   b. What is the probability three or fewer passengers arrive in a 1-minute period?
   c. What is the probability of no arrivals in a 15-second period?
   d. What is the probability of at least one arrival in a 15-second period?

39. Williams Company has observed that calculators fail and need to be replaced at the rate of 3 every 25 days.

   a. What is the expected number of calculators that will fail in 30 days?
   b. What is the probability that at least two will fail in 50 days?
   c. What is the probability that exactly three will fail in 10 days?

5.6 ■ THE HYPERGEOMETRIC PROBABILITY DISTRIBUTION

The hypergeometric probability distribution is closely related to the binomial probability distribution. It also provides the probability of obtaining x successes in n trials when there are two possible outcomes (success and failure) on each trial. The key difference between the two probability distributions is that with the hypergeometric distribution, the trials are not independent; thus the probability of success changes from trial to trial.

To illustrate the hypergeometric distribution, suppose a five-member committee consists of three women and two men. Two of the committee members are expected to represent the group at a meeting in Las Vegas. The committee has decided to choose randomly the two members that will attend the meeting. What is the probability that both persons chosen will be women? Since three of the five committee members are women, the probability that the first person randomly selected is a woman is $3/5 = .60$. However, if the first person chosen is a woman, then the probability of randomly selecting another woman from the four remaining committee members drops to $2/4 = .50$. Therefore, the probability that two women will be selected to attend the meeting is $(.60)(.50) = .30$.

One of the most important applications of the hypergeometric probability distribution involves sampling without replacement from a finite population. The objective is to choose a random sample of n items out of a population of N items, under the condition that once an item has been selected, it is not returned to the population. Thus, on the next selection, the probability of selecting an item of that type goes down.

The usual notation in applications of the hypergeometric probability distribution is to let r denote the number of items in the population that are labeled success and $N - r$ denote the number of items in the population that are labeled failure. The hypergeometric probability function is used to compute the probability that in a random sample of n items, selected without replacement, we will obtain x items labeled success and $n - x$ items labeled failure. Note that for this to occur, we must obtain x successes from the r successes in the population and $n - x$ failures from the $N - r$ failures. The following hypergeometric probability function provides $f(x)$, the probability of obtaining x successes in a sample of size n.

Hypergeometric Probability Function

$$f(x) = \frac{\binom{r}{x}\binom{N-r}{n-x}}{\binom{N}{n}} \tag{5.14}$$

where

$f(x)$ = the probability of x successes in n trials

n = the number of trials

N = number of elements in the population

r = number of elements in the population labeled success

Note that $\binom{N}{n}$, defined in (5.7), simply represents the number of ways a sample of size n can be selected from a population of size N; $\binom{r}{x}$ represents the number of ways x successes can be selected from a total of r successes in the population; and $\binom{N-r}{n-x}$ represents the number of ways $n - x$ failures can be selected from a total of $N - r$ failures in the population. To illustrate the computations involved in using (5.14), let us reconsider the problem of selecting two of five committee members to send to Las Vegas.

Recall that the objective is to select two members from the five-member committee consisting of three women and two men. To determine the probability of obtaining a sample that consists of two women, we can use (5.14) with $N = 5$, $r = 3$, and $x = 2$.

$$f(2) = \frac{\binom{3}{2}\binom{2}{0}}{\binom{5}{2}} = \frac{\dfrac{3!}{2!1!}\dfrac{2!}{2!0!}}{\dfrac{5!}{3!2!}} = \frac{3}{10} = .30$$

Note that this is the same answer we obtained previously. Suppose, however, that we learn that three committee members will be allowed to make the trip. The probability that two of the three members will be women is

$$f(2) = \frac{\binom{3}{2}\binom{2}{1}}{\binom{5}{2}} = \frac{\dfrac{3!}{1!2!}\dfrac{2!}{1!1!}}{\dfrac{5!}{3!2!}} = \frac{6}{10} = .60$$

As another illustration, suppose a population consists of ten items, four of which are classified as defective and six of which are classified as acceptable. What is the probability that a random sample of size three will contain two defective items? For this problem we can think of obtaining a defective item as a "success" and obtaining an acceptable item as a "failure." Thus, $N = 10$, $r = 4$, $n =$

3, and $x = 2$. Using (5.14) we can compute $f(2)$ as follows:

$$f(2) = \frac{\binom{4}{2}\binom{6}{1}}{\binom{10}{3}} = \frac{\frac{4!}{2!2!}\frac{6!}{1!5!}}{\frac{10!}{3!7!}} = \frac{36}{120} = .30$$

NOTES AND COMMENTS

The hypergeometric distribution is used to determine the probability of obtaining a certain sample when sampling without replacement. When sampling with replacement, each item that is selected is returned to the population before another item is selected. The binomial probability distribution is then used to compute the probability of x successes.

■■ EXERCISES

40. Suppose $N = 10$ and $r = 3$. Compute the hypergeometric probabilities for the following values of n and x.

   a. $n = 4, x = 1$
   b. $n = 2, x = 2$
   c. $n = 2, x = 0$
   d. $n = 4, x = 2$

41. Suppose $N = 15$ and $r = 4$. What is the probability of $x = 5$ for $n = 10$?

42. What is the probability of being dealt three aces in a seven-card poker hand?

43. There are 25 students (14 boys and 11 girls) in the sixth-grade class at St. Andrew School. Five students were absent Thursday.

   a. What is the probability 2 were girls?
   b. What is the probability 2 were boys?
   c. What is the probability all were boys?
   d. What is the probability none were boys?

44. Axline Computers manufactures personal computers at two plants; one is in Las Vegas, the other in Hawaii. There are 40 employees at the Las Vegas plant and 20 in Hawaii. A random sample of 10 different employees is to be asked to fill out a benefits' questionnaire.

   a. What is the probability none will be from the plant in Hawaii?
   b. What is the probability 1 will be from the plant in Hawaii?
   c. What is the probability 2 or more will be from the plant in Hawaii?
   d. What is the probability 9 will be from the plant in Las Vegas?

■■ SUMMARY

The concept of a random variable was introduced in order to provide a numerical description of the outcome of an experiment. We saw that the probability distribution for a random variable describes how the probabilities are distributed over the values the random variable can take on. For any discrete random variable x, the probability distribution is defined by a probability function, denoted by $f(x)$, which provides the probability associated with each value of the random variable. Once the probability function has been defined, we can then compute the expected value and the variance for the random variable.

The binomial probability distribution can be used to determine the probability of x successes in n trials whenever the experiment has the following properties:

1. The experiment consists of a sequence of n identical trials.

2. Two outcomes are possible on each trial, one called success and the other failure.

3. The probability of a success, p, does not change from trial to trial. Consequently, the probability of failure, $1 - p$, does not change from trial to trial.

4. The trials are independent.

When the above conditions hold, a binomial probability function, or a table of binomial probabilities, can be used to determine the probability of x successes in n trials. Formulas were also presented for the mean and variance of the binomial probability distribution.

The Poisson probability distribution is used when it is desired to determine the probability of obtaining x occurrences over an interval of time or space. The following assumptions are required for the Poisson distribution to be applicable:

1. The probability of an occurrence of the event is the same for any two intervals of equal length.

2. The occurrence or nonoccurrence of the event in any interval is independent of the occurrence or nonoccurrence in any other interval.

A third discrete probability distribution, the hypergeometric, was introduced in Section 5.6. Like the binomial, it is used to compute the probability of x successes in n trials. But, unlike the binomial, the probability of success changes from trial to trial.

■■■ GLOSSARY

Random variable   A numerical description of the outcome of an experiment.

Discrete random variable   A random variable that can assume only a finite or infinite sequence of values.

Continuous random variable   A random variable that may assume all values in an interval or collection of intervals.

Probability distribution   A description of how the probabilities are distributed over the values the random variable can take on.

Probability function   A function, denoted by $f(x)$, that for a discrete random variable, provides the probability that x takes on a particular value.

Expected value   A measure of the mean, or central location, value of a random variable.

Variance   A measure of the dispersion, or variability, of a random variable.

Standard deviation   The positive square root of the variance.

Binomial experiment   A probability experiment possessing the four properties stated in Section 5.4.

Binomial probability distribution   A probability distribution showing the probability of x successes in n trials of a binomial experiment.

Binomial probability function   The function used to compute probabilities in a binomial experiment.

Poisson probability distribution   A probability distribution showing the probability of x occurrences of an event over a specified interval of time or space.

Poisson probability function   The function used to compute Poisson probabilities.
Hypergeometric probability function   The function used to compute the probability of
x successes in n trials when the trials are dependent.

■■ KEY FORMULAS

Expected Value of a Discrete Random Variable

$$E(x) = \mu = \Sigma \, xf(x) \tag{5.3}$$

Variance of a Discrete Random Variable

$$\text{Var}(x) = \sigma^2 = \Sigma \, (x - \mu)^2 f(x) \tag{5.4}$$

Expected Value of the Sum of Two Random Variables

$$E(x + y) = E(x) + E(y) \tag{5.5}$$

Variance of the Sum of Two Independent Random Variables

$$\text{Var}(x + y) = \text{Var}(x) + \text{Var}(y) \tag{5.6}$$

Number of Experimental Outcomes Providing Exactly x Successes in n Trials

$$\binom{n}{x} = \frac{n!}{x!(n - x)!} \tag{5.7}$$

Binomial Probability Function

$$f(x) = \binom{n}{x} p^x (1 - p)^{(n-x)} \tag{5.10}$$

Expected Value for the Binomial Probability Distribution

$$E(x) = \mu = np \tag{5.11}$$

Variance for the Binomial Probability Distribution

$$\text{Var}(x) = \sigma^2 = np(1 - p) \tag{5.12}$$

Poisson Probability Function

$$f(x) = \frac{\mu^x e^{-\mu}}{x!} \tag{5.13}$$

Hypergeometric Probability Function

$$f(x) = \frac{\binom{r}{x}\binom{N - r}{n - x}}{\binom{N}{n}} \tag{5.14}$$

■ SUPPLEMENTARY EXERCISES

45. Which of the following are and which are not probability distributions? Explain.

x	f(x)	y	f(y)	z	f(z)
0	.20	0	.25	−1	.20
1	.30	2	.05	0	.50
2	.25	4	.10	1	−.10
3	.35	6	.60	2	.40

46. An automobile agency located in Beverly Hills specializes in the rental of luxury automobiles. Assume that the probability distribution of daily demand at their agency is as follows.

x	f(x)
0	.15
1	.30
2	.40
3	.10
4	.05

a. Compute the expected value of daily demand.
b. If the daily rental cost for an automobile is $75, what is the expected value of daily automobile rental?

47. At a large university, the number of student problems handled by the dean for student affairs varies from semester to semester. Assume that the number of student problems (x) handled by the dean has the following probability distribution.

x	f(x)
0	.10
1	.15
2	.30
3	.25
4	.10
5	.10

What are the mean and variance of the number of student problems handled by the dean each semester?

48. The number of weekly lost-time injuries at a particular plant (x) has the following probability distribution.

x	f(x)
0	.05
1	.20
2	.40
3	.20
4	.15

a. Compute the expected value.
b. Compute the variance.

49. Assume that the plant in Exercise 48 initiated a safety training program and that the

number of lost-time injuries during the 20 weeks following the training program was as follows.

Number of Injuries	Number of Weeks
0	2
1	8
2	6
3	3
4	$\frac{1}{20}$

a. Construct a probability distribution for weekly lost-time injuries based on these data.

b. Compute the expected value and the variance and use both to evaluate the effectiveness of the safety training program.

50. The Hub Real Estate Investment stock is currently selling for $16 per share. An investor plans to buy shares and hold the stock for 1 year. Let x be the random variable indicating the price of the stock after 1 year. The probability distribution for x is shown.

Price of Stock (x)	$f(x)$
16	.35
17	.25
18	.25
19	.10
20	.05

a. Show that the above probability distribution possesses the properties of all probability distributions.

b. What is the expected price of the stock after 1 year?

c. What is the expected gain per share of the stock over the 1-year period? What percent return on the investment is reflected by this expected value?

d. What is the variance in the price of the stock over the 1-year period?

e. Another stock with a similar expected return has a variance of 3. Which stock appears to be the better investment in terms of minimizing risk or uncertainty associated with the investment? Explain.

51. The budgeting process for a midwestern college resulted in expense forecasts for the coming year (in 1,000,000s) of $9, $10, $11, $12, and $13. Since the actual expenses are unknown, the following respective probabilities are assigned: .3, .2, .25, .05, and .2.

a. Show the probability distribution for the expense forecast.

b. What is the expected value of the expenses for the coming year?

c. What is the variance in the expenses for the coming year?

d. If income projections for the year are estimated at $12 million, comment on the financial position of the college.

52. Exercise 9 provided a probability function for x, the hours required to change over a production system, as follows.

$$f(x) = \frac{x}{10} \quad \text{for } x = 1, 2, 3, \text{ or } 4$$

a. What is the expected value of the changeover time?

b. What is the variance of the changeover time?

53. Students who complete a statistics course are required to take a midterm exam and a

final exam. The expected value or average on the midterm is 72 with a standard deviation of 12. The expected value or average on the final is 68 with a standard deviation of 14.

a. What is the expected value of the total number of points scored on the two exams?
b. What are the variance and standard deviation of the total number of points scored on the two exams? Assume that the number of points scored on the exams are independent of one another.

54. Customers entering a fast-food store average spending $2.28 per person. The standard deviation in the amount spent is $.75.

a. Compute the mean, variance, and standard deviation of the total food bill for 10 randomly selected customers. Assume that the customers order independently.
b. Repeat part (a) for 100 randomly selected customers.

55. The police department of a major midwestern city makes arrests on 40% of its reported robberies. Assume that we are interested in the number of arrests that will be made in the next 20 reported robberies. Describe whether or not you feel the properties of a binomial experiment are satisfied.

56. In October 1986, *Better Homes and Gardens* published the results of a reader survey. Over 30,000 readers responded. Findings indicated that 34% of the women who responded worked full time outside home and 24% worked part-time outside the home.

a. For a random sample of five women who are *Better Homes and Gardens* readers, what is the probability that three work full time outside the home?
b. For a random sample of five women readers, what is the probability that two are part-time workers?
c. Suppose a random sample of five women (not necessarily *Better Homes and Gardens* readers) was taken. Would your answers to parts a and b change? Why or why not?

57. Refer again to the *Better Homes and Gardens* survey in Exercise 56. In 65% of the two-parent households, the wife does most of the child care; in 1%, the husband does most. For a sample of four respondents from two-parent households answer the following:

a. What is the probability that none of the respondents will say the wife does most?
b. What is the probability that none of the respondents will say the husband does most?
c. What is the probability that three or more of the respondents will say the wife does most?
d. What is the probability that one of the respondents will say the husband does most?

58. A new clothes-washing compound is found to remove excess dirt and stains satisfactorily on 88% of the items washed. Assume that 10 items are to be washed with the new compound.

a. What is the probability of satisfactory results on all 10 items?
b. What is the probability at least 2 items are found with unsatisfactory results?

59. In an audit of a company's billings, an auditor randomly selects 5 bills. If 3% of all bills contain an error, what is the probability that the auditor will find the following?

a. exactly one bill in error
b. at least one bill in error

60. Many companies use a quality-control technique referred to as *acceptance sampling* in order to monitor incoming shipments of parts, raw materials, and so on. In the electronics industry, it is common to have component parts shipped from suppliers in large lots. Inspection of a sample of n components can be viewed as the n trials of a binomial experiment. The outcome for each component tested (trial) will be that the component is good or defective. Reynolds Electronics accepts lots from a particular supplier as long as

the percent defective in the lot is not greater than 1%. Suppose a random sample of five items from a recent shipment has been tested.

    a. Assume that 1% of the shipment is defective. Compute the probability that no items in the sample are defective.

    b. Assume that 1% of the shipment is defective. Compute the probability that exactly one item in the sample is defective.

    c. What is the probability of observing one or more defective items in the sample if 1% of the shipment is defective?

    d. Would you feel comfortable accepting the shipment if one item was found defective? Why or why not?

61. Cars arrive at a carwash at the average rate of 15 cars per hour. If the number of arrivals per hour follows a Poisson distribution, what is the probability of 20 or more arrivals during any given hour of operation? Use the Poisson probability table.

62. A new automated production process has been experiencing an average of 1.5 breakdowns per day. Because of the cost associated with a breakdown, management is concerned about the possibility of having three or more breakdowns during a given day. Assume that the number of breakdowns per day follows a Poisson distribution. What is the probability of observing three or more breakdowns?

63. A regional director responsible for business development in Pennsylvania is concerned about the number of businesses that end as failures. If the average number of failures per month is ten, what is the probability that exactly four businesses will fail during a given month? Assume that the number of businesses failing per month follows a Poisson distribution.

64. The arrivals of customers at a bank follow the Poisson distribution. Answer the following questions assuming a mean arrival rate of three per minute.

    a. What is the probability of exactly three arrivals in a 1-minute period?

    b. What is the probability of at least three arrivals in a 1-minute period?

65. During the registration period at a local university, students consult advisors with questions about course selection. A particular advisor noted that during the registration period an average of eight students per hour ask questions, although the exact arrival times of the students were random in nature. Use the Poisson distribution to answer the following questions:

    a. What is the probability that exactly eight students come in for consultation during a particular 1-hour period?

    b. What is the probability that three students come in for consultation during a particular ½-hour period?

APPLICATION

Xerox Corporation*

Stamford, Connecticut

Xerox Corporation is in the information products and systems business worldwide. As a major part of this business, it develops, makes, and markets xerographic copiers and duplicators; facsimile transceivers; electrostatic printers; processor memory disks, drives and high-speed terminals; electronic typewriters; information-processing products, office information systems; electronic printing systems; automatic labeling, binding, and mailing machines, and xeroradiographic devices.

▃▃ MULTINATIONAL DOCUMENTATION & TRAINING SERVICES

Multinational Documentation & Training Services (MD&TS) provides customers with timely, cost effective, and high quality communication services. In this regard, MD&TS provides four basic services:

1. *Documentation.* Quality documentation assures that the intent of service or product design is upheld, marketing and service goals are maintained, and that the entire organization is striving toward the same end result. MD&TS provides a variety of documentation services such as: Operation Manuals; Installation Instructions; Preventive Maintenance Procedures; Testing and Troubleshooting Procedures.

2. *Training.* MD&TS has the experience and the facilities to provide training programs that use a variety of proven instructional methods. These programs can be implemented at any of our 28 modern training facilities, the customer location, or through correspondence.

3. *Translation.* Documentation and training requirements that are multinational must also be multilingual. MD&TS can provide translation services for technical English materials into a number of different foreign languages. Our highly skilled linguistic specialists ensure grammatical and connotative correctness, as well as technical accuracy.

4. *Publishing.* MD&TS offers complete publishing capabilities with over 100 communication and media specialists. The quality of the publishing service provided is assured by following a policy whereby every document is scrutinized and validated for technical accuracy, consistency, grammar, and print quality.

Our professionals have many years of experience in the design, development and implementation of multinational documentation and training materials. We believe this

*The authors are indebted to Soterios M. Flouris, Manager, Systems Development and Maintenance, Webster, New York, for providing this application.

experience enables us to provide the most flexible, efficient, and cost-effective documentation and training services available today.

■■■ PERFORMANCE TEST SIMULATION

The professional writers and translators working for MD&TS use an on-line computerized publication system. Management of MD&TS was interested in determining the effect of different system configurations (for example, type of computer, maximum number of on-line users, and so forth) on system performance. Specifically, for a given system configuration, management was interested in determining the following:

1. The probability of a user being refused access by the system because of an excess number of users.

2. The probability of any specific number of users being on the system simultaneously.

To determine the above probabilities, a computer simulation model was developed. The purpose of the computer simulation model was to provide a representation of different system configurations. Through a series of computer runs the behavior of the simulation model—and hence each system configuration—was studied. The operating characteristics of the simulation models were then used to make inferences about the operating characteristics of different system configurations.

In order to build the simulation model it was necessary to determine a probability distribution for the following two random variables:

1. The length of time a user is on the system, referred to as the on time per session.

2. The length of time between one user session and the next user session, referred to as the idle time per session.

Based upon a survey of users, the probability distribution of on time per session was approximated* as shown in Table 5A.1 and Figure 5A.1, where x is the random variable indicating on time in minutes. Using the data in Table 5A.1 the expected value and variance of on time per session were calculated with $E(x) = 48.8$ and $Var(x) = 336.2$. The

TABLE 5A.1   Probability Distribution of on Time per Session

x	$f(x)$
10	.05
20	.06
30	.08
40	.20
50	.25
60	.20
70	.08
80	.06
90	.02

*The actual distribution used in the simulation study has been modified to protect proprietary information and to simplify the discussion.

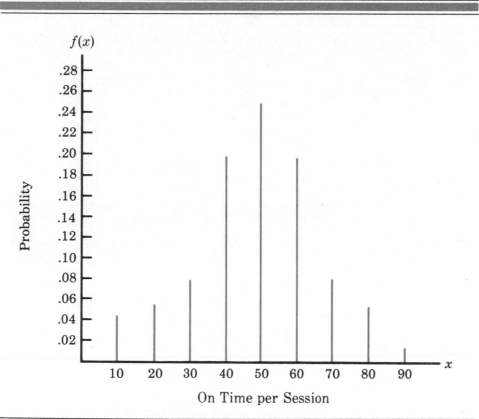

■■■ FIGURE 5A.1   **Graphical Representation of the Probability Distribution of On Time Per Session**

expected value shows that the mean or "average" length of time a user spends on the system is 48.8 minutes. Another probability distribution was developed for the random variable indicating idle time per session.

In the computer simulation model, whenever the simulation indicated that a user was to begin operation on the system, it was necessary to generate a value of the random variable from the on time per session distribution. Similarly, whenever the simulation indicated that a user had completed a session, it was necessary to generate a value of the random variable from the idle time distribution. Although the details of the simulation model are somewhat complex and beyond the scope of this application, the concept of a random variable and its probability distribution was an important part of this simulation study.

■■■ RESULTS

The probability distributions for on time per session and idle time per session were key components in the simulation model developed to investigate the effect of different system configurations. The results from the simulation study helped MD&TS to determine a system configuration that would ensure a near-zero probability that a user would be refused access to the system.

CHAPTER 6

Continuous Probability Distributions

I n the previous chapter we discussed discrete random variables and their probability distributions. In this chapter we turn our attention to the study of continuous random variables. Specifically, we discuss three continuous probability distributions: the uniform, the normal, and the exponential probability distributions.

In order better to understand the difference between discrete and continuous random variables, first recall that for a discrete random variable we can compute the probability of the random variable taking on a particular value. For continuous probability distributions, the situation is much different. A continuous random variable may assume any value in an interval or a collection of intervals. Since there are an infinite number of values in any interval, it is no longer possible to talk about the probability that the random variable will take on a specific value; instead, we must think in terms of the probability that a continuous random variable will lie within a specific interval.

In our discussion of discrete probability distributions, we introduced the concept of a probability function $f(x)$. Recall that this function provided the probability that the random variable x assumed some specific value. In the continuous case the counterpart of the probability function is the *probability density function*, also denoted by $f(x)$. For a continuous random variable, the probability density function provides the height or value of the function at any particular value of x; it does not directly provide the probability of the random variable taking on some specific value. However, the area under the graph of $f(x)$ corresponding to some interval provides the probability that the continuous random variable will take on a value in that interval. In Section 6.1 we demonstrate these concepts for a continuous random variable that has a uniform probability distribution.

195

6.1 ■ THE UNIFORM DISTRIBUTION

Consider the random variable x that represents the total flight time of an airplane traveling from Chicago to New York. Assume that the flight time can be any value in the interval from 120 minutes to 140 minutes (e.g., 124 minutes, 125.48 minutes, and so on). Since the random variable x can take on all values in the interval from 120 to 140 minutes, x is a continuous rather than a discrete random variable. Let us assume that sufficient actual flight data are available to conclude that the probability of a flight time between 120 and 121 minutes is the same as the probability of a flight time within any other 1-minute interval up to and including 140 minutes. With every 1-minute interval being equally likely, the random variable x is said to have a *uniform probability distribution*. The *probability density function*, which defines the uniform probability distribution for the flight time random variable, is

$$f(x) = \begin{cases} 1/20 & \text{for } 120 \leq x \leq 140 \\ 0 & \text{elsewhere} \end{cases}$$

A graph of this probability density function is shown in Figure 6.1. In general, the uniform probability density function for a random variable x is as follows:

Uniform Probability Density Function

$$f(x) = \begin{cases} \dfrac{1}{b-a} & \text{for } a \leq x \leq b \\ 0 & \text{elsewhere} \end{cases} \tag{6.1}$$

In the flight-time example, $a = 120$ and $b = 140$.

The graph of the probability density function, $f(x)$, provides the height or value of the function at any particular value of x. Note that for a *uniform*

■ FIGURE 6.1   **Uniform Probability Density Function for Flight Time**

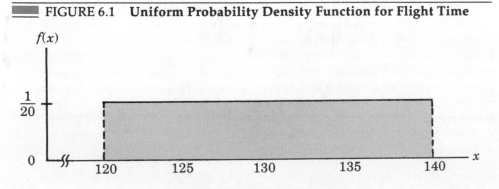

Flight Time in Minutes

probability density function the height or value of the function is the same for each value of x. For example, in the flight-time example, $f(x) = \frac{1}{20}$ for all values of x between 120 and 140. *In general, the probability density function $f(x)$, unlike the probability function for a discrete random variable, does not represent probability. Rather it simply provides the height of the function at any particular value of x.*

For a continuous random variable, we consider probability only in terms of the likelihood that a random variable has a value within a *specified interval*. In our flight-time problem an acceptable probability question is, "What is the probability that the flight time is between 120 and 130 minutes?" That is, what is $P(120 \leq x \leq 130)$? Since the flight time must be between 120 and 140 minutes and since the probability was described as being uniform over this interval, we feel comfortable saying $P(120 \leq x \leq 130) = .50$. In the following subsection we will show that this probability can be computed as the area under the graph of $f(x)$ from 120 to 130.

Area as a Measure of Probability

Let us make an observation about the graph shown in Figure 6.2. Consider the *area under the graph of $f(x)$* in the interval from 120 to 130. The region is rectangular in shape and the area of a rectangle is simply the width times the height. With the width of the interval equal to $130 - 120 = 10$ and the height equal to the value of probability density function $f(x) = \frac{1}{20}$, we have area = width $\times$ height $= 10(\frac{1}{20}) = \frac{10}{20} = .50$.

What observation can you make about the area under the graph of $f(x)$ and probability? They are identical! Indeed, this is true for all continuous random variables. Namely, once a probability density function $f(x)$ has been identified, the probability that x takes on a value between some lower value x_1 and some higher value x_2 can be found by computing the *area* under the graph of $f(x)$ over the interval x_1 to x_2.

Once we have the appropriate probability distribution and accept the interpretation of area as probability, we can answer any number of probability questions. For example, what is the probability of a flight time between 128 and

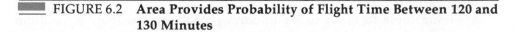

FIGURE 6.2   **Area Provides Probability of Flight Time Between 120 and 130 Minutes**

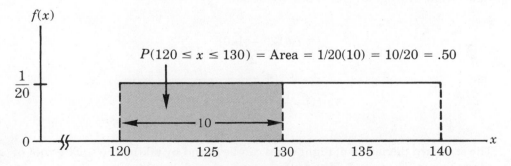

$$P(120 \leq x \leq 130) = \text{Area} = 1/20(10) = 10/20 = .50$$

Flight Time in Minutes

136 minutes? The width of the interval is $136 - 128 = 8$. With the uniform height of $\frac{1}{20}$, we see that $P(128 \le x \le 136) = \frac{8}{20} = .40$.

Note that $P(120 \le x \le 140) = 20(\frac{1}{20}) = 1$. That is, the total area under the graph of $f(x)$ is equal to 1. This property holds for all continuous probability distributions and is the analog of the condition that the sum of the probabilities has to equal one for a discrete probability function. For a continuous probability density function, we must also require that $f(x) \ge 0$ for all values of x. This is the analog of the requirement that $f(x) \ge 0$ for discrete probability functions.

When we deal with continuous random variables and probability distributions, two major differences stand out as compared to the treatment of their discrete counterparts:

1. We no longer talk about the probability of the random variable taking on a particular value. Instead we talk about the probability of the random variable taking on a value within some given interval.

2. The probability of the random variable taking on a value within some given interval from x_1 to x_2 is defined to be the area under the graph of the probability density function between x_1 and x_2. *This implies that the probability that a continuous random variable takes on any particular value exactly is zero, since the area under the graph of $f(x)$ at a single point is zero.*

The calculation of the mean and variance for a continuous random variable is analogous to that for a discrete random variable. However, since the computational procedure involves integral calculus, we leave the derivation of the appropriate formulas to more advanced texts.

For the uniform continuous probability distribution introduced in this section the formulas for the expected value and variance are

Expected Value

$$E(x) = \frac{a+b}{2}$$

Variance

$$Var(x) = \frac{(b-a)^2}{12}$$

In these formulas a is the smallest value and b is the largest value that the random variable may take on.

Applying these formulas to the uniform probability distribution for flight times from Chicago to New York, we obtain

$$E(x) = \frac{(120 + 140)}{2} = 130$$

$$Var(x) = \frac{(140 - 120)^2}{12} = 33.33$$

The standard deviation of flight times can be found by taking the square root of the variance. Thus $\sigma = 5.77$ minutes.

▬▬ NOTES AND COMMENTS ⋀ ▬▬

1. Since for any continuous random variable the probability of any particular value is zero, we have $P(a \leq x \leq b) = P(a < x < b)$. This shows that the probability of a random variable assuming a value in any interval is the same whether or not the endpoints are included.

2. To see more clearly why the height of a probability density function is not a probability, think about a random variable with the following uniform probability distribution:

$$f(x) = \begin{cases} 2 & \text{for } 0 \leq x \leq .5 \\ 0 & \text{elsewhere} \end{cases}$$

The height of the probability density function is 2 for values of x between 0 and .5. But, we know probabilities can never be greater than 1.

▬▬ EXERCISES

1. The random variable x is known to be uniformly distributed between 1.0 and 1.5.
   a. Show the graph of the probability density function.
   b. Find $P(x = 1.25)$.
   c. Find $P(1.0 \leq x \leq 1.25)$.
   d. Find $P(1.20 < x < 1.5)$.

2. The random variable x is known to be uniformly distributed between 10 and 20.
   a. Show the graph of the probability density function.
   b. Find $P(x < 15)$.
   c. Find $P(12 \leq x \leq 18)$.
   d. Find $E(x)$.
   e. Find $\text{Var}(x)$.

3. Delta Airlines quotes a flight time of 1 hour, 52 minutes for its flights from Cincinnati to Tampa. Suppose that we believe actual flight times are uniformly distributed between the quoted time and 2 hours, 10 minutes.
   a. Show the graph of the probability density function for flight times.
   b. What is the probability the flight will be no more than 5 minutes late?
   c. What is the probability the flight will be more than 10 minutes late?
   d. What is the expected flight time?

4. Most computer languages have a function that can be used to generate random numbers. In Microsoft's QuickBASIC, the RND function can be used to generate random numbers between 0 and 1. If we let x denote the random number generated, then x is a continuous random variable with the following probability density function:

$$f(x) = \begin{cases} 1 & \text{for } 0 \leq x \leq 1 \\ 0 & \text{elsewhere} \end{cases}$$

   a. Graph the probability density function.
   b. What is the probability of generating a random number between .25 and .75?

c. What is the probability of generating a random number with values less than or equal to .30?

d. What is the probability of generating a random number with value greater than .60?

5. The total time to process a loan application is uniformly distributed between 3 and 7 days.

a. Give a mathematical expression for the probability density function.

b. What is the probability that the loan application will be processed in fewer than 3 days?

c. Compute the probability that a loan application will be processed in 5 days or less.

d. Find the expected processing time and the standard deviation.

6. The label on a bottle of liquid detergent shows contents to be 12 ounces per bottle. The production operation fills the bottle uniformly according to the following probability density function:

$$f(x) = \begin{cases} 8 & \text{for } 11.975 \le x \le 12.10 \\ 0 & \text{elsewhere} \end{cases}$$

a. What is the probability that a bottle will be filled with between 12 and 12.05 ounces?

b. What is the probability that a bottle will be filled with 12.02 or more ounces?

c. Quality control accepts production that is within .02 ounces of the number of ounces shown on the container label. What is the probability that a bottle of this liquid detergent will fail to meet the quality control standard?

6.2 ▰ THE NORMAL DISTRIBUTION

Perhaps the most important probability distribution used to describe a continuous random variable is the *normal probability distribution*. The normal probability distribution has been applied in a wide variety of practical applications in which the random variables involved are heights and weights of people, IQ scores, scientific measurements, amounts of rainfall, and so on. In order to use this probability distribution, the random variable must be continuous. However, as we shall see, a continuous normal random variable can also used as an approximation in situations involving discrete random variables.

▰▰▰ FIGURE 6.3 **Bell-Shaped Curve for the Normal Probability Distribution**

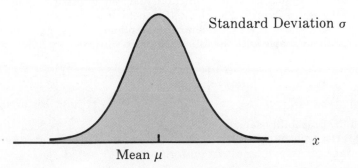

The Normal Curve

The form, or shape, of the normal probability distribution is illustrated by the bell-shaped curve shown in Figure 6.3. The probability density function that defines the bell-shaped curve of the normal probability distribution is as follows:

Normal Probability Density Function

$$f(x) = \frac{1}{\sqrt{2\pi}\sigma} e^{-(x-\mu)^2/2\sigma^2} \tag{6.2}$$

where μ is the mean
σ is the standard deviation
$\pi = 3.14159$
and $e = 2.71828$

We make some observations about the characteristics of the normal probability distribution:

1. There is an entire family of normal probability distributions with each specific normal distribution being differentiated by its mean μ and its standard deviation σ.

2. The highest point on the normal curve occurs at the mean, which is also the median and mode of the distribution.

3. The mean of the distribution can be any numerical value: negative, zero, or positive. Three normal curves with the same standard deviation but three different means (-10, 0, and 20) are shown.

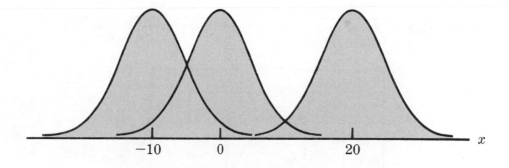

4. The normal probability distribution is symmetric, with the shape of the curve to the left of the mean a mirror image of the shape of the curve to the right of the mean. The tails of the curve extend to infinity in both directions and theoretically never touch the horizontal axis.

5. The standard deviation determines the width of the curve. Larger values of the standard deviation result in wider, flatter curves, showing more disper-

sion in the data. Two normal distributions with the same mean but with different standard deviations are shown.

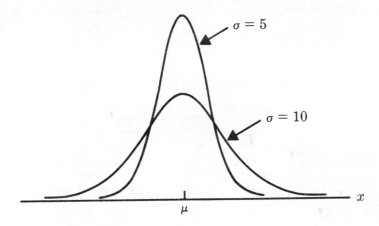

6. The total area under the curve for the normal probability distribution is 1. (This is true for all continuous probability distributions.)

7. Probabilities for the normal random variable are given by areas under the curve. Probabilities for some commonly used intervals are:

   a. 68.26% of the time, a normal random variable assumes a value within plus or minus 1 standard deviation of its mean.

   b. 95.44% of the time, a normal random variable assumes a value within plus or minus 2 standard deviations of its mean.

   c. 99.72% of the time, a normal random variable assumes a value within plus or minus 3 standard deviations of its mean. Figure 6.4 shows properties (a), (b), and (c) graphically.

■■■■ FIGURE 6.4 **Areas Under the Curve for Any Normal Probability Distribution**

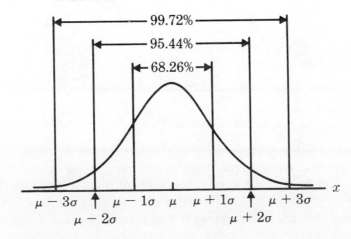

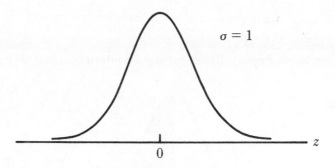

■■■ FIGURE 6.5   **The Standard Normal Probability Distribution**

The Standard Normal Probability Distribution

A random variable that has a normal distribution with a mean of 0 and a standard deviation of 1 is said to have a *standard normal probability distribution.* The letter z is commonly used to designate this particular normal random variable. The graph of the standard normal probability distribution is shown in Figure 6.5. This too is a normal probability distribution; hence it has the same general appearance as other normal distributions but with the special properties of $\mu = 0$ and $\sigma = 1$.

As with other continuous random variables, probability calculations with any normal probability distribution are made by computing areas under the graph of the probability density function. Thus, to find the probability that a normal random variable lies within any specific interval, we must compute the area under the normal curve over that interval. For the standard normal probability distribution, areas under the normal curve have been computed and are available in tables that can be used in computing probabilities. Table 6.1 is such a table; it is also available as Table 1 of Appendix B and inside the back cover of the text. The reason for interest in this table is that it can be used to compute probabilities for any normal distribution.

Let us show how the table of areas under the curve for the standard normal probability distribution (Table 6.1) can be used to find probabilities by considering some examples. Later we will see how this same table can be used to compute probabilities for any normal distribution. To begin with let us see how we can compute the probability that the z value for the standard normal random variable will be between 0.00 and 1.00; that is, $P(0.00 \leq z \leq 1.00)$. The shaded region in the following graph shows this area or probability.

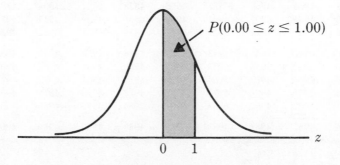

TABLE 6.1   Areas, or Probabilities, for the Standard Normal Distribution

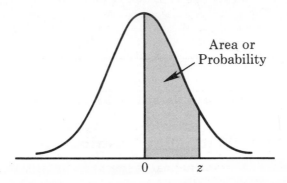

z	.00	.01	.02	.03	.04	.05	.06	.07	.08	.09
.0	.0000	.0040	.0080	.0120	.0160	.0199	.0239	.0279	.0319	.0359
.1	.0398	.0438	.0478	.0517	.0557	.0596	.0636	.0675	.0714	.0753
.2	.0793	.0832	.0871	.0910	.0948	.0987	.1026	.1064	.1103	.1141
.3	.1179	.1217	.1255	.1293	.1331	.1368	.1406	.1443	.1480	.1517
.4	.1554	.1591	.1628	.1664	.1700	.1736	.1772	.1808	.1844	.1879
.5	.1915	.1950	.1985	.2019	.2054	.2088	.2123	.2157	.2190	.2224
.6	.2257	.2291	.2324	.2357	.2389	.2422	.2454	.2486	.2518	.2549
.7	.2580	.2612	.2642	.2673	.2704	.2734	.2764	.2794	.2823	.2852
.8	.2881	.2910	.2939	.2967	.2995	.3023	.3051	.3078	.3106	.3133
.9	.3159	.3186	.3212	.3238	.3264	.3289	.3315	.3340	.3365	.3389
1.0	.3413	.3438	.3461	.3485	.3508	.3531	.3554	.3577	.3599	.3621
1.1	.3643	.3665	.3686	.3708	.3729	.3749	.3770	.3790	.3810	.3830
1.2	.3849	.3869	.3888	.3907	.3925	.3944	.3962	.3980	.3997	.4015
1.3	.4032	.4049	.4066	.4082	.4099	.4115	.4131	.4147	.4162	.4177
1.4	.4192	.4207	.4222	.4236	.4251	.4265	.4279	.4292	.4306	.4319
1.5	.4332	.4345	.4357	.4370	.4382	.4394	.4406	.4418	.4429	.4441
1.6	.4452	.4463	.4474	.4484	.4495	.4505	.4515	.4525	.4535	.4545
1.7	.4554	.4564	.4573	.4582	.4591	.4599	.4608	.4616	.4625	.4633
1.8	.4641	.4649	.4656	.4664	.4671	.4678	.4686	.4693	.4699	.4706
1.9	.4713	.4719	.4726	.4732	.4738	.4744	.4750	.4756	.4761	.4767
2.0	.4772	.4778	.4783	.4788	.4793	.4798	.4803	.4808	.4812	.4817
2.1	.4821	.4826	.4830	.4834	.4838	.4842	.4846	.4850	.4854	.4857
2.2	.4861	.4864	.4868	.4871	.4875	.4878	.4881	.4884	.4887	.4890
2.3	.4893	.4896	.4898	.4901	.4904	.4906	.4909	.4911	.4913	.4916
2.4	.4918	.4920	.4922	.4925	.4927	.4929	.4931	.4932	.4934	.4936
2.5	.4938	.4940	.4941	.4943	.4945	.4946	.4948	.4949	.4951	.4952
2.6	.4953	.4955	.4956	.4957	.4959	.4960	.4961	.4962	.4963	.4964
2.7	.4965	.4966	.4967	.4968	.4969	.4970	.4971	.4972	.4973	.4974
2.8	.4974	.4975	.4976	.4977	.4977	.4978	.4979	.4979	.4980	.4981
2.9	.4981	.4982	.4982	.4983	.4984	.4984	.4985	.4985	.4986	.4986
3.0	.4986	.4987	.4987	.4988	.4988	.4989	.4989	.4989	.4990	.4990

The entries in Table 6.1 give the area under the standard normal curve between the mean, $z = 0$, and a specified positive value of z. In this case we are interested in the area between $z = 0$ and $z = 1.00$. Thus we must find the entry in the table corresponding to $z = 1.00$. To do this, we first find 1.0 in the left-hand column of the table and then find .00 in the top row of the table. Then by looking in the body of the table we find that the 1.0 row of the table and the .00 column of the table intersect at the value of .3413. We have found the desired probability; $P(.00 \leq z \leq 1.00) = .3413$. A portion of Table 6.1 showing these steps is shown below.

z	.00	.01	.02
.			
.			
.9	.3159	.3186	.3212
1.0	.3413	.3438	.3461
1.1	.3643	.3665	.3686
1.2	.3849	.3869	.3888
.	.		
.	.	$P(0.00 \leq z \leq 1.00)$	

Using the same approach we can find $P(.00 \leq z \leq 1.25)$. We first locate the 1.2 row and then move across to the .05 column. Doing so, we find $P(0.00 \leq z \leq 1.25) = .3944$.

As a third example of the use of the table of areas for the standard normal distribution we compute the probability of obtaining a z value between $z = -1.00$ and $z = 1.00$; that is, $P(-1.00 \leq z \leq 1.00)$.

First note that we have already used Table 6.1 to show that the probability of a z value between $z = 0.00$ and $z = 1.00$ is .3413. Recall now that the normal probability distribution is *symmetric*. Thus the probability of a z value between $z = .00$ and $z = -1.00$ is the *same* as the probability of a z value between $z = .00$ and $z = +1.00$. Hence the probability of a z value between $z = -1.00$ and $z = +1.00$ is

$$P(-1.00 \leq z \leq 0.00) + P(0.00 \leq z \leq 1.00) = .3413 + .3413 = .6826$$

This area is shown graphically as follows.

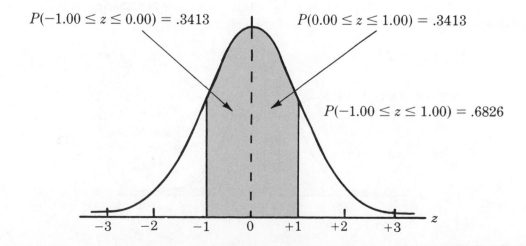

$P(-1.00 \leq z \leq 0.00) = .3413$ $P(0.00 \leq z \leq 1.00) = .3413$

$P(-1.00 \leq z \leq 1.00) = .6826$

In a similar manner we can use the values in Table 6.1 to show that the probability of a z value between -2.00 and $+2.00$ is $.4772 + .4772 = .9544$ and that the probability of a z value between -3.00 and $+3.00$ is $.4986 + .4986 = .9972$. Since we know that the total probability or total area under the curve for any continuous random variable must be 1.0000, the probability .9972 tells us that the value of z will almost always fall between -3.00 and $+3.00$.

Next, we compute the probability of obtaining a z value of at least 1.58; that is, $P(z \geq 1.58)$. First, we use the $z = 1.5$ row and the .08 column of Table 6.1 to find that $P(.00 \leq z \leq 1.58) = .4429$. Now, since the normal probability distribution is symmetric and the total area under the curve equals 1, we know that 50% of the area must be above the mean (i.e., $z = 0$) and 50% of the area must be below the mean. Since .4429 is the area between the mean and $z = 1.58$, the area or probability corresponding to $z \geq 1.58$ must be $.5000 - .4429 = .0571$. This probability is shown in the following figure.

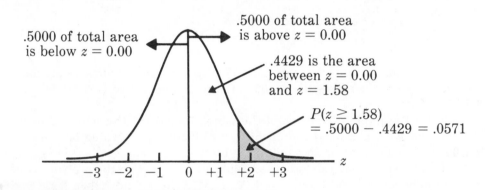

As another illustration, consider the probability the random variable z assumes a value of $-.50$ or larger; that is, $P(z \geq -.50)$. To make this computation, we note that the probability we are seeking can be written as the sum of two probabilities: $P(z \geq -.50) = P(-.50 \leq z \leq .00) + P(z \geq 0.00)$. We have previously seen that $P(z \geq .00) = .50$. Also, we know that since the normal distribution is symmetric, $P(-.50 \leq z \leq .00) = P(.00 \leq z \leq .50)$. Referring to Table 6.1 we find that $P(.00 \leq z \leq .50) = .1915$. Therefore $P(-.50 \leq z \leq .00) = .1915$. Thus $P(z \geq -.50) = P(-.50 \leq z \leq .00) + P(z \geq .00) = .1915 + .5000 = .6915$. The graph shows this area.

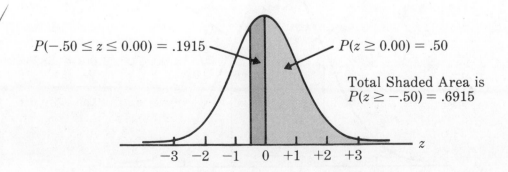

Next, we compute the probability of obtaining a z value between 1.00 1.58; that is, $P(1.00 \leq z \leq 1.58)$. From our previous examples we know that there is a .3413 probability of a z value between $z = 0.00$ and $z = 1.00$ and that there is a .4429 probability of a z value between $z = 0.00$ and $z = 1.58$. Thus there must be a $.4429 - .3413 = .1016$ probability of a z value between $z = 1.00$ and $z = 1.58$. Thus $P(1.00 \leq z \leq 1.58) = .1016$. This situation is shown graphically in the following figure.

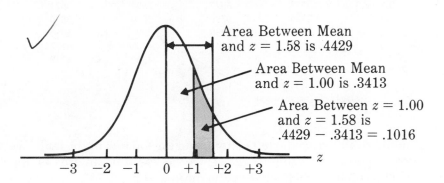

As a final illustration, let us find a z value such that the probability of obtaining a larger z value is only .10. This situation is shown graphically as follows:

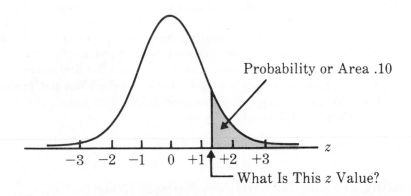

This problem is somewhat different from the examples we have considered thus far. The difference is that previously we specified the z value of interest and then found the corresponding probability, or area. In this example we are given the probability, or area, information and asked to find the corresponding z value. This can be found by using the table of areas for the standard normal probability distribution (Table 6.1) a little differently.

Recall that the body of Table 6.1 gives the area under the curve between the mean and a particular z value. In the above example we are given the information that the area in the upper tail of the curve is .10. Thus we must determine how much of the area is between the mean and the z value of interest. Since we know .5000 of the area is above the mean, $.5000 - .1000 = .4000$ must be the area under the curve *between* the mean and the desired z value. Scanning the body of the table, we find .3997 as the probability value closest to .4000. The section of the table providing this result is shown next:

z	.06	.07	.08	.09
.				
.				
.				
1.0	.3554	.3577	.3599	.3621
1.1	.3770	.3790	.3810	.3830
1.2	.3962	.3980	.3997	.4015
1.3	.4131	.4147	.4162	.4177
1.4	.4729	.4292	.4306	.4319
.				

Area value in body
of table closest to .4000

Reading the z value from the left column and the top row of the table, we find that the corresponding z value is 1.28. Thus there will be an area of approximately .4000 (actually .3997) between the mean and $z = 1.28$*. In terms of the question originally asked, there is approximately a .10 probability of a z value larger than 1.28.

The examples illustrate that the table of areas for the standard normal probability distribution can be used to find probabilities associated with values of the standard normal random variable z. Two types of questions can be asked. The first type of question specifies a value, or values, for z and asks us to use the table to determine the corresponding areas, or probabilities. The second type of question provides an area, or probability, and asks us to use the table to determine the corresponding z value. Thus we need to remain flexible in terms of using the standard normal probability table to answer the desired probability question. In most cases, sketching a graph of the standard normal probability distribution and shading the appropriate area helps to visualize the situation and aids in determining the correct answer.

Computing Probabilities for Any Normal Distribution

The reason that we have been discussing the standard normal distribution so extensively is that probabilities for all normal distributions are computed using the standard normal distribution. That is, when we have a normal distribution with any mean μ and any standard deviation σ, we answer probability questions about the distribution by first converting to the standard normal distribution. Then we can use Table 6.1 and the appropriate z values to find the desired probabilities. The formula used to convert any normal random variable x with

---

*We could use interpolation in the body of the table to get a better approximation of the z value that cuts off an area of .4000. Doing so to provide one more decimal place of accuracy would yield a z value of 1.282. However, in most practical situations sufficient accuracy is obtained by simply using the table value closest to the desired probability.

mean μ and standard deviation σ to the standard normal distribution is as follows.

Converting to the Standard Normal Distribution

$$z = \frac{(x - \mu)}{\sigma} \tag{6.3}$$

A value of x equal to its mean μ results in $z = (\mu - \mu)/\sigma = 0$. Thus we see that a value of x equal to its mean μ corresponds to a value of z at its mean 0. Now suppose that x is one standard deviation above its mean; that is, $x = \mu + \sigma$. Applying equation 6.3 we see that the corresponding z value is $z = [(\mu + \sigma) - \mu]/\sigma = \sigma/\sigma = 1$. Thus a value of x that is one standard deviation above its mean yields, $z = 1$. In other words, we can interpret the z value as *the number of standard deviations that the normal random variable, x, is from its mean μ.*

To see how this conversion enables us to compute probabilities for any normal distribution, suppose we have a normal distribution with $\mu = 10$ and $\sigma = 2$. What is the probability that the random variable, x, is between 10 and 14? Using (6.3) we see that at $x = 10$, $z = (x - \mu)/\sigma = (10 - 10)/2 = 0$ and that at $x = 14$, $z = (14 - 10)/2 = 4/2 = 2$. Thus the answer to our question about the probability of x being between 10 and 14 is given by the equivalent probability that z is between 0 and 2 for the standard normal distribution. In other words, the probability that we are seeking is the probability that the random variable x is between its mean and two standard deviations above the mean. Using $z = 2.00$ and Table 6.1, we see that the probability is .4772. Hence the probability that x is between 10 and 14 is .4772.

The Grear Tire Company Problem

Let us look at an application of the use of the normal probability distribution. Suppose that the Grear Tire Company has just developed a new steel-belted radial tire that will be sold through a national chain of discount stores. Since the tire is a new product, Grear's management believes that the mileage guarantee offered with the tire will be an important factor in the acceptance of the product. Before finalizing the tire mileage guarantee policy, Grear's management would like some probability information concerning the number of miles the tires will last.

From actual road tests with the tires, Grear's engineering group has estimated the mean tire mileage at $\mu = 36,500$ miles and the standard deviation at $\sigma = 5000$. In addition, the data collected indicate that a normal distribution is a reasonable assumption.

Using the normal distribution, what percentage of the tires can be expected to last more than 40,000 miles? In other words, what is the probability that the tire mileage will exceed 40,000? This question can be answered by finding the area of the shaded region in Figure 6.6.

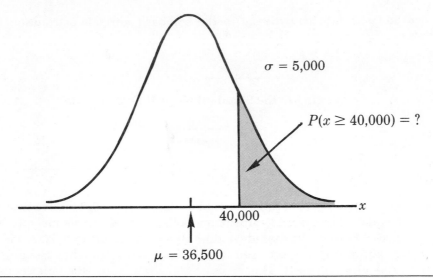

$\sigma = 5{,}000$

$P(x \geq 40{,}000) = ?$

x

40,000

$\mu = 36{,}500$

◼◼ FIGURE 6.6   **Grear Tire Company Mileage Distribution**

At $x = 40{,}000$ we have

$$z = \frac{x - \mu}{\sigma} = \frac{40{,}000 - 36{,}500}{5000} = \frac{3500}{5000} = .70$$

Using Table 6.1 we see that the area between the mean and $z = .70$ is .2580. Thus .5000 − .2580 = .2420 is the probability that x will exceed 40,000. We can conclude that about 24.2% of the tires will exceed 40,000 in mileage.

Let us now assume that Grear is considering a guarantee that will provide a discount on a new set of tires if the original tires do not exceed the mileage stated in the guarantee. What should the guarantee mileage be if Grear would like no more than 10% of the tires to be eligible for the discount guarantee? This question is interpreted graphically in Figure 6.7. According to Figure 6.7, 40% of the area must be between the mean and the unknown guarantee mileage. We look up .4000 in the body of Table 6.1 and see that this area occurs at approximately 1.28 standard deviations *below the mean*. That is, $z = -1.28$. To find the mileage (x) corresponding to $z = -1.28$ we have

$$z = \frac{x - \mu}{\sigma} = -1.28$$

$$x - \mu = -1.28\sigma$$

$$x = \mu - 1.28\sigma$$

or, with $\mu = 36{,}500$ and $\sigma = 5000$,

$$x = 36{,}500 - 1.28(5000) = 30{,}100$$

Thus a guarantee of 30,100 miles will meet the requirement that approxi-

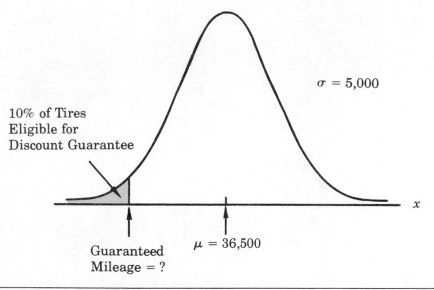

$\sigma = 5,000$

10% of Tires
Eligible for
Discount Guarantee

Guaranteed
Mileage = ?

$\mu = 36,500$

x

FIGURE 6.7   Grear's Discount Guarantee

mately 10% of the tires will be eligible for the guarantee. Perhaps with this information the firm will set its tire mileage guarantee policy at 30,000 miles.

Again we see the important role that probability distributions play in providing decision-making information. Namely, once a probability distribution is established for a particular problem situation, it can be used to rather quickly and easily provide probability data about the problem. While the data do not make a decision recommendation directly, they do provide information that helps the decision maker better understand the problem. Ultimately this information may assist the decision maker in reaching a good decision.

Normal Approximation of Binomial Probabilities

As discussed in the previous chapter, binomial probability tables for large values of n usually are not available. A normal distribution approximation of binomial probabilities is considered acceptable when $np \geq 5$ and $n(1 - p) \geq 5$.

When using the normal approximation to the binomial we set $\mu = np$ and $\sigma = \sqrt{np (1 - p)}$ in the definition of the normal curve. Let us illustrate the normal approximation to the binomial by supposing that a particular company has a history of making errors in 10% of its invoices. A sample of 100 invoices has been taken, and we would like to compute the probability that 12 invoices contain errors. That is, we would like to find the binomial probability of 12 successes in 100 trials. Since the binomial tables in Appendix B are not tabulated for values of n greater than 20, we will use the normal approximation to compute the desired probability.

In applying the normal approximation to the binomial we set $\mu = np = (100)(.1) = 10$ and $\sigma = \sqrt{np (1 - p)} = \sqrt{(100)(.1)(.9)} = 3$. A normal distribution with $\mu = 10$ and $\sigma = 3$ is shown in Figure 6.8.

Recall that with a continuous probability distribution probabilities are computed as areas under the probability density function. As a result the probability of any single value for the random variable is zero. Thus to approximate the

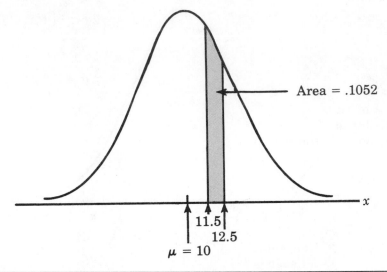

Area = .1052

11.5
12.5
$\mu = 10$

x

**Normal Approximation to a Binomial Probability
Distribution with n = 100 and p = .10 Showing the
Probability of 12 Errors**

binomial probability of 12 successes we must compute the area under the
corresponding normal curve between 11.5 and 12.5. The .5 that we added and
subtracted from 12 is called a *continuity correction factor*. It is introduced because a
continuous distribution is being used to approximate a discrete distribution.
Thus $P(x = 12)$ for the *discrete* binomial distribution is approximated by $P(11.5 \leq
x \leq 12.5)$ for the *continuous* normal distribution.

Converting to the standard normal distribution in order to compute $P(11.5 \leq
x \leq 12.5)$, we have

$$z = \frac{x - \mu}{\sigma} = \frac{12.5 - 10.0}{3} = .83 \qquad \text{at } x = 12.5$$

$$z = \frac{x - \mu}{\sigma} = \frac{11.5 - 10.0}{3} = .50 \qquad \text{at } x = 11.5$$

From Table 6.1 we find the area under the curve (in Figure 6.8) between 10 and
12.5 is .2967. Similarly, the area under the curve between 10 and 11.5 is .1915.
Therefore, the area between 11.5 and 12.5 is .2967 − .1915 = .1052. The normal
approximation to the probability of 12 successes in 100 trials thus is .1052.

For another illustration, suppose that we want to compute the probability of
13 or fewer errors in the sample of 100 invoices. Figure 6.9 shows the area under
the normal curve which approximates this probability. Note that the use of the
continuity correction factor results in the value of 13.5 being used to compute
the desired probability. The z value corresponding to $x = 13.5$ is

$$z = \frac{13.5 - 10.0}{3.0} = 1.17$$

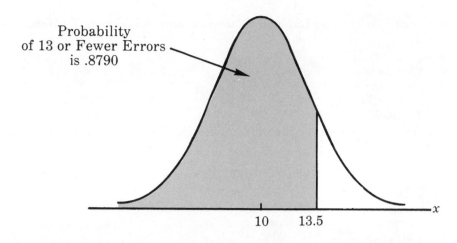

Probability
of 13 or Fewer Errors
is .8790

x

10   13.5

■■■ FIGURE 6.9   **Normal Approximation to a Binomial Probability
Distribution with n = 100 and p = .10 Showing the
Probability of 13 or Fewer Errors**

Table 6.1 shows that the area under the curve between 10 and 13.5 is .3790.
Hence the shaded portion of the graph in Figure 6.9 represents an area, or
probability, of .3790 + .5000 = .8790.

■■■ EXERCISES

7. Using Figure 6.4 as a guide, sketch a normal curve for a random variable x that has a
mean $\mu = 100$ and a standard deviation $\sigma = 10$. Label the horizontal axis with values of 70,
80, 90, 100, 110, 120 and 130.

8. The length of time required to complete a college examination is normally distributed
with a mean of $\mu = 50$ minutes and a standard deviation of $\sigma = 5$ minutes.

   a. Sketch a normal curve for the length of the examination. Label the horizontal axis
   with values of 35, 40, 45, 50, 55, 60 and 65 minutes. Figure 6.4 shows that the normal
   curve almost touches the horizontal line at three standard deviations below and at
   three standard deviations above the mean (in this case at 35 and 65).
   b. What is the probability that a student will take between 45 and 55 minutes to
   complete the exam?
   c. What is the probability that a student will take between 40 and 60 minutes to
   complete the exam?

9. Given that z is the standard normal random variable, sketch the standard normal
curve. Label the horizontal axis at values of $-3, -2, -1, 0, 1, 2,$ and 3. Then use the table
of probabilities for the standard normal distribution to compute the following probabilities.

   a. $P(0 \leq z \leq 1)$
   b. $P(0 \leq z \leq 1.5)$
   c. $P(0 < z < 2)$
   d. $P(0 < z < 2.5)$

10. Given that z is the standard normal random variable, compute the following probabilities.

   a. $P(-1 \leq z \leq 0)$
   b. $P(-1.5 \leq z \leq 0)$
   c. $P(-2 < z < 0)$
   d. $P(-2.5 \leq z \leq 0)$
   e. $P(-3 < z \leq 0)$

11. Given that z is the standard normal random variable, compute the following probabilities.

   a. $P(0 \leq z \leq .83)$
   b. $P(-1.57 \leq z \leq 0)$
   c. $P(z > .44)$
   d. $P(z \geq -.23)$
   e. $P(z < 1.20)$
   f. $P(z \leq -.71)$

12. Given that z is the standard normal random variable, compute the following probabilities.

   a. $P(-1.98 \leq z \leq .49)$
   b. $P(.52 \leq z \leq 1.22)$
   c. $P(-1.75 \leq z \leq -1.04)$

13. Given that z is the standard normal random variable, find z for each situation.

   a. The area between 0 and z is .4750.
   b. The area between 0 and z is .2291.
   c. The area to the right of z is .1314.
   d. The area to the left of z is .6700.

14. Given that z is the standard normal random variable, find z for each situation.

   a. The area to the left of z is .2119.
   b. The area between $-z$ and z is .9030.
   c. The area between $-z$ and z is .2052.
   d. The area to the left of z is .9948.
   e. The area to the right of z is .6915.

15. Given that z is the standard normal random variable, find z for each situation.

   a. The area to the right of z is .01.
   b. The area to the right of z is .025.
   c. The area to the right of z is .05.
   d. The area to the right of z is .10.

16. The demand for a new product is assumed to be normally distributed with $\mu = 200$ and $\sigma = 40$. Letting x be the number of units demanded, find the following:

   a. $P(180 \leq x \leq 220)$
   b. $P(x \geq 250)$
   c. $P(x \leq 100)$
   d. $P(225 \leq x \leq 250)$

17. The Webster National Bank is reviewing its service charges and interest-paying policies on checking accounts. The bank has found that the average daily balance on personal checking accounts is $550.00, with a standard deviation of $150.00. In addition, the average daily balances have been found to be normally distributed.

   a. What percentage of personal checking account customers carry average daily balances in excess of $800.00?
   b. What percentage of the bank's customers carry average daily balances below $200.00?
   c. What percentage of the bank's customers carry average daily balances between $300.00 and $700.00?

d. The bank is considering paying interest to customers carrying average daily balances in excess of a certain amount. If the bank does not want to pay interest to more than 5% of its customers, what is the minimum average daily balance it should be willing to pay interest on?

18. General Hospital's patient account division has compiled data on the age of accounts receivables. The data collected indicate that the age of the accounts follows a normal distribution, with $\mu = 28$ days and $\sigma = 8$ days.

a. What portion of the accounts are between 20 and 40 days old? $(P (20 \le x \le 40))$?

b. The hospital administrator is interested in sending reminder letters to the oldest 15% of accounts. How many days old should an account be before a reminder letter is sent?

c. The hospital administrator would like to give a discount to those accounts that pay their balance by the 21st day. What percentage of the accounts will receive the discount?

19. The time required to complete a final examination in a particular college course is normally distributed, with a mean of 80 minutes and a standard deviation of 10 minutes. Answer the following questions:

a. What is the probability of completing the exam in 1 hour or less?

b. What is the probability a student will complete the exam in more than 60 minutes but less than 75 minutes?

c. Assume that the class has 60 students and that the examination period is 90 minutes in length. How many students do you expect will be unable to complete the exam in the allotted time?

20. The useful life of a computer terminal at a university computer center is known to be normally distributed, with a mean of 3.25 years and a standard deviation of .5 years.

a. Historically 22% of the terminals have had a useful life less than the manufacturer's advertised life. What is the manufacturer's advertised life for the computer terminals?

b. What is the probability that a computer terminal will have a useful life of at least 3 but less than 4 years?

21. From past experience, the management of a well-known fast-food restaurant estimates that the number of weekly customers at a particular location is normally distributed, with a mean of 5000 and a standard deviation of 800 customers.

a. What is the probability that on a given week the number of customers will be 4760 to 5800?

b. What is the probability of more than 6500 customers?

c. For 90% of the weeks the number of customers should exceed what amount?

22. In order to obtain cost savings, a company is considering offering an early retirement incentive for its older management personnel. The consulting firm that designed the early retirement program has found that approximately 22% of the employees qualifying for the program will select early retirement during the first year of eligibility. Assume that the company offers the early retirement program to 50 of its management personnel.

a. What is the expected number of employees who will elect early retirement in the first year?

b. What is the probability at least 8 but not more than 12 employees will elect early retirement in the first year?

c. What is the probability that 15 or more employees will select the early retirement option in the first year?

d. For the program to be judged successful, the company believes that it should entice at least 10 management employees to elect early retirement in the first year. What is the probability that the program is successful?

23. Suppose that 54% of a large population of registered voters favor the Democratic candidate for state senator. A public opinion poll uses randomly selected samples of voters and asks each person in the sample his or her preference: the Democratic candidate or the Republican candidate. The weekly poll is based on the response of 100 voters.

 a. What is the expected number of voters who will favor the Democratic candidate?
 b. What is the variance in the number of voters who will favor the Democratic candidate?
 c. What is the probability that 49 or fewer individuals in the sample express support for the Democratic candidate?

24. Thirty percent of the students at a particular university attended Catholic high schools. A random sample of 50 of this university's students has been taken. Use the normal approximation to the binomial probability distribution to answer the following questions:

 a. What is the probability that exactly 10 of the students selected attended Catholic high schools?
 b. What is the probability that 20 or more of the students attended Catholic high schools?
 c. What is the probability that the number of students from Catholic high schools is between 10 and 20 inclusively?

25. A Myrtle Beach resort hotel has 120 rooms. In the spring months, hotel room occupancy is approximately 75%. Use the normal approximation to the binomial distribution to answer the following questions:

 a. What is the probability that at least half the rooms are occupied on a given day?
 b. What is the probability that 100 or more rooms are occupied on a given day?
 c. What is the probability that 80 or fewer rooms are occupied on a given day?

26. It is known that 30% of all customers of a major national charge card pay their bills in full before any interest charges are incurred. Use the normal approximation to the binomial distribution to answer the following questions for a group of 150 credit card holders:

 a. What is the probability that between 40 and 60 customers pay their account charges before any interest charges are incurred? That is, find $P(40 \leq x \leq 60)$.
 b. What is the probability that 30 or fewer customers pay their account charges before any interest charges are incurred?

6.3 ■ THE EXPONENTIAL DISTRIBUTION

A continuous probability distribution that is often useful in describing the time it takes to complete a task is the *exponential probability distribution*. The exponential random variable can be used to describe such things as the time between arrivals at a carwash, the time required to load a truck, the distance between major defects in a highway, and so on. The exponential probability density function is as follows:

Exponential Probability Density Function

$$f(x) = \frac{1}{\mu} e^{-x/\mu} \qquad \text{for } x \geq 0, \mu > 0 \tag{6.4}$$

To provide an example of the exponential probability distribution, assume that the time it takes to load a truck at the Schips loading dock follows an exponential probability distribution. If the mean, or average, time to load a truck is 15 minutes ($\mu = 15$), then the appropriate probability density function is

$$f(x) = \frac{1}{15} e^{-x/15}$$

The graph of this density function is shown in Figure 6.10.

Computing Probabilities for the Exponential Distribution

As with any continuous probability distribution, the area under the curve corresponding to some interval provides the probability that the random variable takes on a value in that interval. For example, for the Schips loading dock example the probability that it takes 6 *minutes or less* ($x \leq 6$) to load a truck is defined to be the area under the curve from $x = 0$ to $x = 6$. Similarly, the probability that a truck is loaded in 18 *minutes or less* ($x \leq 18$) is the area under the curve from $x = 0$ to $x = 18$. Note also that the probability that it takes between 6 minutes and 18 minutes ($6 \leq x \leq 18$) to load a truck is given by the area under the curve from $x = 6$ to $x = 18$.

In order to compute exponential probabilities such as those described above, we make use of the following formula, which provides the probability of obtaining a value for the exponential random variable of less than or equal to some specific value of x, denoted by x_0:

Exponential Distribution Probabilities

$$P(x \leq x_0) = 1 - e^{-x_0/\mu} \tag{6.5}$$

Thus for the Schips loading dock example, (6.5) can be written as

$$P(\text{loading time} \leq x_0) = 1 - e^{-x_0/15}$$

Hence, the probability that it takes 6 minutes or less ($x \leq 6$) to load a truck is

$$P(\text{loading time} \leq 6) = 1 - e^{-6/15} = .3297$$

Note also the probability that it takes 18 minutes or less ($x \leq 18$) to load a truck is

$$P(\text{loading time} \leq 18) = 1 - e^{-18/15} = .6988$$

Thus we see that the probability that it takes between 6 minutes and 18 minutes to load a truck is equal to $.6988 - .3297 = .3691$. Probabilities for any other interval can be computed in a similar manner.

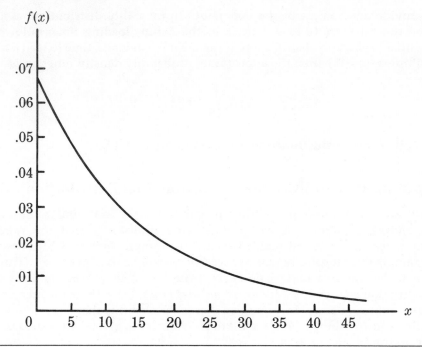

▩▩▩ FIGURE 6.10 **Exponential Probability Distribution for the Schips Loading Dock Example**

Relationship Between the Poisson and Exponential Distributions

In Section 5.5 we introduced the Poisson distribution as a discrete probability distribution that is often useful when dealing with the number of occurrences of an event over a specified interval of time or space. Recall that the Poisson probability function is

$$f(x) = \frac{\mu^x e^{-\mu}}{x!}$$

where

μ = expected value or mean number of occurrences in an interval.

The continuous exponential probability distribution is related to the discrete Poisson distribution in that if the Poisson distribution provides an appropriate description of the number of occurrences per interval, then the exponential distribution provides a description of the length of the interval between occurrences.

To provide an example that illustrates this relationship, suppose that the number of cars that arrive at a car wash during 1 hour is described by a Poisson probability distribution with a mean of 10 cars per hour. Thus the Poisson probability function that provides the probability of x arrivals per hour is

$$f(x) = \frac{10^x e^{-10}}{x!}$$

Since the average number of arrivals is 10 cars per hour, the average time between cars arriving is

$$\frac{1 \text{ hour}}{10 \text{ cars}} = .1 \text{ hour/car}$$

Thus the corresponding exponential distribution that describes the time between the arrival of cars has a mean of $\mu = .1$ hours per car; the appropriate exponential probability density function is

$$f(x) = \frac{1}{.1} e^{-x/.1} = 10 e^{-10x}$$

■■■ EXERCISES

27. Answer the following questions concerning the given exponential probability distribution

$$f(x) = \frac{1}{8} e^{-x/8} \qquad \text{for } x \geq 0$$

  a. Find $P(x \leq 6)$.
  b. Find $P(x \leq 4)$.
  c. Find $P(x \geq 6)$.
  d. Find $P(4 \leq x \leq 6)$.

28. There were 34 traffic fatalities in Clermont County, Ohio, during 1987 (The *Cincinnati Enquirer*, December 8, 1988). Assume that given an average of 34 fatalities per year an exponential distribution accurately describes the time between fatalities.

  a. What is the probability that the time between fatalities is one month or less?
  b. What is the probability that the time between fatalities is one week or more?

29. The time between arrivals of vehicles at a particular intersection follows an exponential probability distribution with a mean of 12 seconds.

  a. Sketch this exponential probability distribution.
  b. What is the probability that the arrival time between vehicles is 12 seconds or less?
  c. What is the probability that the arrival time between vehicles is 6 seconds or less?
  d. What is the probability that there will be 30 or more seconds between arriving vehicles?

30. The lifetime (hours) of an electronic device is a random variable with the following exponential probability density function:

$$f(x) = \frac{1}{50} e^{-x/50} \qquad \text{for } x \geq 0$$

  a. What is the mean lifetime of the device?
  b. What is the probability the device fails in the first 25 hours of operation?
  c. What is the probability the device operates 100 or more hours before failure?

31. A new automated production process has been averaging 2 breakdowns per day, where the number of breakdowns per day follows a Poisson probability distribution.

a. What is the mean time between breakdowns, assuming 8 hours of operation per day?

b. Show the exponential probability density function that can be used for the time between breakdowns.

c. What is the probability the process will run 1 hour or more before another breakdown?

d. What is the probability that the process can run a full 8-hour shift without a breakdown?

32. The time in minutes for a student using a computer terminal at the computer center of a major university follows an exponential probability distribution with a mean of 36 minutes. Assume a second student arrives at the terminal just as another student is beginning to work on the terminal.

a. What is the probability the wait for the second student will be 15 minutes or less?

b. What is the probability the wait for the second student will be between 15 minutes and 45 minutes?

c. What is the probability the second student will have to wait an hour or more?

■■■ SUMMARY

This chapter extended the discussion of probability distributions to the case of continuous random variables. The major conceptual difference between discrete and continuous probability distributions is in the method of computing probabilities. With discrete distributions the probability function, $f(x)$, provides the probability that the random variable, x, assumed various values. With continuous probability distributions, we associate a probability density function, denoted by $f(x)$. The difference is that the probability density function does not provide probability values for a continuous random variable directly. Probabilities are given by areas under the curve or graph of the probability density function, $f(x)$. Since the area under the curve above a single point is zero, we observe that the probability of any particular value is zero for a continuous random variable.

Three continuous probability distributions, the uniform, normal, and exponential, were treated in detail. The normal probability distribution is used widely in statistical inference and will be used extensively in the remainder of the text.

■■■ GLOSSARY

Uniform probability distribution A continuous probability distribution where the probability that the random variable will assume a value in any interval of equal length is the same for each interval.

Probability density function The function that defines the probability distribution of a continuous random variable.

Normal probability distribution A continuous probability distribution. Its probability density function is bell shaped and determined by the mean μ and standard deviation σ.

Standard normal distribution A normal distribution with a mean of 0 and a standard deviation of 1.

Continuity correction factor A value of .5 that is added and/or subtracted from a value of x when the continuous normal probability distribution is used to approximate the discrete binomial probability distribution.

Exponential probability distribution A continuous probability distribution that is useful in computing probabilities for the time, or space, between occurrences of an event.

■■■ KEY FORMULAS

Uniform Probability Density Function

$$f(x) = \begin{cases} \dfrac{1}{b-a} & \text{for } a \le x \le b \\ \\ 0 & \text{elsewhere} \end{cases} \tag{6.1}$$

Normal Probability Density Function

$$f(x) = \frac{1}{\sqrt{2\pi}\sigma} e^{-(x-\mu)^2/2\sigma^2} \tag{6.2}$$

Converting to the Standard Normal Distribution

$$z = \frac{x-\mu}{\sigma} \tag{6.3}$$

Exponential Probability Density Function

$$f(x) = \frac{1}{\mu} e^{-x/\mu} \qquad \text{for } x \ge 0, \mu > 0 \tag{6.4}$$

Exponential Distribution Probabilities

$$P(x \le x_0) = 1 - e^{-x_0/\mu} \tag{6.5}$$

■■■ SUPPLEMENTARY EXERCISES

33. In an office building the waiting time for an elevator is found to be uniformly distributed between 0 minutes and 5 minutes.

    a. What is the probability density function, $f(x)$, for this uniform distribution?
    b. What is the probability of waiting longer than 3.5 minutes?
    c. What is the probability that the elevator arrives in the first 45 seconds?
    d. What is the probability of a waiting time between 1 and 3 minutes?
    e. What is the expected waiting time?

34. The time required to complete a particular assembly operation is uniformly distributed between 30 and 40 minutes.

    a. What is the mathematical expression for the probability density function?
    b. Compute the probability that the assembly operation will require more than 38 minutes to complete.
    c. If management wants to set a time standard for this operation, what time should be selected such that 70% of the time the operation will be completed within the time specified?
    d. Find the expected value and standard deviation for the assembly time.

35. A particular make of automobile is listed as weighing 4000 pounds. Because of weight differences due to the options ordered with the car, the actual weight varies uniformly between 3900 and 4100 pounds.

    a. What is the mathematical expression for the probability density function?
    b. What is the probability that the car will weigh less than 3950 pounds?

36. Given that z is a standard normal random variable, compute the following probabilities.

    a. $P(-.72 \leq z \leq 0)$
    b. $P(-.35 \leq z \leq .35)$
    c. $P(.22 \leq z \leq .87)$
    d. $P(z \leq -1.02)$

37. Given that z is a standard normal random variable, compute the following probabilities.

    a. $P(z \geq -.88)$
    b. $P(z \geq 1.38)$
    c. $P(-.54 \leq z \leq 2.33)$
    d. $P(-1.96 \leq z \leq 1.96)$

38. Given that z is a standard normal random variable, find z if it is known that

    a. the area between $-z$ and z is .90
    b. the area to the right of z is .20
    c. the area between -1.66 and z is .25
    d. the area to the left of z is .40
    e. the area between z and 1.80 is .20

39. In 1985 the average household income for Americans was $23,618 (*Louis Rukeyser's Business Almanac*, Simon and Schuster, New York, 1988).

    a. It was noted that 7.7% of the households earned less than $5000. Assuming that household income is normally distributed, what is the standard deviation of household income?
    b. It was also noted that 14.8% of households earned more than $50,000. Does this seem reasonable given the standard deviation computed in part a? Explain.
    c. In part a we said to assume that household income is normally distributed. Does this assumption appear to be reasonable? Explain.

40. A soup company markets eight varieties of homemade soup throughout the Eastern states. The standard-size soup can holds a maximum of 11 ounces, while the label on each can advertises contents of $10\frac{3}{4}$ ounces. The extra $\frac{1}{4}$ ounce is to allow for the possibility of the automatic filling machine placing more soup than the company actually wants in a can. Past experience shows that the number of ounces placed in a can is approximately normally distributed, with a mean of $10\frac{3}{4}$ ounces and a standard deviation of .1 ounce. What is the probability that the machine will attempt to place more than 11 ounces in a can, causing an overflow to occur?

41. The sales of High-Brite Toothpaste are believed to be approximately normally distributed, with a mean of 10,000 tubes per week and a standard deviation of 1500 tubes per week.

    a. What is the probability that more than 12,000 tubes will be sold in any given week?
    b. In order to have a .95 probability that the company will have sufficient stock to cover the weekly demand, how many tubes should be produced?

42. Points scored by the winning team in NCAA college football games are approximately normally distributed, with a mean of 24 and a standard deviation of 6.

   a. What is the probability that a winning team in a football game scores between 20 and 30 points; that is, $P(20 \leq x \leq 30)$?
   b. How many points does a winning team have to score to be in the highest 20% of scores for college football games?

43. Ward Doering Auto Sales is considering offering a special service contract that will cover the total cost of any service work required on leased vehicles. From past experience the company manager estimates that yearly service costs are approximately normally distributed, with a mean of $150 and a standard deviation of $25.

   a. If the company offers the service contract to customers for a yearly charge of $200, what is the probability that any one customer's service costs will exceed the contract price of $200?
   b. What is Ward's expected profit per service contract?

44. The attendance at football games at a certain stadium is normally distributed, with a mean of 45,000 and a standard deviation of 3000.

   a. What percentage of the time should attendance be between 44,000 and 48,000?
   b. What is the probability of the attendance exceeding 50,000?
   c. Eighty percent of the time the attendance should be at least how many?

45. Assume that the test scores from a college admissions test are normally distributed, with a mean of 450 and a standard deviation of 100.

   a. What percentage of the people taking the test score between 400 and 500?
   b. Suppose that someone receives a score of 630. What percentage of the people taking the test score better? What percentage score worse?
   c. If a particular university will not admit anyone scoring below 480, what percentage of the persons taking the test would be acceptable to the university?

46. The Office Products Group of the former Burroughs Corporation manufacturers plastic credit cards used in automatic bank teller machines. Any card with a length of less than 3.365 inches is considered defective. One of the dies used in making the credit cards is producing cards with a mean length of 3.367 inches. The lengths are normally distributed with a standard deviation of .001 inch.

   a. What is the probability of obtaining a defective card using this die?
   b. The company does not want to see any die that produces more than 1% defective cards. What should the company do in this instance?
   c. Assuming the standard deviation stays at .001 inch, what is the smallest acceptable mean length for cards manufactured? (*Hint:* For what mean length will no more than 1% of the card be shorter than 3.365 inches?)

47. A machine fills containers with a particular product. The standard deviation of filling weights is known from past data to be .6 ounces. If only 2% of the containers hold less than 18 ounces, what is the mean filling weight for the machine? That is, what must μ equal? Assume the filling weights have a normal distribution.

48. It is estimated that in criminal trials, the jury will reach the correct decision (guilty or not guilty) 90% of the time. Consider a group of 100 cases that are brought to trial before a jury.

   a. What is the expected number of cases where the jury will reach the correct decision?
   b. What is the probability the jury will judge 95 or more cases correctly?
   c. What is the probability an incorrect decision is reached in 12 or more cases?
   d. Answer the question in (c) if the jury system reaches the correct decision 95% of the time.

49. Consider a multiple-choice examination with 50 questions. Each question has 4 possible answers. Assume that a student who has done the homework and attended lectures has a .75 probability of answering any question correctly.

   a. A student must answer 43 or more questions correctly in order to obtain a grade of A. What percentage of the students who have done their homework and attended lectures will obtain a grade of A on this multiple-choice examination?

   b. A student who answers 35 questions to 39 questions correctly will receive a grade of C. What percentage of students who have done their homework and attended lectures will obtain a grade of C on this multiple-choice examination?

   c. A student must answer 30 or more questions correctly in order to pass the examination. What percentage of the students who have done their homework and attended lectures will pass the examination?

   d. Assume that a student has not attended class and has not done the homework for the course. Furthermore, assume that the student will simply guess at the answer to each question. What is the probability that this student answers 30 or more questions correctly and passes the examination?

50. The time (in minutes) between telephone calls at an insurance claims office has the following exponential probability distribution:

$$f(x) = .50e^{-.50x} \qquad \text{for } x \geq 0$$

   a. What is the mean time between telephone calls?
   b. What is the probability that there are 30 seconds or less between telephone calls?
   c. What is the probability that there is 1 minute or less between telephone calls?
   d. What is the probability of going 5 or more minutes without a telephone call?

51. The time (in minutes) a checkout lane is idle between customers at a supermarket follows an exponential probability distribution with a mean of 1.2 minutes.

   a. Show the probability density function for this distribution.
   b. What is the probability the next customer arrives between .5 to 1.0 minutes after a customer is served?
   c. What is the probability of the checkout lane being idle for more than a minute between customers?

APPLICATION

Procter & Gamble*

Cincinnati, Ohio

Procter & Gamble (P&G) is in the consumer-products business worldwide. The company competes in over 36 product categories, such as detergents, disposable diapers, over-the-counter pharmaceuticals, dentifrices, bar soaps, mouthwashes, and paper towels. P&G has the leading brand in more categories than any other consumer products company. The company develops and manufactures its products internally; also, raw materials are processed internally for paper products, detergents, and bar soaps.

MATHEMATICAL AND INFORMATION SERVICES

Mathematical and Information Services (MIS) provides quantitative and information systems support to the Corporate Engineering Division. MIS is a section comprised of people with diverse academic backgrounds: engineering, statistics, operations research, and business, with an emphasis on quantitative methods. The major quantitative technologies MIS provides support for and leadership in are as follows:

1. *Probabilistic Decision and Risk Analysis.* This technology has been applied corporately since 1980. It is composed of a five-step process:

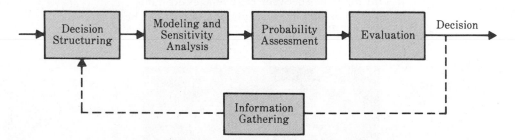

2. *Advanced Simulation.* MIS is pioneering P&G's use of graphic animated simulation to model manufacturing operating policies, test new equipment proposals, and determine the impact of product changes on system capacity. Probability distributions are important components of the simulation models being developed.

3. *Quality Improvement.* Quality improvement techniques, often based on probability and statistical considerations, are in use to improve business performance. MIS both provides staff consulting on major projects and trains personnel in the philosophy and techniques of quality improvement. The area has trained people from every P&G product division and geographic location.

*The authors are indebted to Mr. Joel Kahn of Procter & Gamble for providing this application.

4. *Quantitative Methods.* Linear programming is used for blend and capacity problems. Regression analysis and test design are used for modeling processes, and probability theory is used for capacity analysis. These are only a few of the quantitative methods employed to support Engineering.

▬ PROBABILISTIC DECISION AND RISK ANALYSIS APPLIED TO MANUFACTURING STRATEGY

P&G's Industrial Chemicals Division is a major supplier of fatty alcohols derived from natural substances such as coconut oil. Competitive products are petroleum based derivatives. The company uses both.

The Industrial Chemicals Division wanted to know the economic risks and opportunities of expanding its fatty-alcohol production facilities. MIS consulted with this area using probabilistic decision and risk analysis. After structuring and modeling the problem, it was determined that the key to profitability was the cost difference between the petroleum and coconut-based raw materials. Both costs were unknown in the future and needed to be represented as continuous random variables. Since we questioned profitability over the next 10 years, the two random variables were defined as follows:

$$x = \text{the average coconut oil price} \\ \text{per pound of fatty alcohol}$$

and

$$y = \text{the average petroleum based raw} \\ \text{material price per pound of fatty alcohol.}$$

Since the key to profitability was the difference between these two random variables, a third random variable $d = x - y$ was developed, where

$$d = \text{the difference between the average coconut} \\ \text{oil and petroleum prices per pound fatty alcohol}$$

Tide is well recognized as one of Procter & Gamble's leading household products.

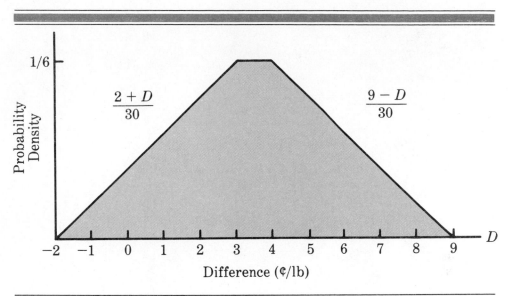

▉▉▉ FIGURE 6A.1 **Distribution of the Difference Between Coconut Oil and Petroleum Prices per Pound of Fatty Alcohol**

Interviews with experts were used to determine the probability distribution for x and y. This information was used to develop the special continuous probability distribution for the difference d, as shown in Figure 6A.1. Using the continuous probability distribution, it was observed that there was a .90 probability that the price difference would be 6.55¢ or less and that there was a .50 probability that the price difference would be 3.50¢ or less. In addition, there was only a .10 probability that the price difference would be 0.45¢ or less.

▉▉▉ SUMMARY

The Industrial Chemicals Division thought that being able to quantify the impact of raw material price differences was key to reaching a consensus. We used these numbers in a sensitivity analysis on the raw material price difference. This analysis yielded sufficient insight to form the basis for a recommendation to management. (*Note:* Data for Figure 6A.1 have been modified to protect proprietary data.)

CHAPTER 7

Sampling and Sampling Distributions

I n Chapter 1, we defined a *population* and a *sample* as two important aspects of a statistical study. The definitions are restated here:

A *population* is the collection of all the elements of interest.

A *sample* is a subset of the population.

As also stated in Chapter 1, the purpose of *statistical inference* is to provide information about a population based upon information contained in a sample. Let us begin by citing two situations where sampling is conducted in order to provide a manager or decision maker with information about a population.

1. A tire manufacturer is considering producing a new tire, which is designed to provide an increase in mileage over the firm's current line of tires. In order to determine whether or not to market the new tire, management needs an estimate of the mean number of miles provided by the new tires. The manufacturer selects a sample of 120 new tires for testing. The test results in a sample mean tire life of 36,500 miles. Thus 36,500 miles is used as an estimate of the mean tire life for the population of new tires.

2. Members of a political party are considering supporting a particular candidate for election to the United States Senate. In order to decide whether or not to enter the candidate in the upcoming primary election, party leaders need an estimate of the proportion of registered voters favoring the candidate. The time and cost associated with contacting every individual in the population of registered voters are prohibitive. Thus a sample of 400 registered voters is selected. If 160 of the 400 voters indicate a preference for the candidate, an estimate of the proportion of the population of registered voters favoring the candidate is 160/400 = .40.

The preceding examples show how sampling and the subsequent sample results can be used to develop estimates of population characteristics. Note that in the tire mileage example, collecting the data on tire life results in wearing out each tire tested. As a result, it is not feasible to test every tire in the population; a sample is the only realistic way to obtain the desired tire mileage data. In the example involving the primary election, it is theoretically possible to contact every registered voter in the population; however, the time and cost involved in doing so are prohibitive. Thus a sample of registered voters is preferred.

The above examples point out some of the reasons for using samples. However, it is important to realize that sample results only provide *estimates* of the values of the population characteristics. That is, we do not expect the sample mean of 36,500 miles to *exactly equal* the mean mileage for all tires in the population; neither do we expect *exactly* 40% of the population of registered voters to favor the candidate. We cannot expect this simply because the sample only contains a portion of the population. However, we believe that with proper sampling methods the sample results will provide "good" estimates of the population characteristics. But how "good" can we expect the sample results to be? Fortunately, statistical procedures are available for answering this question.

In this chapter we show how simple random sampling can be used to select a sample from a population. We then show how data obtained from a simple random sample can be used to compute estimates of a population mean, a population standard deviation, and/or a population proportion. In addition, we introduce the important concept of a sampling distribution. As we shall show, knowledge of the appropriate sampling distribution is what enables us to make statements about the "goodness" of the sample results.

7.1 ▪ THE ELECTRONICS ASSOCIATES SAMPLING PROBLEM

Electronics Associates, Inc. (EAI) is an international company that manufactures a diverse line of products in plants located throughout the United States, Canada, and Europe. The firm's director of personnel has been assigned the task of developing a profile of the company's 2500 managers. This group includes department heads, plant superintendents, and division managers. The characteristics that are to be identified include the mean annual salary for the managers and the proportion of managers having completed the company's management training program.

With the 2500 managers considered the population for this study, we can find the annual salary and the training program status for each individual in the population by referring to the firm's personnel records. Let us assume that this has been done and that we have obtained the corresponding annual salary and management training program participation information for all 2500 managers in the population.

Using the formulas for a population mean and a population standard deviation that were presented in Chapter 3, we can compute the mean and standard deviation of annual salary for the population. Assume that these calculations have been performed with the following results for annual salary:

$$\text{Population mean:} \quad \mu = \frac{\Sigma x_i}{2500} = \$41,800$$

$$\text{Population standard deviation:} \quad \sigma = \sqrt{\frac{\Sigma(x_i - \mu)^2}{2500}} = \$4000$$

Furthermore, assume that a review of the 2500 records shows that 1500 managers have completed the training program. Letting p denote the proportion of the population having completed the training program, we see that $p = 1500/2500 = .60$.

Whenever a numerical characteristic of a population is calculated using observations from all elements of the population, the numerical characteristic is called a *parameter* of the population. Thus, the population mean annual salary ($\mu = \$41,800$), the population standard deviation of annual salary ($\sigma = \$4000$), and the population proportion having completed the training program ($p = .60$) are parameters of the population of EAI managers.

The question that we would now like to consider is how the firm's director of personnel could have obtained estimates of these population parameters by using a sample of managers rather than using all 2500 managers in the population. Assume that a sample of 30 managers will be used. Clearly the time and the cost required to develop a profile for 30 managers would be substantially less than those for the entire population. If the personnel director could be assured that a sample of 30 managers would provide adequate information about the population of 2500 managers, working with a sample would be preferred to working with the entire population. Let us explore the possibility of using a sample for the EAI study by first considering how we could identify a sample of 30 managers.

7.2 ■ SIMPLE RANDOM SAMPLING

There are several methods that can be used to select a sample from a population; one of the most common of these is *simple random sampling*. The definition of a simple random sample and the process of selecting a simple random sample depends upon whether the population is *finite* or *infinite* in size. Since the EAI sampling problem introduced in the previous section involves a finite population of 2500 managers, we first consider sampling from finite populations.

Sampling from Finite Populations

A simple random sample of size n from a finite population of size N is defined as follows:

> #### Simple Random Sample (Finite Population)
>
> A simple random sample of size n from a finite population of size N is a sample selected such that each possible sample of size n has the same probability of being selected.

Fortunately, there is a relatively easy and straightforward procedure for identifying a simple random sample from a finite population. This method of selecting a simple random sample enables us to choose, or select, the elements for the sample *one at a time*. At each selection we make sure that each of the elements remaining in the population has the *same probability* of being selected for the sample. Sampling n elements in this fashion will satisfy the definition of a simple random sample from a finite population.

Let us demonstrate the selection of a simple random sample from a finite population by referring to the EAI problem. First, we will assume that the 2500 EAI managers have been numbered sequentially (i.e., 1, 2, 3, . . . , 2499, 2500) in the order that their names appear in the EAI personnel file. We could then write the numbers from 1 to 2500 on equal sized pieces of paper. The 2500 pieces of paper could then be placed in a hat and mixed thoroughly. We would begin the process of identifying managers for the sample by reaching into the hat and selecting one piece of paper *randomly*. The number on the chosen piece of paper would correspond to one of the numbered managers in the file of 2500 managers; thus, that manager would be selected for the sample. The remaining 2499 pieces of paper would be thoroughly mixed again, after which another piece of paper would be selected. This second number corresponds to another EAI manager to be included in the sample. The process continues until 30 managers have been selected from the population. The 30 managers identified in this manner form a simple random sample from the population.

In this procedure we took care not to place a selected (sampled) piece of paper back into the hat after it was drawn. Thus we are selecting a simple random sample *without replacement*. Certainly we could have followed the sampling procedure of *replacing* each sampled element before selecting subsequent elements. This form of sampling, referred to as sampling *with replacement*, would have made it possible for some elements to appear in the sample more than once. While sampling with replacement is a valid way of identifying a simple random sample, sampling *without replacement* is the sampling procedure used most often. Whenever we refer to simple random sampling, we will assume that the sampling is done without replacement.

This procedure for selecting a simple random sample of 30 EAI managers requires the labeling of 2500 pieces of paper. In practice, tables of random numbers can be used to provide the same results much more easily. Tables of random numbers are available from a variety of handbooks* that contain page after page of random numbers. We have included one such page of random numbers in Table 8 of Appendix B. A portion of this page of random numbers is shown in Table 7.1. The first line of this table of random numbers begins as follows:

63271   59986   71744   51102   15141   80714

The digit appearing in any one of the above positions is a random selection of the digits 0, 1, . . . , 9 with each digit having an equal chance of occurring. The grouping of the numbers into sets of 5 is simply for the convenience of making the table easy to read.

Let us see how the numbers in this random number table can be used to select a simple random sample of 30 EAI managers. As we did using the pieces of paper, we want to select numbers from 1 to 2500 such that every number has an equal chance of being selected. Since the largest number in the EAI population, 2500, has 4 digits, we will select random numbers from the table in sets or groups of 4 digits. While we could select 4-digit numbers from any portion of

---

*For example, The Rand Corporation, *A Million Random Digits with 100,000 Normal Deviates*. New York: The Free Press, 1983.

TABLE 7.1   **Random Numbers**

63271	59986	71744	51102	15141	80714	58683	93108	13554	79945
88547	09896	95436	79115	08303	01041	20030	63754	08459	28364
55957	57243	83865	09911	19761	66535	40102	26646	60147	15702
46276	87453	44790	67122	45573	84358	21625	16999	13385	22782
55363	07449	34835	15290	76616	67191	12777	21861	68689	03263
69393	92785	49902	58447	42048	30378	87618	26933	40640	16281
13186	29431	88190	04588	38733	81290	89541	70290	40113	08243
17726	28652	56836	78351	47327	18518	92222	55201	27340	10493
36520	64465	05550	30157	82242	29520	69753	72602	23756	54935
81628	36100	39254	56835	37636	02421	98063	89641	64953	99337
84649	48968	75215	75498	49539	74240	03466	49292	36401	45525
63291	11618	12613	75055	43915	26488	41116	64531	56827	30825
70502	53225	03655	05915	37140	57051	48393	91322	25653	06543
06426	24771	59935	49801	11082	66762	94477	02494	88215	27191
20711	55609	29430	70165	45406	78484	31639	52009	18873	96927
41990	70538	77191	25860	55204	73417	83920	69468	74972	38712
72452	36618	76298	26678	89334	33938	95567	29380	75906	91807
37042	40318	57099	10528	09925	89773	41335	96244	29002	46453
53766	52875	15987	46962	67342	77592	57651	95508	80033	69828
90585	58955	53122	16025	84299	53310	67380	84249	25348	04332
32001	96293	37203	64516	51530	37069	40261	61374	05815	06714
62606	64324	46354	72157	67248	20135	49804	09226	64419	29457
10078	28073	85389	50324	14500	15562	64165	06125	71353	77669
91561	46145	24177	15294	10061	98124	75732	00815	83452	97355
13091	98112	53959	79607	52244	63303	10413	63839	74762	50289

the random number table, suppose we start by using the first row of random numbers appearing in Table 7.1. The 4-digit grouping of the first 28 random numbers in the first row provides

6327   1599   8671   7445   1102   1514   1807

Since the numbers in the table are random, the preceding 4-digit numbers are all equally probable, or equally likely.

We can now use the equally likely 4-digit random numbers to give each element in the population an equal chance of being included in the sample. The first number, 6327, is greater than 2500. It does not correspond to an element in the population, and thus it is discarded. The second number, 1599, is between 1 and 2500. Thus the first individual selected for the sample is manager 1599 on the list of EAI managers. Continuing the process, we ignore 8671 and 7445 before identifying individuals 1102, 1514, and 1807 as the next managers to be included in the sample. This process of selecting managers continues until the desired simple random sample of size 30 has been obtained. We note that with this random number procedure for simple random sampling, a random number previously used to identify an element for the sample may reappear in the random number table. Since we want to select the simple random sample *without replacement*, previously used random numbers are ignored because the corresponding element is already included in the sample.

As a final comment, note that random numbers can be selected from anywhere in the random number table. We chose to use the first row of the table in the above example. However, we could have started at any other point in the table and continued in any direction. Once the arbitrary starting point is selected, it is recommended that a predetermined systematic procedure, such as reading across rows or down columns, be used to determine the subsequent random numbers.

Sampling from Infinite Populations

To this point we have restricted our attention to selecting a simple random sample from a finite population. Although many sampling situations in business and economics involve finite populations, there are situations in which the population is either infinite or so large that for practical purposes it must be treated as infinite. In sampling from an infinite population we must provide a new definition of a simple random sample. In addition, since the elements in an infinite population cannot be numbered, we must use a different process for selecting elements for the sample.

Let us consider an example requiring a simple random sample from an infinite population. Suppose we want to estimate the average time between placing an order and receiving food for customers at a fast-food restaurant during the 11:30 A.M. to 1:30 P.M. lunch period. If we consider the population as being all possible customer visits, we see that it would not be feasible to specify a finite limit on the number of possible visits. In fact, if we define the population as being all customer visits that could *conceivably* occur during the lunch period, we can consider the population as being infinite. Our task is now to select a simple random sample of n customers from this population. With this situation in mind we now state the definition of a simple random sample from an infinite population:

Simple Random Sample (Infinite Population)

A simple random sample from an infinite population is a sample selected such that the following conditions are satisfied:

1. Each element selected comes from the same population.
2. Each element is selected independently.

For the problem of selecting a simple random sample of customer visits at a fast-food restaurant, we find that the first condition defined above is satisfied by any customer visit occurring during the 11:30 A.M. to 1:30 P.M. lunch period while the restaurant is operating with its regular staff under "normal" operating conditions. The second condition is satisfied by ensuring that the selection of a particular customer does not influence the selection of any other customer. That is, the customers are selected independently.

A well known fast-food restaurant has implemented a simple random sampling procedure for just such a situation. The sampling procedure is based on the fact that some customers will present discount coupons which provide special prices on sandwiches, drinks, french fries, and so on. Whenever a customer presents a discount coupon, the *next* person served is selected for the sample. Since the customers present discount coupons in a random and independent fashion, the firm is satisfied that the sampling plan satisfies the two conditions for a simple random sample from an infinite population.

NOTES AND COMMENTS

1. Finite populations are often defined by lists such as organization membership rosters, enrolled students, credit-card accounts, inventory product numbers, and so on. Infinite populations are often defined by an ongoing process where the elements of the population consist of items generated if the process were to operate indefinitely under the same conditions; in such cases, it is impossible to obtain a list of all items in the population. For example, populations consisting of all possible parts to be manufactured, all possible customer visits, all possible bank transactions, and so on can be classified as infinite populations.

2. The number of different simple random samples of size n that can be selected from a finite population of size N is:

$$\frac{N!}{n!(N-n)!}$$

In this formula, $N!$ and $n!$ refer to the factorial computations discussed in Chapter 5. For the EAI problem with $N = 2500$ and $n = 30$, this expression can be used to show that there are approximately 2.75×10^{69} different simple random samples of 30 EAI managers.

3. Simple random sampling from a finite population is generally done without replacement. However, if a simple random sample is selected from a finite population *with replacement*, the elements are selected from the same population and the elements are selected independently. Since these conditions satisfy the requirements of a simple random sample from an infinite population, simple random sampling with replacement from a finite population is equivalent to simple random sampling from an infinite population.

EXERCISES

1. Assume that the simple random sample for the EAI study had been based on the seventh column of the 5-digit random numbers shown in Table 7.1. Ignoring the first digit in the column and moving down the column, identify the first 5 EAI manager numbers that will be selected for the simple random sample. Note that this procedure begins with the 4-digit random number 8683.

2. Based on sales (*Wall Street Journal*, October 13, 1988), the top 10 athletic footwear manufacturers are

1.	Reebok	2.	Nike
3.	Converse	4.	Avia
5.	Adidas	6.	L. A. Gear
7.	Etonic/Tretorn	8.	New Balance
9.	ASICS Tiger	10.	British Knights

a. Beginning with the first random digit in Table 7.1 (6) and reading down (8, 5, 4, and so on), use single-digit random numbers from the first column to select a simple random sample of five footwear manufacturers from the population of the top ten footwear manufacturers.

b. If the random number 1 corresponds to Reebok, 2 corresponds to Nike, and so on, what single-digit random digit would have to appear in the first column in order to select British Knights for the sample?

c. How many different simple random samples of size 5 can be selected from this population of ten manufacturers?

3. A student government organization is interested in estimating the proportion of students who favor a mandatory "pass-fail" grading policy for elective courses. A list of names and addresses of the 645 students enrolled during the current quarter is available from the registrar's office. Using row 10 of Table 7.1 and moving across the row from left to right, identify the first ten students who would be selected by simple random sampling. When every digit in row 10 is used the three-digit random numbers begin with 816, 283, and 610.

4. The *County and City Data Book*, published by the Bureau of Census, lists information on 3139 counties throughout the United States. Assume that a national study will collect data from 30 randomly selected counties. Use 4-digit random numbers from the last column of Table 7.1 to identify the numbers corresponding to the first 5 counties selected for the sample. Ignore the first digit in the last column and begin with the 4-digit random numbers 9945, 8364, 5702, and so on.

5. Assume that we wish to identify a simple random sample of 12 of the 372 doctors located in a particular city. The doctors' names are available from a local medical organization. Use the eighth column of 5-digit random numbers in Table 7.1 to identify the 12 doctors for the sample. Ignore the first 2 random digits in each 5-digit grouping of the random numbers. This process begins with random number 108 and proceeds down the column of random numbers.

6. *Business Week*, February 20, 1989, provided detailed information for 640 mutual funds available to investors. Data about the funds included assets, fees, return on investment, price/earnings ratios, and more. Assume that you would like to do a statistical study of the financial characteristics for the population of 640 mutual funds by using a simple random sample of 12 mutual funds. Use the third column of 5-digit random numbers in Table 7.1, beginning with 71744. Ignore the 44 and only use the first three-digits, 717. Reading down the column of three-digit random numbers, identify the numbers corresponding to the 12 mutual funds to be included in the sample.

7. Schuster's Interior Design, Inc., specializes in a variety of home decorating services for its clients. During the previous year the firm provided major decorating consultation for 875 homes. Schuster's management was interested in obtaining information about customer satisfaction 6 to 12 months after the project was complete. To obtain this information, the firm decided to sample 30 of the 875 clients and interview the group to learn about client satisfaction and ways that Schuster might improve its service. Using the last three digits in column 10 of Table 7.1, and moving down the column, the random number sequence would be 945, 364, 702, and so on. Use this procedure to identify the

first 10 clients that would be included in the sample. Assume that the 875 clients are numbered sequentially in the order in which the decorating projects were conducted.

8. Haskell Public Opinion Poll, Inc., conducts telephone surveys concerning a variety of political and general public interest issues. The households included in the survey are identified by taking a simple random sample from telephone directories in selected metropolitan areas. The telephone directory for a major Midwest area contains 853 pages with 400 lines per page.

　　a. Describe a two-stage random selection procedure that could be used to identify a simple random sample of 200 households. The selection process should involve first selecting a page at random (Stage 1) and then selecting a line on the sampled page (Stage 2). Use the random numbers in Table 7.1 to illustrate this process. Select your own arbitrary starting point in the table.

　　b. What would you do if the line selected in part (a) was clearly inappropriate for the study (that is, the line provided the phone number of a business, restaurant, etc.)?

9. Read the Kings Island consumer profile sample survey application at the end of Chapter 1.

　　a. Assume that the Kings Island research group treats the population of consumer visits as an infinite population. Is this acceptable? Explain.

　　b. Assume that immediately after completing an interview with a consumer, the interviewer returns to the entrance gate and begins counting individuals as they enter the park. The 25th individual counted is selected as the next person to be sampled for the survey. After completing this interview, the interviewer returns to the entrance and again selects the 25th individual entering the park. Does this sampling process appear to provide a simple random sample? Explain.

10. Indicate whether the populations listed below should be considered finite or infinite:

　　a. All the registered voters in the state of California.

　　b. All the television sets that could be produced by the Allentown, Pennsylvania plant of the TV-M Company.

　　c. All orders that could be processed by a mail-order firm.

　　d. All emergency telephone calls that could come into a local police station.

　　e. All items that were produced on the second shift on May 17.

7.3 ■■■ POINT ESTIMATION

Now that we have described how to select a simple random sample, let us return to the EAI problem. We will assume that a simple random sample of 30 managers has been selected and that the corresponding data on annual salary and management training program participation are as shown in Table 7.2. The notation x_1, x_2, and so on, is used to denote the annual salary of the first manager in the sample, the second manager in the sample, and so on. Participation in the management training program is indicated by a "yes" in the management training program column.

In order to estimate the value of a population parameter, we compute a corresponding characteristic of the sample, referred to as a *sample statistic*. For example, in order to estimate the population mean, μ, and the population standard deviation, σ, for the annual salary of EAI managers, we simply use the data in column 1 of Table 7.2 to calculate the corresponding sample statistics: the sample mean $\bar{x}$ and the sample standard deviation, s. Using the formulas for a

TABLE 7.2　**Annual Salary and Training Program Status for a Simple Random Sample of 30 Managers**

Annual Salary (dollars)	Management Training Program?	Annual Salary (dollars)	Management Training Program?
x_1 = 39,094.30	Yes	x_{16} = 41,766.00	Yes
x_2 = 43,263.90	Yes	x_{17} = 42,541.30	No
x_3 = 39,643.50	Yes	x_{18} = 34,980.00	Yes
x_4 = 39,894.90	Yes	x_{19} = 41,932.60	Yes
x_5 = 37,621.60	No	x_{20} = 42,973.00	Yes
x_6 = 45,924.00	Yes	x_{21} = 35,120.90	Yes
x_7 = 39,092.30	Yes	x_{22} = 41,753.00	Yes
x_8 = 41,404.40	Yes	x_{23} = 44,391.80	No
x_9 = 40,957.70	Yes	x_{24} = 40,164.20	No
x_{10} = 45,109.70	Yes	x_{25} = 42,973.60	No
x_{11} = 35,922.60	Yes	x_{26} = 40,241.30	No
x_{12} = 47,268.40	No	x_{27}= 42,793.90	No
x_{13} = 45,688.80	Yes	x_{28} = 40,979.40	Yes
x_{14} = 41,564.70	No	x_{29} = 45,860.90	Yes
x_{15} = 46,188.20	No	x_{30} = 47,309.10	No

sample mean and a sample standard deviation as presented in Chapter 3, the sample mean is

$$\bar{x} = \frac{\Sigma x_i}{n} = \frac{1,254,420}{30} = \$41,814.00$$

and the sample standard deviation is

$$s = \sqrt{\frac{\Sigma(x_i - \bar{x})^2}{n - 1}} = \sqrt{\frac{325,009,260}{29}} = \$3347.72$$

In addition, by computing the proportion of managers in the sample who have responded yes, we can estimate the proportion of managers in the population who have completed the management training program. Column 2 of Table 7.2 shows that 19 of the 30 managers in the sample have completed the training program. Thus the sample proportion, denoted by $\bar{p}$, is given by

$$\bar{p} = \frac{19}{30} = .63$$

This value is used as the estimate of the population proportion p.

By making the preceding computations, we have completed the statistical procedure called *point estimation*. In point estimation we use the data from the sample to compute a value of a sample statistic that serves as an estimate of a population parameter. Using the terminology of point estimation, we would refer to $\bar{x}$ as the *point estimator* of the population mean μ, s as the *point estimator* of the population standard deviation σ, and $\bar{p}$ as the *point estimator* of the population

TABLE 7.3 **Summary of Point Estimates Obtained from a Simple Random Sample of 30 EAI Managers**

Population Parameter	Parameter Value	Point Estimator	Point Estimate
μ = Population mean annual salary	$41,800.00	$\bar{x}$ = Sample mean annual salary	$41,814.00
σ = Population standard deviation for annual salary	$4,000.00	s = Sample standard deviation for annual salary	$ 3,347.72
p = Population proportion having completed the management training program	.60	$\bar{p}$ = Sample proportion having completed the management training program	.63

proportion p. The actual numerical value obtained for $\bar{x}$, s, or $\bar{p}$ in a particular sample is called the *point estimate* of the parameter. Thus, based upon the sample of 30 EAI managers, $41,814.00 is the point estimate of μ, $3,347.72 is the point estimate of σ, and .63 is the point estimate of p. Table 7.3 provides a summary of the sample results and compares the point estimates to the actual values of the population parameters.

■■■ EXERCISES

11. A simple random sample of 5 months of sales data provides the following:

Month	1	2	3	4	5
Units Sold	94	100	85	94	92

a. What is a point estimate of the mean number of units sold per month?
b. What is a point estimate of the standard deviation for the population?

12. Forty-nine television commercials were shown during the televising of the 1989 Super Bowl Game. A sample of 60 viewers was used to rate each of the commercials (*USA Today*, January 23, 1989). Each viewer in the sample provided a commercial rating on a scale of 1 to 10, with a higher rating indicating a more preferred commercial. Based on the sample mean ratings, American Express and Diet Pepsi provided the two best-liked commercials shown during the Super Bowl game. Suppose that the following data represent a portion of the ratings obtained for the American Express commercial.

10 9 10 9 7 8 10 10 9 6 10 8 7 10 9 10

Assume these ratings are from a simple random sample of 16 viewers selected from a population of all viewers of the Super Bowl game.

a. What is the point estimate of the mean rating of the American Express commercial for the population?
b. What is the point estimate of the standard deviation of the American Express commercial rating for the population? (.31

13. The California Highway Patrol maintains records showing the time between an accident report being received and an officer arriving at the accident scene. A simple random sample of ten records shows the following times in minutes:

12.6 3.4 4.8 5.0 6.8 2.3 3.6 8.1 2.5 10.3

   a. What is a point estimate of the population mean time between accident report and officer arrival?
   b. What is a point estimate of the population standard deviation of time between accident report and officer arrival?

14. An official for United Airlines reported that out of 104 United Airlines flights arriving at Chicago's O'Hare Airport, 3 flights arrived more than 15 minutes late (*Wall Street Journal*, November 7, 1988). Assuming that the 104 flights are a random sample of all United Airline flights into O'Hare, what is the point estimate of the proportion of all United Airline flights into O'Hare that are more than 15 minutes late?

15. A survey of 400 women college students was conducted to determine future plans concerning career, marriage, and family. The following results were recorded:

   ■■■ 310 women answered yes to the question, Do you plan to begin a full-time career immediately following graduation?
   ■■■ 225 women answered yes to the question, Do you plan to marry before the age of 30?
   ■■■ 175 women answered yes to the question, Do you plan to have children?

Use the survey results to provide point estimates of each of the following.

   a. The proportion of college women planning to begin full-time careers immediately following graduation
   b. The proportion of college women planning to marry before the age of 30
   c. The proportion of college women planning to have children

7.4 ■■■ INTRODUCTION TO SAMPLING DISTRIBUTIONS

In the previous section we saw how a simple random sample of 30 EAI managers could be used to develop point estimates of the mean and standard deviation of annual salary for the population of all EAI managers as well as the proportion of the managers in the population that have completed the company's management training program. In this section we consider the point estimates that would be observed if different simple random samples, each of size 30, were selected. The resulting analysis will introduce the important concept of a *sampling distribution*.

Suppose we were to select another simple random sample of 30 EAI managers. Let us assume that this has been done and that an analysis of the data from the second simple random sample provides the following:

$$\text{Sample Mean } \bar{x} = \$42,669.70$$

$$\text{Sample Standard Deviation } s = \$4,239.07$$

$$\text{Sample Proportion } \bar{p} = .70$$

These results show that different values of $\bar{x}$, s, and $\bar{p}$ have been obtained with the second sample. In general, this is to be expected because this second simple

TABLE 7.4   Values of $\bar{x}$, s, and $\bar{p}$ From 500 Simple Random Samples of 30 EAI Managers

Sample Number	Sample Mean $\bar{x}$	Sample Standard Deviation s	Sample Proportion $\bar{p}$
1	$41,814.00	$3,347.72	.63
2	$42,669.70	$4,239.07	.70
3	$41,780.30	$4,433.43	.67
4	$41,587.90	$3,985.32	.53
.	.	.	.
.	.	.	.
.	.	.	.
500	$41,752.00	$3,857.82	.50

random sample will most likely not contain the same 30 managers that were in the first sample. Let us imagine carrying out the same process of selecting a new simple random sample of 30 managers over and over again, each time computing values of $\bar{x}$, s, and $\bar{p}$. In this way, we could begin to identify the variety of values that these point estimators can take on. To illustrate this, we repeated the simple random sampling process for the EAI problem until we obtained 500 samples of 30 managers each and their corresponding $\bar{x}$, s, and $\bar{p}$ values. A portion of the results are shown in Table 7.4. Table 7.5 shows the frequency distribution for the 500 $\bar{x}$ values. Figure 7.1 shows the relative frequency histogram for the $\bar{x}$ results.

Recall that in Chapter 5 we defined a random variable as a numerical description of the outcome of an experiment. If we consider the simple random sampling process as an experiment, the sample mean $\bar{x}$ is the numerical description of the outcome of the experiment. Thus the sample mean $\bar{x}$ is a random variable. As a result $\bar{x}$, just like other random variables, has a mean or expected value, a variance, and a probability distribution. Since the various possible values of $\bar{x}$ are the result of different simple random *samples*, the probability distribution of $\bar{x}$ is called the *sampling distribution of $\bar{x}$*. Knowledge of

TABLE 7.5   Frequency Distribution of $\bar{x}$ From 500 Simple Random Samples of 30 EAI Managers

Mean Annual Salary ($)	Frequency	Relative Frequency
$39,500.00–39,999.99	2	.004
$40,000.00–40,499.99	16	.032
$40,500.00–40,999.99	52	.104
$41,000.00–41,499.99	101	.202
$41,500.00–41,999.99	133	.266
$42,000.00–42,499.99	110	.220
$42,500.00–42,999.99	54	.108
$43,000.00–43,499.99	26	.052
$43,500.00–43,999.99	6	.012
Totals	500	1.000

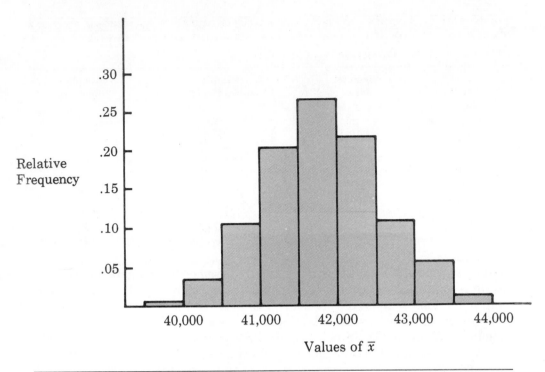

Relative
Frequency

Values of $\bar{x}$

**■■■ FIGURE 7.1   Relative Frequency Histogram of $\bar{x}$ Values for 500 Simple
Random Samples of Size 30 Each**

this sampling distribution and its properties will enable us to make probability
statements about how close the sample mean $\bar{x}$ is to the population mean μ.

Let us return to Figure 7.1. We would need to enumerate every possible
sample of 30 managers and compute each sample mean in order to determine
completely the sampling distribution of $\bar{x}$. However, the histogram of 500 $\bar{x}$
values gives an approximation of this sampling distribution. From this approximation we observe the bell-shaped appearance of the distribution. We also note
that the mean of the 500 $\bar{x}$ values is near the population mean μ = \$41,800. We
will describe the properties of the sampling distribution of $\bar{x}$ more fully in the
next section.

The 500 values of the sample standard deviation s and the 500 values of the
sample proportion $\bar{p}$ are summarized by the relative frequency histograms in
Figures 7.2 and 7.3. As in the case of $\bar{x}$, both s and $\bar{p}$ are random variables that
provide numerical descriptions of the outcome of a simple random sampling
process. If every possible sample of size 30 were selected from the population
and if a value of s and a value of $\bar{p}$ were computed for each sample, the resulting
probability distributions would be called the sampling distribution of s and the
sampling distribution of $\bar{p}$, respectively. The relative frequency histograms of
the 500 sample values shown in Figures 7.2 and 7.3 provide a general idea of the
appearance of these two sampling distributions.

In closing this section let us note that in practice we select *only one simple
random sample* from the population. The reason that we repeated the sampling
process 500 times in this section was to illustrate that many different samples are
possible and that the different samples generate a variety of values for the
sample statistics $\bar{x}$, s and p. The probability distribution of any particular sample

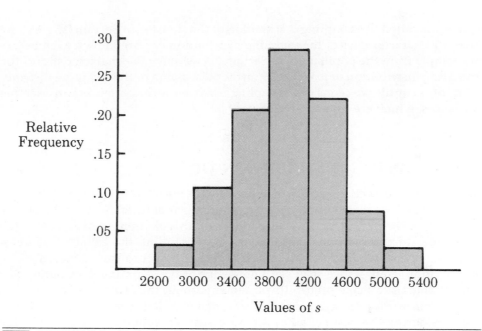

■■ FIGURE 7.2   **Relative Frequency Histogram of s Values for 500 Simple Random Samples of Size 30 Each**

■■ FIGURE 7.3   **Relative Frequency Histogram of $\bar{p}$ Values for 500 Simple Random Samples of Size 30 Each**

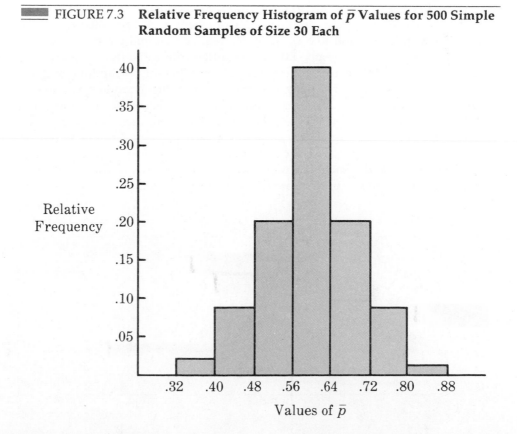

statistic is called the sampling distribution of the statistic. In Section 7.5 we show that the characteristics of the sampling distribution of $\bar{x}$ are known even before we sample from the population. In Section 7.6 we show the characteristics of the sampling distribution of $\bar{p}$. We defer further discussion of the sampling distribution of s until we consider sampling distributions pertaining to sample variances, which are covered in Chapter 11.

7.5 ■■■ SAMPLING DISTRIBUTION OF $\bar{x}$

In the previous section we saw that the sample mean, $\bar{x}$, is a random variable. The probability distribution of this random variable is referred to as the sampling distribution of $\bar{x}$. The purpose of this section is to describe the properties of the sampling distribution of $\bar{x}$, including the expected value or mean of $\bar{x}$, the standard deviation of $\bar{x}$, and the shape or form of the sampling distribution itself. As we shall see, knowledge of the sampling distribution of $\bar{x}$ will enable us to make probability statements about the error involved when a sample mean $\bar{x}$ is used as a point estimator of a population mean μ. Let us begin by considering the mean of all possible $\bar{x}$ values, or, simply, the expected value of $\bar{x}$.

Expected Value of $\bar{x}$

As we saw in the EAI sampling problem, different simple random samples result in a variety of values for the sample mean (for example, $41,814.00, $42,669.70, $41,780.30, $41,587.90, and so on). When we realize that many different values of the random variable $\bar{x}$ are possible, we are often interested in the mean of all possible values of $\bar{x}$ that can be generated by the various simple random samples. This can be provided by realizing that the mean of the $\bar{x}$ random variable is simply the expected value of $\bar{x}$. As a result, we are able to use concepts from sampling theory to calculate the expected value of $\bar{x}$; we need not actually go through the process of computing all possible values of $\bar{x}$ and then computing their mean. Let $E(\bar{x})$ represent the expected value of $\bar{x}$, or simply the mean of all possible $\bar{x}$ values, and μ equal the population mean. It can be shown that when using simple random sampling these two values are the same.

Expected Value of $\bar{x}$

$$E(\bar{x}) = \mu \qquad (7.1)$$

where

$E(\bar{x})$ = the expected value of the random variable $\bar{x}$

μ = the population mean

This result, which is derived in the appendix to this chapter, shows that with simple random sampling, the expected value or mean for $\bar{x}$ is equal to the mean of the population. Refer to the EAI study and recall that in Section 7.1 we saw that the mean annual salary for the population of EAI managers was $\mu = \$41,800$. Thus according to (7.1) the mean of all possible sample means for the EAI study also is $\$41,800$.

Standard Deviation of $\bar{x}$

As we have stated, various simple random samples can be expected to generate a variety of $\bar{x}$ values. Let us now explore what sampling theory tells us about the standard deviation of all possible $\bar{x}$ values. We will use the following notation:

$\sigma_{\bar{x}}$ = the standard deviation of all possible $\bar{x}$ values

σ = the population standard deviation

n = the sample size

N = the population size

It can be shown that with simple random sampling the standard deviation of $\bar{x}$ depends upon whether the population is finite or infinite. The two expressions for the standard deviation of $\bar{x}$ are as follows:

$$
\begin{array}{cc}
\textbf{Standard Deviation of } \bar{x} & \\
\textit{Finite Population} & \textit{Infinite Population} \\
\sigma_{\bar{x}} = \sqrt{\dfrac{N-n}{N-1}}\left(\dfrac{\sigma}{\sqrt{n}}\right) & \sigma_{\bar{x}} = \dfrac{\sigma}{\sqrt{n}}
\end{array} \tag{7.2}
$$

A derivation of the formulas for $\sigma_{\bar{x}}$ is discussed in the appendix to this chapter. In comparing the two expressions in (7.2) we see that the factor $\sqrt{(N-n)/(N-1)}$ is required for the finite population but not for the infinite population case. This factor is commonly referred to as the *finite population correction factor*. In many practical sampling situations we find that the population involved, although finite, is "large," whereas the sample size is relatively "small." In such cases the finite population correction factor $\sqrt{(N-n)/(N-1)}$ is close to 1. As a result the difference between the values of the standard deviation of $\bar{x}$ for the finite and infinite population cases becomes negligible. When this occurs, $\sigma_{\bar{x}} = \sigma/\sqrt{n}$ becomes a very good approximation to the standard deviation of $\bar{x}$ even though the population is finite. As a general guideline or rule of thumb for computing the standard deviation of $\bar{x}$, we state the following:

Use the following expression to calculate the standard deviation of $\bar{x}$

$$\sigma_{\bar{x}} = \frac{\sigma}{\sqrt{n}} \qquad (7.3)$$

whenever

1. The population is infinite, or
2. The population is finite *and* the sample size is less than or equal to 5% of the population size; that is, $n/N \leq .05$

In cases where $n/N > .05$, the finite population version of (7.2) should be used in the computation of $\sigma_{\bar{x}}$. Note however that unless specifically noted, throughout the text we will be assuming that the population size is "large," the finite population correction factor is unnecessary, and (7.3) can be used to compute $\sigma_{\bar{x}}$.

Now let us return to the EAI study and determine the standard deviation of all possible sample means that can be generated with samples of 30 EAI managers. Recall that in Section 7.1 we identified the population standard deviation for the annual salary data to be $\sigma = 4000$. In this case the population is finite, with $N = 2500$. However, with a sample size of 30, we have $n/N = 30/2500 = .012$. Following the rule of thumb given earlier, we can ignore the finite population correction factor and use (7.3) to compute the standard deviation of $\bar{x}$:

$$\sigma_{\bar{x}} = \frac{\sigma}{\sqrt{n}} = \frac{4000}{\sqrt{30}} = 730.30$$

Later we will see that the value of $\sigma_{\bar{x}}$ is helpful in determining how far the sample mean may be from the population mean. Because of the role that $\sigma_{\bar{x}}$ plays in computing possible estimation errors, $\sigma_{\bar{x}}$ is referred to as the *standard error of the mean*.

Central Limit Theorem

The final step in identifying the characteristics of the sampling distribution of $\bar{x}$ is to determine the form of the probability distribution of $\bar{x}$. We consider two cases: one where the population distribution is unknown and one where the population distribution is known to be normal.

For the situation where the population distribution is unknown, we rely on one of the most important theorems in statistics—the *central limit theorem*. A statement of the central limit theorem as it applies to the sampling distribution of $\overline{x}$ is as follows:

Central Limit Theorem

In selecting simple random samples of size n from a population with mean μ and standard deviation σ, the sampling distribution of the sample mean $\overline{x}$ approaches a *normal probability distribution* with mean μ and standard deviation $\sigma / \sqrt{n}$ as the sample size becomes large.

Figure 7.4 shows how the central limit theorem works for three different populations; in each case the population clearly is not normal. However, note what begins to happen to the sampling distribution of $\overline{x}$ as the sample size is increased. When the samples are of size 2 we see that the sampling distribution of $\overline{x}$ begins to take on an appearance different than the population distribution. For samples of size 5 we see all three sampling distributions beginning to take on a bell-shaped appearance. Finally, the samples of size 30 show all three sampling distributions to be approximately normal. General statistical practice is to assume that regardless of the population distribution, the sampling distribution of $\overline{x}$ can be approximated by a normal probability distribution whenever the *sample size is 30 or more*. In effect the sample size of 30 is the rule of thumb that allows us to assume that the large sample conditions of the central limit theorem have been satisfied. This observation about the sampling distribution of $\overline{x}$ is so important that we restate it:

The sampling distribution of $\overline{x}$ can be approximated by a normal probability distribution whenever the sample size is large. The large-sample-size condition can be assumed for simple random samples of size 30 or more.

The central limit theorem is the key to identifying the form of the sampling distribution of $\overline{x}$ whenever the population distribution is unknown. However, we may encounter some sampling situations where the population is assumed or believed to have a normal distribution. When this condition occurs, the following result identifies the form of the sampling distribution of $\overline{x}$:

Whenever the population has a normal probability distribution, the sampling distribution of $\overline{x}$ is a normal probability distribution for any sample size.

In summary, whenever we are using a large simple random sample (rule of thumb: $n \geq 30$), the central limit theorem enables us to conclude that the sampling distribution of $\overline{x}$ can be approximated by a normal probability

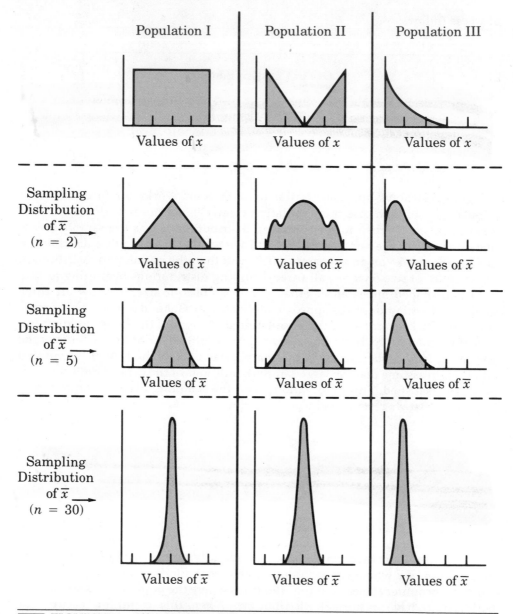

Population I   Population II   Population III

Values of x Values of x Values of x

Sampling Distribution of $\bar{x}$ ($n = 2$)

Values of $\bar{x}$ Values of $\bar{x}$ Values of $\bar{x}$

Sampling Distribution of $\bar{x}$ ($n = 5$)

Values of $\bar{x}$ Values of $\bar{x}$ Values of $\bar{x}$

Sampling Distribution of $\bar{x}$ ($n = 30$)

Values of $\bar{x}$ Values of $\bar{x}$ Values of $\bar{x}$

■■■ FIGURE 7.4   **Illustration of the Central Limit Theorem for Three Populations**

distribution. In cases where the simple random sample is small ($n < 30$), the sampling distribution of $\bar{x}$ can be considered normal only if we assume that the population has a normal probability distribution.

Sampling Distribution of $\bar{x}$ for the EAI Problem

Let us draw upon our knowledge of the sampling distribution of $\bar{x}$ to determine the properties of the sampling distribution of $\bar{x}$ for the EAI study. Previously, we have shown that $E(\bar{x}) = 41,800$ and $\sigma_{\bar{x}} = 730.30$. Since we are using a simple

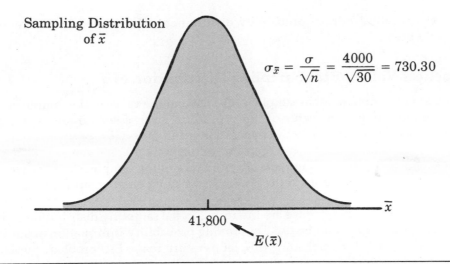

Sampling Distribution
of $\bar{x}$

$$\sigma_{\bar{x}} = \frac{\sigma}{\sqrt{n}} = \frac{4000}{\sqrt{30}} = 730.30$$

41,800

$E(\bar{x})$

$\bar{x}$

▬▬ FIGURE 7.5   **Sampling Distribution of $\bar{x}$ for the Mean Annual Salary of a Simple Random Sample of 30 Managers**

random sample of 30 managers, the central limit theorem enables us to conclude that the sampling distribution of $\bar{x}$ is approximately normal. Thus, the sampling distribution of $\bar{x}$ for the EAI problem is as shown in Figure 7.5. Although we do not have actual data available for all possible $\bar{x}$ values, we do have the 500 values of $\bar{x}$ that were obtained from the 500 simple random samples referred to in Section 7.4 (see Figure 7.1). In Figure 7.6 we compare the theoretical sampling distribution of $\bar{x}$ with the relative frequency histogram of the 500 $\bar{x}$ values that we have actually observed. Note how close the distribution of the 500 $\bar{x}$ values is

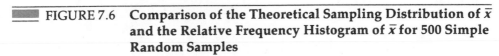

▬▬ FIGURE 7.6   **Comparison of the Theoretical Sampling Distribution of $\bar{x}$ and the Relative Frequency Histogram of $\bar{x}$ for 500 Simple Random Samples**

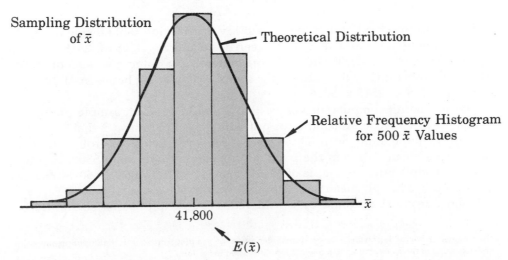

Sampling Distribution
of $\bar{x}$

Theoretical Distribution

Relative Frequency Histogram
for 500 $\bar{x}$ Values

41,800

$E(\bar{x})$

$\bar{x}$

to the theoretical normal probability distribution, as indicated by the central limit theorem.

Practical Value of the Sampling Distribution of $\bar{x}$

Whenever a simple random sample is selected and the value of the sample mean $\bar{x}$ is used to estimate the value of the population mean μ, we cannot expect the sample mean to *equal exactly* the population mean. The difference between the value of the sample mean $\bar{x}$ and the value of the population mean μ is called the *sampling error*. Before putting faith or confidence in the sample mean, we would like to know something about the potential sampling error when $\bar{x}$ is used as an estimator of μ.

The practical reason that we are interested in the sampling distribution of $\bar{x}$ is that this distribution can be used to provide probability information about the sampling error.* To demonstrate this, let us return to the EAI problem. Suppose that the personnel director believes the sample mean will be an acceptable estimate of the population mean if the sample mean is within $500 of the population mean. In probability terms, the personnel director is really concerned with the following question: What is the probability that the sample mean we obtain from a simple random sample of 30 EAI managers will be within $500 of the population mean?

Since we have identified the sampling distribution for $\bar{x}$ (see Figure 7.5), we will use this distribution to answer the probability question. Refer to the sampling distribution of $\bar{x}$ shown again in Figure 7.7. The personnel director is asking about the probability that the sample mean is between $41,300 and $42,300. If the value of the sample mean $\bar{x}$ is in this interval, the value of $\bar{x}$ will be within $500 of the population mean. The probability is given by the shaded area of the sampling distribution shown in Figure 7.7. Since the sampling distribution is normal, with mean 41,800 and standard deviation 730.30, we can use the standard normal distribution table to find the area or probability. At $\bar{x} = 41,300$, we have

$$z = \frac{41,300 - 41,800}{730.30} = -.68$$

Referring to the standard normal probability distribution table inside the back cover of the text, we find an area between $z = 0$ and $z = -.68$ of .2518. Similar calculations for $\bar{x} = 42,300$ show an area between $z = 0$ and $z = +.68$ of .2518. Thus the probability of the value of the sample mean being between 41,300 and 42,300 is .2518 + .2518 = .5036.

The preceding computations show that a simple random sample of 30 EAI managers has a .5036 probability of providing a sample mean $\bar{x}$ that is within $500 of the population mean. Thus, there is a $1 - .5036 = .4964$ probability that the sample mean will miss the population mean by more than $500. In other words, a simple random sample of 30 EAI managers has roughly a 50–50 chance of providing a sample mean within the allowable $500 margin of error. Perhaps a larger sample size should be considered. Let us explore this possibility by

---

*In Chapter 8 we will provide a more formal definition of sampling error; this definition requires an understanding of the concept of unbiasedness, a topic discussed in Section 7.7.

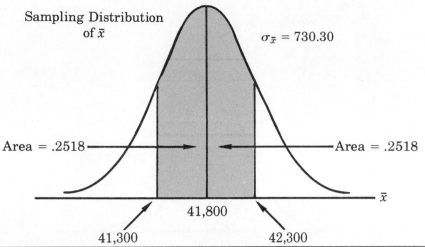

Sampling Distribution of $\bar{x}$

$\sigma_{\bar{x}} = 730.30$

Area = .2518

Area = .2518

$\bar{x}$

41,800

41,300

42,300

■■■ FIGURE 7.7   **Shaded Area is Probability of a Sample Mean Being Within $500 of the Population Mean**

considering the relationship between the sample size and the sampling distribution of $\bar{x}$.

The Relationship Between the Sample Size and the Sampling Distribution of $\bar{x}$

Suppose that in the EAI sampling problem, we selected a simple random sample of 100 EAI managers instead of the 30 EAI managers originally considered. Intuitively, it would seem that with more data provided by the larger sample size, the sample mean based on $n = 100$ should tend to provide a better estimate of the population mean than the sample mean based on $n = 30$. To see that this is true, let us consider the relationship between the sample size and the sampling distribution of $\bar{x}$.

First note that $E(\bar{x}) = \mu$ regardless of the sample size. Thus, the mean of all possible values of $\bar{x}$ is equal to the population mean μ regardless of the sample size n. However, note that the standard error of the mean, $\sigma_{\bar{x}} = \sigma/\sqrt{n}$, is inversely proportional to the sample size. That is, whenever the sample size is increased, the standard error of the mean, $\sigma_{\bar{x}}$, is decreased. With $n = 30$, the standard error of the mean for the EAI problem was 730.30. However, with the increase in the sample size to $n = 100$, the standard error of the mean is decreased to

$$\sigma_{\bar{x}} = \frac{\sigma}{\sqrt{n}} = \frac{4000}{\sqrt{100}} = 400$$

The sampling distributions of $\bar{x}$ with $n = 30$ and $n = 100$ are shown in Figure 7.8. Since the sampling distribution with $n = 100$ has a smaller standard error, the various values of $\bar{x}$ have less variation and tend to fall closer to the population mean than do the values of $\bar{x}$ with $n = 30$.

We can use the sampling distribution of $\bar{x}$ for the case with $n = 100$ to compute the probability that a simple random sample of 100 EAI managers will provide a sample mean that is within $500 of the population mean. Since the sampling distribution is normal, with mean 41,800 and standard deviation 400, we can use

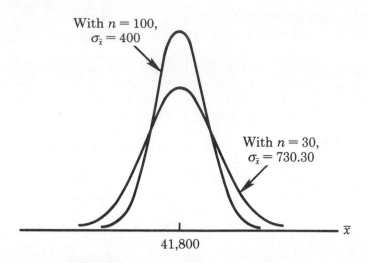

With $n = 100$,
$\sigma_{\bar{x}} = 400$

With $n = 30$,
$\sigma_{\bar{x}} = 730.30$

$\bar{x}$

41,800

▦ FIGURE 7.8 **A Comparison of the Sampling Distributions of $\bar{x}$ for Simple Random Samples of $n = 30$ and $n = 100$ EAI Managers**

the standard normal distribution table to find the area or probability. At $\bar{x} = 41,300$ (Figure 7.9), we have

$$z = \frac{41,300 - 41,800}{400} = -1.25$$

Referring to the standard normal probability distribution table, we find an area between $z = 0$ and $z = -1.25$ of .3944. With a similar calculation for $\bar{x} = 42,300$,

▦ FIGURE 7.9 **Shaded Area is Probability of a Sample Mean Being Within \$500 of the Population Mean when a Simple Random Sample of 100 EAI Managers is Used**

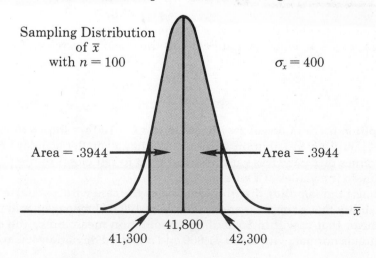

Sampling Distribution
of $\bar{x}$
with $n = 100$

$\sigma_x = 400$

Area = .3944 ⟶   ⟵ Area = .3944

$\bar{x}$

41,300   41,800   42,300

we see that the probability of the value of the sample mean being between 41,300 and 42,300 is .3944 + .3944 = .7888. Thus by increasing the sample size from 30 to 100 EAI managers, the probability of obtaining a sample mean within $500 of the population mean has increased from .5036 to .7888.

Perhaps even a larger sample size should be considered. However, the important point in this discussion is that as the sample size is increased, the standard error of the mean is decreased. As a result, the sampling distribution of $\bar{x}$ will have less variation. In effect, the larger sample size will provide a higher probability that the value of the sample mean is within a specified distance of the population mean.

NOTES AND COMMENTS

1. In presenting the sampling distribution of $\bar{x}$ for the EAI problem we took advantage of the fact that the population mean $\mu = 41,800$ and the population standard deviation $\sigma = 4000$ were provided in the discussion of Section 7.1. However, in general, the values of the population mean μ and the population standard deviation σ that are needed to determine the sampling distribution of $\bar{x}$ will be unknown. In Chapter 8 we will show how the sample mean $\bar{x}$ and the sample standard deviation s from a simple random sample are used when μ and σ are unknown.

2. The expression for $\sigma_{\bar{x}}$ in the finite population case is based on sampling without replacement. Although sampling with replacement is rarely conducted, sampling with replacement from a finite population possesses the same properties as sampling from an infinite population. Thus if a population is finite and if sampling is done with replacement, the expression of $\sigma_{\bar{x}}$ for the infinite population case would be appropriate.

3. The theoretical proof of the central limit theorem requires independent observations or items in the sample. This condition exists for infinite populations and for finite populations where sampling is done with replacement. Although the central limit theorem does not directly address sampling without replacement from finite populations, general statistical practice has been to apply the findings of the central limit theorem in this situation provided that the population size is large.

▬ EXERCISES

16. Assume we have a population with mean $\mu = 32$ and standard deviation $\sigma = 5$. Furthermore, assume that the population has 1000 items and that a simple random sample of 30 items is used to obtain information about this population. Let $\bar{x}$ denote the sample mean that will be used to estimate the value of the population mean.

   a. What is the expected value of $\bar{x}$?
   b. What is the standard deviation of $\bar{x}$?

17. Refer to the EAI sampling problem. Suppose that the simple random sample had contained 60 managers.

a. Sketch the sampling distribution of $\bar{x}$ when simple random samples of size 60 are used.
b. What happens to the sampling distribution of $\bar{x}$ if simple random samples of size 120 are used?
c. What general statement can you make about what happens to the sampling distribution of $\bar{x}$ as the sample size is increased? Does this seem logical? Explain.

18. In the EAI sampling problem, we showed that with a sample size of $n = 30$ there was .5036 probability of selecting a simple random sample having a sample mean within plus or minus $500 of the population mean.

a. What is the probability of an $\bar{x}$ value being within $500 of the population mean if a sample of size 60 is used?
b. Answer part (a) for a sample of size 120.

19. What important role does the central limit theorem serve whenever a sample mean $\bar{x}$ is used as a point estimator of a population mean μ?

20. Statistics help computer scientists and data-processing specialists understand the operating characteristics of computer systems. Statistical information includes waiting time, running time, central processor time, number of disk accesses, and so on. For a particular class of jobs run on the Amdahl 5880 mainframe computer (*Technical Update*, University of Cincinnati, Spring 1989), the population mean running time is 12.55 minutes per job. The population standard deviation is 4.0 minutes. Assume that a simple random sample of 40 jobs will be used to monitor the running time for the jobs.

a. Show the sampling distribution of $\bar{x}$, where $\bar{x}$ is the sample mean running time.
b. What is the probability that a simple random sample of 40 jobs will provide a sample mean within 1 minute of the population mean?
c. What is the probability that a simple random sample of 40 jobs will provide a sample mean within 30 seconds of the population mean?

21. A statistics class has 80 students. The mean score on the midterm exam was $\mu = 72$ and the standard deviation was $\sigma = 12$. Assume that a simple random sample of 20 students will be selected and the sample mean exam score $\bar{x}$ will be computed. What is the expected value and standard deviation of $\bar{x}$?

22. Weights for males between the ages of 20 and 30 have a mean $\mu = 170$ pounds with a standard deviation of $\sigma = 28$ pounds. If a simple random sample of 40 males in this age group is to be selected and the sample mean weight $\bar{x}$ computed, what are the values of $E(\bar{x})$ and $\sigma_{\bar{x}}$?

23. Consider a population of 1000 items. Assume the population standard deviation is $\sigma = 25$. Use (7.2) to compute the standard error of the mean, $\sigma_{\bar{x}}$, for sample sizes of 50, 100, 150, and 200. What can you say about the size of the standard error of the mean as the sample size is increased?

24. *Money* (February 1989) listed the national mean interest rate for new auto loans as 11.28. This was up from the 10.81 mean interest rate in 1988. Assume that the mean interest rate for the population of all new auto loans is 11.28 and that the population standard deviation is 1.5. Suppose that a simple random sample of 50 loans will be used to monitor interest rates of auto loans.

a. What is the probability that a simple random sample of 50 loans will provide a mean interest rate within .2 of the population mean?
b. What is the probability that a simple random sample of 50 loans will provide a mean interest rate within .1 of the population mean?

25. In a study of the growth rate of a certain plant, a botanist is planning to use a simple random sample of 25 plants for data-collection purposes. After analyzing the data on plant growth rate, the botanist believes that the standard error of the mean is too large. What size simple random sample should the botanist use in order to reduce the standard error to one-half its current value?

26. A simple random sample of size 50 is to be selected from a population with $\sigma = 10$. Find the value of the standard error of the mean in each of the following cases; use the finite population correction factor if appropriate.

    a. The population size is infinite.
    b. The population size is $N = 50,000$.
    c. The population size is $N = 5000$.
    d. The population size is $N = 500$.

27. The mean and standard deviation for the number of calories in a 12-ounce can of light beer are as follows: $\mu = 105$ and $\sigma = 3$. A simple random sample of 30 cans will be selected and a laboratory test conducted to determine the number of calories present in each of the 30 cans. The sample mean $\bar{x}$ will be computed.

    a. What is the expected value of $\bar{x}$?
    b. What is the standard deviation of $\bar{x}$?
    c. What probability distribution can be used to approximate the sampling distribution of $\bar{x}$?
    d. Sketch a graph of the sampling distribution of $\bar{x}$.

28. The length of time of long-distance telephone calls has a mean $\mu = 18$ minutes and a standard deviation $\sigma = 4$ minutes. Sketch the sampling distribution of $\bar{x}$ if a simple random sample of 50 telephone calls will be used to estimate the mean length of long-distance telephone calls.

29. A population has a mean $\mu = 400$ and a standard deviation $\sigma = 50$. The probability distribution of the population is unknown.

    a. A research study will use simple random samples of either 10, 20, 30, or 40 items to collect data about the population. In which of these sample-size alternatives will we be able to use a normal probability distribution to describe the sampling distribution of $\bar{x}$? Explain.
    b. Sketch the sampling distribution of $\bar{x}$ for the instances where the normal probability distribution is appropriate.

30. The mean wage rate for workers at General Motors is $14.25 per hour (*Detroit Daily News*, April 14, 1989). Assume that the population standard deviation is $2.00. Answer the following questions if a simple random sample of 50 workers is selected from the population of workers at General Motors.

    a. Show the sampling distribution of $\bar{x}$, where $\bar{x}$ is the sample mean hourly wage rate for the sample of 50 workers.
    b. What is the probability that the sample mean $\bar{x}$ is at least $13.80 per hour?
    c. What is the probability that the sample mean $\bar{x}$ is within $.25 of the population mean of $14.25 per hour?
    d. Answer parts (b) and (c) if the sample size is increased to 100 workers.

31. An automatic machine used to fill cans of soup has the following characteristics: $\mu = 15.9$ ounces and $\sigma = .5$ ounces.

    a. Show the sampling distribution of $\bar{x}$, where $\bar{x}$ is the sample mean for 40 cans selected randomly by a quality control inspector.
    b. What is the probability of finding a sample of 40 cans with a mean $\bar{x}$ greater than 16 ounces?

32. In a population of 4000 employees, a simple random sample of 40 employees is selected in order to estimate the mean age for the population.

    a. Would you use the finite population correction factor in calculating the standard error of the mean? Explain.

    b. If the population standard deviation is $\sigma = 8.2$ years, compute the standard error both with and without using the finite population correction factor. What is the rationale behind ignoring the finite population correction factor whenever $n/N \leq .05$?

33. What is the probability that the sample mean age of the employees in Exercise 32 will be within (plus or minus) 2 years of the population mean age? Use the population standard deviation of $\sigma = 8.2$ years and the sample size of 40.

34. A library checks out an average of $\mu = 320$ books per day, with a standard deviation of $\sigma = 75$ books. Consider a sample of 30 days of operation, with $\bar{x}$ being the sample mean number of books checked out per day.

    a. Show the sampling distribution of $\bar{x}$.

    b. What is the standard deviation of $\bar{x}$.

    c. What is the probability that the sample mean for the 30 days will be between 300 and 340 books?

    d. What is the probability the sample mean will show 325 or more books checked out?

7.6 ■ SAMPLING DISTRIBUTION OF $\bar{p}$

As we observed in Section 7.4, the sample proportion $\bar{p}$ is a random variable. In order to determine how close the sample proportion $\bar{p}$ is to the population proportion p, we need to understand the properties of the sampling distribution of $\bar{p}$: the expected value of $\bar{p}$, the standard deviation of $\bar{p}$, and the shape of the sampling distribution of $\bar{p}$.

Expected Value of $\bar{p}$

It can be shown that the expected value of $\bar{p}$—that is, the mean of all possible values of $\bar{p}$—is as follows:

Expected Value of $\bar{p}$

$$E(\bar{p}) = p \qquad (7.4)$$

where

$$E(\bar{p}) = \text{the expected value of the random variable } \bar{p}$$
$$p = \text{the population proportion}$$

Equation (7.4) shows that the mean of all possible $\bar{p}$ values is equal to the population proportion p. Recall that in Section 7.1 we showed that $p = .60$ for the EAI population, where p was the proportion of the population of managers who had participated in the company's management-training program. Thus the expected value of the random variable $\bar{p}$ for the EAI sampling problem is .60.

Standard Deviation of $\bar{p}$

Different simple random samples generate a variety of values for $\bar{p}$. We now are interested in determining the standard deviation of $\bar{p}$, which is referred to as the *standard error of the proportion*. Just as we found for the sample mean $\bar{x}$, the standard deviation of $\bar{p}$ depends upon whether the population is finite or infinite. The two expressions for the standard deviation of $\bar{p}$ are as follows:

Standard Deviation of $\bar{p}$

Finite Population	Infinite Population	
$\sigma_{\bar{p}} = \sqrt{\dfrac{N-n}{N-1}}\sqrt{\dfrac{p(1-p)}{n}}$	$\sigma_{\bar{p}} = \sqrt{\dfrac{p(1-p)}{n}}$	(7.5)

Comparing the two expressions in (7.5), we see that the only difference is the use of the finite population correction factor $\sqrt{(N-n)/(N-1)}$.

As was the case with the sample mean $\bar{x}$, we find that the difference between the above expressions for the finite population and the infinite population becomes negligible if the size of the finite population is large compared to the sample size. We follow the same rule of thumb that we recommended for the sample mean. That is, if the population is finite with $n/N \leq .05$, we will use $\sigma_{\bar{p}} = \sqrt{p(1-p)/n}$. However, if the population is finite and if $n/N > .05$, the finite population correction factor should be used, as shown in (7.5). Again, unless specifically noted, throughout the text we will be assuming that the population size is large relative to the sample size and that the finite population correction factor is unnecessary.

For the EAI study the population proportion of managers that have participated in the management training program is $p = .60$. With $n/N = 30/2500 = .012$, we can ignore the finite population correction factor when we compute the standard deviation of the $\bar{p}$ values. For the simple random sample of 30 managers, $\sigma_{\bar{p}}$ is

$$\sigma_{\bar{p}} = \sqrt{\frac{p(1-p)}{n}} = \sqrt{\frac{.60(1-.60)}{30}} = \sqrt{.008} = .0894$$

Form of the Sampling Distribution of $\bar{p}$

Now that we know the mean and standard deviation of $\bar{p}$, we want to consider the form of the sampling distribution of $\bar{p}$. Applying the central limit theorem as it relates to the $\bar{p}$ random variable, we have the following:

The sampling distribution of $\bar{p}$ can be approximated by a normal probability distribution whenever the sample size is large.

With $\bar{p}$, the sample size can be considered large whenever the following two conditions are satisfied:

$$np \geq 5$$
$$n(1 - p) \geq 5$$

Recall that for the EAI sampling problem we know that the population proportion of managers having participated in the training program is $p = .60$. Let us use this value to determine the characteristics of the sampling distribution of $\bar{p}$ for the EAI study. With a simple random sample of size 30, we have $np = 30(.60) = 18$ and $n(1 - p) = 30(.40) = 12$. Thus according to the above rule of thumb, the sampling distribution of $\bar{p}$ can be approximated by a normal probability distribution. Figure 7.10 shows this sampling distribution for the EAI study. Figure 7.11 shows the close agreement between this theoretical sampling distribution of $\bar{p}$ and the histogram of the 500 $\bar{p}$ values we obtained from the 500 repeated samples of size 30 (see Figure 7.3).

Practical Value of the Sampling Distribution of $\bar{p}$

Whenever a simple random sample is selected and the value of the sample proportion $\bar{p}$ is used to estimate the value of the population proportion p, we anticipate some sampling error. In this case, the sampling error is the difference between the value of the sample proportion $\bar{p}$ and the value of the population proportion p. The practical value of the sampling distribution of $\bar{p}$ is that it can be used to provide probability information about the sampling error.

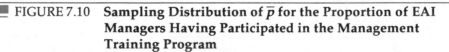

■■■■ FIGURE 7.10   **Sampling Distribution of $\bar{p}$ for the Proportion of EAI Managers Having Participated in the Management Training Program**

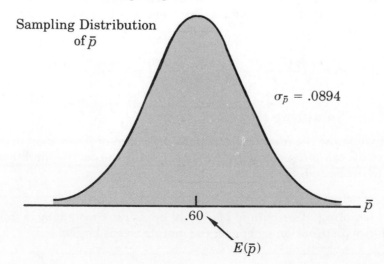

Sampling Distribution of $\bar{p}$

$\sigma_{\bar{p}} = .0894$

.60

$E(\bar{p})$

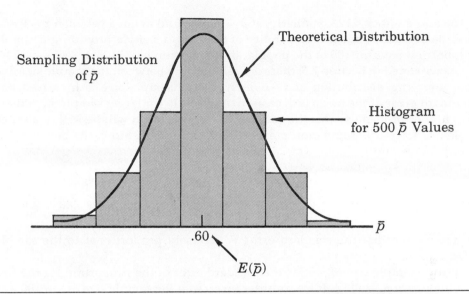

FIGURE 7.11   Comparison of the Theoretical Sampling Distribution of $\bar{p}$ and the Relative Frequency Histogram of $\bar{p}$ for 500 Simple Random Samples

Suppose, in the EAI problem, the personnel director wanted to know the probability of obtaining a value of $\bar{p}$ that is within .05 of the population proportion of EAI managers who have participated in the training program. That is, what is the probability of obtaining a sample with a sample proportion $\bar{p}$ between .55 and .65? The shaded area in Figure 7.12 shows this probability. Using the fact that the sampling distribution of $\bar{p}$ is normal with mean .60 and standard deviation $\sigma_{\bar{p}} = .0894$, the standard normal random variable corresponding to $\bar{p} = .55$ has a value of $z = (.55 - .60)/.0894 = -.56$. Referring to the standard normal probability distribution table, we find an area between $z =$

FIGURE 7.12   Sampling Distribution of $\bar{p}$ for the EAI Sampling Problem

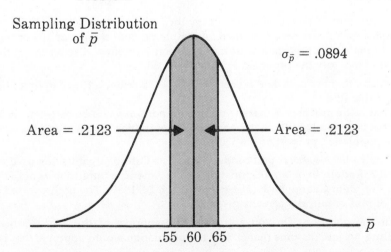

$-.56$ and $z = 0$ of .2123. Similarly, at $\bar{p} = .65$ we find an area between $z = 0$ and $z = .56$ of .2123. Thus the probability of selecting a sample providing a sample proportion $\bar{p}$ within .05 of the population proportion p is $.2123 + .2123 = .4246$.

As we stated in Section 7.5, there is a relationship between the sample size and the sampling distribution of $\bar{x}$—namely, as the sample size is increased, the standard error of the mean is decreased. As a result, for larger sample sizes there is a higher probability that the value of the sample mean will be within a stated distance from the population mean. A similar relationship exists between the sample size and the sampling distribution of $\bar{p}$. The formula for the standard error of the proportion is given by

$$\sigma_{\bar{p}} = \sqrt{\frac{p(1 - p)}{n}}$$

Thus, we see that the standard error is inversely proportional to the sample size.

With a sample size of $n = 30$, the standard error of the proportion for the EAI problem was $\sigma_{\bar{p}} = .0894$. If we consider increasing the sample size to $n = 100$, the standard error of the proportion becomes

$$\sigma_{\bar{p}} = \sqrt{\frac{.60(1 - .60)}{100}} = \sqrt{.0024} = .0490$$

With a sample size of 100 EAI managers, the probability of the sample proportion having a value within .05 of the population proportion can now be computed. Since the sampling distribution is normal, with mean .60 and standard deviation .0490, we can use the standard normal probability distribution table to find the area or probability. At $\bar{p} = .55$, we have $z = (.55 - .60)/.0490 = -1.02$. Referring to the standard normal probability distribution table, we find the area between $z = -1.02$ and 0 is .3461. Similarly, at .65 we find the area between $z = 0$ and $z = 1.02$ is .3461. Thus if the sample size is increased from 30 to 100, the probability that the sample proportion $\bar{p}$ is within .05 of the population proportion p will be increased from .4246 to $.3461 + .3461 = .6922$.

◼◼◼ EXERCISES

35. The president of Doerman Distributors, Inc., believes that 30% of the firm's orders come from new or first-time customers. A simple random sample of 100 orders will be used to estimate the proportion of new or first-time customers. The results of the sample will be used to verify the president's claim of $p = .30$.

    a. Assume that the president is correct and $p = .30$. What is the sampling distribution of $\bar{p}$ for this study?
    b. What is the probability that the sample proportion $\bar{p}$ will be between .20 and .40?
    c. What is the probability that the sample proportion will be within plus or minus .05 of the population proportion $p = .30$?

36. A March 1989 *Newsweek* poll conducted by The Gallup Organization used a national sample of 756 adults to obtain information on the American public views of the job being done by President George Bush (*Newsweek*, March 20, 1989). The poll reported a margin of error of plus or minus 4 percentage points.

    a. For the question 'Do you approve or disapprove of the way George Bush is handling his job?' assume that 60% of the population would approve; that is, $p = .60$.

What is the probability the sample proportion $\bar{p}$ from the sample of 756 adults would be within 4 percentage points of this population proportion?

b. What is the probability the sample proportion from the sample of 756 adults would be within 2 percentage points of the population proportion?

c. Why did the Newsweek poll quote a 4% margin of error rather than a 2% margin of error?

37. A particular county in West Virginia has a 9% unemployment rate. A monthly survey of 800 individuals is conducted by a state agency. This study provides the basis for monitoring the unemployment rate of the county.

a. Assume that $p = .09$. What is the sampling distribution of $\bar{p}$ when a sample of size 800 is used?

b. What is the probability that a sample proportion $\bar{p}$ of at least .08 will be observed?

38. Surveys of subscribers of the *Wall Street Journal* and *Investor's Daily* in 1988 provided reader profile information for these two business publications (*Investor's Daily*, March 23, 1989). The profile information include median reader age, percentage of college graduates, percentage in top management, mean income, and mean net worth. Based on data provided by these surveys, assume that the 80% of the population of all subscribers of these business publications are college graduates. Answer the following questions about the results of simple random sample of subscribers. The statistic of interest is the sample proportion of subscribers found to be college graduates.

a. Show the sampling distribution of $\bar{p}$ if a simple random sample of 400 subscribers is used. Assume $p = .80$.

b. With a sample size of 400, what is the probability that the margin of error will be 3% or less? That is, what is the probability the sample proportion of college graduates among the 400 subscribers will be within .03 of the population proportion $p = .80$?

c. Answer part (b) if the size of the simple random sample is increased to 750 subscribers.

39. What is the probability that the EAI estimate of the proportion of managers having completed the firm's training program, $\bar{p}$, is within $\pm.05$ of the population proportion $p = .60$? Use samples of size 60 and 120.

40. Assume that 15% of the items produced in an assembly line operation are defective, but that the firm's production manager is not aware of this situation. Assume further that 50 parts are tested by the quality assurance department in order to determine the quality of the assembly operation. Let $\bar{p}$ be the sample proportion defective found by the quality assurance test.

a. Show the sampling distribution for $\bar{p}$.

b. What is the probability that the sample proportion will be within $\pm.03$ of the population proportion defective?

c. If the test shows $\bar{p} = .10$ or more, the assembly line operation will be shut down to check for the cause of the defects. What is the probability that the sample of 50 parts will lead to the conclusion that the assembly line should be shut down?

41. Baskin Robbins offers frozen yogurt at 900 of its 2500 outlets (*Advertising Age*, April 10, 1989).

a. What is the population proportion of Baskin Robbins outlets offering frozen yogurt?

b. Assume that a simple random sample of 40 Baskin Robbins outlets will be used and that the sample proportion $\bar{p}$ will be used to estimate the population proportion of Baskin Robbins outlets offering frozen yogurt. What is the probability that the sample proportion will be within $\pm.05$ of the population proportion?

c. What is the probability that the sample proportion will be within $\pm.05$ of the population proportion if the sample size is increased to 120 outlets?

7.7 ■■■ PROPERTIES OF POINT ESTIMATORS

In this chapter we have shown how sample statistics such as a sample mean, $\bar{x}$, a sample standard deviation, s, and a sample proportion, $\bar{p}$, can be used as point estimators of their corresponding population parameters μ, σ, and p. It is intuitively appealing that each sample statistic should be the point estimator of its corresponding population parameter. However, before using a sample statistic as a point estimator, statisticians check to see whether or not the sample statistic has some of the properties associated with "good" point estimators. In this section we discuss three properties of good point estimators referred to as unbiasedness, efficiency, and consistency.

Unbiasedness

If the expected value of the sample statistic is equal to the population parameter being estimated, the sample statistic is said to be an *unbiased* estimator of the population parameter. If we denote the sample statistics $\hat{\theta}$ and the population parameter as θ, the property of unbiasedness is defined as follows:

Unbiasedness

The sample statistic $\hat{\theta}$ is an unbiased estimator of the population parameter θ if

$$E(\hat{\theta}) = \theta \qquad (7.6)$$

where

$$E(\hat{\theta}) = \text{expected value of the sample statistic } \hat{\theta}$$

In words, the unbiasedness property states that the expected value, or mean, of all possible values of the sample statistic is equal to the population parameter being estimated.

Figure 7.13 shows the cases of unbiased and biased point estimators. In the illustration showing the unbiased estimator, the mean of the sampling distribution is equal to the value of the population parameter. The sampling errors balance out in this case, since sometimes the value of the point estimator $\hat{\theta}$ may be less than θ and other times it may be greater than θ. In the case of the biased estimator, the mean of the sampling distribution is less than or greater than the value of the population parameter. In the illustration shown in Figure 7.13(b), $E(\hat{\theta}) > \theta$; thus, the sample statistic has a high probability of overestimating the value of the population parameter. This tendency to overestimate the parameter is the biased property of the estimator. The amount of the bias is shown in the figure.

In discussing the sampling distributions of the sample mean and the sample

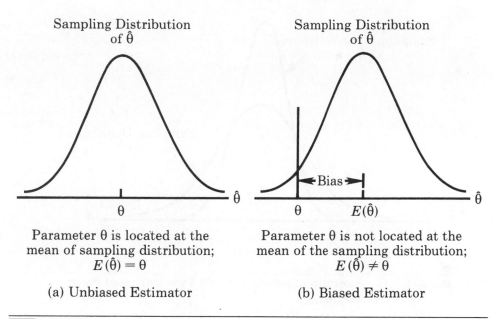

(a) Unbiased Estimator | (b) Biased Estimator

FIGURE 7.13   **Examples of Unbiased and Biased Point Estimators**

proportion, we stated that $E(\bar{x}) = \mu$ and $E(\bar{p}) = p$. Thus both $\bar{x}$ and $\bar{p}$ are unbiased estimators of their corresponding population parameters μ and p.

We delay a more complete discussion of the sampling distribution of the sample standard deviation s and the sample variance s^2 until Chapter 11. However, it can be shown that $E(s^2) = \sigma^2$. Thus we conclude that the sample variance s^2 is an unbiased estimator of the population variance σ^2. In fact, when we first presented the formulas for the sample variance and the sample standard deviation in Chapter 3, $n - 1$ rather than n was used in the denominators. The reason for using $n - 1$ rather than n is to make the sample variance an unbiased estimator of the population variance. If we had used n in the denominator, the sample variance would have been a biased estimator, tending to slightly underestimate the population variance.

Efficiency

Assume that a simple random sample of n elements can be used to provide two unbiased point estimators of the same population parameter. In this situation, we would prefer to use the point estimator with the smaller variance, since it tends to provide estimates closer to the population parameter. The point estimator with the smaller variance is said to have greater *relative efficiency* than the other.

Figure 7.14 shows the sampling distributions of two unbiased point estimators, $\hat{\theta}_1$ and $\hat{\theta}_2$. Note that the variance of $\hat{\theta}_1$ is less than the variance of $\hat{\theta}_2$; thus, values of $\hat{\theta}_1$ have a greater chance of being close to the parameter θ than do values of $\hat{\theta}_2$. Since the variance of point estimator $\hat{\theta}_1$ is less than the variance of point estimator $\hat{\theta}_2$, $\hat{\theta}_1$ is relatively more efficient than $\hat{\theta}_2$; thus $\hat{\theta}_1$ is the preferred point estimator.

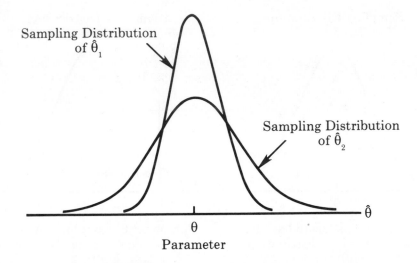

Sampling Distributions of Two Unbiased Point Estimators

Consistency

A third property associated with good point estimators is *consistency.* Loosely speaking, a point estimator is consistent if the values of the point estimator tend to become closer to the population parameter as the sample size becomes larger. In other words, a larger sample size tends to provide a better point estimate than does a smaller sample size. Note here that for the sample mean $\bar{x}$, we showed that the standard deviation of $\bar{x}$ is given by $\sigma_{\bar{x}} = \sigma/\sqrt{n}$. Thus, we observed that $\sigma_{\bar{x}}$ is inversely proportional to the sample size. As a result, larger samples sizes provide smaller values for $\sigma_{\bar{x}}$. Thus we conclude that a larger sample size tends to provide point estimates closer to the population mean μ. In this sense, we would say that the sample mean $\bar{x}$ is a consistent estimator of the population mean μ. Using a similar rationale, we also could conclude that the sample proportion $\bar{p}$ is a consistent estimator of the population proportion p.

NOTES AND COMMENTS

In Chapter 3, we stated that the mean and the median are two measures of central location. In this chapter we discussed only the mean. The reason for this is that the sample mean is a more efficient estimator than the sample median. It can be shown that in sampling from a normal population, where the population mean and population median are identical, the standard error of the median is approximately 25% larger than the standard error of the mean. Recall for the EAI problem with $n = 30$ that the standard error of the mean is $\sigma_{\bar{x}} = 730.30$. The standard error of the median for this problem would have been approximately $1.25 \times (730.30) = 913$. As a result, the more relatively efficient sample mean will have a higher probability of being within a specified distance of the population mean.

7.8 ▬ OTHER SAMPLING METHODS

In this chapter we have described how simple random sampling can be used to estimate population parameters. It is important to realize that simple random sampling is not the only sampling method available. Sampling methods such as stratified simple random sampling, cluster sampling, systematic sampling, convenience sampling, and judgment sampling offer alternatives to simple random sampling. We briefly describe each of these other sampling methods.

Stratified Simple Random Sampling

Using stratified simple random sampling, the population is first divided into groups of elements called *strata*, such that each element in the population belongs to one and only one stratum. Figure 7.15 shows a diagram with a population divided into H strata. For example, suppose a college of business administration conducts an annual survey of its graduating seniors in order to learn about starting salaries and the job market for new graduates. Before selecting a sample of graduating seniors for the survey, the population of graduating seniors is divided into strata based on each student's major. Thus, there is a separate stratum for accounting majors, finance majors, management majors, marketing majors, and so on. At this point, the method of simple random sampling is used to select a separate simple random sample of graduates from *each stratum*. The sample data are collected and summarized for each stratum.

To illustrate how the data for each stratum is used to estimate the parameter of interest in the population, assume that for the annual survey of graduating seniors we wanted to estimate the mean annual starting salary for the population. First, the sample mean annual starting salary would be computed for each major (stratum). The results for the individual strata are then combined to estimate the mean of the population. The formula used to estimate the population mean μ for a stratified simple random sample involving H strata is as

---

▬ FIGURE 7.15 **Diagram for Stratified Simple Random Sampling**

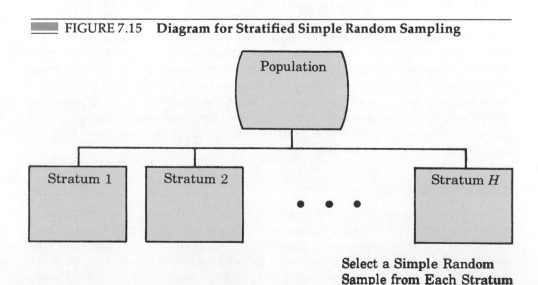

Select a Simple Random Sample from Each Stratum

follows:

Point Estimate of μ Based on Stratified Simple Random Sampling

$$\bar{x}_{st} = \frac{\Sigma N_h \bar{x}_h}{N} \tag{7.7}$$

where

$\bar{x}_{st}$ = point estimator of the population mean

$\bar{x}_h$ = sample mean for stratum h, $h = 1, 2, \ldots, H$

N_h = the number of elements in stratum h, $h = 1, 2, \ldots, H$

N = the number of elements in the population

For example, assume that a problem involving three strata provided the following sample means for the three strata: $\bar{x}_1 = 12$, $\bar{x}_2 = 19$, and $\bar{x}_3 = 20$. The strata sizes were $N_1 = 100$, $N_2 = 150$, and $N_3 = 250$; thus, the number of elements in the population was $N = 100 + 150 + 250 = 500$. Using (7.7), the stratified simple random sampling estimate of the population mean μ is

$$\bar{x}_{st} = \frac{100(12) + 150(19) + 250(20)}{500} = \frac{9050}{500} = 18.1$$

Formulas are available for approximating the sampling distribution of $\bar{x}_{st}$ in order to obtain probability information about how close $\bar{x}_{st}$ is the population mean μ. The details regarding how to use these formulas are beyond the scope of this text.

The basis for forming the various strata is up to the designer of the sample. Some criteria for stratifying include department, location, age, product type, and industry type. The value of stratified simple random sampling depends upon how homogeneous the elements are within the strata. If units within the strata are *alike* (homogeneity), the strata will have low variances. Thus relatively small sample sizes can be used to obtain good estimates of the strata characteristics. If homogeneous strata can be developed, the stratified simple random sampling procedure will provide results similar to those obtained using simple random sampling; the advantage is that with stratified simple random sampling, the total sample size will be smaller.

Cluster Sampling

Cluster sampling requires the population to be divided into groups of elements called clusters, such that each element in the population belongs to one and only one cluster; thus, in this regard, cluster sampling is similar to stratified simple random sampling. Figure 7.16 shows a population divided into K clusters. To provide an illustration of how cluster sampling can be applied, consider a study

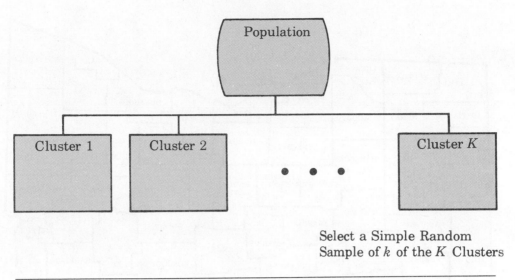

Select a Simple Random
Sample of k of the K Clusters

▬▬▬ FIGURE 7.16   **Diagram for Cluster Sampling**

that is being conducted of registered voters in the state of Ohio. Rather than attempt to list all registered voters in the population, we might begin with a list of the 88 counties in the state. The counties identify the clusters for the study in that each county consists of a group of registered voters and each registered voter in the population belongs to one and only one cluster. A map of Ohio with its 88 counties (clusters) is shown in Figure 7.17.

In cluster sampling we begin by selecting a simple random sample of the clusters. For example, we could randomly select 12 of the 88 counties in Ohio. At this point, we may choose to collect data from all registered voters in the sampled clusters or we may proceed by selecting a simple random sample of registered voters from each of the sampled clusters. In either case, formulas are available for using the results from the sampled clusters to develop point estimates of population parameters such as the population mean μ.

Cluster sampling tends to provide better results when the elements within the clusters are heterogeneous (not alike). In the ideal case, each cluster would be a representative small-scale version of the entire population. In this case, sampling a small number of clusters would provide good information about the characteristics of the entire population.

One of the primary applications of cluster sampling is area sampling, where the clusters are counties, townships, city blocks, or other well-defined sections of the population area. Since data are collected from only a sample of the sections or clusters, time and cost savings result due to the fact that when a data collector is sent to a sampled cluster, sample data can be obtained from the cluster in a relatively short time. As a result, a larger sample may be obtained with relatively low time and cost per element. Thus, even if a larger total sample size is required, cluster sampling may be less costly than either simple random sampling or stratified simple random sampling. In addition, cluster sampling can minimize the time and cost associated with organizing and listing the elements to be sampled, since cluster sampling requires this organizing and listing only in the sampled clusters and not the entire population.

■■■ FIGURE 7.17   **Counties of the State of Ohio Used as Clusters of Registered Voters**

Systematic Sampling

In some sampling situations, especially those with large populations, it is often time-consuming to select a simple random sample by first finding a random number and then counting or searching through the list of population items until the corresponding element is found. An alternative to simple random sampling is a systematic sampling procedure. For example, if a sample size of 50 is desired from a population containing 5000 elements, we might sample one element for every $5000/50 = 100$ elements in the population. A systematic sample for this case would involve selecting randomly 1 of the first 100 elements from the population list. Other sample elements are identified by starting with the first sampled element and then selecting every 100th element that follows in the population list. In effect the sample of 50 is identified by moving systematically through the population and identifying every 100th element after the first randomly selected element. The sample of 50 usually will be easier to identify in this manner than would be the case if simple random sampling were used. Since the first element selected is a random choice, the assumption is usually made that a systematic sample has the properties of a simple random sample. This assumption is especially applicable when the list of the population elements is believed to be a random ordering of the elements.

Convenience Sampling

All the sampling methods discussed thus far fall under the heading of *probability sampling* techniques. By this we mean that elements selected from the population have a known probability of being included in the sample. The advantage of probability sampling is that the sampling distribution of the appropriate sample statistic generally can be identified. Formulas such as the ones for simple random sampling presented earlier in this chapter can be used to determine the properties of the sampling distribution. Then the sampling distribution can be used to make probability statements about possible sampling errors associated with the sample results.

Convenience sampling falls under the heading of *nonprobability sampling* techniques. As the name implies, the sample is identified primarily by convenience. Elements are included in the sample without prespecified or known probabilities of being selected. For example, a professor conducting research at a university may ask student volunteers to constitute a sample for the study simply because they are readily available and will often participate as subjects for little or no cost. In another example, a shipment of oranges may be sampled by an inspector who selects oranges haphazardly from among several crates. Labeling each orange and using a probability method of sampling would be impractical. Samples such as wildlife captures and volunteer panels for consumer research would also have to be considered convenience samples.

Convenience samples have the advantage of relatively easy sample selection and data collection; however, it is impossible to evaluate the "goodness" of the sample in terms of its ability to estimate characteristics of the population. A convenience sample may provide good results, or it may not. However, there is no statistically justified procedure that will allow a probability analysis and inference about the quality of the sample results. Nevertheless, at times you will see statistical methods designed for probability samples applied to a convenience sample. The researcher argues that the convenience sample may well provide a sample which can be treated as if it were a random sample. However, this argument should be questioned, and we should be very cautious in interpreting convenience samples that are used to make inferences about populations.

Judgment Sampling

One additional nonprobability sampling technique is judgment sampling. In this situation the person most knowledgeable on the subject of the study selects individuals or other elements of the population that he or she feels are most representative of the population. Often this is a relatively easy way of selecting a sample. However, the quality of the sample results are dependent on the judgment of the person selecting the sample. Again great caution must be used in drawing conclusions based on judgment samples used to make inferences about populations.

NOTES AND COMMENTS

We recommend sampling using one of the probability sampling methods: simple random sampling, stratified simple random sampling, cluster sampling, or systematic sampling. By using these methods, formulas are available for evaluating the "goodness" of the sample results in terms of how close the results may be to the population characteristics being estimated. An evaluation of the goodness cannot be made with convenience or judgment sampling. Thus, great care should be used in interpreting the results when either of these nonprobability sampling methods have been used to obtain statistical information.

■ SUMMARY

In this chapter we presented the important concepts of simple random sampling and sampling distributions. We demonstrated how a simple random sample can be selected and how the data collected for the sample can serve as the basis for developing point estimates of population parameters. Since different simple random samples provided a variety of different values for the point estimators, we saw that point estimators, such as $\bar{x}$ and $\bar{p}$, are random variables. The probability distribution of such a random variable is called a sampling distribution. In particular, we described the sampling distributions of the sample mean, $\bar{x}$, and the sample proportion, $\bar{p}$.

In considering the characteristics of the sampling distributions of $\bar{x}$ and $\bar{p}$, we stated that $E(\bar{x}) = \mu$ and $E(\bar{p}) = p$. After developing the standard deviation or standard error formulas for these estimators, we showed how the central limit theorem provided the basis for using a normal distribution to approximate these sampling distributions in the large-sample case. Rules of thumb were provided for determining when "large" sample size conditions were satisfied.

We then discussed some properties of point estimators, including unbiasedness, efficiency, and consistency. We concluded the chapter by describing sampling methods other than simple random sampling, including stratified, cluster, systematic, convenience, and judgment sampling.

■ GLOSSARY

Parameter   A population characteristic, such as a population mean, μ, a population standard deviation σ, a population proportion, p, and so on.

Simple random sampling   Finite population: a sample selected such that each possible sample of size n has the same probability of being selected. Infinite population: a sample selected such that each element comes from the same population and the successive elements are selected independently.

Sampling without replacement   Once an item from the population has been included in the sample it is removed from further consideration and thus cannot be selected a second time.

Sampling with replacement As each item is selected for the sample, it is returned to the population. It is possible that a previously selected item may be selected again and therefore appear in the sample more than once.

Sample statistic A sample characteristic, such as a sample mean, $\bar{x}$, a sample standard deviation, s, a sample proportion, $\bar{p}$, and so on. The value of the sample statistic is used to estimate the value of the population parameter.

Sampling distribution A probability distribution consisting of all possible values of a sample statistic.

Point estimate A single numerical value used as an estimate of a population parameter.

Point estimator The sample statistic, such as $\bar{x}$, s, $\bar{p}$, etc., that provides the point estimate of the population parameter.

Finite population correction factor The term $\sqrt{(N - n)/(N - 1)}$ that is used in the formulas for $\sigma_{\bar{x}}$ and $\sigma_{\bar{p}}$ whenever a finite population, rather than an infinite population, is being sampled. The generally accepted rule of thumb is to ignore the finite population correction factor whenever $n/N \leq .05$.

Standard error The standard deviation of a point estimator.

Central limit theorem A theorem that allows us to use the normal probability distribution to approximate the sampling distribution of $\bar{x}$ and $\bar{p}$ whenever the sample size is large.

Unbiasedness A property of a point estimator that occurs whenever the expected value of the point estimator is equal to the population parameter it estimates.

Relative efficiency Given two unbiased point estimators of the same population parameter, the point estimator with the smaller variance is said to have greater relative efficiency than the other.

Consistency A property of a point estimator that occurs whenever larger sample sizes tend to provide point estimates closer to the population parameter.

Probability sample A sample selected such that each element in the population has a known probability of being included in the sample. Simple random sampling, stratified simple random sampling, cluster sampling, and systematic sampling are probability samples.

Nonprobability sample A sample selected such that the probability of each element being included in the sample is unknown. Convenience and judgment samples are nonprobability samples.

▬ KEY FORMULAS

Expected Value of $\bar{x}$

$$E(\bar{x}) = \mu \tag{7.1}$$

Standard Deviation of $\bar{x}$

Finite Population *Infinite Population*

$$\sigma_{\bar{x}} = \sqrt{\frac{N - n}{N - 1}} \left(\frac{\sigma}{\sqrt{n}} \right) \qquad \sigma_{\bar{x}} = \frac{\sigma}{\sqrt{n}} \tag{7.2}$$

Expected Value of $\bar{p}$

$$E(\bar{p}) = p \tag{7.4}$$

Standard Deviation of $\bar{p}$

Finite Population　　　　*Infinite Population*

$$\sigma_{\bar{p}} = \sqrt{\frac{N-n}{N-1}}\sqrt{\frac{p(1-p)}{n}} \qquad \sigma_{\bar{p}} = \sqrt{\frac{p(1-p)}{n}} \qquad (7.5)$$

■ SUPPLEMENTARY EXERCISES

42. Nationwide Supermarkets has 4800 retail stores located in 32 states. At the end of each year a sample of 35 stores is selected for physical inventories. Results from the inventory samples are used in annual tax reports. Assume that the retail stores are listed sequentially on a computer printout. Begin at the bottom of the second column of the random numbers shown in Table 7.1. Using four-digit random numbers beginning with 8112, read *up* the column to identify the first five stores to be included in the simple random sample.

43. A study of the time from computer program submission until program return (i.e., "turnaround" time) was conducted at a university computer center. Assume that under standard operating conditions the population mean is 120 minutes, with a population standard deviation of 40 minutes.

    a. Future studies of turnaround time are to be based on simple random samples of 30 programs. Show the sampling distribution of the sample mean turnaround time.

    b. What is the probability that the sample mean for 30 programs will be less than 100 minutes? Over 125 minutes?

44. *Survey and Analysis of Salary Trends*, 1988, from the research department of the American Federation of Teachers, lists the average annual salary for elementary/secondary school teachers as $28,085. Answer the following questions using this value as a population mean and 3,200 as the population standard deviation.

    a. What is the probability that a simple random sample of 100 teachers will provide a sample mean annual salary within $200 of the population mean?

    b. What is the probability that a simple random sample of 300 teachers will provide a sample mean annual salary within $200 of the population mean?

    c. Assuming that we would like a 90% chance of having a sample mean annual salary within $200 of the population mean, how many teachers should be included in the simple random sample?

45. An electrical component is designed to provide a mean service life of 3000 hours, with a standard deviation of 800 hours. A customer purchases a batch of 50 components; assume that this batch can be considered a simple random sample of the population of components. What is the probability that the mean life for the group of 50 components will be at least 2750 hours? At least 3200 hours?

46. An automatic machine used to fill cans of soft drink has the following characteristics: $\mu = 12.0$ ounces and $\sigma = .4$ ounces.

    a. Show the sampling distribution of $\bar{x}$, where $\bar{x}$ is the sample mean for 50 cans selected randomly by a quality control inspector.

    b. What is the probability of finding a sample of 50 cans with a sample mean, $\bar{x}$, greater than 12 ounces?

47. The time it takes a fire department to respond to a request for emergency aid has a mean of $\mu = 14$ minutes with a standard deviation of $\sigma = 4$ minutes. Suppose we randomly

sample 50 emergency requests over a 2-month period. Records of aid-request times and arrival times will be used to compute a sample mean response time for the 50 requests.

a. Show the sampling distribution of $\bar{x}$.
b. What role does the central limit theorem play in identifying this sampling distribution?
c. What is the probability that the sample mean will be 15 minutes or less?
d. What is the probability that the sample mean will be within ±.5 minutes of the mean time for the population?

48. The speed of automobiles on a section of I-75 in northern Florida has a mean of $\mu = 67$ miles per hour with a standard deviation of $\sigma = 6$ miles per hour. Answer the following questions if the population can be assumed to have a *normal distribution* and if a sample of 16 automobiles will be selected to compute a sample mean automobile speed.

a. What is the expected value of $\bar{x}$?
b. What is the value of the standard error of the mean?
c. Show the sampling distribution of $\bar{x}$.
d. What is the probability that the value of the sample mean will be 65 miles per hour or more?
e. What is the probability that the value of the sample mean will be between 66 and 68 miles per hour?

49. Consider a population of size $N = 500$ with a mean $\mu = 200$ and a standard deviation $\sigma = 40$. Assume that a simple random sample of size $n = 100$ will be selected from this population.

a. Should the finite population correction factor be used in computing the standard error of the mean?
b. What is the value of the standard error of the mean for this problem?
c. What is the probability of selecting a simple random sample that provides a value of $\bar{x}$ that is within ±5 of the population mean μ?

50. In a population of 5000 students, a simple random sample of 50 students is selected in order to estimate the mean grade point average for the population.

a. Would you use the finite population correction factor in calculating the standard error of the mean? Explain.
b. If the population standard deviation is $\sigma = .4$ years, compute the standard error of the mean, first with and then without the finite population correction factor. What is the rationale for ignoring the finite population correction factor whenever $n/N \leq .05$?
c. What is the probability that the sample grade point average for 50 students will be within ±.10 of the population mean grade point average?

51. During a complete review of 2 months of billings an accountant found the following values for the mean and standard deviation of the dollar amounts per billing: $\mu = \$22.00$ and $\sigma = \$7.00$. The company's controller believes that the accountant could have obtained very good estimates of the mean billing amount by taking a simple random sample of 50 billings. Assume that a simple random sampling procedure was conducted.

a. Explain how the sample mean billing $\bar{x}$ would have a sampling distribution.
b. Show the sampling distribution of $\bar{x}$.
c. What is the standard error of the mean?
d. What would happen to the sampling distribution of $\bar{x}$ if a sample size of 100 was considered?

52. In the EAI study the population of managers had annual salaries with $\mu = \$41,800$ and $\sigma = \$4,000$. Samples of size 30 provided a .5036 probability of selecting a sample with $\bar{x}$ within ±\$500 of the population mean. How large a sample should be selected if the personnel director wishes the probability of a sample mean $\bar{x}$ being within ±\$500 of μ to be .95?

53. Three firms have inventories that vary in size. Firm A has a population of 2000 items, firm B has a population of 5000 items, and firm C has a population of 10,000 items. The population standard deviation for the cost of the items is $\sigma = 144$. A statistical consultant recommends that each firm take a sample of 50 items from their respective populations in order to provide statistically valid estimates of the average cost per item. Management of the small firm states that since it has the smallest population, it should be able to obtain the data from a much smaller sample size than required by the larger firms. However, the consultant states that in order to obtain the same standard error and thus the same precision in the sample results, all firms should take the same sample size regardless of population size.

   a. Using the finite population correction factor, compute the standard error for each of the three firms given a sample of size 50.
   b. For each firm what is the probability that the sample mean $\bar{x}$ will be within ± 25 of the population mean, μ?

54. A survey reports its results by stating that the standard error of the mean is 20. The population standard deviation is 500.

   a. How large is the sample used in this survey?
   b. What is the probability that the estimate would be within ± 25 of the population mean?

55. A production process is checked periodically by a quality control inspector. The inspector selects simple random samples of 30 finished products and computes the sample mean product weights, $\bar{x}$. If test results over a long period of time show that 5% of the $\bar{x}$ values are over 2.1 pounds and 5% are under 1.9 pounds, what are the mean and the standard deviation for the population produced with this process?

56. The grade point average for all juniors at Strausser College has a standard deviation of .50.

   a. A random sample of 20 students is to be used to estimate the population mean grade point average. What assumption is necessary in order to compute the probability of obtaining a sample mean within plus or minus .2 of the population mean?
   b. Provided that this assumption can be made, what is the probability of $\bar{x}$ being within plus or minus .2 of the population mean?
   c. If this assumption cannot be made, what would you recommend doing?

57. Assume that the proportion of persons having a college degree is $p = .35$.

   a. Explain how the sampling distribution of $\bar{p}$ results from random samples of size 80 being used to estimate the proportion of individuals having a college degree.
   b. Show the sampling distribution for $\bar{p}$ in this case.
   c. If the sample size is increased to 200, what happens to the sampling distribution of $\bar{p}$? Compare the standard error for the $n = 80$ and $n = 200$ alternatives.

58. CinemaScore is a movie research firm that uses a sample of film audiences to learn why members of the audience chose to attend a particular motion picture (*USA Today*, April 11, 1989). Possible responses include the subject matter, the cast, and other reasons. Assume that for a particular new motion picture, 65% of the audience population chose the movie due to subject matter. In addition, assume that a simple random sample will be used to estimate the proportion of the population that will choose the movie due to subject matter.

   a. What is the probability that a simple random sample of 100 people will provide a sample proportion within .04 of the population proportion?
   b. What is the probability that a simple random sample of 200 people will provide a sample proportion within .04 of the population proportion?
   c. Assume that we would like to select a simple random sample that will provide a 90% chance of obtaining a sample proportion within .04 of the population proportion. How many people should be included in the simple random sample?

59. A market research firm conducts telephone surveys with a 40% historical response rate. What is the probability that in a new sample of 400 telephone numbers at least 150 individuals will cooperate and respond to the questions? In other words, what is the probability of a sample proportion $\bar{p} \geq 150/400 = .375$?

60. A production run is not acceptable for shipment to customers if a sample of 100 items contains 5% or more defective items. If a production run has a population proportion defective of $p = .10$, what is the probability that $\bar{p}$ will be at least .05?

61. The proportion of individuals insured by the All-Driver Automobile Insurance Company that have received at least one traffic ticket during a 5-year period is .15.

a. Show the sampling distribution of $\bar{p}$ if a random sample of 150 insured individuals is used to estimate the proportion having received at least one ticket.

b. What is the probability that the sample proportion will be within plus or minus .03 of the population proportion?

62. Historical records show that .50 of all orders placed at Big Burger fast-food restaurants include a soft drink. With a simple random sample of 40 orders, what is the probability that between .45 and .55 of the sampled orders will include a soft drink?

63. Lori Jeffrey is a successful sales representative for a major publisher of college textbooks. Historically, Lori obtains a book adoption on 25% of her sales calls. Viewing her sales calls for 1 month as a sample of all possible sales calls, a statistical analysis of the data yields a standard error of the proportion of .0625.

a. How large was the sample used in this analysis? That is, how many sales calls did Lori make during the month?

b. Let $\bar{p}$ indicate the sample proportion of book adoptions obtained during the month. Show the sampling distribution $\bar{p}$.

c. Using the sampling distribution of $\bar{p}$, compute the probability that Lori will obtain book adoptions on 30% or more of her sales calls during the 1 month period?

64. Comment on why each of the following samples do not constitute a simple random sample. What kind of samples are they?

a. To obtain consumer reaction to a new product, a firm contacts women's groups at several local churches and offers to pay the organization for each person participating in the study.

b. A psychology professor uses a freshman class in Psychology 101 as a sample of subjects for a research project.

c. After a television debate for presidential candidates, viewers are encouraged to phone the television station to indicate the candidate of their preference.

APPENDIX

The Expected Value and Standard Deviation of $\bar{x}$

In this appendix we present the mathematical basis for the expressions for the expected value of $\bar{x}$, $E(\bar{x})$, as given by (7.1) and the standard deviation of $\bar{x}$, $\sigma_{\bar{x}}$, as given by (7.2).

Expected Value of $\bar{x}$

Assume a population with mean μ and variance σ^2. A simple random sample of size n is selected with individual observations denoted $x_1, x_2, \ldots, x_n$. A sample mean $\bar{x}$ is computed as follows:

$$\bar{x} = \frac{\Sigma x_i}{n}$$

With repeated simple random samples of size n, $\bar{x}$ is a random variable that takes on different numerical values depending upon the specific n items selected. The expected value of the random variable $\bar{x}$, or the mean of all possible $\bar{x}$ values, is as follows:

$$\text{Mean of } \bar{x} = E(\bar{x}) = E\left(\frac{\Sigma x_i}{n}\right)$$

$$= \frac{1}{n}[E(x_1 + x_2 + \cdots + x_n)]$$

$$= \frac{1}{n}[E(x_1) + E(x_2) + \cdots + E(x_n)]$$

Since for any x_i we have $E(x_i) = \mu$, we can write

$$E(\bar{x}) = \frac{1}{n}(\mu + \mu + \cdots + \mu)$$

$$= \frac{1}{n}(n\mu) = \mu$$

The above expression shows that the mean of all possible $\bar{x}$ values is the same as the population mean μ. That is, $E(\bar{x}) = \mu$.

Standard Deviation of $\bar{x}$

Again assume a population with mean μ, variance σ^2, and a sample mean given by

$$\bar{x} = \frac{\Sigma x_i}{n}$$

With repeated simple random samples of size n, we know that $\bar{x}$ is a random variable that takes on different numerical values depending upon the specific n items selected. Shown below is the derivation of the expression for the standard deviation of the $\bar{x}$ values, $\sigma_{\bar{x}}$, for the case where the population is infinite. The derivation of the expression for $\sigma_{\bar{x}}$ for a finite population when sampling is done without replacement is more difficult and is beyond the scope of this text.

Returning to the infinite population case, recall that a simple random sample from an infinite population consists of observations $x_1, x_2, \ldots, x_n$ that are independent. The following two expressions are general formulas concerning the variance of random variables:

$$\text{Var}(ax) = a^2 \text{Var}(x) \tag{A.1}$$

where a is a constant and x is a random variable, and

$$\text{Var}(x + y) = \text{Var}(x) + \text{Var}(y) \tag{A.2}$$

where x and y are *independent* random variables. Using (A.1) and (A.2), we can develop the expression for the variance of the random variable $\bar{x}$ as follows:

$$\text{Var}(\bar{x}) = \text{Var}\left(\frac{\Sigma x_i}{n}\right) = \text{Var}\left(\frac{1}{n}\Sigma x_i\right)$$

Using (A.1) with $1/n$ viewed as the constant, we have

$$\text{Var}(\bar{x}) = \left(\frac{1}{n}\right)^2 \text{Var}(\Sigma x_i)$$

$$= \left(\frac{1}{n}\right)^2 \text{Var}(x_1 + x_2 + \cdots + x_n)$$

With the infinite population case, the random variables $x_1, x_2, \ldots, x_n$ are independent. Thus (A.2) enables us to write

$$\text{Var}(\bar{x}) = \left(\frac{1}{n}\right)^2 \left[\text{Var}(x_1) + \text{Var}(x_2) + \cdots + \text{Var}(x_n)\right]$$

Since for any x_i we have $\text{Var}(x_i) = \sigma^2$, we have

$$\text{Var}(\bar{x}) = \left(\frac{1}{n}\right)^2 \underbrace{\left(\sigma^2 + \sigma^2 + \cdots + \sigma^2\right)}_{n \text{ items}}$$

With n values of σ^2 in this expression, we have

$$\text{Var}\,(\bar{x}) = \left(\frac{1}{n}\right)^2 (n\sigma^2) = \frac{\sigma^2}{n}$$

Taking the square root provides the formula for the standard deviation of $\bar{x}$ for the infinite population case:

$$\sigma_{\bar{x}} = \sqrt{\text{Var}\,(\bar{x})} = \frac{\sigma}{\sqrt{n}}$$

This expression, which appeared as (7.2), shows that as the sample size n is increased the standard deviation of the various sample means $\bar{x}$ will decrease. In effect, the larger samples tend to provide better estimates of the population mean.

APPLICATION

Mead Corporation*

Dayton, Ohio

Mead is a diversified paper and forest products company that manufactures paper, pulp, and lumber and converts paperboard into shipping containers and beverage carriers. The company's strong distribution capability is used to market many of its own products, including paper, school supplies, and stationery. Mead's Advanced Systems group develops business applications for the future. These involve storing, retrieving, and reproducing data through the innovative application of digital technology.

DECISION ANALYSIS AT MEAD CORPORATION

Decision Analysis is an internal consulting group located in a larger department known as Decision Support Applications (DSA). The principal thrust of the DSA department is to increase the productivity of Mead's human and computer resources by providing products and training which will enable Mead employees worldwide to use the computer to do their jobs more efficiently and/or effectively. DSA focuses on providing timely and relevant responses that will satisfy staff/line business needs in the following areas:

- Word processing/office automation
- Operations research/statistical analysis
- Financial/ planning/modeling
- Hotline assistance and training in the use of user-friendly computer products
- Identification of end-user software
- Consulting on efficient use of end-user-developed applications

To accomplish these goals, DSA is divided into four departments: Office Systems, Decision Analysis, Information Center, and Financial Modeling Coordination.

The major role of the Decision Analysis department is to provide quantitative support to decision makers throughout the corporation, including both line and staff management in the functional areas of operations, finance, accounting, and human resources. Consulting projects include not only analysis for one-time decision situations, but also the development of support tools for periodic and recurring decision problems. The activities of the department can be divided into 6 major categories: financial analysis, data and statistical analysis, resource allocation modeling, simulation modeling, forecasting, and user-friendly planning models. Decision Analysis is also responsible for monitoring and disseminating to appropriate Mead management new developments in operations research and decision support methodologies. The emphasis in this effort is to identify those developments that would result in significant productivity benefits to the organization.

*The authors are indebted to Dr. Edward P. Winkofsky, Mead Corporation, Dayton, Ohio for providing this application.

▬ AN APPLICATION USING SAMPLING

Mead maintains a continuous forest inventory (CFI) system which provides information concerning the stocking of a large portion of its timberland holdings. The CFI system consists of permanent plots of $\frac{1}{5}$ or $\frac{1}{7}$ acre, which are systematically located throughout the Mead forest. On a periodic basis the trees on certain CFI plots are measured and general site information is gathered on each plot. These data are used to estimate the present volume and past growth of the forest and to project the future growth of the forest.

To identify the plots for measurement, a sampling technique was used. First, the forest was divided into three sections. Then, based upon previous measurements, estimates of the variance of the volume of each section were developed. Using the estimated variance and the accuracy specified, the sample size needed in each section was computed.

Once the sample size was determined, a random sample of plots within each section was selected. In order to determine whether CFI plots had been treated differently from the surrounding forest, for each CFI plot selected, two temporary plots—each located in close proximity to the permanent CFI plot—were located at randomly selected points. Decision Analysis provided tables of computer generated pseudorandom numbers to be used in the selection process. Specifically, these numbers were used to identify which permanent plots would be included in the sample and to define the location of the associated temporary plots.

Foresters throughout the organization participated in the field measurements. They gathered information on each tree of every selected plot, as well as general plot information. These data were entered in the field on computer generated plot reports. These reports provided space to enter the new data and showed the data from the previous measurements for the permanent plots. The data were collected by several two person teams, and an additional two persons acted as an audit team. As the plot information was collected, it was entered into a data base by CRT operators.

Decision Analysis generated a number of frequency distributions from the data base. These reports were used in part to edit the data base but also to estimate the proportions of the forest in each of several species groups. The proportions for each section were first determined and then weighted to provide an estimate for the overall forest. Since certain species groups are more valuable than others, these estimates provided an initial indication of the value of Mead's land holdings. A more accurate measure was then determined through volume estimates.

The volume of each tree was determined using several species-specific formulas

A paper machine known as the "Spirit of Escanaba" is one of the largest and most modern paper machines in the world

common to forestry. Next, the volume estimates were compiled for each section and then weighted to provide the forest estimates. These reports provided management with the necessary information to evaluate the Mead timberlands. Growth estimates are now in the process of being developed from these data. These estimates will be used in a large scale linear program which will develop long term harvest schedules.

▬▬ ADVANTAGES TO MEAD

With sampling techniques, a small number of measurements were used to estimate the value of the total Mead forest ownership. Decision Analysis assisted in the sample selection and played a major role in the development of the programs necessary to compute the required statistical and summary information. In addition, the group has developed a user-friendly front end to the data to allow Corporate and the divisional woodlands to make their own report requests. This system allows the user to select from over ten prespecified reports and plots to be included in the estimates, thus providing a facility to develop estimates for any subset of the total ownership. The information provided by these reports is important input for decisions involving the management of Mead's woodland assets.

CHAPTER 8

Interval Estimation

In Chapter 7 we showed that the value of the sample mean, $\bar{x}$, provides a point estimate of the population mean, μ, and that the value of the sample proportion, $\bar{p}$, provides a point estimate of the population proportion, p. However, in any point-estimation process, we cannot expect the value of a point estimate to be *exactly* equal to the corresponding population parameter. Some degree of error due to sampling is anticipated.

Point estimates of population parameters do not provide information about the *precision*, or magnitude of the sampling error, present in the estimation process. Interval estimates of population parameters have an advantage over point estimates in that interval estimates provide the desired precision information. Often the precision information is essential in evaluating and interpreting the sample results. For example, assume that a sample is used to estimate the mean annual starting salary for recent college graduates with degrees in business administration. Suppose the sample mean for this study is $\bar{x} = \$20,500$. The sample mean of \$20,500 provides the point estimate of the population mean annual starting salary. If the margin of error associated with the estimate is $\pm\$10,000$, the interval estimate of \$10,500 to \$30,500 shows that the point estimate has limited use due to the wide margin of error. On the other hand, if the margin of error is $\pm\$100$, the information is extremely useful in that the population mean annual salary for the graduates is provided by the interval estimate of \$20,400 to \$20,600.

In this chapter, we will make use of the sampling distributions of $\bar{x}$ and $\bar{p}$ presented in Chapter 7 to develop interval estimates of the population mean μ and the population proportion p. Let us introduce the procedure for interval estimation of a population mean by considering a sampling study conducted by Statewide Insurance Company.

8.1 ▪ INTERVAL ESTIMATION OF A POPULATION MEAN — σ KNOWN

The Statewide Insurance Company provides a variety of life, health, disability, and business insurance policies for customers located throughout the United States. As part of an annual review of life insurance policies, a simple random sample of 36 Statewide policyholders is selected. The corresponding 36 life insurance policies are reviewed in terms of the amount of the coverage, the cash value of the policy, the disability options, and so on. For the current policy review study the project manager has requested information on the ages of the life insurance policyholders. Table 8.1 shows the age data collected from the simple random sample of 36 policyholders. Let us use these data to develop an interval estimate of the mean age of the population of life insurance policyholders covered by Statewide.

The estimation procedure that we develop in this section is based on the assumption that the value of the population standard deviation is *known*. For the Statewide study, previous studies on policyholder ages permit us to use a known population standard deviation of $\sigma = 7.2$ years. In Section 8.2 we will show the interval estimation procedure for a population mean when the value of the population standard deviation is unknown.

Let x_1 indicate the age of the first policyholder in the sample, x_2 the age of the second policyholder, and so on. The sample mean, $\bar{x}$, provides a point estimate of the population mean μ. Using the data in Table 8.1, we obtain

$$\bar{x} = \frac{\Sigma x_i}{n} = \frac{1422}{36} = 39.5$$

Thus the point estimate of the mean age of the population of Statewide life insurance policyholders is 39.5 years.

In discussing the practical value of the sampling distribution of $\bar{x}$ in Chapter 7, we indicated that we cannot expect the value of a sample mean $\bar{x}$ to equal *exactly*

TABLE 8.1 **Ages of Life Insurance Policyholders from a Simple Random Sample of 36 Statewide Policyholders**

Policyholder	Age	Policyholder	Age	Policyholder	Age
1	32	13	39	25	23
2	50	14	46	26	36
3	40	15	45	27	42
4	24	16	39	28	34
5	33	17	38	29	39
6	44	18	45	30	34
7	45	19	27	31	35
8	48	20	43	32	42
9	44	21	54	33	53
10	47	22	36	34	28
11	31	23	34	35	49
12	36	24	48	36	39

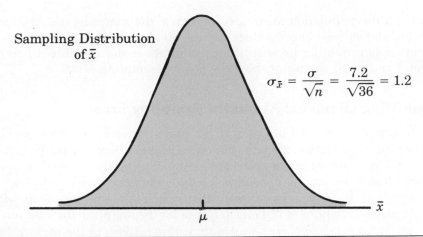

Sampling Distribution
of $\bar{x}$

$$\sigma_{\bar{x}} = \frac{\sigma}{\sqrt{n}} = \frac{7.2}{\sqrt{36}} = 1.2$$

FIGURE 8.1 Sampling Distribution of the Sample Mean Age ($\bar{x}$) from Simple Random Samples of 36 Statewide Insurance Company Policyholders

the value of the population mean μ. Thus anytime a sample mean is used to provide a point estimate of a population mean, someone may ask, "How good is the estimate?" The "how good" question is a way of asking about the error involved when the value of $\bar{x}$ is used as a point estimate of the population mean μ. In general, we shall refer to the magnitude of the difference between an unbiased point estimator and the population parameter as the *sampling error*. For the case of a sample mean estimating a population mean, the sampling error would be expressed as follows:

$$\text{Sampling error} = |\bar{x} - \mu| \tag{8.1}$$

Note, however, that even after a sample is selected and the sample mean is computed ($\bar{x} = 39.5$ in the Statewide example), we will not be able to use (8.1) to find the value of the sampling error because the population mean μ is unknown. However, we shall see that the sampling distribution of $\bar{x}$ developed in Chapter 7 can be used to make probability statements about the sampling error.

From the central limit theorem we know that whenever the sample size is large ($n \geq 30$) the sampling distribution of $\bar{x}$ can be approximated by a normal probability distribution with a mean μ and a standard deviation* $\sigma_{\bar{x}} = \sigma/\sqrt{n}$. For the Statewide Insurance study, with $\sigma = 7.2$ years and $n = 36$, this theorem enables us to conclude that the sampling distribution of $\bar{x}$ is approximately normal with a mean μ and a standard deviation $\sigma_{\bar{x}} = 7.2/\sqrt{36} = 1.2$ years. This sampling distribution is shown in Figure 8.1.

---

*An unbiased estimate of the standard error of the proportion is given by $\sqrt{\bar{p}(1 - \bar{p})/(n - 1)}$. The bias introduced by using n in the denominator does not cause any difficulty because large samples are used in making estimates concerning population proportions and the numerical difference between the results using n and $n - 1$ is negligible.

Although the population mean μ is unknown, the sampling distribution in Figure 8.1 shows how the $\bar{x}$ values are distributed around μ. In effect, this distribution is providing us with information about the possible differences between $\bar{x}$ and μ and as a result about the possible sampling error.

Probability Statements About the Sampling Error

Since the sampling distribution of $\bar{x}$ can be approximated by a normal probability distribution, we can use the table of normal probabilities to make probability statements about the sampling error. For example, using the table of areas for a standard normal probability distribution (inside the back cover of the text), we find that 95% of the values of a normally distributed random variable lie within ±1.96 standard deviations of the mean. Hence for the sampling distribution of $\bar{x}$ shown in Figure 8.1, 95% of all $\bar{x}$ values are within ±1.96$\sigma_{\bar{x}}$ of the mean μ. Since $1.96\sigma_{\bar{x}} = 1.96(1.2) = 2.35$, we can state that 95% of all sample means lie within plus or minus 2.35 years of the population mean μ. The location of all the sample means that provide a sampling error of 2.35 years or less is shown in Figure 8.2. It is possible for a sample mean to fall in one of the two tails of the sampling distribution, which would result in a sampling error greater than 2.35 years. However, we see from Figure 8.2 that the probability of this occurring is only $1 - .95 = .05$. Thus our knowledge of the sampling distribution of $\bar{x}$ enables us to make the following probability statement about the sampling error whenever a simple random sample of 36 Statewide policyholders is used to provide a point estimate of the mean age of the population:

There is a .95 probability that the sample mean will provide a sampling error of 2.35 years or less.

■■■■ FIGURE 8.2 **Sampling Distribution of $\bar{x}$ Showing the Location of Sample Means that Provide a Sampling Error of 2.35 Years or Less**

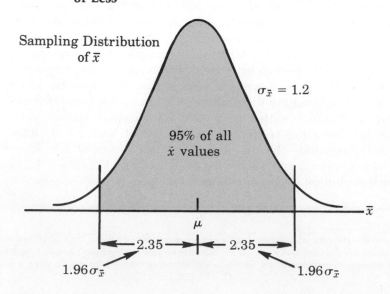

Sampling Distribution of $\bar{x}$

$\sigma_{\bar{x}} = 1.2$

95% of all $\bar{x}$ values

μ

$\bar{x}$

$\longleftarrow$ 2.35 $\longrightarrow$ $\longleftarrow$ 2.35 $\longrightarrow$

$1.96\sigma_{\bar{x}}$ $1.96\sigma_{\bar{x}}$

The above probability statement about the sampling error is a statement of the *precision* of the estimate. If the project manager is not satisfied with this degree of precision, a larger sample size will be necessary. We shall discuss a procedure for determining the sample size necessary to obtain a desired precision in Section 8.3.

Note that in the above analysis the .95 probability used in the statement about the sampling error was arbitrary. Although a .95 probability is frequently used in making such statements, other probability values can be selected. Probabilities of .90 and .99 are popular alternatives. Let us consider what would have happened to the precision statement if a probability of .99 had been selected. Figure 8.3 shows the location of 99% of the sample means for the Statewide Insurance sampling problem. From the standard normal probability distribution table, we find that 99% of the $\bar{x}$ values lie within ±2.575 standard deviations of the mean μ. Since $2.575\sigma_{\bar{x}} = 2.575(1.2) = 3.09$, we can make the following statement about the sampling error whenever a simple random sample of 36 Statewide policyholders is used to provide a point estimate of the mean age of the population:

There is a .99 probability that the sample mean will provide a sampling error of 3.09 years or less.

A similar calculation with a .90 probability shows that there is a .90 probability that the sample mean will provide a sampling error of $1.645\sigma_{\bar{x}} = 1.645(1.2) = 1.97$ years or less.

These results show that there are various probability statements that can be made about the sampling error. They also show that there is a trade-off between the probability specified and the stated limit on the sampling error. In particu-

---

■■■■ FIGURE 8.3 **Sampling Distribution of $\bar{x}$ Showing the Location of 99% of the $\bar{x}$ Values**

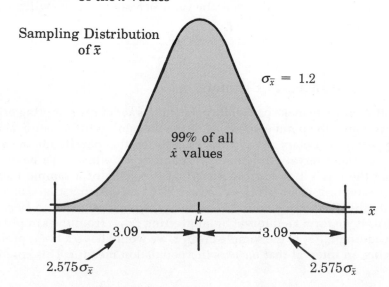

lar, note that the higher probability statements possess larger values for the sampling error.

Let us now generalize the procedure we are using to make probability statements about the sampling error whenever a sample mean is used to provide a point estimate of a population mean. We will use the Greek letter α ("alpha") to indicate the probability that a sampling error is *larger* than the sampling error mentioned in the precision statement. Refer to Figure 8.4. We see that $\alpha/2$ will be the area or probability in each tail of the distribution, and $1 - \alpha$ will be the area or probability that a sample mean will provide a sampling error *less than or equal to* the sampling error contained in the precision statement.

Refer to the Statewide Insurance example. The statement that there is a .95 probability that the value of a sample mean will provide a sampling error of 2.35 years or less is based on $\alpha = .05$ and $1 - \alpha = .95$. The area in each tail of the sampling distribution is $\alpha/2 = .025$ (see Figure 8.2).

Using the z notation for the standard normal random variable, we will place a subscript on the z value to denote the *area in the upper tail* of the probability distribution. Thus $z_{.025}$ will correspond to the z value with .025 of the area in the upper tail of the probability distribution. As can be found in the standard normal probability distribution table, $z_{.025} = 1.96$. If we desired a .99 probability statement, $\alpha = .01$, we would be interested in an area of $\alpha/2 = .005$ in the upper tail of the distribution and hence, $z_{.005} = 2.575$.

Let $z_{\alpha/2}$ denote the value for the standard normal random variable corresponding to an area of $\alpha/2$ in the upper tail of the distribution. Also let $\sigma_{\bar{x}}$ denote the standard deviation of the sampling distribution of $\bar{x}$ (also called the standard error of the mean). We now have the following general procedure for making a probability statement about the sampling error whenever $\bar{x}$ is used to estimate μ:

Probability Statement About the Sampling Error

There is a $1 - \alpha$ probability that the value of a sample mean will provide a sampling error of $z_{\alpha/2}\sigma_{\bar{x}}$ or less.

Calculating an Interval Estimate

We have the ability to make probability statements about the sampling error. We now can combine the point estimate with the probability information about the sampling error to obtain an *interval estimate* of the population mean. The rationale for the interval estimation procedure is as follows: We have already stated that there is a $1 - \alpha$ probability that the value of a sample mean will provide a sampling error of $z_{\alpha/2}\sigma_{\bar{x}}$ or less. This means that there is a $1 - \alpha$ probability that the sample mean *will not miss* the population mean *by more than* $z_{\alpha/2}\sigma_{\bar{x}}$. Thus if we form an interval by subtracting $z_{\alpha/2}\sigma_{\bar{x}}$ from the sample mean $\bar{x}$ and then adding $z_{\alpha/2}\sigma_{\bar{x}}$ to the sample mean $\bar{x}$, we would have a $1 - \alpha$ probability of obtaining an interval that *includes* the population mean μ. This condition is

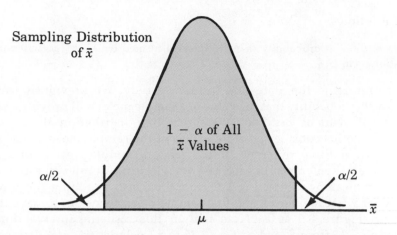

FIGURE 8.4 **Areas of a Sampling Distribution of $\bar{x}$ Used to Make Probability Statements About the Sampling Error**

FIGURE 8.5 **Intervals Formed from Selected Sample Means at Locations $\bar{x}_1$, $\bar{x}_2$, and $\bar{x}_3$**

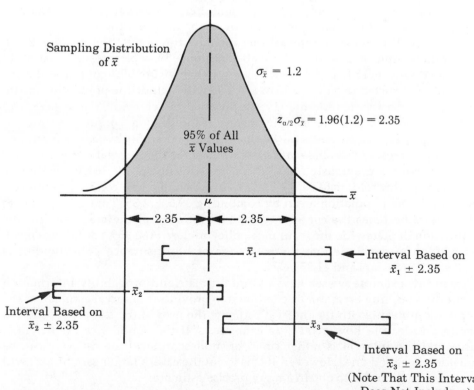

stated as follows:

There is a $1 - \alpha$ probability that the interval formed by $\bar{x} \pm z_{\alpha/2}\sigma_{\bar{x}}$ will contain the population mean μ.

Let us return to the Statewide Insurance study. We previously stated that there is a .95 probability that the value of a sample mean will provide a sampling error of 2.35 years or less. Look at the sampling distribution of $\bar{x}$ as shown in Figure 8.5. Let us consider possible values of the sample mean $\bar{x}$ that could be obtained from three different simple random samples, each containing 36 policyholders. Remember, in each case we will form an interval estimate of the population mean by subtracting 2.35 from $\bar{x}$ and also by adding 2.35 to $\bar{x}$.

Consider what happens if the first sample mean turns out to have the value shown in Figure 8.5 as $\bar{x}_1$. Note that in this case the interval formed by subtracting 2.35 from $\bar{x}_1$ and adding 2.35 to $\bar{x}_1$ includes the population mean μ. Now consider what happens if the sample mean turns out to have the value shown in Figure 8.5 as $\bar{x}_2$. While this next sample mean is different from the first sample mean, we see that the interval based on $\bar{x}_2$ also includes the population mean μ. However, the interval based on the third sample mean, denoted by $\bar{x}_3$, does not include the population mean. The reason for this is that the sample mean $\bar{x}_3$ lies in a tail of the probability distribution at a distance further than 2.35 from μ. Thus subtracting and adding 2.35 to $\bar{x}_3$ forms an interval that does not include μ.

Now think of repeating the sampling process many times, each time computing the value of the sample mean and then forming an interval from $\bar{x} -$ 2.35 to $\bar{x} + 2.35$. Any sample mean ($\bar{x}$) that falls between the vertical lines in Figure 8.5 will provide an interval that includes the population mean μ. Since 95% of the sample means are in the shaded region, 95% of all intervals that could be formed will include μ. As a result, we say that we are 95% confident that an interval constructed from $\bar{x} -$ 2.35 to $\bar{x} +$ 2.35 will include the population mean. In common statistical terminology the interval is referred to as a *confidence interval*. With 95% of the sample means leading to a confidence interval including μ, we say the interval is established at the 95% *confidence level*. The value .95 is referred to as the *confidence coefficient* for the interval estimate.

Recall that we previously found the sample mean age for 36 Statewide Life Insurance policyholders to be $\bar{x} = 39.5$. We can obtain a 95% confidence interval estimate of the population mean by computing 39.5 ± 2.35. Thus 37.15 years to 41.85 years becomes the confidence interval estimate of the mean age for the population of Statewide life insurance policyholders. At a 95% confidence level, Statewide can conclude that the mean age of life insurance policyholders is between 37.15 years and 41.85 years.

From this example we see how a point estimate and probability information about the sampling error can be combined to provide an interval estimate of the population mean. With the interval estimate the user of the sample results can obtain an idea of "how good" the estimate is. If the confidence coefficient is large and the interval is rather small or modest in size, the estimate can be considered a good one. However, if this is not the case, a larger sample size will be necessary in order to obtain a more precise estimate.

Let us now state the general procedure for computing an interval estimate of a population mean. As previously noted, there is a $1 - \alpha$ probability that the interval formed by $\bar{x} \pm z_{\alpha/2}\sigma_{\bar{x}}$ includes the population mean μ. Using the fact that $\sigma_{\bar{x}} = \sigma/\sqrt{n}$, the general procedure for the *interval estimate* of population mean can be written as follows:

Interval Estimate of a Population Mean

$$\bar{x} \pm z_{\alpha/2} \frac{\sigma}{\sqrt{n}} \qquad (8.2)$$

where $1 - \alpha$ is the confidence coefficient and $z_{\alpha/2}$ is the z value providing an area of $\alpha/2$ in the upper tail of the standard normal probability distribution.

An Example: Estimating Television-Viewing Time

Nielsen Media Research used a sample to estimate that the mean time a person spends watching television per day is 6 hours 59 minutes (*USA Today*, December 13, 1988). Assume that this estimate was based on a sample of 600 television viewers and that the population standard deviation of viewing time is $\sigma = 120$ minutes. Develop a 90% confidence interval estimate for the mean viewing time per day for the population.

At 90% confidence, the confidence coefficient is $1 - \alpha = .90$. Thus, $\alpha = .10$ and $\alpha/2 = .10/2 = .05$. Using the standard normal probability distribution table we find the z value corresponding to an area or probability of .05 in the upper tail of the standard normal curve is 1.645. Thus, $z_{.05} = 1.645$. With the sample mean $\bar{x}$ given to be 6 hours 59 minutes, the population standard deviation $\sigma = 120$, and the sample size $n = 600$, equation (8.2) can be used to compute the interval estimate of the population mean as follows:

$$6 \text{ hours } 59 \text{ minutes} \pm 1.645 \frac{120}{\sqrt{600}}$$

or

$$6 \text{ hours } 59 \text{ minutes} \pm 8 \text{ minutes}$$

Thus, the 90% confidence interval for the population mean viewing time per day is 6 hours 51 minutes to 7 hours 7 minutes. Given these data, we can be 90% confident that the population mean viewing time per day is contained within this interval.

Using a Small Sample Size ($n < 30$)

The central limit theorem plays an important role in the development of the interval estimation procedure shown in (8.2). Specifically, with a large sample size ($n \geq 30$) the central limit theorem allowed us to conclude that the sampling distribution of $\bar{x}$ can be approximated by a normal probability distribution. This nomal distribution approximation is what makes the use of $z_{\alpha/2}$ appropriate in (8.2). However, what happens to the interval estimation procedure when the sample size is small ($n < 30$) and we cannot use the central limit theorem to justify the use of a normal sampling distribution? Recall that in Chapter 7 we pointed out that *if the population has a normal distribution*, the sampling distribution of $\bar{x}$ is normal regardless of the sample size. Thus if we are faced with a small sample size situation ($n < 30$), we consider the possibility of the population having a normal distribution. If this appears to be a reasonable assumption, the sampling distribution of $\bar{x}$ will be normal regardless of the sample size. Thus with σ known, (8.2) can be used to compute an interval estimate of a population mean for the small sample size case. However, if we are unwilling to make the assumption of a normal population, the only alternative is to increase the sample size to $n \geq 30$ and rely on the central limit theorem as the basis for the interval estimation given by (8.2).

Finally, note that whether the sample size is large or small, the interval estimation procedure we have been using in this section is based on the assumption that the population standard deviation σ is known. The procedure for computing an interval estimate of a population mean when σ is unknown is the topic of the next section.

▬ NOTES AND COMMENTS ▬

1. In using (8.2) to develop an interval estimate of the population mean, we specify the desired confidence coefficient ($1 - \alpha$) before selecting the sample. Thus, prior to selecting the sample, we conclude that there is a $1 - \alpha$ probability that the confidence interval we eventually compute will contain the population mean μ. However, once the sample is taken, the sample mean $\bar{x}$ is computed, and the particular interval estimate is determined, the resulting interval *may or may not* contain μ. However, if $1 - \alpha$ is reasonably large, we can be confident that the resulting interval contains μ because we know that if we use this procedure in the long run, $100(1 - \alpha)$ percent of all possible intervals developed in this manner will contain μ.

2. Note that the sample size n appears in the denominator of the interval estimation expression (8.2). Thus if a particular sample size provides too wide an interval to be of any practical use, we may want to consider increasing the sample size. With n in the denominator, a larger sample size will reduce the margin of error, resulting in a narrower interval and a greater precision. The procedure for determining the size of a simple random sample required to obtain a desired precision is discussed in Section 8.3.

■ EXERCISES

1. In an effort to estimate the mean amount spent per customer for dinner meals at a major Atlanta restaurant, data were collected for a sample of 49 customers over a 3-week period.

   a. Assume a population standard deviation of $2.50. What is the standard error of the mean?
   b. With a .95 probability, what statement can be made about the sampling error?
   c. If the sample mean is $12.60, what is the 95% confidence interval estimate of the population mean?

2. The mean annual income of U.S. factory workers is $24,000 (*Barron's*, April 10, 1989). Assume that this estimate was based on a sample of 250 U.S. factory workers and that the population standard deviation was $\sigma = \$5,000$.

   a. Compute the 90% confidence interval for the population mean.
   b. Compute the 95% confidence interval for the population mean.
   c. Compute the 99% confidence interval for the population mean.
   d. Discuss what happens to the width of the interval estimate as the confidence level is increased. Why does this seem reasonable?

3. A production filling operation has a historical standard deviation of 5.5 ounces. A quality control inspector periodically selects 36 containers at random and uses the sample mean filling weight to estimate the population mean filling weight for the production process.

   a. What is the standard error of the mean, $\sigma_{\bar{x}}$?
   b. With .75, .90, and .99 probabilities, what statements can be made about the sampling error? What happens to the statement about the sampling error when the probability is increased? Why does this happen?
   c. What is the 99% confidence interval estimate for the population mean filling weight for the process if a sample mean is 48.6 ounces?

4. A survey of readers of *Money* magazine found that the sample mean age of men was 47 years and the sample mean age of women was 44 years (*Money Extra*, Fall 1988). All together, 454 people were included in the reader poll—340 men and 114 women. Assume that the population standard deviation of age for both men and women is 8 years.

   a. Develop a 95% confidence interval estimate for the mean age of the population of men who read *Money* magazine.
   b. Develop a 95% confidence interval estimate for the mean age of the population of women who read *Money* magazine.
   c. Compare the widths of the two interval estimates from parts (a) and (b). Did the estimate of the mean age of men or the mean age of women have the better precision? Why?

5. E. Lynn and Associates is an energy research firm that provides estimates of monthly heating costs for new homes based on style of house, square footage, insulation, and so on. The firm's service is used by both builders and potential buyers of new homes who wish advance information on heating costs. For winter months the standard deviation in the home heating bills for residential homes in a certain area is $100. Assume that a sample of 36 homes in a particular subdivision will be used to estimate the mean monthly heating bill for the population all homes in this type of subdivision.

   a. What is the standard error of the mean, $\sigma_{\bar{x}}$?
   b. Show the sampling distribution for the sample mean heating bill.

c. At an 80% probability, what can be said about the sampling error? Show this probability on the graph of the sampling distribution in part (b).

d. What is the 98% confidence interval for the population mean monthly heating bill if the sample mean is $196.50?

6. *Consumer Research*, April 1989, reports information on the time required for caffeine from products such as coffee and soft drinks to leave the body after consumption. Assume that the 95% confidence interval estimate of the population mean time for adults is 5.6 hours to 6.4 hours.

a. What is the point estimate of the mean time for caffeine to leave the body of adults after consumption?

b. If the population standard deviation is 2 hours, how large a sample was used to provide the interval estimate?

8.2 ■ INTERVAL ESTIMATION OF A POPULATION MEAN — σ UNKNOWN

In the previous section we showed that an interval estimate of a population mean μ is given by

$$\bar{x} \pm z_{\alpha/2} \frac{\sigma}{\sqrt{n}} \tag{8.3}$$

A difficulty in using this expression is that in many sampling situations the value of the population standard deviation σ is *unknown*. In these instances we simply use the value of the sample standard deviation s as the point estimate of the population standard deviation σ. The estimator of the standard deviation of $\bar{x}$ can then be computed as follows:

$$\text{Estimator of } \sigma_{\bar{x}} = s_{\bar{x}} = \frac{s}{\sqrt{n}} \tag{8.4}$$

The $s_{\bar{x}}$ notation is used for the estimator of $\sigma_{\bar{x}}$ just as s is used as the estimator of σ. At this point the interval estimation procedure depends upon whether the sample size is large or small. Let us first consider the interval estimation of a population mean when the sample size is large.

Large-Sample Case

As a working rule, we shall consider the large-sample-size case appropriate whenever the size of the simple random sample is 30 *or more*. Whenever the sample size is $n \geq 30$ and the sample standard deviation s is used as the estimator of the population standard deviation σ, the interval estimate of a population mean is as follows:

Interval Estimate of a Population Mean (Large-Sample Case with σ Unknown)

$$\bar{x} \pm z_{\alpha/2} \frac{s}{\sqrt{n}} \tag{8.5}$$

where $1 - \alpha$ is the confidence coefficient, $z_{\alpha/2}$ is the z value providing an area of $\alpha/2$ in the upper tail of the standard normal probability distribution, and s is the sample standard deviation.

Thus with a large sample size the procedure for developing an interval estimate of a population mean with σ unknown follows the interval estimation procedure of Section 8.1. The only difference is that the sample standard deviation s is used to estimate the population standard deviation σ.

For example, the *1988 Information Please Almanac* lists more than 1700 accredited senior colleges and universities in the United States. A simple random sample of 50 colleges and universities from this population was used to estimate the mean annual room and board expenses associated with attending college. The sample of 50 provided a sample mean of $\bar{x} = \$2999$ per year and a sample standard deviation of $s = \$678$. Since the sample size is large and the population standard deviation σ is unknown, the 95% confidence interval estimate of the population mean annual cost can be found using (8.5). At 95% confidence, $z_{\alpha/2} = z_{.025} = 1.96$. Thus we have

$$\bar{x} \pm z_{.025} \frac{s}{\sqrt{n}}$$

$$2999 \pm 1.96 \frac{678}{\sqrt{50}}$$

$$2999 \pm 188$$

As a result, we can be 95% confident that the interval \$2811 to \$3187 contains the mean annual room and board cost for the population of accredited senior colleges and universities.

Small-Sample Case

If the sample size is small ($n < 30$), we can develop an interval estimate of a population mean *only if* the population has a normal probability distribution. If the sample standard deviation s is used as an estimator of the population standard deviation σ and if the population has a normal distribution, interval estimation of the population mean can be based upon a probability distribution known as the *t distribution*.

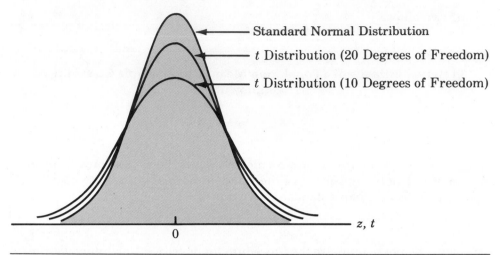

■ FIGURE 8.6   **Comparison of the Standard Normal Distribution with**
t-Distributions Having 10 and 20 Degrees of Freedom

The t distribution is actually a family of similar probability distributions, with a specific t distribution depending upon a parameter known as the *degrees of freedom*. That is, there is a unique t distribution with 1 degree of freedom, with 2 degrees of freedom, with 3 degrees of freedom, and so on. As the number of degrees of freedom increases, the difference between the t distribution and the standard normal distribution becomes smaller and smaller. Figure 8.6 shows t distributions with 10 and 20 degrees of freedom and their relationship to the standard normal probability distribution. Note that a t distribution with more degrees of freedom has less dispersion and more closely resembles the standard normal distribution.

We will use a subscript for t to indicate the area in the upper tail of the t

■ FIGURE 8.7   ***t*-Distribution with $\alpha/2$ Area or Probability in the Upper Tail**

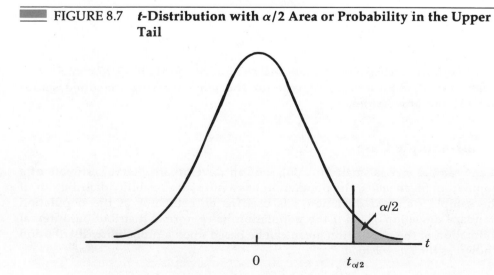

distribution. For example, just as we used $z_{.025}$ to indicate the z value providing a .025 area in the upper tail of a standard normal probability distribution, we will use $t_{.025}$ to indicate a .025 area in the upper tail of the t distribution. In general, we will use the notation $t_{\alpha/2}$ to represent a t value with an area of $\alpha/2$ in the upper tail of the t distribution. See Figure 8.7.

A table for the t distribution is provided in Table 8.2. This table is also shown inside the back cover of the text. Note, for example, that for a t distribution with 10 degrees of freedom, $t_{.025} = 2.228$. Similarly, for a t distribution with 20 degrees of freedom, $t_{.025} = 2.086$. As the degrees of freedom continue to increase, $t_{.025}$ approaches $z_{.025}$.

Now that we have an idea of what the t distribution is, let us show how it is used to develop an interval estimate of a population mean. Assume that the population has a normal probability distribution and that the sample standard deviation s is used as a point estimate of the population standard deviation σ. The following interval estimation procedure is applicable:

Interval Estimate of a Population Mean (Small-Sample Case with σ Unknown)

$$\bar{x} \pm t_{\alpha/2} \frac{s}{\sqrt{n}} \qquad (8.6)$$

where $1 - \alpha$ is the confidence coefficient, $t_{\alpha/2}$ is the t value providing an area of $\alpha/2$ in the upper tail of a t distribution with $n - 1$ *degrees of freedom*, and s is the sample standard deviation.

The reason the number of degrees of freedom associated with the t value in (8.6) is $n - 1$ has to do with the use of s as an estimate of the population standard deviation σ. The expression for the sample standard deviation is

$$s = \sqrt{\frac{\Sigma(x_i - \bar{x})^2}{n - 1}}$$

Degrees of freedom here refers to the number of independent pieces of information that go into the computation of $\Sigma(x_i - \bar{x})^2$. The pieces of information involved in computing $\Sigma(x_i - \bar{x})^2$ are $x_1 - \bar{x}, x_2 - \bar{x}, \ldots, x_n - \bar{x}$. In Section 3.2 we indicated that $\Sigma(x_i - \bar{x}) = 0$ for any data set. Thus only $n - 1$ of the $x_i - \bar{x}$ values are independent; if we know $n - 1$ of the values, the remaining value can be determined exactly using the condition that all of the $x_i - \bar{x}$ values must sum to 0. Thus $n - 1$ is the number of degrees of freedom for the t distribution used in (8.6).

Let us demonstrate the use of the above interval estimate by considering the training program evaluation conducted by Scheer Industries. Scheer's director of manufacturing is interested in a computer assisted training program that can be used to train the firm's maintenance employees for machine repair operations. It is anticipated that the computer assisted training will reduce training

TABLE 8.2 *t* Distribution Table for Areas in the Upper Tail.
Example: with 10 Degrees of Freedom $t_{.025} = 2.228$

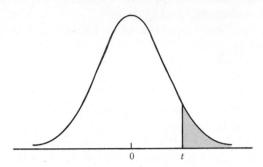

Degrees of Freedom	Upper-Tail Area (Shaded)				
	.10	.05	.025	.01	.005
1	3.078	6.314	12.706	31.821	63.657
2	1.886	2.920	4.303	6.965	9.925
3	1.638	2.353	3.182	4.541	5.841
4	1.533	2.132	2.776	3.747	4.604
5	1.476	2.015	2.571	3.365	4.032
6	1.440	1.943	2.447	3.143	3.707
7	1.415	1.895	2.365	2.998	3.499
8	1.397	1.860	2.306	2.896	3.355
9	1.383	1.833	2.262	2.821	3.250
10	1.372	1.812	**2.228**	2.764	3.169
11	1.363	1.796	2.201	2.718	3.106
12	1.356	1.782	2.179	2.681	3.055
13	1.350	1.771	2.160	2.650	3.012
14	1.345	1.761	2.145	2.624	2.977
15	1.341	1.753	2.131	2.602	2.947
16	1.337	1.746	2.120	2.583	2.921
17	1.333	1.740	2.110	2.567	2.898
18	1.330	1.734	2.101	2.552	2.878
19	1.328	1.729	2.093	2.539	2.861
20	1.325	1.725	2.086	2.528	2.845
21	1.323	1.721	2.080	2.518	2.831
22	1.321	1.717	2.074	2.508	2.819
23	1.319	1.714	2.069	2.500	2.807
24	1.318	1.711	2.064	2.492	2.797
25	1.316	1.708	2.060	2.485	2.787
26	1.315	1.706	2.056	2.479	2.779
27	1.314	1.703	2.052	2.473	2.771
28	1.313	1.701	2.048	2.467	2.763
29	1.311	1.699	2.045	2.462	2.756
30	1.310	1.697	2.042	2.457	2.750
40	1.303	1.684	2.021	2.423	2.704
60	1.296	1.671	2.000	2.390	2.660
120	1.289	1.658	1.980	2.358	2.617
∞	1.282	1.645	1.960	2.326	2.576

time and training costs for Scheer employees. In order to evaluate the training method, the director of manufacturing has requested an estimate of the mean training time required using the computer assisted training technique.

Suppose that management has agreed to train 15 employees with the new approach. The data on training days required for each employee in the sample are shown in Table 8.3. The sample mean and sample standard deviation for these data are as follows:

$$\bar{x} = \frac{\Sigma x_i}{n} = \frac{808}{15} = 53.87 \text{ days}$$

$$s = \sqrt{\frac{\Sigma(x_i - \bar{x})^2}{n - 1}} = \sqrt{\frac{651.73}{14}} = 6.82 \text{ days}$$

The point estimate of the mean training time for the population of employees is 53.87 days. We can obtain information about the precision of this estimate by developing an interval estimate of the population mean. Since the population standard deviation is unknown, we will use the sample standard deviation $s = 6.82$ days as the point estimate of σ. With the small sample size, $n = 15$, we will use (8.6) to develop the interval estimate. Assuming that the population of training times has a normal probability distribution, the t distribution with $n - 1 = 14$ degrees of freedom is the appropriate probability distribution for the interval-estimation procedure. Selecting a .95 confidence coefficient, we see from Table 8.2 that with 14 degrees of freedom, $t_{\alpha/2} = t_{.025} = 2.145$. Using (8.6), we have

$$\bar{x} \pm t_{.025} \frac{s}{\sqrt{n}}$$

$$53.87 \pm (2.145) \frac{6.82}{\sqrt{15}}$$

$$53.87 \pm 3.78$$

Thus the 95% confidence interval estimate of the population mean training time is 50.09 days to 57.65 days.

Strictly speaking, the above approach for interval estimation based on the use of the t distribution is applicable whenever the population standard deviation is

TABLE 8.3 Machine Repair Training Time in Days for the Computer-Assisted Instruction Program

Employee	Time	Employee	Time	Employee	Time
1	52	6	59	11	54
2	44	7	50	12	58
3	55	8	54	13	60
4	44	9	62	14	62
5	45	10	46	15	63

unknown and the population being sampled has a normal probability distribution. However, statistical research has shown that (8.6) is applicable even if the population being sampled is not quite normal. That is, confidence intervals based on the t distribution provide good results provided that the population distribution does not differ extensively from a normal probability distribution. The fact that the t distribution can be used to provide satisfactory results for many possible population distributions is referred to as the *robustness* property of the t distribution.

NOTES AND COMMENTS

We would like to point out that the t distribution is not restricted to the small-sample situation. Actually the t distribution is applicable whenever the population is normal or near normal and whenever the sample standard deviation is used to estimate the population standard deviation. If these conditions exist, the t distribution can be used for any sample size. However, (8.5) shows that with a large sample ($n \geq 30$), interval estimation of a population mean can be based on the standard normal distribution and the value $z_{\alpha/2}$. Thus with (8.5) available for the large-sample-size case, we generally do not consider the use of the t distribution until we encounter a small-sample-size case.

■■■ EXERCISES

7. During a water shortage a water company randomly sampled residential water meters in order to monitor daily water consumption. On one particular day, a sample of 50 meters showed a sample mean of $\bar{x} = 240$ gallons and a sample standard deviation of $s = 45$ gallons. Provide a 90% confidence interval estimate of the mean water consumption for the population.

8. Researchers at the University of Illinois used a sample of 70 students from five high schools in a large metropolitan area to learn about performance of students on the American College Test (ACT) (*Journal of College Student Development*, February 1989). Of the students in the sample, 47 went on to college and 23 did not. ACT scores for the sample of students who went on to college had a mean $\bar{x} = 25.23$ and a standard deviation $s = 3.21$.

    a. Provide a 95% confidence interval estimate of the population mean ACT score for students going on to college.
    b. Provide a 98% confidence interval estimate of the population mean ACT score for students going on to college.

9. Data were collected on the golf-ball driving distances by professional golfers in a recent tournament. Using data on the first drives of 30 randomly selected golfers, it was found that the sample mean distance was 250 yards and the sample standard deviation was 10 yards. Develop a 95% confidence interval for the mean driving distance for the population.

10. For a t distribution with 12 degrees of freedom, find the area, or probability, that lies in each region.

    a. to the left of 1.782
    b. to the right of -1.356

c. to the right of 2.681
d. to the left of -1.782
e. between -2.179 and $+2.179$
f. between -1.356 and $+1.782$

11. Find the t value for each of the following.

a. upper tail area of .05 with 18 degrees of freedom
b lower tail area of .10 with 22 degrees of freedom
c. upper tail area of .01 with 5 degrees of freedom
d. 90% of the area is between these two t values with 14 degrees of freedom
e. 95% of the area is between these two t values with 28 degrees of freedom

12. A sample of 12 cab fares in New York City shows a sample mean of $\bar{x}$ = $8.50 and a sample standard deviation of s = $2.40. Develop a 90% confidence interval estimate of the mean cab fares in New York City. Assume the population of cab fares has a normal distribution.

13. In the testing of a new production method, 18 employees were randomly selected and asked to try the new method. The sample mean production rate for the 18 employees was 80 parts per hour. The sample standard deviation was 10 parts per hour. Provide 90% and 95% confidence interval estimates for the mean production rate for the new method assuming the population has a normal distribution.

14. The Money & Investing section of the *Wall Street Journal* contains a summary of the daily investment performances for the New York Stock Exchange, the American Stock Exchange, overseas markets, options, commodities, futures and so on. In the New York Stock Exchange section, information is provided on each stock's 52-week high price per share, 52-week low price per share, dividend rate, yield, P/E ratio, daily volume, daily high price per share, daily low price per share, closing price per share, and daily net change. The P/E ratio for each stock is determined by dividing the price of a share of stock by the earnings per share reported by the company for the most recent four quarters. A sample of ten stocks taken from the *Wall Street Journal*, May 19, 1989, provided the following data on P/E ratios:

5   7   9   10   14   23   20   15   3   26

a. What is the point estimate of the mean P/E ratio for the population of all stocks listed on the New York Exchange?
b. What is the point estimate of the standard deviation of the P/E ratios for the population of all stocks listed on the New York Stock Exchange?
c. Using 95% confidence, what is the interval estimate of the mean P/E ratio for the population of all stocks listed on the New York Stock Exchange?
d. Comment on the precision of the results.

15. The following data are family sizes from a simple random sample of households in a new test market area:

Household	Family Size	Household	Family Size
1	4	7	3
2	3	8	2
3	2	9	3
4	2	10	6
5	4	11	3
6	5	12	2

Provide a 95% confidence interval estimate for the mean family size for the population.

16. Sales personnel for Skillings Distributors are required to submit weekly reports listing the customer contacts made during the week. A sample of 61 weekly contact reports showed a mean of 22.4 customer contacts per week for the sales personnel. The sample standard deviation was 5 contacts.

a. Develop a 95% confidence interval estimate for the mean number of weekly customer contacts for the population of sales personnel.

b. Assume that the population of weekly contact data has a normal distribution. The *t* distribution can also be used to develop an interval estimate of the population mean. Use the *t* distribution with 60 degrees of freedom to develop a 95% confidence interval for the mean number of weekly customer contacts.

c. Compare your answers for parts (a) and (b). Comment on why in the large-sample case it is permissible to base interval estimates on the procedure used in part (a) even though the *t* distribution may also be applicable.

17. Researchers at the University of Georgia used a sample of college students to study psychological issues associated with attending college (*Journal of College Student Development*, January 1989). One variable of interest was how the students progressed in terms of working on autonomy. The autonomy score was lowest in the fall quarter, better in the winter quarter, and highest in the spring quarter. A sample of 16 students provided the following autonomy statistics for the fall quarter: sample mean $\bar{x} = 51$ and sample standard deviation $s = 10.18$.

a. Using 95% confidence, provide an interval estimate of the population mean autonomy score for the fall quarter.

b. What assumption about the population was necessary in order to obtain an answer to part (a)?

c. Assume that it was desirable to estimate the population mean autonomy score with a sampling error of ± 3 points. Does the statistical data provide this desired level of precision? What action, if any, would you recommend be taken?

18. Shown are the duration (in minutes) for a sample of 20 flight reservation telephone calls:

2.1	4.8	5.5	10.4
3.3	3.5	4.8	5.8
5.3	5.5	2.8	3.6
5.9	6.6	7.8	10.5
7.5	6.0	4.5	4.8

a. What is the point estimate of the population mean time for flight reservation phone calls?

b. Assuming that the population has a normal distribution, develop a 95% confidence interval estimate of the population mean time.

8.3 ■ DETERMINING THE SIZE OF THE SAMPLE

Recall that in Section 8.1 we were able to make the following probability statement about the sampling error whenever a sample mean was used to provide a point estimate of a population mean:

There is a $1 - \alpha$ probability that the value of the sample mean will provide a sampling error of $z_{\alpha/2}\sigma_{\bar{x}}$ or less.

Note that $\sigma_{\bar{x}} = \sigma/\sqrt{n}$. We now can rewrite this statement as follows:

There is a $1 - \alpha$ probability that the value of the sample mean will provide a sampling error of $z_{\alpha/2}(\sigma/\sqrt{n})$ or less.

From this statement we see that the values of $z_{\alpha/2}$, σ, and the sample size n combine to determine the sampling error mentioned in the precision statement. Once we select a confidence coefficient or probability of $1 - \alpha$, $z_{\alpha/2}$ can be determined. Given values for $z_{\alpha/2}$ and σ, we can adjust the sample size n to provide any sampling error value desired. The formula used to compute the required sample size n is developed as follows.

Let E = the sampling error mentioned in the statement about the desired precision. We now have

$$E = z_{\alpha/2} \frac{\sigma}{\sqrt{n}} \tag{8.7}$$

Using (8.7) to solve for $\sqrt{n}$, we have

$$\sqrt{n} = \frac{z_{\alpha/2}\sigma}{E}$$

Squaring both sides of this equation, we obtain the following equation for the sample size.

Sample Size for An Interval Estimate of a Population Mean

$$n = \frac{(z_{\alpha/2})^2\sigma^2}{E^2} \tag{8.8}$$

This sample size n will provide a precision statement with a $1 - \alpha$ probability that the sampling error will be E *or less*.

To see how (8.8) can now be used to determine a recommended sample size, let us return to the Scheer Industries problem. We showed in Section 8.2 that for a sample of 15 employees the 95% confidence interval was 53.87 days ± 3.78 days. Assume that after considering the interval estimate of 50.09 days to 57.65 days, Scheer's director of manufacturing is not satisfied with this degree of precision. Furthermore, suppose that the director makes the following statement about the desired precision: "I would like a .95 probability that the value of the sample mean will provide a sampling error of 2 days or less." From the above statement we have E = 2 days. In addition, with $1 - \alpha$ specified at .95, α = .05 and $z_{\alpha/2} = z_{.025} = 1.96$. Note that we are using the z value in our calculations of the sample size even though the original computations for the Scheer problem employed the t distribution. The reason for this is that since the sample size is yet to be determined, we are anticipating that it will be larger than 30. Thus (8.8) is based on a z value rather than a t value.

Refer to (8.8). We see that with E = 2 and $z_{.025}$ = 1.96 we need a value for σ in order to compute the sample size that will provide the desired precision. Do we have a value of σ that we could use for the Scheer problem? Although σ is unknown, let us take advantage of the sample results reported in Section 8.2, where we found a sample standard deviation $s = 6.82$ days. Using $s = 6.82$ days as a planning value for the population standard deviation σ, (8.8) shows that the following sample size should provide the director's desired precision:

$$n = \frac{(z_{\alpha/2})^2 \sigma^2}{E^2} = \frac{(1.96)^2 (6.82)^2}{2^2} = 44.67$$

In cases where the computed n is a fraction, we round up to the next integer value; thus the recommended sample size for the Scheer problem is 45 employees. Since Scheer already has test data for 15 employees, an additional $45 - 15 = 30$ employees should be tested if the director wishes to obtain the desired precision of ± 2 days at a 95% confidence level.

Note that in (8.8) the values of $z_{\alpha/2}$ and E follow directly from the statement about the desired precision. However, the value of the population standard deviation σ may or may not be known. In cases where σ is known we have no problem in determining the sample size. However, in cases where σ is unknown we see that we must at least have a preliminary or *planning value* for σ in order to compute the sample size. In the Scheer Industries example, we were fortunate to have a sample of 15 employees which provided a point estimate for σ of 6.82 days. In instances where this initial or preliminary sample is unavailable, we may be able to obtain a good approximation of σ from past data on "similar" studies. Without such past data we may have to use a judgment or "best-guess" value for σ. In any case, regardless of the source, we see from (8.8) that we must have a planning value for σ in order to determine a recommended sample size.

▬ EXERCISES

19. What sample size would have been recommended for the Scheer Industries problem if the director of manufacturing had specified a .95 probability for a sampling error of 1.5 days or less? How large a sample would have been necessary if the precision statement had specified a .90 probability for a sampling error of 2 days or less?

20. In Section 8.1 the Statewide Insurance Company used a simple random sample of 36 policyholders to estimate the mean age of the population of policyholders. The resulting precision statement was reported to have a .95 probability that the value of the sample mean provided a sampling error of 2.35 years or less. This statement was based on a known population standard deviation of 7.2 years.

   a. How large a simple random sample would have been necessary to reduce the sampling error to 2 years or less? To 1.5 years or less? To 1 year or less?

   b. Would you recommend that Statewide attempt to estimate the mean age of the policyholders with $E = 1$ year? Explain.

21. Starting annual salaries for college graduates with business administration degrees are believed to have a standard deviation of approximately $2000.00. Assume that a 95% confidence interval estimate of the mean annual starting salary is desired. How large a sample size should be taken if the size of the sampling error in the precision statement is

   a. $500.00?
   b. $200.00?
   c. $100.00?

22. Refer to Exercise 14, which showed P/E ratios for 10 stocks listed on the New York Stock Exchange (*Wall Street Journal*, May 19, 1989). Assume that we are interested in estimating the population mean P/E ratio for all stocks listed on the New York Stock Exchange. How many stocks should be included in the sample if we would like a .95 probability that the sampling error is 2 or less?

23. Refer to Exercise 2, which provided the mean annual income for U.S. factory workers (*Barron's*, April 10, 1989). The population standard deviation of annual income was given to be $\sigma = \$5000$. If we wanted to estimate the mean annual income for the population of U.S. factory workers with a $500 margin of error, what sample size should be used. Assume 95% confidence.

24. A national survey research firm has past data that indicate that the interview time for a consumer opinion study has a standard deviation of 6 minutes.
    a. How large a sample should be taken if the firm desires a .98 probability of estimating the mean interview time to within 2 minutes or less?
    b. Assume that the simple random sample you recommended in part (a) is taken and that the mean interview time for the sample is 32 minutes. What is the 98% confidence interval estimate for the mean interview time for the population of interviews?

25. A gasoline service station shows a standard deviation of $6.25 for the charges made by the credit card customers. Assume that the station's management would like to estimate the population mean gasoline bill for its credit card customers to within ±$1.00. For a 95% confidence level, how large a sample would be necessary?

8.4 ▆ INTERVAL ESTIMATION OF A POPULATION PROPORTION

In Section 8.2 we presented the Scheer Industries problem, which involved estimating the mean employee training time for a new machine-repair training program. In order to evaluate the program from a different perspective, management has requested that some measure of program quality be developed. In the past the degree of success of the training program has been measured by the score the employee obtains on a standard examination given at the end of the training program. From past experience the company has found that an individual scoring 75 or better on the examination has an excellent chance of high performance on the job. After some discussion management has agreed to evaluate the program quality for the new training method based on the proportion of the employees that score 75 or better on the examination. Let us assume that Scheer implemented the sample size recommendation of the preceding section. Thus we now have a sample of 45 employees which can be used to develop an interval estimate for the proportion of the population that score 75 or better on the examination.

In Chapter 7 we learned that a sample proportion $\bar{p}$ is an unbiased estimator of a population proportion p and that the large sample approximation of the sampling distribution of $\bar{p}$ is normal as shown in Figure 8.8. Recall that the use of the normal distribution as an approximation of the sampling distribution of $\bar{p}$ is based on the condition that both np and $n(1 - p)$ are 5 or more. We will be using our knowledge of the sampling distribution of $\bar{p}$ to make probability statements about the sampling error whenever a sample proportion $\bar{p}$ is used to estimate a population proportion p. In this case the sampling error is defined as the magnitude of the difference between $\bar{p}$ and p, written $|\bar{p} - p|$.

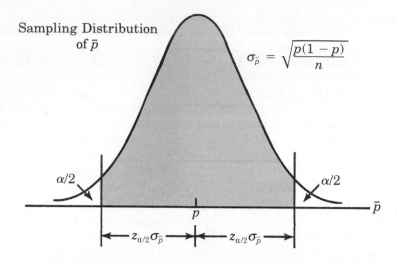

Sampling Distribution of $\bar{p}$

$$\sigma_{\bar{p}} = \sqrt{\frac{p(1-p)}{n}}$$

$\alpha/2$

$\alpha/2$

p

$\bar{p}$

$z_{\alpha/2}\sigma_{\bar{p}}$

$z_{\alpha/2}\sigma_{\bar{p}}$

▬▬ FIGURE 8.8 **Normal Approximation of the Sampling Distribution of $\bar{p}$ When Both $np \geq 5$ and $n(1 - p) \geq 5$**

The probability statements that we can make about the sampling error for the proportion take the following form:

There is a $1 - \alpha$ probability that the value of the sample proportion will provide a sampling error of $z_{\alpha/2}\sigma_{\bar{p}}$ or less.

The rationale for the preceding statement is the same as we used when the value of a sample mean was used as an estimate of a population mean. Namely, since we know that the sampling distribution of $\bar{p}$ can be approximated by a normal probability distribution, we can use the value of $z_{\alpha/2}$ and the value of the standard error of the proportion, $\sigma_{\bar{p}}$, to make the probability statement about the sampling error.

Once we see that the probability statement concerning the sampling error is based on $z_{\alpha/2}\sigma_{\bar{p}}$, we can subtract this value from $\bar{p}$ and add it to $\bar{p}$ in order to obtain an interval estimate of the population proportion. Such an interval estimate is given by

$$\bar{p} \pm z_{\alpha/2}\sigma_{\bar{p}} \tag{8.9}$$

where $1 - \alpha$ is the confidence coefficient. Since $\sigma_{\bar{p}} = \sqrt{p(1-p)/n}$, we can rewrite (8.9) as follows:

$$\bar{p} \pm z_{\alpha/2}\sqrt{\frac{p(1-p)}{n}} \tag{8.10}$$

However, note that in using (8.10) to develop an interval estimate of a population proportion p, the value of p would have to be *known*. Since the value of p is *unknown*, we simply substitute the sample proportion $\bar{p}$ for p. As a result, the general expression for a confidence interval estimate of a population proportion is as follows:*

Interval Estimate of a Population Proportion

$$\bar{p} \pm z_{\alpha/2} \sqrt{\frac{\bar{p}(1 - \bar{p})}{n}} \tag{8.11}$$

where $1 - \alpha$ is the confidence coefficient and $z_{\alpha/2}$ is the z value providing an area of $\alpha/2$ in the upper tail of the standard normal probability distribution.

Let us return to the Scheer Industries problem. Assume that in the sample of 45 employees who completed the new training program, 36 scored 75 or above on the examination. Thus the point estimate of the proportion in the population that score 75 or above on the examination is $\bar{p} = 36/45 = .80$. Using (8.11) and a .95 confidence coefficient, the interval estimate for the population proportion is given by

$$\bar{p} \pm z_{.025} \sqrt{\frac{\bar{p}(1 - \bar{p})}{n}}$$

$$.80 \pm 1.96 \sqrt{\frac{.80(1 - .80)}{45}}$$

$$.80 \pm .12$$

Thus we see that at the 95% confidence level, the interval estimate of the population proportion is .68 to .92.

Determining the Sample Size

Let us consider the question of how large the sample size should be in order to obtain an estimate of a population proportion at a specified level of precision. The rationale for the sample-size determination in developing interval estimates of p is very similar to the rationale used in Section 8.3 to determine the sample size for estimating a population mean.

Earlier in this section we provided the following probability statement about the sampling error:

There is a $1 - \alpha$ probability that the value of the sample proportion will provide a sampling error of $z_{\alpha/2}\sigma_{\bar{p}}$ or less.

---

*In this chapter we will be assuming that $n/N \leq .05$. Thus the finite population correction factor is not needed in the computation of $\sigma_{\bar{x}}$.

With $\sigma_{\bar{p}} = \sqrt{p(1-p)/n}$, the sampling error in this statement is based on the values of $z_{\alpha/2}$, the population proportion p, and the sample size n. For a given confidence coefficient $1 - \alpha$, $z_{\alpha/2}$ can be determined. Then, since the value of the population proportion is fixed, the sampling error mentioned in the precision statement is determined by the sample size n. Larger sample sizes again provide better precision.

Let E = the sampling error mentioned in the statement about the desired precision. We then have

$$E = z_{\alpha/2} \sqrt{\frac{p(1-p)}{n}}$$

Solving the preceding equation for n provides the following formula for the sample size.

Sample Size for An Interval Estimate of a Population Proportion

$$n = \frac{(z_{\alpha/2})^2 p(1-p)}{E^2} \tag{8.12}$$

Assume that the manager in the Scheer study has requested a sampling error of .10 or less with a .95 probability. This degree of precision specifies $z_{.025} = 1.96$ and $E = .10$. In order to use (8.12) to find the necessary sample size, we need a planning value for the population proportion p. Obviously, p will never be known exactly, since it is the population parameter we are trying to estimate. Thus we will use past data, a preliminary sample, or judgment to determine a planning value for p. For the Scheer Industries example, we can use the sample proportion of .80 from the sample of 45 employees as the planning value for the population proportion p. Substituting this value for p into (8.12) provides the following sample size:

$$n = \frac{(1.96)^2 .80(1 - .80)}{(.10)^2} = 61.47, \quad \text{or} \quad 62$$

Thus a sample size of 62, or 17 more than the current sample of 45, would be necessary to meet the precision requirement of $\pm .10$ at a 95% confidence level.

In this example we were fortunate to have the sample of 45 and the associated planning value of .80 available for p. In other cases it may be more difficult to determine an appropriate planning value for p. However, note that the numerator of (8.12) shows that the sample size is proportional to the quantity $p(1 - p)$. Table 8.4 shows some possible values for this quantity. To be on the safe or conservative side we have to consider the largest possible value for $p(1 - p)$. Thus, whenever there is any question regarding an appropriate planning value for p, we suggest using $p = .50$, since this value provides the largest sample size recommendation. If the proportion is different than the .50 planning value, the precision statement will be better than anticipated. However, in using the .50

TABLE 8.4 **Some Possible Values for** $p(1 - p)$

p	$p(1 - p)$
.10	$(.10)(.90) = .09$
.30	$(.30)(.70) = .21$
.40	$(.40)(.60) = .24$
.50	$(.50)(.50) = .25 \leftarrow$ Largest Value for $p(1 - p)$
.60	$(.60)(.40) = .24$
.70	$(.70)(.30) = .21$
.90	$(.90)(.10) = .09$

planning value, at least we have guaranteed that the required level of precision will be obtained.

In the Scheer Industries example, a planning value of $p = .50$ would have provided the following recommended sample size:

$$n = \frac{(1.96)^2.50(1 - .50)}{(.10)^2} = 96$$

The larger recommended sample size reflects the caution we took in using the conservative planning value for the population proportion.

NOTES AND COMMENTS

The desired margin of error for estimation of a population proportion is almost always .10 or less. In national public opinion polls conducted by organizations such as Gallup and Harris, a .03 or .04 margin of error is generally reported. The use of these values of error (E in equation 8.12) will generally provide a sample size that is large enough to satisfy the central limit theorem requirements of $np \geq 5$ and $n(1 - p) \geq 5$.

■■■ EXERCISES

26. What is the public opinion toward the U.S. Supreme Court ruling on abortion which stated that women may end a pregnancy during the first 3 months? A survey of 1227 adults in 12 southern states found that only 429 adults favored the Supreme Court ruling (*Atlanta Constitution*, April 13, 1989). What is the 95% confidence interval for the proportion of adults in the 12 southern states favoring the Supreme Court ruling? What is your interpretation of this interval estimate?

27. In a telephone followup survey of a new advertising campaign, 45 of 150 individuals contacted could recall the new advertising slogan associated with the product. Develop a 90% confidence interval estimate of the proportion in the population that will recall the advertising slogan.

28. Medical researchers at Cornell University studied the effect of tight neckties on the flow of blood to the head and the possible decrease in the brain's ability to respond to visual information (*Medical Self Care*, July–August 1988). Results of a sample of business-men found that 67% wear their ties too tight. Assuming a sample size of 250 businessmen, what is the 98% percent confidence interval estimate of the proportion of the population of businessmen who wear their ties too tight?

29. In an election campaign, a campaign manager requests that a sample of voters be polled to determine the support for the candidate. From a sample of 120 voters, 64 express plans to support the candidate.

   a. What is the point estimate of the proportion of the voters in the population who will support the candidate?

   b. Develop and interpret the 95% confidence interval for the proportion of voters in the population who will support the candidate.

   c. From the result from part (b), is the campaign manager justified in feeling confident that the candidate has the support of at least 50% of the voters? Explain.

   d. How many voters should be sampled if we want to estimate the population proportion with a sampling error of 5% or less? Continue to use the 95% confidence level.

30. It is estimated that 29% of all minimum-wage workers are teenagers (*Boston Globe*, April 12, 1989). Using a 95% confidence, provide an interval estimate of the proportion of the population of minimum-wage workers who are teenagers if the estimate is based on

   a. a simple random sample of 200

   b. a simple random sample of 600

   c. a simple random sample of 1000

   d. In general, what happens to the interval estimate of a population proportion as the sample size is increased?

31. The Tourism Institute for the State of Florida plans to sample visitors at major beaches throughout the state in order to estimate the proportion of beach visitors that are not residents of Florida. Preliminary estimates are that 55% of the beach visitors are not residents of Florida.

   a. How large a sample should be taken to estimate the proportion of out-of-state visitors to within ±3% of the actual value? Use a 95% confidence level.

   b. How large a sample should be taken if the error is increased to ±6%?

32. Where do people place their investment dollars? A sample conducted by the *Wall Street Journal* showed that 47% of all investors own at least some real estate (*Wall Street Journal*, December 2, 1988).

   a. Compute a 95% confidence interval estimate for the proportion of the population of investors who own at least some real estate. Assume that a sample size of 250 investors was used in the preceding study.

   b. How large a sample of investors should be used if it is desired to be 95% confident that the sampling error is 5% or less?

33. A firm provides national survey and interview services designed to estimate the proportion of the population that have certain beliefs or preferences. Typical questions seek to find the proportion favoring gun control, the proportion favoring abortion, the proportion favoring a particular political candidate, and so on. Assume that all interval estimates of population proportions are conducted at the 95% confidence level. How large a sample size would you recommend if the firm desired the sampling error to be

   a. 3% or less?

   b. 2% or less?

   c. 1% or less?

34. A Gallup poll asked the following question: "Would you favor or oppose federal legislation banning the manufacture, sale and possession of semiautomatic assult guns,

such as the AK-47"? (*Wall Street Journal*, April 7, 1989). If it is believed approximately 75% of the population would favor such legislation, how many individuals should be sampled for each of the following margins of error? Use a 95% confidence level.

    a. 10% margin of error
    b. 7.5% margin of error
    c. 5% margin of error
    d. 3% margin of error
    e. In general, what happens to the sample size as the margin of error decreases?

▬ SUMMARY

In this chapter we presented confidence interval estimation procedures for a population mean μ and a population proportion p. The purpose of developing an interval estimate of a population parameter is to provide the user of the sample results with a better understanding of the sampling error that may exist. If the width of an interval estimate is

▬ **FIGURE 8.9** **Summary of Interval Estimation Procedures for a Population Mean**

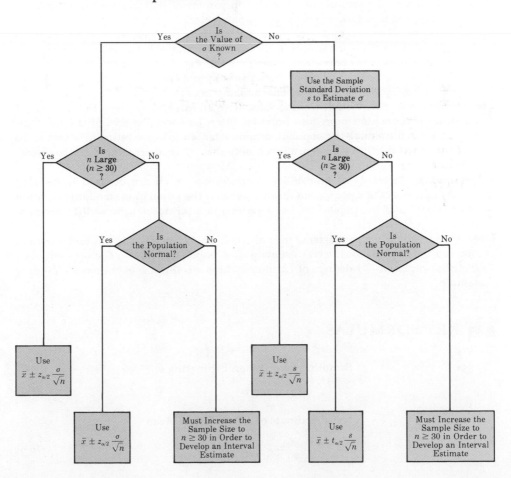

considered to be too large, the sample size can be increased in order to improve the precision of the estimate.

Figure 8.9 summarizes the interval estimation procedures for a population mean. The figure shows that the expression used to compute an interval estimate depends on whether or not the population standard deviation σ is known, whether the sample size is large ($n \geq 30$) or small ($n < 30$), and in some cases, whether or not the population has a normal probability distribution. If the sample size is large, no assumption is required about the distribution of the population and $z_{\alpha/2}$ is always used in the computation of the interval estimate. If the sample size is small, the population must be normally distributed in order to compute an interval estimate of μ. If the sample size is small and the assumption of a normally distributed population is inappropriate, the sample size must be increased to $n \geq 30$ in order to develop an interval estimate of μ. Finally, note that the use of the t distribution is reserved for the case where the population standard deviation σ is unknown, the sample size is small, and the population has a 'normal probability distribution.

In addition, we showed how to determine the sample size so that the interval estimates of μ and p would possess a specified or desired level of precision. In practice, the sample sizes required for interval estimates of a population proportion are generally large. Thus we restricted our discussion to the large sample case ($np \geq 5$ and $n(1 - p) \geq 5$).

■ GLOSSARY

Interval estimate An estimate of a population parameter that provides an interval of values believed to contain the value of the parameter.

Sampling error The magnitude of the difference between the value of an unbiased point estimator, such as the sample mean $\bar{x}$, and the value of the population parameter it estimates, such as the population mean μ. In this case the sampling error is $|\bar{x} - \mu|$. In the case of the population proportion, the sampling error is $|\bar{p} - p|$.

Precision A probability statement about the sampling error.

Confidence level The confidence associated with an interval estimate. For example, if an interval estimation procedure provides intervals such that 95% of the intervals developed will include the population parameter, an interval estimate is said to be constructed at the 95% confidence level; note that .95 is referred to as the *confidence coefficient*.

t distribution A family of probability distributions which can be used to develop interval estimates of a population mean whenever the population standard deviation is unknown and the population has a normal or near-normal probability distribution.

Degrees of freedom A parameter of the t distribution. When the t distribution is used in the computation of an interval estimate of a population mean, the appropriate t distribution has $n - 1$ degrees of freedom, where n is the size of the simple random sample.

■ KEY FORMULAS

Sampling Error When Estimating μ

$$|\bar{x} - \mu| \tag{8.1}$$

Interval Estimate of a Population Mean

$$\bar{x} \pm z_{\alpha/2} \frac{\sigma}{\sqrt{n}} \tag{8.2}$$

Estimator of $\sigma_{\bar{x}}$

$$s_{\bar{x}} = \frac{s}{\sqrt{n}} \qquad\qquad (8.4)$$

Interval Estimate of a Population Mean (Large-Sample Case with σ Unknown)

$$\bar{x} \pm z_{\alpha/2}\frac{s}{\sqrt{n}} \qquad\qquad (8.5)$$

Interval Estimate of a Population Mean (Small-Sample Case with σ Unknown)

$$\bar{x} \pm t_{\alpha/2}\frac{s}{\sqrt{n}} \qquad\qquad (8.6)$$

Sample Size for An Interval Estimate of a Population Mean

$$n = \frac{(z_{\alpha/2})^2\sigma^2}{E^2} \qquad\qquad (8.8)$$

Interval Estimate of a Population Proportion

$$\bar{p} \pm z_{\alpha/2}\sqrt{\frac{\bar{p}(1-\bar{p})}{n}} \qquad\qquad (8.11)$$

Sample Size for An Interval Estimate of a Population Proportion

$$n = \frac{(z_{\alpha/2})^2 p(1-p)}{E^2} \qquad\qquad (8.12)$$

▬ SUPPLEMENTARY EXERCISES

35. The North Carolina Savings and Loan Association would like to develop an estimate of the mean size of home improvement loans granted by its member institutions. A sample of 100 loans granted by member institutions resulted in a sample mean of $3,400 and a sample standard deviation of $650. With these data develop a 98% confidence interval for the population mean dollar amount of home improvement loans.

36. A sample of 1033 recreational fishermen was used in a study reported in the *Journal of Marketing Research*, November 1987. Data on temperature were collected in order to study the relationship between temperature and a fisherman's decision to go fishing. The sample mean temperature was 55.6° and the sample standard deviation was 7.37°. Construct a 95% confidence interval for the population mean temperature for the recreational fishermen considered in this study.

37. Dailey Paints, Inc. implemented a long-term painting test study designed to check the wear resistance of its major brand of paint. The test consisted of painting eight houses in various parts of the United States and observing the number of months until signs of peeling were observed. The following data were obtained from a normal population:

House	1	2	3	4	5	6	7	8
Months Until Signs of Peeling	60	51	64	45	48	62	54	56

a. What is a point estimate of the mean number of months until signs of peeling are observed?

b. Develop a 95% confidence interval to estimate the population mean number of months until signs of peeling are observed.

c. Develop a 99% confidence interval for the population mean.

38. What is the mean annual salary/bonus paid to the chief executive officers of the largest firms in the United States? A random sample of seven firms provided the following annual salary/bonus data (*Business Week*, May 1, 1989).

Firm	Annual Salary/Bonus (000s)
Bank of Boston	1200
Citicorp	1798
Du Pont	1611
Abbott Laboratories	1920
Teledyne	860
Emerson Electric	1681
Conagra	1312

a. What is the point estimate of the population mean annual salary/bonus for chief executives?

b. What is the point estimate of the population standard deviation for annual salaries?

c. What is the 95% confidence interval estimate of the population mean annual salary/bonus for chief executives?

39. The Atlantic Fishing and Tackle Company has developed a new synthetic fishing line. In order to estimate the breaking strength of this line, testers subjected six lengths of line to breakage testing. The following data were obtained from a normal population.

Line	1	2	3	4	5	6
Breaking Strength (pounds)	18	24	19	21	20	18

Develop a 95% confidence interval estimate of the mean breaking strength of the new line.

40. Sample assembly times for a particular manufactured part were 8, 10, 10, 12, 15, and 17 minutes. If the mean of the sample is to be used to estimate the mean of the population of assembly times, provide a point estimate and a 90% confidence interval estimate of the population mean. Assume that the population has a normal distribution.

41. A utility company finds that a sample of 100 delinquent accounts yields an average amount owed of $131.44, with a sample standard deviation of $16.19. Develop a 90% confidence interval for the population mean amount owed.

42. In Exercise 41 the utility company sampled 100 delinquent accounts in order to estimate the mean amount owed by these accounts. The sample standard deviation was $16.19. How large a sample should be taken if the company wants to be 90% confident that the estimate of the population mean will have a sampling error of $1 or less?

43. Consider the Atlantic Fishing and Tackle Company problem presented in Exercise 39. How large a sample would be necessary in order to estimate the mean breaking strength of the new line with a .99 probability of a sampling error of 1 pound or less?

44. Mileage tests are conducted for a particular model of automobile. If the desired precision is stated such that there is to be .98 probability of a sampling error of 1 mile per gallon or less, how many automobiles should be used in the test? Assume that preliminary mileage tests indicate a standard deviation in miles per gallon for the automobiles to be 2.6 miles per gallon.

45. In developing patient appointment schedules, a medical center desires to estimate

the mean time that a staff member spends with each patient. How large a sample should be taken if the precision of the estimate is to be ±2 minutes at a 95% level of confidence? How large a sample should be taken for a 99% level of confidence? Use a planning value for the population standard deviation of 8 minutes.

46. Exercise 38 provided annual salary/bonus data for chief executives of firms in the United States (*Business Week,* May 1, 1989). The sample standard deviation was $375, with data provided in thousands of dollars. How many chief executives should be in the sample if we would like to estimate the population mean annual salary/bonus with a margin of error of $100,000? Note that the margin of error should be stated as $100 since the data were provided in thousands of dollars. Use a 95% confidence level.

47. The New Orleans Beverage Company has been experiencing problems with the automatic machine that places labels on bottles. The company desires an estimate of the percentage of bottles that have improperly applied labels. A simple random sample of 400 bottles resulted in 18 bottles with improperly applied labels. Using these data, develop a 90% confidence interval estimate of the population proportion of bottles with improperly applied labels.

48. H. G. Forester and Company is a distributor of lumber supplies throughout the Southwest United States. Management of H. G. Forester would like to check a shipment of over 1 million pine boards in order to determine if excessive warpage exists for the boards. A sample of 50 boards resulted in the identification of 7 boards with excessive warpage. With these data develop a 95% confidence interval estimate of the proportion of boards defective in the whole shipment.

49. Consider the H. G. Forester and Company problem represented in Exercise 48. How large a sample size would be required to estimate the proportion of boards with warpage to within ±.01 at a 95% confidence level?

50. A well-known bank credit card firm is interested in estimating the proportion of credit card holders that carry a nonzero balance at the end of the month and incur an interest charge. Assume that the desired precision for the proportion estimate is ±3% at a 98% confidence level.

    a. How large a sample should be recommended if it is anticipated that roughly 70% of the firm's cardholders carry a nonzero balance at the end of the month?
    b. How large a sample would be recommended if no planning value for the population proportion can be specified?

51. A sample of 200 people were asked to identify their major source of news information; 110 stated that their major source was television news coverage.

    a. Construct a 95% confidence interval for the proportion of the people in the population that consider television their major source of news information.
    b. How large a sample would be necessary to estimate the population proportion with a sampling error of .05 or less at a 95% confidence level?

52. A Gallup poll of labor-management negotiations found that 55% of the cases studied showed that labor is taking a more aggressive bargaining stance (*Barron's,* April 10, 1989). If a sample of 500 firms was used in the poll, what is the 90% confidence interval for the proportion on negotiations where labor is taking a more aggressive bargaining stance?

53. A survey is to be taken to estimate the proportion of high-school graduates in a particular school district that plan to attend college. How large a sample of students should be selected in order to provide a 95% confidence of reporting a sample proportion that is within ±.025 of the population proportion? Use $p = .35$ as a planning value for the proportion of high-school students who plan to attend college.

54. A survey of executives will be used to learn about how people in business view the quality of education provided by public school systems (*Fortune,* March, 1989). Using a

90% confidence level, the study is to be designed to have a 4% margin of error for questions where the population proportion is approximately .60.

a. How many executives should be included in the sample?

b. Using your sample size, what is the 90% confidence interval estimate for the population proportion rating public schools "fair/poor" if 313 executives in the sample respond "fair/poor"?

c. Using your sample size, what is the 90% confidence interval estimate for the population proportion responding "public schools have deteriorated in the last 10 years" if 260 executives respond yes to this statement?

d. Do the interval estimates in parts (b) and (c) provide the desired 4% margin of error? Explain.

■■■ COMPUTER EXERCISE

A consumer research organization has been studying the repair history of diesel automobiles produced by a major Detroit manufacturer. Of particular concern has been the performance of diesel engines used in the manufacturer's full-sized cars. Preliminary evidence shows that owners of these cars have experienced relatively early failures in the car's transmission system. Part of the consumer research organization's study has uncovered the fact that the transmission used by the manufacturer may be too small for the diesel engines. To aid in the investigation, the research organization sent question-naires to owners of the diesel automobiles. Data were collected from 50 owners who had experienced a transmission failure. The following data show the number of miles the vehicle had been driven at the time of the failure.

85,092	32,609	59,465	77,437	32,534	64,090	32,464	59,902
39,323	89,641	94,219	116,803	92,857	63,436	65,605	85,861
64,342	61,978	67,998	59,817	101,769	95,774	121,352	69,568
74,276	66,998	40,001	72,069	25,066	77,098	69,922	35,662
74,425	67,202	118,444	53,500	79,294	64,544	86,813	116,269
37,831	89,341	73,341	85,288	138,114	53,402	85,586	82,256
77,539	88,798						

■■■ QUESTIONS

1. Use appropriate descriptive statistics to summarize this data.

2. Develop a 95% confidence interval for the mean number of miles driven until transmission failure for the population of vehicles that have experienced transmission failure.

3. Discuss the implications of your statistical findings in terms of the claim that some owners of the diesel automobiles have been experiencing early transmission failures.

APPLICATION

Dollar General Corporation

Nashville, Tennessee

Dollar General Corporation was founded in 1939 by J.L. Turner and his son Cal Turner as a dry goods wholesale company. After World War II the Turners began opening retail locations in rural southcentral Kentucky. Today Dollar General Corporation operates more than 1300 neighborhood stores in 23 states. Serving predominately low and middle income customers, Dollar General markets soft goods, health, beauty and cleaning supplies, and hard goods at low, everyday prices.

Being in an inventory-intense business with approximately 17,000 different products, Dollar General Corporation made the decision to adopt the LIFO (last in-first out) method of inventory valuation. Under this accounting practice, the inventory on hand at the close of an accounting period is valued at the first price paid regardless of fluctuation affecting the actual cost of the total inventory. The reasons for adopting this method of valuation are many; the significant reasons are as follows:

1. To better match current costs against current revenues, thereby minimizing the effect of radical price changes and their influence on profit or loss results.

2. To reduce income and thereby income taxes during periods of inflation. This in turn brings disposable cash generated from operations more in line with current income and allows for replacement of inventory at current costs.

▬▬ ESTABLISHING AN ANNUAL LIFO INDEX

The LIFO computations require the company to establish a LIFO index for its inventory over the base year. The desired LIFO index is based on two components, as shown in Figure 8A.1. Component 1 is referred to as the base cost for the January 31 fiscal year end inventory, and component 2 is referred to as the current cost for the same inventory. For example, a base cost for a given mix of inventory items might be $210 million. However, using current costs, which are usually higher because they reflect the year's inflationary effect, this same mix of inventory items might cost $220 million. The LIFO index would be $220 million/$210 million = 1.048. Interpretation of this value is that the company's inventory at current costs contains a 4.8% increase in value due to the inflation occurring during the one year period.

To use the LIFO method of inventory valuation, the company needs a method for computing the annual LIFO index. The straightforward approach is to determine the actual number of units in inventory for every item (a census) carried at Dollar General's 1300 locations. Then, using the current cost per unit, as well as the preceding year's base

*The authors are indebted to Mr. Robert S. Knaul, Controller, Dollar General Corporation for providing this application.

Actual Inventory Units
January 31 of
Current Year

Component 1

Dollar Value of
This Inventory Using
Costs Effective
January 31 of
Preceding Year

Component 2

Dollar Value of
This Inventory Using
Costs Effective
January 31 of
Current Year

■■■ FIGURE 8A.1 **Components of the LIFO Index for the Dollar General Corporation**

cost per unit, the two components of the index can be determined. However, with over 17,000 products and over 1300 retail locations, the time and cost of conducting this census would be prohibitive. This is where sampling and the statistical estimation procedures you have been studying play an essential role at Dollar General.

One of the more than 1300 Dollar General Stores

■ DEVELOPING AN INTERVAL ESTIMATE OF THE LIFO INDEX

Rather than collecting data for every product in over 1300 retail locations and 3 warehouses, a random sample 800 items from 75 retail locations and the 3 warehouses is selected. The physical inventory counts for the sample items are taken during the last week of January. A clerical employee identifies the current cost per unit and base cost per unit for each item in the sample. Component 1 of the LIFO index is the dollar value of the sample inventory using the base costs, while component 2 is the dollar value of the sample inventory using current costs. The ratio of the sample current cost to the sample base cost provides the point estimate for the company's LIFO index.

For example, for a given year the sample LIFO index was 1.070. However, since this index is only an estimate of the population's LIFO index, a statement about the sampling error and the associated interval estimate are essential in determining the goodness of the sample index. Using the sample results and a 95% confidence level, the maximum sampling error was .006. Thus the interval of 1.064 to 1.076 provided the 95% confidence interval estimate for the population LIFO index. The precision of the sample was judged acceptable, and the point estimate of the index, 1.070, was used in the LIFO computations.

The sample of 800 items provided the time and cost savings that made the use of the LIFO inventory policy acceptable to Dollar General. Without sampling and the interval estimate showing the goodness of the sample index, Dollar General would have been unable to obtain the advantages of the LIFO policy.

CHAPTER 9

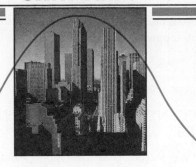

Hypothesis Testing

I n Chapters 7 and 8 we showed how a sample could be used to develop point and internal estimates of population parameters. In this chapter we continue the discussion of statistical inference by showing how *hypothesis testing* can be used to test claims about population parameters.

In hypothesis testing we begin by making a tentative assumption about a population parameter. This tentative assumption is called the *null hypothesis* and is denoted by H_0. We then define another hypothesis, called the *alternative hypothesis*, which is the opposite of what is stated in the null hypothesis. This alternative hypothesis is denoted by H_a. The hypothesis testing procedure involves using data from a sample to test the two competing claims indicated by H_0 and H_a.

The situation encountered in hypothesis testing is similar to the one encountered in a criminal trial. In a criminal trial the assumption is that the defendant is innocent. Thus, the null hypothesis is one of innocence. The opposite of the null hypothesis is the alternative hypothesis—that the defendant is guilty. Thus the hypotheses for a criminal trial would be written

H_0: The defendant is innocent

H_a: The defendant is guilty

To test these competing claims, or hypotheses, a trial is held. The testimony and evidence obtained during the trial provide the sample information. If the sample information is not inconsistent with the assumption of innocence, the null hypothesis that the defendant is innocent cannot be rejected. However, if the sample information is inconsistent with the assumption of innocence, the null hypothesis will be rejected. In this case, action will be taken based upon the alternative hypothesis that the defendant is guilty.

The purpose of this chapter is to show how hypothesis tests can be conducted about a population mean and a population proportion. We begin by providing examples that illustrate approaches to developing null and alternative hypotheses.

9.1 ■ DEVELOPING NULL AND ALTERNATIVE HYPOTHESES

In the introduction we stated that in hypothesis testing we develop two competing claims, or hypotheses, about a population parameter. One claim is called the null hypothesis, H_0, and the other claim is called the alternative hypothesis, H_a. In some applications it may not be obvious how the null and alternative hypotheses should be formulated. Care must be taken to be sure that the hypotheses are structured appropriately and that the hypothesis-testing conclusion provides the information that the researcher or decision maker desires.

Guidelines for establishing the null and alternative hypotheses will be given for three types of situations that frequently employ hypothesis-testing procedures. A discussion of each of these situations follows.

Testing Research Hypotheses

Consider a particular model automobile that currently obtains an average of 24 miles per gallon. A product-research group has developed a new carburetor designed specifically to increase the miles-per-gallon performance. In order to evaluate the new design, several new carburetors will be manufactured, installed in automobiles, and subjected to research-controlled driving tests. Note that the product-research group is looking for evidence to enable them to conclude that the new design *increases* the mean number of miles per gallon. In this case, the research hypothesis is that the new carburetor will provide a mean miles per gallon exceeding 24; that is, $\mu > 24$. As a general guideline, a research hypothesis such as this should be formulated as the *alternative hypothesis*. Thus, the appropriate null and alternative hypotheses for the study are as follows:

$$H_0: \quad \mu \leq 24$$
$$H_a: \quad \mu > 24$$

If the sample results indicate that H_0 cannot be rejected, we will not be able to conclude that the new carburetor is better. Perhaps more research and subsequent testing should be conducted. However, if the sample results indicate H_0 can be rejected, the inference can be made that $H_a: \mu > 24$ is true. With this conclusion, the researcher has the statistical support necessary to conclude that the new carburetor increases the mean number of miles per gallon.

In research studies such as these, the null and alternative hypotheses should be formulated so that the rejection of H_0, and hence the inference that H_a is true, will provide the researcher with the conclusion and action being sought. Thus, the research hypothesis should be expressed as the alternative hypothesis.

Testing the Validity of Assumption

Examples of testing the validity of assumptions can be found when companies make claims about their products. For example, a manufacturer of soft drinks states that 2-liter containers of its products have an average of at least 67.6 fluid ounces. A sample of 2-liter containers will be selected, and the contents will be measured in order to test the manufacturer's statement. In this type of hypothesis-testing situation, we generally follow the rationale suggested by the criminal trial analogy. That is, the manufacturer's statement should be assumed true (innocent) unless the sample evidence proves otherwise (guilty). Using this approach for the soft-drink example, the null and alternative hypotheses would be stated as follows:

$$H_0: \quad \mu \geq 67.6$$

$$H_a: \quad \mu < 67.6$$

If the sample results indicate H_0 cannot be rejected, the manufacturer's claim cannot be challenged. However, if the sample results indicate H_0 can be rejected, the inference will be made that $H_a: \mu < 67.6$ is true. With this conclusion, statistical evidence indicates that the manufacturer's statement is incorrect and that the soft-drink containers are being filled with a mean less than the claimed 67.6 ounces. Appropriate action against the manufacturer may be considered.

In any situation which involves testing the validity of some assumption, the null hypothesis is generally formulated based on the assumption. The alternative hypotheses is then formulated so that rejection of H_0 will provide the statistical evidence that the stated assumption is incorrect. Action to correct the claim or assumption should be considered whenever H_0 is rejected.

Decision Making

A hypothesis-testing situation involving decision making occurs when a decision maker must choose between two courses of action, one associated with the null hypothesis and another associated with the alternative hypothesis. For example, on the basis of a sample of parts from a shipment that has just been received, a quality-control inspector must decide whether to accept the entire shipment or to return the shipment to the supplier because it does not meet specifications. Assume that specifications for a particular part indicate a mean length of 2 inches per part is required. If the average length of the parts is greater or less than the 2-inch standard, the parts will cause quality problems in the assembly operation. In this case the null and alternative hypotheses would be formulated as follows:

$$H_0: \quad \mu = 2$$

$$H_a: \quad \mu \neq 2$$

If the sample results indicate H_0 cannot be rejected, the quality control inspector will have no reason to doubt that the shipment meets specifications, and thus the shipment will be accepted. However, if the sample results indicate that H_0

should be rejected, the conclusion can be made that the parts do not meet specifications. In this case, the quality-control inspector has sufficient evidence to return the shipment to the supplier.

A Summary of Forms for Null and Alternative Hypotheses

The three hypothesis-testing examples that we have discussed in this section all concern the value of a population mean. Let μ_0 denote the numerical value being considered in the hypotheses. In general, a hypothesis test concerning the value of a population mean must take one of the following three forms:

$$H_0: \quad \mu \geq \mu_0 \qquad H_0: \quad \mu \leq \mu_0 \qquad H_0: \quad \mu = \mu_0$$
$$H_a: \quad \mu < \mu_0 \qquad H_a: \quad \mu > \mu_0 \qquad H_a: \quad \mu \neq \mu_0$$

The following exercises are designed to provide practice in choosing the proper form for a hypothesis test.

▭ EXERCISES

1. The manager of the Danvers-Hilton Resort Hotel has stated that the mean guest bill for a weekend is \$400 or less. A member of the hotel's accounting staff has noticed that the total charges for guest bills have been increasing in recent months. The accountant will use a sample of weekend guest bills to test the manager's claim.

   a. Which of the forms of hypotheses should be used to test the manager's claim? Explain.

$$H_0: \quad \mu \geq 400 \qquad H_0: \quad \mu \leq 400 \qquad H_0: \quad \mu = 400$$
$$H_a: \quad \mu < 400 \qquad H_a: \quad \mu > 400 \qquad H_a: \quad \mu \neq 400$$

   b. Comment on the conclusion when H_0 cannot be rejected.
   c. Comment on the conclusion when H_0 can be rejected.

2. The manager of an automobile dealership is considering a new bonus plan that is designed to increase sales volume. Currently, the mean sales volume is 14 automobiles per month. The manager would like to conduct a research study to see if there is evidence that the new bonus plan increases sales volume. In order to collect data on the plan, a sample of sales personnel will be allowed to sell under the new bonus plan for a 1-month period.

   a. Develop the null and alternative hypotheses that are most appropriate for this research situation.
   b. Comment on the conclusion when H_0 cannot be rejected.
   c. Comment on the conclusion when H_0 can be rejected.

3. A production-line operation is supposed to fill cartons of laundry detergent with a mean weight of 32 ounces. A sample of cartons is periodically selected and weighed in order to determine if underfilling or overfilling exists. If the sample data enable the conclusion of underfilling or overfilling, the production line will be shut down and adjusted to obtain proper filling.

   a. Formulate the null and alternative hypotheses that will help make the decision of whether or not to shut down and adjust the production line.

b. Comment on the conclusion and the decision when H_0 cannot be rejected.

c. Comment on the conclusion and the decision when H_0 can be rejected.

4. Because of high production-changeover time and costs, a director of manufacturing must convince management that a proposed manufacturing method reduces costs before the new method can be implemented. The current production method operates with a mean cost of $220 per hour. A research study will be conducted, with the cost of the new method measured over a sample production period.

a. Develop the null and alternative hypotheses that are most appropriate for this study.

b. Comment on the conclusion when H_0 cannot be rejected.

c. Comment on the conclusion when H_0 can be rejected.

9.2 ■■■ TYPE I AND TYPE II ERRORS

The null and alternative hypotheses are competing claims about the true state of nature. Either the null hypotheses, H_0, is true or the alternative hypothesis, H_a, is true, but not both. Ideally the hypothesis testing procedure should lead to the acceptance of H_0 when H_0 is the true state of nature and the rejection of H_0 when H_a is the true state of nature. Unfortunately, these results are not always possible. Since hypothesis tests are based upon sample information, we must allow for the possibility of errors. Table 9.1 provides a summary of the correct decisions and the errors in hypothesis testing.

The first row of Table 9.1 shows what can happen if we make the decision to accept H_0. If H_0 is the true state of nature, this is the correct decision. However, if H_a is the true state of nature, we have made a *Type II error*; that is, we have accepted H_0 when it is false.

The second row of Table 9.1 shows what can happen if we make the decision to reject H_0. If the true state of nature is H_0, we have made a *Type I error*; that is, we rejected H_0 when it is true. However, if H_a is the true state of nature, then rejecting H_0 is the correct decision.

Although we cannot eliminate the possibility of errors in hypothesis testing, we can consider the probability of their occurrence. Using common statistical notation, we denote the probabilities of making the two errors as follows:

$$\alpha = \text{the probability of making a Type I error}$$

$$\beta = \text{the probability of making a Type II error}$$

TABLE 9.1 **Errors and Correct Decisions in Hypothesis Testing**

		State of Nature	
		H_0 *True*	H_a *True*
Decision	Accept H_0	Correct Decision	Type II Error
	Reject H_0	Type I Error	Correct Decision

In most applications of hypothesis testing, the probability of making a Type I error is controlled, but the probability of making a Type II error is not. In practice, the person conducting the hypothesis test specifies the maximum allowable probability of making a Type I error, called the *level of significance* for the test. Common choices for the level of significance are .05 and .01. Referring to the second row of Table 9.1, note that the decision to *reject* H_0 indicates that either a Type I error or a correct decision has been made. Thus, if the probability of making a Type I error is controlled for by selecting a low value for the level of significance, we have a high degree of confidence that the decision to reject H_0 is correct. In such cases we have statistical support to conclude that H_0 is false and H_a is true. Any action suggested by the alternative hypothesis, H_a, is appropriate.

As we stated previously, however, although applications of hypothesis testing control the probability of making a Type I error, they do not always control for the probability of making a Type II error. Thus, whenever we decide to accept H_0 without taking into consideration the probability of making a Type II error, we cannot determine how confident we can be with the decision to accept H_0. Because of the uncertainty associated with making a Type II error, statisticians often recommend that we use the statement *do not reject H_0* instead of *accept H_0*. Using the statement *do not reject H_0* carries the recommendation to withhold both judgment and action. In effect, by never directly concluding that H_0 is true, the statistician avoids the risk of making a Type II error. Whenever the probability of making a Type II error has not been determined and controlled, we will not make the decision to accept H_0. In such cases, only two conclusions are possible: *do not reject H_0* or *reject H_0*.

Note however, that just because it is not common to control for the Type II error in hypothesis testing, this does not mean that it is not possible to do so. In fact, as we will show in Sections 9.7 and 9.8, procedures exist for determining and controlling for the probability of making a Type II error. If proper controls have been established for this error, it can be appropriate to take action based on the decision to accept H_0.

NOTES AND COMMENTS

As we indicated in this section, the statement *do not reject H_0* is used to avoid the conclusion that H_0 is true, and thus eliminate the possibility of making a Type II error. As a result, the statement *do not reject H_0* means we do not conclude that H_0 is true or that H_a is true. In other words, the statement *do not reject H_0* indicates that the test results are inconclusive. Since we are unable to conclude H_0 is true or H_a is true, the only action that should be taken is to recommend that further study be conducted in order to clarify the situation.

▨▨▨ EXERCISES

5. The average American buys 6.08 books per year (*Louis Rukeyser's Business Almanac*, 1988). A University of Iowa researcher believes that adults in Des Moines purchase books

at an annual rate higher than the national average. The following null and alternative hypotheses have been formulated by the researcher.

$$H_0: \quad \mu \le 6.08$$

$$H_a: \quad \mu > 6.08$$

a. What is the Type I error in this situation? What are the consequences of making this error?

b. What is the Type II error in this situation? What are the consequences of making this error?

6. The label on a 3 quart container of orange juice claims that the orange juice contains, on the average, 1 gram of fat or less. Answer the following questions for a hypothesis test that could be used to test the claim on the label.

a. Develop the appropriate null and alternative hypotheses.

b. What is the Type I error in this situation? What are the consequences of making this error?

c. What is the Type II error in this situation? What are the consequences of making this error?

7. Carpetland salespersons have been selling an average of $8000 of carpeting per week. Steve Contois, the firm's vice president, has a proposed a compensation plan with new selling incentives. Steve hopes to use a trial selling period to be able to conclude that the compensation plan increases the average sales per salesperson.

a. Develop the appropriate null and alternative hypotheses.

b. What is the Type I error in this situation? What are the consequences of making this error?

c. What is the Type II error in this situation? What are the consequences of making this error?

8. Refer to Exercise 4. Suppose that the new production method will be implemented if a hypothesis test supports the conclusion that the new method reduces the mean operating cost per hour.

a. State the appropriate null and alternative hypotheses if the mean cost for the current production method is $220 per hour.

b. What is the Type I error in this situation? What are the consequences of making this error?

c. What is the Type II error in this situation? What are the consequences of making this error?

9.3 ■■■ ONE-TAILED HYPOTHESIS TESTS ABOUT A POPULATION MEAN

The Federal Trade Commission (FTC) periodically conducts studies designed to test the claims manufacturers make about their products. For example, the label on a large can of Hilltop Coffee states that the can contains at least 3 pounds of coffee. Suppose that it is desired to test this claim using hypothesis testing.

The first step is to develop the null and the alternative hypotheses. We begin by tentatively assuming that the manufacturer's claim is correct. Note that if the population of coffee cans has a mean weight of 3 or more pounds per can, Hilltop's claim about its product is correct. However, if the population of coffee cans has a mean weight less than 3 pounds per can, Hilltop's claim is invalid; in this case, a charge of underfilling should be made against Hilltop.

With μ denoting the mean weight for the population, the null and the alternative hypotheses are formulated as follows:

$$H_0: \quad \mu \geq 3$$

$$H_a: \quad \mu < 3$$

If the sample data indicate that H_0 cannot be rejected, the statistical evidence does not support the conclusion that a label violation has occurred. Thus, no action would be taken against Hilltop. However, if sample data indicate that H_0 can be rejected, we will conclude that the alternative hypothesis, $H_a: \mu < 3$, is true. In this case, an FTC claim of underfilling and a charge of a label violation would be appropriate.

Suppose that a random sample of 36 cans of coffee will be taken to test these hypotheses for the Hilltop study. Note that if the sample mean filling weight, $\bar{x}$, for the 36 cans is less than 3 pounds, the sample results will begin to cast doubt on the null hypothesis, $H_0: \mu \geq 3$. But how much less than 3 must $\bar{x}$ be before we would be willing to risk making a Type I error and falsely accuse the company of a label violation?

To answer this question, let us tentatively assume that the null hypothesis is true with $\mu = 3$. From our study of sampling distributions in Chapter 7, we know that whenever the sample size is large ($n \geq 30$), the sampling distribution of $\bar{x}$ can be approximated by a probability distribution normal with mean μ and standard deviation given by $\sigma_{\bar{x}} = \sigma/\sqrt{n}$. Figure 9.1 shows the sampling distribution of $\bar{x}$ when the null hypothesis is true at $\mu = 3$.

The value of $z = (\bar{x} - 3)/\sigma_{\bar{x}}$ gives the number of standard deviations $\bar{x}$ is from $\mu = 3$. For hypothesis tests about a population mean, we will use z as a *test statistic* to determine whether or not $\bar{x}$ deviates enough from $\mu = 3$ to justify rejecting the null hypothesis. Note that a value of $z = -1$ means that $\bar{x}$ is 1 standard deviation below $\mu = 3$, a value of $z = -2$ means that $\bar{x}$ is 2 standard deviations below $\mu = 3$,

■■■ FIGURE 9.1   **Sampling Distribution of $\bar{x}$ for the Hilltop Coffee Study when the Null Hypothesis is True ($\mu = 3$)**

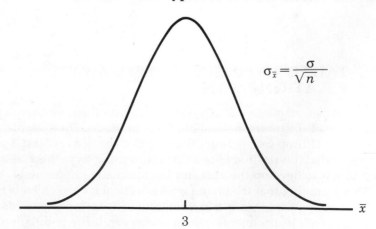

and so on. Obtaining a value of $z < -3$ is very unlikely if the null hypothesis is true. The key question is: How small must the test statistic, z, be before we have enough evidence to reject the null hypothesis?

Figure 9.2 shows that the probability of observing a value of $\bar{x}$ more than 1.645 standard deviations below the mean of $\mu = 3$ is .05. Thus, if we were to reject the null hypothesis whenever the value of the test statistic, $z = (\bar{x} - 3)/\sigma_{\bar{x}}$, is less than -1.645, the probability of making a Type I error would be .05. If this were a reasonable probability of a Type I error, then we would reject the null hypothesis whenever the test statistic indicated that the sample mean was more than 1.645 standard deviations below $\mu = 3$. Thus, we would reject H_0 whenever $z < -1.645$.

The methodology of hypothesis testing requires that we specify the maximum allowable probability of a Type I error. As noted in the previous section, this maximum probability is called the level of significance for the test; it is denoted by α, and it represents the probability of making a Type I error when the null hypothesis is true as an equality. Specifying the level of significance is a job for the manager. It requires an evaluation of the seriousness of making a Type I error. If the cost of making a Type I error is high, then a small value should be chosen for the level of significance. If the cost is not too great, a larger value may be appropriate.

In the Hilltop Coffee study, the director of the weight-testing program has made the following statement: "If the company is meeting its weight specifications exactly ($\mu = 3$), I would like a 99% chance of not taking any action against the company. While I do not want to accuse the company wrongly of underfilling its product, I am willing to live with a 1% chance of making this error."

From the director's statement, the maximum probability of a Type I error is .01. Thus, the level of significance for the hypothesis test is $\alpha = .01$. Figure 9.3 shows both the sampling distributions of $\bar{x}$ and $z = (\bar{x} - \mu)/\sigma_{\bar{x}}$ for the Hilltop Coffee example. Note that when the null hypothesis is true at $\mu = 3$, the

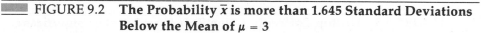

FIGURE 9.2   The Probability $\bar{x}$ is more than 1.645 Standard Deviations Below the Mean of $\mu = 3$

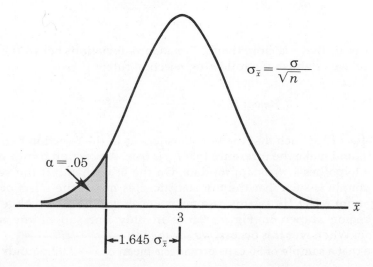

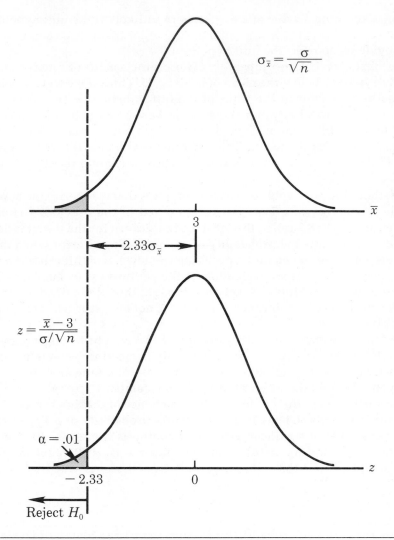

$$\sigma_{\bar{x}} = \frac{\sigma}{\sqrt{n}}$$

$$z = \frac{\bar{x} - 3}{\sigma/\sqrt{n}}$$

$\alpha = .01$

-2.33 0 z

Reject H_0

■■■■ FIGURE 9.3   **Hilltop Coffee Rejection Rule has a Level of Significance of $\alpha = .01$**

probability is .01 that $\bar{x}$ is more than 2.33 standard deviations below the mean of 3. Therefore, we establish the following rejection rule:

$$\text{Reject } H_0 \text{ if } z = \frac{\bar{x} - \mu}{\sigma_{\bar{x}}} < -2.33$$

If the value of $\bar{x}$ is such that the test statistic, z, is in the rejection region, then we reject H_0 and make the inference that H_a is true. Any action indicated by the alternative hypothesis H_a is appropriate. On the other hand, if the value of $\bar{x}$ from the sample is such that the test statistic, z, is not in the rejection region, then we cannot reject H_0. In this case no action should be taken. Note that the rejection region shown in Figure 9.3 is in only one tail of the sampling distribution. Whenever this occurs, we say the test is a *one-tailed* test.

Suppose that a sample of 36 cans provides a mean of $\bar{x} = 2.92$ pounds and that it is known from previous studies that the population standard deviation is $\sigma = .18$. With $\sigma_{\bar{x}} = \sigma/\sqrt{n}$, the value of the test statistic is given by

$$z = \frac{\bar{x} - 3}{\sigma/\sqrt{n}} = \frac{2.92 - 3}{.18/\sqrt{36}} = -2.67$$

Figure 9.4 shows that the value of the test statistic is in the rejection region. We are now justified in making the inference that $\mu < 3$ at a .01 level of significance. The director now has the statistical justification to take action against Hilltop Coffee for underfilling its product.

Suppose, instead, that the sample of 36 cans provides a sample mean of $\bar{x} = 2.97$. In this case, the value of the test statistic would be

$$z = \frac{\bar{x} - 3}{\sigma/\sqrt{n}} = \frac{2.97 - 3}{.18/\sqrt{36}} = -1.00$$

Since $z = -1.00$ is greater than -2.33, the value of the test statistic is not in the rejection region (see Figure 9.5). Hence we cannot reject the null hypothesis. No further inference can be made and no statistical justification has been provided to take action against Hilltop Coffee.

The value of z that establishes the boundary of the rejection region is called the *critical value*. In establishing the critical value, we tentatively assume the null hypothesis is true. But, for Hilltop Coffee, the null hypothesis is true whenever $\mu \geq 3$, and we considered only the case when $\mu = 3$. What about the case when $\mu > 3$? If $\mu > 3$, the probability of making a Type I error will be less than it is when $\mu = 3$; that is, in this case, it is even less likely that we will find a value of the test statistic that is in the rejection region. Since the objective of the hypothesis testing procedure is to limit the maximum probability of making a Type I error, the critical value for the test is established using $\mu = 3$.

■■■ FIGURE 9.4   **Value of the Test Statistic for $\bar{x} = 2.92$ is in the Rejection Region**

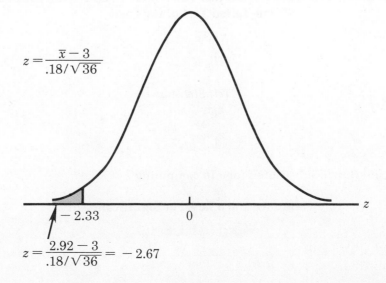

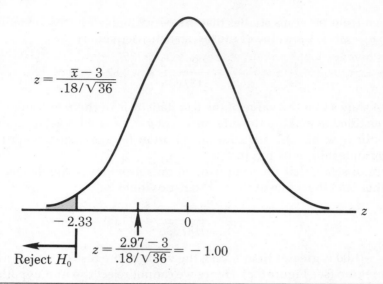

$$z = \frac{\bar{x} - 3}{.18/\sqrt{36}}$$

-2.33 0 z

$$z = \frac{2.97 - 3}{.18/\sqrt{36}} = -1.00$$

Reject H_0

■ FIGURE 9.5   **Value of the Test Statistic for $\bar{x} = 2.97$ is not in the Rejection Region**

Summary: One-Tailed Tests About a Population Mean

Let us generalize the hypothesis-testing procedure for one-tailed tests about a population mean. We restrict ourselves here to the large-sample case ($n \geq 30$) where the central limit theorem permits us to assume a normal sampling distribution for $\bar{x}$. In this large-sample case when σ is unknown, we simply substitute the sample standard deviation s for σ in computing the test statistic. The general form of a lower-tailed test is as follows (see Figure 9.3), where μ_0 is a stated value for the population mean.

Large-Sample ($n \geq 30$) Hypothesis Test About a Population Mean for a One-Tailed Test of the Form

$$H_0: \mu \geq \mu_0$$

$$H_a: \mu < \mu_0$$

Test Statistic:

$$z = \frac{\bar{x} - \mu_0}{\sigma/\sqrt{n}}$$

If σ is unknown, substitute s for σ in computing z.

Rejection Rule at a Level of Significance of α

Reject H_0 if $z < -z_\alpha$ (9.1)

A second form of the one-tailed test rejects the null hypothesis when the test statistic is in the upper tail of the sampling distribution. This one-tailed test and rejection rule are summarized next (see Figure 9.6). Again, we are considering the large-sample case; when σ is unknown, s may be substituted for σ in the computation of the test statistic, z.

Large-Sample ($n \geq 30$) Hypothesis Test About a Population Mean for a One-Tailed Test of the Form

$$H_0: \mu \leq \mu_0$$

$$H_a: \mu > \mu_0$$

Test Statistic:

$$z = \frac{\bar{x} - \mu_0}{\sigma/\sqrt{n}}$$

If σ is unknown, substitute s for σ in computing z.

Rejection Rule at a Level of Significance of α

$$\text{Reject } H_0 \text{ if } z > z_\alpha \tag{9.2}$$

The Use of *p*-Values

We now consider another approach that is sometimes used in hypothesis testing. This approach is based upon what is called a *p-value*. We will show how the

---

▆▆▆ FIGURE 9.6   **Rejection Region for an Upper-Tailed Hypothesis Test about a Population Mean**

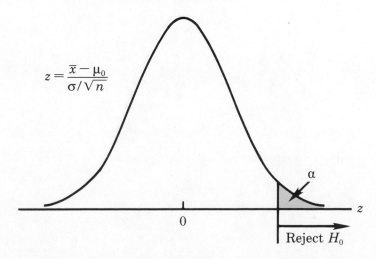

p-value for a sample can be computed from the value of the test statistic z. The p-value can also be used to make the decision whether or not to reject H_0.

Assuming that the null hypothesis is true, the p-value is the probability of obtaining a sample result that is more unlikely than what is observed. In the Hilltop Coffee example, the rejection region is in the lower tail; therefore the p-value is the probability of observing a sample mean smaller (more unlikely given the null hypothesis) than what was observed.

Let us compute the p-value associated with the sample mean $\bar{x} = 2.92$ in the Hilltop Coffee example. The p-value in this case is the probability of obtaining a value for the sample mean that is smaller than the observed value of $\bar{x} = 2.92$, given the hypothesized value for the population mean of $\mu = 3$. Previously we showed that the test statistic, $z = -2.67$, corresponded to $\bar{x} = 2.92$. Thus, as shown in Figure 9.7, the p-value is the area in the tail of the standard normal probability distribution less than $z = -2.67$. Using the standard normal probability distribution table, we find that the area between the mean and $z = -2.67$ is .4962. Thus there is a $.5000 - .4962 = .0038$ probability of obtaining a sample mean that is smaller than the observed $\bar{x} = 2.92$. Thus the p-value is .0038. This p-value shows us that there is a very small probability of obtaining a sample mean smaller than $\bar{x} = 2.92$ when sampling from a population with $\mu = 3$.

The p-value can be used to make the decision for a hypothesis test by noting that if the *p-value is less than the level of significance, α*, the value of the test statistic must be in the *rejection region*. Similarly, if the p value is *greater than or equal to α*, the value of the test statistic is not in the rejection region. For the Hilltop Coffee example, the fact that the p-value of .0038 is less than the level of significance, $\alpha = .01$, indicates that the null hypothesis should be rejected. Given the stated level of significance, α, for any hypothesis test, the decision of whether or not to reject H_0 can be made in terms of the p-value as follows:

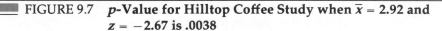

FIGURE 9.7 ▬ ***p*-Value for Hilltop Coffee Study when $\bar{x} = 2.92$ and $z = -2.67$ is .0038**

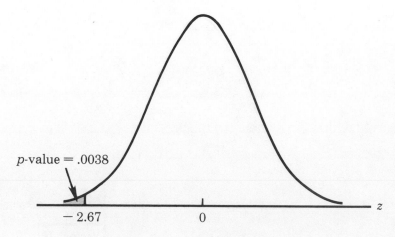

p-value $= .0038$

-2.67 0 z

> ## *p*-Value Criterion for Hypothesis Testing
>
> Reject H_0 if the *p*-value $< \alpha$

The *p*-value and the corresponding test statistic will always provide the same hypothesis-testing conclusion at a chosen level of significance α. When the rejection region is in the lower tail of the sampling distribution, the *p*-value is the area under the curve to the left of the test statistic. When the rejection region is in the upper tail of the sampling distribution, the *p*-value is the area under the curve to the right of the test statistic. A small *p*-value thus indicates a sample result that is unusual given the assumption that H_0 is true. Thus small *p*-values lead to rejection of H_0. On the other hand, a relatively large *p*-value indicates that the sample results are not inconsistent with what should be expected when the null hypothesis is true.

Refer again to the Hilltop Coffee example, but suppose that the sample result was $\bar{x} = 2.97$. The corresponding value of the test statistic is $z = -1.00$. Using the standard normal probability distribution table, we find an area of $.5000 - .3413 = .1587$ in the lower tail of the sampling distribution. Thus, the *p*-value $= .1587$; since the *p*-value is larger than $\alpha = .01$, we can conclude that the test statistic is not in the rejection region. As a result, the null hypothesis $\mu \geq 3$ cannot be rejected.

▬▬ **NOTES AND COMMENTS** ▬▬

The *p*-value is often called the *observed level of significance* for the test. It is a measure of how unlikely the sample results are assuming the null hypothesis is true. The smaller the *p*-value, the less likely the sample results. Most statistical software packages print the *p*-value associated with a hypothesis test.

▬▬ EXERCISES

9. According to *Business Week* (February 6, 1989), the mean cost of a heart-bypass operation is $26,100; in addition, approximately 230,000 operations are performed annually. A sample of 36 bypass operations in a particular city showed a mean cost of $\bar{x} = \$25,000$ and $s = \$2400$.

   a. Develop appropriate hypotheses to test whether or not the mean bypass operation cost is less than $26,000 in this city.

   b. Use $\alpha = .05$. What is your conclusion?

   c. What is the *p*-value for this test?

10. Fightmaster and Associates Real Estate, Inc., advertises that the mean selling time of a residential home is 40 days or less after it is listed with the company. A sample of 50

recently sold residential homes shows a sample mean selling time of 45 days and a sample standard deviation of 20 days. Using a .02 level of significance, test the validity of the company's claim.

11. Fowle Marketing Research, Inc., bases charges to a client on the assumption that telephone surveys can be completed with a mean time of 15 minutes or less. If a greater mean survey time is required, a premium rate is charged the client. Suppose that a sample of 35 surveys shows a sample mean of 17 minutes and a sample standard deviation of 4 minutes. Is the premium rate justified? Test at the α = .01 level of significance.

12. New tires manufactured by a company in Findlay, Ohio, are designed to provide a mean of at least 28,000 miles. Tests with 30 tires show a sample mean of 27,500 miles with a sample standard deviation of 1000 miles. Using a .05 level of significance, test whether or not there is sufficient evidence to reject the claim of a mean of at least 28,000 miles. What is the p-value?

13. A company currently pays its production employees a mean wage of $15.00 per hour. The company is planning to build a new factory, and several locations are being considered. The availability of labor at a rate less than $15.00 per hour is a major factor in the location decision. For one location, a sample of 40 workers showed a current mean hourly wage of $\bar{x}$ = $14.00 and a sample standard deviation of s = $2.40.

   a. Using a .10 level of significance, does the sample data indicate that the location has a mean wage rate significantly below the $15.00 per hour rate?
   b. What is the p-value?

14. A new diet program claims that participants will lose on average at least 8 pounds during the first week of the program. A random sample of 40 people participating in the program showed a sample mean weight loss of 7 pounds. The sample standard deviation was 3.2 pounds.

   a. What is the rejection rule with α = .05?
   b. What is your conclusion about the claim made by the diet program?
   c. What is the p-value?

9.4 ■ TWO-TAILED TESTS ABOUT A POPULATION MEAN

Two-tailed hypothesis tests differ from one-tailed tests in that the rejection region is placed in both the lower and the upper tails of the sampling distribution. Let us introduce an example to show how and why two-tailed tests are conducted.

The United States Golf Association (USGA) has established rules which manufacturers of golf equipment must meet in order to have their products acceptable for use in USGA events. One of the rules regarding the manufacture of golf balls states that "A brand of golf ball, when tested on apparatus approved by the USGA on the outdoor range at the USGA Headquarters . . . shall not cover an average distance in carry and roll exceeding 280 yards. . . ." Suppose that Superflight, Inc. has recently developed a high-technology manufacturing method which can produce golf balls that have an average distance in carry and roll of 280 yards.

Superflight realizes, however, that if the new manufacturing process goes out of adjustment, the process may produce balls with an average distance less than

280 yards or with an average distance greater than 280 yards. In the former case, Superflight may experience a downturn in sales as a result of marketing an inferior product, and in the latter case Superflight may have their golf balls rejected by the USGA. As a result, management of Superflight have instituted a quality control program in order to carefully monitor the new manufacturing process.

As part of the quality control program, an inspector periodically selects a sample of balls from the production line and subjects them to tests which are equivalent to those performed by the USGA. With no good reason to doubt that the manufacturing process is functioning correctly, the "innocent until proven guilty" analogy suggests establishing the following null and alternative hypotheses:

$$H_0: \quad \mu = 280$$

$$H_a: \quad \mu \neq 280$$

As usual, we make the tentative assumption that the null hypothesis is true. A rejection region must be established for the test statistic, z. We want to reject the claim that $\mu = 280$ when the z-value indicates that the sample mean, $\bar{x}$, is significantly less than 280 yards or when the z-value indicates that the sample mean is significantly greater than 280 yards. Thus, H_0 should be rejected for values of the test statistic in either the lower tail or the upper tail of the sampling distribution. As a result the test will be referred to as a *two-tailed* test.

Following the hypothesis-testing procedure developed in the previous sections, we first specify a level of significance by determining a maximum allowable probability of making a Type I error. Suppose we choose $\alpha = .05$ as the level of significance. This means that there will be a .05 probability of concluding that the mean distance is not 280 yards when in fact it is. The test statistic is

$$z = \frac{x - \mu}{\sigma/\sqrt{n}}$$

Figure 9.8 shows the sampling distribution of z with the two-tailed rejection region for $\alpha = .05$. With two-tailed hypothesis tests, we will always determine the rejection region by placing an area or probability of $\alpha/2$ in each tail of the distribution. The values of z that provide an area of .025 in each tail can be found from the standard normal probability distribution table. We see in Figure 9.8 that $-z_{.025} = -1.96$ identifies an area of .025 in the lower tail and $z_{.025} = +1.96$ identifies an area of .025 in the upper tail. Referring to Figure 9.8, we can establish the following rejection rule:

Reject H_0 if $z < -1.96$ or if $z > 1.96$

Suppose that a sample of 36 golf balls provides a sample mean distance of $\bar{x} = 278.5$ yards and a sample standard deviation of $s = 12$ yards. Using the value of μ from the null hypothesis and the sample standard deviation of $s = 12$ as an

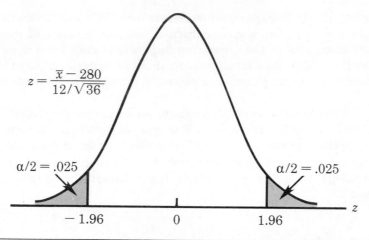

$$z = \frac{\bar{x} - 280}{12/\sqrt{36}}$$

$\alpha/2 = .025$

$\alpha/2 = .025$

-1.96 0 1.96 z

■■■ FIGURE 9.8 **FIGURE 9.8   Rejection Region (Shaded Area) for the Two-Tailed Hypothesis Test of the Superflight, Inc. Golf Ball Claim**

estimate of the population standard deviation, σ, the value of the test statistic is

$$z = \frac{\bar{x} - \mu}{\sigma/\sqrt{n}} = \frac{278.5 - 280}{12/\sqrt{36}} = -0.75$$

According to the rejection rule, H_0 cannot be rejected. The sample results do not indicate that the quality control manager has reason to doubt that the manufacturing process is producing golf balls with a mean distance of 280 yards.

p-Values for Two-Tailed Tests

To compute the *p*-value for a two-tailed test we must first calculate the area in the tail of the sampling distribution corresponding to the observed value of $\bar{x}$. The *p*-value for the two-tailed test is then computed by doubling this area or probability. For instance, for the Superflight golf ball example, $\bar{x} = 278.5$; the corresponding z value was -0.75. The table for the standard normal probability distribution shows that the area between the mean and $z = -0.75$ is .2734. Thus, the area in the lower tail is $.5000 - .2734 = .2266$. The *p*-value for this test is double this value: *p*-value $= 2(.2266) = .4532$. Since the *p*-value is greater than the level of significance ($\alpha = .05$), the *p*-value also indicates that the null hypothesis cannot be rejected.

Summary: Two-Tailed Tests About a Population Mean

Let μ_0 represent the value of the mean as claimed in the null hypothesis. The general form of the two-tailed hypothesis test about a population mean is

---

Large-Sample ($n \geq 30$) Hypothesis Test About a Population Mean for a Two-Tailed Test of the Form

$$H_0: \quad \mu = \mu_0$$

$$H_a: \quad \mu \neq \mu_0$$

Test Statistic:

$$z = \frac{\bar{x} - \mu_0}{\sigma/\sqrt{n}}$$

If σ is unknown, substitute s for σ in computing z.

Rejection Rule at a Level of Significance of α

Reject H_0 if $z < -z_{\alpha/2}$ or if $z > z_{\alpha/2}$ $\qquad\qquad$ (9.3)

The Relationship Between Interval Estimation and Hypothesis Testing

In Chapters 8 and 9 we have discussed statistical procedures that can be used to make inferences about the value of a population mean. In Chapter 8 we discussed interval estimation, and in Chapter 9 we have focused on hypothesis testing. In the case of interval estimation, the population mean μ was unknown. Once the sample was selected and the sample mean $\bar{x}$ computed, we developed an interval around the value of $\bar{x}$ that had a good chance of including the value of the parameter μ. The interval estimate computed was referred to as a confidence interval with $1 - \alpha$ defined as the confidence coefficient. In the large-sample case, the formula for interval estimation of a population mean was given by

$$\bar{x} \pm z_{\alpha/2} \frac{\sigma}{\sqrt{n}} \qquad\qquad (9.4)$$

Conducting a hypothesis test requires us first to make an assumption about the value of a population parameter. In the case of the population mean, the two-tailed hypothesis test has the form

$$H_0: \quad \mu = \mu_0$$

$$H_a: \quad \mu \neq \mu_0$$

where μ_0 is the hypothesized value for the population mean. Using the decision rule provided by (9.1), we see that the region over which we do not reject includes all values of the sample mean $\bar{x}$ that are within $-z_{\alpha/2}$ and $+z_{\alpha/2}$ standard errors of μ_0. Thus the following expression provides the do-not-reject region for the sample mean $\bar{x}$ in a two-tailed hypothesis test with a level of significance of α:

$$\mu_0 \pm z_{\alpha/2} \frac{\sigma}{\sqrt{n}} \qquad (9.5)$$

A close look at (9.4) and (9.5) will provide insight into the relationship between the estimation and hypothesis-testing approaches to statistical inference. Note in particular that both procedures require the computation of the values $z_{\alpha/2}$ and $\sigma/\sqrt{n}$. Focusing on α, we see that a confidence coefficient of $(1 - \alpha)$ for interval estimation corresponds to a level of significance of α in hypothesis testing. For example, a 95% confidence interval for estimation corresponds to a .05 level of significance for hypothesis testing. Furthermore, (9.4) and (9.5) show that since $z_{\alpha/2}$ $(\sigma/\sqrt{n})$ is the plus or minus value for both expressions, if $\bar{x}$ falls in the do-not-reject region defined by (9.5), the hypothesized value μ_0 will be in the confidence interval defined by (9.4). Conversely, if the hypothesized value μ_0 falls in the confidence interval defined by (9.4), the sample mean $\bar{x}$ will be in the do-not-reject region for the hypothesis $H_0: \mu = \mu_0$. These observations lead to the following procedure for using confidence interval results to draw hypothesis-testing conclusions:

A Confidence Interval Approach to Testing a Hypothesis of the Form

$$H_0: \quad \mu = \mu_0$$

$$H_a: \quad \mu \neq \mu_0$$

1. Select a simple random sample from the population and use the value of the sample mean $\bar{x}$ to develop the confidence interval

$$\bar{x} \pm z_{\alpha/2} \frac{\sigma}{\sqrt{n}}$$

2. If the confidence interval contains the hypothesized value μ_0, do not reject H_0. Otherwise, reject H_0.

Let us return to the Superflight golf ball study discussed earlier to demonstrate the use of a confidence interval for hypothesis testing. The Superflight golf ball study resulted in the following two-tailed test:

$$H_0: \mu = 280$$

$$H_a: \mu \neq 280$$

In order to test this hypothesis with a level of significance of $\alpha = .05$, we sampled 36 golf balls and found a sample mean distance of $\bar{x} = 278.5$ yards and a sample standard deviation of $s = 12$ yards. Using these results with $z_{.025} = 1.96$, the 95%

confidence interval estimate of the population mean becomes

$$\bar{x} \pm z_{.025} \frac{\sigma}{\sqrt{n}}$$

$$278.5 \pm (1.96) \left(\frac{12}{\sqrt{36}} \right)$$

$$278.5 \pm 3.92$$

or

$$274.58 \text{ to } 282.42$$

This finding enables the quality control manager to conclude with 95% confidence that the mean distance for the population of golf balls is between 274.58 yards and 282.42 yards. Since the hypothesized value for the population mean, $\mu_0 = 280$, is in this interval, the hypothesis-testing conclusion is that the null hypothesis, $H_0: \mu = 280$, cannot be rejected.

Note that this discussion and example have been devoted to two-tailed hypothesis tests about a population mean. However, the same confidence interval and hypothesis-testing relationship exists for other population parameters as well. In addition, the relationship can be extended to make one-tailed tests about population parameters. However, this requires the development of one-sided confidence intervals.

NOTES AND COMMENTS

1. The p-value is called the observed level of significance; it depends only on the sample outcome. One does not need to know the level of significance to compute the p-value. But, it is necessary to know whether the hypothesis test being investigated is one-tailed or two-tailed. Given the value of $\bar{x}$ in a sample, the p-value for a two-tailed test will always be *twice* the area in the tail of the sampling distribution at the value of $\bar{x}$.

2. The interval estimation approach to hypothesis testing helps to highlight the role of the sample size. From (9.4), it can be seen that larger sample sizes, n, lead to more narrow confidence intervals. Thus, for a given level of significance (α), a larger sample is less likely to lead to an interval containing μ_0 when the null hypothesis is false. That is, the larger sample size will provide a higher probability of rejecting H_0 when H_0 is false.

▪▪▪ EXERCISES

15. The U.S. Bureau of the Census reported mean hourly earnings in the wholesale trade industry of $9.70 per hour in 1987. A sample of 49 wholesale trade workers in a particular

city showed a sample mean hourly wage of $\bar{x}$ = $9.30 with a standard deviation of s = $1.05. Use $H_0: \mu = 9.70$ and $H_a: \mu \neq 9.70$.

a. Test to see if wage rates in the city differ significantly from the reported $9.70. Use $\alpha = .05$.

b. What is the p-value for this test?

16. A study of the operation of a city-owned parking garage shows a historical mean parking time of 220 minutes per car. The garage area has recently been remodeled and the parking charges have been increased. The city manager would like to know if these changes have had any effect on the mean parking time. Test the hypotheses $H_0: \mu = 220$ and $H_a: \mu \neq 220$ at a .05 level of significance.

a. What is your conclusion if a sample of 50 cars showed $\bar{x}$ = 208 and s = 80?

b. What is the p-value?

17. A production line operates with a filling weight standard of 16 ounces per container. Overfilling or underfilling is a serious problem, and the production line should be shut down if either occurs. From past data σ is known to be .8 ounces. A quality-control inspector samples 30 items every 2 hours and at that time makes the decision of whether or not to shut the line down for adjustment.

a. With a .05 level of significance, what is the rejection rule for the hypothesis testing procedure?

b. If a sample mean of $\bar{x}$ = 16.32 ounces occurs, what action would you recommend?

c. If $\bar{x}$ = 15.82 ounces, what action would you recommend?

d. What is the p-value for parts (b) and (c)?

18. An automobile assembly-line operation has a scheduled mean completion time of 2.2 minutes. Because of the effect of completion time on both earlier and later assembly operations it is important to maintain the 2.2-minute standard. A random sample of 45 times shows a sample mean completion time of 2.39 minutes, with a sample standard deviation of .20 minutes. Use a .02 level of significance and test whether or not the operation is meeting its 2.2-minute standard.

19. Historically, evening long-distance phone calls from a particular city have averaged 15.20 minutes per call. In a random sample of 35 calls, the sample mean time was 14.30 minutes per call, with a sample standard deviation of 5 minutes. Use this sample information to test whether or not there has been a change in the mean duration of long distance phone calls. Use a .05 level of significance. What is the p-value?

20. The mean salary for full professors at public universities is $45,300 (*Statistical Abstract of the United States*, 1988). A sample of 36 full professors at business colleges showed $\bar{x}$ = $52,000 and s = $5,000. Choose $H_0: \mu = 45,300$ and $H_a: \mu \neq 45,300$.

a. Develop a 95% confidence interval for the mean salary of business college professsors using the sample data.

b. Use the confidence interval to conduct the hypothesis test. What is your conclusion?

21. At Western University the historical mean scholarship examination score of entering students has been 900, with a standard deviation of 180. Each year a sample of applications is taken to see if the examination scores are at the same level as in previous years. The null hypothesis tested is $H_0: \mu = 900$. A sample of 200 students in this year's class shows a sample mean score of 935. Use a .05 level of significance.

a. Use a confidence interval estimation procedure to conduct a hypothesis test.

b. Use a test statistic to test this hypothesis.

c. What is the p-value for this test?

22. An industry pays an average wage rate of $9.00 per hour. A sample of 36 workers from one company showed a mean wage of $\bar{x}$ = $8.50 and a sample standard deviation of s = $.60.

a. A one-sided confidence interval uses the sample results to establish either an upper limit or a lower limit for the value of the population parameter. For this exercise establish an upper 95% confidence limit for the hourly wage rate paid by the company. The form of this one-sided confidence interval requires that we be 95% confident that the population mean is this value or less. What is the 95% confidence statement for this one-sided confidence interval?

b. Use the one-sided confidence interval result to test the hypothesis $H_0: \mu \geq 9$. What is your conclusion? Explain.

9.5 ■■■ HYPOTHESIS TESTS ABOUT A POPULATION MEAN—SMALL-SAMPLE CASE

The methods of hypothesis testing that we have discussed thus far have required sample sizes of at least 30 items. The reason for this is that in the large-sample situation ($n \geq 30$), the central limit theorem enables us to approximate the sampling distribution of $\bar{x}$ with a normal probability distribution. Thus

$$z = \frac{\bar{x} - \mu_0}{\sigma/\sqrt{n}} \tag{9.6}$$

is a standard normal random variable and is used as the test statistic. The sample standard deviation, s, can be used in (9.6) when the population standard deviation, σ, is unknown.

If the sample size is small ($n < 30$), we no longer use (9.6). However, if it is reasonable to assume the population has a normal distribution, the t distribution can be used to make inferences about the value of a population mean. In using the t distribution for hypothesis tests about a population mean, the test statistic is

$$t = \frac{\bar{x} - \mu_0}{s/\sqrt{n}} \tag{9.7}$$

This test statistic has a t distribution with $n - 1$ degrees of freedom when the null hypothesis is true. Noting the similarities of the test statistics defined by (9.6) and (9.7), it should not be surprising that the small-sample procedure for hypothesis tests about a population mean is very similar to the large-sample procedure. The difference is that in the small-sample case, the t distribution is used to establish the rejection region, and the t value given in (9.7) is used as the test statistic.

Consider the following example: A company that specializes in products for home gardening has developed a new plant food that has been designed to increase the growing height of plants. The new plant food was tested on a sample of 12 plants of a type known to have a mean growing height of 18 inches. Results showed a sample mean height of 19.4 inches and a sample standard deviation of 3 inches. Assume that the heights of the plants are normally distributed.

Using a .10 level of significance, is there reason to believe that the new plant food *increases* plant height? The null and alternative hypotheses are as follows:

$$H_0: \quad \mu \le 18$$
$$H_a: \quad \mu > 18$$

The new plant food will be judged to increase plant height if H_0 can be rejected.

The rejection region is located in the upper tail of the sampling distribution. With $n - 1 = 12 - 1 = 11$ degrees of freedom, Table 2 of Appendix B shows that $t_{.10} = 1.363$. Thus the rejection rule is

$$\text{Reject } H_0 \text{ if } t > 1.363$$

Using the sample results ($\bar{x} = 19.4$ and $s = 3$) and (9.7), we have the following value for the test statistic:

$$t = \frac{\bar{x} - \mu_0}{s/\sqrt{n}} = \frac{19.4 - 18}{3/\sqrt{12}} = 1.62$$

Since 1.62 is greater than 1.363, the null hypothesis is rejected. At the .10 level of significance, it can be concluded that the mean plant height exceeds 18 inches when the new plant food is used. Figure 9.9 shows that the value of the test statistic is in the rejection region.

The *p*-value for this example can be determined by using the observed *t* value of 1.62. However, due to the format of the *t* distribution table, *p*-values are slightly more difficult to determine than they were in Sections 9.3 and 9.4. For example, the *t* distribution in the plant food test has 11 degrees of freedom.

■■ FIGURE 9.9 **Value of Test Statistic ($t = 1.62$) is in the Rejection Region; Conclude the New Plant Food Increases Growing Height**

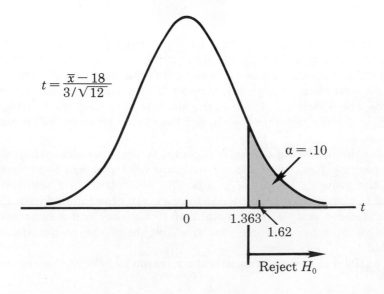

Referring to the row for 11 degrees of freedom in Table 2 of Appendix B, we see that 1.62 is between 1.363, occurring at a p-value of .10, and 1.796, occurring at a p-value of .05. Interpolation between these two values shows that $t = 1.62$ corresponds to a p-value of approximately .08. With a p-value of .08, which is less than the level of significance, $\alpha = .10$, the p-value criterion also indicates the null hypothesis should be rejected.

As an example of a two-tailed hypothesis test about a population mean using a small sample, consider the following production problem. A production process is designed to fill containers with a mean filling weight of $\mu = 16$ ounces. An undesirable condition exists if the process is underfilling containers and the consumer is not receiving the amount of product indicated on the container label. In addition, an equally undesirable condition exists if the process is overfilling containers; in this case the firm is losing money since the process is placing more product in the container than is required. Quality-assurance personnel periodically select a simple random sample of eight containers and test the following two-tailed hypotheses:

$$H_0: \quad \mu = 16$$

$$H_a: \quad \mu \neq 16$$

If H_0 is rejected, the production manager will request that the production process be stopped and that the mechanism for regulating filling weights be readjusted to provide a mean filling weight of 16 ounces. If the sample provides data values of 16.02, 16.22, 15.82, 15.92, 16.22, 16.32, 16.12, and 15.92 ounces, what action should be taken at a .05 level of significance? Assume that the population of filling weights is normally distributed.

Since the data have not been summarized, we must first compute the sample mean and sample standard deviation. Doing so provides the following results:

$$\bar{x} = \frac{\Sigma x_i}{n} = \frac{128.56}{8} = 16.07 \text{ ounces}$$

and

$$s = \sqrt{\frac{\Sigma(x_i - \bar{x})^2}{n - 1}} = \sqrt{\frac{.22}{7}} = .18 \text{ ounces}$$

With a two-tailed test and a level of significance of $\alpha = .05$, $-t_{.025}$ and $t_{.025}$ determine the rejection region for the test. Using the table for the t distribution, we find that with $n - 1 = 8 - 1 = 7$ degrees of freedom, $-t_{.025} = -2.365$ and $t_{.025} = +2.365$. Thus the rejection rule is written

Reject H_0 if $t < -2.365$ or if $t > 2.365$

Using $\bar{x} = 16.07$ and $s = .18$, we have

$$t = \frac{\bar{x} - \mu_0}{s/\sqrt{n}} = \frac{16.07 - 16.00}{.18/\sqrt{8}} = 1.10$$

Since $t = 1.10$ is not in the rejection region, the null hypothesis $\mu = 16$ ounces cannot be rejected. There is not enough evidence to stop the production process.

Using Table 2 of Appendix B and the row for 7 degrees of freedom, we see that the computed t value of 1.10 has an upper tail area of more than .10. Although the format of the t table prevents us from being more specific, we can at least conclude that the two-tailed p-value is greater than $2(.10) = .20$. Since this is greater than the .05 level of significance, we see that the p-value leads to the same conclusion; that is, do not reject H_0.

■ EXERCISES

23. City Homes bought 101 badly deteriorated Baltimore rowhouses and turned them into housing for the poor. The mean monthly rental for these houses was $200 (*Financial World*, November 29, 1988). Nine hundred homes in another city were built and turned over to a foundation to rent. A sample of 10 of these homes showed the following monthly rental rates:

$220 $190 $250 $230 $185 $210 $240 $260 $200 $195

a. Test at the $\alpha = .05$ level to see if the mean rental rate for the population of 900 homes exceeds the $200 per month mean rental rate in Baltimore. What is your conclusion?
b. Approximate the p-value for this test.

24. The manager of a long-distance trucking firm believes that the mean weekly loss due to damaged shipments is $2000 or less. A sample of 15 weeks of operations shows a sample mean weekly loss of $2200, with a sample standard deviation of $500. Using a .05 level of significance, test the assumption that the mean weekly loss is $2000 or less.

25. It is estimated that, on the average, a housewife with a husband and two children works 55 hours or less per week on household related activities. Shown below are the hours worked during a week for a sample of eight housewives.

58   52   64   63   59   62   62   55

a. Use $\alpha = .05$ to test the hypotheses $H_0: \mu \leq 55$, $H_a: \mu > 55$. What is your conclusion about the mean number of hours worked per week?
b. What is the p-value?

26. A study of a drug designed to reduce blood pressure used a sample of 25 men between the ages of 45 and 55. With μ indicating the mean change in blood pressure for the population of men receiving the drug, the hypotheses in the study were written: $H_0: \mu \geq 0$ and $H_a: \mu < 0$. Rejection of H_0 shows that the mean change is negative, indicating that the drug is effective in lowering blood pressure.

a. At a .05 level of significance, what conclusion should be drawn if $\bar{x} = -10$ and $s = 15$?
b. What is the p-value?

27. Last year the number of lunches served at an elementary-school cafeteria was normally distributed with a mean of 300 lunches per day. At the beginning of the current year, the price of a lunch was raised by 25¢. A sample of 6 days during the months of September, October, and November provided the following number of children being

served lunches: 290, 275, 310, 260, 270, and 275. Do these data indicate that the mean number of lunches per day has dropped compared to last year? Test the hypothesis H_0: $\mu \geq 300$ against the alternative hypothesis H_a: $\mu < 300$ at a .05 level of significance.

28. Joan's Nursery specializes in custom designed landscaping for residential areas. The labor cost associated with a particular landscaping proposal is estimated based on the number of plantings of trees, shrubs, and so on associated with the project. For cost-estimating purposes, management figures 2 hours of labor time for the planting of a medium-size tree. Actual times from a sample of 10 plantings during the past month are as follows (times in hours):

1.9 1.7 2.8 2.4 2.6 2.5 2.8 3.2 1.6 2.5

Using a .05 level of significance, test to see if the mean tree planting time exceeds 2 hours. What is your conclusion, and what recommendations would you consider making to management?

9.6 ■■■ HYPOTHESIS TESTS ABOUT A POPULATION PROPORTION

Using the symbol p to denote the population proportion and p_0 to denote a particular hypothesized value for the population proportion, the three forms for a hypothesis test about a population proportion are as follows:

$$H_0: \quad p \geq p_0 \qquad H_0: \quad p \leq p_0 \qquad H_0: \quad p = p_0$$
$$H_a: \quad p < p_0 \qquad H_a: \quad p > p_0 \qquad H_a: \quad p \neq p_0$$

The first two forms are one-tailed tests, whereas the third form is a two-tailed test. The specific form used depends upon the application of interest.

Hypothesis tests about a population proportion are based on the difference between the sample proportion, $\bar{p}$, and the hypothesized value, p_0. The methods used to conduct the tests are very similar to the procedures used for hypothesis tests about a population mean. The only difference is that we use the sample proportion, $\bar{p}$, and its standard deviation, $\sigma_{\bar{p}}$, in developing the test statistic. We begin by formulating null and alternative hypotheses about the value of the population proportion. Then, using the value of the sample proportion, $\bar{p}$, and its standard deviation, $\sigma_{\bar{p}}$, we compute a value for the test statistic, z. Comparing the value of the test statistic to the critical value enables us to determine whether or not the null hypothesis should be rejected.

Let us illustrate hypothesis testing for a population proportion by considering the situation faced by Pine Creek golf course. Over the past few weeks, 20% of the players at Pine Creek have been female. In an effort to increase the proportion of female players, course management used a special promotion for female players. After one month, a random sample of 400 players showed 300 male and 100 female players. Course management would like to determine if the data support the conclusion that there has been an increase in the proportion of female players at Pine Creek.

Since we want to determine if the effect of the promotion is significant, the null hypothesis is that the promotion has no effect; the alternative hypothesis is

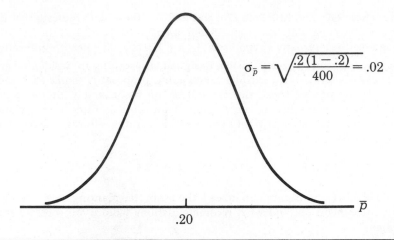

$$\sigma_{\bar{p}} = \sqrt{\frac{.2(1 - .2)}{400}} = .02$$

.20

$\bar{p}$

■■ FIGURE 9.10   Sampling Distribution of $\bar{p}$ for the Proportion of Female Golfers at Pine Creek Golf Course

that the promotion has increased the proportion of female players. The appropriate hypotheses are:

$$H_0: \quad p \leq .20$$

$$H_a: \quad p > .20$$

As usual, we tentatively begin the hypothesis-testing procedure by assuming that H_0 is true and hence that $p = .20$. Using the sample proportion $\bar{p}$ to estimate p, we next consider the sampling distribution of $\bar{p}$. Since $\bar{p}$ is an unbiased estimator of p, we know that if $p = .20$, the mean of the sampling distribution of $\bar{p}$ is .20. In addition, we know from Chapter 7 that the standard deviation of $\bar{p}$ is given by

$$\sigma_{\bar{p}} = \sqrt{\frac{p(1 - p)}{n}}$$

With the assumed value of $p = .20$ and a sample size of $n = 400$, the standard deviation of $\bar{p}$ for the Pine Creek golf course problem is

$$\sigma_{\bar{p}} = \sqrt{\frac{.20(1 - .20)}{400}} = .02$$

In Chapter 7 we saw that the sampling distribution of $\bar{p}$ can be approximated by a normal probability distribution if both np and $n(1 - p)$ are greater than or equal to 5. In the Pine Creek case, $np = 400(.20) = 80$ and $n(1 - p) = 400(.80) = 320$; thus the normal distribution approximation is appropriate. The sampling distribution of $\bar{p}$ for the Pine Creek problem is shown in Figure 9.10.

Since the sampling distribution of $\bar{p}$ is approximately normal, the following test statistic can be used:

Test Statistic for Hypothesis Tests about Proportions

$$z = \frac{\bar{p} - p_0}{\sigma_{\bar{p}}} \tag{9.8}$$

where

$$\sigma_{\bar{p}} = \sqrt{\frac{p_0(1 - p_0)}{n}} \tag{9.9}$$

Let us assume that $\alpha = .05$ has been selected as the level of significance for the test. With $z_{.05} = 1.645$, the upper-tail rejection region for the hypothesis test (see Figure 9.11) provides the following rejection rule:

$$\text{Reject } H_0 \text{ if } z > 1.645$$

Once the rejection rule has been determined, it is necessary to collect the data, compute the value of the point estimate, $\bar{p}$, and compute the corresponding value of the test statistic, z. By comparing the value of z to the critical value ($z_{.05} = 1.645$), a decision whether or not to reject the null hypothesis can be made.

Since 100 of the 400 players during the promotion were female, we obtain $\bar{p} = 100/400 = .25$. With $\sigma_{\bar{p}} = .02$, the value of the test statistic is

$$z = \frac{\bar{p} - p_0}{\sigma_{\bar{p}}} = \frac{.25 - .20}{.02} = 2.5$$

▬ FIGURE 9.11 **Rejection Region for Hypothesis Test Concerning the Proportion of Female Golfers at Pine Creek**

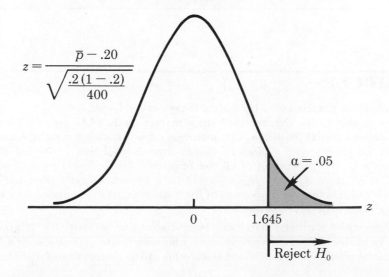

Thus the rejection rule tells us to reject H_0. As a result, Pine Creek would conclude that there has been an increase in the proportion of female players.

Using the table of areas for the standard normal probability distribution, we find that the p-value for the test can also be computed. For example, with $z = 2.50$, the table of areas shows a .4938 area or probability between the mean and $z = 2.50$. Thus the p-value for the test is $.5000 - .4938 = .0062$. With a p-value less than α, the rule for hypothesis tests using p-values shows that the null hypothesis should be rejected.

We see that hypothesis tests about a population proportion and a population mean are similar; the primary difference being that the test statistic is based on the sampling distribution of $\bar{x}$ when the hypothesis test involves a population mean and on the sampling distribution of $\bar{p}$ when the hypothesis test involves a population proportion. The tentative assumption that the null hypothesis is true, the use of the level of significance to establish the critical value, and the comparison of the test statistic to the critical value are identical for both testing procedures. A summary of the decision rules for hypothesis tests about a population proportion is presented in Figure 9.12. We assume the large-sample case ($np \geq 5$ and $n(1 - p) \geq 5$) where the normal distribution is a good approximation to the sampling distribution of $\bar{p}$.

NOTES AND COMMENTS

1. We have not shown the procedure for small-sample hypothesis tests involving population proportions. In the small-sample case the sampling distribution of $\bar{p}$ follows the binomial distribution and thus the normal approximation is not applicable. More advanced texts show how hypothesis tests are conducted for this situation. However, in practice small-sample tests are rarely conducted for a population proportion.
2. The relationship of interval estimation to hypothesis testing for the two-tailed test about a population proportion is the same as for the population mean. If the interval estimate computed using $\bar{p}$ does not contain p_0, the null hypothesis can be rejected.

■ EXERCISES

29. The American Association of Individual Investors conducted a survey to learn about member responses to the October 1987 stock market crash. (*AAII Journal*, May 1988). Results indicated that 11.3% of the AAII members owned stock on margin. A sample of 200 individual clients with a large brokerage firm showed that 29 owned stock on margin. Let p be the proportion of all the brokerage firm's clients owning stock on margin. Use H_0: $p \leq .113$ and H_a: $p > .113$ in a hypothesis test. Does the sample information indicate that the proportion of the brokerage firm's clients owning stock on margin is greater than .113? Use $\alpha = .05$.

30. The director of a college placement office claims that at least 80% of graduating seniors have made employment commitments 1 month prior to graduation. At a .05 level of significance, what is your conclusion if a sample of 100 seniors shows that 75 actually

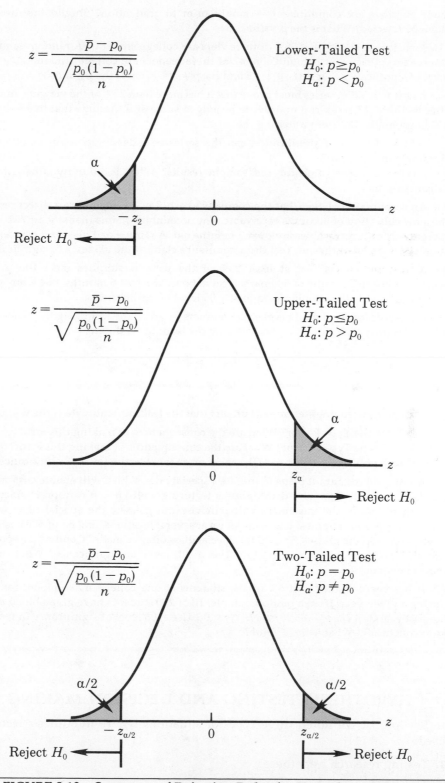

$$z = \frac{\bar{p} - p_0}{\sqrt{\dfrac{p_0(1 - p_0)}{n}}}$$

Lower-Tailed Test
$H_0: p \geq p_0$
$H_a: p < p_0$

α

$-z_2$ 0 z

Reject H_0 ◀

$$z = \frac{\bar{p} - p_0}{\sqrt{\dfrac{p_0(1 - p_0)}{n}}}$$

Upper-Tailed Test
$H_0: p \leq p_0$
$H_a: p > p_0$

α

0   z_α z

Reject H_0 ▶

$$z = \frac{\bar{p} - p_0}{\sqrt{\dfrac{p_0(1 - p_0)}{n}}}$$

Two-Tailed Test
$H_0: p = p_0$
$H_a: p \neq p_0$

$\alpha/2$ $\alpha/2$

$-z_{\alpha/2}$ 0 $z_{\alpha/2}$ z

Reject H_0 ◀   ▶ Reject H_0

▬▬ FIGURE 9.12   **Summary of Rejection Rules for Hypothesis Tests about Population Proportions**

made employment commitments 1 month prior to graduation? Should the director's claim be rejected? What is the p-value?

31. A magazine claims that 25% of its readers are college students. A random sample of 200 readers is taken. It is found that 42 of these readers are college students. Use a .10 level of significance to test $H_0: p = .25$ and $H_a: p \neq .25$. What is the p-value?

32. A new television series must prove that it has more than 25% of the viewing audience after its initial 13-week run in order to be judged successful. Assume that in a sample of 400 households, 112 were watching the series.

   a. At a .10 level of significance, can the series be judged successful based on the sample information?
   b. What is the p-value for the sample results? What is your hypothesis-testing conclusion?

33. An accountant believes that the company's cash-flow problems are a direct result of the slow collection of accounts receivable. The accountant claims that at least 70% of the current accounts receivables are over 2 months old. A sample of 120 accounts receivable shows 78 over 2 months old. Test the accountant's claim at the .05 level of significance.

34. A supplier claims that at least 96% of the parts it supplies meet the product specifications. In a sample of 500 parts received over the past 6 months, 464 were within specification. Test the supplier's claim at a .05 level of significance.

35. A television station predicts election winners based on the following hypothesis test, where p is the proportion of voters selecting the leading candidate.

$$H_0: \quad p \leq .50$$

$$H_a: \quad p > .50$$

If H_0 can be rejected, the station will predict that the leading candidate is the winner.

   a. What is the Type I error? What are the consequences of making this error?
   b. What is the Type II error? What are the consequences of making this error?
   c. Which value, $\alpha = .05$ or $\alpha = .01$, makes more sense as the level of significance?

36. A fast-food restaurant plans to initiate a special offer, which will enable customers to purchase specially designed drink glasses featuring well-known cartoon characters. If more than 15% of the customers will purchase the glasses, the special offer will be initiated. A preliminary test has been set up at several locations, and 88 of 500 customers have purchased the glasses. Should the special glass offer be made? Conduct a hypothesis test that will support your decision. Use a .01 level of significance. What is your recommendation?

37. It has been claimed that 5% of the students at one college make blood donations during a given year. If, in a random sample, 10 of 250 students have given blood during the past year, test $H_0: p = .05$ versus $H_a: p \neq .05$. Use a .05 level of significance in reaching your conclusion. What is the p value?

9.7 ▬ HYPOTHESIS TESTING AND DECISION MAKING

In Section 9.1 we noted three types of situations where hypothesis testing is used:

1. Testing research hypotheses
2. Testing the validity of assumptions
3. Decision making

In the first two situations, action is taken only when the null hypothesis H_0 is rejected and hence the alternative hypothesis H_a concluded to be true. In the third situation—decision making—it is appropriate to take action when the null hypothesis is not rejected as well as when it is rejected.

The hypothesis-testing procedures presented thus far have limited applicability in the decision-making situation because it is not considered appropriate to accept H_0 and take action based on the conclusion that H_0 is true. The reason given for not taking action when the test results indicate *do not reject* H_0 is that the decision to accept H_0 exposes the decision maker to the risk of making a Type II error; that is, accepting H_0 when it is false. With the hypothesis-testing procedures described in the previous sections, the probability of a Type I error was controlled by establishing a level of significance for the test. However, the probability of making the Type II error was not controlled.

Clearly, in decision-making situations the decision maker may want—and in some cases may be forced—to take action under both the conclusion do not reject H_0 and the conclusion reject H_0. A good illustration of this situation is lot-acceptance sampling, a topic we will discuss in more depth in Chapter 20. For example, a quality-control manager must decide to accept a shipment of batteries from a supplier or to return the shipment due to poor quality. Assume that design specifications require batteries from the supplier to have a mean useful life of at least 120 hours. In order to evaluate the quality of an incoming shipment, a sample of 36 batteries will be selected and tested. On the basis of the sample, a decision must be made to accept the shipment of batteries or return the shipment to the supplier because of poor quality. Let μ denote the mean hours of useful life for batteries in the shipment (population). The null and alternative hypotheses about the population mean are as follows:

$$H_0: \quad \mu \geq 120$$

$$H_a: \quad \mu < 120$$

If H_0 is rejected, the alternative hypothesis is concluded to be true. This conclusion indicates that the appropriate decision is to return the shipment to the supplier. However, if H_0 is not rejected, the decision maker must still make a decision regarding what action should be taken concerning the shipment. Thus, without directly concluding that H_0 is true, the decision maker will make the decision to accept the shipment as being of satisfactory quality.

In decision-making situations such as these, it is recommended that the hypothesis-testing procedure be extended to include a consideration of the probability of making a Type II error. Since a decision will be made and action taken when we do not reject H_0, knowledge of the probability of making a Type II error will determine the confidence we can have in the decision to accept that H_0 is true. Sections 9.8 and 9.9 are devoted to computing the probability of making a Type II error and to understanding how the sample size can be adjusted to help control the probability of making a Type II error.

9.8 ■ CALCULATING THE PROBABILITY OF TYPE II ERRORS

In this section we show how to calculate the probability of making a Type II error for a hypothesis test concerning a population mean. We will illustrate the procedure using the lot-acceptance example described in Section 9.7. The null and alternative hypotheses about the mean hours of useful life for a shipment of batteries were written as follows:

$$H_0: \quad \mu \geq 120$$

$$H_a: \quad \mu < 120$$

If H_0 is rejected, the decision will be made to return the shipment to the supplier because the mean hours of useful life is less than the specified 120 hours. If H_0 is not rejected, the decision will be made to accept the shipment.

Suppose that a level of significance of $\alpha = .05$ is used to conduct the hypothesis test. The test statistic is

$$z = \frac{\bar{x} - \mu}{\sigma/\sqrt{n}} = \frac{\bar{x} - 120}{\sigma/\sqrt{n}}$$

With $z_{.05} = 1.645$, the rejection rule for the lower-tailed test becomes

$$\text{Reject } H_0 \text{ if } z < -1.645$$

Suppose that a sample of 36 batteries will be selected, and that it is known from previous testing that the standard deviation for the population is $\sigma = 12$ hours. The preceding rejection rule indicates we will reject H_0 when

$$z = \frac{\bar{x} - 120}{12/\sqrt{36}} < -1.645$$

Solving for $\bar{x}$ in the preceding expression indicates that we will reject H_0 whenever

$$\bar{x} < 120 - 1.645\left(\frac{12}{\sqrt{36}}\right) = 116.71$$

Rejecting H_0 when $\bar{x} < 116.71$ means that we will make the decision to accept the shipment whenever

$$\bar{x} \geq 116.71$$

Based on this information, we are now ready to compute probabilities associated with making a Type II error. First, recall that we make a Type II error whenever the true shipment mean is less than 120 hours and we make the decision to accept $H_0: \mu \geq 120$. Thus, in order to compute the probability of making a Type II error, we must select a value of μ less than 120 hours. For example, suppose that the shipment is considered to be of poor-quality if the batteries have a mean life of $\mu = 112$ hours. If $\mu = 112$ is really true, what is the probability of accepting $H_0: \mu \geq 120$, and hence committing a Type II error? To find this probability, note that it is the probability that the sample mean, $\bar{x}$, is greater than or equal to 116.71 when $\mu = 112$.

Figure 9.13 shows the sampling distribution of $\bar{x}$ when the mean is $\mu = 112$. The shaded area in the upper tail shows the probability of obtaining $\bar{x} \geq 116.71$. Using the standard normal distribution calculation based on Figure 9.13, we see that

$$z = \frac{\bar{x} - \mu}{\sigma/\sqrt{n}} = \frac{116.71 - 112}{12/\sqrt{36}} = 2.36$$

The standard normal probability distribution table shows that with $z = 2.36$, the area in the upper tail is $.5000 - .4909 = .0091$. This is the probability of accepting H_0 when $\mu = 112$. In other words, this is the probability of making a Type II error when $\mu = 112$. Denoting the probability of making a Type II error as β, we see that when $\mu = 112$, $\beta = .0091$. Therefore, we can conclude that *if the mean of the population is* 112 *hours*, the probability of making a Type II error is only .0091.

We can repeat these calculations for other values of μ less than 120. Doing this will show that there is a different probability of making a Type II error for each value of μ less than 120. For example, suppose that we consider the case where the shipment of batteries has a mean useful life of $\mu = 115$ hours. Using $\bar{x} \geq$ 116.71 to accept H_0, the z-value for $\mu = 115$ is given by

$$z = \frac{\bar{x} - \mu}{\sigma/\sqrt{n}} = \frac{116.71 - 115}{12/\sqrt{36}} = 0.86$$

From the standard normal probability distribution table, we find that the area in the upper tail of the standard normal probability distribution for $z = .86$ is

---

FIGURE 9.13   Probability of a Type II Error when $\mu = 112$

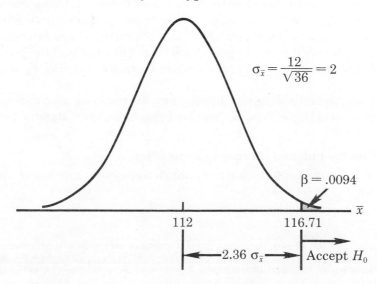

.5000 − .3051 = .1949. Thus the probability of making a Type II error is β = .1949 when the true mean is μ = 115.

In Table 9.2 we show the probability of making a Type II error for a variety of values of μ less than 120. Note that as μ increases toward 120, the probability of making a Type II error increases toward an upper bound of .95. However, as μ decreases to values further below 120, the probability of making a Type II error becomes smaller and smaller. This is what we should expect. When the true population mean μ is close to the null hypothesis value of μ = 120, there is a high probability that we will accept H_0 and make a Type II error. However, when the true population mean μ is far below the null hypothesis value of μ = 120, there is a low probability that we will accept H_0 and make a Type II error.

TABLE 9.2 Probabilities of Making a Type II Error for the Lot-Acceptance Hypothesis Test

Value of μ	$z = \dfrac{116.71 - \mu}{12/\sqrt{36}}$	Probability of Making a Type II Error (β)	Power $(1 - \beta)$
112	2.36	.0091	.9909
114	1.36	.0869	.9131
115	.86	.1949	.8051
116.71	.00	.5000	.5000
117	−.15	.5596	.4404
118	−.65	.7422	.2578
119.999	−1.645	.9500	.0500

The probability of correctly rejecting H_0 when it is false is called the *power* of the test. For any particular value of μ, the power is $1 - \beta$. That is, the probability of correctly rejecting the null hypothesis is 1 minus the probability of making a Type II error. Values of power are also shown in Table 9.2. Using these values, the power associated with each value of μ is shown graphically in Figure 9.14. Such a graph is called a *power curve*. Note that the power curve extends over the values of μ for which the null hypothesis is false. The height of the power curve at any value of μ provides the probability of correctly rejecting H_0 when H_0 is false.*

In summary, the following step-by-step procedure can be used to compute the probability of making a Type II error in hypothesis tests about a population mean.

1. Formulate the null and alternative hypotheses.

2. Use the level of significance α to establish a rejection rule based on the test statistic.

---

*Another graph, called the *operating characteristic curve*, is sometimes used to provide information about the probability of making a Type II error. The operating characteristic curve is a graph of β for the values of μ for which the null hypothesis is false. The probabilities of making Type II errors can be read directly from this graph.

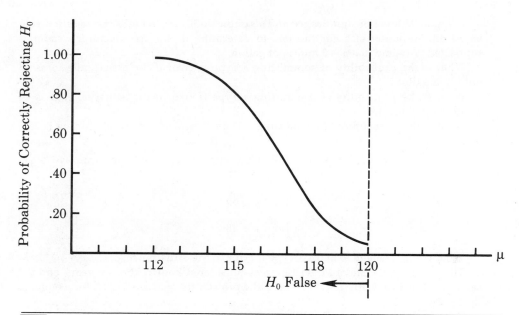

▬ FIGURE 9.14 **Power Curve for the Lot-Acceptance Hypothesis Test**

3. Using the rejection rule, solve for the value of the sample mean that identifies the rejection region for the test.

4. Use the results from step 3 to state the values of the sample mean that lead to the acceptance of H_0; this defines the acceptance region for the test.

5. Using the sampling distribution of $\bar{x}$ for any value of μ from the alternative hypothesis, and the acceptance region from step 4, compute the probability that the sample mean will be in the acceptance region. This probability is the probability of making a Type II error at the chosen value of μ.

6. Repeat step 5 for other values of μ from the alternative hypothesis.

▬ **EXERCISES**

38. In Exercise 11 Fowle Marketing Research, Inc. based charges to a client on the assumption that telephone surveys can be completed within 15 minutes or less. If more time is required, a premium rate is charged to the client. Using a sample of 35 surveys, a standard deviation of 4 minutes, and a level of significance of .01, the sample mean will be used to test the null hypothesis $H_0: \mu \le 15$.

   a. What is your interpretation of the Type II error for this problem? What is its impact on the firm?

   b. What is the probability of making a Type II error when the actual mean time is $\mu = 17$ minutes?

   c. What is the probability of making a Type II error when the actual mean time is $\mu = 18$ minutes?

   d. Sketch the general shape of the power curve for this test.

39. From the results of mileage testing, an automobile manufacturer claims that a new economy model will get at least 25 miles per gallon of gasoline.

a. Using a .02 level of significance and a sample of 30 cars, what is the rejection rule based on the value of $\bar{x}$ for the test to determine if the manufacturer's claim is supported? Assume that σ is 3 miles per gallon.
b. What is the probability of committing a Type II error if the actual mileage is 23 miles per gallon?
c. What is the probability of committing a Type II error if the actual mileage is 24 miles per gallon?
d. What is the probability of committing a Type II error if the actual mileage is 25.5 miles per gallon?

40. *Young Adult* magazine states the following hypotheses about the mean age of its subscribers:

$$H_0: \quad \mu = 28$$

$$H_a: \quad \mu \neq 28$$

a. What would it mean to make a Type II error in this situation?
b. The population standard deviation was considered known at $\sigma = 6$ years, and the sample size is 100. With $\alpha = .05$, what is the probability of accepting H_0 for μ equal to 26, 27, 29, and 30?
c. What is the power at $\mu = 26$? What does this tell you?

41. The production line operation referred to in Exercise 17 was tested for filling weight accuracy with the following hypotheses:

Hypothesis	Conclusion and Action
$H_0: \mu = 16$	Filling okay; keep running
$H_a: \mu \neq 16$	Filling off standard; stop and adjust machine

The sample size is 30 and the population standard deviation is considered known at $\sigma = .8$. Use $\alpha = .05$.
a. What would a Type II error mean in this situation?
b. What is the probability of making a Type II error when the machine is overfilling by .5 ounces?
c. What is the power of the statistical test when the machine is overfilling by .5 ounces?
d. Show the power curve for this hypothesis test. What information does it contain for the production manager?

42. Refer to Exercise 38. Repeat parts (b) and (c) if the firm selects a sample of 50 surveys. What observation can you make about how increasing the sample size affects the probability of making a Type II error?

43. Sparr Investments, Inc., specializes in tax-deferred investment opportunities for its clients. Recently Sparr has offered a payroll deduction investment program for the employees of a particular company. Sparr estimates that the employees are currently averaging $100 or less per month in tax deferred investments. A sample of 40 employees will be used to test Sparr's hypothesis about the current level of investment activity among the population of employees. Assume that the monthly employee tax deferred investment amounts have a standard deviation of $75 and that a .05 level of significance will be used in the hypothesis test.
a. What is the Type II error in this situation?
b. What is the probability of the Type II error if the actual mean employee monthly investment is $120?
c. What is the probability of the Type II error if the actual mean employee monthly investment is $130?

d. Repeat parts (b) and (c) if a sample of size 80 employees is used.

9.9 ■■ DETERMINING SAMPLE SIZES FOR HYPOTHESIS TESTING

The sample size for a hypothesis test can be selected in order to control both the probabilities of making a Type I error and a Type II error. To determine the sample size, the person conducting the hypothesis test must specify allowable probabilities for both the Type I and Type II errors. Using the lot-acceptance example discussed in Sections 9.7 and 9.8, the statements about the error probabilities might appear as follows:

Type I error statement: If the mean life of the batteries in the shipment is $\mu = 120$, we are willing to risk an $\alpha = .05$ probability of rejecting the shipment.

Type II error statement: If the mean life of the batteries in the shipment is 5 hours under the specification (that is, $\mu = 115$), we are willing to risk a $\beta = .10$ probability of accepting the shipment.

These statements are based on the judgment of the person conducting the test. Someone else might specify different restrictions on the probabilities. However, statements about the allowable probabilities of both errors must be made before the sample size can be determined.

In the two statements just shown, the statement about the probability of making a Type I error is a restatement of the level of significance for the test. As we have seen in the previous sections, the level of significance must be known in order to establish the rejection region for the test. The statement about the probability of making a Type II error is the additional statement that must be provided in order to determine the sample size.

Given the two statements, the sample size required for a one-tailed test can be determined using the following formula:

Recommended Sample Size for a One-Tailed Hypothesis Test

$$n = \frac{(z_\alpha + z_\beta)^2 \sigma^2}{(\mu_0 - \mu_a)^2} \tag{9.10}$$

where

z_α = z value providing an area of α in the upper tail of a standard normal distribution

z_β = z value providing an area of β in the upper tail of a standard normal distribution

σ = the population standard deviation

μ_0 = the value of the popualtion mean in the null hypothesis

μ_a = the value of the population mean in the statement about the Type II error

Note: In a two-tailed hypothesis test, use (9.10) with $z_{\alpha/2}$ replacing z_α.

For the lot-acceptance example, $\alpha = .05$ and $\beta = .10$. Thus, using the standard normal probability distribution, we have $z_{.05} = 1.645$ and $z_{.10} = 1.28$. From the statements about the error probabilities we note that null hypothesis, $\mu_0 = 120$ and $\mu_a = 115$. Finally the population standard deviation was assumed known at a value of $\sigma = 12$. Using (9.10), the recommended sample size for the lot-acceptance problem is

$$n = \frac{(1.645 + 1.28)^2(12)^2}{(120 - 115)^2} = 49.3$$

Rounding up, the recommended sample size is 50.

Since both the Type I and Type II error probabilities have been controlled at allowable levels with $n = 50$, the quality-control manager is now justified in using the accept H_0 and reject H_0 statements for the hypothesis test. The accompanying inferences of H_0 true or H_a true are made with known and allowable probabilities of error.

Before closing this section, let us make some comments about the relationship among α, β, and the sample size n.

1. Once two of the three values are known, the other can be computed.
2. For a given level of significance α, increasing the sample size will reduce β.
3. For a given sample size, decreasing α will increase β, whereas increasing α will decrease β.

The third observation is something to keep in mind when the probability of a Type II error is not being controlled. It suggests that one should not choose unnecessarily small values for the level of significance. For a given sample size, choosing a smaller level of significance α means more exposure to a Type II error. Inexperienced users of hypothesis testing often think that smaller values of α are always better. This is true if we are concerned only about the Type I error. However, smaller values of α have the disadvantage of increasing the risk of making a Type II error.

■■■ EXERCISES

44. Suppose that the project director for the Hilltop Coffee study (see Section 9.3) had asked for a .10 probability of not claiming that Hilltop was in violation when it really was underfilling by 1 ounce ($\mu = 2.9375$ pounds). What sample size would have been recommended?

45. A special industrial battery must have a life of at least 400 hours. A hypothesis test is to be conducted with a .02 level of significance. If the batteries from a particular production run have an actual mean use life of 385 hours, the production manager would like a sampling procedure that erroneously concludes the batch is acceptable only 10% of the time. What sample size is recommended for the hypothesis test? Use 30 hours as an estimate of the population standard deviation.

46. For the *Young Adult* magazine study in Exercise 40, H_0: $\mu = 28$ was tested at a .05 level of significance. If the manager conducting the test will permit only a .15 probability of

making a Type II error when the true mean age is 29, what sample size should be selected? Assume $\sigma = 6$.

47. The automobile mileage study in Exercise 39 tested the following hypotheses:

Hypothesis	Conclusion
$H_0: \mu \geq 25$	Manufacturer's claim supported
$H_1: \mu < 25$	Manufacturer's claim rejected; average mileage per gallon less than stated

For $\sigma = 3$ and a .02 level of significance, what sample size would be recommended if the tester desired an 80% chance of detecting that μ was less than 25 miles per gallon when it was actually $\mu = 24$?

▆ SUMMARY

Hypothesis testing is a statistical inference process that uses sample data to test claims about a population and its parameters. The claims, which come from a variety of sources, must be stated in two competing hypotheses: a null hypothesis, H_0, and an alternative hypothesis, H_a. In some applications it is not obvious how the null and alternative hypotheses should be formulated. We suggested guidelines for developing hypotheses based on three types of situations most frequently encountered.

1. Testing research hypotheses: The research hypothesis should be formulated as the alternative hypothesis. The null hypothesis is based on an established theory or the statement that the research treatment will have no effect. Whenever the sample data contradict the null hypothesis, the null hypothesis is rejected. In this case, the alternative, or research, hypothesis is supported and can be claimed true.

2. Testing the validity of assumptions: Generally, this situation corresponds to the "innocent until proven guilty" analogy. The assumption or claim made is chosen as the null hypothesis; the challenge to the assumption or claim is chosen as the alternative hypothesis. Action against the assumption will be taken whenever the sample data contradict the null hypothesis. When this occurs, the challenge implied by the alternative hypothesis can be claimed true.

3. Decision making: This type of situation occurs when a decision maker must choose between two courses of action, one associated with the null hypothesis and one associated with the alternative hypothesis. In this situation it is often suggested that the hypotheses be formulated such that the Type I error is the more serious error. However, whenever a decision must be made based on the decision to accept H_0, the hypothesis-testing procedure should be extended to control the probability of making a Type II error. These procedures were discussed in Sections 9.8 and 9.9.

We discussed in detail hypothesis-testing procedures concerning a population mean. In the large-sample case ($n \geq 30$) the central limit theorem is applicable and the test statistic

$$z = \frac{\bar{x} - \mu_0}{\sigma/\sqrt{n}}$$

is used when the population standard deviation, σ, is known. When σ is unknown, the sample standard deviation, s, is substituted for σ in computing the above test statistic.

In the small-sample case ($n < 30$), we must be able to assume that the population has a normal distribution in order to conduct hypothesis tests about a population mean. In this case, we use the sample standard deviation and the t distribution to develop the following test statistic:

$$t = \frac{\bar{x} - \mu_0}{s/\sqrt{n}}$$

Hypothesis tests concerning a population proportion were developed for the large-sample case ($np \geq 5$ and $n(1 - p) \geq 5$). The test statistic used is

$$z = \frac{\bar{p} - p_0}{\sqrt{\dfrac{p_0(1 - p_0)}{n}}}$$

The rejection rule for all the hypothesis-testing procedures involves comparing the value of the test statistic with a critical value. For lower-tail tests, the null hypothesis is rejected if the value of the test statistic is less than the critical value. For upper-tail tests, the null hypothesis is rejected if the test statistic is greater than the critical value. For two-tailed tests, the null hypothesis is rejected for values of the test statistic in either tail of the sampling distribution.

We also saw that p-values could be used for hypothesis testing. The p-value yields the probability, when the null hypothesis is true, of obtaining a sample result more unlikely than what is observed. When p-values are used to conduct a hypothesis test, the decision rule calls for rejecting the null hypothesis whenever the p-value is less than α. The p-value is often called the observed level of significance because the null hypothesis will be rejected for any value of α larger than the p-value.

Extensions of the hypothesis-testing procedure in order to control the probability of making a Type II error were also presented. In Section 9.8, we showed how to compute the probability of making a Type II error. In Section 9.9 we showed how to determine a sample size that would enable us to control the probabilities of making both a Type I error and Type II error.

■■■ GLOSSARY

Null hypothesis   The hypothesis tentatively assumed true in the hypothesis-testing procedure.

Alternative hypothesis   The hypothesis concluded to be true if the null hypothesis is rejected.

Type I error   The error of rejecting H_0 when it is true.

Type II error   The error of accepting H_0 when it is false.

Critical value   A value that is compared with the test statistic to determine whether or not H_0 should be rejected.

Level of significance   The maximum allowable probability of a Type I error.

One-tailed test   A hypothesis test in which rejection of the null hypothesis occurs for values of the test statistic in one tail of the sampling distribution.

Two-tailed test   A hypothesis test in which rejection of the null hypothesis occurs for
values of the test statistic in either tail of the sampling distribution.

p-value   The probability, when the null hypothesis is true, of obtaining a sample result
that is more unlikely than what is observed. It is often called the observed level of
significance.

Power curve   A graph of the probability of rejecting H_0 for all possible values of the
population parameter not satisfying the null hypothesis. The power curve provides
the probability of correctly rejecting the null hypothesis.

■■■ KEY FORMULAS

Test Statistic – Population Mean (Large-Sample Case)

$$z = \frac{\bar{x} - \mu_0}{\sigma/\sqrt{n}} \tag{9.6}$$

When σ is unknown, substitute s.

Test Statistic – Population Mean (Small-Sample Case)

$$t = \frac{\bar{x} - \mu_0}{s/\sqrt{n}} \tag{9.7}$$

Test Statistic – Population Proportion

$$z = \frac{\bar{p} - p_0}{\sigma_{\bar{p}}} \tag{9.8}$$

where

$$\sigma_{\bar{p}} = \sqrt{\frac{p_0(1 - p_0)}{n}} \tag{9.9}$$

■■■ SUPPLEMENTARY EXERCISES

48. The Graduate Management Admission Council conducted an extensive survey of
first-year students in MBA programs ("An Overview of Demographic and Family
Characteristics of First-Year Students in U.S. MBA Programs," *GMAC Occasional Papers*,
April 1988). This study reported that the mean age of first-year MBA students is 27 years.
A sample of 45 first-year MBA students at a particular Eastern school has a mean age of
28.2 years and a standard deviation of $s = 1.5$ years.

   a. Conduct a hypothesis test to see if the mean age for the population of students at the school is greater than the GMAC reported mean age. Use $\alpha = .01$. What is your conclusion?

b. What is the p-value for the hypothesis test?

c. Construct a 95% confidence interval estimate of the mean age for MBA students admitted to this school.

49. The manager of the Keeton Department Store has assumed that the mean annual income of the store's credit-card customers is at least $28,000 per year. A sample of 58 credit card-customers shows a sample mean of $27,200 and a sample standard deviation of $3000. At the .05 level of significance should this assumption be rejected? What is the p-value?

50. The chamber of commerce of a Florida Gulf Coast community advertises area residential property available at a mean cost of $25,000 or less per lot. Using a .05 level of significance, test the validity of this claim. Suppose a sample of 32 properties provided a sample mean of $26,000 per lot and a sample standard deviation of $2500. What is the p-value?

51. A bath soap manufacturing process is designed to produce a mean of 120 bars of soap per batch. Quantities over or under this standard are undesirable. A sample of ten batches shows the following numbers of bars of soap:

108   118   120   122   119   113   124   122   120   123

Using a .05 level of significance, test to see if the sample results indicate that the manufacturing process is functioning properly.

52. The monthly rent for a two-bedroom apartment in a particular city is reported to averge $350. Suppose we would like to test the hypothesis $H_0 = 350$ versus the hypothesis $H_a: \mu \neq 350$. A sample of 36 two-bedroom apartments is selected. The sample mean turns out to be $\bar{x} = \$362$, with a sample standard deviation of $s = \$40$.

a. Conduct this hypothesis test with a .05 level of significance.

b. Compute the p-value.

c. Use the sample results to construct a 95% confidence interval for the population mean. What hypothesis-testing conclusion would you draw based on the confidence interval result?

53. Stout Electric Company operates a fleet of trucks which provide electrical service to the construction industry. Monthly mean maintenance costs have been $75 per truck. A random sample of 40 trucks shows a sample mean maintenance cost of $82.50 per month, with a sample standard deviation of $30. Management would like a test to determine whether or not the mean monthly maintenance cost has increased.

a. With a .05 level of significance, what is the rejection rule for this test?

b. What is your conclusion based on the sample mean of $82.50?

c. What is the p-value associated with this sample result? What is your conclusion based on the p-value?

54. In making bids on building projects, Sonneborn Builders, Inc. assumes construction workers are idle no more than 15% of the time. For a normal 8-hour shift, this means that the mean idle time per worker should be 72 minutes or less per day. A sample of 30 construction workers provided a mean idle time of 80 minutes per day. The sample standard deviation was 20 minutes. Suppose a hypothesis test is to be designed to test the validity of the company's assumption.

a. What is the p-value associated with the sample result?

b. Using a .05 level of significance and the p-value, test the hypothesis $H_0: \mu \leq 72$. What is your conclusion?

55. Sixty percent of Americans believe that business profits are distributed unfairly (*General Social Surveys*, National Opinion Research Center, University of Chicago). Suppose a sample of 40 Midwesterners showed that 27 believe business profits are distributed unfairly.

    a. Do these results justify making the inference that a larger proportion of Midwesterners believe that business profits are distributed unfairly? Use $\alpha = .05$.

    b. What is the p-value?

56. In the past the Dumont Clothing Store has recorded 72% charge purchases and 28% cash purchases. A sample of 200 recent purchases shows that 160 were charge purchases. Does this suggest a change in the paying practices of the Dumont customers? Test with $\alpha = .05$. What is the p-value?

57. The manager of K-Mark Supermarkets assumes that at least 30% of the Saturday customers purchase the price-reduced special advertised in the Friday newspaper. Use $\alpha = .05$ and test the validity of the manager's assumption if a sample of 250 Saturday customers show that 60 purchased the advertised special.

58. The filling machine for a production operation must be adjusted if more than 8% of the items being produced are underfilled. A random sample of 80 items from the day's production contained 9 underfilled items. Does the sample evidence indicate that the filling machine should be adjusted? Use $\alpha = .02$. What is the p-value?

59. A radio station in a major resort area announced that at least 90% of the hotels and motels would be full for the Memorial Day weekend. The station went on to advise listeners to make reservations in advance if they planned to be in the resort over the weekend. On Saturday night a sample of 58 hotels and motels showed 49 with a no-vacancy sign and 9 with vacancies. What is your reaction to the radio station's claim based on the sample evidence? Use $\alpha = .05$ in making this statistical test. What is the p-value for the sample results?

60. It is assumed that at least 90% of juvenile first-time criminals are given probation upon the admission of guilt. Test this hypothesis at a .02 level of significance, if a sample of 92 juvenile criminal convictions shows 78 juveniles receiving probation. What is the p-value?

61. Refer again to Exercise 54.

    a. What is the probability of making a Type II error when the population mean idle time is 80 minutes?

    b. What is the probability of making a Type II error when the population mean idle time is 75 minutes?

    c. What is the probability of making a Type II error when the population mean idle time is 70?

    d. Sketch the power curve for this problem.

62. A federal funding program is available to low-income neighborhoods. To qualify for the funding a neighborhood must have a mean household income of less than $7000 per year. Neighborhoods with mean annual household incomes of $7000 or more do not qualify. Funding decisions are based on a sample of residents in the neighborhood. A hypothesis test with a .02 level of significance is conducted. If the funding guidelines call for a maximum probability of .05 of not funding a neighborhood with a mean annual household income of $6500, what sample size should be used in the funding decision study? Use $\sigma = \$2000$ as a planning value.

63. The bath soap production process in Exercise 51 uses the hypothesis $H_0: \mu = 120$ and $H_a: \mu \neq 120$ in order to test whether or not the production process is meeting the standard output of 120 bars per batch. Use a .05 level of significance for the test and a planning value of 5 for the standard deviation.

    a. If the mean output drops to 117 bars per batch, the firm would like a 98% chance of concluding that the standard production output is not being met. How large a sample should be selected?

b. Using your sample size from part (a), what is the probability of concluding that the process is operating satisfactorily for each of the following actual mean outputs: 117, 118, 119, 121, 122, and 123 bars per batch? That is, what is the probability of a Type II error in each case?

APPLICATION

Harris Corporation[*]

Melbourne, Florida

Harris Corporation/RF Communications Division is a major manufacturer of point-to-point radio communications equipment. It is a horizontally integrated manufacturing company with a multi-plant facility. Harris' primary factory for catalog products specializes in medium to high volume assembly; factory operations include printed circuit assembly, final assembly and testing.

A HYPOTHESIS-TESTING APPLICATION

One of the higher volume items at the catalog products factory involves an assembly called an RF deck. The assembly consists of sixteen electronic components soldered to a machined casting which forms the plated surface of the deck. During a manufacturing run a problem developed in the soldering process; the flow of solder onto the deck did not meet the criteria established. A manufacturing engineer was summoned to solve the problem.

A preliminary investigation conducted by the manufacturing engineer uncovered the following factors which could cause the soldering problem:

1. impure flux or solder
2. improper temperature of the soldering iron
3. operator training
4. contaminated RF deck surface
5. defective plating from supplier

After considering these factors, the engineer made a preliminary determination that the soldering problem was most likely due to defective plating from the supplier. But, before drawing a final conclusion, the engineer decided to set up an experiment to test this hypothesis. The objective of the experiment was to generate data that could be used for a hypothesis test on a population proportion.

The question raised by the engineer was: Did the proportion of defective platings exceed that which could be expected when the supplier's operation conforms to design specifications? Letting p indicate the proportion of defective platings in the Harris inventory and p_0 indicate the proportion of defective platings when the supplier

---

[*]The authors are indebted to Richard A. Marshall of the Harris Corporation for providing this application.

operation conforms with design specifications, the following hypothesis test was conducted:

$$H_0: \quad p \le p_0$$
$$H_a: \quad p > p_0$$

H_0 indicates that the Harris inventory has a defective plating proportion less than or equal to the design specification, while H_a indicates that the Harris inventory has a defective plating proportion greater than the design specification.

Tests made on a sample of the Harris inventory showed a sample proportion defective of $\bar{p} = .15$. This proportion defective resulted in the rejection of H_0; the current inventory had a higher defective proportion than would exist if the supplier's operation conformed to design specifications.

As a result the manufacturing engineer concluded that the plating was defective and pressure was applied to purchasing management to have the plating supplier held responsible for both rejected parts and damages.

■■■ FURTHER ANALYSIS AND RECONSIDERATION

The conclusion that the plating supplier was responsible for the soldering problem resulted in additional meetings with management at various levels of the organization. Finally, a meeting was called with the plant manager and the materials manager to review the data and decide on a course of action.

At this meeting several questions were raised concerning the experimental process that resulted in the conclusion that defective plating was the underlying problem. For example, one of the defective pieces had only one of sixteen components that did not accept solder; further investigation of this piece showed that the plating was not at fault, but that the component itself was contaminated. Another defective piece examined was scorched and bleeded. In this case, further investigation revealed high soldering temperatures as the most likely cause of the defect.

After reflection on the experimental process, the plant manager and materials manager concluded that, although the hypothesis test correctly indicated an undesirably high proportion of defective platings, it *did not prove* defective plating was the cause of the soldering problem. Additional experimentation was requested in order to identify other

A business executive uses a radio communication car telephone manufactured by the Harris Corporation

possible causes. This further investigation ultimately led to the conclusion that the true underlying problem was shelf contamination, not plating. Management thus recognized that the supplier was not at fault and accordingly did not press a complaint.

▬ CONCLUSION

The hypothesis test described in this application pointed to an undesirable level of defective platings in the Harris inventory. However, the statistical evidence did not prove the supplier was the cause of the soldering problem. In this case, the fact that management did not blindly accept "official looking" statistics prevented a wrong decision from being made.

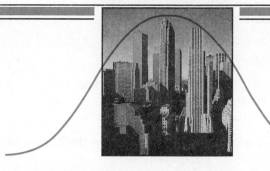

CHAPTER 10

Statistical Inference About Means and Proportions with Two Populations

I n Chapters 8 and 9 we presented statistical methods for developing interval estimates and conducting hypothesis tests for population means and population proportions. However, the statistical procedures we have discussed thus far have considered only single-population situations. In this chapter we consider cases where *two populations* are involved. For example, we may want to compare a population of men versus a population of women, a population of parts from supplier A versus a population of parts from supplier B, a population of oil-industry firms versus a population of steel-industry firms, and so on. In cases such as these we will be using simple random sampling to develop interval estimates and test hypotheses about the difference between the means and/or the difference between the proportions of two populations.

10.1 ▰ ESTIMATION OF THE DIFFERENCE BETWEEN THE MEANS OF TWO POPULATIONS—INDEPENDENT SAMPLES

In many practical situations we are faced with two separate populations where the difference between the means of the two populations is of prime importance. We know from Chapters 7 and 8 that we can take a simple random sample from a single population and use the sample mean $\bar{x}$ as a point estimator of the population mean. In the two-population case we will select two separate and independent simple random samples, one from population 1 and another from population 2. Let

$$\mu_1 = \text{mean of population 1}$$

$$\mu_2 = \text{mean of population 2}$$

$\bar{x}_1$ = sample mean for the simple random sample taken from population 1

$\bar{x}_2$ = sample mean for the simple random sample taken from population 2

The difference between the two population means is $\mu_1 - \mu_2$. The point estimator of $\mu_1 - \mu_2$ is as follows:

Point Estimator of the Difference Between the Means of Two Populations

$$\bar{x}_1 - \bar{x}_2 \tag{10.1}$$

Thus we see that the point estimator of the difference between two population means is the difference between the sample means of the two independent simple random samples.

Sampling Distribution of $\bar{x}_1 - \bar{x}_2$

In the study of the difference between the means of two populations, $\bar{x}_1 - \bar{x}_2$ is the point estimator of interest. This point estimator, just like the point estimators discussed previously, has its own sampling distribution. If we can identify the sampling distribution of $\bar{x}_1 - \bar{x}_2$, we can use it to develop an interval estimate of the difference between the two population means in much the same way that we used the sampling distribution of $\bar{x}$ for interval estimation about a single population mean. The properties of the sampling distribution of $\bar{x}_1 - \bar{x}_2$ are as follows:

Sampling Distribution of $\bar{x}_1 - \bar{x}_2$

$$\text{Mean:} \quad E(\bar{x}_1 - \bar{x}_2) = \mu_1 - \mu_2 \tag{10.2}$$

$$\text{Standard Deviation:} \quad \sigma_{\bar{x}_1 - \bar{x}_2} = \sqrt{\frac{\sigma_1^2}{n_1} + \frac{\sigma_2^2}{n_2}} \tag{10.3}$$

where

σ_1 = standard deviation of population 1

σ_2 = standard deviation of population 2

n_1 = sample size for the simple random sample selected from population 1

n_2 = sample size for the simple random sample selected from population 2

Distribution form: Provided that the sample sizes are both *large* ($n_1 \geq 30$ and $n_2 \geq 30$), the sampling distribution of $\bar{x}_1 - \bar{x}_2$ can be approximated by a normal probability distribution.

Figure 10.1 shows the sampling distribution of $\bar{x}_1 - \bar{x}_2$ and its relationship to the individual sampling distributions of $\bar{x}_1$ and $\bar{x}_2$.

Let us use the sampling distribution of $\bar{x}_1 - \bar{x}_2$ to develop an interval estimate of the difference between the means of 2 populations. We shall consider 2 cases, one where the sample sizes are large ($n_1 \geq 30$ and $n_2 \geq 30$) and the other where one or both sample sizes are small ($n_1 < 30$ and/or $n_2 < 30$). Let us consider the large-sample-size case first.

■■■ FIGURE 10.1   **Sampling Distribution of $\bar{x}_1 - \bar{x}_2$ and Its Relationship to the Individual Sampling Distributions of $\bar{x}_1$ and $\bar{x}_2$**

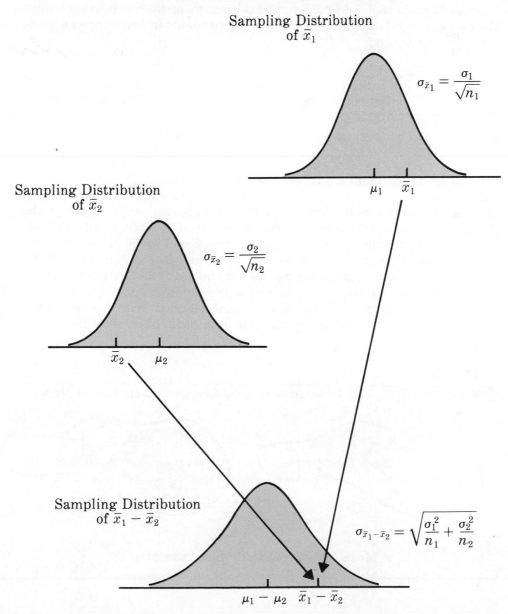

Large-Sample Case

Greystone Department Stores, Inc., operates two stores in Buffalo, New York, one located in the inner city and one located in a suburban shopping center. Over a period of time the regional manager has noticed that products that sell extremely well in one store do not always sell well in the other. One plausible explanation for the differences in sales at the two stores is that there are differences between the customers who shop at the two locations. Customer differences may be noticeable in age, education, income, and so on. While we could elect to study any of these characteristics, let us assume that the manager has asked about the difference between the average or mean ages of the customers who shop at the two stores. Figure 10.2 provides an illustration of this two population situation.

Let us suppose that Greystone conducts a survey of customers at each store. The customer age data collected from two independent random samples provide the following results:

Store	Number of Customers Sampled	Sample Mean Age	Sample Standard Deviation
Inner city	36	$\bar{x}_1 = 40$ years	$s_1 = 9$ years
Suburban	49	$\bar{x}_2 = 35$ years	$s_2 = 10$ years

Using (10.1), the point estimate of the difference between mean ages of the 2 populations is $\bar{x}_1 - \bar{x}_2 = 40 - 35 = 5$ years. Thus, we are led to believe that the customers at the inner-city store have a mean age approximately 5 years greater than the mean age of the suburban store customers. However, as with all point estimates, we know that 5 years is only an approximate value of the difference between mean ages of the 2 populations. Thus we will compute a confidence interval estimate of the difference between the means of the two populations.

In the large-sample-size case we know that the sampling distribution of $\bar{x}_1 - \bar{x}_2$ can be approximated by a normal probability distribution. With this approximation we can use the following expression to develop an interval estimate of the

---

■■■ FIGURE 10.2   **Two Populations for the Greystone Department Store**

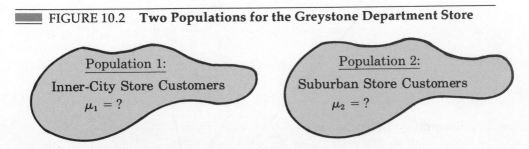

where

$\mu_1 =$ Mean Age of Inner-City Store Customers.

$\mu_2 =$ Mean Age of Surburban Store Customers

difference between the means of the 2 populations:

Interval Estimate of the Difference Between the Means of Two Populations
(Large-Sample Case with $n_1 \geq 30$ and $n_2 \geq 30$)

$$\bar{x}_1 - \bar{x}_2 \pm z_{\alpha/2}\sigma_{\bar{x}_1-\bar{x}_2} \tag{10.4}$$

where $1 - \alpha$ is the confidence coefficient.

Let us use (10.4) to develop a confidence interval estimate of the difference between the mean ages for the two populations in the Greystone Department Store study. Since the population standard deviations σ_1 and σ_2 are unknown, we cannot use (10.3) to calculate $\sigma_{\bar{x}_1-\bar{x}_2}$. However, we can use the sample standard deviations as estimates of the population standard deviations and estimate $\sigma_{\bar{x}_1-\bar{x}_2}$ as follows:

Estimate of $\sigma_{\bar{x}_1-\bar{x}_2}$

$$s_{\bar{x}_1-\bar{x}_2} = \sqrt{\frac{s_1^2}{n_1} + \frac{s_2^2}{n_2}} \tag{10.5}$$

With large sample sizes, $s_{\bar{x}_1-\bar{x}_2}$ can be accepted as a good estimate of $\sigma_{\bar{x}_1-\bar{x}_2}$.

Using (10.5) to estimate $\sigma_{\bar{x}_1-\bar{x}_2}$ for the Greystone Department Store problem, we have

$$s_{\bar{x}_1-\bar{x}_2} = \sqrt{\frac{(9)^2}{36} + \frac{(10)^2}{49}} = \sqrt{4.29} = 2.07$$

With this value as the estimate of $\sigma_{\bar{x}_1-\bar{x}_2}$ and with $z_{\alpha/2} = z_{.025} = 1.96$, (10.4) provides the following 95% confidence interval estimate:

$$5 \pm (1.96)(2.07)$$

or

$$5 \pm 4.06$$

Thus at a 95% level of confidence the interval estimate for the difference between the mean ages of the two Greystone populations is .94 years to 9.06 years.

Small-Sample Case

Let us now consider the interval-estimation procedure for the difference between the means of two populations whenever one or both sample sizes are less than 30—that is, $n_1 < 30$ and/or $n_2 < 30$. This will be referred to as the small-sample case.

In Chapter 8 we presented a procedure for interval estimation of the mean for a single population whenever a small sample was used. Recall that the procedure required the assumption that the population had a normal distribution. With the sample standard deviation s used as an estimate of the population standard deviation σ, the t distribution was used to develop an interval estimate of the population mean.

To develop interval estimates for the two-population small-sample case, we will make some assumptions about the two populations and the samples selected from the two populations. The specific assumptions required are as follows:

1. Both populations have normal distributions.
2. The variances of the populations are equal ($\sigma_1^2 = \sigma_2^2 = \sigma^2$). *not necessary but assume this*
3. Independent random samples are selected from the two populations.

Given these assumptions, the sampling distribution of $\bar{x}_1 - \bar{x}_2$ is normal regardless of the sample sizes involved. The mean of the sampling distribution is $\mu_1 - \mu_2$. Because of the equal variances assumption, equation (10.3) can be written

$$\sigma_{\bar{x}_1 - \bar{x}_2} = \sqrt{\frac{\sigma^2}{n_1} + \frac{\sigma^2}{n_2}} = \sqrt{\sigma^2 \left(\frac{1}{n_1} + \frac{1}{n_2}\right)} \qquad (10.6)$$

The sampling distribution of $\bar{x}_1 - \bar{x}_2$ is shown in Figure 10.3.

Since the value of σ^2 is unknown, the two sample variances, s_1^2 and s_2^2, can be

---

■ FIGURE 10.3   **Sampling Distribution of $\bar{x}_1 - \bar{x}_2$ When the Populations Have Normal Distributions with Equal Variances**

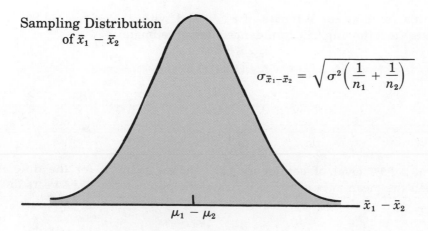

combined to compute the following estimate of σ^2:

Estimate of σ^2

$$s^2 = \frac{(n_1 - 1)s_1^2 + (n_2 - 1)s_2^2}{n_1 + n_2 - 2} \qquad (10.7)$$

The process of combining the results of the two independent random samples to provide one estimate of σ^2 is referred to as *pooling*. The sample variance, s^2, is referred to as the *pooled estimator* of σ^2.

With s^2 as the estimator of σ^2 and using (10.6), the following estimate of the standard deviation of $\bar{x}_1 - \bar{x}_2$ can be obtained:

Estimate of $\sigma_{\bar{x}_1 - \bar{x}_2}$ when $\sigma_1^2 = \sigma_2^2$

$$s_{\bar{x}_1 - \bar{x}_2} = \sqrt{s^2\left(\frac{1}{n_1} + \frac{1}{n_2}\right)} \qquad (10.8)$$

The t distribution can now be used to compute an interval estimate of the difference between the means of the two populations. Since there are $n_1 - 1$ degrees of freedom associated with the random sample from population 1 and $n_2 - 1$ degrees of freedom associated with the random sample from population 2, the t distribution will have $n_1 + n_2 - 2$ degrees of freedom. The interval-estimation procedure is as follows:

Interval Estimate of the Difference Between the Means of Two Populations
(Small-Sample Case with $n_1 < 30$ and/or $n_2 < 30$)

$$\bar{x}_1 - \bar{x}_2 \pm t_{\alpha/2}s_{\bar{x}_1 - \bar{x}_2} \qquad (10.9)$$

where the t value is based on a t distribution with $n_1 + n_2 - 2$ degrees of freedom and where $1 - \alpha$ is the confidence coefficient.

Let us demonstrate the interval-estimation procedure for the sampling study conducted by the Clearview National Bank. Independent random samples of checking account balances for customers at two Clearview branch banks show the following results:

Branch Bank	Number of Checking Accounts	Sample Mean Balance	Sample Standard Deviation
Cherry Grove	12	$\bar{x}_1 = \$1000$	$s_1 = \$150$
Beechmont	10	$\bar{x}_2 = \$\ 920$	$s_2 = \$120$

Let us use these data to develop a 90% confidence interval estimate of the difference between the mean checking account balances for the two branch banks. Using (10.7), the pooled estimate of the population variance becomes

$$s^2 = \frac{(n_1 - 1)s_1^2 + (n_2 - 1)s_2^2}{n_1 + n_2 - 2} = \frac{(11)(150)^2 + (9)(120)^2}{12 + 10 - 2} = 18,855$$

The corresponding estimate of the standard deviation of $\bar{x}_1 - \bar{x}_2$ is

$$s_{\bar{x}_1 - \bar{x}_2} = \sqrt{s^2 \left(\frac{1}{n_1} + \frac{1}{n_2} \right)} = \sqrt{18,855 \left(\frac{1}{12} + \frac{1}{10} \right)} = 58.79$$

The appropriate t distribution for the interval estimation procedure has $n_1 + n_2 - 2 = 12 + 10 - 2 = 20$ degrees of freedom. With $\alpha = .10$, $t_{\alpha/2} = t_{.05} = 1.725$. Thus using (10.9) the interval estimate becomes

$$\bar{x}_1 - \bar{x}_2 \pm t_{.05} s_{\bar{x}_1 - \bar{x}_2}$$
$$1000 - 920 \pm (1.725)(58.79)$$
$$80 \pm 101.41$$

At a 90% level of confidence the interval estimate of the difference between the mean account balances at the 2 branch banks is $-\$21.41$ to $\$181.41$. The fact that the interval includes a negative range of values indicates that the actual difference between the two means, $\mu_1 - \mu_2$, may be negative. Thus μ_2 could actually be larger than μ_1, indicating that the population mean balance could be greater for the Beechmont branch even though the results show a greater sample mean balance at the Cherry Grove branch. The fact that the confidence interval contains the value 0 can be interpreted as indicating that we do not have sufficient evidence to conclude that the population mean account balances differ at the 2 branches.

■■ NOTES AND COMMENTS ■■

The t distribution is not restricted to the small-sample situation. Whenever we are interested in the difference between the means of two populations and are able to conclude that the populations are normally distributed with equal variances, the t distribution can be used to develop the appropriate interval estimate. However, (10.4) shows how to determine an interval estimate of the difference between the means of two populations whenever the sample sizes are large. In this case the use of the t distribution and its corresponding assumptions are not required. As a result, we do not need to refer to the t distribution until we have a small-sample-size situation.

■ EXERCISES

1. A college admissions board is interested in estimating the difference between the mean grade point averages of students from two high schools. Independent simple random samples of students at the two high schools provide the following results:

Mt. Washington	Country Day
$n_1 = 46$	$n_2 = 33$
$\bar{x}_1 = 3.02$	$\bar{x}_2 = 2.72$
$s_1 = .38$	$s_2 = .45$

a. What is the point estimate of the difference between the means of the two populations?

b. Develop a 90% confidence interval estimate of the difference between the two population means.

c. Answer part (b) using a 95% confidence interval estimate.

2. *Working Woman* (January 1989) provided salary survey information for women in a variety of occupations. The mean entry level salary for women accountants in large public accounting firms was $23,750. The mean entry level for women accountants in large corporations was $21,000. Assume that the following sample sizes and sample standard deviations were available:

Public Accounting	Corporations
$n_1 = 220$	$n_2 = 250$
$s_1 = 1200$	$s_2 = 1000$

Provide a 95% confidence interval estimate of the difference between the mean entry level salaries for women in large public accounting firms and women in large corporations.

3. The Butler County Bank and Trust Company is interested in estimating the difference between the mean credit card balances at two of its branch banks. Independent samples of credit card customers provide the following results:

Branch 1	Branch 2
$n_1 = 32$	$n_2 = 36$
$\bar{x}_1 = \$500$	$\bar{x}_2 = \$375$
$s_1 = \$150$	$s_2 = \$130$

a. Develop a point estimate of the difference between the mean balances at the two branches.

b. Develop a 99% confidence interval estimate of the difference between mean balances.

4. An urban-planning group is interested in estimating the difference between mean household income for two neighborhoods in a large metropolitan area. Independent samples of households in the neighborhoods provide the following results:

Neighborhood 1	Neighborhood 2
$n_1 = 8$	$n_2 = 12$
$\bar{x}_1 = \$15,700$	$\bar{x}_2 = \$14,500$
$s_1 = \$700$	$s_2 = \$850$

a. Develop a point estimate of the difference between mean incomes in the two neighborhoods.

b. Develop a 95% confidence interval estimate of the difference between mean incomes in the two neighborhoods.

c. What assumptions were made in order to compute the interval estimates in part (b)?

5. Production quantities for two assembly-line workers are shown below. Each data value indicates the amount produced during a randomly selected 1-hour period.

Worker 1	Worker 2
20	22
18	18
21	20
22	23
20	24

a. Develop a point estimate of the difference between the mean hourly production rates of the two workers. Which worker appears to have the higher mean production rate?

b. Develop a 90% confidence interval estimate of the difference between the mean production rates of the two workers. Consider the confidence interval estimate. Does the result provide conclusive evidence that the worker having the higher sample mean production rate is actually the worker with the overall higher production rate? Explain.

6. A sample of 15 graduates from Eastern University showed that the mean time until they received their first job promotion was 5.2 years, with a sample standard deviation of 1.4 years. A sample of 12 graduates from Midwestern University showed a sample mean of 2.7 years, with a standard deviation of 1.1 years.

a. What is the pooled estimate of the population variance σ^2?

b. Using a 95% confidence interval, estimate the difference between the mean time until the first job promotion for the populations of Eastern and Midwestern University graduates.

10.2 ■ HYPOTHESIS TESTS ABOUT THE DIFFERENCE BETWEEN THE MEANS OF TWO POPULATIONS—INDEPENDENT SAMPLES

In this section we present procedures that can be used to test hypotheses about the difference between the means of two populations. The methodology is again divided into large-sample ($n_1 > 30$, $n_2 > 30$) and small-sample ($n_1 < 30$ and/or $n_2 < 30$) cases.

Large-Sample Case

As part of a study to evaluate differences between the educational quality of two training centers, a company administers a standardized examination to individuals who were trained at the two centers. The examination scores are a major factor in assessing any quality differences between the centers.

Let

μ_1 = the mean examination score for the population
of individuals trained at center A

μ_2 = the mean examination score for the population
of individuals trained at center B

We begin with the tentative assumption that there is no difference between the training quality provided at the two centers. Thus, in terms of the mean examination scores, the null hypothesis is that $\mu_1 - \mu_2 = 0$. If sample evidence leads to the rejection of this hypothesis, we will conclude that the mean examination scores differ for the two populations. This conclusion indicates that a quality differential exists between the two centers and that a follow-up study investigating the reasons for the differential may be warranted. The null and alternative hypotheses are written as follows:

$$H_0: \quad \mu_1 - \mu_2 = 0$$

$$H_a: \quad \mu_1 - \mu_2 \neq 0$$

Following the hypothesis-testing procedure from Chapter 9, we will make the tentative assumption that H_0 is true. Using the difference between the sample means as the point estimator of the difference between the population means, we consider the sampling distribution of $\bar{x}_1 - \bar{x}_2$ when H_0 is true. Assuming the large-sample case, this distribution is as shown in Figure 10.4. Since the sampling distribution is approximately normal, the following test statistic can be used:

$$z = \frac{(\bar{x}_1 - \bar{x}_2) - (\mu_1 - \mu_2)}{\sqrt{\sigma_1^2/n_1 + \sigma_2^2/n_2}} \tag{10.10}$$

■ FIGURE 10.4 **Sampling Distribution of $\bar{x}_1 - \bar{x}_2$ with $H_0: \mu_1 - \mu_2 = 0$**

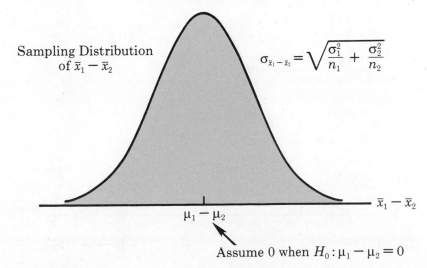

Sampling Distribution
of $\bar{x}_1 - \bar{x}_2$

$\sigma_{\bar{x}_1 - \bar{x}_2} = \sqrt{\dfrac{\sigma_1^2}{n_1} + \dfrac{\sigma_2^2}{n_2}}$

$\bar{x}_1 - \bar{x}_2$

$\mu_1 - \mu_2$

Assume 0 when $H_0 : \mu_1 - \mu_2 = 0$

In this case, z can be interpreted as the number of standard deviations $\bar{x} - \bar{x}_2$ is from the value of $\mu_1 - \mu_2$ specified in H_0. For $\alpha = .05$ and thus $z_{\alpha/2} = z_{.025} = 1.96$, the rejection region for the two-tailed hypothesis test is shown in Figure 10.5. The rejection rule is as follows:

$$\text{Reject } H_0 \text{ if } z < -1.96 \text{ or if } z > +1.96$$

Let us assume that independent random samples of individuals trained at the two centers provide the following examination score results:

Training Center A	Training Center B
$n_1 = 30$	$n_2 = 40$
$\bar{x}_1 = 82.5$	$\bar{x}_2 = 78$
$s_1 = 8$	$s_2 = 10$

Using the sample standard deviations to estimate the population standard deviations and using (10.10), the test statistic for the null hypothesis $H_0 : \mu_1 - \mu_2 = 0$ becomes

$$z = \frac{(\bar{x}_1 - \bar{x}_2) - (\mu_1 - \mu_2)}{\sqrt{\sigma_1^2/n_1 + \sigma_2^2/n_2}} = \frac{(82.5 - 78) - 0}{\sqrt{(8)^2/30 + (10)^2/40}} = 2.09$$

With this value of z, the conclusion is to reject H_0. Thus the sample leads the firm to suspect that educational quality differs at the two centers.

With $z = 2.09$, the standard normal probability distribution table can be used to compute the p-value for this two-tailed test. With an area of .4817 between the mean and $z = 2.09$, the p-value becomes $2(.5000 - .4817) = .0366$.

In this hypothesis test we were interested in determining if a difference exists between the means of the two populations. Since we did not have a prior belief about one mean being specifically greater than or less than the other, the

---

▬▬ FIGURE 10.5 **Rejection Region for the Two-Tailed Hypothesis Test with $\alpha = .05$**

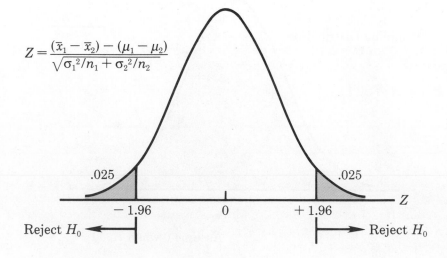

hypotheses $H_0: \mu_1 - \mu_2 = 0$ and $H_a: \mu_1 - \mu_2 \neq 0$ were appropriate. In other hypothesis tests about the difference between the means of two populations, we may want to test a belief about one of the means being greater than or perhaps less than the other mean. In these cases a one-tailed testing procedure would be appropriate. The two forms of a one-tailed test about the difference between two population means are as follows:

$$H_0: \mu_1 - \mu_2 \leq 0 \qquad H_0: \mu_1 - \mu_2 \geq 0$$
$$H_a: \mu_1 - \mu_2 > 0 \qquad H_a: \mu_1 - \mu_2 < 0$$

These hypotheses are tested with use of the test statistic z given by (10.10) with the rejection region determined in a manner identical to the one-tailed test rules presented in Chapter 9.

Small-Sample Case

Let us now consider hypothesis tests about the difference between the means of two populations for the small-sample case; i.e., where $n_1 < 30$ and/or $n_2 < 30$. The procedure we will use is based on the t distribution with $n_1 + n_2 - 2$ degrees of freedom. As discussed in Section 10.1, the assumptions that both populations have normal distributions and that the variances of the populations are equal must be made for the procedure to be valid. We present the hypothesis-testing methodology through the following illustration.

A new computer software package has been developed to help systems analysts in developing computer-based information systems. The objective of the new software package is to reduce the time required to design, develop, and implement an information system. In order to evaluate the benefits of the new software package, a random sample of 20 systems analysts was selected. Each analyst was given specifications for a hypothetical information system. Ten of the analysts were instructed to design, develop, and implement the information system using current technology. The other ten analysts were first trained in the use of the new software package and then instructed to design, develop, and implement the information system using the new software package.

In this study, there are two populations: a population of systems analysts using the current technology and a population of systems analysts using the new software package. In terms of time required to complete the information systems design project, the population means are as follows:

μ_1 = the mean project completion time for analysts using
    current technology

μ_2 = the mean project completion time for analysts using
    the new software package

The researcher in charge of the new software-evaluation project hopes to show that the new software package will provide a smaller mean project-completion time. Thus the researcher is looking for evidence to conclude that μ_2 is less than μ_1; in this case, the difference between the two population means, $\mu_1 - \mu_2$, will be greater than zero. In formulating the hypotheses, the research hypothesis

$\mu_1 - \mu_2 > 0$ is stated as the alternative hypothesis:

$$H_0: \quad \mu_1 - \mu_2 \le 0$$
$$H_a: \quad \mu_1 - \mu_2 > 0$$

In tentatively assuming H_0 is true, we are taking the position that using the new software package takes the same time or perhaps even longer than the current technology. The researcher is looking for evidence to reject H_0 and conclude that the new software package possess a smaller mean completion time.

Let us assume that the 20 analysts completed the study with the following results:

Current Technology	New Software Package
$n_1 = 10$	$n_2 = 10$
$\bar{x}_1 = 325$ hours	$\bar{x}_2 = 295$ hours
$s_1 = 40$ hours	$s_2 = 44$ hours

Under the assumption that the variances of the populations are the same, (10.7) is used to compute the pooled estimate of σ^2.

$$s^2 = \frac{(n_1 - 1)s_1^2 + (n_2 - 1)s_2^2}{(n_1 + n_2 - 2)} = \frac{9(40)^2 + 9(44)^2}{(10 + 10 - 2)} = 1768$$

The test statistic for the small-sample case is

$$t = \frac{(\bar{x}_1 - \bar{x}_2) - (\mu_1 - \mu_2)}{\sqrt{s^2 \left(\frac{1}{n_1} + \frac{1}{n_2}\right)}} \tag{10.11}$$

In the case of two independent random samples of sizes n_1 and n_2, the t distribution will have $n_1 + n_2 - 2$ degrees of freedom. For $\alpha = .05$, the t distribution table shows that with $10 + 10 - 2 = 18$ degrees of freedom, $t_\alpha = t_{.05} = 1.734$. Thus the rejection region for the one-tailed test is as follows:

Reject H_0 if $t > 1.734$

The sample data and (10.11) provide the following value for the test statistic:

$$t = \frac{(325 - 295) - 0}{\sqrt{1768 \left(\frac{1}{10} + \frac{1}{10}\right)}} = 1.60$$

Checking the rejection region, we see that $t = 1.60$ does not allow the rejection of H_0 at the .05 level of significance. Thus, the sample results do not enable the researcher to conclude that the use of the new software package has a lower mean completion time. If the new software package is better, the sample information has not detected it. Since the test is inconclusive, additional research into the merits of the new software package is suggested.

The preceding example leads to a cautionary note concerning hypothesis testing with small samples. With small sample sizes, the probability of making a type II error tends to be high. That is, there is a high probability that the small samples will not reject H_0 when, it fact, it should be rejected. Thus if small samples fail to reject H_0, further study with addition data is frequently recommended.

NOTES AND COMMENTS

In hypothesis tests about the difference between the means of two populations, the null hypothesis almost always contains the condition that there is no difference between the means. Thus null hypotheses of the form

$$H_0: \mu_1 - \mu_2 = 0 \qquad H_0: \mu_1 - \mu_2 \leq 0 \qquad H_0: \mu_1 - \mu_2 \geq 0$$

are possible choices. In some instances, we may wish to determine if a nonzero difference, D_0, exists between the population means. The specific value chosen for D_0 depends upon the application under study. However, in this case, the null hypothesis may be of the following forms:

$$H_0: \mu_1 - \mu_2 = D_0 \qquad H_0: \mu_1 - \mu_2 \leq D_0 \qquad H_0: \mu_1 - \mu_2 \geq D_0$$

The hypothesis-testing computations remain the same with the exception that D_0 is used for the value of $\mu_1 - \mu_2$ in (10.10) and (10.11).

▬ EXERCISES

7. The Greystone Department Store study in Section 10.1 provided the following data on customer ages from independent random samples taken at two store locations:

Inner-City Store	Suburban Store
$n_1 = 36$	$n_2 = 49$
$\bar{x}_1 = 40$ years	$\bar{x}_2 = 35$ years
$s_1 = 9$ years	$s_2 = 10$ years

For $\alpha = .05$, test the hypothesis $H_0: \mu_1 - \mu_2 = 0$ against the alternative hypothesis $H_a: \mu_1 - \mu_2 \neq 0$. What is your conclusion about the mean ages of the populations of customers at the two stores?

8. The Educational Testing Service conducted a study to investigate difference between the scores of males and females on the Scholastic Aptitude Test (*Journal of Educational Measurement,* Spring 1987). The study identified a random sample of 562 females and 852 males who had achieved the same high score on the mathematics portion of the test. That is, both the females and males were viewed as having similar high abilities in mathemat-

ics. The SAT verbal scores for the two samples are summarized as follows:

Females	Males
$\bar{x}_1 = 547$	$\bar{x}_2 = 525$
$s_1 = 83$	$s_2 = 78$

Do the data support the conclusion that given a population of females and a population of males with similar high mathematical abilities, the females will have a significantly higher verbal ability? Test at a .02 level of significance. What is your conclusion?

9. A firm is studying the delivery times for two raw material suppliers. The firm is basically satisfied with supplier A and is prepared to stay with this supplier provided that the mean delivery times are the same as or less than those of supplier B. However, if the firm finds that the mean delivery times from supplier B are less than those of supplier A, it will begin making raw material purchases from supplier B.

 a. What are the null and alternative hypotheses for this situation?

 b. Assume that independent samples show the following delivery time characteristics for the two suppliers:

Supplier A	Supplier B
$n_1 = 50$	$n_2 = 30$
$\bar{x}_1 = 14$ days	$\bar{x}_2 = 12.5$ days
$s_1 = 3$ days	$s_2 = 2$ days

Using $\alpha = .05$, what is your conclusion for the hypotheses from part (a)? What action do you recommend in terms of supplier selection?

10. In a wage discrimination case involving male and female employees, independent samples of male and female employees with 5 years or more experience show the following hourly wage results:

Male Employees	Female Employees
$n_1 = 44$	$n_2 = 32$
$\bar{x}_1 = \$6.25$	$\bar{x}_2 = \$5.70$
$s_1 = \$1.00$	$s_2 = \$.80$

The null hypothesis is stated such that male employees have a mean hourly wage less than or equal to that of the female employees. Rejection of H_0 leads to the conclusion that male employees have a mean hourly wage exceeding the female employee wages. Test the hypothesis with $\alpha = .01$. Does wage discrimination appear to exist in this case?

11. A production line is designed on the assumption that the difference between mean assembly times for two operations is 5 minutes. Independent tests for the two assembly operations show the following results:

Operation A	Operation B
$n_1 = 100$	$n_2 = 50$
$\bar{x}_1 = 14.8$ minutes	$\bar{x}_2 = 10.4$ minutes
$s_1 = .8$ minutes	$s_2 = .6$ minutes

For $\alpha = .02$, test the hypothesis that the difference between the mean assembly times is $\mu_1 - \mu_2 = 5$ minutes.

12. Salary surveys of marketing and management majors show the following starting annual salary data:

Marketing Majors	Management Majors
$n_1 = 14$	$n_2 = 16$
$\bar{x}_1 = \$19{,}800$	$\bar{x}_2 = \$19{,}300$
$s_1 = \$1000$	$s_2 = \$1400$

The null hypothesis is that the mean annual salaries are the same for both majors.

a. What assumptions must be made in order to test the hypothesis?

b. Assume that these assumptions are appropriate. What is the pooled estimate of the variance σ^2?

c. For $\alpha = .05$, can you conclude that a difference exists between the mean annual salaries of the two majors?

10.3 ■■■ INFERENCES ABOUT THE DIFFERENCE BETWEEN THE MEAN OF TWO POPULATIONS — MATCHED SAMPLES

Suppose that a manufacturing company has two methods available for employees to perform a production task. In order to maximize production output, the company would like to identify the method with the smallest mean completion time per unit. Let μ_1 denote the mean completion time for production method 1 and μ_2 denote the mean completion time for production method 2. With no preliminary indication of the preferred production method, we begin by tentatively assuming that the production methods have the same mean completion time. Thus, the null hypothesis becomes H_0: $\mu_1 - \mu_2 = 0$. If this hypothesis is rejected, we can conclude a difference between mean completion times exists. In this case, the method providing the smaller mean completion time would be recommended. The null and alternative hypotheses are written as follows:

$$H_0: \quad \mu_1 - \mu_2 = 0$$

$$H_a: \quad \mu_1 - \mu_2 \neq 0$$

In designing the sampling procedure that will be used to collect production time data and test the above hypotheses, we consider two alternative designs. One design is based on *independent samples*, and the other design is based on *matched samples*. The designs are described as follows:

1. *Independent-sample design:* A simple random sample of workers is selected and uses method 1. A second independent simple random sample of workers is selected and uses method 2. The test of the difference between means is based on the procedures of Section 10.2.

2. *Matched-sample design:* One simple random sample of workers is selected with the workers first using one method and then using the other method. The order of the two methods is assigned randomly to the workers, with some workers performing method 1 first and others performing method 2 first. Each worker provides a pair of data values, one value for method 1 and another value for method 2.

Our interest in the matched-sample design is that since both production methods are tested under similar conditions (i.e., same workers), this design often leads to a smaller sampling error than the independent sample design. The primary reason for this is that each worker in a matched sample design provides data first under one method and then under the other method. Thus variation between workers is eliminated as a source of the sampling error.

Let us demonstrate the analysis of a matched-sample design by assuming that this method is used to test the difference between the two production methods. A random sample of 6 workers is used. The data on completion times for the 6 workers are shown in Table 10.1. Note that each worker provides a pair of data values, one for each production method. Also note that the last column contains the difference in completion times (d_i) for each worker in the sample.

The key to the analysis of the matched-sample design is to realize that we consider only the column of differences in the two methods. As a result we have six data values (.6, −.2, .5, .3, .0, and .6) that will be used in the analysis of the difference between means of the two production methods.

Let μ_d = the mean of the *difference* values for the population of workers. With this notation the null and alternative hypotheses are rewritten as follows:

$$H_0: \quad \mu_d = 0$$

$$H_a: \quad \mu_d \neq 0$$

If H_0 can be rejected, we can conclude that a difference between the mean completion times exists.

The d notation is a reminder that the matched sample provides *difference* data. The sample mean and sample standard deviation for the 6 difference values in

TABLE 10.1 **Task-Completion Times for a Matched Sample of 6 Workers**

Worker	Completion Time, Method 1 (minutes)	Completion Time, Method 2 (minutes)	Difference in Completion Times (d_i)
1	6.0	5.4	.6
2	5.0	5.2	−.2
3	7.0	6.5	.5
4	6.2	5.9	.3
5	6.0	6.0	.0
6	6.4	5.8	.6

Table 10.1 are as follows:

$$\bar{d} = \frac{\Sigma d_i}{n} = \frac{1.8}{6} = .30$$

$$s_d = \sqrt{\frac{\Sigma(d_i - \bar{d})^2}{n - 1}} = \sqrt{\frac{.56}{5}} = \sqrt{.112} = .335$$

In Chapter 9 we found that if the population can be assumed normal the t distribution with $n - 1$ degrees of freedom can be used to test the null hypothesis about a population mean. With difference data, the test statistic becomes

$$t = \frac{\bar{d} - \mu_d}{s_d/\sqrt{n}} \qquad (10.12)$$

With $\alpha = .05$ and with $n - 1 = 5$ degrees of freedom ($t_{.025} = 2.571$), the rejection rule for the two-tailed test becomes

$$\text{Reject } H_0 \text{ if } t < -2.571 \text{ or if } t > 2.571$$

With $\bar{d} = .30$, $s_d = .335$, and $n = 6$, the value of the test statistic for the null hypothesis H_0: $\mu_d = 0$ becomes

$$t = \frac{\bar{d} - \mu_d}{s_d/\sqrt{n}} = \frac{.30 - 0}{.335/\sqrt{6}} = 2.19$$

Thus, the sample data do not provide sufficient evidence to reject H_0. It is recommended that the firm take a larger sample and study the issue further.

With the above sample results, an interval estimate of the difference between the two population means can be based on the single-population methodology of Chapter 8. Doing so provides the following:

$$0.3 \pm t_{\alpha/2} \frac{s_d}{\sqrt{n}}$$

$$0.3 \pm 2.571 \frac{.335}{\sqrt{6}}$$

$$0.3 \pm .35$$

Thus the 95% confidence interval estimate of the difference between the means of the two production methods is $-.05$ to $.65$ minutes.

NOTES AND COMMENTS

1. In this example, workers performed the production task using first one method and then the other method. This is an example of a matched-sample design, where each sampled item (worker) provides a pair of data values. Although this is often the procedure used in the matched-samples analysis, it is possible to use different but "similar" items to provide the pair of data values. In this sense, a worker at one location could be matched with a similar worker at another location (similarity based on age, education, sex, experience, etc.). The pairs of workers would provide the difference data that could be used in the matched-sample analysis.

2. Since a matched-sample procedure for inferences about two population means generally provides better precision than the independent-sample approach, it is the recommended design. However, in some applications the matching cannot be achieved, or perhaps the time and cost associated with matching is excessive. In these cases the independent-sample design should be used.

3. The example presented in this section employed a sample size of 6 workers. As such, the small-sample case existed, and the t distribution was used in both the test of hypothesis and interval estimation computations. If the sample size is large ($n \geq 30$), the use of the t distribution is unnecessary, and the statistical inferences can be based on the z values of the standard normal probability distribution.

▬ EXERCISES

13. A market research firm used a sample of individuals to rate the potential to purchase for a particular product before and after the individuals saw a new television commercial about the product. The potential to purchase ratings were based on a 0 to 10 scale, with higher values indicating a higher potential to purchase. The null hypothesis stated that the mean rating "after" would be less than or equal to the mean rating "before." Rejection of this hypothesis would provide the conclusion that the commercial improved the mean potential to purchase rating. Use $\alpha = .05$ and the following data to test the hypothesis and comment on the value of the commercial:

Individual	Purchase Rating Before Seeing Commercial	Purchase Rating After Seeing Commercial
1	5	6
2	4	6
3	7	7
4	3	4
5	5	3
6	8	9
7	5	7
8	6	6

14. Earnings per share for 1987 and 1988 are presented below for a sample of 10 corporations (*Business Week,* 1989 Bonus Issue).

Corporation	1987	1988
IBM	8.72	9.27
Sears Roebuck	4.35	2.72
Chevron	3.65	5.17
Walt Disney	2.85	3.80
American Express	1.20	2.31
McDonalds	2.89	3.43
Anheuser-Busch	2.04	2.45
Kellogg	3.20	3.90
J. C. Penny	4.11	6.02
Motorola	2.39	3.43

a. Based on the preceding data, is it appropriate to conclude that mean earnings per share were significantly higher in 1988? Test at a .05 level of significance.

b. Provide a 95% confidence interval estimate of the mean increase in earnings per share for the 1-year period.

15. A manufacturer produces both a deluxe and a standard model automatic sander designed for home use. Selling prices obtained from a sample of retail outlets are as follows:

Retail Outlet	Price, Deluxe Model	Price, Standard Model	Difference
1	$39	$27	12
2	39	28	11
3	45	35	10
4	38	30	8
5	40	30	10
6	39	34	5
7	35	29	6

a. The manufacturer's suggested retail prices for the two models show a $10 differential in prices. Use a .05 level of significance and test to see if the mean difference between prices of the two models is $10.

b. What is the 95% confidence interval estimate of the difference between the mean prices for the two models?

16. A company attempts to evaluate the potential for a new bonus plan by selecting a random sample of 5 salespersons to use the bonus plan for a trial period. The weekly sales volumes before and after implementing the bonus plan are shown below:

Salesperson	Weekly Sales Before	Weekly Sales After
1	15	18
2	12	14
3	18	19
4	15	18
5	16	18

a. Use $\alpha = .05$ and test to see if the bonus plan will result in an increase in the mean weekly sales.

b. Provide a 90% confidence interval estimate for the mean *increase* in weekly sales that can be expected if a new bonus plan is implemented.

17. Word-processing systems are often justified on the basis of improved efficiencies for a secretarial staff. Shown below are typing rates in words per minute for 7 secretaries who previously used electronic typewriters and who are now using computer-based word processors. Test at the .05 level of significance to see if there has been an increase in the mean typing rate due to the word-processing system.

Secretary	Electronic Typewriter	Word Processor
1	72	75
2	68	66
3	55	60
4	58	64
5	52	55
6	55	57
7	64	64

10.4 ■ INFERENCES ABOUT THE DIFFERENCE BETWEEN THE PROPORTIONS OF TWO POPULATIONS

A tax preparation firm is interested in comparing the quality of work at 2 of its regional offices. By randomly selecting samples of tax returns prepared at each office and having the sample returns verified for accuracy, the firm will be able to estimate the proportion of erroneous returns prepared at each office. Of particular interest is the difference between the proportions of erroneous returns existing at the two offices.

Let

p_1 = proportion of erroneous returns for population 1 (office 1)
p_2 = proportion of erroneous returns for population 2 (office 2)
$\bar{p}_1$ = sample proportion for a simple random sample taken from population 1 (i.e., the point estimator of p_1)
$\bar{p}_2$ = sample proportion for a simple random sample taken from population 2 (i.e., the point estimator of p_2)

The difference between the two population proportions is given by $p_1 - p_2$. The point estimator of $p_1 - p_2$ is as follows:

Point Estimator of the Difference Between the Proportions of Two Populations

$$\bar{p}_1 - \bar{p}_2$$

Thus, the point estimator of the difference between two population proportions is the difference between the sample proportions of two independent simple random samples.

Sampling Distribution of $\bar{p}_1 - \bar{p}_2$

In the study of the difference between two population proportions, $\bar{p}_1 - \bar{p}_2$ is the point estimator of interest. As we have seen in several previous cases, the sampling distribution of the point estimator is a key factor in developing interval estimates and in testing hypotheses about the parameters of interest. The properties of the sampling distribution of $\bar{p}_1 - \bar{p}_2$ are as follows:

Sampling Distribution of $\bar{p}_1 - \bar{p}_2$

$$\text{Mean:} \quad E(\bar{p}_1 - \bar{p}_2) = p_1 - p_2 \qquad (10.13)$$

$$\text{Standard deviation:} \quad \sigma_{\bar{p}_1 - \bar{p}_2} = \sqrt{\frac{p_1(1 - p_1)}{n_1} + \frac{p_2(1 - p_2)}{n_2}} \qquad (10.14)$$

where

n_1 = sample size for the simple random sample selected from population 1

n_2 = sample size for the simple random sample selected from population 2

Distribution form: If the sample sizes are large (that is, $n_1 p_1$, $n_1(1 - p_1)$, $n_2 p_2$, and $n_2(1 - p_2)$ are all greater than or equal to 5), the sampling distribution of $\bar{p}_1 - \bar{p}_2$ can be approximated by a normal probability distribution.

Figure 10.6 shows the sampling distribution of $\bar{p}_1 - \bar{p}_2$.

▪▪▪ FIGURE 10.6   **Sampling Distribution of $\bar{p}_1 - \bar{p}_2$**

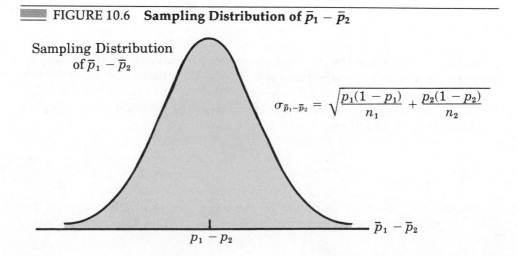

Sampling Distribution of $\bar{p}_1 - \bar{p}_2$

$$\sigma_{\bar{p}_1 - \bar{p}_2} = \sqrt{\frac{p_1(1 - p_1)}{n_1} + \frac{p_2(1 - p_2)}{n_2}}$$

$p_1 - p_2$

$\bar{p}_1 - \bar{p}_2$

Interval Estimation of $p_1 - p_2$

Let us assume that independent simple random samples of tax returns from the two offices show the following:

Office 1	Office 2
$n_1 = 250$	$n_2 = 300$
Number of returns with errors = 35	Number of returns with errors = 27

The sample proportions for the two offices are as follows:

$$\bar{p}_1 = \frac{35}{250} = .14$$

$$\bar{p}_2 = \frac{27}{300} = .09$$

Using (10.12), the point estimate of the difference between the proportion of erroneous tax returns for the two populations is $\bar{p}_1 - \bar{p}_2 = .14 - .09 = .05$. Specifically, we are led to believe that office 1 possesses a 5% greater error rate than office 2. However, as with all point estimates, we know that the .05 difference is only one of many possible sample values for the difference between the two population proportions. The following expression can be used to develop an interval estimate of the difference between the proportions of the two populations:

Interval Estimate of the Difference Between the Proportions of Two Populations (Large-Sample Case with $n_1 p_1$, $n_1(1 - p_1)$, $n_2 p_2$, and $n_2(1 - p_2)$ All Greater Than or Equal to 5)

$$\bar{p}_1 - \bar{p}_2 \pm z_{\alpha/2} \sigma_{\bar{p}_1 - \bar{p}_2} \qquad (10.15)$$

where $1 - \alpha$ is the confidence coefficient.

Let us use this procedure to develop an interval estimate of the difference between the population proportions of erroneous tax returns existing at the two offices. Since p_1 and p_2 are unknown, we cannot use (10.14) to calculate $\sigma_{\bar{p}_1 - \bar{p}_2}$. However, using $\bar{p}_1$, the point estimator of p_1, and $\bar{p}_2$, the point estimator of p_2, we can estimate $\sigma_{\bar{p}_1 - \bar{p}_2}$ as follows:

Estimate of $\sigma_{\bar{p}_1 - \bar{p}_2}$

$$s_{\bar{p}_1 - \bar{p}_2} = \sqrt{\frac{\bar{p}_1(1 - \bar{p}_1)}{n_1} + \frac{\bar{p}_2(1 - \bar{p}_2)}{n_2}} \qquad (10.16)$$

The above expression provides an estimate of $\sigma_{\bar{p}_1 - \bar{p}_2}$ and can be used in (10.15) to obtain an interval estimate of $p_1 - p_2$.

Let us make these calculations. Using (10.16) we have

$$s_{\bar{p}_1 - \bar{p}_2} = \sqrt{\frac{.14(.86)}{250} + \frac{.09(.91)}{300}} = .0275$$

With a 90% confidence interval, $z_{\alpha/2} = z_{.05} = 1.645$. Expression (10.15) provides the following interval estimate:

$$(.14 - .09) \pm 1.645(.0275)$$
$$.05 \pm .045$$

Thus the 90% confidence interval estimate of the difference in error rates at the two offices is .005 to .095.

Hypothesis Tests About $p_1 - p_2$

As an example of hypothesis tests concerning the difference between the proportions of two populations, let us consider the data collected in the preceding example and assume that the firm had been attempting to determine whether or not a difference exists in the error proportions at the two offices. Let us illustrate the statistical analysis we could use to test the following hypotheses:

$$H_0: \quad p_1 - p_2 = 0$$
$$H_a: \quad p_1 - p_2 \neq 0$$

Figure 10.7 shows the sampling distribution of $\bar{p}_1 - \bar{p}_2$ based on the assumption

---

■■■ FIGURE 10.7   Sampling Distribution of $\bar{p}_1 - \bar{p}_2$ with H_0: $p_1 - p_2 = 0$

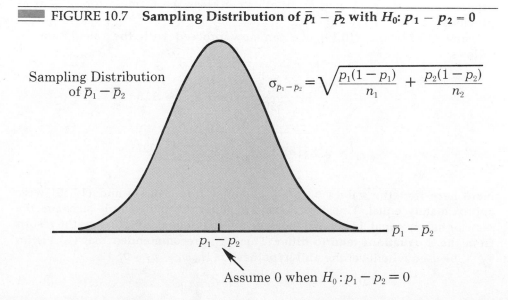

Sampling Distribution of $\bar{p}_1 - \bar{p}_2$

$$\sigma_{\bar{p}_1 - \bar{p}_2} = \sqrt{\frac{p_1(1 - p_1)}{n_1} + \frac{p_2(1 - p_2)}{n_2}}$$

$\bar{p}_1 - \bar{p}_2$

$p_1 - p_2$

Assume 0 when $H_0: p_1 - p_2 = 0$

that there is no difference between the two population proportions. That is, $p_1 - p_2 = 0$. With the sampling distribution approximately normal, the test statistic for the difference between two population proportions can be written

$$z = \frac{(\bar{p}_1 - \bar{p}_2) - (p_1 - p_2)}{\sigma_{\bar{p}_1 - \bar{p}_2}} \tag{10.17}$$

Using $\alpha = .10$ and $z_{\alpha/2} = z_{.05} = 1.645$, the rejection rule is

Reject H_0 if $z < -1.645$ or if $z > 1.645$

We may be tempted to use (10.16) to estimate the standard error of the difference between proportions $\sigma_{\bar{p}_1 - \bar{p}_2}$. However, we generally adjust the formula for $s_{\bar{p}_1 - \bar{p}_2}$ for the *special case* of a hypothesis test of the form H_0: $p_1 - p_2 = 0$. That is, (10.16) is modified to reflect the fact that when we assume the null hypothesis to be true we are assuming that $p_1 = p_2$. Whenever this is the case, there is no need to use the individual values for $\bar{p}_1$ and $\bar{p}_2$ in the formula for $s_{\bar{p}_1 - \bar{p}_2}$. Only one value, $\bar{p}$, is needed, since it can be used as an estimate for both p_1 and p_2. Thus we will combine or *pool* the two sample results to provide one estimate of the population proportion. The pooled estimate is as follows:

$$\bar{p} = \frac{n_1 \bar{p}_1 + n_2 \bar{p}_2}{n_1 + n_2} \tag{10.18}$$

Since $\bar{p}$ can be used as an estimate of both p_1 and p_2, (10.16) is revised to

$$s_{\bar{p}_1 - \bar{p}_2} = \sqrt{\bar{p}(1 - \bar{p})\left(\frac{1}{n_1} + \frac{1}{n_2}\right)} \tag{10.19}$$

Remember, however, that (10.18) and (10.19) are only used whenever the null hypothesis for a statistical test about the difference between two population proportions is H_0: $p_1 - p_2 = 0$. In all other cases the formula for $s_{\bar{p}_1 - \bar{p}_2}$ is as previously shown in (10.16).

Using (10.18) and (10.19), we can now proceed with the calculations as follows:

$$\bar{p} = \frac{250(.14) + 300(.09)}{550} = \frac{62}{550} = .113$$

$$s_{\bar{p}_1 - \bar{p}_2} = \sqrt{(.113)(.887)\left(\frac{1}{250} + \frac{1}{300}\right)} = .0271$$

Note here that the values of $s_{\bar{p}_1 - \bar{p}_2}$ obtained from (10.16) and (10.19) were approximately equal. This case exists whenever the two sample sizes are the same or nearly equal. In general, however, with unequal sample sizes the results from the 2 equations tend to differ. Thus, it is recommended that (10.19) for $s_{\bar{p}_1 - \bar{p}_2}$ be used whenever the null hypothesis is H_0: $p_1 - p_2 = 0$.

Using (10.17), the value of the test statistic becomes

$$z = \frac{(\bar{p}_1 - \bar{p}_2) - (p_1 - p_2)}{s_{\bar{p}_1 - \bar{p}_2}} = \frac{(.14 - .09) - 0}{.0271} = 1.85$$

Since $1.85 > 1.645$, at the .10 level of significance the null hypothesis is rejected. The sample evidence indicates that there is a difference between the error proportions at the two offices.

As we saw with the hypothesis tests about differences between two population means, one-tailed tests can also be developed for the difference between two population proportions. The one-tailed rejection regions are established in a manner similar to the one-tailed hypothesis testing procedures for a single-population proportion.

■ EXERCISES

18. A sample of 400 items produced by supplier A contained 70 defective items. A sample of 300 items produced by supplier B contained 40 defective items. Compute a 90% confidence interval estimate for the difference in the proportion defective from the two suppliers.

 During the primary elections of 1988 a particular presidential candidate showed the following preelection voter support in Wisconsin and Illinois:

State	Voters Surveyed	Voters Favoring the Candidate
Wisconsin	500	270
Illinois	360	162

Compute a 95% confidence interval estimate for the difference between the proportion of voters favoring the candidate in the two states.

20. Leo J. Shapiro & Associates, a Chicago market-research firm, surveys consumers on a variety of issues (*Wall Street Journal*, October 17, 1988). In late summer and early fall of each year, the firm surveys consumers to learn about spending plans for the forthcoming holiday season. In 1988, 36% of the consumers surveyed indicated they planned to "spend less" during the 1988 season. In 1987, 28% of the consumers surveyed indicated they planed to "spend less" during the 1987 season. Assume that 400 consumers were surveyed each year.

    a. Compare the results for the two years. Use $\alpha = .05$ to see if there has been a significant increase in the proportion of consumers indicating they will spend less during the 1988 season. What is your conclusion? What is the *p*-value?

    b. Assume you are a retailer in October 1988. Comment on the value of the 2-year study and what the results might mean to your business.

    c. Use estimation procedures to develop an interval estimate of the 1-year increase in the proportion of consumers who indicate they will spend less during the 1988 season. Use a 95% confidence level.

21. Two loan officers at the North Ridge National Bank show the following data for defaults on loans that they have approved (the data are based on samples of loans granted over the past 5 years):

	Loans Reviewed in the Sample	Defaulted Loans
Loan Officer A	60	9
Loan Officer B	80	6

Use $\alpha = .05$ and test using the hypothesis that the default rates are the same for the two loan officers.

22. A Media General/Associated Press Poll (*USA Today*, April 27, 1989) reported that 16% of men and 5% of women would want to be president of the United States. Assume 500 men and 500 women participated in the poll. Test $H_0: p_1 - p_2 = 0$ versus $H_a: p_1 - p_2 \neq 0$. Use $\alpha = .05$. What is your conclusion about the proportions of men and women wanting to be president of the United States?

23. A survey firm conducts door-to-door surveys on a variety of issues. Some individuals cooperate with the interviewer and complete the interview questionnaire, while others do not. The following sample data are available (showing the response data for men and women):

	Sample Size	Number Cooperating with the Survey
Men	200	110
Women	300	210

a. Use $\alpha = .05$ and test the hypothesis that the response rate is the same for both men and women.
b. Compute the 95% confidence interval estimate for the difference between the proportion of men and the proportion of women that cooperate with the survey.

24. In a test of the quality of two television commercials, each commercial was shown in a separate test area 6 times over a 1-week period. The following week a telephone survey was conducted to identify individuals who had seen the commercials. The individuals who had seen the commercials were asked to state the primary message in the commercial. The following results were recorded:

	Number Reporting Having Seen the Commercial	Number Recalling Primary Message
Commercial A	150	63
Commercial B	200	60

a. Use $\alpha = .05$ and test the hypothesis that there is no difference in the recall proportions for the two commercials.
b. Compute a 95% confidence interval for the difference between the recall proportions for the two populations.

■ SUMMARY

In this chapter we have discussed procedures for interval estimation and hypothesis testing involving two populations. First, we showed how to make inferences about the differences between the means of two populations when independent simple random samples are selected. We considered both the large-sample and small-sample cases. The z values from the standard normal probability distribution are used for inferences about the difference between two population means when the sample sizes are large. The t distribution is used for the inferences when the populations are assumed normal with equal variances. This permits inferences in the small-sample case.

Inferences about the difference between the means of two populations were then discussed for the matched-sample design. In the matched-sample design each element provides a pair of data values, one from each population. The difference between the paired data values is then used in the statistical analysis. The matched-sample design is generally preferred over the independent-sample design because the matched-sample procedure reduces variability, thus tending to reduce the sampling error and to improve the precision of the estimate.

Finally, interval estimation and hypothesis testing about the difference between two population proportions were discussed. Statistical procedures for analyzing the difference between proportions for two populations are similar to the procedures for analyzing the difference between means for two populations.

■ GLOSSARY

Pooled variance An estimate of the variance of a population based on the combination of two (or more) sample results. The pooled variance estimate is appropriate whenever the variances of two (or more) populations are assumed equal. For the two-population case, the pooled estimator of the variance is computed as follows:

$$s^2 = \frac{(n_1 - 1)s_1^2 + (n_2 - 1)s_2^2}{n_1 + n_2 - 2}$$

Independent samples Samples selected from two (or more) populations where the elements making up one sample are chosen independently of the elements making up the other sample(s).

Matched samples Samples where each data value in one sample is matched with a corresponding data value in the other sample.

■ KEY FORMULAS

Mean of $\bar{x}_1 - \bar{x}_2$

$$E(\bar{x}_1 - \bar{x}_2) = \mu_1 - \mu_2 \tag{10.2}$$

Standard Deviation of $\bar{x}_1 - \bar{x}_2$

$$\sigma_{\bar{x}_1 - \bar{x}_2} = \sqrt{\frac{\sigma_1^2}{n_1} + \frac{\sigma_2^2}{n_2}} \tag{10.3}$$

Interval Estimate of the Difference Between the Means of Two Populations (Large-Sample Case with $n_1 \geq 30$ and $n_2 \geq 30$)

$$\bar{x}_1 - \bar{x}_2 \pm z_{\alpha/2}\sigma_{\bar{x}_1 - \bar{x}_2} \tag{10.4}$$

Estimator of $\sigma_{\bar{x}_1 - \bar{x}_2}$

$$s_{\bar{x}_1 - \bar{x}_2} = \sqrt{\frac{s_1^2}{n_1} + \frac{s_2^2}{n_2}} \tag{10.5}$$

Standard Deviation of $\bar{x}_1 - \bar{x}_2$ When $\sigma_1^2 = \sigma_2^2$,

$$\sigma_{\bar{x}_1-\bar{x}_2} = \sqrt{\frac{\sigma^2}{n_1} + \frac{\sigma^2}{n_2}} = \sqrt{\sigma^2\left(\frac{1}{n_1} + \frac{1}{n_2}\right)} \qquad (10.6)$$

Pooled Estimator of σ^2

$$s^2 = \frac{(n_1 - 1)s_1^2 + (n_2 - 1)s_2^2}{n_1 + n_2 - 2} \qquad (10.7)$$

Estimator of $\sigma_{\bar{x}_1-\bar{x}_2}$ When $\sigma_1^2 = \sigma_2^2$

$$s_{\bar{x}_1-\bar{x}_2} = \sqrt{s^2\left(\frac{1}{n_1} + \frac{1}{n_2}\right)} \qquad (10.8)$$

Interval Estimate of the Difference Between the Means of Two Populations (Small-Sample Case with $n_1 < 30$ and/or $n_2 < 30$)

$$\bar{x}_1 - \bar{x}_2 \pm t_{\alpha/2}s_{\bar{x}_1-\bar{x}_2} \qquad (10.9)$$

Test Statistic for Hypothesis Tests About the Difference Between the Means of Two Populations

$$z = \frac{(\bar{x}_1 - \bar{x}_2) - (\mu_1 - \mu_2)}{\sqrt{\sigma_1^2/n_1 + \sigma_2^2/n_2}} \qquad (10.10)$$

Sample Mean — Matched Samples

$$\bar{d} = \frac{\Sigma d_i}{n}$$

Sample Standard Deviation — Matched Samples

$$s_d = \sqrt{\frac{\Sigma(d_i - \bar{d})^2}{n - 1}}$$

Test Statistic — Matched Samples

$$t = \frac{\bar{d} - \mu_d}{s_d/\sqrt{n}} \qquad (10.12)$$

Mean of $\bar{p}_1 - \bar{p}_2$

$$E(\bar{p}_1 - \bar{p}_2) = p_1 - p_2 \qquad (10.13)$$

Standard Deviation of $\bar{p}_1 - \bar{p}_2$

$$\sigma_{\bar{p}_1-\bar{p}_2} = \sqrt{\frac{p_1(1 - p_1)}{n_1} + \frac{p_2(1 - p_2)}{n_2}} \qquad (10.14)$$

Interval Estimate of the Difference Between the Proportions of Two Populations (Large-Sample Case with $n_1 p_1$, $n_1(1 - p_1)$, $n_2 p_2$, and $n_2(1 - p_2)$ All Greater Than or Equal to 5)

$$\bar{p}_1 - \bar{p}_2 \pm z_{\alpha/2}\sigma_{\bar{p}_1-\bar{p}_2} \qquad (10.15)$$

Estimator of $\sigma_{\bar{p}_1 - \bar{p}_2}$

$$s_{\bar{p}_1 - \bar{p}_2} = \sqrt{\frac{\bar{p}_1(1 - \bar{p}_1)}{n_1} + \frac{\bar{p}_2(1 - \bar{p}_2)}{n_2}} \tag{10.16}$$

Test Statistic for Hypothesis Tests About the Difference Between Proportions of Two Populations

$$z = \frac{(\bar{p}_1 - \bar{p}_2) - (p_1 - p_2)}{\sigma_{\bar{p}_1 - \bar{p}_2}} \tag{10.17}$$

Pooled Estimator of the Population Proportion

$$\bar{p} = \frac{n_1 \bar{p}_1 + n_2 \bar{p}_2}{n_1 + n_2} \tag{10.18}$$

Estimator of $\sigma_{\bar{p}_1 - \bar{p}_2}$ When $p_1 = p_2$

$$s_{\bar{p}_1 - \bar{p}_2} = \sqrt{\bar{p}(1 - \bar{p})\left(\frac{1}{n_1} + \frac{1}{n_2}\right)} \tag{10.19}$$

■ SUPPLEMENTARY EXERCISES

25. Starting annual salaries for individuals with master's and bachelor's degrees in business were collected with use of two independent random samples. Use the data shown to develop a 90% confidence interval estimate of the increase in starting salary that can be expected upon completion of the master's degree:

Master's Degree	Bachelor's Degree
$n_1 = 60$	$n_2 = 80$
$\bar{x}_1 = \$23,000$	$\bar{x}_2 = \$21,000$
$s_1 = \$2500$	$s_2 = \$2000$

26. Samples of dinner and luncheon receipts at a major downtown restaurant show the following results:

Dinner Receipts	Luncheon Receipts
$n_1 = 70$	$n_2 = 55$
$\bar{x}_1 = \$32.65$	$\bar{x}_2 = \$12.80$
$s_1 = \$7.20$	$s_2 = \$3.60$

Provide a 98% confidence interval estimates of the difference between mean receipt amounts for the two meals.

27. Safegate Foods, Inc., is redesigning the checkout lanes in its supermarkets throughout the country. Two designs have been suggested. Tests on customer checkout times have been collected at two stores where the two new systems have been installed. The sample data are as follows:

Times for Checkout System A	Times for Checkout System B
$n_1 = 120$	$n_2 = 100$
$\bar{x}_1 = 4.1$ minutes	$\bar{x}_2 = 3.3$ minutes
$s_1 = 2.2$ minutes	$\bar{s}_2 = 1.5$ minutes

Test at the .05 level of significance to determine if there is a difference between the mean checkout times for the two systems. Which system is preferred?

28. Samples of final examination scores for two statistics classes with different instructors showed the following results:

Instructor A's Class	Instructor B's Class
$n_1 = 12$	$n_2 = 15$
$\bar{x}_1 = 72$	$\bar{x}_2 = 78$
$s_1 = 8$	$s_2 = 10$

With $\alpha = .05$, test whether or not these data are sufficient to conclude that the mean grades differ for the two classes.

29. In a study of job attitudes and job satisfaction, a sample of 50 men and 50 women were asked to rate their overall job satisfaction on a 1 to 10 scale. A high rating indicates a higher degree of job satisfaction. Using the sample results shown below, does there appear to be a significant difference between the levels of job satisfaction for men and women? Use $\alpha = .05$.

Men	Women
$\bar{x}_1 = 7.2$	$\bar{x}_2 = 6.4$
$s_1 = 1.7$	$s_2 = 1.4$

30. Figure Perfect, Inc., is a women's figure salon that specializes in weight-reduction programs. Weights for a sample of clients before and after a 6-week introductory program are as follows:

Client	Weight Before	Weight After
1	140	132
2	160	158
3	210	195
4	148	152
5	190	180
6	170	164

Use $\alpha = .05$ and test to determine if the introductory program provides a statistically significant weight loss.

31. A cable television firm is considering submitting bids for rights to operate in two regions of the state of Florida. Surveys of the two regions provide the following data on customer acceptance of the cable television service:

Region I	Region II
$n_1 = 500$	$n_2 = 800$
Number indicating an intent to purchase = 175	Number indicating an intent to purchase = 360

Develop a 99% confidence interval estimate of the difference between population proportions of customers who have an intent to purchase in the two regions.

32. A group of physicians from Denmark conducted a study on the effectiveness of nicotine chewing gum in helping people stop smoking (*New England Journal of Medicine*, 1988). The 113 people who participated in the study were all smokers. Of these, 60 were given chewing gum with 2 milligrams of nicotine and 53 were given identical appearing

chewing gum that had no nicotine content. No one in the study knew which type of gum he or she had been given. All were told to use the gum and refrain from smoking.

a. Define the null and alternative hypothesis that would be appropriate if the researchers hoped to show that the group given chewing with nicotine had a higher proportion of nonsmokers 1 year after the study began.

b. Results 1 year later found that 23 of the smokers given nicotine chewing gum had remained nonsmokers for the 1-year period. In addition, 12 of the smokers given the chewing gum with no nicotine content had remained nonsmokers for the 1-year period. Do these results support the conclusion that nicotine gum can help stop smoking? Test using $\alpha = .05$. What is the p-value?

33. A large automobile-insurance company selected samples of single and married male policyholders and recorded the number who had made an insurance claim over the past 3-year period:

Single Policyholders	Married Policyholders
$n_1 = 400$	$n_2 = 900$
Number making claims = 76	Number making claims = 90

a. Test with $\alpha = .05$ to determine if the claim rates differ between single and married male policyholders.

b. Provide a 95% confidence interval estimate of the difference between the claim proportions for the two populations.

34. A political opinion survey shows that of 200 Republicans surveyed 80 opposed the building of nuclear power plants. Similar results for a sample of 300 Democrats showed that 150 opposed the building of such nuclear plants. Do these results indicate that there is a significant difference between the attitudes of Republicans and Democrats toward building nuclear power plants? Use $\alpha = .05$.

▬▬ COMPUTER EXERCISE

Par, Inc., is a major manufacturer of golf equipment. The research group at Par has been investigating a new golf ball designed to resist cuts and yet still offer good driving distances. In tests with the new golf balls, 40 balls of the new model and 40 balls of the current model were subjected to distance tests. The testing was performed with a mechanical hitting machine in ideal weather conditions, so that if a difference existed between the mean distance of the two models it could be attributed to a real difference in design performance. The results of the test are shown. The distance data are measured to the nearest yard.

Current Model	New Model	Current Model	New Model
242	274	248	269
239	266	265	256
245	260	267	261
250	263	258	277
236	259	250	271
261	248	253	278
236	259	243	273
244	286	238	266
237	283	256	265
248	261	253	259
241	271	259	280
242	263	252	247

Current Model	New Model	Current Model	New Model
262	259	251	250
241	268	241	257
238	257	253	267
261	278	245	260
233	247	257	258
250	260	252	252
244	275	254	260
246	261	240	276

▬ QUESTIONS

1. Develop numerical and graphical measures to summarize the given data.

2. Develop interval estimates for the mean distance traveled for both types of balls.

3. What statistical conclusion can you reach regarding the mean distances for the two models? What are your recommendations?

APPLICATION

Pennwalt Corporation*

Rochester, New York

Pennwalt Corporation was founded in 1850 as a chemical company manufacturing caustic soda. The company steadily expanded its products to a variety of chemical and other markets, such as fruit sizing, dental products, agricultural needs, food processing, and pharmaceuticals. Today, Pennwalt is an international organization that employs over 9000 people, with revenues of approximately $1 billion. The techniques discussed in this application pertain to operations in the Pharmaceutical Division and involve statistical analyses that will provide conclusions about the differences between population means.

One of the major responsibilities of the Pharmaceutical Division involves testing new drugs. This process consists of three stages: (1) initial preclinical testing; (2) testing for longterm usage and safety; and (3) actual clinical testing. At each successive stage the chances that the drug survives the rigorous testing decreases; however, the cost of testing increases geometrically. Thus it is important to weed out unsuccessful new drugs in the early stages of the testing process. The statistical techniques that are discussed in this chapter play an important role in helping both to identify new drugs that should be rejected and to ensure that potential good candidates are not incorrectly rejected.

▦ INITIAL PRECLINICAL TESTING

Once it has been shown that a new drug is chemically stable, it is sent to the pharmacology department for testing of efficacy (i.e., its capacity to produce the desired effect) in experimental animal models and to check for any untoward (i.e., unfavorable) effects. As part of this process, a statistician is asked to design an experiment that can be used to test for significant effect; this design must specify the number of observations to be collected, the methods of statistical analyses, and a statement of the statistical requirements that must be met. The efficacy of this new drug (population 1) is checked by comparing its average efficacy with that of a standard drug (population 2). Based upon its intended use, the new drug may be tested in various subdisciplines (e.g., central nervous, cardiovascular, immunology and/or biochemical). In each discipline the chemical will be repeatedly tested in different settings; typically, a total of more than 100 different tests are performed. Each test involves the comparison of the new drug with a standard drug in its category. Thus the actual statistical analysis involves testing for the difference between the means of two populations. The new drug can be rejected at any level of testing because of lack of efficacy or undesirable effects.

▦ TESTING FOR LONGTERM USAGE AND SAFETY

The next stage of testing is in the Drug Safety Evaluation section. The new drug is tested here for its cumulative long-term effect. With the use of at least two different animal

*The authors are indebted to Dr. M. C. Trivedi, Pennwalt Corporation, Pharmaceutical Division for providing this application.

species and with the application of repeated doses, the new drug is tested for 15 days, 30 days, 90 days, and/or longer. For each experiment a protocol is prepared that is approved by persons in different disciplines, including the statistician for preclinical testing. The major statistical analyses involve comparisons of two population means based on procedures discussed in this chapter. If the chemical survives the initial testing, an application is made to the Food and Drug Administration (FDA) for permission to test the new drug in humans. This is known as an Investigational New Drug (IND) application. If the application is not rejected clinical testing may begin.

■■ CLINICAL TESTING

The third stage of testing involves hospitals, physicians, nurses, and either volunteers or patients, depending upon the phase of testing. Each volunteer or patient is informed of the experimental nature of the drug, and a signed consent form is obtained before the person is admitted into the study. The first phase involves volunteers who are intensively monitored by the participating physician according to the protocol to detect any untoward effect. The second phase involves a selected set of patients to evaluate efficacy of the drug compared to a standard drug.

A battery of efficacy and safety parameters, such as blood pressure, heart rate, body weight, and blood and urine chemistries, are monitored to detect any untoward effects. The person's sex may be a contributing factor. A person may have hypertension complicated by other diseases, or subjects may be in different phases of the life cycle (e.g., childhood, youth, pregnancy, old age); thus the drug has to be tested for each combination in order to determine its true effect in each case and in order to make sure that untoward effects, if any, are identified. During this phase the drug is tested in several hundred subjects. The statistical analysis involves comparing the average efficacy of the new drug with that of the standard drug.

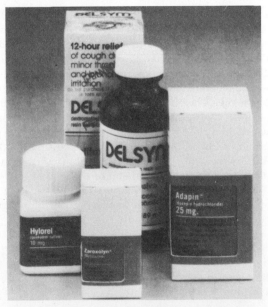

Some of the drug products manufactured by the pharmaceutical division of Pennwalt Corporation

Another phase of clinical experiments involves evaluating the onset of drug effect and its duration. In these studies the new drug is compared with a standard drug, if applicable, and the average time to onset of the effect for the new drug is compared with that for the standard drug. Similarly, duration of effect is also compared (or studied by itself when there is no standard drug on the market).

If the new drug meets the requirements of the FDA, a New Drug Application (NDA) is filed with the FDA. Although the guidelines call for approval or disapproval within 6 months, the FDA usually takes longer to respond.

Besides being tested to meet the requirements of the FDA, the new drug is also tested to meet the needs of the marketing division. That is, the new drug may be good enough as a "me, too" or even a little bit better than the standard drug. However, the marketing division may not be able to sell it if it is not substantially better than the standard drug on the market. In this testing, one-sided statistical hypotheses are used when comparing average efficacies of the drugs.

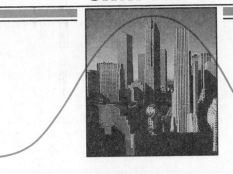

CHAPTER 11

Inferences About Population Variances

I n the previous four chapters we focused on methods of statistical inference involving population means and population proportions. In this chapter we expand the discussion to situations involving inferences about population variances. As an example of a case where a variance can provide important decision-making information, consider the production process of filling containers with a liquid detergent product. The filling mechanism for the process is adjusted so that the mean filling quantity is 16 ounces per container. While a mean of 16 ounces is desired, the variance of the fillings is also critical. That is, even with the filling mechanism properly adjusted for the mean of 16 ounces, we cannot expect every container to have exactly 16 ounces. By selecting a sample of containers we can compute a sample variance for the filling quantities. This value will serve as an estimate of the variance for the population of containers being filled by the production process. If the sample variance is modest, the production process will be continued. However, if the sample variance is excessive, overfilling and underfilling may exist even though the mean may be correct at 16 ounces. In this case the filling mechanism will be readjusted in an attempt to reduce the filling variance for the containers.

In the following section we consider inferences about the variance of a single population. Later we will discuss procedures that can be used to make inferences about the variances of two different populations.

11.1 ▄▄ INFERENCES ABOUT A POPULATION VARIANCE

In earlier chapters we used the sample variance

$$s^2 = \frac{\Sigma(x_i - \bar{x})^2}{n - 1} \tag{11.1}$$

as the point estimator of the population variance σ^2. In using the sample variance as a basis for making inferences about a population variance, the sampling distribution of the quantity $(n - 1)s^2/\sigma^2$ is very helpful. This sampling distribution is described as follows:

Sampling Distribution of $(n - 1)s^2/\sigma^2$

Whenever a simple random sample of size n is selected from a *normal population*, the quantity

$$\frac{(n - 1)s^2}{\sigma^2} \qquad\qquad (11.2)$$

has a *chi-square distribution* with $n - 1$ degrees of freedom, where s^2 is the sample variance and σ^2 is the population variance.

A graph of the sampling distribution of $(n - 1)s^2/\sigma^2$ is shown in Figure 11.1.

In previous chapters we have shown that knowledge of a sampling distribution is essential for computing interval estimates and conducting hypothesis tests about a population parameter. For inferences about a population variance we find it convenient to work with the sampling distribution of $(n - 1)s^2/\sigma^2$. The reason for this is that with normal populations the sampling distribution of $(n - 1)s^2/\sigma^2$ is known to have a chi-square probability distribution with $n - 1$ degrees of freedom. Since tables of areas or probabilities are readily available for

---

■■■ FIGURE 11.1   **Examples of the Sampling Distribution of $(n - 1)s^2/\sigma^2$ (a Chi-Square Distribution) with 2, 5, and 10 Degrees of Freedom**

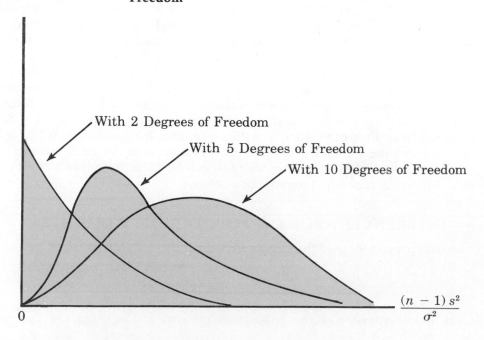

the chi-square distribution (see Table 3 of Appendix B), it becomes relatively easy to use the chi-square distribution to make interval estimates and hypothesis tests about a population variance.

From the three chi-square distributions shown in Figure 11.1, we see that the chi-square distribution is not symmetric and that the shape of a particular chi-square distribution depends upon the number of degrees of freedom. Also note that chi-square values can never be negative.

Interval Estimation of σ^2

Let us show how the chi-square distribution can be used to develop a confidence interval estimate of a population variance σ^2. As an illustration of this process, let us assume that we are interested in estimating the population variance for the production filling process mentioned at the beginning of this chapter. A sample of 20 containers is taken, and the sample variance for the filling quantities is found to be $s^2 = .0025$. However, we know we cannot expect the variance of a sample of 20 containers to provide the exact value of the variance for the population of containers filled by the production process. Thus our interest will be in developing an interval estimate for the population variance.

We will use the notation χ_α^2 to denote the value for the chi-square distribution that provides an area or probability of α to the *right* of the stated χ_α^2 value. For example, as noted in Figure 11.2, the chi-square distribution with 19 degrees of freedom is shown with $\chi_{.025}^2 = 32.85$ indicating that 2.5% of the chi-square values are to the right of 32.85, and $\chi_{.975}^2 = 8.91$ indicating that 97.5% of the chi-square values are to the right of 8.91. Refer to Table 3 of Appendix B and verify that these chi-square values with 19 degrees of freedom (19th row of the table) are correct.

From the graph in Figure 11.2 we see that .95, or 95%, of the chi-square (χ^2) values are between $\chi_{.975}^2$ and $\chi_{.025}^2$. That is, there is a .95 probability of obtaining a χ^2 value such that

$$\chi_{.975}^2 \leq \chi^2 \leq \chi_{.025}^2$$

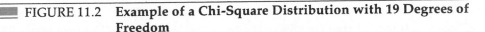

■■■ FIGURE 11.2 **Example of a Chi-Square Distribution with 19 Degrees of Freedom**

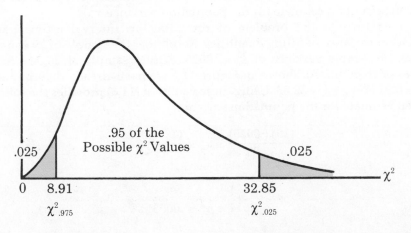

However, since we stated in (11.2) that $(n - 1)s^2/\sigma^2$ follows a chi-square distribution, we can substitute $(n - 1)s^2/\sigma^2$ for the above χ^2 and write

$$\chi^2_{.975} \leq \frac{(n - 1)s^2}{\sigma^2} \leq \chi^2_{.025} \tag{11.3}$$

In effect the above expression provides an interval estimate for the quantity $(n - 1)s^2/\sigma^2$ in that .95, or 95%, of all possible values for $(n - 1)s^2/\sigma^2$ will be in the interval $\chi^2_{.975}$ to $\chi^2_{.025}$. We now want to do some algebraic manipulations with (11.3) in order to establish an interval estimate for the population variance σ^2. Working with the leftmost inequality in (11.3), we have

$$\chi^2_{.975} \leq \frac{(n - 1)s^2}{\sigma^2}$$

Thus

$$\sigma^2 \chi^2_{.975} \leq (n - 1)s^2$$

or

$$\sigma^2 \leq \frac{(n - 1)s^2}{\chi^2_{.975}} \tag{11.4}$$

Performing similar algebraic manipulations with the rightmost inequality in (11.3) provides

$$\frac{(n - 1)s^2}{\chi^2_{.025}} \leq \sigma^2 \tag{11.5}$$

Finally, the results of (11.4) and (11.5) can be combined to provide

$$\frac{(n - 1)s^2}{\chi^2_{.025}} \leq \sigma^2 \leq \frac{(n - 1)s^2}{\chi^2_{.975}} \tag{11.6}$$

Since (11.3) will be true for 95% of the $(n - 1)s^2/\sigma^2$ values, (11.6) provides a 95% confidence interval estimate for the population variance σ^2.

Let us return to the problem of providing an interval estimate of the population variance of filling quantities. Recall that the sample of 20 containers provided a sample variance of $s^2 = .0025$. With a sample of 20, we have 19 degrees of freedom. As shown in Figure 11.2 we have already determined that $\chi^2_{.975} = 8.91$ and $\chi^2_{.025} = 32.85$. Using these values in (11.6) provides the following interval estimate for the population variance:

$$\frac{(19)(.0025)}{32.85} \leq \sigma^2 \leq \frac{(19)(.0025)}{8.91}$$

or

$$.0014 \leq \sigma^2 \leq .0053$$

Taking the square root of the above terms provides the following 95% confidence interval estimate for the population standard deviation:

$$.0374 \leq \sigma \leq .0728$$

Thus we have illustrated the process of using the chi-square distribution to establish interval estimates of a population variance and a population standard deviation. Note specifically that since $\chi^2_{.975}$ and $\chi^2_{.025}$ were used, the interval estimate has a .95 confidence coefficient. Extending (11.6) to the general case of any confidence coefficient, we have the following interval estimate of a population variance:

Interval Estimate of a Population Variance

$$\frac{(n-1)s^2}{\chi^2_{\alpha/2}} \leq \sigma^2 \leq \frac{(n-1)s^2}{\chi^2_{(1-\alpha/2)}} \tag{11.7}$$

where the χ^2 values are based on a chi-square distribution with $n-1$ degrees of freedom and where $1-\alpha$ is the confidence coefficient.

Hypothesis Testing

Let us now consider an example and the statistical methodology necessary to test hypotheses concerning the value of a population variance. The St. Louis Metro Bus Company has recently made a concerted effort to promote an image of reliability by encouraging its drivers to maintain consistent schedules. As a standard policy the company expects arrival times at various bus stops to have low variability. In terms of the variance of arrival times, the company standard specifies an arrival time variance of 4 or less with arrival times measured in minutes. Periodically, the company collects arrival time data at various bus stops in order to determine if the variability guideline is being maintained. The sample results are used to test the following hypotheses:

$$H_0: \quad \sigma^2 \leq 4$$

$$H_a: \quad \sigma^2 > 4$$

In tentatively assuming H_0 true, we are assuming that the variance for arrival times is within the company guidelines. We will reject H_0 only if the sample evidence indicates that the guidelines are not being maintained. In this sense, rejection of H_0 suggests that followup steps are necessary in order to reduce the arrival time variability.

Assume for the moment that a random sample of 10 bus arrivals will be taken at a particular downtown intersection. If the population of arrival times has a normal distribution, we know from (11.1) that the quantity $(n-1)s^2/\sigma^2$ has a chi-square distribution with $n-1$ degrees of freedom. With the null hypothesis assumed true at $\sigma^2 = 4$ and with a sample size of $n = 10$, we can conclude that

$$\frac{(n-1)s^2}{\sigma^2} = \frac{9s^2}{4}$$

has a chi-square distribution with $n - 1 = 9$ degrees of freedom. Thus once the sample data are obtained and the sample variance s^2 computed the following equation will provide an observed chi-square (χ^2) value

$$\chi^2 = \frac{9s^2}{4} \qquad (11.8)$$

The chi-square distribution showing this rejection region for this one-tailed test is shown in Figure 11.3. Note that we will reject H_0 only if the sample variance s^2 leads to a large χ^2 value. With $\alpha = .05$, Table 3 of Appendix B shows that with 9 degrees of freedom $\chi^2_{.05} = 16.92$. With this as the critical value for the test, the rejection rule is as follows:

Reject H_0 if $\chi^2 > 16.92$

Let us assume that the sample of arrival times for 10 buses shows a sample variance of $s^2 = 4.8$. Is this sample evidence sufficient to reject H_0 and conclude that the buses are not meeting the company's arrival time variance guideline? With $s^2 = 4.8$ and (11.8) we obtain the following χ^2 value:

$$\chi^2 = \frac{9(4.8)}{4} = 10.80$$

Since $\chi^2 = 10.80$ is less than 16.92, we cannot reject H_0. Thus the sample variance of $s^2 = 4.8$ is insufficient evidence to conclude that the bus-arrival-time variance is not meeting the company standard.

One-tailed tests concerning the value of a population variance as just demonstrated are in practice perhaps the most frequently encountered tests about population variances. That is, in situations involving arrival times, production times, filling weights, part dimensions, and so on, low variances generally are desired, whereas large variances tend to be unacceptable. With a statement about the maximum allowable variance, we frequently test the null hypothesis that the variance is less than or equal to this value against the alternative hypothesis that the variance is greater than this value. We now outline the decision rule for making this one-tailed test about a population variance:

One-Tailed Test About a Population Variance

$H_0: \quad \sigma^2 \leq \sigma_0^2$

$H_a: \quad \sigma^2 > \sigma_0^2$

Test Statistic

$$\chi^2 = \frac{(n - 1)s^2}{\sigma_0^2}$$

Rejection Rule

Reject H_0 if $\chi^2 > \chi_\alpha^2$

where σ_0^2 is the hypothesized value for the population variance, α is the level of significance for the test, and the value of χ_α^2 is based on a chi-square distribution with $n - 1$ degrees of freedom.

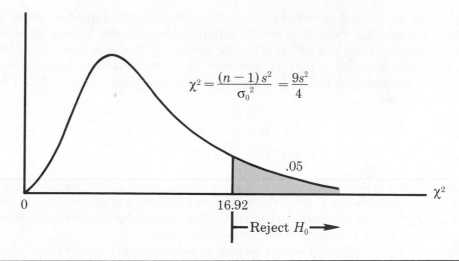

$$\chi^2 = \frac{(n-1)\,s^2}{\sigma_0^{\,2}} = \frac{9s^2}{4}$$

.05

0   16.92 χ^2

├─ Reject H_0 ⟶

───

■■■ FIGURE 11.3   **Rejection Region for the St. Louis Metro Bus Test with** $\alpha = .05$**.**

However, just as we saw with population means and population proportions, other forms of the hypotheses can be developed. The one-tailed test for H_0: $\sigma^2 \geq \sigma_0^2$ is similar to the test shown above with the exception that the one-tailed rejection region occurs in the lower tail at a critical value of $\chi^2_{(1-\alpha)}$. The two-tailed test with H_0: $\sigma^2 = \sigma_0^2$, as with other two-tailed tests, places an area of $\alpha/2$ in each tail in order to establish the two critical values for the test. The decision rule for conducting a two-tailed test about the value of a population variance is summarized as follows:

Two-Tailed Test About a Population Variance

$$H_0: \quad \sigma^2 = \sigma_0^2$$

$$H_a: \quad \sigma^2 \neq \sigma_0^2$$

Test Statistic

$$\chi^2 = \frac{(n-1)s^2}{\sigma_0^2}$$

Rejection Rule

Reject H_0 if $\chi^2 < \chi^2_{(1-\alpha/2)}$ or if $\chi^2 > \chi^2_{\alpha/2}$

where σ_0^2 is the hypothesized value for the population variance, α is the level of significance for the test, and the values of $\chi^2_{(1-\alpha/2)}$ and $\chi^2_{\alpha/2}$ are based on a chi-square distribution with $n - 1$ degrees of freedom.

Let us demonstrate the use of the chi-square distribution to conduct a two-tailed test about a population variance by considering a situation faced by a bureau of motor vehicles. Historically, the variance in test scores for individuals applying for drivers' licenses has been $\sigma^2 = 100$. A new examination with new test questions has been developed. Administrators of the bureau of motor vehicles believe that it is desirable for the variance in the test scores for the new examination to remain at the historical level. In order to evaluate the variance in the new examination test scores, the following two-tailed hypothesis test has been proposed:

$$H_0: \quad \sigma^2 = 100$$

$$H_a: \quad \sigma^2 \neq 100$$

Rejection of H_0 will indicate that some questions in the new examination may need revision in order to make the variance of the new test scores similar to the variance of the old test scores. A sample of 30 applicants for driver's licenses will be given the new version of the driver's examination.

The chi-square distribution can be used to conduct this two-tailed test. With a .05 level of significance, the critical regions will be based on $\chi^2_{.975}$ and $\chi^2_{.025}$. With $n - 1 = 29$ degrees of freedom, Table 3 of Appendix B shows that $\chi^2_{.975} = 16.05$ and $\chi^2_{.025} = 45.72$. The rejection rule for the two-tailed test becomes

$$\text{Reject } H_0 \text{ if } \chi^2 < 16.05 \text{ or if } \chi^2 > 45.72$$

What is the appropriate conclusion if the sample of 30 driver-examination scores provides a sample variance of $s^2 = 64$? With $H_0: \sigma^2 = 100$, the value of the χ^2 statistic is computed to be

$$\chi^2 = \frac{(n-1)s^2}{\sigma_0^2} = \frac{29(64)}{100} = 18.56$$

With this value of χ^2, we are not able to reject H_0. There is no statistical evidence that the variance in the new examination scores differ from the historical variance in examination scores.

■■■ EXERCISES

1. In the St. Louis Metro Bus Company example, the sample of 10 bus arrivals showed a sample variance of $s^2 = 4.8$.

   a. Provide a 95% confidence interval estimate of the variance for the population of arrival times.

   b. Assume that the sample variance of $s^2 = 4.8$ had been obtained from a sample of 25 bus arrivals. Provide a 95% confidence interval estimate of the variance for the population of arrival times.

   c. What effect does a larger sample size have on the interval estimate of a population variance? Does this seem reasonable?

2. The variance in drug weights is very critical in the pharmaceutical industry. For a specific drug, with weights measured in grams, a sample of 18 units provided a sample variance of $s^2 = .36$.

   a. Construct a 90% confidence interval estimate of the population variance for the weights of this drug.

   b. Construct a 90% confidence interval estimate of the population standard deviation.

3. A sample of cans of soups produced by Carle Foods shows the following weights, measured in ounces:

$$12.2 \quad 11.9 \quad 12.0 \quad 12.2$$
$$11.7 \quad 11.6 \quad 11.9 \quad 12.0$$
$$12.1 \quad 12.3 \quad 11.8 \quad 11.9$$

Provide 95% confidence interval estimates of the variance and the standard deviation of the population.

4. A study of worker attitudes about their jobs was conducted for the airline industry (*Industrial Relations*, Winter 1988). A sample of airline employees provided a sample mean age of 40 years and a sample standard deviation of 9.5 years. For purposes of illustration, compute the 95% confidence interval estimate of the population standard deviation under the assumption the following sample sizes were used.

   a. $n = 30$
   b. $n = 51$
   c. $n = 101$
   d. What happens to the interval estimate of the population standard deviation as the sample size increases?

5. A certain part must be machined to very close tolerances or it is not acceptable to customers. Production specifications call for a maximum variance in the lengths of the parts of .0004. The sample variance for 30 parts turns out to be $s^2 = .0005$. Using $\alpha = .05$, test to see if the production specifications are being violated.

6. City Trucking, Inc., claims consistent delivery times for its routine customer deliveries. A sample of 22 truck deliveries shows a sample variance of 1.5. Test to determine if H_0: $\sigma^2 \leq 1$ can be rejected. Use $\alpha = .10$.

7. The variance in the filling amounts of cups of soft drink from an automatic drink machine is an important consideration to the owner of the soft-drink service. If the variance is too large, overfilling and underfilling of cups will cause customer dissatisfaction with the service. An acceptable variance in filling amounts is $\sigma^2 = .25$ when filling amounts are measured in ounces. In a test of filling amounts for a particular machine, a sample of 18 cups showed a sample variance of .40.

   a. Do the sample results indicate that the filling mechanism on the machine should be replaced due to a large variance in filling amounts? Use a .05 level of significance.
   b. Provide a 90% confidence interval of the variance in the filling amounts for this machine.

8. From a sample of 9 days over the past 6 months, a dentist has seen the following number of patients: 22, 25, 20, 18, 15, 22, 24, 19, and 26. Assuming that the number of patients seen per day is normally distributed, would an analysis of this sample data reject the hypothesis that the variance in the number of patients seen per day is equal to 10? Use a .10 level of significance. What is your conclusion?

11.2 ▬ INFERENCES ABOUT THE VARIANCES OF TWO POPULATIONS

In some statistical applications it is desirable to compare the variances of two populations. For instance, we might want to compare the variances in product quality resulting from two different production processes, the variances in assembly times for two assembly methods, or the variances in temperatures for two heating devices. In addition, recall that in Chapter 10 we developed a pooled variance estimate based on the assumption that two populations had equal variances. Thus we might want to compare the variances of two populations to determine if the equal variance assumption can be justified.

In making comparisons about the variances of two populations we will be using data collected from two independent random samples, one from population 1 and another from population 2. In using the two sample variances, s_1^2 and s_2^2, as a basis for making inferences about the two population variances, σ_1^2 and σ_2^2, the sampling distribution of the ratio of the two sample variances, s_1^2/s_2^2, is very helpful. The sampling distribution of this ratio is described as follows:

Sampling Distribution of s_1^2/s_2^2

Whenever independent simple random samples of sizes n_1 and n_2 are selected from normal populations with equal variances, the ratio

$$\frac{s_1^2}{s_2^2} \qquad (11.9)$$

has an F distribution with $n_1 - 1$ degrees of freedom for the numerator and $n_2 - 1$ degrees of freedom for the denominator; s_1^2 is the sample variance for the random sample of n_1 items from population 1 and s_2^2 is the sample variance for the random sample of n_2 items from population 2.

A graph of the F distribution with 20 degrees of freedom for both the numerator and denominator is shown in Figure 11.4. As can be seen from this graph, the F distribution is not symmetric and the F values can never be negative. The actual shape of any particular F distribution depends upon its corresponding numerator and denominator degrees of freedom.

■ FIGURE 11.4   *F* **Distribution with 20 Degrees of Freedom for the Numerator and 20 Degrees of Freedom for the Denominator**

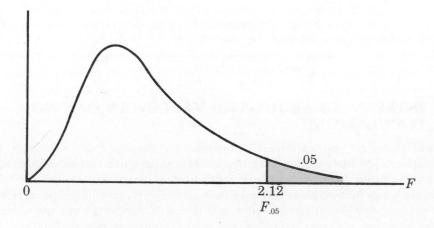

We will use the notation F_α to denote the value for the F distribution that provides an area or probability of α to the *right* of the stated F_α value. For example, as noted in Figure 11.4, $F_{.05}$ denotes the upper 5% of the F values for an F distribution with 20 degrees of freedom for the numerator and 20 degrees of freedom for the denominator. Table 4 of Appendix B shows that for this particular F distribution, $F_{.05} = 2.12$. Let us show how the F distribution can be used for a hypothesis test concerning the variances of two populations.

Dullus County Schools is renewing its school bus service contract for the coming year and must select one of the two bus companies, the Milbank Company or the Gulf Park Company. We will be interested in using the variance of the arrival or pickup/delivery times as a primary measure of the quality of the bus service. Low variance values indicate the more consistent and higher-quality service. If the variances of arrival times associated with the two services are equal, Dullus School administrators will select the company offering the better financial terms. However, if the sample data on bus arrival times for the two companies indicate that a significant difference exists between the variances, the administrators may want to give special consideration to the company with the better or lower-variance service. The appropriate hypotheses are as follows:

$$H_0: \quad \sigma_1^2 = \sigma_2^2$$

$$H_a: \quad \sigma_1^2 \neq \sigma_2^2$$

If H_0 can be rejected, the conclusion of unequal service qualities is appropriate. In this case, the company with the lower sample variance would be preferred.

Assume that the above hypothesis test will be conducted with $\alpha = .10$. Furthermore, assume that we obtain samples of arrival times from school systems currently using the two school bus services. A sample of 25 arrival times is available for the Milbank service (population 1) and a sample of 16 arrival times is available for the Gulf Park service (population 2). The graph of the F distribution with $n_1 - 1 = 24$ degrees of freedom for the numerator and $n_2 - 1 = 15$ degrees of freedom for the denominator is shown in Figure 11.5. Note that the two-tailed rejection region is indicated by the critical values at $F_{.95}$ and $F_{.05}$.

Let us assume that the 2 samples of bus arrival times resulted in sample variances of $s_1^2 = 48$ for the Milbank service and $s_2^2 = 20$ for the Gulf Park service. What conclusion is now appropriate concerning the quality of the two bus services? Assume that the 2 populations of arrival times have normal distributions, and assume that H_0 is true with $\sigma_1^2 = \sigma_2^2$. The F distribution now can be used to reach a decision. Specifically, we compute $F = s_1^2/s_2^2$ and use the rejection region shown in Figure 11.5. Thus we find

$$F = \frac{s_1^2}{s_2^2} = \frac{48}{20} = 2.40$$

Using Table 4 of Appendix B to determine the critical value for the test, we find the upper-tail critical value with 24 numerator degrees of freedom and 15 denominator degrees of freedom is $F_{.05} = 2.29$. While the appendix does not provide $F_{.95}$ values, note that the determination of this lower-tail critical value is not necessary. We can already observe that $F = 2.40$ exceeds $F_{.05} = 2.29$. Thus at

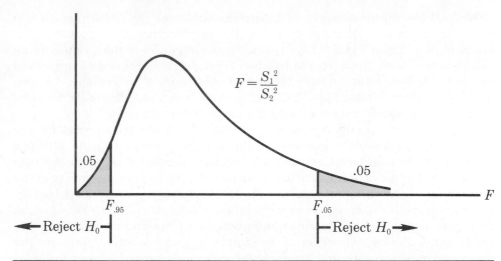

Rejection Region for the Dullus County School Bus Problem with $\alpha = .10$

the .10 level of significance H_0 is rejected. This leads us to the conclusion that the two bus services differ in terms of pickup/delivery time variances. Specifically, it is recommended that the Dullus School administrators give special consideration to the better, or lower-variance, service offered by the Gulf Park Company.

In these statistical computations, you might feel that we were lucky in carrying out the test because the lower-tail critical value, $F_{.95}$, which could not be found in the table, was not even necessary. Thus we were able to draw the appropriate conclusion without knowing the $F_{.95}$ value. While lower-tail F values can be determined, it is quite common to find only upper-tail F values in statistical references simply because F tables would require several pages if very many F_α values were provided. Thus in order to limit the size of the appendices many statistical texts present only the upper-tail values, and then for only a relatively few α values. However, in hypothesis-testing situations involving $H_0: \sigma_1^2 = \sigma_2^2$, the lack of availability of the lower-tail critical values is not as restrictive as you might think. We can get around this apparent problem simply by denoting the population with the *larger sample variance* as population 1. That is, which population we denote population 1 and which we denote 2 is arbitrary. By labeling the population with the larger sample variance as population 1, we guarantee that $F = s_1^2/s_2^2$ will be greater than or equal to 1. Thus if a rejection of H_0 is to occur, it can occur *only* in the *upper tail*. While the lower-tail critical value still exists and could be correctly labeled $F_{.95}$, we do not need to know its value simply because the convention of using the population generating the largest sample variance as population 1 always places the ratio s_1^2/s_2^2 in the upper-tail direction. In the Dullus Schools example, population 1, the Milbank bus service, possessed the largest sample variance. Thus we proceeded directly with the test. If the Gulf Park bus service had provided the largest sample variance, we simply would have denoted Gulf Park as population 1 and followed the same statistical testing procedure. A summary of the two-tailed test for the equality of two population variances with this procedure is as follows:

Two-Tailed Test About Variances of Two Populations

$$H_0: \quad \sigma_1^2 = \sigma_2^2$$

$$H_a: \quad \sigma_1^2 \neq \sigma_2^2$$

Denote the population providing the *largest sample variance* as population 1.

Test Statistic	*Rejection Rule*
$F = \dfrac{s_1^2}{s_2^2}$	Reject H_0 if $F > F_{\alpha/2}$

where α is the level of significance for the test and the value of $F_{\alpha/2}$ is based on an F distribution with $n_1 - 1$ degrees of freedom for the numerator and $n_2 - 1$ degrees of freedom for the denominator.

One-tailed tests involving two population variances are also possible. Again the F distribution is used, with the one-tailed rejection region enabling us to conclude whether or not one population variance is significantly greater or significantly less than the other. Only upper-tail F values are needed. For any one-tailed test we set up the null hypothesis so that the rejection region is in the upper tail. This can be accomplished by labeling the population with the larger variance in H_a as population 1. For instance, suppose that we want to test the null hypothesis that the variance of population B is less than or equal to the variance of population A versus the alternative hypothesis that the variance of population B is greater than the variance of population A. In this case, we would label population B population 1. The general procedure is as follows:

One-Tailed Test About Variances of Two Populations

$$H_0: \quad \sigma_1^2 \leq \sigma_2^2$$

$$H_a: \quad \sigma_1^2 > \sigma_2^2$$

Test Statistic	*Rejection Rule*
$F = \dfrac{s_1^2}{s_2^2}$	Reject H_0 if $F > F_{\alpha}$

where α is the level of significance for the test and the value of F_{α} is based on an F distribution with $n_1 - 1$ degrees of freedom for the numerator and $n_2 - 1$ degrees of freedom for the denominator.

Let us demonstrate the use of the F distribution to conduct a one-tailed test about the variances of two populations by considering the following public opinion survey. Samples of 31 men and 41 women will be used to study attitudes about current political issues. The researcher conducting the study would like to test to see if the sample data indicate that women demonstrate a greater variation in attitude on policital issues than men. Using the form of the one-tailed hypothesis test shown earlier, women will be denoted as population 1 and men will be denoted as population 2. The hypothesis test will be stated

$$H_0: \sigma^2_{\text{women}} \leq \sigma^2_{\text{men}}$$

$$H_a: \sigma^2_{\text{women}} > \sigma^2_{\text{men}}$$

If H_0 is rejected, the researcher will have the statistical support necessary to conclude that women demonstrate a greater variation in attitude on political issues.

With the sample variance for women in the numerator and the sample variance for men in the denominator, the F distribution with $41 - 1 = 40$ degrees of freedom in the numerator and $31 - 1 = 30$ degrees of freedom in the denominator will be used to conduct the one-tailed test. Using a .05 level of significance, the rejection region is based on $F_{.05}$. Using Table 4 of Appendix B, we find that $F_{.05} = 1.79$. Thus the rejection rule becomes

$$\text{Reject } H_0 \text{ if } F > 1.79$$

where F is computed from the ratio of the 2 sample variances, s^2_1/s^2_2.

Assume that the survey shows a sample variance of $s^2_1 = 120$ for the 41 women and a sample variance of $s^2_2 = 80$ for the 31 men. What is the appropriate statistical conclusion? The F statistic becomes

$$F = \frac{s^2_1}{s^2_2} = \frac{120}{80} = 1.50$$

Since 1.50 is less than 1.79, H_0 cannot be rejected. Thus the sample results do not support the conclusion that women have significantly greater variation in attitudes on political issues.

▆▆▆ EXERCISES

9. A manager is interested in estimating the difference between the mean checking account balances at two branches of the Clearview National Bank. The data collected from two independent random samples are as follows:

Branch Bank	Number of Checking Accounts	Sample Mean Balance	Sample Standard Deviation
Cherry Grove	12	$\bar{x}_1 = \$1000$	$s_1 = \$150$
Beechmont	10	$\bar{x}_2 = \$920$	$s_2 = \$120$

In using the t distribution to estimate the difference between means (see Chapter 10) we make the assumption that the variances of the two populations were equal. This assumption is the basis for developing a pooled variance estimate. Use $\alpha = .10$ and conduct a test for the equality of the population variances. Do your results justify the use of a pooled variance estimate of σ^2?

10. The Educational Testing Service has conducted studies designed to identify differences between the scores of males and females on the Scholastic Aptitude Test (*Journal of Educational Measurement,* Spring 1987). For a sample of females, the standard deviation was 83 on the verbal portion of the SAT. For a sample of males, the standard deviation was 78 on the same test. Assume that standard deviations were based on random samples of 121 females and 121 males. Do the data indicate there are differences between the variances of females and males on the verbal portion of the SAT? Test using $\alpha = .05$.

11. Independent random samples of parts manufactured by 2 suppliers show the following results:

Supplier	Sample Size	Sample Variance of Part Sizes
Durham Electric	41	$s_1^2 = 3.8$
Raleigh Electronics	31	$s_2^2 = 2.0$

The firm making the supplier-selection decision is prepared to use the Durham supplier unless the test results show that the Raleigh supplier provides significantly lower variance in part sizes. Use $\alpha = .05$ and conduct the statistical test that will help the firm select a supplier. Which supplier do you recommend?

12. The following sample data have been collected from two independent random samples:

Population	Sample Size	Sample Mean	Sample Variance
A	$n_A = 25$	$\bar{x}_A = 40$	$s_A^2 = 5$
B	$n_B = 21$	$\bar{x}_B = 50$	$s_B^2 = 11$

In a test for the difference between the two population means, the statistical analyst is considering using a pooled estimate of the population variance based on the assumption the variances of the two populations are equal. Is pooling appropriate in this case? Use $\alpha = .10$ for your test.

13. Two secretaries are each given eight typing assignments of equal difficulty. The sample standard deviations of the completion times were 3.8 minutes and 5.2 minutes, respectively. Do the data suggest that there is a difference in the variability of completion times for the two secretaries? Test the hypothesis at a .10 level of significance.

14. A research hypothesis is that the variance of stopping distances of automobiles on wet pavement is substantially greater than the variance of stopping distances of automobiles on dry pavement. In a research study, 16 automobiles traveling at the same speeds are tested with respect to stopping distances on wet pavement and then tested with respect to stopping distances on dry pavement. On wet pavement, the standard deviation of stopping distances was 32 feet. On dry pavement, the standard deviation was 16 feet.

a. At a .05 level of significance, do the sample data justify the conclusion that the variance in stopping distances on wet pavement is greater than the variance in stopping distances on dry pavement?

b. What are the implications of your statistical conclusions in terms of driving safety recommendations?

■ SUMMARY

In this chapter we have presented statistical procedures that can be used to make inferences about population variances. In the process we have introduced two new probability distributions: the chi-square distribution and the F distribution. The chi-square distribution can be used as the basis for interval estimation and hypothesis tests concerning the variance of a normal population. In particular, we showed that for simple random samples of size n selected from a normal population, the quantity $(n-1)s^2/\sigma^2$ has a chi-square distribution with $n-1$ degrees of freedom.

We illustrated the use of the F distribution in making hypothesis tests concerning the variances of two normal populations. In particular, we showed that with independent simple random samples of sizes n_1 and n_2 selected from two normal populations with equal variances, $\sigma_1^2 = \sigma_2^2$, the sampling distribution of the ratio of the two sample variances, s_1^2/s_2^2, has an F distribution with n_1-1 degrees of freedom for the numerator and n_2-1 degrees of freedom for the denominator.

▬ KEY FORMULAS

Interval Estimate of a Population Variance

$$\frac{(n-1)s^2}{\chi_{\alpha/2}^2} \le \sigma^2 \le \frac{(n-1)s^2}{\chi_{(1-\alpha/2)}^2} \tag{11.7}$$

Sampling Distribution of s_1^2/s_2^2 When $\sigma_1^2 = \sigma_2^2$

$$F = \frac{s_1^2}{s_2^2} \tag{11.9}$$

▬ SUPPLEMENTARY EXERCISES

15. Because of staffing decisions, management of the Gibson-Marimont Hotel is interested in the variability for the number of rooms occupied per day during a particular season of the year. A sample of 20 days of operation shows a sample mean of 290 rooms occupied per day and a sample standard deviation of 30 rooms.

  a. What is the point estimate of the population variance?
  b. Provide a 90% confidence interval estimate of the population variance.
  c. Provide a 90% confidence interval estimate of the population standard deviation.

16. Initial public offerings (IPOs) of stocks that go public are on average underpriced. However, in some cases the IPOs are actually overpriced (*Financial Analysts Journal*, December 1987). The standard deviation measures the dispersion or variation in the underpricing-overpricing indicators. A sample of 13 Canadian IPOs that were subsequently traded on the Toronto Stock Exchange had a standard deviation of 14.95. Develop a 95% confidence interval estimate of the population standard deviation for the underpricing-overpricing indicator.

17. Historical delivery times for Buffalo Trucking, Inc., have had a mean of 3 hours and a standard deviation of .5 hours. A sample of 22 deliveries over the past month provides a sample mean of 3.1 hours and a sample standard deviation of .75 hours.

  a. Use a test of hypothesis to determine if the sample results lead to rejection the historical delivery variance of H_0: $\sigma^2 = (.5)^2 = .25$. Use $\alpha = .05$.
  b. Provide 95% confidence interval estimates of the population variance and the population standard deviation.

18. Part variability is very critical in the manufacturing of ball bearings. Large variances in the size of the ball bearings cause bearing failure and rapid wearout. Production standards call for a maximum variance of .0001 when the bearing sizes are measured in inches. A sample of 15 bearings shows a sample standard deviation of .014 inches.

a. Using $\alpha = .10$, determine if the sample indicates the standard maximum variance is being exceeded.

b. Provide a 90% confidence interval estimate of the variance of the ball bearings in the population.

19. The filling variance for boxes of cereal is designed to be .02 or less. A sample of 41 boxes of cereal shows a sample standard deviation of .16 ounces. Using $\alpha = .05$, determine if the variance in the cereal box fillings is exceeding the standard.

20. A sample standard deviation for the number of passengers taking a particular airline flight is 8. A 95% confidence interval estimate of the standard deviation is 5.86 to 12.62.

a. Was a sample size of 10 or 15 used in the above statistical analysis?

b. If the sample standard deviation of $s = 8$ had been based on a sample of 25 flights, what change would you expect in the confidence interval for the population standard deviation? Compute a 95% confidence interval for σ if a sample of size 25 had been used.

21. A firm gives a mechanical aptitude test to all job applicants. A sample of 20 male applicants shows a sample variance of 80 for the test scores. A sample of 16 female applicants shows a sample variance of 220. Using $\alpha = .05$, determine if the test score variances differ for male and female job applicants. If a difference in variances exists, which group has the higher variance in mechanical aptitude?

22. The grade point averages of 352 students who completed a college course in financial accounting had a standard deviation of .940 (*The Accounting Review*, January 1988). The grade point averages of 73 students who dropped out of the same course has a standard deviation of .797. Do the data indicate that there is a difference between the variances of grade point averages for students who complete financial accounting courses and students who drop out? Use a .05 level of significance. *Note:* $F_{.025}$ with 351 and 72 degrees of freedom is approximately 1.45.

23. The accounting department analyzes the variance of the weekly unit costs reported by two production departments. A sample of 16 cost reports for each of the two departments shows cost variances of 2.3 and 5.4, respectively. Is this sample sufficient to conclude that the two production departments differ in terms of unit cost variances? Use $\alpha = .10$.

24. In using the t distribution to estimate the difference between two population means an analyst is interested in computing a pooled estimate of the variance of the populations. Pooling is justified only if it appears reasonable to assume that the two populations have equal variances. Use the following data to determine if pooling is appropriate for this situation (test with $\alpha = .10$):

| Sample 1 | 80 | 72 | 75 | 90 | 78 | 75 | 72 | 85 |
| Sample 2 | 50 | 48 | 45 | 60 | 65 | 66 | 70 | 54 |

If pooling is appropriate, what is the pooled estimate of σ^2?

25. Two new assembly methods are tested with the following variances in assembly times.

Method	Sample Size	Sample Variance
A	31	$s_1^2 = 25$
B	25	$s_2^2 = 12$

Using $\alpha = .10$, test for equality of the two population variances.

■■■ COMPUTER EXERCISE

An Air Force introductory course in electronics is currently taught using computer-assisted instruction, with each student in the course working individually at a computer terminal. It has been proposed that a better approach to teaching the course would be to have a pair of students work together at each computer terminal. In addition to the fact that a greater number of students could be taught at the same time, the proposed method may have the positive effect of reducing overall training time due to the fact that the students can help each other. In order to test the proposed method, an entering class of 120 students was randomly assigned to two groups of 60 students each. One group of 60 was taught using the current method; the second group was taught using the new method. The time in hours was recorded for each student in the sample. The data are shown.

Proposed Method	Current Method	Proposed Method	Current Method
75	73	75	79
76	77	71	79
75	72	75	74
73	76	72	79
74	65	77	72
73	72	78	75
75	71	74	75
73	73	76	71
74	67	75	74
76	82	77	80
77	68	76	82
77	72	71	76
76	69	77	76
72	76	74	77
77	79	82	81
75	76	75	74
76	80	73	82
77	73	72	75
72	79	76	72
75	77	73	76
74	75	77	63
75	77	75	79
70	70	73	75
74	73	78	74
78	78	75	76
80	73	74	77
72	76	72	74
76	72	77	70
76	61	76	72
73	76	78	78

■■■ QUESTIONS

1. Develop numerical and graphical measures to summarize the data.

2. Does there appear to be any difference between the mean training times for the two methods? Explain.

3. Does variance in the training times appear to be a significant factor in terms of the difference between the two methods?

4. What conclusion can you reach about any differences in the two methods? What is your recommendation? Explain.

APPLICATION

U.S. General Accounting Office*
Washington, D.C.

The United States General Accounting Office (GAO) is an independent, nonpolitical audit organization in the Legislative branch of the federal government. GAO was created by the Budget and Accounting Act of 1921 and has three basic purposes:

- To assist Congress, its committees, and its members carry out their legislative and oversight responsibilities, consistent with its role as an independent, nonpolitical agency.
- To audit and evaluate the programs, activities, and financial operations of federal departments and agencies, and to make recommendations toward more efficient and effective operations.
- To carry out financial control and other functions with respect to federal government programs and operations including accounting, legal, and claims settlement work.

GAO evaluators determine the effectiveness of existing or proposed federal programs and the efficiency, economy, legality, and effectiveness with which federal agencies carry out their responsibilities. These evaluations culminate in reports to the Congress and to the heads of federal departments and agencies. Such reports typically include recommendations to Congress concerning the need for legislation and suggestions to agencies concerning the need for changes to improve economy, efficiency, and effectiveness.

GAO evaluators analyze policies and practices, and the use of resources within and among federal programs; identify problem areas and deficiencies in meeting program goals; develop and analyze alternative solutions to problems of program execution; and develop and recommend changes to enable the programs to better conform to Congressional goals and legislative intent. To carry out their duties effectively evaluators must be proficient in interviewing, data processing, records review, operations research, legislative research, and statistical analysis techniques.

STATISTICAL ANALYSIS AT GAO

GAO evaluators perform statistical analysis and conduct audits of the validity of statistical analyses conducted by other governmental agencies and private organizations. Sampling procedures are frequently employed to collect data for studies. GAO evaluators use probability samples wherever possible. However, at times they must work with whatever data is available, including data from judgment samples. Care must be exercised to ensure that the sample selected is representative of the population to which inferences

*The authors are indebted to Mr. Art Foreman and Mr. Dale Ledman of the U.S. General Accounting Office for providing this application.

are being made. Otherwise legislators and others might be led to draw invalid conclusions.

Regression and correlation analysis is another area in which GAO becomes involved. For instance a regression model might be developed to predict demand for hospital intensive care units. Such a model can have a great impact on government policies for medicare and medicaid reimbursement. GAO conducts careful evaluations of such models to ensure the validity of the results.

Other areas of statistical analysis have involved nonparametric studies of the consistency of parole determinations, probabilistic analysis of the disposition of disability applications, and time series analyses. In the following we describe an application where an hypothesis test concerning a population variance was used.

▬ AN AUDIT OF A SEWAGE TREATMENT FACILITY

A few years ago a program was established by the Department of Interior to clean up the nation's rivers and lakes. As a part of the program, federal grants were made to small cities scattered throughout the United States. Congress asked GAO to determine how effectively the program was operating. To make this determination GAO examined records and conducted site visits at several plants.

One objective of these audits was to ensure that the effluent (treated sewage) at the plants met certain standards. Among other things, the following characteristics of the effluents were examined:

1. Oxygen content

2. pH level

3. Amount of suspended solids

4. Amount of soluble solids

This application overviews an interesting finding that resulted from the audit of one particular plant.

A requirement of the program grants was that a variety of tests be taken and recorded daily with the records periodically sent to the state engineering department. GAO's

General Accounting Office Headquarters in Washington D.C.

investigation of the plant in question began with an examination of a sample of the records submitted to the state. Initially GAO auditors were concerned with determining whether or not various characteristics of the effluent were within acceptable limits. In this case the measurements were within acceptable limits, but the auditors noted that there was very little variability in the data. This led to further investigation along another line.

The pH level of water is 7. A certain variance is normal and expected for samples taken from different sources at different times. The apparent low variance in the sample data caused the auditors to conduct the following hypothesis test concerning the variance in pH level for the population:

$$H_0: \quad \sigma^2 \geq \sigma_0^2$$
$$H_a: \quad \sigma^2 < \sigma_0^2$$

where σ_0^2 is the population variance in pH level found at other properly functioning plants. The hypothesis test led to rejection of H_0 and the conclusion that the variance in pH level at the plant in question was significantly less than normal.

The auditors then made a plant visit to examine the measuring equipment and discuss their findings with the plant operator. They found that the measuring equipment was not even being used because the plant operator did not know how to operate it. Instead, the operator was told by an engineer what an acceptable pH level was and had simply recorded numbers within the acceptable range each day without actually conducting the required tests.

■■ BENEFITS OF STUDY

This particular case caused GAO evaluators to investigate further the ability of sewage treatment plant operators to use the effluent measuring equipment as well as other equipment and chemicals used to treat the sewage. It was found that even though the plants and equipment purchased by the grant were modern and of high quality, many of the operators did not have adequate training to operate the equipment. This particular study and followup investigations led to a GAO recommendation that states receiving funds from the program be required to establish training programs for plant operators.

CHAPTER 12

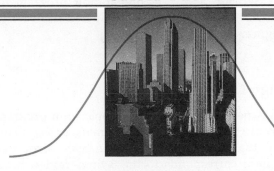

Experimental Design and Analysis of Variance

I n Chapter 10 we showed how to test whether or not the means of two populations are equal. Recall that the test involved the selection of an independent random sample from each population. In this chapter we present a statistical procedure for determining whether or not the means of more than two populations are equal. This technique is called the *analysis of variance (ANOVA) procedure*. We also discuss the process of designing an experiment that results in the collection of data; this process is referred to as *experimental design*.

In order to introduce the concepts of experimental design and analysis of variance, let us consider the situation facing Dr. Edward A. Johnson, Dean of the College of Business at a major Texas university.

12.1 ▬ THE GMAT EXPERIMENT

The Graduate Management Admissions Test (GMAT) is a standardized test used by graduate schools of business to evaluate an applicant's ability to pursue successfully a graduate program in that field. Scores on the GMAT range from 200 to 800, with higher scores implying a higher aptitude.

In an attempt to improve the performance of undergraduate students on the exam, Dr. Johnson is considering the use of a GMAT preparation program. The following three preparation programs have been proposed:

1. A 3-hour review session covering the types of questions generally asked on the GMAT.

2. A 1-day program covering relevant exam material, along with the taking and grading of a sample exam.

3. An intensive 10-week course involving the identification of each student's weaknesses and the setting up of individualized programs for improvement.

Before making a final decision as to the preparation program to adopt, Dr. Johnson has requested that further study be conducted in order to determine how the proposed programs affect GMAT scores.

Experimental Design Considerations

In the GMAT study, or experiment, the GMAT preparation program is referred to as a *factor*. Since there are three preparation programs, or *levels*, corresponding to this factor, we say that there are three *treatments* associated with the experiment: one treatment corresponds to the 3-hour review, another to the 1-day program, and the third to the 10-week course. In general, a treatment corresponds to a level of a factor.*

The three GMAT preparation programs or treatments define the three populations of interest in this experiment. One population corresponds to all students who take the 3-hour review session, another corresponds to those who take the 1-day program, and the third to those who take the 10-week course. Note that for each population the random variable of interest is the GMAT score, and the primary statistical objective for the experiment is to determine whether or not the mean GMAT score is the same for all three populations. In experimental design terminology the random variable of interest is referred to as the *dependent variable*, the *response variable*, or simply the *response*.

Now that we have defined the response, factor, and treatments for the GMAT experiment, let us turn to the method of assigning the treatments. To begin with, let us assume that a random sample of three students has been selected from the population of seniors considering attending graduate school. In experimental design terminology, the students are referred to as the *experimental units*.

The experimental design that we will use for the GMAT experiment is referred to as a *completely randomized design*. This type of design requires that each of the three preparation programs or treatments be randomly assigned to one of the experimental units or students. For example, the 3-hour review session might be randomly assigned to the second student, the 1-day program to the first student, and the 10-week course to the third student. The concept of randomization, as illustrated in this example, is an important principle of all experimental designs.

Note that the experiment as described above would only result in one measurement or GMAT score for each treatment. In other words we have a sample size of 1 corresponding to each treatment. Thus to obtain additional data for each preparation program we must repeat, or *replicate*, the basic experimental process. For example, suppose that instead of selecting just 3 students at random we had selected 15 students and then randomly assigned *each* of the three treatments to 5 of the students. Since each preparation program is assigned 5 students, we say that 5 replicates have been obtained. The concept of *replication* is another important principle of experimental design. Figure 12.1 shows the completely randomized design for the GMAT study.

*The term *treatment* was originally used in experimental design because many of the applications were in agriculture, where the treatments often corresponded to different types of fertilizers applied to selected agricultural plots. Today the term is used in a more general context.

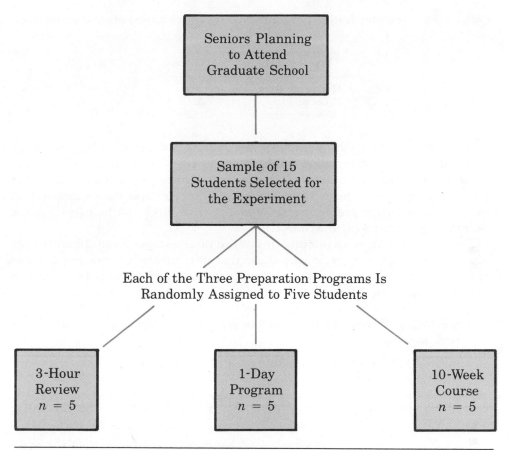

Completely Randomized Design for Evaluating the GMAT Preparation Programs

Data Collection

Once we are satisfied with the experimental design, we proceed by carrying out the study and collecting the data. In this case the students would prepare for the GMAT exam under their assigned preparation program. The students would then take the exam and have their GMAT scores recorded. Suppose this has been done and that the GMAT scores for the 15 students in the study are as shown in Table 12.1. Using these data we calculated the sample mean GMAT score for

TABLE 12.1   **GMAT Scores for 15 Students**

		Observation				
Treatment	3-Hour Review	491	579	451	521	503
	1-Day Program	588	502	550	520	470
	10-Week Course	533	628	501	537	561

each of the three preparation programs; the results obtained are shown below:

Type of Program	Mean GMAT Score
3-hour review	509
1-day program	526
10-week course	552

From these data it appears that the 10-week course may result in higher GMAT scores than either of the other methods. However, before we make any final recommendations we must remember that each of these means is based upon the test results of just 5 students. Thus we are looking at 3 sample means drawn from the 3 populations representing all students who might participate in these programs prior to taking the GMAT.

The real issue, then, is whether or not the three sample means observed are different enough for us to conclude that the means of the populations corresponding to the three preparation programs are different. To write this question in statistical terms we introduce the following notation:

μ_1 = mean GMAT score for the population of all students who take the 3-hour review session

μ_2 = mean GMAT score for the population of all students who take the 1-day program

μ_3 = mean GMAT score for the population of all students who take the 10-week course

Letting $\bar{x}_1$, $\bar{x}_2$, and $\bar{x}_3$ denote the corresponding sample means, the experimental results for the three programs yielded $\bar{x}_1 = 509$, $\bar{x}_2 = 526$, and $\bar{x}_3 = 552$. Although we will never know the actual values of μ_1, μ_2, and μ_3, what we want to do is use the sample means to test the following hypotheses:

Hypotheses for Analysis of Variance

H_0: $\mu_1 = \mu_2 = \mu_3$

H_a: Not all the programs have equal means

In the next section we show how the ANOVA procedure can be used to determine if there is a difference among the three population means.

12.2 ▬ THE ANALYSIS OF VARIANCE PROCEDURE FOR COMPLETELY RANDOMIZED DESIGNS

The analysis of variance (ANOVA) procedure is designed to test the following hypotheses:

$$H_0: \mu_1 = \mu_2 = \cdots = \mu_k$$

H_a: Not all the means are equal

where

μ_i = mean of the ith population

k = number of populations or treatments

We assume that a simple random sample of size n has been selected from each of the k populations and that the sample means $\bar{x}_1, \bar{x}_2, \ldots, \bar{x}_k$ have been computed. In general, we will refer to the sample mean corresponding to the ith population as $\bar{x}_i$ and the overall sample mean for all the treatments as $\bar{\bar{x}}$. That is,

$$\bar{x}_i = \frac{\sum_j x_{ij}}{n} \tag{12.1}$$

$$\bar{\bar{x}} = \frac{\sum_i \sum_j x_{ij}}{n_T} \tag{12.2}$$

where

x_{ij} = jth observation corresponding to the ith treatment

n_T = total sample size for the experiment

Note that since the sample size is n for each of the k treatments, $n_T = kn$. For the Dr. Johnson GMAT experiment, where $k = 3$ and $n = 5$, $n_T = (3)(5) = 15$.

Assumptions for Analysis of Variance

The ANOVA procedure is based upon the following two assumptions:

1. The random variable of interest for each population has a normal probability distribution. *Implication:* In the GMAT study this assumption would require that the random variable of interest, GMAT score, be normally distributed for each of the three programs under study.

2. The variance associated with the random variable denoted by σ^2, must be the same for each population. *Implication:* In the GMAT study this assumption would require that the variance of GMAT scores be the same for students participating in each of the three programs.

The logic behind the ANOVA procedure is based upon the development of two independent estimates of the common population variance (σ^2). One estimate of σ^2 is based upon the differences *between* the treatment means and the overall sample mean, and the other estimate is based upon the differences of observations *within* each treatment from the corresponding treatment mean. By comparing these two estimates of σ^2 we will be able to answer the question about whether or not the population means are equal. We note that the approach presented in this section applies only for situations in which the sample sizes are the same for each treatment. In Section 12.3 we present a general approach that applies in situations where the sample sizes are not necessarily the same.

Between-Treatments Estimate of Population Variance

The between treatments estimate of σ^2 is based on the assumption that the null hypothesis of equal population means is true. In this case, if we let μ denote the common population mean (that is, $\mu_1 = \mu_2 = \cdots = \mu_k = \mu$), then all of the n_T sample observations would represent data values drawn from the same normal probability distribution with mean μ and variance σ^2. If we let $\bar{x}$ denote the mean of a simple random sample of size n selected from this probability distribution, then the sampling distribution of $\bar{x}$ would be normal with mean μ and variance $\sigma_{\bar{x}}^2 = \sigma^2/n$. Figure 12.2 illustrates such a sampling distribution. Thus, under the null hypothesis, we can think of each $\bar{x}_i$ as a value drawn at random from this sampling distribution.

An estimate of the mean μ of this sampling distribution can be obtained by computing $\bar{\bar{x}}$, the overall sample mean. For the GMAT scores in Table 12.1 we use all the data to find

$$\sum_i \sum_j x_{ij} = (491 + 579 + 451 + \cdots + 537 + 561) = 7935$$

Then using (12.2) with $n_T = 15$ we compute an overall sample mean

$$\bar{\bar{x}} = \frac{7935}{15} = 529$$

To estimate the variance of the sampling distribution of $\bar{x}$ (that is, $\sigma_{\bar{x}}^2$), we can use the variance of the individual sample means about the overall sample mean $\bar{\bar{x}}$. For the case where each sample is the same size, the estimated variance, denoted $s_{\bar{x}}^2$, is

$$s_{\bar{x}}^2 = \frac{\sum_i (\bar{x}_i - \bar{\bar{x}})^2}{k - 1} \tag{12.3}$$

where $\bar{x}_i$ = mean of the ith sample.

▬▬ FIGURE 12.2 **Sampling Distribution of $\bar{x}$ Given the Null Hypothesis $\mu_1 = \mu_2 = \cdots = \mu_k = \mu$ and a Sample of Size n**

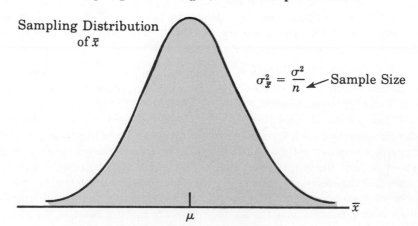

Since the three sample means are $\bar{x}_1 = 509$, $\bar{x}_2 = 526$, and $\bar{x}_3 = 552$, we have

$$\sum_i (\bar{x}_i - \bar{\bar{x}})^2 = (509 - 529)^2 + (526 - 529)^2 + (552 - 529)^2$$
$$= 400 + 9 + 529 = 938$$

With $k = 3$ populations or treatments, $k - 1 = 2$. Thus using (12.3) the estimate of the variance of the sampling distribution of $\bar{x}$ is

$$s_{\bar{x}}^2 = \frac{938}{2} = 469$$

Since $\sigma_{\bar{x}}^2 = \sigma^2/n$, solving for σ^2 gives

$$\sigma^2 = n\,\sigma_{\bar{x}}^2 \tag{12.4}$$

where n is the sample size involved in computing $\bar{x}$. Hence an estimate of the population variance σ^2 can be obtained from (12.4) by multiplying n by the estimate of $\sigma_{\bar{x}}^2$:

Between-Treatments Estimate of σ^2

$$\text{Estimate of } \sigma^2 = n\,(\text{Estimate of } \sigma_{\bar{x}}^2) = n s_{\bar{x}}^2 \tag{12.5}$$

With all samples of size $n = 5$ and with $s_{\bar{x}}^2 = 469$, for the GMAT study we have

$$\text{Estimate of } \sigma^2 = 5\,(469) = 2345$$

This estimate of σ^2 is given the name *mean square between treatments* and is denoted by MSTR.

From (12.3) and (12.5) we see that MSTR can be written

$$\text{MSTR} = \frac{n \displaystyle\sum_i (\bar{x}_i - \bar{\bar{x}})^2}{k - 1} \tag{12.6}$$

The numerator of (12.6) is called the *sum of squares between treatments* and is denoted by SSTR. The denominator, $k - 1$, represents the *degrees of freedom* corresponding to SSTR. Thus

---

Mean Square Between Treatments (MSTR)

$$\text{MSTR} = \frac{\text{SSTR}}{k - 1} \tag{12.7}$$

where

$$\text{SSTR} = \text{sum of squares between treatments}$$
$$= n \Sigma\, (\bar{x}_i - \bar{\bar{x}})^2$$

Using the previous calculations we have the following values for SSTR, degrees of freedom, and MSTR for the GMAT study:

$$\text{SSTR} = n \sum_i (\bar{x}_i - \bar{\bar{x}})^2 = 5\,(938) = 4690, \qquad k - 1 = 3 - 1 = 2$$

and thus

$$\text{MSTR} = \frac{\text{SSTR}}{k-1} = \frac{4690}{2} = 2345$$

Within-Treatments Estimate of Population Variance

We will now develop a second estimate of σ^2 that is not based on the null hypothesis assumption that the population means are equal. It is instead based upon the variation of the sample observations "within" each treatment. This estimate of σ^2 is called the mean square due to error and is denoted MSE. It is also referred to as the *mean square within treatments*.

If each of the k samples is a simple random sample, then each of the k sample variances provides an estimate of σ^2. For each sample the sample variance is computed in the usual fashion:

$$\text{Variance of sample } i = s_i^2 = \frac{\sum_j (x_{ij} - \bar{x}_i)^2}{n - 1} \tag{12.8}$$

For example, in considering the GMAT scores for the five students taking the 3-hour review, we have $n = 5$ and $\bar{x}_1 = 509$. Using (12.8) the variance estimate from this first sample becomes

$$s_1^2 = \frac{(491 - 509)^2 + (579 - 509)^2 + (451 - 509)^2 + (521 - 509)^2 + (503 - 509)^2}{5 - 1}$$

$$= \frac{324 + 4900 + 3364 + 144 + 36}{4} = \frac{8768}{4} = 2192$$

The variance estimates for the two remaining samples can be similarly computed:

$$s_2^2 = \frac{\sum_j (x_{2j} - \bar{x}_2)^2}{n - 1} = \frac{8168}{4} = 2042$$

$$s_3^2 = \frac{\sum_j (x_{3j} - \bar{x}_3)^2}{n - 1} = \frac{9044}{4} = 2261$$

Rather than working with three separate estimates of the population variance we combine, or *pool*, the results to obtain MSE. To do so we must first compute the sum of the numerators and the sum of the denominators of the three within-treatment variance estimates. Dividing the numerator total by the denominator total provides MSE, the within-treatment estimate of the population variance. Thus for the GMAT data we obtain

$$MSE = \frac{8768 + 8168 + 9044}{4 + 4 + 4} = \frac{25,980}{12} = 2165$$

Applying the procedure described above in a general context for the case of k treatments allows us to write MSE as follows:

$$MSE = \frac{\sum_{j}(x_{1j} - \bar{x}_1)^2 + \sum_{j}(x_{2j} - \bar{x}_2)^2 + \cdots + \sum_{j}(x_{kj} - \bar{x}_k)^2}{(n-1) + (n-1) + \cdots + (n-1)}$$

$$= \frac{\sum_{i}\sum_{j}(x_{ij} - \bar{x}_i)^2}{kn - k}$$

$$= \frac{\sum_{i}\sum_{j}(x_{ij} - \bar{x}_i)^2}{n_T - k} \qquad (12.9)$$

The numerator in the MSE computation is given the name *sum of squares within* or *sum of squares due to error* and is denoted by SSE. Thus from the above result we see that SSE = 25,980 for the GMAT experiment. The denominator of MSE is referred to as the *degrees of freedom* associated with the within-treatment variance estimate. In the above case the MSE value is based on 12 degrees of freedom, since $n_T = 15$, $k = 3$, and $n_T - k = 15 - 3 = 12$.

In general, we can compute MSE as follows:

Mean Square Within Treatments (MSE)

$$MSE = \frac{SSE}{n_T - k} \qquad (12.10)$$

where

$$SSE = \text{sum of squares within or due to error}$$

$$= \sum_{i}\sum_{j}(x_{ij} - \bar{x}_i)^2 \qquad (12.11)$$

Comparing the Variance Estimates: The *F* Test

We have now developed two estimates of σ^2. The first estimate (MSTR) is based upon the variation between treatment means, and the second estimate (MSE) is based upon the variation within the treatments. Recall that in order to compute MSTR we had to assume that the null hypothesis was true (that is, $\mu_1 = \mu_2 = \cdots = \mu_k$). However, in computing MSE this assumption was not required.

In fact, regardless of whether or not the means of the k populations are equal, MSE always provides an unbiased estimate of σ^2.

It can be shown that MSTR provides an unbiased estimate of σ^2 when H_0 is true. In this case, MSTR and MSE are approximately equal and thus the ratio MSTR/MSE is near 1. However, if the means of the k populations are not equal (that is, H_a is true), MSTR is not an unbiased estimate of σ^2. In fact, in this case MSTR overestimates σ^2. This is the key to the analysis of variance procedure: that is, we test the hypotheses

$$H_0: \quad \mu_1 = \mu_2 = \cdots = \mu_k$$

$$H_a: \quad \text{Not all } \mu_i \text{ are equal}$$

by comparing the two estimates of the population variance MSTR and MSE. If MSTR is much larger than MSE, such that the ratio MSTR/MSE is much larger than 1, we reject H_0 and conclude that the means are not all equal.

To obtain a better intuitive feel for why larger values of MSTR are obtained when the null hypothesis is false, recall the formula for computing MSTR:

$$\text{MSTR} = \frac{\text{SSTR}}{k-1} = \frac{n\Sigma\,(\bar{x}_i - \bar{\bar{x}})^2}{k-1} \tag{12.12}$$

Note that the numerator (SSTR) is based on the dispersion of the k sample means, $\bar{x}_i$'s, around the overall sample mean $\bar{\bar{x}}$. Now consider the two situations shown in Figure 12.3. In part A we have the sampling distribution of $\bar{x}$ under the assumption that H_0 is true with $\mu_1 = \mu_2 = \mu_3 = \mu$. In this case each sample comes from the same population, and there is only one sampling distribution. Although the three sample means $\bar{x}_1$, $\bar{x}_2$, and $\bar{x}_3$ are not the same, they are "close" to one another. In this case MSTR as computed in (12.12) will provide an unbiased estimate of σ^2.

The second diagram, part B of Figure 12.3, shows a situation where H_0 is false and the three means μ_1, μ_2, and μ_3 are not the same. What happens to the value of SSTR in this case? Note that since the sample means $\bar{x}_1$, $\bar{x}_2$, and $\bar{x}_3$ are coming from distributions with different μ's, they have different sampling distributions and show a much larger dispersion than when H_0 is true. In this case the value of SSTR in (12.12) will be larger, causing the value of MSTR to overestimate the population variance σ^2.

Let us assume for the moment that the null hypothesis is true and that $\mu_1 = \mu_2 = \cdots = \mu_k$. In this case it can be shown that MSTR and MSE provide two independent unbiased estimates of σ^2. Recall from Chapter 11 that the sampling distribution of the ratio of two independent estimates of σ^2 for a normal population follows an F probability distribution. Thus if the null hypothesis is true and the ANOVA assumptions are valid, the sampling distribution of $F = $ MSTR/MSE is an F distribution with numerator degrees of freedom equal to $k - 1$ and denominator degrees of freedom equal to $n_T - k$.

On the other hand, if the means of the k populations are not all equal, the value of MSTR/MSE will be inflated because MSTR overestimates σ^2. Hence we will reject H_0 if the resulting value of $F = $ MSTR/MSE appears to be too "large" to have been selected at random from an F distribution with degrees of freedom $k - 1$ in the numerator and $n_T - k$ in the denominator. The value of F that will cause us to reject H_0 depends upon α, the level of significance. Once α is selected

A. H_0 True

$\mu_1 = \mu_2 = \mu_3 = \mu$

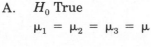

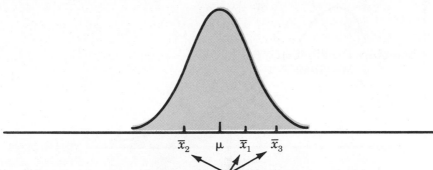

$\bar{x}_2 \quad \mu \quad \bar{x}_1 \quad \bar{x}_3$

Sample Means Are "Close
Together," Since There Is Only
One Sampling Distribution
When H_0 Is True

B. H_0 False

μ_i's Not All Equal

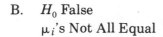

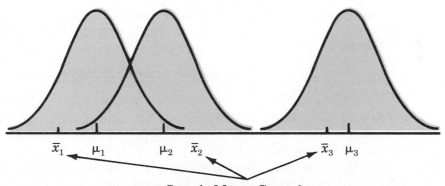

$\bar{x}_1 \quad \mu_1 \qquad \mu_2 \quad \bar{x}_2 \qquad\qquad \bar{x}_3 \quad \mu_3$

Sample Means Come from
Different Sampling Distributions
and Are Not as Close Together When
H_0 Is Not True

▦ FIGURE 12.3 **Examples of the Sampling Distributions of $\bar{x}$ for the
Cases of H_0 True (Part A) and H_0 False (Part B)**

a critical value of F can be determined. Figure 12.4 shows the sampling
distribution of MSTR/MSE and the rejection region associated with a level of
significance equal to α. Note that F_α denotes the critical value.

Let us finish the ANOVA procedure for the GMAT experiment. Assume that
Dr. Johnson was willing to accept a Type I error probability of $\alpha = .05$. From
Table 4 of Appendix B we can determine the critical value by locating the value
corresponding to numerator degrees of freedom equal to $k - 1 = 3 - 1 = 2$ and

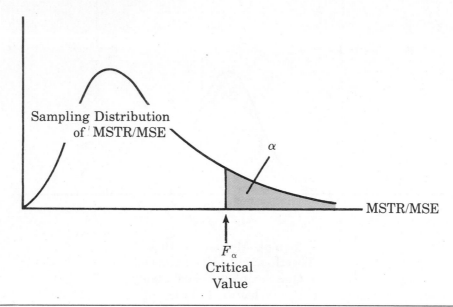

Sampling Distribution
of MSTR/MSE

α

MSTR/MSE

F_α
Critical
Value

<hr/>

■■■■ FIGURE 12.4   **Sampling Distribution of MSTR/MSE; The Critical Value for Rejecting the Null Hypothesis of Equality of Means is F_α**

denominator degrees of freedom equal to $n_T - k = 15 - 3 = 12$. Thus we obtain the value $F_{.05} = 3.89$. Hence the appropriate rejection rule for the GMAT problem is written

$$\text{Reject } H_0 \text{ if MSTR/MSE} > 3.89$$

Since MSTR/MSE = 2345/2165 = 1.08 is less than the critical value, $F_{.05} = 3.89$, there is not sufficient statistical evidence to reject the null hypothesis that the means of the three programs are the same.

Although the three sample means observed were different, there is not sufficient evidence to conclude that a statistically significant difference exists in the three population means at the .05 level. Actually, this is not too surprising considering the relatively few observations that were available and the size of the variance that is associated with GMAT scores actually observed. If Dr. Johnson believed that additional investigation was still warranted, any further experiments should seek to increase the number of observations available in each sample.

<hr/>

■■■■ EXERCISES

1. The Jacobs Chemical Company wants to estimate the mean time (minutes) required to mix a batch of material on machines produced by three different manufacturers. In order to limit the cost of testing, four batches of material were mixed on machines produced by each of the three manufacturers. The time needed to mix the material was recorded. The times in minutes are shown:

Manufacturer 1	Manufacturer 2	Manufacturer 3
20	28	20
26	26	19
24	31	23
22	27	22

Use these data and test to see if the mean time needed to mix a batch of material is the same for each manufacturer. Use $\alpha = .05$.

2. Four different paints are advertised as having the same drying time. In order to check the manufacturers' claims, five paint samples were tested for each make of paint. The time in minutes until the paint was dry enough for a second coat to be applied was recorded. The following data were obtained:

Paint 1	Paint 2	Paint 3	Paint 4
128	144	133	150
137	133	143	142
135	142	137	135
124	146	136	140
141	130	131	153

At the $\alpha = .05$ level of significance test to see if the mean drying time is the same for each type of paint.

3. A study investigated the perception of corporate ethical values among individuals specializing in marketing (*Journal of Marketing Research*, July 1989). Suppose that the data shown below were obtained in a similar study (higher scores indicate higher ethical values). Using $\alpha = .05$, test to see if there are significant differences in perception for the three groups of specialists.

Marketing Managers	Marketing Research	Advertising
6	5	6
5	5	7
4	4	6
5	4	5
6	5	6
4	4	6

4. Three top-of-the-line intermediate-sized automobiles manufactured in the United States have been test driven and compared on a variety of criteria by a well-known automotive magazine. In the area of gasoline mileage performance, five automobiles of each brand were each test driven 500 miles; the miles per gallon data obtained are shown:

	Miles per Gallon Data				
Automobile A	19	21	20	19	21
Automobile B	19	20	22	21	23
Automobile C	24	26	23	25	27

Use the analysis of variance procedure with $\alpha = .05$ to determine if there is a significant difference in the mean miles per gallon for the three types of automobiles.

5. Three brands of paper towels were tested for their ability to absorb water. Equal-sized towels were used, with four sections of towels tested per brand. The absorbency rating data are given. Using a .05 level of significance, does there appear to be a difference in the ability of the brands to absorb water?

Brand	Absorbency Rating			
X	91	100	88	89
Y	99	96	94	99
Z	83	88	89	76

12.3 ▬ THE ANOVA TABLE AND OTHER CONSIDERATIONS

A convenient way to summarize the computations and results of the analysis of variance procedure involves the development of an *analysis of variance (ANOVA) table*. Before presenting the ANOVA table for the GMAT study, however, we must perform some additional calculations with the data collected (see Table 12.1). Treating the entire data set as one sample of 15 observations, we have already found the overall sample mean to be $\bar{\bar{x}} = 529$. Using the entire data set as one sample, let us now compute an overall sample variance:

$$s^2 = \frac{\sum_i \sum_j (x_{ij} - \bar{\bar{x}})^2}{n_T - 1} \qquad (12.13)$$

The numerator of (12.13) is called the total sum of squares about the mean (SST), and the denominator represents the degrees of freedom associated with this total sum of squares. Check for yourself that the 15 values in Table 12.1 provide the following value for SST:

$$SST = \sum_i \sum_j (x_{ij} - \bar{\bar{x}})^2$$

$$= (491 - 529)^2 + (579 - 529)^2 + \cdots + (537 - 529)^2 + (561 - 529)^2 = 30,670$$

We can now present the ANOVA table for a completely randomized design. Table 12.2 shows a general ANOVA table, and Table 12.3 shows the ANOVA table for the GMAT experiment.

TABLE 12.2   Anova Table for a Completely Randomized Design

Source of Variation	Sum of Squares	Degrees of Freedom	Mean Square	F
Between treatments	SSTR	$k - 1$	$MSTR = \dfrac{SSTR}{k - 1}$	$\dfrac{MSTR}{MSE}$
Error	SSE	$n_T - k$	$MSE = \dfrac{SSE}{n_T - k}$	
Total	SST	$n_T - 1$		

TABLE 12.3   Anova Table for the Dr. Johnson GMAT Experiment

Source of Variation	Sum of Squares	Degrees of Freedom	Mean Square	F
Between treatments	4690	2	$\dfrac{4690}{2} = 2345$	$\dfrac{2345}{2165} = 1.08$
Error	25,980	12	$\dfrac{25,980}{12} = 2165$	
Total	30,670	14		

Note that the rows of the table provide information concerning the two sources of variation: "between treatments," which refers to the between-group variation, and "error," which refers to the within-group variation. The "Sum of Squares" column and the "Degrees of Freedom" column provide the corresponding values as defined in the previous section. The column labeled "Mean Square" is simply the sum of squares divided by the corresponding degrees of freedom. Recall that in Section 12.2 we showed that when we divided the sum of squares by the corresponding degrees of freedom we obtained two estimates of the population variance (MSTR and MSE). The ANOVA table simply summarizes these values in the "Mean Square" column.

Finally, the last column in the table contains the F value corresponding to MSTR/MSE. Since the variance estimates are found in the "Mean Square" column, the F value is computed by dividing the mean square treatments (MSTR) by the mean square error (MSE). That is, $F = $ MSTR/MSE.

What observation can you make from the "Sum of Squares" columns in Tables 12.2 and 12.3? Note in particular that the following condition holds:

$$SST = SSTR + SSE \qquad (12.14)$$

or

$$\sum_i \sum_j (x_{ij} - \bar{\bar{x}})^2 = n \sum_i (\bar{x}_i - \bar{\bar{x}})^2 + \sum_i \sum_j (x_{ij} - \bar{x}_i)^2 \qquad (12.15)$$

Thus we see that SST can be partitioned into two sums—one called the sum of squares between treatments and one called the sum of squares due to error. This is known as *partitioning* the sum of squares.

Note also that the degrees of freedom corresponding to SST ($n_T - 1 = 14$) can be partitioned into the degrees of freedom corresponding to SSTR ($k - 1 = 2$) and the degrees of freedom corresponding to SSE ($n_T - k = 12$). The analysis of variance procedure can be viewed as the process of partitioning the sum of squares and degrees of freedom into their corresponding sources: treatments and error. Dividing the sum of squares by the appropriate degrees of freedom provides the variance estimates and the F value used to test the hypothesis of equal population or treatment means.

Unbalanced Designs

The formula presented previously for computing SSTR applies only to *balanced designs*. Balanced designs are ones in which the sample size is the same for each treatment. Any experimental design for which the sample size is not the same for each treatment is said to be *unbalanced*. Although we would prefer a balanced design,* in some cases we must work with unequal sample sizes. For unbalanced designs the analysis of variance procedure described above may be used with the following modification of the formula for SSTR:

$$\text{SSTR} = \sum_i n_i (\bar{x}_i - \bar{\bar{x}})^2 \tag{12.16}$$

where

$$n_i = \text{sample size for the } i\text{th treatment}$$

In such cases

$$n_T = n_1 + n_2 + \cdots + n_k$$

$$= \sum_i n_i$$

Note that (12.16) yields the same formula we used earlier when the sample size is the same for each treatment, that is, $n_i = n$.

Computer Results for Analysis of Variance

The computational aspect of the analysis of variance procedure is devoted primarily to computing the appropriate sums of squares. When a hand calculator is used to compute the sum of squares, some computational help can be obtained by using alternate forms of the sums-of-squares formulas. In Appendix 12.1 we provide a step-by-step procedure that uses these revised formulas to compute the sums of squares for a completely randomized design.

Because of the widespread availability of statistical computer packages, problems involving large sample sizes and/or large numbers of treatments can be routinely solved. In Figure 12.5 we show part of the output obtained from the Minitab computer package when it was used to solve the GMAT problem.

We see that the computer output contains the familiar ANOVA table format. It should prove easy for you to interpret. Comparing Figure 12.5 with Table 12.3, we see that the same information is available, although some of the headings are a little different. "SOURCE" is the heading used for the source of variation column, "FACTOR" identifies the between-treatments row, and the sum of squares and degrees of freedom columns are interchanged. Note also that below the ANOVA table the computer output contains all of the sample means, their standard deviations, and respective sample sizes. The "F" column contains the F statistic used for the hypothesis test. Next to the F statistic is the p-value showing the level of significance of the sample results.

---

*With a balanced design the F statistic is less sensitive to small departures from the assumption that σ^2 is the same for each population. In addition choosing equal sample sizes maximizes the power of the test.

```
ANALYSIS OF VARIANCE
SOURCE      DF        SS        MS         F        p
FACTOR       2      4690      2345      1.08    0.369
ERROR       12     25980      2165
TOTAL       14     30670

LEVEL       N      MEAN     STDEV
3-HOUR       5    509.00     46.82
1-DAY        5    526.00     45.19
10-WEEK      5    552.00     47.55
```

■■■■ FIGURE 12.5   **Computer Output for GMAT Problem**

■■■■ **EXERCISES**

6. In a completely randomized experimental design, 7 experimental units were used for each of the 5 levels of the factor. Complete the ANOVA table shown.

Source of Variation	Sum of Squares	Degrees of Freedom	Mean Squares	F
Between treatments	300			
Error				
Total	480			

7. In an experiment designed to test the output levels of three different machines, the following results were obtained: SST = 400, SSTR = 150, $n_T = 18$. Set up the ANOVA table and test for any significant difference between the mean output levels of the three machines. Use $\alpha = .05$.

8. Managers at all levels of an organization need to have the information necessary to perform their respective tasks. A recent study investigated the effect the source has on the dissemination of the information (*Journal of Management Information Systems*, Fall 1988). In the particular study the sources of information were a superior, a peer, and a subordinate. In each case a measure of dissemination was obtained with higher values indicating greater dissemination of the information. Using $\alpha = .05$, and the data shown below, test whether or not the source of information significantly affects dissemination. What is your conclusion and what does this suggest about the use and dissemination of information?

Superior	Peer	Subordinate
8	6	6
5	6	5
4	7	7
6	5	4
6	3	3
7	4	5
5	7	7
5	6	5

9. Auditors must make judgments concerning various aspects of an audit based on their own direct experience, indirect experience, or a combination of the two. In a study, auditors were asked to make judgments about the frequency of errors to be found in an

audit (*Journal of Accounting Research*, Autumn 1988). The judgments by the auditors were then compared to the actual results. Suppose the following data were obtained from a similar study; lower scores indicate better judgments. Using $\alpha = .05$, test to see if the basis for the judgment affects the quality of the judgment. What is your conclusion?

Direct Experience	Indirect Experience	Combination
17.0	16.6	25.2
18.5	22.2	24.0
15.8	20.5	21.5
18.2	18.3	26.8
20.2	24.2	27.5
16.0	19.8	25.8
13.3	21.2	24.2

10. Three different methods for assembling a product were proposed by an industrial engineer. To investigate the number of units assembled correctly using each method, 30 employees were randomly selected and randomly assigned to the three proposed methods such that 10 workers were associated with each method. The number of units assembled correctly was recorded, and the analysis of variance procedure was applied to the resulting data set. The following results were obtained: SST = 10,800, SSTR = 4560.
   a. Set up the ANOVA table for this problem.
   b. Using $\alpha = .05$, test for any significant difference in the means for the three assembly methods.

11. In an experiment designed to test the breaking strength of four types of cables, the following results were obtained: SST = 85.05, SSTR = 61.64, $n_T = 24$. Set up the ANOVA table and test for any significant difference in the mean breaking strength of the four cables. Use $\alpha = .05$.

12. To test for any significant difference in the time between breakdowns for four machines, the following data were obtained:

Time (hours)

Machine 1	6.4	7.8	5.3	7.4	8.4	7.3
Machine 2	8.7	7.4	9.4	10.1	9.2	9.8
Machine 3	11.1	10.3	9.7	10.3	9.2	8.8
Machine 4	9.9	12.8	12.1	10.8	11.3	11.5

At the $\alpha = .05$ level of significance, is there any difference in the mean time between breakdowns among the four machines?

13. In order to study the effect of temperature upon yield in a chemical process, 5 batches were produced under each of 3 temperature levels. The results are given. Construct an analysis of variance table. Using a .05 level of significance, test to see if the temperature level appears to have an effect upon the mean yield of the process.

Temperature (°C)		Yield			
50	34	24	36	39	32
60	30	31	34	23	27
70	23	28	28	30	31

14. In an unbalanced experimental design, 12 experimental units were used for the first treatment, 15 experimental units were used for the second treatment, 20 experimental units were used for the third treatment, and 10 experimental units were used for the

fourth treatment. Complete the analysis of variance table shown. Using a .05 level of significance, is there a significant difference between the treatments?

Source	Sum of Squares	Degrees of Freedom	Mean Square	F
Between treatments	1200			
Error				
Total	1800			

15. Develop the analysis of variance computations for the following unbalanced experimental design. Using $\alpha = .05$, is there a significant difference between the treatment means?

Treatment	Value of Observation									
A	136	120	113	107	131	114	129	102		
B	107	114	120	104	107	109	97	114	104	94
C	92	82	85	101	89	117	110	120	98	106

12.4 ■■■ RANDOMIZED BLOCK DESIGN

Thus far we have considered the completely randomized experimental design. Recall that in order to test for a difference among treatment means we computed an F value using the ratio

$$F = \frac{\text{MSTR}}{\text{MSE}} \qquad (12.17)$$

A problem can arise whenever differences due to extraneous factors (ones not considered in the experiment) cause the MSE term in this ratio to become large. In such cases the F value in (12.17) can become small, signaling no difference among treatment means when in fact such a difference exists.

In this section we present an experimental design referred to as a *randomized block design*. The purpose of this design is to control some of the extraneous sources of variation by removing such variation from the MSE term. This design tends to provide a better estimate of the true error variance, and leads to a more powerful hypothesis test in terms of the ability to detect differences among treatment means. To illustrate, let us consider a stress study for air traffic controllers.

Air Traffic Controllers Stress Test

A study directed at measuring the fatigue and stress on air traffic controllers has resulted in proposals for modification and redesign of the controller's work station. After consideration of several designs for the work station, three specific alternatives have been selected as having the best potential for reducing controller stress. The key question is: To what extent do the three alternatives differ in terms of their effect on controller stress? To answer this question we need to design an experiment that will provide measurements of air traffic controller stress under each alternative.

In a completely randomized design a random sample of controllers would be assigned to each work station alternative. However, it is believed that controllers differ substantially in terms of their ability to handle stressful situations. What is high stress to one controller might be only moderate or even low stress to another. Thus when considering the within-group source of variation (MSE) we must realize that this variation includes both random error and error due to individual controller differences. In fact, for this study management expected controller variability to be a major contributor to the MSE term.

One way to separate the effect of the individual differences is to use the randomized block design. This design will identify the variability stemming from individual controller differences and remove it from the MSE term. The randomized block design calls for a single sample of controllers. Each controller in the sample is tested using each of the three work station alternatives. In experimental design terminology, the work station is the *factor of interest,* and the controllers are referred to as the *blocks.* The three treatments or populations associated with the work station factor correspond to the three work station alternatives. For simplicity, we will refer to these alternatives as System A, System B, and System C.

The *randomized* aspect of the randomized block design refers to the fact that the order in which the treatments (systems) are assigned to the controllers is chosen randomly. If every controller were to test the three systems in the same order, any observed difference in systems might be due to the order of the test rather than to true differences in the systems.

To provide the necessary data, the three types of work stations were installed at the Cleveland Control Center in Oberlin, Ohio. Six controllers were selected at random and assigned to operate each of the systems. A follow-up interview and a medical examination of each controller participating in the study provided a measure of the stress for each controller on each system. The data are shown in Table 12.4.

A summary of the stress data collected is shown in Table 12.5. In this table we have included column totals (blocks) and row totals (treatments) as well as some sample means that will be helpful in making the sum of squares computations for the ANOVA procedure. Since lower stress values are viewed as better, the sample data available would seem to favor System B with its mean stress rating of 13. However, the usual question remains: Do the sample results justify the conclusion that the mean stress levels for the three systems differ? That is, are the differences statistically significant? An analysis of variance computation similar to the one performed for the completely randomized design can be used to answer this statistical question.

The ANOVA Procedure for a Randomized Block Design

The ANOVA procedure for the randomized block design requires us to partition the sum of squares total (SST) into three groups: sum of squares between treatments, sum of squares due to blocks, and sum of squares due to error. The formula for this partitioning is as follows:

$$SST = SSTR + SSB + SSE \qquad (12.18)$$

TABLE 12.4 A Randomized Block Design for the Air Traffic Controllers Stress Test

Treatments	Block 1 (Controller 1)	Block 2 (Controller 2)	Block 3 (Controller 3)	Block 4 (Controller 4)	Block 5 (Controller 5)	Block 6 (Controller 6)
System A	15	14	10	13	16	13
System B	15	14	11	12	13	13
System C	18	14	15	17	16	13

Stress Value (arrow pointing to the 18 of System C, Block 1)

TABLE 12.5 Summary of Stress Data for the Air Traffic Controllers Stress Test

	Blocks						Row or Treatment Totals	Treatment Means ($\bar{x}_{i.}$)
Treatments	Controller 1	Controller 2	Controller 3	Controller 4	Controller 5	Controller 6		
System A	15	14	10	13	16	13	81	$\bar{x}_{1.} = 81/6 = 13.5$
System B	15	14	11	12	13	13	78	$\bar{x}_{2.} = 78/6 = 13.0$
System C	18	14	15	17	16	13	93	$\bar{x}_{3.} = 93/6 = 15.5$
Column or block totals	48	42	36	42	45	39	252	$\bar{\bar{x}} = \dfrac{252}{18} = 14.0$
Block Means	$\bar{x}_{.1} = \dfrac{48}{3}$ $= 16.0$	$\bar{x}_{.2} = \dfrac{42}{3}$ $= 14.0$	$\bar{x}_{.3} = \dfrac{36}{3}$ $= 12.0$	$\bar{x}_{.4} = \dfrac{42}{3}$ $= 14.0$	$\bar{x}_{.5} = \dfrac{45}{3}$ $= 15.0$	$\bar{x}_{.6} = \dfrac{39}{3}$ $= 13.0$		

This sum of squares partition is summarized in the ANOVA table for the randomized block design as shown in Table 12.6. The notation used in this table is as follows.

$$k = \text{the number of treatments}$$
$$b = \text{the number of blocks}$$
$$n_T = \text{the total sample size } (n_T = kb)$$

Note that the ANOVA table in Table 12.6 also shows how the $n_T - 1$ total degrees of freedom are partitioned such that $k - 1$ go to treatments, $b - 1$ go to blocks, and $(k - 1)(b - 1)$ go to the error term. The mean square column shows the sum of squares divided by the degrees of freedom, and $F = \text{MSTR}/\text{MSE}$ is the F ratio used to test for a significant difference among the treatment means. The primary contribution of the randomized block design is that by including blocks we have removed the individual controller differences from the MSE term and obtained a more powerful test for the stress differences in the three work station alternatives.

Computations and Conclusions

To compute the F statistic needed to test for a difference among treatment means using a randomized block design, we need to compute MSTR and MSE. To calculate these two mean squares, we must first compute SSTR and SSE; in doing so, we will also compute SSB and SST. To simplify the presentation, we will perform the calculations using four steps. In addition to k, b, and n_T as previously defined, the following notation is used:

$$x_{ij} = \text{value of the observation under treatment } i \text{ in block } j$$
$$\bar{x}_{i.} = \text{sample mean of the } i\text{th treatment}$$
$$\bar{x}_{.j} = \text{sample mean for the } j\text{th block}$$
$$\bar{\bar{x}} = \text{overall sample mean}$$

Step 1. Compute the total sum of squares (SST):

$$\text{SST} = \sum_i \sum_j (x_{ij} - \bar{\bar{x}})^2 \tag{12.19}$$

TABLE 12.6 **ANOVA Table for the Randomized Block Design with k Treatments and b Blocks**

Source of Variation	Sum of Squares	Degrees of Freedom	Mean Square	F
Between treatments	SSTR	$k - 1$	$\text{MSTR} = \dfrac{\text{SSTR}}{k - 1}$	$\dfrac{\text{MSTR}}{\text{MSE}}$
Between blocks	SSB	$b - 1$	$\text{MSB} = \dfrac{\text{SSB}}{b - 1}$	
Error	SSE	$(k - 1)(b - 1)$	$\text{MSE} = \dfrac{\text{SSE}}{(k - 1)(b - 1)}$	
Total	SST	$n_T - 1$		

Step 2. Compute the sum of squares between treatments (SSTR):

$$\text{SSTR} = b \sum_i (\bar{x}_{i\cdot} - \bar{\bar{x}})^2 \qquad (12.20)$$

Step 3. Compute the sum of squares due to blocks (SSB):

$$\text{SSB} = k \sum_j (\bar{x}_{\cdot j} - \bar{\bar{x}})^2 \qquad (12.21)$$

Step 4. Compute the sum of squares due to error (SSE):

$$\text{SSE} = \text{SST} - \text{SSTR} - \text{SSB} \qquad (12.22)$$

For the air traffic controller data in Table 12.5, these steps lead to the following sum of squares:

Step 1. $\text{SST} = (15 - 14)^2 + (14 - 14)^2 + (10 - 14)^2 + \cdots + (13 - 14)^2 = 70$

Step 2. $\text{SSTR} = 6[(13.5 - 14)^2 + (13.0 - 14)^2 + (15.5 - 14)^2] = 21$

Step 3. $\text{SSB} = 3[(16 - 14)^2 + (14 - 14)^2 + (12 - 14)^2 + (14 - 14)^2 + (15 - 14)^2 + (13 - 14)^2] = 30$

Step 4. $\text{SSE} = 70 - 21 - 30 = 19$

These sums of squares divided by their degrees of freedom provide the corresponding mean square values shown in Table 12.7. The F ratio used to test for differences between treatment means is $\text{MSTR}/\text{MSE} = 10.5/1.9 = 5.53$. Checking the F values in Table 4 of Appendix B, we find that the critical F value at $\alpha = .05$ (2 numerator degrees of freedom and 10 denominator degrees of freedom) is 4.10. With $F = 5.53$, we reject the null hypothesis $H_0 : \mu_1 = \mu_2 = \mu_3$ and conclude that the work station designs differ in terms of their mean stress effects on air traffic controllers.

Before leaving this section let us make some general comments about the randomized block design. The blocking as described in this section is referred to as a *complete* block design; the word "complete" indicates that each block is

TABLE 12.7 **ANOVA Table for the Air Traffic Controller Stress Test**

Source of Variation	Sum of Squares	Degrees of Freedom	Mean Square	F
Between treatments	21	2	10.5	$\frac{10.5}{1.9} = 5.53$
Between blocks	30	5	6.0	
Error	19	10	1.9	
Total	70	17		

subjected to all k treatments. That is, all controllers (blocks) were tested using all three systems (treatments). Experimental designs employing blocking where some but not all treatments are applied to each block are referred to as *incomplete* block designs. A discussion of incomplete block designs is outside the scope of this text.

In addition, note that in the air traffic controller stress test each controller in the study was required to use all 3 systems. While this guarantees a complete block design, in some cases blocking is carried out with "similar" experimental units in each block. For example, assume that in a pretest of air traffic controllers the population of controllers was divided into groups ranging from extremely high stress individuals to extremely low stress individuals. The blocking could still have been accomplished by having three controllers from each of the stress classifications participate in the study. Each block would then be formed from 3 controllers in the same stress class. The randomized aspect of the block design would be conducted by randomly assigning the 3 controllers in each block to the 3 systems.

Finally, note that the ANOVA table shown in Table 12.7 provides an F value to test for treatment effects but *not* for blocks. The reason is that the experiment was designed to test a single factor—work station design. The blocking based on individual stress differences was conducted in order to remove this variation from the MSE term. However, the study was not designed to test specifically for individual differences in stress.

Some analysts compute $F = MSB/MSE$ and use this statistic to test for significance of the blocks. Then they use the result as a guide to whether or not this type of blocking would be desired in future experiments. However, if individual stress difference is to be a factor in the study, a different experimental design should be used. A test of significance on blocks should not be performed to attempt to draw such a conclusion about a second factor.

In closing this section, we also note that alternate formulas for SST, SSTR, and SSB can be developed that can ease the computational burden when using hand calculation. In Appendix 12.2 we have included a step-by-step procedure that illustrates the use of these alternate formulas.

■ EXERCISES

16. An automobile dealer conducted a test to determine if the time needed to complete a minor engine tuneup depends upon whether a computerized engine analyzer or an electronic analyzer is used. Because tuneup time varies among compact, intermediate, and full-sized cars, the three types of cars were used as blocks in the experiment. The data obtained are shown below:

		Car		
		Compact	*Intermediate*	*Full-size*
Analyzer	Computerized	50	55	63
	Electronic	42	44	46

Time (minutes)

Using $\alpha = .05$, test for any significant differences.

17. An important factor in selecting software for word processing and data base management systems is the time required to learn how to use a particular system. In order to evaluate 3 file management systems, a firm designed a test involving 5 different word processing operators. Since operator variability was believed to be a significant factor, each of the 5 operators was trained on each of the 3 file management systems. The data obtained are shown below.

		Operator				
		1	*2*	*3*	*4*	*5*
	A	16	19	14	13	18
System	B	16	17	13	12	17
	C	24	22	19	18	22

Time (hours)

Using $\alpha = .05$, test to see if there is any difference in training time for the three software packages.

18. The following data were obtained for a randomized block design involving 5 treatments and 3 blocks: SST = 430, SSTR = 310, SSB = 85. Set up the ANOVA table and test for any significant differences. Use $\alpha = .05$.

19. Five different auditing procedures were compared with respect to total audit time. To control for possible variation due to the person conducting the audit, 4 accountants were selected randomly and treated as blocks in the experiment. The following values were obtained using the ANOVA procedure: SST = 100, SSTR = 45, SSB = 36. Using $\alpha = .05$, test to see if there is any significant difference in total audit time stemming from the auditing procedure used.

20. An experiment has been conducted for 4 treatments using 8 blocks. Complete the following analysis of variance table.

Source	Sum of Squares	Degrees of Freedom	Mean Square	*F*
Between Treatments	900			
Blocks	400			
Error				
Total	1800			

Using $\alpha = .05$, test for any significant differences.

21. Consider the following experimental results of a randomized block design. Make the calculations necessary to set up the analysis of variance table, and using $\alpha = .05$, test for any significant differences.

		Blocks				
		1	*2*	*3*	*4*	*5*
	A	10	12	18	20	8
Treatment	B	9	6	15	18	7
	C	8	5	14	18	8

12.5 ■ INFERENCES ABOUT INDIVIDUAL TREATMENT MEANS

Suppose that in carrying out an ANOVA procedure we reject the null hypothesis and conclude that the population or treatment means are not all the same. Sometimes we may be satisfied with this conclusion, but in other cases we will want to go a step further and determine where the differences occur. The purpose of this section is to show how to conduct statistical comparisons between pairs of treatment means.

Recall that in the GMAT study (Sections 12.1 to 12.3) we could not reject the null hypothesis that the three GMAT preparation programs led to the same mean test scores. Further tests comparing specific programs were unwarranted because there was not sufficient statistical evidence to conclude that any differences in means existed. However, in the air traffic controllers stress test we rejected the null hypothesis and concluded the 3 work station systems differed in terms of their effects on controller stress. In this case the follow-up question is, "We believe that the systems differ; where do the differences occur?"

Let us show the details of a procedure that could be used to test for the equality of two treatment means. For example, let us test to see if there is a significant difference between the means of Systems A and B. From the data in Table 12.5 we found the treatment means to be $\bar{x}_1 = 13.5$ for System A and $\bar{x}_2 = 13$ for System B; thus System B shows the better (lower) stress level. But is the sample information sufficient to justify the conclusion that a difference in stress level exists between System A and System B?

A simple test for a difference between two treatment means is based on the use of the t distribution (as presented for the two population case in Chapter 10). The mean square error (MSE) is used as an unbiased estimate of the population variance σ^2. A confidence interval estimate of the difference between the population means μ_1 and μ_2 is given by the following expression:

Interval Estimate of $\mu_1 - \mu_2$

$$\bar{x}_1 - \bar{x}_2 \pm t_{\alpha/2} \sqrt{\text{MSE}\left(\frac{1}{n_1} + \frac{1}{n_2}\right)} \qquad (12.23)$$

where

$\bar{x}_1$ = sample mean for the first treatment

$\bar{x}_2$ = sample mean for the second treatment

$t_{\alpha/2}$ = t value based upon the degrees of freedom for error (see ANOVA table "Degrees of Freedom" column)

MSE = mean square error (see ANOVA table)

n_1 = sample size for the first treatment

n_2 = sample size for the second treatment

If the confidence interval in (12.23) includes the value 0, we have to conclude that there is no significant difference between the treatment means. In this case we cannot reject the hypothesis that no difference exists between the treatment means. However, if the confidence interval does not include the value 0, we conclude that there is a difference between the treatment means.

Let us return to the air traffic controllers stress test and compare the stress levels for System A and System B at the .05 level of significance. The following data are needed:

$$\bar{x}_1 = 13.5 \quad \text{(System A)} \qquad \bar{x}_2 = 13 \quad \text{(System B)}$$

With $\alpha = .05$, $t_{.025} = 2.228$ (*Note:* 10 degrees of freedom for error)

$$\text{MSE} = 1.9$$

$$n_1 = n_2 = 6$$

Using these data and (12.23) we have

$$13.5 - 13 \pm 2.228\sqrt{1.9(\tfrac{1}{6} + \tfrac{1}{6})}$$

$$.5 \pm 2.228(.7958)$$

$$.5 \pm 1.77$$

Thus a 95% confidence interval for the difference between System A and System B is given by -1.27 to 2.27. Since this interval includes 0, we are unable to reject the hypothesis that the systems provide the same level of stress.

At this point, analysis of variance has told us that the stress levels for the three systems are not all the same. However, the above result tells us that Systems A and B appear to have the same level of stress. With $\bar{x}_3 = 15.5$, System C appears to have the highest level of stress. Thus we should feel comfortable in concluding that the difference in population means is due to System C.

A word of caution is needed at this point: This test of an individual difference should be applied only if we reject the null hypothesis of equal population means. That is, while it may appear natural to use (12.23) to compare all possible pairs of treatment means (System A versus System B, System A versus System C, System B versus System C), statistical problems can occur with this sequential approach. If a null hypothesis is true (that is, the two population means are equal) and a test is conducted using (12.23) with a Type I error probability of $\alpha = .10$, there is a probability of .10 of rejecting the null hypothesis when it is really true; hence the probability of making a correct decision is .90. If two tests are conducted in this manner, the probability that a correct decision is made on both tests is $(.90)(.90) = .81$. Thus the probability that *at least* one of the tests would result in rejecting a true null hypothesis is $1 - .81 = .19$.* Thus we see that the actual Type I error probability using a sequential testing procedure to test two hypotheses at the .10 significance level is .19 and not .10. For three tests at the .10 level of significance the Type I error probability is actually $1 - (.90)(.90)(.90) =$

---

*This assumes that the two tests are independent and hence the joint probability of the two events can be obtained simply by multiplying the individual probabilities. In fact the two tests are not independent (MSE is used in each test), and hence the error involved is even greater than that shown.

$1 - .729 = .271$. Note that the probability of making a Type I error increases rapidly as the number of multiple tests increases.

A simple procedure to adjust (approximately) for this increasing probability of making the Type I error is to reduce the α level for each separate test. The aim is to reduce to a satisfactory level the α level for all tests taken together. With an α level of significance desired and m tests to be made, we would use α/m as the probability of making a Type I error on any one test. Then the overall probability of making a Type I error on any one of the tests will approximately equal α.

Because of the difficulty of the Type I error increasing when making multiple tests of difference between individual means, a variety of specialized tests have been developed. Often these tests carry the name of the developer. The better known tests for multiple comparisons include the Duncan multiple range test, the Newman-Kuels test, Tukey's test, and Scheffe's method. References in the bibliography provide details for these methods. It is recommended that these tests be considered whenever multiple comparisons among treatment means are expected to be a major concern in the study.

■■■ EXERCISES

22. Refer to Exercise 1. Use the procedure described in this section to test for the equality of the mean for manufacturers 1 and 3. Use $\alpha = .05$. What conclusion can you make after carrying out this test?

23. Problem 9 described three sources of information that auditors use in making judgments (*Journal of Accounting Research*, Autumn 1988). Compare the direct and indirect sources of information. Does there appear to be a significant difference? Use $\alpha = .05$.

24. Refer to Exercise 4. Use the procedure described in this section to test for the equality of the mean mileage for automobiles A and B. What general conclusion can you make after carrying out this test? Use $\alpha = .05$.

25. Refer to Exercise 5. Use the procedure described in this section to test for the equality of brands Y and Z. What general conclusions can you make after carrying out this test? Use $\alpha = .05$.

12.6 ■■■ FACTORIAL EXPERIMENTS

The experimental designs we have considered thus far enable statistical conclusions to be drawn about one factor. However, in some experiments we want to draw conclusions about more than one variable or factor. *Factorial experiments* and their corresponding ANOVA computations are valuable designs when simultaneous conclusions are required about two or more factors. The term "factorial" is used because the experimental conditions include all possible combinations of the factors involved. For example, if there are a levels of factor A and b levels of factor B, the experiment will involve collecting data on ab treatment combinations. In this section we will show the analysis of a 2-factor factorial experiment. This basic approach can be extended to experiments involving more than two factors.

As an illustration of a 2-factor factorial experiment, let us return to the GMAT study (Sections 12.1 to 12.3). One factor in this study is the GMAT preparation program, which has three treatments: 3-hour review, 1-day program, and 10-week course. Let us assume that Dr. Johnson, Dean of the College of Business, has talked to the deans of the College of Engineering and the College of Arts and Sciences and that these deans are interested in the GMAT preparation programs because a number of their students are also considering attending graduate business school. Thus, also of interest in this experiment is whether or not a student's undergraduate college affects the GMAT score and a second factor, undergraduate college, is defined. This factor also has three treatments: business, engineering, and arts and sciences. The factorial design for this experiment with 3 treatments corresponding to factor A, the preparation program, and 3 treatments corresponding to factor B, the undergraduate college, will have a total of $3 \times 3 = 9$ treatment combinations. These treatment combinations or experimental conditions are summarized in Table 12.8.

Assume that a sample of 2 students will be selected corresponding to each of the 9 treatment combinations shown in Table 12.8: two business students will take the 3-hour review, 2 will take the 1-day program, and 2 will take the 10-week course. In addition, 2 engineering students and 2 arts and sciences students will take each of the three preparation programs. In experimental design terminology the sample size of 2 for each treatment combination indicates that we have 2 replications. Additional replications and an increased sample size could easily be made, but we elected not to do so in order to minimize the computational aspects for this illustration.

This experimental design requires that 6 students who plan to attend graduate school be randomly selected from *each* of the 3 undergraduate colleges. Then 2 students from each college should be assigned randomly to each preparation program, resulting in a total of 18 students being used in the study.

Let us assume that the students have been randomly selected, have participated in the preparation program, and have taken the GMAT. The scores obtained are shown in Table 12.9.

The analysis of variance computations using the data in Table 12.9 will provide answers to the following questions:

■ **Main effect (factor A):** Do the preparation programs differ in terms of effect on GMAT scores?

TABLE 12.8 **Nine Treatment Combinations for the 2-Factor GMAT Experiment**

| | Treatments ——→ | Factor B: College | | |
		Business	Engineering	Arts and Sciences
Factor A:	3-hour review	1	2	3
Preparation	1-day program	4	5	6
Program	10-week course	7	8	9

TABLE 12.9   **GMAT Scores for the Two Factor Experiment**

		Factor B: College		
		Business	Engineering	Arts and Sciences
	3-hour review	500	540	480
		580	460	400
Factor A: Preparation Program	1-day program	460	560	420
		540	620	480
	10-week course	560	600	480
		600	580	410

■■■ **Main effect (factor B):** Do the undergraduate colleges differ in terms of student ability to perform on the GMAT?

■■■ **Interaction effect (factors A and B):** Do students in some colleges do better on one type of preparation program while others do better on a different type of preparation program?

The term *interaction* refers to a new effect that we can now study because we have used a factorial experiment. If the interaction effect has a significant impact on the GMAT scores, it will mean that the effect of the type of preparation program depends on the undergraduate college.

The ANOVA Procedure

The ANOVA procedure for the 2-factor factorial experiment is similar to the completely randomized experiment and the randomized block experiment in that we once again partition the sum of squares and the degrees of freedom into their respective sources. The formula for partitioning the sum of squares for the 2-factor factorial experiments is as follows:

$$SST = SSA + SSB + SSAB + SSE \qquad (12.24)$$

The partitioning of the sum of squares and degrees of freedom is summarized in Table 12.10. The following notation is used:

a = number of levels of factor A

b = number of levels of factor B

r = number of replications

n_T = total number of observations taken in the experiment; $n_T = abr$

TABLE 12.10   ANOVA Table for the 2-Factor Factorial Experiment with r Replications

Source of Variation	Sum of Squares	Degrees of Freedom	Mean Square	F
Factor A treatments	SSA	$a - 1$	$MSA = \dfrac{SSA}{a - 1}$	$\dfrac{MSA}{MSE}$
Factor B treatments	SSB	$b - 1$	$MSB = \dfrac{SSB}{b - 1}$	$\dfrac{MSB}{MSE}$
Interaction	SSAB	$(a - 1)(b - 1)$	$MSAB = \dfrac{SSAB}{(a - 1)(b - 1)}$	$\dfrac{MSAB}{MSE}$
Error	SSE	$ab(r - 1)$	$MSE = \dfrac{SSE}{ab(r - 1)}$	
Total	SST	$n_T - 1$		

Computations and Conclusions

To compute the F statistics needed to test for the significance of factor A, factor B, and interaction, we need to compute MSA, MSB, MSAB, and MSE. To calculate these four mean squares, we must first compute SSA, SSB, SSAB, and SSE; in doing so we will also compute SST. To simplify the presentation, we will perform the calculations using five steps. In addition to a, b, r, and n_T as previously defined, the following notation is used:

x_{ijk} = observation corresponding to the kth replicate taken from treatment i of factor A and treatment j of factor B

$\bar{x}_{i.}$ = sample mean for the observations in treatment i (factor A)

$\bar{x}_{.j}$ = sample mean for the observations in treatment j (factor B)

$\bar{x}_{ij}$ = sample mean for the observations corresponding to the combination of treatment i (factor A) and treatment j (factor B)

$\bar{\bar{x}}$ = overall sample mean of all n_T observations

Step 1. Compute the total sum of squares:

$$SST = \sum_i \sum_j \sum_k (x_{ijk} - \bar{\bar{x}})^2 \tag{12.25}$$

Step 2. Compute the sum of squares for factor A:

$$SSA = br \sum_i (\bar{x}_{i.} - \bar{\bar{x}})^2 \tag{12.26}$$

Step 3. Compute the sum of squares for factor B:

$$SSB = ar \sum_j (\bar{x}_{.j} - \bar{\bar{x}})^2 \tag{12.27}$$

TABLE 12.11　GMAT Summary Data for the 2-Factor Experiment

Treatment Combination Totals

		Factor B: College				
Factor A: Preparation Program		Business	Engineering	Arts and Sciences	Row Totals	Factor A Means
3-hour review		500 580 $\overline{1080}$ $\bar{x}_{11} = \frac{1080}{2} = 540$	540 460 $\overline{1000}$ $\bar{x}_{12} = \frac{1000}{2} = 500$	480 400 $\overline{880}$ $\bar{x}_{13} = \frac{880}{2} = 440$	2960	$\bar{x}_{1.} = \frac{2960}{6} = 493.33$
1-day program		460 540 $\overline{1000}$ $\bar{x}_{21} = \frac{1000}{2} = 500$	560 620 $\overline{1180}$ $\bar{x}_{22} = \frac{1180}{2} = 590$	420 480 $\overline{900}$ $\bar{x}_{23} = \frac{900}{2} = 450$	3080	$\bar{x}_{2.} = \frac{3080}{6} = 513.33$
10-week course		560 600 $\overline{1160}$ $\bar{x}_{31} = \frac{1160}{2} = 580$	600 580 $\overline{1180}$ $\bar{x}_{32} = \frac{1180}{2} = 590$	480 410 $\overline{890}$ $\bar{x}_{33} = \frac{890}{2} = 445$	3230	$\bar{x}_{3.} = \frac{3230}{6} = 538.33$
Column totals		3240	3360	2670	9270　⟶　Overall Total	
Factor B Means		$\bar{x}_{.1} = \frac{3240}{6}$ $= 540$	$\bar{x}_{.2} = \frac{3360}{6}$ $= 560$	$\bar{x}_{.3} = \frac{2670}{6}$ $= 445$		$\bar{x} = \frac{9270}{18} = 515$

Step 4. Compute the sum of squares for interaction:

$$\text{SSAB} = r \sum_i \sum_j (\bar{x}_{ij} - \bar{x}_{i.} - \bar{x}_{.j} + \bar{\bar{x}})^2 \tag{12.28}$$

Step 5. Compute the sum of squares due to error:

$$\text{SSE} = \text{SST} - \text{SSA} - \text{SSB} - \text{SSAB} \tag{12.29}$$

Table 12.11 shows the data collected in the experiment, along with the various sums that will help us with the sum of squares computations. Using (12.26) to (12.29) we have the following sum of squares for the GMAT 2-factor factorial experiment:

Step 1. SST = $(500 - 515)^2 + (580 - 515)^2 + (540 - 515)^2 + \cdots +$
$(410 - 515)^2 = 82{,}450$

Step 2. SSA = $(3)(2)[(493.33 - 515)^2 + (513.33 - 515)^2 + (538.33 - 515)^2] =$
6100

Step 3. SSB = $(3)(2)[540 - 515)^2 + (560 - 515)^2 + (445 - 515)^2] = 45{,}300$

Step 4. SSAB = $2[(540 - 493.33 - 540 + 515)^2 + (500 - 493.33 - 560 + 515)^2 +$
$\cdots + (445 - 538.33 - 445 + 515)^2] = 11{,}200$

Step 5. SSE = $82{,}450 - 6100 - 45{,}300 - 11{,}200 = 19{,}850$

These sums of squares divided by their corresponding degrees of freedom, as shown in Table 12.12, provide the appropriate mean square values for testing the two main effects (preparation program and undergraduate college) and the interaction effect. The F ratio used to test for differences among preparation programs is 1.38. The critical F value at $\alpha = .05$ (with 2 numerator degrees of freedom and 9 denominator degrees of freedom) is 4.26. With $F = 1.38$, we cannot reject the null hypothesis and must conclude that there is not a significant difference in the preparation provided by the 3 preparation programs. However, for the undergraduate college effect, $F = 10.27$ exceeds the critical F value of 4.26. Thus the analysis of variance results allow us to conclude that there is a difference in GMAT test scores among the three undergraduate colleges; that is, the three undergraduate colleges do not provide the same preparation for performance on the GMAT. Finally, the interaction F value of $F = 1.27$ (critical F value = 3.63 at $\alpha = .05$) means that we cannot identify a significant interaction effect. Thus there is no reason to believe that the 3

TABLE 12.12 ANOVA Table for the 2-Factor GMAT Study

Source of Variation	Sum of Squares	Degrees of Freedom	Mean Square	F
Factor A treatments	6,100	2	3,050	3050/2206 = 1.38
Factor B treatments	45,300	2	22,650	22,650/2206 = 10.27
Interaction	11,200	4	2,800	2800/2206 = 1.27
Error	19,850	9	2,206	
Total	82,450	17		

```
ANALYSIS OF VARIANCE   GMAT

SOURCE         DF      SS       MS
FACTOR A       2      6100     3050
FACTOR B       2     45300    22650
INTERACTION    4     11200     2800
ERROR          9     19850     2206
TOTAL         17     82450
```

■■■ FIGURE 12.6   **Computer Output for Two-Factor Design**

preparation programs differ in their ability to prepare students from the different colleges for the GMAT.

Undergraduate college was found to be a significant factor. Checking the calculations in Table 12.11 we see that the sample means are as follows: business students $\bar{x}_1 = 540$, engineering students $\bar{x}_2 = 560$, and arts and sciences students $\bar{x}_3 = 445$. Tests on individual treatment means can be conducted; yet after reviewing the 3 sample means we would anticipate no difference in preparation for business and engineering graduates. However, the arts and sciences students appear to be significantly less prepared for the GMAT than students in the other colleges. Perhaps this observation will lead the Dean of the College of Arts and Sciences to consider other options for assisting these students in preparing for graduate management admission tests.

Because of the computational effort involved in any modest to large-size factorial experiment, the computer usually plays an important role in making and summarizing the analysis of variance computations. The computer printout for the analysis of variance of the GMAT 2-factor factorial experiment is shown in Figure 12.6.

■■■ EXERCISES

26. A mail-order catalog firm designed a factorial experiment to test the effect of the size of a magazine advertisement and the advertisement design on the number of catalog requests received (1000s). Three advertising designs and two different-size advertisements were considered. The data obtained are shown:

		Size of Advertisement	
		Small	*Large*
	A	8 12	12 8
Design	B	22 14	26 30
	C	10 18	18 14

Use the ANOVA procedure for factorial designs to test for any significant effects due to type of design, size of advertisement, or interaction. Use $\alpha = .05$.

27. An amusement park has been studying methods for decreasing the waiting time on rides by loading and unloading riders more efficiently. Two alternative loading/ unloading methods have been proposed. To account for potential differences due to the type of ride and the possible interaction between the method of loading and unloading and the type of ride, a factorial experiment was designed. Using the data shown below, test for any significant effect due to the loading and unloading method, the type of ride, and interaction. Use $\alpha = .05$.

| | **Type of Ride** | | |
	Roller Coaster	*Screaming Demon*	*Log Flume*
Method 1	41	52	50
	43	44	46
Method 2	49	50	48
	51	46	44

Waiting Time (minutes)

28. The calculations for a factorial experiment involving 4 levels of factor A, 3 levels of factor B, and 3 replications resulted in the following data: SST = 280, SSA = 26, SSB = 23, SSAB = 175. Set up the ANOVA table and test for any significant main effect and any interaction effect. Use $\alpha = .05$.

■■■ SUMMARY

In this chapter we have discussed several types of experimental designs that can be used to test for differences among means of several populations or treatments. Specifically, we introduced the single-factor completely randomized, the randomized block, and the 2-factor factorial experimental designs. The completely randomized design and the randomized block designs were used to draw conclusions about differences in the means of a single factor. The primary purpose of the blocking in the randomized block design was to remove extraneous sources of variation from the error term. This blocking provided a better estimate of the error variance and a better test to determine whether or not the population or treatment means of the single factor differed significantly.

Although the various experimental designs required different formulas and computations, we showed that the basis for the statistical tests used is the development of independent estimates of the population variance σ^2. In the single factor case one estimator, MSTR, is based upon the variation between the treatments. This value provides an unbiased estimate of σ^2 only if the means $\mu_1, \mu_2, \ldots, \mu_k$ are all equal. A second estimator, MSE, is based upon the variation of the observations within each sample; this estimator will always provide an unbiased estimate of σ^2. By computing $F = \text{MSTR}/\text{MSE}$ and using the F distribution, we developed a rejection rule for determining whether or not to reject the null hypothesis that the treatment means are equal. In all the experimental designs considered, the partitioning of the sum of squares and degrees of freedom into their various sources enabled us to compute the appropriate values for making the analysis of variance calculations and tests.

Whenever an analysis of variance conclusion results in the rejection of the equal-means hypothesis (H_0), we may want to consider testing for a difference between the individual treatment means. We showed how the t distribution test could be used to compare two treatment means. However, we warned against indiscriminate use of this testing procedure because of the increasing probability of making a Type I error. By simultaneously making several tests for individual differences, the probability of erroneously claiming that a difference exists increases. Several specialized tests, discussed in more advanced tests, are available if multiple comparisons of the treatment means are to be considered.

■■ GLOSSARY

Analysis of variance (ANOVA) procedure A statistical approach for determining whether or not the means of several different populations are equal.

Factor Another word for the variable of interest in an ANOVA procedure.

Treatment Different levels of a factor.

Single-factor experiment An experiment involving only one factor with k populations or treatments.

Experimental units The objects of interest in the experiment.

Completely randomized design An experimental design where the treatments are randomly assigned to the experimental units.

Mean square The sum of squares divided by its corresponding degrees of freedom. This quantity is used in the F ratio to determine if significant differences among means exist or not.

ANOVA table A table used to summarize the analysis of variance computations and results. It contains columns showing the source of variation, the degrees of freedom, the sum of squares, the mean squares, and the F values.

Partitioning The process of allocating the total sum of squares and degrees of freedom into the various components.

Blocking The process of using the same or similar experimental units for all treatments. The purpose of blocking is to remove a source of variation from the error term and hence provide a more powerful test for a difference in population or treatment means.

Randomized block design An experimental design employing blocking. The experimental unit(s) within a block are randomly ordered for the treatments.

Factorial experiments An experimental design that permits statistical conclusions about two or more factors. All levels of each factor are considered with all levels of the other factors in order to specify the experimental conditions for the experiment.

Replication The number of times each experimental condition is repeated in an experiment. It is the sample size associated with each treatment combination.

Interaction The response produced when the levels of one factor interact with the levels of another factor in influencing the response variable.

■■ KEY FORMULAS

Completely Randomized Designs

Sum of Squares About the Mean

$$SST = \sum_i \sum_j (x_{ij} - \bar{\bar{x}})^2 \tag{12.15}$$

Sum of Squares Between Treatments

$$SSTR = \sum_i n_i (\bar{x}_i - \bar{\bar{x}})^2 \tag{12.16}$$

Sum of Squares Due to Error

$$SSE = SST - SSTR \tag{12.14}$$

The F Value

$$F = \frac{\text{MSTR}}{\text{MSE}} \tag{12.17}$$

Randomized Block Designs

Total Sum of Squares

$$\text{SST} = \sum_i \sum_j (x_{ij} - \bar{\bar{x}})^2 \tag{12.19}$$

Sum of Squares Between Treatments

$$\text{SSTR} = b \sum_i (\bar{x}_{i.} - \bar{\bar{x}})^2 \tag{12.20}$$

Sum of Squares Due to Blocks

$$\text{SSB} = k \sum_j (\bar{x}_{.j} - \bar{\bar{x}})^2 \tag{12.21}$$

Sum of Squares Due to Error

$$\text{SSE} = \text{SST} - \text{SSTR} - \text{SSB} \tag{12.22}$$

Factorial Experiments

Total Sum of Squares

$$\text{SST} = \sum_i \sum_j \sum_k (x_{ijk} - \bar{\bar{x}})^2 \tag{12.25}$$

Sum of Squares for Factor A

$$\text{SSA} = br \sum_i (\bar{x}_{i.} - \bar{\bar{x}})^2 \tag{12.26}$$

Sum of Squares for Factor B

$$\text{SSB} = ar \sum_j (\bar{x}_{.j} - \bar{\bar{x}})^2 \tag{12.27}$$

Sum of Squares for the Interaction

$$\text{SSAB} = r \sum_i \sum_j (\bar{x}_{ij} - \bar{x}_{i.} - \bar{x}_{.j} + \bar{\bar{x}})^2 \tag{12.28}$$

Sum of Squares Due to Error

$$\text{SSE} = \text{SST} - \text{SSA} - \text{SSB} - \text{SSAB} \tag{12.29}$$

■■ SUPPLEMENTARY EXERCISES

29. In your own words explain what the ANOVA procedure is used for.

30. What has to be true in order for MSTR to provide a good estimate of σ^2? Explain.

31. Why do we assume that the populations sampled all have the same variance when we apply the ANOVA procedure?

32. Explain why MSTR and MSE provide two independent estimates of σ^2.

33. Explain why MSTR provides an inflated estimate of σ^2 when the population means are not the same.

34. A simple random sample of the asking price ($1000s) of four houses currently for sale in each of two residential areas resulted in the following data:

Area 1	Area 2
92	90
89	102
98	96
105	88

a. Use the procedure developed in Chapter 10 and test if the mean asking price is the same in both areas. Use $\alpha = .05$.

b. Use the ANOVA procedure to test if the mean asking price is the same. Compare your analysis with part (a). Use $\alpha = .05$.

35. Suppose that in Exercise 34 data were collected for another residential area. The asking prices for the simple random sample from the third area were as follows: $81,000, $86,000, $75,000, and $90,000. Is the mean asking price for all three areas the same? Use $\alpha = .05$.

36. An analysis of the number of units sold by 10 salespersons in each of 4 sales territories resulted in the following data:

	Sales Territory			
	1	*2*	*3*	*4*
Number of salespersons	10	10	10	10
Average number sold ($\bar{x}$)	130	120	132	114
Sample variance (s^2)	72	64	69	67

Test at the $\alpha = .05$ level if there is any significant difference in the mean number of units sold in the 4 sales territories.

37. Suppose that in Exercise 36 the number of salespersons in each territory was as follows: $n_1 = 10$, $n_2 = 12$, $n_3 = 10$, and $n_4 = 15$. Using the same data for $\bar{x}$ and s^2 as given in Exercise 36, test at the $\alpha = .05$ level if there is any significant difference in the mean number of units sold in the four sales territories.

38. Three different assembly methods have been proposed for a new product. In order to determine which assembly method results in the greatest number of parts produced per hour, 30 workers were randomly selected and assigned to use one of the proposed methods. The number of units produced by each worker is given below:

Method A	97	73	93	100	73	91	100	86	92	95
Method B	93	100	93	55	77	91	85	73	90	83
Method C	99	94	87	66	59	75	84	72	88	86

Use these data and test to see if the mean number of parts produced with each method is the same. Use $\alpha = .05$.

39. In order to test to see if there is any significant difference in the mean number of units produced per week by each of three production methods, the following data were collected:

Method 1	Method 2	Method 3
58	52	48
64	63	57
55	65	59
66	58	47
67	62	49

At the $\alpha = .05$ level of significance is there any difference in the means for the three methods?

40. Pappashales Restaurant is considering introducing a new specialty sandwich. For a determination of the effect of sandwich price on sales, the new sandwich was test marketed at three prices in selected company restaurants. The following data, in terms of the number of sandwiches sold per day, were obtained:

$1.49	$1.79	$1.99
925	910	860
850	845	935
930	905	820
955	860	845

At the $\alpha = .05$ level of significance is there any difference in the mean number of sandwiches sold per day for the 3 prices? What should management of Pappashales do?

41. Hargreaves Automotive Parts, Inc., would like to compare the mileage for 4 different types of brake linings. Thirty linings of each type were produced and placed on a fleet of rental cars. The number of miles that each brake lining lasted until it no longer met the required federal safety standard was recorded, and an average value was computed for each type of lining. The following data were obtained:

	Sample Size	Sample Mean	Standard Deviation
Type A	30	32,000	1450
Type B	30	27,500	1525
Type C	30	34,200	1650
Type D	30	30,300	1400

Use these data and test to see if the corresponding population means are equal. Use =.05.

42. A manufacturer of batteries for electronic toys and calculators is considering 3 new battery designs. An attempt was made to determine if the mean lifetime in hours is the same for each of the 3 designs. The following data were collected:

Design A	Design B	Design C
78	112	115
98	99	101
88	101	100
96	116	120

Test to see if the population means are equal. Use $\alpha = .05$.

43. A study was conducted to investigate the browsing activity by shoppers (*Journal of the Academy of Marketing Science*, Winter 1989). Each shopper was initially classified as

non-browser, light browser, or heavy browser. For each shopper in the study a measure was obtained to determine how comfortable the shopper was in a store. Higher scores indicated greater comfort. Suppose the following data is from a related study.

Non-Browser	Light Browswer	Heavy Browser
4	5	5
5	6	7
6	5	5
3	4	7
3	7	4
4	4	6
5	6	5
4	5	7

a. Using $\alpha = .05$, test for differences among comfort levels for the three types of browsers.

b. Use the procedure described in Section 12.5 for testing for differences between individual treatment means to compare the comfort levels of non-browsers and light browsers. Use $\alpha = .05$. What is your conclusion.

44. In Exercise 40, would it make sense to do a test on individual treatment means as described in Section 12.5? Explain.

45. A research firm tests the miles per gallon characteristics of three brands of gasoline. Because of different gasoline performance characteristics in different brands of automobiles, five brands of automobiles are selected and treated as blocks in the experiment. That is, each brand of automobile is tested with each type of gasoline. The results of the experiment are shown:

		Blocks: Automobiles				
		A	B	C	D	E
	I	18	24	30	22	20
Gasoline Brands	II	21	26	29	25	23
	III	20	27	34	24	24

Miles per gallon

With $\alpha = .05$, is there a significant difference in the mean miles per gallon characteristics of the 3 brands of gasoline?

46. Analyze the experimental data provided in Exercise 45 using the ANOVA procedure for completely randomized designs. Compare your findings with those obtained in Exercise 45. What is the advantage of attempting to remove the block effect?

47. The following data were obtained for a randomized block design involving three treatments and four blocks: SST = 148, SSTR = 84, SSB = 50. Set up the ANOVA table and test for any significant differences. Use $\alpha = .05$.

48. Three different road-repair compounds were tested at four different highway locations. At each location, three sections of the road were repaired, with each section using one of the three compounds. Data were then collected on the number of days of traffic usage before additional repair was required. These data are as follows.

		Location			
		1	2	3	4
	A	99	73	85	103
Compound	B	82	72	85	97
	C	81	79	82	86

With $\alpha = .01$ test to see if there is a significant difference in the compounds.

49. A factorial experiment was designed to test for any significant differences in the time needed to perform English to foreign language translations using two computerized language translators. Since the type of language translated was also considered a significant factor, translations were made using both systems for three different languages: Spanish, French, and German. Use the data shown:

	Language		
	Spanish	*French*	*German*
System 1	8	10	12
	12	14	16
System 2	6	14	16
	10	16	22

Translation Time (hours)

Test for any significant differences due to language translator, type of language, and interaction. Use $\alpha = .05$.

50. A manufacturing company designed a factorial experiment in order to determine if the number of defective parts produced by two machines differed. Since it was believed that the number of defective parts produced also depends upon whether or not raw material needed by the machine was loaded manually or using an automatic feed system, use the following data to test for any significant effect due to machine, loading system, and interaction. Use $\alpha = .05$.

	Loading System	
	Manual	*Automatic*
Machine 1	30	30
	34	26
Machine 2	20	24
	22	28

Number of Defective Parts

▬ COMPUTER EXERCISE

As part of a long-term study of individuals 65 years of age or older, sociologists and physicians at Upstate Medical Center recently conducted an experiment to study the relationship between geographic location and depression levels. A sample of 60 individuals was selected; 20 of the individuals were lifetime residents of Florida, 20 were lifetime residents of New York, and 20 were lifetime residents of North Carolina. To account for any possible effects that might be due to the health status of the individual, 50% of the subjects that were sampled from each state had a chronic health condition (arthritis, hypertension, hearing loss, or heart ailment), and 50% did not have a chronic health condition.

Each individual who participated in the study was given a standardized test in order to measure depression. Higher test scores indicate higher levels of depression. The data are shown. A code of 1 is used to indicate that the subject had a chronic health condition and a code of 2 is used to indicate that the subject did not have a chronic health condition.

Florida		New York		North Carolina	
Score	Condition	Score	Condition	Score	Condition
7	2	9	2	14	1
10	1	19	1	14	1
12	1	10	1	20	1
19	1	11	2	15	1
11	1	8	2	6	2
2	2	17	1	5	2
4	2	15	1	12	2
13	1	11	1	8	2
14	1	9	2	12	1
17	1	6	2	9	2
8	2	7	2	16	1
5	2	12	1	12	1
11	1	20	1	7	2
18	1	10	2	13	1
7	2	13	1	6	2
4	2	10	2	15	1
3	2	12	1	4	2
4	2	15	1	6	2
1	2	4	2	6	2
9	1	10	2	10	1

▬ QUESTIONS

1. Use numerical and graphical measures to summarize the data.

2. Ignoring the data on health condition, does there appear to be a significant difference in depression levels for residents of the three states?

3. Considering health condition as a block, does there appear to be a significant difference in depression levels for residents of the three states? Explain.

APPENDIX 12.1

Computational Procedure for a Completely Randomized Design

▬▬▬

The following step-by-step procedure is designed to ease the burden in computing the appropriate sums of squares for completely randomized designs. The formulas shown below can be applied to both balanced and unbalanced designs.

Notation

x_{ij} = value of the jth observation under treatment i

T_i = sum of all observations for treatment i

T = sum of all observations

n_i = sample size for the ith treatment

n_T = total sample size for the experiment

Procedure

Step 1. Compute the sum of squares about the mean (SST):

$$SST = \sum_i \sum_j x_{ij}^2 - \frac{T^2}{n_T}$$

Step 2. Compute the sum of squares due to treatments (SSTR):

$$SSTR = \sum_i \frac{T_i^2}{n_i} - \frac{T^2}{n_T}$$

Step 3. Compute the sum of squares due to error (SSE):

$$SSE = SST - SSTR$$

Example

Using this computational procedure with the GMAT data in Table 12.1, we obtain the following results:

Step 1. $SST = 4,228,285 - \dfrac{(7935)^2}{15} = 30,670$

Step 2. $SSTR = \dfrac{(2545)^2}{5} + \dfrac{(2630)^2}{5} + \dfrac{(2760)^2}{5} - \dfrac{(7935)^2}{15} = 4690$

Step 3. $SSE = SST - SSTR = 30,670 - 4690 = 25,980$

APPENDIX 12.2

Computational Procedure for a Randomized Block Design

The following step-by-step procedure is designed to help in computing the appropriate sums of squares for randomized block designs.

Notation

x_{ij} = value of the observation under treatment i in block j

$T_{i\cdot}$ = the total of all observations in treatment i

$T_{\cdot j}$ = the total of all observations in block j

T = the total of all observations

k = the number of treatments

b = the number of blocks

n_T = the total sample size ($n_T = kb$)

Procedure

Step 1. Compute the total sum of squares (SST):

$$\text{SST} = \sum_i \sum_j x_{ij}^2 - \frac{T^2}{n_T}$$

Step 2. Compute the sum of squares between treatments (SSTR):

$$\text{SSTR} = \frac{\sum_i T_{i\cdot}^2}{b} - \frac{T^2}{n_T}$$

Step 3. Compute the sum of squares due to blocks (SSB):

$$\text{SSB} = \frac{\sum_j T_{\cdot j}^2}{k} - \frac{T^2}{n_T}$$

Step 4. Compute the sum of squares due to error (SSE):

$$\text{SSE} = \text{SST} - \text{SSTR} - \text{SSB}$$

Example

For the air traffic controller data in Table 12.5, these steps lead to the sum of squares:

Step 1. $\quad \text{SST} = 3598 - \dfrac{(252)^2}{18} = 70$

Step 2. $\quad \text{SSTR} = \dfrac{(81)^2 + (78)^2 + (93)^2}{6} - \dfrac{(252)^2}{18} = 21$

Step 3. $\quad \text{SSB} = \dfrac{(48)^2 + (42)^2 + \cdots + (39)^2}{3} - \dfrac{(252)^2}{18} = 30$

Step 4. $\quad \text{SSE} = 70 - 21 - 30 = 19$

APPENDIX 12.3

Computational Procedure for Two-Factor Factorial Design

Notation

x_{ijk} = observation corresponding to the kth replicate taken from treatment i of factor A and treatment j of factor B

$T_{i.}$ = total of all observations in treatment i (factor A)

$T_{.j}$ = total of all observations in treatment j (factor B)

T_{ij} = total of all observations in the combination of treatment i (factor A) and treatment j (factor B)

T = total of all observations

a = number of levels of factor A

b = number of levels of factor B

r = number of replications

n_T = total number of observations; $n_T = abr$

Procedure

Step 1. Compute the total sum of squares:

$$SST = \sum_i \sum_j \sum_k x_{ijk}^2 - \frac{T^2}{n_T}$$

Step 2. Compute the sum of squares for factor A:

$$SSA = \frac{\sum_i T_{i.}^2}{br} - \frac{T^2}{n_T}$$

Step 3. Compute the sum of squares for factor B:

$$SSB = \frac{\sum_j T_{.j}^2}{ar} - \frac{T^2}{n_T}$$

Step 4. Compute the sum of squares for the interaction:

$$\text{SSAB} = \frac{\displaystyle\sum_i \sum_j T_{ij}^2}{r} - \frac{T^2}{n_T} - \text{SSA} - \text{SSB}$$

Step 5. Compute the sum of squares due to error:

$$\text{SSE} = \text{SST} - \text{SSA} - \text{SSB} - \text{SSAB}$$

Example

For the GMAT data in Table 12.11, these steps lead to the following sum of squares:

Step 1. $\text{SST} = 4{,}856{,}500 - \dfrac{(9270)^2}{18} = 82{,}450$

Step 2. $\text{SSA} = \dfrac{(2960)^2 + (3080)^2 + (3230)^2}{6} - \dfrac{(9270)^2}{18} = 6100$

Step 3. $\text{SSB} = \dfrac{(3240)^2 + (3360)^2 + (2670)^2}{6} - \dfrac{(9270)^2}{18} = 45{,}300$

Step 4. $\text{SSAB} = \dfrac{(1080)^2 + (1000)^2 + \cdots + (890)^2}{2} - \dfrac{(9270)^2}{18}$

$\qquad\qquad - 6100 - 45{,}300 = 11{,}200$

Step 5. $\text{SSE} = 82{,}450 - 6100 - 45{,}300 - 11{,}200 = 19{,}850$

APPLICATION

Burke Marketing
Services, Inc.*

Cincinnati, Ohio

For half a century Burke Marketing Services has been a leader in solving marketing problems; during this time, Burke has become the preeminent, most experienced custom survey research firm in the industry. Burke writes more proposals, on more projects, every day than any other research company in the world. Supported by state-of-the-art technology, four Burke divisions offer a wide range of research capabilities, providing answers to nearly every marketing question. The divisions are BASES, Test Marketing Group, Burke Marketing Research, and Consulting and Analytical Services.

The BASES system translates key attitudinal measures into sales estimates at four critical decision points in the new product development process: when the product is just a concept; when a prototype has been developed; at the stage when final packaging and so on has been developed; and in the test market or rollout stage. In effect, BASES answers the toughest new product question of them all: "Does this product have the potential to meet the company's minimum business requirements?"

The Test Marketing Group consists of two operating units: AdTel and Market Audits. AdTel conducts advertising tests using one-of-a kind cable TV systems in 6 test markets across the nation. The unique AdTel construction allows a company to develop an alternative media plan and test it alongside its current plan. This service allows a company to evaluate new product viability, test new packaging, compare strategies and schedules, and test weights and promotions.

Market Audits provides consumer products manufacturers with a complete record of their product's performance from which they may make marketing decisions. Each year Market Audits conducts over 100 in-store tests, primarily in food, drug, and mass merchandiser retail outlets. Market Audits offers a full complement of in-store research: minimarket testing, controlled store tests, standard retail sales audits, distribution checks, and observation studies.

Burke Marketing Research provides a complete range of ad hoc research services—television and radio commercial tests, magazine and newspaper advertising tests, concept generation and evaluation studies, package evaluations, claim support, and tracking studies, to name just a few. The Consulting and Analytical Services Division of Burke provides consulting in matters of marketing problems, research design, and interpretation of results to a wide range of clients. Consulting and Analytical Services provides analytic support to all Burke divisions as well as to Burke clients.

*The authors are indebted to Dr. Ronald Tatham of Burke Marketing Services for providing this application.

■■■ STATISTICAL ANALYSIS AT BURKE MARKETING SERVICES, INC.

A wide range of statistical and other analytical techniques are used to facilitate research design and data interpretation. A simple division of research into experimental and survey research facilitates discussion.

■■ *Experimental research:* BASES, Test Marketing Group, and Burke Marketing Research all conduct experiments and use regression, analysis of variance, and analysis of covariance to estimate the effect of the experiment. This is about 30% of the research conducted by the company.

■■ *Survey research:* The results of survey research are often examined and interpretation facilitated through the use of such techniques as multivariate exploratory methods, regression, analysis of variance, and statistical tests of differences.

Almost all research at Burke involves statistical analysis of some type or the provision of statistics (standard errors, etc.) to the client to facilitate later analysis.

■■■ AN EXPERIMENTAL DESIGN APPLICATION

A major manufacturer of children's dry cereal continually strives for formula improvements that offer a better tasting product. To maintain confidentiality we will refer to the manufacturer as the Anon Company. Burke's Consulting and Analytical Services Division was retained by Anon Company to evaluate potential new formulations designed to improve the taste of the cereal. The 4 key ingredients that Anon Company thought would enhance taste were

1. Ratio of wheat to corn in the cereal flake (2 levels)

2. Types of sweetness (3 types)

3. Flavor bits (present or absent)

4. Manufacturing cooking time (short, long)

The sweeteners were sugar, honey, and an artificial sweetener additive. The flavor bits were small particles with a fruit taste that could be included in the cereal. The cooking time influences the "crunchiness" of the cereal.

Using all combinations of the four ingredients would lead to $2 \times 3 \times 2 \times 2 = 24$ different cereal formulations. If the 24 cereal formulations were tested independently,

Burke's in-store research provides valuable information for clients.

TABLE 12A.1 **Experimental Design for Cereal Taste Test. There are 18 Blocks Each Containing a Different Set of 3 Cereals**

	Cereal Formulations								
Blocks	C1	C2	C3	C4	C5	C6	C7	C8	C9
1	x			x			x		
2	x			x				x	
3	x				x				x
4	x				x		x		
5	x					x		x	
6	x					x			x
7		x		x			x		
8		x		x				x	
9		x			x				x
10		x			x		x		
11		x				x		x	
12		x				x			x
13			x	x			x		
14			x	x				x	
15			x		x				x
16			x		x		x		
17			x			x		x	
18			x			x			x

the sample sizes would have to be very large to provide sufficient power to measure the effects of the various ingredients on taste perception. In order to use fewer respondents, thus saving time and money, a variation of the factorial experiment was used; it is called a fractional factorial design. Only 9 of the 24 cereal formulations were used, but they were selected in such a way that the best possible measure of the independent effect of the 4 ingredients could be obtained.

The nine cereal formulations were then placed in blocks of size 3. This meant that each child in the sample would taste test 3 different cereals (the order of tasting was randomized for each child). There were 18 blocks, each involving a different combination of three of the 9 cereals in the test. Table 12A.1 shows this experimental design. Each block of three cereals was then tested by six children, resulting in an overall sample size of $6 \times 18 = 108$ children. To maintain control over the sampling Burke prepackaged the cereals into 108 packages of 3 cereals each. This ensured that the random sample of 108 children would have the desired characteristics.

■ RESULTS

An analysis of the sample results indicated the following:

■ The flake composition and sweetener type were *very* influential in taste evaluation.
■ The flavor bits actually *detracted* from the perception of the taste of the cereal.
■ The cooking time had *no* impact on the taste.

Note that managerial considerations such as time, cost, and control over the experimenters influenced the choice of the experimental design. These are often as important as statistical considerations.

CHAPTER 13

Simple Linear Regression and Correlation

M anagerial decisions are often based upon the relationship between two or more variables. For example, after considering the relationship between advertising expenditures and subsequent sales volume, a marketing manager might attempt to predict sales volume from a known level of advertising expenditures. In another case, public utilities often use the relationship between temperature and electricity use to predict demand for electricity. Sometimes, a manager will rely on intuition to judge how two variables are related, but a more objective approach is to collect data on the two variables and then use statistical procedures to determine how the variables are related.

Regression analysis is a statistical procedure that can be used to develop a mathematical equation showing how variables are related. In regression terminology the variable which is being predicted by the mathematical equation is called the *dependent* variable. The variable or variables being used to predict the value of the dependent variable are called the *independent* variables. In the sales-advertising expenditures example, the marketing manager's desire to predict sales volume would suggest making sales volume the dependent variable for the analysis. Advertising expenditure would be the independent variable used to predict the sales volume. Common statistical notation is to use y to denote the dependent variable and x to denote the independent variable.

In this chapter we consider the simplest type of regression: situations involving one independent and one dependent variable for which the relationship between the variables is approximated by a straight line. This is called *simple linear regression*. Regression analysis involving two or more independent variables is called multiple regression analysis; multiple regression is covered in Chapters 14 and 15.

Another topic discussed in this chapter is correlation. In correlation analysis we are not concerned with identifying a mathematical equation relating an independent and dependent variable; we are concerned only with determining the extent to which the variables are linearly related. Correlation analysis is a procedure for making this determination and, if such a relationship exists, providing a measure of the relative strength of the relationship.

We caution the reader before beginning this chapter that neither regression nor correlation analysis can be interpreted as establishing cause-effect relationships. Regression and correlation analyses can indicate only how or to what extent variables are associated with each other. Any conclusions about a cause and effect relationship must be based on the *judgment of the analyst.*

13.1 ■ THE LEAST SQUARES METHOD

In this section we show how the least squares method can be used to develop a linear equation relating two variables. As an illustration, let us consider the problem currently being faced by the management of Armand's Pizza Parlors, a chain of Italian-food restaurants located in a five-state area. One of the most successful types of locations for Armand's has been near a college campus.

Prior to opening a new restaurant Armand's management requires an estimate of annual sales revenues. Such an estimate is used in planning the appropriate restaurant capacity, making initial staffing decisions, and deciding whether or not the potential revenues justify the cost of the operation.

Suppose management believes that the size of the student population on the nearby campus is related to the annual sales revenues. On an intuitive basis, management believes restaurants located near larger campuses generate more revenue than those near small campuses. To evaluate the relationship between student population (x) and annual sales (y), Armand's collected data from a sample of 10 of its restaurants located near college campuses. The data for these 10 observations are summarized in Table 13.1. For the ith observation, x_i represents the value of the independent variable and y_i represents the value of the dependent variable. We see that restaurant 1, with $x_1 = 2$ and $y_1 = 58$, is located near a campus with 2000 students and has annual sales of $58,000;

TABLE 13.1 Data on Student Population and Annual Sales for 10 Armand's Restaurants

Restaurant i	Student Population (1000s) x_i	Annual Sales ($1000s) y_i
1	2	58
2	6	105
3	8	88
4	8	118
5	12	117
6	16	137
7	20	157
8	20	169
9	22	149
10	26	202

restaurant 2 is located near a campus with 6000 students and has annual sales of $105,000; and so on.

Figure 13.1 shows graphically the data presented in Table 13.1. The size of the student population is shown on the horizontal axis, with annual sales on the vertical axis. A graph such as this is known as a *scatter diagram*. Values for the independent variable are shown on the horizontal axis and the corresponding values for the dependent variable are shown on the vertical axis. The scatter diagram provides an overview of the data and enables us to draw preliminary conclusions about a possible relationship between the variables.

What preliminary conclusions can we draw from Figure 13.1? It appears that low sales volumes are associated with small student populations and higher sales volumes are associated with larger student populations. It also appears that the relationship between the two variables can be approximated by a straight line. In Figure 13.2 we have drawn a straight line through the data that appears to provide a good linear approximation of the relationship between the variables. The equation for this line is $y = 50 + 5.5\,x$. The y-intercept, the point at which the line intersects the y-axis, is 50 and the slope, the amount of change in y per unit change in x, is 5.5.

Clearly, there are many different straight lines that we could have drawn in Figure 13.2 to represent the relationship between x and y. The question is, Which of the straight lines that could be drawn "best" represents the relationship?

▉▉ FIGURE 13.1 **Scatter Diagram of Annual Sales versus Student Population**

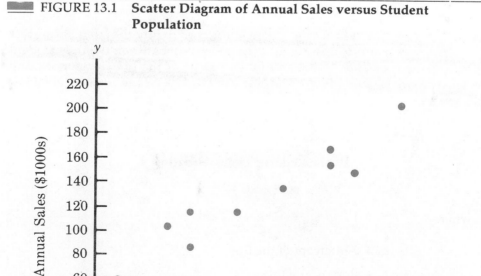

Student Population (1000s)

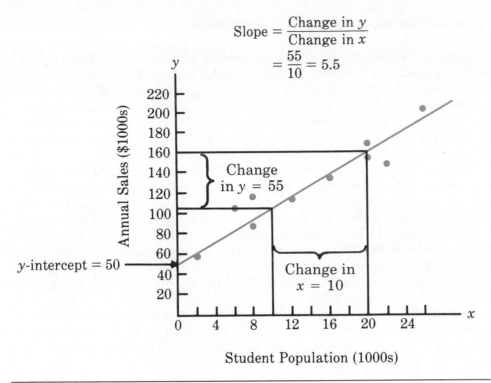

$$\text{Slope} = \frac{\text{Change in } y}{\text{Change in } x}$$

$$= \frac{55}{10} = 5.5$$

▬ FIGURE 13.2 **Straight-Line Approximation**

The *least squares method* is a procedure that is used to find the straight line that provides the best approximation for the relationship between the independent and dependent variables. We refer to the equation of the line developed using the least squares method as the *estimated regression line,* or the *estimated regression equation.*

Estimated Regression Equation

$$\hat{y} = b_0 + b_1 x \tag{13.1}$$

where

b_0 = y-intercept of the line

b_1 = slope of the line

$\hat{y}$ = estimated value of the dependent variable

For any particular value of the independent variable, x_i, the corresponding value on the estimated regression line is denoted by $\hat{y}_i = b_0 + b_1 x_i$.

Application of the least squares method provides the values of b_0 and b_1 that make the sum of the squared deviations between the observed values of the dependent variable (y_i) and the estimated values of the dependent variable ($\hat{y}_i$) a minimum. The criterion for the least squares method is given by (13.2).

Least Squares Criterion

$$\min \Sigma \, (y_i - \hat{y}_i)^2 \qquad\qquad (13.2)$$

where

y_i = observed value of the dependent variable for the ith observation

$\hat{y}_i$ = estimated value of the dependent variable for the ith observation.

Using differential calculus, it can be shown (see the appendix to this chapter) that the values of b_0 and b_1 that minimize (13.2) can be found using (13.3) and (13.4).

Slope and y-Intercept for the Estimated Regression Equation

$$b_1 = \frac{\Sigma \, (x_i - \bar{x})(y_i - \bar{y})}{\Sigma \, (x_i - \bar{x})^2} = \frac{\Sigma x_i y_i - (\Sigma x_i \Sigma y_i)/n}{\Sigma x_i^2 - (\Sigma x_i)^2/n} \qquad\qquad (13.3)$$

$$b_0 = \bar{y} - b_1 \bar{x} \qquad\qquad (13.4)$$

where

x_i = value of the independent variable for the ith observation

y_i = value of the dependent variable for the ith observation

$\bar{x}$ = mean value for the independent variable

$\bar{y}$ = mean value for the dependent variable

n = total number of observations

The second form of (13.3) is normally used for computing b_1 with a calculator because it avoids the tedious calculations involving the computation of each ($x_i - \bar{x}$) and ($y_i - \bar{y}$). However, to avoid rounding errors, it is best to carry as many significant digits as possible in the calculation; we recommend carrying at least four significant digits.

Some of the calculations necessary to develop the least squares estimated regression equation for the Armand's Pizza problem are shown in Table 13.2. In this example there are 10 restaurants, or observations; hence $n = 10$. Using (13.3) and (13.4), we can now compute the slope and intercept of the estimated

TABLE 13.2   **Calculations Necessary to Develop the Least Squares Estimated Regression Equation for Armand's Pizza**

Restaurant i	x_i	y_i	$x_i y_i$	x_i^2
1	2	58	116	4
2	6	105	630	36
3	8	88	704	64
4	8	118	944	64
5	12	117	1404	144
6	16	137	2192	256
7	20	157	3140	400
8	20	169	3380	400
9	22	149	3278	484
10	26	202	5252	676
Totals	140	1300	21,040	2528
	Σx_i	Σy_i	$\Sigma x_i y_i$	Σx_i^2

regression equation for Armand's Restaurants. The calculation of the slope (b_1) proceeds as follows:

$$b_1 = \frac{\Sigma x_i y_i - (\Sigma x_i \Sigma y_i)/n}{\Sigma x_i^2 - (\Sigma x_i)^2/n}$$

$$= \frac{21,040 - (140)(1300)/10}{2528 - (140)^2/10}$$

$$= \frac{2840}{568}$$

$$= 5$$

The calculation of the y intercept (b_0) is as follows:

$$\bar{x} = \frac{\Sigma x_i}{n} = \frac{140}{10} = 14$$

$$\bar{y} = \frac{\Sigma y_i}{n} = \frac{1300}{10} = 130$$

$$b_0 = \bar{y} - b_1 \bar{x}$$

$$= 130 - 5(14)$$

$$= 60$$

Thus the estimated regression equation found by using the method of least

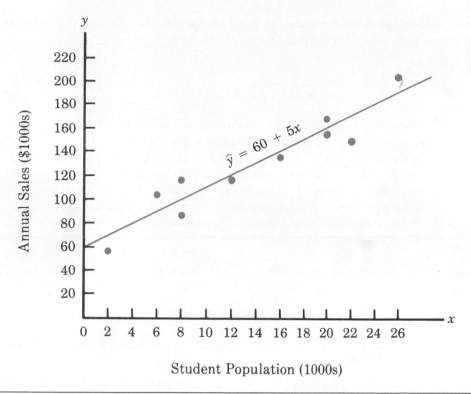

Graph of the Estimated Regression Equation for Armand's Pizza: $\hat{y} = 60 + 5x$

squares is

$$\hat{y} = 60 + 5x$$

In Figure 13.3 we show the graph of this equation.

The slope of the estimated regression equation ($b_1 = 5$) is positive, implying that as student population increases, annual sales increase. In fact we can conclude (since sales are measured in $1000s and student population in 1000s) that an increase in the student population of 1000 is associated with an increase of $5000 in expected annual sales; that is, sales are expected to increase by $5.00 per student.

If we believe that the least squares estimated regression equation adequately describes the relationship between x and y, then it would seem reasonable to use the estimated regression equation to predict the value of y for a given value of x. For example, if we wanted to predict annual sales for a restaurant to be located near a campus with 16,000 students we would compute

$$\hat{y} = 60 + 5\,(16)$$

$$= 140$$

Hence we would predict sales of $140,000 per year. In the following sections we will discuss methods for assessing the appropriateness of using the estimated regression equation for estimation and prediction.

NOTES AND COMMENTS

The least squares method provides an estimated regression equation that minimizes the sum of squared deviations between the observed values of the dependent variable (y_i) and the estimated values of the dependent variable ($\hat{y}_i$). This is the least squares criterion for choosing the equation that provides the best fit. If some other criterion were used, such as minimizing the sum of the absolute deviations between y_i and $\hat{y}_i$, a different equation would be obtained. In practice, the least squares method is the most widely used approach.

▬ EXERCISES

1. The following data were collected regarding the height (inches) of women swimmers and their weight (pounds).

Height	68	64	62	65	66
Weight	132	108	102	115	128

a. Develop a scatter diagram for these data with height as the independent variable.
b. What does the scatter diagram developed in part a indicate about the relationship between the two variables?
c. Try to approximate the relationship between height and weight by drawing a straight line through the data.
d. Develop the estimated regression equation by computing the values of b_0 and b_1 using (13.3) and (13.4).
e. If a swimmer's height is 63 inches, what would you estimate her weight to be?

2. Given are five observations taken for two variables, x and y.

Observation i	x_i	y_i
1	2	25
2	3	25
3	5	20
4	1	30
5	8	16

a. Develop a scatter diagram for these data.
b. Use the method of least squares to compute an estimated regression equation for the data.

3. The following data were collected regarding the monthly starting salaries and the

grade point averages (GPA) for undergraduate students who had obtained a degree in political science.

GPA	Monthly Salary ($)
2.6	1100
3.4	1400
3.6	1800
3.2	1300
3.5	1600
2.9	1200

a. Develop a scatter diagram for these data with GPA as the independent variable.
b. What does the scatter diagram developed in (a) indicate about the relationship between the two variables?
c. Draw a straight line through the data to approximate a linear relationship between GPA and salary.
d. Use the least squares method to develop the estimated regression equation.
e. Predict the monthly starting salary for a student with a 3.0 GPA and for a student with a 3.5 GPA.

4. Eddie's Restaurants collected the following data on the relationship between advertising and sales at a sample of seven restaurants.

Advertising Expenditures ($1000s)	Sales ($1000s)
1.0	19.0
2.0	32.0
4.0	44.0
6.0	40.0
10.0	52.0
14.0	53.0
20.0	54.0

Develop a scatter diagram for these data with advertising expenditures as the independent variable and sales as the dependent variable. Does there appear to be a linear relationship?

5. Shown are some data that a sales manager has collected concerning annual sales and years of experience.

Salesperson	Years of Experience	Annual Sales ($1000s)
1	1	80
2	3	97
3	4	92
4	4	102
5	6	103
6	8	111
7	10	119
8	10	123
9	11	117
10	13	136

a. Develop a scatter diagram for these data with years of experience as the independent variable.

b. Use the method of least squares to compute an estimated regression equation for the relationship between years of experience and annual sales.

6. Tyler Realty collected the following data regarding the selling price of new homes and the size of the homes measured in terms of square footage of living space.

Square Footage	Selling Price
2500	$124,000
2400	$108,000
1800	$ 92,000
3000	$146,000
2300	$110,000

a. Develop a scatter diagram for these data with square footage on the horizontal axis.
b. Try to approximate the relationship between square footage and selling price by drawing a straight line through the data.
c. Does there appear to be a linear relationship?
d. Develop an estimated regression equation using the least squares method.
e. Predict the selling price for a home with 2700 square feet.

7. The following data show the percentage of women working in each company (x) and the percentage of management jobs held by women in that company (y); the data shown represent companies in retailing and trade (*Louis Rukeyser's Business Almanac*).

Company	x	y
Federated Department Stores	72	61
Kroger	47	16
Marriott	51	32
McDonald's	57	46
Sears, Roebuck	55	36

a. Develop a scatter diagram for these data.
b. What does the scatter diagram developed in part a indicate about the relationship between x and y?
c. Use the method of least squares to develop an estimated regression equation for the data.
d. Predict the percentage of management jobs held by women in a company that has 60% women employees.
e. Use the estimated regression equation to predict the percentage of management jobs held by women in a company where 55% of the jobs are held by women. How does this predicted value compare to the 36% value observed for Sears, Roebuck, a company for which 55% of its employees are women?

8. A university medical center has developed a test designed to measure a patient's stress level. The test is designed so that higher scores on the test correspond to higher levels of stress. As part of a research study, the blood pressure (low reading) of patients who took the test was recorded. The following results were obtained.

Stress Test Score	Blood Pressure
53	70
94	91
64	78
73	78
82	85
90	84

a. Develop a scatter diagram for these data with stress test score on the horizontal axis. Does a linear relationship between the two variables appear to be appropriate?

b. Develop the estimated regression equation for these data.

c. Estimate an individual's blood pressure if he or she scored 85 on the stress test.

9. The cash flow (millions of dollars) and annual growth rate for nine pharmaceutical companies are shown below (*Forbes,* May 1989)

Firm	Cash Flow ($1,000,000s)	Annual Growth Rate (%)
Pfizer	986	12
Bristol Meyers	957	14
Merck	1412	16
American Home Products	1074	11
Abbott Laboratories	1023	18
Eli Lilly	965	11
SmithKline Beckman	450	3
Upjohn	456	10
Warner-Lambert	437	7

a. Develop a scatter diagram for these data, plotting cash flow on the vertical axis. Does it appear that the two variables are related?

b. Use the least squares method to fit a straight line to the data.

c. Estimate the cash flow for a company that has a 10% growth rate. How does the predicted value compare with the observed results for Upjohn?

13.2 ■■■■ THE COEFFICIENT OF DETERMINATION

In Section 13.1 we showed that the least squares method provides a linear approximation for the relationship between two variables. For example, for the Armand's Pizza Parlors problem, we developed the estimated regression equation $\hat{y} = 60 + 5x$ in order to approximate the linear relationship between the student population (x) and the annual sales (y). A question that might occur to you now is, How good is the fit? In this section we show that the *coefficient of determination* provides a measure of goodness of fit of the estimated regression equation to the data.

Recall that the least squares method is a technique for finding the values of b_0 and b_1 that minimize the sum of squared deviations between the observed values of the dependent variable (y_i) and the predicted values of the dependent variable ($\hat{y}_i$). The difference between y_i and $\hat{y}_i$ actually represents the error in using $\hat{y}_i$ to estimate y_i; the difference for the ith observation is $y_i - \hat{y}_i$. This difference is referred to as the ith *residual*. Thus, the sum of squares minimized by the least squares method is referred to as the sum of squares due to error, or the residual sum of squares. We use SSE to represent this quantity.

Sum of Squares Due to Error

$$SSE = \Sigma\,(y_i - \hat{y}_i)^2 \qquad (13.5)$$

TABLE 13.3   Calculation of SSE for the Armand's Pizza Parlors Problem

Restaurant i	x_i = Student Population (1000s)	y_i = Annual Sales ($1000s)	$\hat{y}_i = 60 + 5x_i$	$y_i - \hat{y}_i$	$(y_i - \hat{y}_i)^2$
1	2	58	70	−12	144
2	6	105	90	15	225
3	8	88	100	−12	144
4	8	118	100	18	324
5	12	117	120	−3	9
6	16	137	140	−3	9
7	20	157	160	−3	9
8	20	169	160	9	81
9	22	149	170	−21	441
10	26	202	190	12	144

SSE ⟶ 1530

Table 13.3 shows the calculations required to compute SSE for the Armand's Pizza Parlors problem. SSE = 1530 is a measure of the error involved in using the estimated regression equation $\hat{y} = 60 + 5x$ to predict the y_i values.

Now suppose that we were asked to develop an estimate of annual sales without using the size of the student population. We could not use the estimated regression equation and would have to use the value of the sample mean, $\overline{y} = 130$, as the best estimate of annual sales. In Table 13.4 we show the errors that would result from using $\overline{y}$ to estimate annual sales at the 10 Armand's Pizza Parlors. The corresponding sum of squared deviations about the mean, denoted by SST, is referred to as the total sum of squares.

Total Sum of Squares

$$SST = \Sigma (y_i - \overline{y})^2 \tag{13.6}$$

For the Armand's Pizza Parlors problem, SST = 15,730.

In Figure 13.4 we show the least squares regression line $\hat{y} = 60 + 5x$ and the line corresponding to $y = \overline{y} = 130$. Note in general that the points cluster more closely around the estimated regression line than they do about the line $\overline{y} = 130$. For example, for the tenth restaurant we see that the error is much larger when $\overline{y}$ is used as an estimate of y_{10} than when $\hat{y}_{10}$ is used. We can think of SST as a measure of how well the observations cluster about the $\overline{y}$ line and SSE as a measure of how well the observations cluster about the $\hat{y}$ line.

To measure how much the predicted values ($\hat{y}$) on the estimated regression line deviate from $\overline{y}$, another sum of squares is computed. This sum of squares is called the *sum of squares due to regression* and is denoted by SSR. The sum of squares due to regression can be written as follows.

Sum of Squares Due to Regression

$$SSR = \Sigma(\hat{y}_i - \overline{y})^2 \qquad (13.7)$$

From the preceding discusion, we should expect that SSE, SST, and SSR are related. Indeed, they are. The relationship among SSE, SST, and SSR is as shown.

Relationship Among SST, SSR, and SSE

$$SST = SSR + SSE \qquad (13.8)$$

where

SST = total sum of squares

SSR = sum of squares due to regression

SSE = sum of squares due to error

Using (13.8), we can conclude that the sum of squares explained by the regression relationship for the Armand's Pizza Parlor problem is

$$SSR = SST - SSE = 15{,}730 - 1530 = 14{,}200$$

Now let us see how these sums of squares can be used to provide a measure of the goodness of fit for the regression relationship. We would have the best possible fit if every observation happened to lie on the least squares line; the

TABLE 13.4 Computation of The Total Sum of Squares

Restaurant i	x_i = Student Population (1000s)	y_i = Annual Sales ($1000s)	$y_i - \overline{y}$	$(y_i - \overline{y})^2$
1	2	58	−72	5184
2	6	105	−25	625
3	8	88	−42	1764
4	8	118	−12	144
5	12	117	−13	169
6	16	137	7	49
7	20	157	27	729
8	20	169	39	1521
9	22	149	19	361
10	26	202	72	5184
				15,730 SST

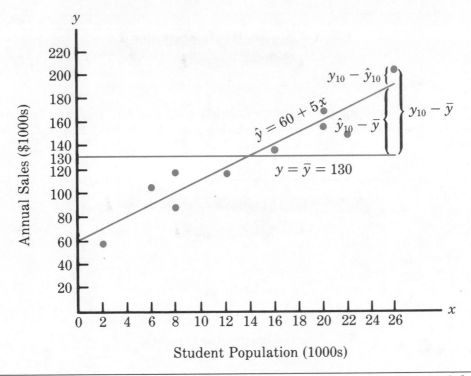

line would pass through each point, and we would have SSE = 0. Hence, for a perfect fit, SSR must equal SST, and thus the ratio SSR/SST = 1. On the other hand, a poorer fit to the observed data results in a larger SSE. Since SST = SSR + SSE, however, the largest SSE (and hence worst fit) occurs when SSR = 0. In this case the estimated regression equation does not help predict y. Thus the worst possible fit yields the ratio SSR/SST = 0.

If we were to use the ratio SSR/SST to evaluate the goodness of fit for the regression relationship, we would have a measure that could take on values between 0 and 1, with values closer to 1 implying a better fit. The fraction SSR/SST is called the *coefficient of determination* and is denoted r^2:

Coefficient of Determination

$$r^2 = \frac{SSR}{SST} \tag{13.9}$$

The value of the coefficient of determination for the Armand's Pizza Parlor Problem is

$$r^2 = \frac{SSR}{SST} = \frac{14{,}200}{15{,}730} = .903$$

To better interpret r^2, we can think of SST as the measure of how good $\bar{y}$ is as a predictor of annual sales volume. After developing the estimated regression equation, we compute SSE as the measure of the goodness of $\hat{y}$ as a predictor of annual sales volume. Thus SSR, the difference between SST and SSE, really measures the portion of SST that is explained by the estimated regression equation. Thus we can think of r^2 as

$$r^2 = \frac{\text{Sum of Squares Explained by Regression}}{\text{Total Sum of Squares (before Regression)}}$$

When it is expressed as a percentage, r^2 can be interpreted as the percentage of the total sum of squares (SST) that can be explained using the estimated regression equation. Statisticians often use r^2 as a measure of the goodness of fit of a regression line to the data. Thus for Armand's Pizza, we conclude that the estimated regression equation has accounted for 90.3% of the total sum of squares. We should be very pleased with such a good fit.

Computational Efficiencies

When using a calculator to compute the value of the coefficient of determination, computational efficiencies can be realized by computing SSR directly using the following formula

$$SSR = \frac{[\Sigma\, x_i y_i - (\Sigma\, x_i\, \Sigma\, y_i)/n]^2}{\Sigma\, x_i^2 - (\Sigma\, x_i)^2/n} \tag{13.10}$$

In addition, we need not compute SST using the expression $\Sigma\,(y_i - \bar{y})^2$; this expression can be algebraically expanded to provide

$$SST = \Sigma\, y_i^2 - \frac{(\Sigma\, y_i)^2}{n} \tag{13.11}$$

For the Armand's Pizza Parlors problem, part of the calculations needed to compute SSR and SST using the above formulas are shown in Table 13.5. Using the values in this table along with (13.10) and (13.11) we can compute SSR and SST as follows:

$$SSR = \frac{[21{,}040 - (140)(1300)/10]^2}{2528 - (140)^2/10}$$

$$= \frac{8{,}065{,}600}{568}$$

$$= 14{,}200$$

$$SST = 184{,}730 - \frac{(1300)^2}{10}$$

$$= 15{,}730$$

TABLE 13.5 **Calculations Used in Computing SSR and SST**

Restaurant i	x_i	y_i	x_iy_i	x_i^2	y_i^2
1	2	58	116	4	3,364
2	6	105	630	36	11,025
3	8	88	704	64	7,744
4	8	118	944	64	13,924
5	12	117	1404	144	13,689
6	16	137	2192	256	18,769
7	20	157	3140	400	24,649
8	20	169	3380	400	28,561
9	22	149	3278	484	22,201
10	26	202	5252	676	40,804
Totals	140	1300	21,040	2528	184,730
	Σx_i	Σy_i	Σx_iy_i	Σx_i^2	Σy_i^2

Note that since these are equivalent formulas we get the same values for SSR and SST that we obtained previously. Thus as before

$$r^2 = \frac{SSR}{SST} = \frac{14,200}{15,730} = .903$$

NOTES AND COMMENTS

1. In developing the least squares estimated regression equation and computing the coefficient of determination, no probabilistic assumptions and no statistical inferences have been made. Larger values of r^2 simply imply that the least squares line provides a better fit to the data; that is, the observations are more closely grouped about the least squares line. But, using only r^2, no conclusion can be made regarding whether or not the relationship between x and y is statistically significant. Such a conclusion must be based on considerations that involve the sample size and the properties of the appropriate sampling distributions of the least squares estimators.

2. As a practical matter, for typical data found in the social sciences, values of r^2 as low as .25 are often considered useful. For data in the physical and medical sciences, r^2 values of .60 or greater are often found; in fact, in some cases r^2 values greater than .90 can be found.

■ EXERCISES

10. Refer again to the data in Exercise 1.
   a. Compute SSE, SST, and SSR using (13.5), (13.6), and (13.8).
   b. Recompute SSR and SST using (13.10) and (13.11). Do you get the same results as in (a)?

c. Compute the coefficient of determination, r^2. Comment on the strength of the relationship.

11. Refer again to the data in Exercise 2.

a. Compute SSR and SST using (13.10) and (13.11).

b. What percentage of the total sum of squares is accounted for by the estimated regression equation?

12. Refer again to the data in Exercise 3.

a. Compute SSE, SST, and SSR using (13.5), (13.6), and (13.8).

b. Recompute SSR and SST using (13.10) and (13.11). Do you get the same results as in (a)?

c. Compute the coefficient of determination, r^2. Comment on the strength of the relationship.

13. Refer again to the data in Exercise 4.

a. Compute SSR and SST using (13.10) and (13.11).

b. What percentage of the total sum of squares can be accounted for by the estimated regression equation?

14. A medical laboratory at Duke University estimates the amount of protein in liver samples through the use of a regression model. A spectrometer emitting light shines through a substance containing the sample, and the amount of light absorbed is used to estimate the amount of protein in the sample. A new estimated regression equation is developed daily because of differing amounts of dye in the solution. On one day six samples with known protein concentrations gave the following absorbence readings.

Absorbence Reading (x)	Milligrams of Protein (y)
.509	0
.756	20
1.020	40
1.400	80
1.570	100
1.790	127

a. Use these data to develop an estimated regression equation relating the light absorbence reading to milligrams of protein present in the sample.

b. Compute r^2. Would you feel comfortable using this regression model to estimate the amount of protein in a sample?

c. In a sample just received the light absorbence reading was .941. Estimate the amount of protein in the sample.

15. A list of the best-selling cars for 1987 whose sales in units varied between 175,000 and 300,000 (rounded to the nearest thousand) is shown in the following table (*The World Almanac*, 1989). The 1988 suggested retail price (in thousands of dollars, rounded to the nearest hundred dollars) is also shown.

Model	Number Sold (1000s)	1988 Price ($1000s)
Hyundai	264	5.4
Oldsmobile Ciera	245	11.4
Nissan Sentra	236	6.4
Ford Tempo	219	9.1
Chev. Corsica/Beretta	214	10.0
Pontiac Grand Am	211	10.3
Toyota Camry	187	11.2
Chevrolet Caprice	177	12.5

a. Use these data to develop an estimated regression equation that could be used to predict the number sold given the price.

b. Compute r^2. Would you feel comfortable using the estimated regression equation to estimate the number sold given the price? Explain.

16. The data from Exercise 2 are repeated here.

Observation i	x_i	y_i
1	2	25
2	3	25
3	5	20
4	1	30
5	8	16

a. Compute $\bar{x}$ and $\bar{y}$.

b. Substitute the values of $\bar{x}$ and $\bar{y}$ for x and $\hat{y}$ in the estimated regression equation. Do these values satisfy the equation?

c. Will the least squares line always pass through $(\bar{x}, \bar{y})$? Why or why not?

13.3 ▬ THE REGRESSION MODEL AND ITS ASSUMPTIONS

An important concept that must be understood before we consider testing for significance in regression analysis involves the distinction between a *deterministic model* and a *probabilistic model*. In a deterministic model the relationship between the dependent variable y and the independent variable x is such that if we specify the value of the independent variable, the value of the dependent variable can be determined *exactly*. For example, if a major oil company leases a service station under a contractual agreement of $500 per month plus 10% of the gross sales, the relationship between the dealer's monthly payment (y) and the gross sales value (x) can be expressed as

$$y = 500 + .10x$$

With this relationship a June gross sales of $6000 would provide a monthly payment of $y = 500 + .10 (6000) = \$1100$, and a July gross sales of $7200 would provide a monthly payment of $y = 500 + .10 (7200) = \$1220$. This type of relationship is deterministic: Once the gross sales value (x) is specified, the monthly payment (y) is determined exactly. Figure 13.5 shows graphically the relationship between gross sales and monthly payment.

To illustrate a relationship between two variables that is probabilistic rather than deterministic, recall the Armand's Pizza Parlors example introduced in Section 13.1; the data for this problem are presented in Table 13.1. Note that restaurants 3 and 4 are both located near college campuses having 8000 students. Restaurant 3 shows annual sales of $88,000; however, restaurant 4 shows annual sales of $118,000. Thus we see that the relationship between y and x cannot be deterministic, since different values of y are observed for the same value of x. Note that this is also the case for restaurants 7 and 8, where a given campus size

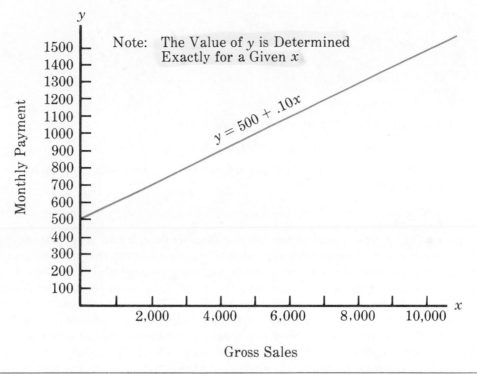

FIGURE 13.5 Illustration of a Deterministic Relationship

generates annual sales of $157,000 for restaurant 7 and $169,000 for restaurant 8. Since the value of y cannot be determined exactly from the value of x, we say that the model relating x and y is *probabilistic*.

Next, let us reconsider Figure 13.1, the scatter diagram for the Armand's Pizza data. We concluded that the relationship between student population (x) and annual sales (y) could be approximated by a straight line. As a result we used the least squares method to develop the following estimated regression equation:

$$\hat{y} = 60 + 5x$$

However, when we graphed this estimated regression equation in Figure 13.3, we saw that the relationship described by this equation was not perfect; that is, none of the observations fell exactly on the regression line.

Since we are unable to guarantee a single value of y for each value of x, the underlying relationship for Armand's Pizza can be explained only with a probabilistic model. Based on the observation that the relationship between student population and annual sales can be approximated by a straight line, we now make the assumption that the following probabilistic model—referred to as the *regression model*—is a realistic representation of the true relationship between the two variables:

Regression Model

$$y = \beta_0 + \beta_1 x + \epsilon \qquad (13.12)$$

where

β_0 = y-intercept of the line given by $\beta_0 + \beta_1 x$

β_1 = the slope of the line given by $\beta_0 + \beta_1 x$

ϵ = the error or deviation of the actual y value from the line given by $\beta_0 + \beta_1 x$

Using (13.12) as a model of the relationship between x and y, we are saying that we believe the two variables are related in such a fashion that the line given by $\beta_0 + \beta_1 x$ provides a good approximation of the y value at each x. However, to identify the exact value of y we must also consider the error term ϵ (the Greek letter epsilon), that corresponds to how far the actual y value is above or below the line $\beta_0 + \beta_1 x$. In the regression model, the independent variable x is treated as being known; the model is used to predict y given knowledge of x. We refer to β_0 (the y-intercept) and β_1 (the slope) as the *parameters* of the model.

The following assumptions are made about the error term ϵ in the regression model $y = \beta_0 + \beta_1 x + \epsilon$:

 Assumptions About the Error Term ϵ in the Regression Model
$$y = \beta_0 + \beta_1 x + \epsilon$$

1. The error term, ϵ, is a random variable with mean or expected value of 0; that is, $E(\epsilon) = 0$.
   Implication: Since β_0 and β_1 are constants, $E(\beta_0) = \beta_0$ and $E(\beta_1) = \beta_1$; thus for a given value of x, the expected value of y is

   Regression equation:
   $$E(y) = \beta_0 + \beta_1 x \qquad (13.13)$$

   Equation (13.13) is referred to as the *regression equation*.
2. The variance of ϵ is denoted by σ^2 and is the same for all values of x.
   Implication: The variance of y equals σ^2 and is the same for all values of x.
3. The values of ϵ are independent.
   Implication: The value of ϵ for a particular value of x is not related to the value of ϵ for any other value of x; thus, the value of y for a particular value of x is not related to the value of y for any other value of x.
4. The error term ϵ is a normally distributed random variable.
   Implication: Since y is a linear function of ϵ, y is also a normally distributed random variable.

Figure 13.6 provides a graphical interpretation of the model assumptions and their implications. As shown in Figure 13.6 the value of $E(y)$ changes according

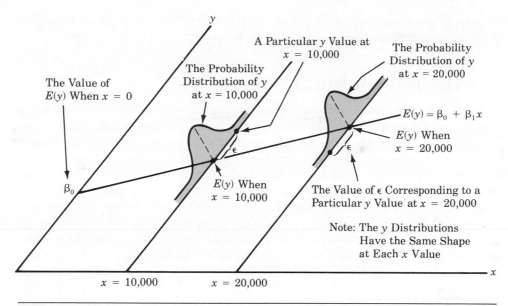

FIGURE 13.6   **Illustration of Assumptions for Regression Model**

to the specific value of x considered. However, note that regardless of the x value, the probability distribution of the errors (ϵ) and hence the probability distribution of y about the regression line form a normal distribution, each with the same shape and hence the same variance. The specific value of the error ϵ at any particular point depends upon whether the actual value of y is greater than or less than $E(y)$.

At this point we must keep in mind that we are also making an assumption or hypothesis about the form of the regression model and the associated regression equation. That is, we have assumed that a straight line represented by $\beta_0 + \beta_1 x$ is the basis for the relationship between the variables. We must not lose sight of the fact that some other model, for instance $\beta_0 + \beta_1 x^2$, may turn out to be a better model for the underlying relationship. After using the sample data to estimate the parameters of the regression model (β_0 and β_1), we will want to conduct further analysis to determine whether or not the specific model that we have assumed appears to be valid.

The Relationship Between the Regression Equation and the Estimated Regression Equation

Recall from Chapter 8 that when data were available for just one variable, the objective was to use a sample statistic (e.g., the sample mean) to make inferences about the corresponding population parameter (e.g., population mean). When we discussed the least squares method in Section 13.1, we presented formulas for computing the y-intercept (b_0) and the slope (b_1) of the estimated regression equation. The value of b_0 is a sample statistic that provides an estimate of the β_0 parameter in the regression model, and the value of b_1 is a sample statistic that provides an estimate of the β_1 parameter. Thus, since the regression equation is $E(y) = \beta_0 + \beta_1 x$, the best estimate of the regression equation is provided by the estimated regression equation $\hat{y} = b_0 + b_1 x$. Hence, $\hat{y}$ provides the estimate of $E(y)$. Figure 13.7 summarizes these concepts.

13.4 ■ TESTING FOR SIGNIFICANCE

In Section 13.2 we saw how the coefficient of determination (r^2) could be used as a measure of the goodness of fit of the estimated regression line. Larger values of r^2 indicated a better fit. However, the value of r^2 does not allow us to conclude whether or not a regression relationship is statistically significant. In order to draw conclusions concerning statistical significance we must, among other things, take the sample size into consideration. It makes intuitive sense that a regression relationship involving 20 sample points would be more statistically significant than one involving 5 sample points even with the same r^2 value. In this section we show how to conduct significance tests that will allow us to draw conclusions about the existence of a regression relationship.

An Estimate of σ^2

In Section 13.2 we used the sum of squares due to error, SSE, as a measure of the variability of the actual observations about the estimated regression line. The mean squared error is used as an estimate of σ^2, the variance of ϵ, and consequently the variance of the y values about the regression line. Recall from

---

■ FIGURE 13.7 **Estimating the Population Regression Equation Using Sample Data**

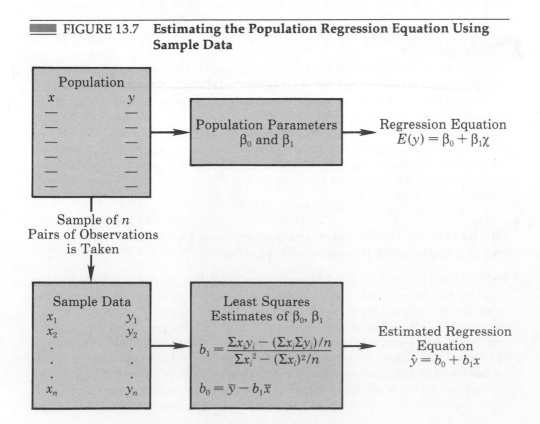

our earlier definition of sample variance that we divided the sum of the squ
deviations about the sample mean by $n - 1$ to obtain an unbiased estimate of the
population variance. We used $n - 1$ instead of n because 1 degree of freedom was
lost when the sample mean was used as an estimate of the population mean in
the computation of the sum of the squared deviations about the mean. In
regression analysis we must estimate the parameters β_0 and β_1 to compute SSE;
that is, we use the estimates b_0 and b_1—obtained from the sample data—to
compute the sum of squares due to error as follows:

$$SSE = \Sigma (y_i - \hat{y}_i)^2 = \Sigma (y_i - b_0 - b_1 x_i)^2$$

For this reason two degrees of freedom are lost; hence we must divide SSE by
$n - 2$ to obtain an unbiased estimate of σ^2. The estimate obtained is called the
mean square due to error, denoted MSE. Since MSE is an estimate of σ^2, the
notation s^2 is also used.

Mean Square Error (Estimate of σ^2)

$$s^2 = MSE = \frac{SSE}{n - 2} \qquad\qquad (13.14)$$

From the data in Table 13.1 we see that SSE = 1530; thus for the Armand's Pizza
example we have

$$s^2 = MSE = \frac{1530}{8} = 191.25$$

In the discussion that follows we will see that this unbiased estimate of σ^2 is used
in computing the F statistic in the test for the significance of the regression
equation.

F Test

Recall that the underlying regression equation is assumed to be $E(y) = \beta_0 + \beta_1 x$.
If there really exists a relationship of this form between x and y, β_1 could not
equal 0. Thus a conclusion regarding the significance of the relationship can be
tested using the following hypotheses:

$$H_0: \quad \beta_1 = 0$$
$$H_a: \quad \beta_1 \neq 0$$

The logic behind the use of the F test for determining whether or not the
relationship between x and y is statistically significant is based upon our being
able to develop two independent estimates of σ^2. We have just seen that MSE
provides an estimate of σ^2. If the null hypothesis $H_0: \beta_1 = 0$ is true, the mean
square due to regression (denoted MSR) provides another *independent* estimate
of σ^2.

To compute MSR we first note that for any sum of squares the mean square is the sum of squares divided by its degrees of freedom. The number of degrees of freedom for the sum of squares due to regression, SSR, is always equal to the number of independent variables. Since in this chapter we are concerned only with models involving one independent variable, the number of regression degrees of freedom is 1. Using DF as an abbreviation for degrees of freedom we can write

$$MSR = \frac{SSR}{\text{Regression DF}} = \frac{SSR}{\text{Number of independent variables}} \qquad (13.15)$$

For the Armand's Pizza problem, we find that MSR = SSR/1 = 14,200.

If the null hypothesis (H_0: $\beta_1 = 0$) is true, MSR and MSE are two independent estimates of σ^2. In this case the sampling distribution of MSR/MSE follows an F distribution with numerator degrees of freedom equal to 1 and denominator degrees of freedom equal to $n - 2$. The test concerning the significance of the regression relationship is based on the following F statistic:

$$F = \frac{MSR}{MSE} \qquad (13.16)$$

Given any sample size, the numerator of the F statistic will increase as more of the variability in y is explained by the regression model and decrease as less is explained. Similarly, the denominator will increase if there is more variability about the estimated regression line and decrease if there is less variability. Thus one would intuitively expect large values of F = MSR/MSE to cast doubt on the null hypotheses and lead us to the conclusion that $\beta_1 \neq 0$ and there is a significant relationship between x and y. Indeed, this is correct; large values of F lead to rejection of H_0 and the conclusion that the relationship is statistically significant.

Let us now conduct the F test for our Armand's Pizza problem. Assume that the level of significance is $\alpha = .01$. From Table 4 of Appendix B we can determine the critical F value by locating the value corresponding to numerator degrees of freedom equal to 1 (the number of independent variables) and denominator degrees of freedom equal to $n - 2 = 10 - 2 = 8$. Thus we obtain $F = 11.26$. Hence the appropriate rejection rule for the Armand's Pizza problem is written

Reject H_0 if MSR/MSE > 11.26

Since MSR/MSE = 14,200/191.25 = 74.25 is greater than the critical value ($F_{.01} = 11.26$), we can reject H_0 and conclude that there is a statistically significant relationship between annual sales and the size of the student population.

We caution here that rejection of H_0 does not yet permit us to conclude that the relationship between x and y is *linear*. However, it is valid to conclude that x and y are related and that a linear approximation explains a significant amount of the variability in y over the range of x values observed in the sample. To illustrate this qualification we call your attention to Figure 13.8, where an F test (on $\beta_1 = 0$) yielded the conclusion that x and y were related. The figure shows that the actual relationship is nonlinear. In the graph we see that the linear

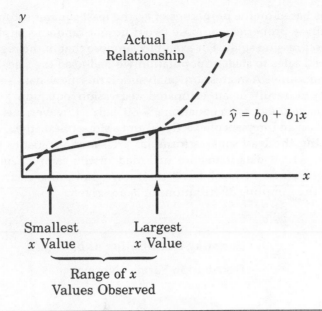

━━━ FIGURE 13.8 **Example of a Linear Approximation of a Nonlinear Relationship**

approximation is very good for the values of x used in developing the least squares line, but it is very bad for larger values of x.

Given a significant relationship, we should feel confident in using the regression equation for predictions corresponding to x values within the range of the x values for the sample. For the Armand's Pizza problem this corresponds to values of x between 2 and 26. But unless there are reasons to believe the model is valid beyond this range, predictions outside the range of the independent variable should be made with caution. For the Armand's Pizza problem, since the regression relationship has been found significant at the .01 level, we should feel confident using it to predict sales whenever the student population is between 2000 and 26,000.

t Test

The F test has been used to test the null hypothesis H_0: $\beta_1 = 0$. A t test also exists for testing this null hypothesis. In regression models with only one independent variable the t test and the F test yield the same results. But with more than one independent variable only the F test can be used to test for a significant relationship between a dependent variable and a set of independent variables. The t test is then used to determine whether or not the coefficient of a particular independent variable is zero. Here we introduce the t test and show that it leads to the same conclusion as the F test. In the next chapter it will be used to test the significance of individual parameters in a multiple regression model.

The hypotheses we will be testing are the same as before:

$$H_0: \quad \beta_1 = 0$$
$$H_a: \quad \beta_1 \neq 0$$

Assumptions being tested w/ t-test.

The t test is based on the properties of b_1, the least squares estimator of β_1. To understand these properties, suppose two different random samples were used for the same regression study. For example, suppose that in the regression study relating annual sales to student population we had used the sales records of 10 different restaurants. A regression analysis of this new data set, or sample, probably would result in an estimated regression equation similar to our previous estimated regression equation $\hat{y} = 60 + 5x$. However, it is doubtful that we would obtain an intercept of exactly 60 and a slope of exactly 5. Thus b_0 and b_1 computed using the least squares formulas are sample statistics whose values depend upon which data items are included in the sample. As a result, we conclude that b_0 and b_1 must have their own sampling distributions. The properties of the sampling distribution for b_1 are given.

Sampling Distribution of b_1

Distribution Form: Normal

$$\text{Mean:} \quad E(b_1) = \beta_1 \tag{13.17}$$

$$\text{Variance:} \quad \sigma_{b_1}^2 = \sigma^2 \frac{1}{\Sigma x_i^2 - (\Sigma x_i)^2 / n} \tag{13.18}$$

Note that the expected value of b_1 is equal to β_1, so b_1 is an unbiased estimator of β_1.

Since we do not know the value of σ^2, we develop an estimate of $\sigma_{b_1}^2$, denoted $s_{b_1}^2$, by first estimating σ^2 with s^2 [see (13.14)]. Thus we obtain the following.

Estimated Variance of b_1

$$s_{b_1}^2 = s^2 \frac{1}{\Sigma x_i^2 - (\Sigma x_i)^2 / n} \tag{13.19}$$

For the Armand's Pizza problem, $s^2 = \text{MSE} = 191.25$. Thus

$$s_{b_1}^2 = 191.25 \frac{1}{2528 - (140)^2 / 10}$$

$$= 191.25 \frac{1}{568}$$

$$= .3367$$

Hence

$$s_{b_1} = \sqrt{.3367} = .5803$$

The t test regarding β_1 is based on the fact that the test statistic

$$\frac{b_1 - \beta_1}{s_{b_1}}$$

follows a t distribution with $n - 2$ degrees of freedom. If the null hypothesis is true, then $\beta_1 = 0$ and we find that b_1/s_{b_1} follows a t distribution with $n - 2$ degrees of freedom. Using b_1/s_{b_1} as the test statistic, we would use the following rejection rule to test H_0: $\beta_1 = 0$ versus H_a: $\beta_1 \neq 0$:

$$\text{Reject } H_0 \text{ if } \quad \frac{b_1}{s_{b_1}} < -t_{\alpha/2}$$

$$\text{or} \quad \frac{b_1}{s_{b_1}} > t_{\alpha/2}$$

For the Armand's Pizza problem we have $b_1 = 5$ and $s_{b_1} = .5803$. Thus we have $b_1/s_{b_1} = 5/.5803 = 8.6162$. From Table 2 of Appendix B we find that the t value corresponding to $\alpha = .01$ and 8 degrees of freedom is $t_{.005} = 3.355$. Since $b_1/s_{b_1} = 8.6162 > 3.355$, we reject H_0 and conclude at the .01 level of significance that β_1 is not equal to zero.

NOTES AND COMMENTS

1. The assumptions made about the error term (Section 13.3) are what permit the tests of statistical significance in this section. The properties of the sampling distribution of b_1 and the subsequent F and t tests directly follow from these assumptions.
2. Do not confuse statistical significance with practical significance. With very large sample sizes it is possible to obtain statistically significant results for small values of b_1; in such cases one must exercise care in concluding that the relationship has practical significance.
3. The reason that the F test and the t test yield the same result *for simple linear regression* is that $F = t^2$. The critical value for the F test is the square of the critical value for the t test, and the test statistic for the F test is the square of the test statistic for the t test.

■ EXERCISES

17. The data from Exercise 2 are repeated here:

Observation i	x_i	y_i
1	2	25
2	3	25
3	5	20
4	1	30
5	8	16

a. Compute an estimate of σ^2.
b. Compute an estimate of the variance of b_1.
c. Use the t test to test the hypotheses ($\alpha = .05$).

$$H_0: \quad \beta_1 = 0$$

$$H_a: \quad \beta_1 \neq 0$$

d. Use the F test to test the hypotheses in part (c) at the $\alpha = .05$ level of significance.

18. Given below are five observations collected in a regression study on two variables.

Observation	x_i	y_i
1	2	2
2	4	3
3	5	2
4	7	6
5	8	4

a. Develop the estimated regression equation for these data.
b. Use the t test to test the hypotheses ($\alpha = .05$).

$$H_0: \quad \beta_1 = 0$$

$$H_a: \quad \beta_1 \neq 0$$

c. Use the F test to test the hypotheses in part (b) at the $\alpha = .05$ level of significance.

19. Refer to Exercise 3, where an estimated regression line relating GPA to monthly starting salaries was developed. SSR and SST were computed in Exercise 12.
a. Compute s^2.
b. Compute $s_{b_1}^2$ and s_{b_1}.
c. Use the t test to determine if GPA and salary are related at the $\alpha = .05$ level of significance.
d. Use the F test to determine if GPA and salary are related at the $\alpha = .05$ level of significance.

20. Refer to Exercise 6, where an estimated regression line relating square footage to selling prices of new homes was developed. Test whether or not selling price and square footage are related at the $\alpha = .01$ level of significance.

21. Refer to Exercise 7, where an estimated regression line relating the percentage of management jobs held by women and the percentage of women employed was developed. Test whether or not these two variables are related at the $\alpha = .05$ level of significance.

22. Refer to Exercise 5, where an estimated regression line relating years of experience and annual sales was developed. At the $\alpha = .05$ level of significance, determine whether or not annual sales and years of experience are related.

23. Refer to Exercise 14, where an estimated regression line relating light absorbence readings and milligrams of protein present in a liver sample was developed. Test whether or not the absorbence readings and amount of protein present are related at the $\alpha = .01$ level of significance.

13.5 ■■■ ESTIMATION AND PREDICTION

For the Armand's Pizza problem we concluded that annual sales (y) and the size of the student population (x) are related. Moreover, the estimated regression equation $\hat{y} = 60 + 5x$ appears to adequately describe the relationship between x and y. Now we can begin to use the estimated regression line to develop interval estimates of annual sales for a given student population.

There are two types of interval estimates to consider. The first is an interval estimate of the mean value of y for a particular value of x. We refer to this type of interval estimate as a _confidence interval estimate_. For instance, we might want a confidence interval estimate of the _expected_ annual sales for an Armand's Pizza Parlor located near any campus with a student population of 10,000. The expected annual sales for campuses with 10,000 students represents the average of the annual sales for all restaurants located near college campuses with 10,000 students.

The second type of interval estimate that we will consider is appropriate in situations where we want to predict an individual value of y corresponding to a given value of x. We refer to this type of interval estimate as a _prediction interval estimate_. For instance, we might be interested in developing a prediction interval estimate for the annual sales of a restaurant site near Talbot College, a school with 10,000 students. In this case our interest is in predicting the annual sales for one particular site, as opposed to predicting the average sales for all restaurants located near campuses with 10,000 students.

Confidence Interval Estimate of the Mean Value of y

Suppose we wanted to develop a confidence interval estimate of the mean or expected value of annual sales for all restaurant sites near campuses with 10,000 students. First, recall that for the Armand's Pizza problem we hypothesized that the expected value of annual sales is given by

$$E(y) = \beta_0 + \beta_1 x$$

Thus when the student population size is 10,000, $x = 10$ and hence $E(y) = \beta_0 + \beta_1(10)$.

The least squares method was used to develop estimates of β_0 and β_1 and hence an estimate of the regression equation; for the Armand's Pizza problem the estimated regression equation was found to be $\hat{y} = 60 + 5x$. Thus for restaurants near a campus with 10,000 students we would obtain $\hat{y} = 60 + 5(10) = 110$. The corresponding estimate of expected annual sales would be $110,000.

In general, the point estimate of $E(y)$ for a particular value of x is the corresponding value of $\hat{y}$ given by the estimated regression equation. We denote the particular value of x by x_p, the mean value of y at x_p by $E(y_p)$, and the estimate of $E(y_p)$ by $\hat{y}_p = b_0 + b_1 x_p$.

Since b_0 and b_1 are only estimates of β_0 and β_1, we cannot expect that the estimated value $\hat{y}_p$ will exactly equal $E(y_p)$. For instance, in our example above we do not expect that the mean annual sales for all sites with 10,000 students to exactly equal $110,000, our estimated value. If we want to make an inference about how close $\hat{y}_p$ is to the true mean value $E(y_p)$, however, we will have to consider the variability that exists when we develop estimates based on the estimated regression equation. Statisticians have developed the following estimate of the variance of $\hat{y}_p$:

$$\star \quad \text{Estimated variance of } \hat{y}_p = s_{\hat{y}_p}^2 = s^2\left(\frac{1}{n} + \frac{(x_p - \bar{x})^2}{\Sigma x_i^2 - (\Sigma x_i)^2/n}\right) \qquad (13.20)$$

For the Armand's Pizza problem, the estimated variance of $\hat{y}_p$ for restaurant sites near campus populations with 10,000 students is

$$s_{\hat{y}_p}^2 = 191.25\left(\frac{1}{10} + \frac{(10 - 14)^2}{2528 - (140)^2/10}\right)$$

$$= 191.25(.1282)$$

$$= 24.52$$

Thus

$$s_{\hat{y}_p} = \sqrt{s_{\hat{y}_p}^2} = \sqrt{24.52} = 4.95$$

The confidence interval estimate of $E(y_p)$ is as follows:

Confidence Interval Estimate of $E(y_p)$

$$\hat{y}_p \pm t_{\alpha/2}\, s_{\hat{y}_p} \qquad (13.21)$$

where the confidence coefficient is $1 - \alpha$ and the t value has $n - 2$ degrees of freedom.

Thus to develop a 95% confidence interval estimate of the expected annual sales for restaurant sites with campus populations of 10,000 students, we need to find the t value from Table 2 of Appendix B corresponding to $n - 2 = 10 - 2 = 8$ degrees of freedom and $\alpha = .05$. Doing so, we find $t_{.025} = 2.306$. Hence the resulting confidence interval is

$$[b_0 + b_1(10)] \pm 2.306\, s_{\hat{y}_p}$$

$$[60 + 5(10)] \pm 2.306\,(4.95)$$

$$110 \pm 11.415$$

In dollars, the 95% confidence interval estimate is $110,000 \pm \$11,415$. Thus we obtain \$98,585 to \$121,415 as a confidence interval estimate of the expected or average sales volume for all restaurant sites near campuses with 10,000 students.

Note that the estimated variance of $\hat{y}_p$ [see (13.20)] is smallest when the given value of $x_p = \bar{x}$. In this case (13.20) becomes

$$s_{\hat{y}_p}^2 = s^2 \left(\frac{1}{n} + \frac{(\bar{x} - \bar{x})^2}{\Sigma x_i^2 - (\Sigma x_i)^2/n} \right)$$

$$= \frac{s^2}{n}$$

which implies that we can expect to make our best estimates at the mean of the independent variable.

Prediction Interval Estimate for an Individual Value of y

In the preceding discussion we showed how to develop a confidence interval estimate of the expected annual sales for all restaurant sites near campuses with 10,000 students. Now we turn to the problem of developing point and interval estimates for an individual value of y corresponding to a particular value of x. Suppose we want to predict annual sales of a restaurant site near Talbot College, a school with 10,000 students.

The point estimate for an individual value of y is given by $\hat{y}_p = b_0 + b_1 x_p$. Hence, the point estimate of annual sales for a site near Talbot College is $\hat{y}_p = 60 + 5(10) = 110$. Note that this is the same as the point estimate of the mean annual sales for all restaurants near campuses with 10,000 students.

In order to develop a prediction interval estimate, we must first determine the variance associated with using $\hat{y}_p$ as an estimate of a particular value of y when $x = x_p$. This variance is made up of the sum of the following two components:

1. The variance of individual y values about the mean $E(y_p)$, an estimate of which is given by s^2.

2. The variance associated with using $\hat{y}_p$ to estimate $E(y_p)$, which we previously found to be $s_{\hat{y}_p}^2$.

Statisticians have shown that the variance of the estimate of an individual value of y_p, which we denote s_{ind}^2, is given by

$$s_{ind}^2 = s^2 + s_{\hat{y}_p}^2$$

$$= s^2 + s^2 \left(\frac{1}{n} + \frac{(x_p - \bar{x})^2}{\Sigma x_i^2 - (\Sigma x_i)^2/n} \right)$$

$$= s^2 \left(1 + \frac{1}{n} + \frac{(x_p - \bar{x})^2}{\Sigma x_i^2 - (\Sigma x_i)^2/n} \right) \qquad (13.22)$$

For the Armand's Pizza problem, the value of (13.22) corresponding to the prediction of annual sales for a particular restaurant site near a campus with 10,000 students is computed as follows:

$$s_{ind}^2 = 191.25 \left(1 + \frac{1}{10} + \frac{(10 - 14)^2}{2528 - (140)^2/10} \right)$$

$$= 191.25 \, (1.1282)$$

$$= 215.77$$

Thus

$$s_{\text{ind}} = \sqrt{s_{\text{ind}}^2} = \sqrt{215.77} = 14.69$$

The prediction interval estimate of y_p is given by (13.23):

Prediction Interval Estimate of y_p

$$\hat{y}_p \pm t_{\alpha/2}\, s_{\text{ind}} \qquad\qquad (13.23)$$

where the confidence coefficient is $1 - \alpha$ and the t value has $n - 2$ degrees of freedom.

Thus a 95% prediction interval for sales of an individual restaurant near a campus with 10,000 students is

$$[b_0 + b_1(10)] \pm t_{\alpha/2}(14.69)$$

$$[60 + 5(10)] \pm 2.306\,(14.69)$$

$$110 \pm 33.875$$

In dollars, the 95% prediction interval for annual sales for an individual restaurant located near a campus with 10,000 students is $76,125 to $143,875. We note that this prediction interval is greater in width than the confidence interval for mean sales of all sites near campuses with 10,000 students ($98,585 to $121,415). This difference simply reflects the fact that we are able to estimate the mean annual sales volume with more precision than we can predict the annual sales for any particular site.

■ EXERCISES

24. As an extension of Exercise 7, develop a 95% confidence interval for estimating the mean percentage of management jobs held by women for companies for which 60% of the employees are women.

25. As an extension of Exercise 3, develop a 95% confidence interval for estimating the mean starting salary for students with a 3.0 GPA.

26. As an extension of Exercise 6, develop a 95% confidence interval for predicting the mean selling price for homes with 2200 square feet of living space.

27. As an extension of Exercise 3, develop a 95% prediction interval for estimating the starting salary of Joe Heller, who has a GPA of 3.0.

28. As an extension of Exercise 6, develop a 95% prediction interval for the selling price of a home on Highland Terrace with 2800 square feet.

29. State in your own words why a smaller interval is obtained when estimating the mean value than when predicting an individual value.

30. A study conducted by a department of transportation regarding driving speed and mileage for midsize automobiles resulted in the following data.

Driving Speed	Mileage
30	28
50	25
40	25
55	23
30	30
25	32
60	21
25	35
50	26
55	25

a. Determine the estimated regression equation that relates mileage to the driving speed.

b. At the $\alpha = .05$ level of significance, determine whether or not mileage and driving speed are related.

c. Did the estimated regression line provide a good fit to the data?

d. Develop a 95% confidence interval for estimating the mean mileage for cars that are driven at 50 miles per hour.

e. If we were interested in one specific car that was driven at 50 miles per hour, how would our estimate of mileage change as compared to the estimate developed in (d)?

13.6 ■■■ COMPUTER SOLUTION OF REGRESSION PROBLEMS

Performing all the computations associated with regression analysis can be quite time-consuming. In this section we discuss how the computational burden can be simplified by using a computer software package. The general procedure followed in using computer packages is for the user to input the data (x and y values for the sample) together with some instructions concerning the types of analyses that are required. The software package performs the analysis and prints the results in an output report. Before discussing the details of this

TABLE 13.6 **General Form of the ANOVA Table for 2-Variable Regression Analysis**

Source of Variation	Sum of Squares	Degrees of Freedom	Mean Square
Regression	SSR	1	$MSR = \dfrac{SSR}{1}$
Error	SSE	$n - 2$	$MSE = \dfrac{SSE}{n - 2}$
Total	SST	$n - 1$	

TABLE 13.7   ANOVA Table for the Armand's Pizza Problem

Source of Variation	Sum of Squares	Degrees of Freedom	Mean Square
Regression	14,200	1	$\frac{14{,}200}{1} = 14{,}200$
Error	1,530	8	$\frac{1530}{8} = 191.25$
Total	15,730	9	

approach, we discuss the use of the analysis of variance (ANOVA) table as a device for summarizing the calculations performed in regression analysis. The ANOVA table is an important component of the output report produced by most software packages.

The ANOVA Table

In Chapter 12 we saw how the ANOVA table could provide a convenient summary of the computational aspects of analysis of variance. In regression analysis a similar table can be developed. Table 13.6 shows the general form of the ANOVA table for 2-variable regression studies and Table 13.7 shows the ANOVA table for the Armand's Pizza problem. It can be seen that the relationship that holds among the sum of squares (that is, SST = SSR + SSE) also holds for the degrees of freedom. That is

$$\text{Total DF} = \text{Regression DF} + \text{Error DF}$$

▬ **FIGURE 13.9   Minitab Output for the Armand's Pizza Problem**

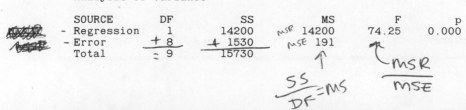

```
The regression equation is
SALES = 60.0 + 5.00 POP

Predictor     Coef      Stdev     t-ratio      p
Constant     60.000     9.226      6.50      0.000
POP           5.0000    0.5803     8.62      0.000

s = 13.83      R-sq = 90.3%      R-sq(adj) = 89.1%

Analysis of Variance

SOURCE        DF        SS        MS         F         p
Regression    1         14200     14200      74.25     0.000
Error         8         1530      191
Total         9         15730
```

Computer Output

In Figure 13.9 we show the Minitab computer output for the Armand's Pizza problem. The dependent variable is labeled "SALES" and the independent variable x is labeled "POP." The interpretation of the output is as follows:

1. Minitab prints the estimated regression equation as SALES = 60.0 + 5.00 POP.

2. A table is printed that shows the values of the coefficients b_0 and b_1, the standard deviation of each coefficient, the t value obtained by dividing each coefficient value by its standard deviation, and the p-value associated with the t-value. Thus, to test H_0: $\beta_1 = 0$ versus H_a: $\beta_1 \neq 0$, we could compare 8.62 (located in the "t-ratio" column) to the appropriate critical value. This is the procedure described in the last part of Section 13.4. Alternatively, we could use the p-value provided by Minitab to perform the same test. Recall from Chapter 9 that the p-value is the probability of obtaining a sample result more unlikely than what is observed. Since the p-value in this case is 0 (to three decimal places), the sample results indicate that the null hypothesis (H_0: $\beta_1 = 0$) should be rejected.

3. Minitab prints the estimate of σ, $s = 13.83$, as well as information regarding the goodness of fit. Note that "R-sq = 90.3%" is the coefficient of determination, which we denoted by r^2. The output "R-sq (adj) = 89.1%" is discussed in Chapter 14.

4. The ANOVA table is printed below the heading "Analysis of Variance." Note that MSR is given as 14,200 and MSE as 191. The ratio of these two values provides the F value of 74.25; in Section 13.4 we showed how the F value can be used to determine if there is a significant statistical relationship between SALES and POP. Minitab also prints the p-value associated with this F test. Since the p-value is 0 (to three decimal places), the relationship is judged statistically significant.

As you can see, the computer output from Minitab is fairly easy to interpret given our current background in regression analysis. Many other computer packages, most with rather cryptic names such as SPSS, SAS, BMDP, and so on, are available for solving regression problems. After a brief period of familiarity with the control language associated with each of these packages, computer output such as that shown in Figure 13.9 can be obtained easily. When large data sets are involved, computer packages provide the only practical means for solving regression problems.

■ EXERCISES

31. The commercial division of a real estate firm is conducting a regression analysis of the relationship between x, annual gross rents ($1,000s), and y, selling price ($1000s) for apartment buildings. Data have been collected on a number of properties recently sold,

and the output has been obtained in a computer run:

X - gross rents
Y - selling prices

```
The regression equation is
Y = 20.0 + 7.21X

Predictor        Coef      Stdev     t-ratio
Constant       20.000     3.2213       6.21
X               7.210     1.3626       5.29

Analysis of Variance

SOURCE        DF          SS
Regression     1       41587.3
Error          7
Total          8       51984.1
```

a. How many apartment buildings were in the sample?
b. Write the estimated regression equation.
c. What is the value of s_{b_1}?
d. Use the F statistic to test the significance of the relationship at an $\alpha = .05$ level of significance.
e. Estimate the selling price of an apartment building with gross annual rents of $50,000.

32. Shown below is a portion of the computer output for a regression analysis relating maintenance expense (dollars per month) to usage (hours per week) of a particular brand of computer terminal:

```
The regression equation is
Y = 6.1092 + .8951X

Predictor        Coef      Stdev
Constant       6.1092     0.9361
X              0.8951     0.1490

Analysis of Variance

SOURCE        DF         SS          MS
Regression     1      1575.76     1575.76
Error          8       349.14       43.64
Total          9      1924.90
```

a. Write the estimated regression equation.
b. Test to see (use a t test) if monthly maintenance expense is related to usage at the .05 level of significance.
c. Use the estimated regression equation to predict monthly maintenance expense for any terminal that is used 25 hours per week.

33. A regression model relating x, number of salespersons at a branch office, to y, annual sales at the office ($1000s), has been developed. Shown below is the computer output

from a regression analysis of the data:

```
The regression equation is
Y = 80.0 + 50.00X

Predictor      Coef      Stdev     t-ratio
Constant       80.0      11.333     7.06
X              50.0       5.482     9.12

Analysis of Variance

SOURCE        DF         SS         MS
Regression     1       6828.6     6828.6
Error         28       2298.8       82.1
Total         29       9127.4
```

a. Write the estimated regression equation.
b. How many branch offices were involved in the study?
c. Compute the F statistic and test the significance of the relationship at an $\alpha = .05$ level of significance.
d. Predict the annual sales at the Memphis branch office. This branch has 12 salespersons.

34. The following data show the dollar value of prescriptions (in thousands) for 13 pharmacies in Iowa and the population of the city served by the given pharmacy ("The Use of Categorical Variables in Data Envelopment Analysis," R. Banker and R. Morey, *Management Science*, December 1986).

Value ($1000s)	Population
61	1410
92	1523
93	1354
45	822
50	746
29	1281
56	1016
45	1070
183	1694
156	1910
120	1745
75	1353
122	1016

a. Use a computer package to develop a scatter diagram for these data; plot population on the horizontal axis.
b. Does there appear to be any relationship between these two variables?
c. Use the computer package to develop the estimated regression line that could be used to predict the dollar value of prescriptions given the population of the city.
d. Test the significance of the relationship at an $\alpha = .05$ level of significance.
e. Predict the dollar value for a particular city with a population of 1500 people. Use $\alpha = .05$.

13.7 ■■■ THE ANALYSIS OF RESIDUALS

For each observation in a regression analysis, there is a residual; it is the difference between the observed value of the dependent variable, y_i, and the value predicted by the regression equation, $\hat{y}_i$. The residual for observation i, $y_i - \hat{y}_i$, is an estimate of the error resulting from using the estimated regression equation to predict the value of y_i.

The analysis of residuals plays an important role in validating the assumptions made in regression analysis. In Section 13.4 we showed how hypothesis testing can be used to determine whether or not a regression relationship is statistically significant. Hypothesis tests concerning regression relationships are based on the assumptions made about the regression model. If the assumptions made are not satisfied, the hypothesis tests are not valid and the estimated regression equation should not be used. However, keep in mind that the regression model is being used only as an approximation of reality, so good judgment must be used to determine whether or not an assumption violation is severe enough to invalidate the model.

There are two key issues in verifying that the assumptions are satisfied in a regression model. Are the four assumptions concerning the error term, ϵ, satisfied and is the form we have assumed for the model appropriate? In this chapter, we restrict our discussion of residual analysis to these two issues concerning model validity. In Chapters 14 and 15, we show some further uses of residual analysis.

Regression analysis begins with an assumption concerning the appropriate form of the regression model. The simple linear regression model assumes the form

$$y = \beta_0 + \beta_1 x + \epsilon$$

With this form, y is a linear function of x. The assumptions regarding the error term (presented in Section 13.3) are as follows:

1. $E(\epsilon) = 0$.
2. The variance of ϵ, denoted by σ^2, is the same for all values of x.
3. The values of ϵ are independent.
4. ϵ is a normally distributed random variable.

Validating the assumptions concerning the error term, ϵ, means using the residuals to check to see if these assumptions seem reasonable.

Recall that the first assumption concerning ϵ implys that the regression equation is

$$E(y) = \beta_0 + \beta_1 x$$

This regression equation shows a linear relationship between x and the expected value of y, $E(y)$. Validating the assumption concerning model form means satisfying ourselves that the relationship between independent and dependent variable is adequately represented by the regression equation. It is possible that the true relationship between x and y is curvilinear and/or that more independent variables should have been included (multiple regression).

We will see how the statistician uses residual analysis to recognize when this assumption concerning model form might be violated.

The residuals, $y_i - \hat{y}_i$, are estimates of ϵ; with n observations in a regression analysis, we have n residuals. Residual plots are graphical presentations of the residuals that help reveal patterns and thus help determine whether or not the assumptions concerning ϵ are satisfied. Three of the most common residual plots are

1. A plot of the residuals against the independent variable, x,
2. A plot of the residuals against the predicted value of the dependent variable, $\hat{y}$, and
3. A standardized plot in which each residual is replaced by its z-score (i.e., the mean is subtracted and the result is divided by the standard error).

Residual Plot Against x

A residual plot against the independent variable, x, is constructed by placing x on the horizontal axis and the residuals on the vertical axis. A residual is plotted for each observation; the first coordinate is x_i, the second is the residual, $y_i - \hat{y}_i$. Figure 13.10 shows some of the patterns statisticians look for when analyzing residuals. Panel A shows the type of plot to expect when the assumptions are satisfied. The patterns shown in Panels B and C indicate violation of one or more assumptions.

If the assumption that the variance of ϵ is the same for all values of x is valid, the residual plot should give an overall impression of a horizontal band of points. Panel A of Figure 13.10 shows the type of pattern to be expected in this case. On the other hand, if the variance of ϵ is not constant—for example, the variability about the regression line is greater for larger values of x—we would observe a pattern such as that of Panel B of Figure 13.10. Finally, if we observe a residual pattern such as that of Panel C of Figure 13.10, we would conclude that the error terms are not independent. The assumption of a linear relationship between x and y is not appropriate, or perhaps a multiple regression model is needed.

A plot of the residuals against the independent variable x for the Armand's Pizza problem is shown in Figure 13.11 (the residuals were computed in Table 13.3). Looking at Figure 13.11, we see that the residuals appear to follow the pattern of Panel A of Figure 13.10. We thus conclude that the assumptions are satisfied and that the simple linear regression model for the Armand's Pizza problem is valid.

Residual Plot Against $\hat{y}$

A residual plot against the predicted value of the independent variable is constructed by placing the predicted value on the horizontal axis and plotting each residual directly above the corresponding value of $\hat{y}_i$. A plot of the residuals against the predicted values for the Armand's Pizza problem (the predicted values were also computed in Table 13.3) is shown in Figure 13.12. Note that the pattern of this residual plot is the same as the pattern of the residual plot against the independent variable x. For simple linear regression, both the residual plot

A

$y - \widehat{y}$

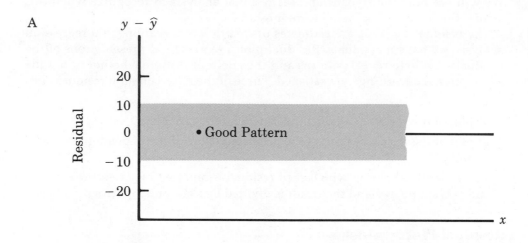

B

$y - \widehat{y}$

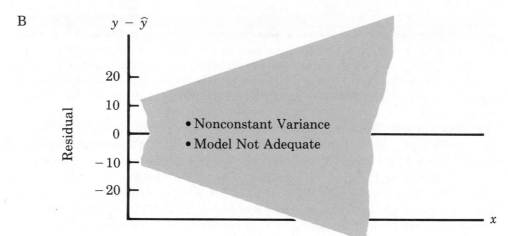

C

$y - \widehat{y}$

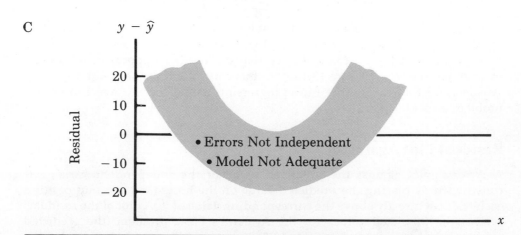

▬ FIGURE 13.10   **Residual Plots From Three Regression Studies**

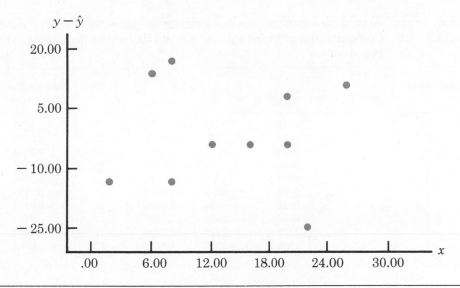

▬ FIGURE 13.11   **Plot of the Residuals Against the Independent Variable**
x for the Armand's Pizza Problem

against x and the plot against $\hat{y}$ provide the same information. With multiple
regression models (more than one independent variable), the residual plot
against $\hat{y}$ is more widely used.

Standardized Residuals

Many of the residual plots provided by computer software packages are a
standardized version of the residuals. As we have seen in earlier chapters, a

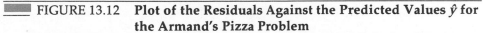

▬ FIGURE 13.12   **Plot of the Residuals Against the Predicted Values $\hat{y}$ for**
the Armand's Pizza Problem

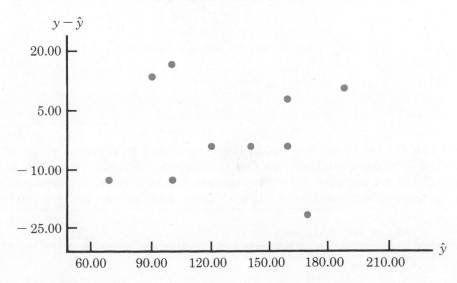

TABLE 13.8 **Computation of Standardized Residuals for the Armand's Pizza Problem**

Restaurant i	$x_i - \bar{x}$	$\dfrac{(x_i - \bar{x})^2}{\Sigma x_i^2 - (\Sigma x_i)^2/10}$	h_i	$s_{y_i - \hat{y}_i}$	$y_i - \hat{y}_i$	Standardized Residual
1	−12	.2535	.3535	11.1193	−12	−1.0792
2	−8	.1127	.2127	12.2709	15	1.2224
3	−6	.0634	.1634	12.6493	−12	−0.9487
4	−6	.0634	.1634	12.6493	18	1.4230
5	−2	.0070	.1070	13.0682	−3	−0.2296
6	2	.0070	.1070	13.0682	−3	−0.2296
7	6	.0634	.1634	12.6493	−3	−0.2372
8	6	.0634	.1634	12.6493	9	0.7115
9	8	.1127	.2127	12.2709	−21	−1.7114
10	12	.2535	.3535	11.1193	12	1.0792

Note: From Table 13.2, we can compute $\bar{x} = 14$ and $\Sigma x_i^2 - (\Sigma x_i^2)/10 = 568$. The values of $y_i - \hat{y}_i$ are given in Table 13.3.

random variable is standardized by subtracting its mean and dividing the result by its standard deviation. With the least squares method, the mean of the residuals is zero. Thus, simply dividing each residual by its standard deviation provides the standardized residual.

It can be shown that the standard deviation of the ith residual depends on $s = \sqrt{MSE}$ and the corresponding value of the independent variable.

Standard Deviation of ith Residual*

$$s_{y_i - \hat{y}_i} = \sqrt{s^2(1 - h_i)} \tag{13.24}$$

where

$$s_{y_i - \hat{y}_i} = \text{standard deviation of residual } i$$

$$h_i = \frac{1}{n} + \frac{(x_i - \bar{x})^2}{\Sigma x_i^2 - (\Sigma x_i)^2/n}$$

Note that (13.24) shows that residuals corresponding to different values of x have different standard deviations. Once the standard deviation of each residual is calculated, we compute the standardized residual by dividing each residual by its corresponding standard deviation. Table 13.8 shows the calculations

*This is actually an estimate of the standard deviation of the ith residual, since s^2 is used instead of σ^2. The value of σ^2 is never known when working with real data and is always estimated by s^2.

Standardized
Residuals

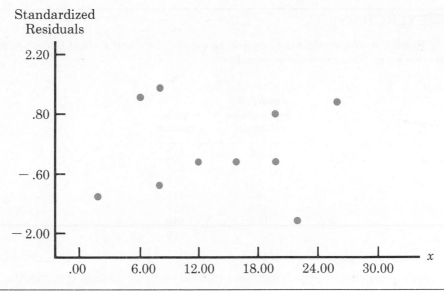

■ FIGURE 13.13   **Plot of the Standardized Residuals Against *x* for the Armand's Pizza Problem**

involved for the Armand's Pizza problem. (Recall that $s^2 = 191.25$ for the Armand's Pizza problem.) Figure 13.13 is a plot of the standardized residuals against *x*. Note that, in this case, the standardized residual plot has the same general pattern as the original residual plot shown in Figure 13.11.

In addition to the information provided by the other residual plots, the standardized residual plot can provide insight concerning the normality assumption for ϵ (assumption 4). If the normality assumption is satisfied, the standardized residuals should appear to come from a standard normal probability distribution.* Thus when looking at a standardized residual plot, we should expect to see approximately 95% of the standardized residuals between -2 and $+2$. Referring to the standardized residuals for the Armand's Pizza problem (Table 13.8), we see that they are all between -2 and $+2$. Thus we conclude that the normality assumption is valid.

The analysis of residuals is the primary method by which statisticians verify that the assumptions are satisfied in a regression model. To validate a model, both the assumption concerning model form (simple linear, in this chapter) and the assumptions on the error term are checked for possible violations. Even if no violations are found, it does not necessarily follow that the model will yield good predictions. However, if in addition r^2 is large and the model is statistically significant, one should obtain good results.

---

*Since s^2 is substituted for σ^2 in (13.24), the probability distribution of the standardized residuals is not technically normal. However, in most regression studies, the sample size is large enough that a normal approximation is very good.

▬ EXERCISES

35. In Exercise 4 data concerning advertising expenditures and sales at Eddie's Restaurants were given. These data are repeated here:

Advertising Expenditures ($1000s)	Sales ($1000s)
1.0	19.0
2.0	32.0
4.0	44.0
6.0	40.0
10.0	52.0
14.0	53.0
20.0	54.0

a. Let x equal advertising expenditures ($1000s) and y equal sales ($1000s). Use the method of least squares to develop a straight line approximation to the relationship between the two variables.

b. Test whether or not sales and advertising expenditures are related at the $\alpha = .05$ level of significance.

c. Prepare a residual plot of $y - \hat{y}$ versus $\hat{y}$. Use the result of part a to obtain the values of $\hat{y}$.

d. What conclusions can you draw from residual analysis? Should this model be used, or should we look for a better one?

36. The following data were used in a regression study:

Observation	x_i	y_i
1	2	4
2	3	5
3	4	4
4	5	6
5	7	4
6	7	6
7	7	9
8	8	5
9	9	11

a. Develop an estimated regression equation for this data.

b. Construct a plot of the residuals. Do the assumptions concerning the error terms seem to be satisfied?

37. Refer to Exercise 5, where an estimated regression equation relating years of experience and annual sales was developed.

a. Compute the residuals and construct a residual plot for this problem.

b. Do the assumptions concerning the error terms seem reasonable in light of the residual plot?

38. The following data show the number of employees and the yearly revenues for the ten largest wholesale bakers (*Louis Rukeyser's Business Almanac*).

Company	Employees	Revenues ($1,000,000s)
Nabisco Brands USA	9,500	1,734
Continental Baking Co.	22,400	1,600
Campbell Taggart, Inc.	19,000	1,044
Keebler Company	8,943	988
Interstate Bakeries Corp.	11,200	704
Flowers Industries, Inc.	10,200	557
Sunshine Biscuits, Inc.	5,000	490
American Bakeries Co.	6,600	461
Entenmann's Inc.	3,734	450
Kitchens of Sara Lee	1,550	405

a. Use a computer package to develop an estimated regression equation relating revenues (y) to the number of employees (x).

b. Construct a residual plot of the standardized residuals against the independent variable.

c. Do the assumptions concerning the error terms and model form seem reasonable in light of the residual plot?

13.8 ■ CORRELATION ANALYSIS

As we indicated in the introduction to this chapter, there are some situations in which the decision maker is not as concerned with the equation that relates two variables as in measuring the extent to which the two variables are related. In such cases a statistical technique referred to as correlation analysis can be used to determine the strength of the relationship between the two variables.* The output of a correlation study is a number referred to as the correlation coefficient. Because of the way in which it is defined, values of the correlation coefficient are always between -1 and $+1$. A value of $+1$ indicates that x and y are perfectly related in a positive linear sense. That is, all the points lie on a straight line that has a positive slope. A value of -1 indicates that x and y are perfectly related in a negative linear sense. That is, all the points lie on a straight line that has a negative slope. Values of the correlation coefficient close to zero indicate that x and y are not linearly related.

To provide an illustration of correlation analysis, we consider the situation of a stereo and sound-equipment store located in San Francisco. Management would like to investigate whether or not there is any relationship between the number of commercials (x) shown on Friday evening television and the resulting sales volume on Saturday (y), measured in hundreds of dollars. The sample data that were obtained are shown in Table 13.9.

In Figure 13.14 we show a scatter diagram of this data. The scatter diagram appears to indicate that there is a positive linear relationship between x and y. To measure the degree of linear association between these two variables, we first define a measure of linear association known as the *covariance*.

---

*In correlation analysis it is assumed that x and y are both random variables.

TABLE 13.9   **Sample Data for the Stero and Sound-Equipment Problem**

Store	x = Number of Commercials	y = Sales Volume ($100s)
1	2	24
2	5	28
3	1	22
4	3	26
5	4	25
6	1	24
7	5	26

Covariance

Sample covariance is defined as follows.

Sample Covariance

$$s_{xy} = \frac{\Sigma (x_i - \bar{x})(y_i - \bar{y})}{n - 1} \qquad (13.25)$$

FIGURE 13.14   **Scatter Diagram for the Stereo and Sound-Equipment Problem**

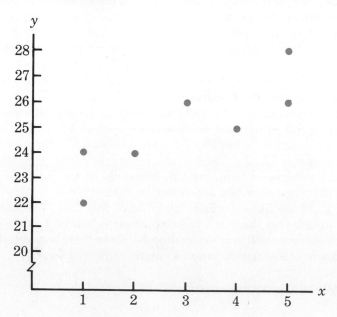

In this formula each x_i value is paired with a y_i value. We then sum the products obtained by multiplying the deviation of each x_i from its sample mean, $\bar{x}$, times the deviation of the corresponding y_i from its sample mean, $\bar{y}$; this sum is then divided by $n - 1$.

To measure the strength of the linear relationship between the number of commercials (x) and the sales volume (y) in the stereo and sound-equipment problem, we can use (13.25) to compute the sample covariance. The calculations shown in Table 13.10 illustrate the computation of $\Sigma (x_i - \bar{x})(y_i - \bar{y})$. Note that $\bar{x} = 21/7 = 3$ and $\bar{y} = 175/7 = 25$. Using (13.25) we obtain

$$s_{xy} = \frac{\Sigma (x_i - \bar{x})(y_i - \bar{y})}{n - 1} = \frac{17}{6} = 2.8333$$

The formula for computing the covariance of a population of size N is similar to (13.25), but we use different notation to indicate that we are dealing with the entire population.

Population Covariance

$$\sigma_{xy} = \frac{\Sigma (x_i - \mu_x)(y_i - \mu_y)}{N} \tag{13.26}$$

In (13.26) we use the notation μ_x for the population mean of the variable x and μ_y for the population mean of the variable y. The sample covariance, s_{xy} is an estimate of the population covariance, σ_{xy} based upon a sample of size n.

Interpretation of the Covariance

To aid in the interpretation of the *sample covariance,* consider Figure 13.15. It is the same as the scatter diagram of Figure 13.14 with a vertical line at $x = 3$ (the value of $\bar{x}$) and a horizontal line at $y = 25$ (the value of $\bar{y}$). Four quadrants have been identified on the graph. Points that fall in quadrant I correspond to x_i

TABLE 13.10 **Calculations for the Sample Covariance**

x_i	y_i	$x_i - \bar{x}$	$y_i - \bar{y}$	$(x_i - \bar{x})(y_i - \bar{y})$
2	24	−1	−1	1
5	28	2	3	6
1	22	−2	−3	6
3	26	0	1	0
4	25	1	0	0
1	24	−2	−1	2
5	26	2	1	2
Totals 21	175	0	0	17

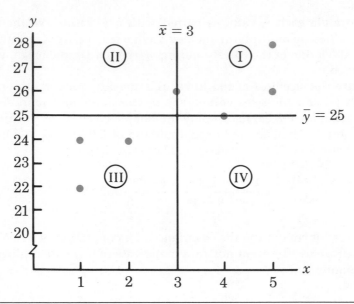

■■■ FIGURE 13.15   **Quadrants I, II, III, and IV for the Stereo and Sound-Equipment Problem**

values greater than $\bar{x}$ and y_i values greater than $\bar{y}$; points that fall in quadrant II correspond to x_i values less than $\bar{x}$ and y_i values greater than $\bar{y}$, and so on. Thus the value of $(x_i - \bar{x})(y_i - \bar{y})$ must be positive for points located in quadrant I, negative for points located in quadrant II, positive for points located in quadrant III and negative for points located in quadrant IV.

If the value of s_{xy} is positive, the points that have had the greatest effect on s_{xy} must lie in quadrants I and/or III. Hence, a positive value for s_{xy} is indicative of a positive linear association between x and y; that is, as the value of x increases, the value of y increases. If the value of s_{xy} is negative, however, the points that have had the greatest effect on s_{xy} lie in quadrants II and/or IV. Hence, a negative value for s_{xy} is indicative of a negative linear association between x and y; that is, as the value of x increases, the value of y decreases. Finally, if the points are evenly distributed across all four quadrants, the value of s_{xy} will be close to 0, indicating no linear association between x and y. Figure 13.16 shows the values of s_{xy} that can be expected with these three different types of scatter diagrams.

From the previous discussion it might appear that a large positive value for the covariance is indicative of a strong positive linear relationship and that a large negative value is indicative of a strong negative linear relationship. However, one problem with using covariance as a measure of the strength of the linear relationship is that the value we obtain for the covariance depends upon the units of measurement for x and y. For example, suppose we were interested in the relationship between height (x) and weight (y) for individuals. If height is measured in inches, we will get much larger numerical values for $(x_i - \bar{x})$ than if it is measured in feet. Thus, with height measured in inches, we would obtain larger values for $\Sigma (x_i - \bar{x})(y_i - \bar{y})$—and hence a larger covariance—when, in fact, there is no difference in the relationship. A measure of relationship that avoids this difficulty is the *correlation coefficient*.

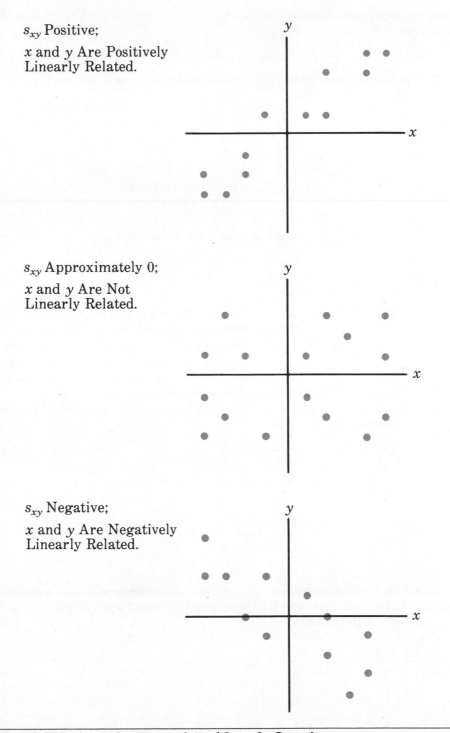

s_{xy} Positive;

x and y Are Positively Linearly Related.

s_{xy} Approximately 0;

x and y Are Not Linearly Related.

s_{xy} Negative;

x and y Are Negatively Linearly Related.

FIGURE 13.16   **Interpretation of Sample Covariance**

Correlation Coefficient

For sample data, the Pearson Product Moment correlation coefficient is defined as follows.

Pearson Product Moment Correlation Coefficient — Sample Data

$$r_{xy} = \frac{s_{xy}}{s_x s_y} \qquad (13.27)$$

where

r_{xy} = sample correlation coefficient

s_{xy} = sample covariance

s_x = sample standard deviation of x

s_y = sample standard deviation of y

Equation (13.27) shows that the Pearson Product Moment correlation coefficient for sample data (commonly referred to more simply as the *sample correlation coefficient*) is computed by dividing the sample covariance by the product of the standard deviation of x and the standard deviation of y. Before we consider further interpretation of the sample correlation coefficient, let us consider the use of (13.27) for the stereo and sound-equipment problem.

Using the data presented in Table 13.10, we can compute the sample correlation coefficient.

$$s_x = \sqrt{\frac{\Sigma (x_i - \overline{x})^2}{n - 1}} = \sqrt{\frac{18}{6}} = 1.7321$$

$$s_y = \sqrt{\frac{\Sigma (y_i - \overline{y})^2}{n - 1}} = \sqrt{\frac{22}{6}} = 1.9149$$

TABLE 13.11 Computations for Using the Alternate Formula for Computing r_{xy}

	x_i	y_i	$x_i y_i$	x_i^2	y_i^2
	2	24	48	4	576
	5	28	140	25	784
	1	22	22	1	484
	3	26	78	9	676
	4	25	100	16	625
	1	24	24	1	576
	5	26	130	25	676
Totals	21	175	542	81	4397

and, since $s_{xy} = 2.8333$, we have

$$r_{xy} = \frac{s_{xy}}{s_x s_y} = \frac{2.8333}{(1.7321)\,(1.9149)} = .854$$

When using a calculator to compute the sample correlation coefficient, the formula given by (13.28) is preferred. Since the computation of each deviation $x_i - \bar{x}$ and $y_i - \bar{y}$ is not necessary, less round-off error is introduced.

Pearson Product Moment Correlation Coefficient — Sample Data Alternate Formula

$$r_{xy} = \frac{\Sigma\, x_i y_i - (\Sigma\, x_i \Sigma\, y_i)/n}{\sqrt{\Sigma\, x_i^2 - (\Sigma\, x_i)^2/n}\,\sqrt{\Sigma\, y_i^2 - (\Sigma\, y_i)^2/n}} \qquad (13.28)$$

Algebraically, equations (13.27) and (13.28) are equivalent. In Table 13.11 we provide the calculations needed to use (13.28). Using these computations and (13.28), we obtain:

$$r_{xy} = \frac{542 - (21)\,(175)/7}{\sqrt{81 - (21)^2/7}\,\sqrt{4397 - (175)^2/7}} = \frac{17}{19.8997} = .854$$

Thus we see that the value obtained for r_{xy} using (13.28) is the same (to three decimal places) as the value obtained using (13.27).

The formula for computing the correlation coefficient of a population, denoted by the Greek letter ρ_{xy} (rho, pronounced "row"), is as follows.

Pearson Product Moment Correlation Coefficient — Population Data

$$\rho_{xy} = \frac{\sigma_{xy}}{\sigma_x \sigma_y} \qquad (13.29)$$

where

$\rho_{xy} = $ population correlation coefficient

$\sigma_{xy} = $ population covariance

$\sigma_x = $ population standard deviation for x

$\sigma_y = $ population standard deviation for y

The sample correlation coefficient, r_{xy}, is an estimate of the population correlation coefficient, ρ_{xy}.

Interpretation of the Correlation Coefficient

First let us consider a simple example that illustrates the concept of perfect positive linear association. The scatter diagram shown in Figure 13.17 depicts the relationship between the following $n = 3$ pairs of points:

x_i	1	2	3
y_i	10	30	50

The straight line drawn through each of the 3 points shows that there is a perfect linear relationship between the two variables x and y. The calculations need to compute r_{xy} are shown in Table 13.12. Using the values in this table we obtain

$$r_{xy} = \frac{\Sigma x_i y_i - (\Sigma x_i \Sigma y_i)/n}{\sqrt{\Sigma x_i^2 - (\Sigma x_i)^2/n} \sqrt{\Sigma y_i^2 - (\Sigma y_i)^2/n}}$$

$$= \frac{220 - (6)(90)/3}{\sqrt{14 - (6)^2/3} \sqrt{3500 - (90)^2/3}} = \frac{40}{40} = 1$$

Thus we see that the value of the sample correlation coefficient for this data set is 1.

In general, it can be shown that if all the points in a data set fall on a straight line having positive slope, then the value of the sample correlation coefficient is $+1$; that is, a sample correlation coefficient of $+1$ corresponds to a perfect positive linear association between x and y. Moreover, if the points in the data set fall on a straight line having negative slope, the value of the sample correlation coefficient is -1; that is, a sample correlation coefficient of -1 corresponds to a perfect negative linear association between x and y.

Let us now suppose that for a certain data set there is a positive linear

---

◼◼ FIGURE 13.17　**Scatter Diagram Depicting a Perfect Positive Linear Association**

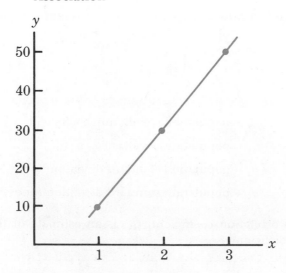

TABLE 13.12 Calculations for Computing r for the Example Used to Illustrate Perfect Positive Linear Association

	x_i	y_i	$x_i y_i$	x_i^2	y_i^2
	1	10	10	1	100
	2	30	60	4	900
	3	50	150	9	2500
Totals	6	90	220	14	3500

association between x and y but that the relationship is not perfect. The value of r_{xy} will be less than $+1$, indicating that the points in the scatter diagram do not all fall on a straight line. As the points in a data set deviate more and more from a perfect positive linear association, the value of r_{xy} becomes smaller and smaller. A value of r_{xy} equal to 0 indicates no linear relationship between x and y and values of r_{xy} near zero indicate a weak relationship.

Recall that for our data set involving the stereo and sound-equipment store, $r_{xy} = +.854$. Since $r_{xy} = +.854$ we conclude that there appears to be a positive linear association between the number of commercials and Saturday sales volume. More specifically, an increase in the number of commercials is associated with an increase in sales volume.

We have stated that values of r_{xy} near $+1$ indicate a strong linear association between two variables and values of r_{xy} near zero indicate little or no linear association between the variables. But we must be careful not to conclude that a value of r near zero means there is no relationship between the variables. The scatter diagram in Figure 13.18 shows a case where $r_{xy} = 0$ and there is no linear relationship; however, in this case, there is a perfect curvilinear relationship between the variables. Table 13.13 provides the calculations needed to compute r_{xy} for this example. Using these calculations, the computation of r_{xy} is as follows:

$$r_{xy} = \frac{210 - (42)(35)/7}{\sqrt{310 - (42)^2/7}\sqrt{231 - (35)^2/7}} = \frac{0}{56.9912} = 0$$

To reiterate, our last example illustrates an important concept regarding the proper interpretation of the sample correlation coefficient. The sample correlation coefficient measures only the degree of *linear association* between the two variables. A value of r_{xy} equal to 0 cannot be interpreted as implying that there is no relationship between the two variables. One should always look at the associated scatter diagram as well as the value of the sample correlation coefficient when attempting to determine if, and how, two variables are related.

In closing this part of the section we caution that while a correlation coefficient near ± 1 does imply a strong linear association between two variables, it does not imply a cause-effect relationship. For instance, it has been noted that as women's hemlines are raised, stock prices go up. There is a positive correlation, but it would be folly to infer a cause-effect relationship. No one

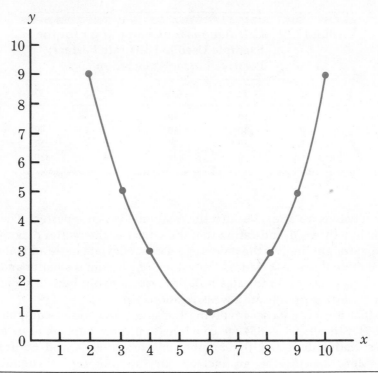

■ FIGURE 13.18   **Even Though $r = 0$, There Is a Perfect Curvilinear Relationship for the Data**

truly believes that changes in the stock market are the result of changes in women's hemlines. Conclusions concerning cause-effect must be based on the judgment of the analyst.

Determining the Sample Correlation Coefficient from the Regression Analysis Output

In this discussion we will assume that the least squares estimated regression equation is $\hat{y} = b_0 + b_1 x$. In such cases the sample correlation coefficient can be

TABLE 13.13   **Calculations for the Example Illustrating $r_{xy} = 0$**

x_i	y_i	$x_i y_i$	x_i^2	y_i^2
2	9	18	4	81
3	5	15	9	25
4	3	12	16	9
6	1	6	36	1
8	3	24	64	9
9	5	45	81	25
10	9	90	100	81
Totals 42	35	210	310	231

computed using one of the following formulas:

Sample Correlation Coefficient

$$r_{xy} = (\text{sign of } b_1) \sqrt{\text{Coefficient of Determination}} = \pm\sqrt{r^2} \qquad (13.30)$$

$$r_{xy} = b_1 \left(\frac{s_x}{s_y}\right) \qquad (13.31)$$

where

$$s_x = \sqrt{\frac{\Sigma (x_i - \bar{x})^2}{(n-1)}} \quad \text{(Sample standard deviation for } x)$$

$$s_y = \sqrt{\frac{\Sigma (y_i - \bar{y})^2}{(n-1)}} \quad \text{(Sample standard deviation for } y)$$

Note that the sign of the sample correlation coefficient is the same as the sign of b_1, the slope of the estimated regression equation. For the Armand's Pizza problem presented earlier in this chapter, $b_1 = 5$, indicating a positive relationship. Thus with $r^2 = .903$ we obtain

$$r_{xy} = \pm \sqrt{0.903}$$
$$= +0.950$$

The sample correlation coefficient is really a point estimator of the population correlation coefficient, a measure of the actual linear relationship between x and y in the population. Recall from our previous discussion that ρ_{xy} denotes the population correlation coefficient. A statistical test for the significance of a linear association between x and y can be performed by testing the following hypotheses:

$$H_0: \quad \rho_{xy} = 0$$
$$H_a: \quad \rho_{xy} \neq 0$$

It can be shown that testing these hypotheses is equivalent to testing the hypotheses regarding the significance of β_1, the slope of the regression equation. Recall that the appropriate hypotheses in this case are

$$H_0: \quad \beta_1 = 0$$
$$H_a: \quad \beta_1 \neq 0$$

Since for the Armand's Pizza problem, we earlier rejected the null hypothesis $H_0: \beta_1 = 0$ (see Section 13.4) we can also reject the null hypothesis $H_a: \rho_{xy} = 0$ and conclude that x and y are correlated. To test for a significant correlation without performing a regression study, statisticians have developed a procedure for testing the following:

$$H_0: \quad \rho_{xy} = 0$$

$$H_a: \quad \rho_{xy} \neq 0$$

It can be shown that if H_0 is true, then the value of

$$r_{xy} \sqrt{\frac{n-2}{1-r_{xy}^2}} \tag{13.32}$$

has a t distribution with $n - 2$ degrees of freedom.

For Armand's Pizza Parlors with $\alpha = .05$ and $n - 2 = 10 - 2 = 8$ degrees of freedom, we see that the appropriate t value from Table 2 of Appendix B is 2.306. Thus if the value of (13.32) exceeds 2.306 or is less than -2.306 we must reject the null hypothesis $H_0: \rho_{xy} = 0$.

With a sample correlation coefficient of $r_{xy} = .950$, the value of (13.32) is

$$.950 \sqrt{\frac{8}{1-.903}} = 8.63$$

Since 8.63 exceeds the t value of 2.306, we reject H_0 and hence conclude that x and y have a significant correlation. We note that this test yields the same result as the previous test on β_1.

■ EXERCISES

39. A high-school guidance counselor collected the following data regarding the grade point average (GPA) and the SAT mathematics test score for 6 seniors.

GPA	2.7	3.5	3.7	3.3	3.6	3.0
SAT	440	560	720	520	640	480

a. Develop a scatter diagram for these data with GPA as the independent variable.

b. Does there appear to be any relationship between the GPA and the SAT mathematics test score? Explain.

c. Compute and interpret the sample covariance for these data.

d. Compute the correlation coefficient for these data using (13.27). What does this value tell us about the relationship between the two variables?

e. Compute the correlation coefficient using (13.28). When using a calculator, why is this formula preferred over (13.27)?

40. Given are five observations taken for two variables.

x	4	6	11	3	16
y	50	50	40	60	30

a. Develop a scatter diagram with x on the horizontal axis.

b. What does the scatter diagram developed in part a indicate about the relationship between the two variables?

c. Compute and interpret the sample covariance for these data.

d. Compute and interpret the sample correlation coefficient for these data.

41. A sociologist collected data regarding the age of the wife and husband when they were married.

Wife's age	19	42	28	25	36
Husband's age	20	32	31	24	33

a. Develop a scatter diagram for these data with the wife's age on the horizontal axis.

b. Does there appear to be a linear association? Explain.

42. Given are five observations taken for two variables.

x	6	11	15	21	27
y	6	9	6	17	12

Comment on the relationship between these 2 variables.

43. The following estimated regression equation has been developed to estimate the relationship between x, the number of units produced per week, and y, the total weekly cost of production ($):

$$\hat{y} = 60 + 3.2x$$

The standard deviation of weekly production is 10 units, and the standard deviation of weekly cost is $35.00. Compute the sample correlation coefficient r_{xy}.

44. Eight observations on two random variables are given.

x_i	y_i
2	11
9	4
6	6
8	5
4	9
7	4
5	9
6	7

a. Compute r_{xy}.

b. Test the hypotheses

$$H_0: \quad \rho_{xy} = 0$$
$$H_a: \quad \rho_{xy} \neq 0$$

at the $\alpha = .01$ level of significance.

45. Use the data in Exercise 3 to compute the sample correlation coefficient between GPA and salary.

46. The data presented in Exercise 30 are shown.

Speed	30	50	40	55	30	25	60	25	50	55
Mileage	28	25	25	23	30	32	21	35	26	25

a. Compute the sample correlation coefficient for these data.

b. Test the hypotheses

$$H_0: \quad \rho_{xy} = 0$$

$$H_a: \quad \rho_{xy} \neq 0$$

at the $\alpha = .01$ level of significance.

47. As more U.S. households receive cable television, the advertising revenue has continued to increase. The following data show the 1988 and 1987 expenditures for the top ten cable television advertisers.

Advertiser	1988 Expenditure ($1,000,000s)	1987 Expenditure ($1,000,000s)
Procter & Gamble Co.	30.2	23.7
Philip Morris Cos.	23.1	20.6
Anheuser-Busch Cos.	21.4	22.9
Time Inc.	21.2	16.4
General Mills Inc.	20.0	18.6
RJR Nabisco Inc.	14.3	14.7
Eastman Kodak	11.0	2.6
Clorox Co.	10.1	6.9
Mars Inc.	10.0	14.9
Chrysler Corp.	9.5	6.1

a. Develop a scatter diagram for the data. Does it appear that the two variables are linearly related?
b. Compute the sample correlation coefficient for these data.
c. Test the hypotheses

$$H_0: \quad \rho_{xy} = 0$$

$$H_a: \quad \rho_{xy} \neq 0$$

at the $\alpha = .01$ level at significance.

■■■ SUMMARY

In this chapter we introduced the topics of regression and correlation analysis. We discussed how regression analysis can be used to develop an equation showing how variables are related and how correlation analysis can be used to determine the strength of the relationship between two variables. Before concluding our discussion, however, we would like to reemphasize a potential misinterpretation of these studies. Regression and correlation analyses can indicate only how or to what extent the variables are associated with each other. These techniques cannot be interpreted directly as showing cause and effect relationships.

■■■ GLOSSARY

Note: The definitions here are all stated with the understanding that simple linear regression and correlation is being considered.

Dependent variable　The variable that is being predicted or explained. It is denoted by y in the regression equation.

Independent variable The variable that is doing the predicting or explaining. It is denoted by x in the regression equation.

Simple linear regression The simplest kind of regression, involving only two variables that are related approximately by a straight line.

Regression equation The mathematical equation relating the independent variable to the expected value of the dependent variable; that is, $E(y) = \beta_0 + \beta_1 x$.

Estimated regression equation The estimate of the regression equation obtained by the least squares method; i.e., $\hat{y} = b_0 + b_1 x$.

Scatter diagram A graph of the available data in which the independent variable appears on the horizontal axis and the dependent variable appears on the vertical axis.

Least squares method The approach used to develop the estimated regression equation which minimizes the sum of squared residuals.

Coefficient of determination (r^2) A measure of the variation explained by the estimated regression equation. It is a measure of how well the estimated regression equation fits the data.

Deterministic model A relationship between an independent variable and a dependent variable whereby specifying the value of the independent variable allows one to compute exactly the value of the dependent variable.

Probabilistic model A relationship between an independent variable and a dependent variable in which specifying the value of the independent variable is not sufficient to allow determination of the value of the dependent variable.

Residual The difference between the observed value of the dependent variable and the value predicted using the estimated regression equation; i.e., $y_i - \hat{y}_i$.

Standardized residual The value obtained by dividing the residual by its standard deviation.

Sample correlation coefficient (r_{xy}) A statistical measure of the linear association between two variables.

■ KEY FORMULAS

Estimated Regression Line

$$\hat{y} = b_0 + b_1 x \tag{13.1}$$

Least Squares Estimates

$$b_1 = \frac{\Sigma(x_i - \bar{x})(y_i - \bar{y})}{\Sigma(x_i - \bar{x})^2} = \frac{\Sigma x_i y_i - (\Sigma x_i \Sigma y_i)/n}{\Sigma x_i^2 - (\Sigma x_i)^2/n} \tag{13.3}$$

$$b_0 = \bar{y} - b_1 \bar{x} \tag{13.4}$$

Sum of Squares Due to Error

$$\text{SSE} = \Sigma(y_i - \hat{y}_i)^2 \tag{13.5}$$

Total Sum of Squares

$$\text{SST} = \Sigma(y_i - \bar{y})^2 \tag{13.6}$$

Sum of Squares Due to Regression

$$\text{SSR} = \Sigma(\hat{y}_i - \bar{y})^2 \tag{13.7}$$

Relationship Among SST, SSR, and SSE

$$SST = SSR + SSE \tag{13.8}$$

Coefficient of Determination

$$r^2 = \frac{SSR}{SST} \tag{13.9}$$

Computational Formula for SSR

$$SSR = \frac{[\Sigma x_i y_i - (\Sigma x_i \, \Sigma y_i)/n]^2}{\Sigma x_i^2 - (\Sigma x_i)^2/n} \tag{13.10}$$

Computational Formula for SST

$$SST = \Sigma y_i^2 - (\Sigma y_i)^2/n \tag{13.11}$$

Regression Model

$$y = \beta_0 + \beta_1 x + \epsilon \tag{13.12}$$

Regression Equation

$$E(y) = \beta_0 + \beta_1 x \tag{13.13}$$

Estimate of σ^2

$$s^2 = MSE = \frac{SSE}{n-2} \tag{13.14}$$

Mean Square Due to Regression

$$MSR = \frac{SSR}{\text{Regression DF}} = \frac{SSR}{\text{No. of independent variables}} \tag{13.15}$$

The F Statistic

$$F = \frac{MSR}{MSE} \tag{13.16}$$

Variance of b_1

$$\sigma_{b_1}^2 = \sigma^2 \left(\frac{1}{\Sigma x_i^2 - (\Sigma x_i)^2/n} \right) \tag{13.18}$$

Estimated Variance of b_1

$$s_{b_1}^2 = s^2 \left(\frac{1}{\Sigma x_i^2 - (\Sigma x_i)^2/n} \right) \tag{13.19}$$

Estimated Variance of $\hat{y}_p$

$$s_{\hat{y}_p}^2 = s^2 \left(\frac{1}{n} + \frac{(x_p - \bar{x})^2}{\Sigma x_i^2 - (\Sigma x_i)^2/n} \right) \tag{13.20}$$

Confidence Interval Estimate of $E(y_p)$

$$\hat{y}_p \pm t_{\alpha/2} s_{\hat{y}_p} \tag{13.21}$$

Estimated Variance When Predicting an Individual Value

$$s_{ind}^2 = s^2 \left(1 + \frac{1}{n} + \frac{(x_p - \bar{x})^2}{\Sigma x_i^2 - (\Sigma x_i)^2/n} \right) \qquad (13.22)$$

Prediction Interval Estimate of y_p

$$\hat{y}_p \pm t_{\alpha/2} s_{ind} \qquad (13.23)$$

Sample Covariance

$$s_{xy} = \frac{\Sigma(x_i - \bar{x})(y_i - \bar{y})}{n - 1} \qquad (13.25)$$

Pearson Product Moment Correlation Coefficient – Sample Data

$$r_{xy} = \frac{s_{xy}}{s_x s_y} \qquad (13.27)$$

Pearson Product Moment Correlation Coefficient – Sample Data, Alternate Formula

$$r_{xy} = \frac{\Sigma x_i y_i - (\Sigma x_i \, \Sigma y_i)/n}{\sqrt{\Sigma x_i^2 - (\Sigma x_i)^2/n} \; \sqrt{\Sigma y_i^2 - (\Sigma y_i)^2/n}} \qquad (13.28)$$

Determining the Sample Correlation Coefficient from the Regression Analysis

$$r_{xy} = (\text{sign of } b_1) \, \sqrt{\text{Coefficient determination}} = \pm \sqrt{r^2} \qquad (13.30)$$

■■■ SUPPLEMENTARY EXERCISES

48. What is the difference between regression analysis and correlation analysis?

49. Does a high value of r^2 imply that two variables are causally related? Explain.

50. In your own words, explain the difference between an interval estimate of the mean value of y for a given x and an interval estimate for an individual value of y for a given x.

51. How do we measure how closely the actual data points are to the estimated regression line? That is, how do we measure the goodness-of-fit of the regression line?

52. What is the purpose of testing whether or not $\beta_1 = 0$?

53. In a manufacturing process the assembly line speed (feet per minute) was thought to affect the number of defective parts found during the inspection process. To test this theory, management devised a situation where the same batch of parts was inspected visually at a variety of line speeds. The following data were collected.

Line Speed	Number of Defective Parts Found
20	21
20	19
40	15
30	16
60	14
40	17

a. Develop the estimated regression equation that relates line speed to the number of defective parts found.

b. At the $\alpha = .05$ level of significance determine whether or not line speed and number of defective parts found are related.

c. Did the estimated regression equation provide a good fit to the data?

d. Develop a 95% confidence interval to predict the mean number of defective parts for a line speed of 50 feet per minute.

54. A study was conducted by Monsanto Company in order to determine the relationship between the percentage of supplemental methionine used in feed and the body weight of the poultry. Using the data collected in this study, regression analysis was used to develop the following estimated regression line.

$$\hat{y} = 0.21 + 0.42x$$

where

$\hat{y}$ = estimated body weight in kilograms

x = percentage of supplemental methionine used in the feed

The coefficient of determination, r^2, was .78, indicating a reasonably good fit for the data.

Suppose it is known that a sample size of 30 was used for the study and that SST = 45.

a. Compute SSR and SSE.

b. Test for a significant regression relationship using $\alpha = .01$.

c. What is the value of the correlation coefficient?

55. The PJH&D Company is in the process of deciding whether or not to purchase a maintenance contract for its new word-processing system. They feel that maintenance expense should be related to usage and have collected the following information on weekly usage (hours) and annual maintenance expense.

Weekly Usage (hours)	Annual Maintenance Expense ($100s)
13	17.0
10	22.0
20	30.0
28	37.0
32	47.0
17	30.5
24	32.5
31	39.0
40	51.5
38	40.0

a. Develop the estimated regression equation that relates annual maintenance expense, in hundreds of dollars, to weekly usage.

b. Test the significance of the relationship in (a) at the $\alpha = .05$ level of significance.

c. PJH&D expects to operate the word processor 30 hours per week. Develop a 95% prediction interval for the company's annual maintenance expense.

d. If the maintenance contract costs $3000 per year, would you recommend purchasing it? Why or why not?

56. A sociologist was hired by a large city hospital to investigate the relationship between the number of unauthorized days that an employee is absent per year and the distance (miles) between home and work for the employees. A sample of ten employees was chosen, and the following data were collected.

Distance to Work (miles)	Number of Days Absent
1	8
3	5
4	8
6	7
8	6
10	3
12	5
14	2
14	4
18	2

a. Develop a scatter diagram for these data. Does a linear relationship appear reasonable? Explain.

b. Develop the least squares estimated regression equation.

c. Is there a significant relationship between the two variables? Use $\alpha = .05$.

d. Did the estimated regression equation provide a good fit? Explain.

e. Use the estimated regression equation developed in (b) to develop a 95% confidence interval estimate of the expected number of days absent for employees living 5 miles from the company.

57. The owner of a chain of fast-food restaurants would like to investigate the relationship between the daily sales volume of a company restaurant and the number of competitor restaurants within a 1-mile radius of the firm's restaurant. The following data have been collected.

Number of Competitors Within 1 Mile	Sales ($)
1	3600
1	3300
2	3100
3	2900
3	2700
4	2500
5	2300
5	2000

a. Develop the least squares estimated regression equation that relates daily sales volume to the number of competitor restaurants within a 1-mile radius.

b. Is there a significant relationship between the two variables? Use $\alpha = .05$.

c. Did the estimated regression equation provide a good fit? Explain.

d. Use the estimated regression line developed in part (a) to develop a 95% interval estimate of the daily sales volume for a particular company restaurant that has four competitors within a 1-mile radius.

58. The regional transit authority for a major metropolitan area would like to determine if there is any relationship between the age of a bus and the annual maintenance cost. A

sample of ten buses resulted in the following data.

Age of Bus (years)	Maintenance Cost ($)
1	350
2	370
2	480
2	520
2	590
3	550
4	750
4	800
5	790
5	950

Compute the sample correlation coefficient for the above data.

59. Reconsider the regional transit authority problem presented in Exercise 58.
    a. Develop the least squares estimated regression equation.
    b. Test to see if the two variables are significantly related at $\alpha = .05$.
    c. Did the least squares line provide a good fit to the observed data? Explain.
    d. Develop a 90% confidence interval estimate of the maintenance cost for a specific bus that is 4 years old.

60. A psychology professor at Givens College is interested in the relation between hours spent studying and total points earned in the course. Data collected on 10 students who took the course last quarter are given below:

Hours Spent Studying	Total Points Earned
45	40
30	35
90	75
60	65
105	90
65	50
90	90
80	80
55	45
75	65

Compute the sample correlation coefficient for these data.

61. Reconsider the Givens College data in Exercise 60.

    a. Develop an estimated regression equation relating total points earned to hours spent studying.
    b. Test the significance of the model at the $\alpha = .05$ level.
    c. Predict the total points earned by Mark Sweeney. He spent 95 hours studying.
    d. Develop a 90% confidence interval for the total points earned by Mark Sweeney.

62. USA Today publishes college basketball computer rankings that are based upon a strength rating computed for each team. The strength ratings can also be used to predict the victory margin for games. For the visiting team, the predicted score is its strength rating. For the home team, the predicted score is the strength rating plus $4\frac{1}{2}$. For each game, the actual victory margin is the winning team's score minus the losing team's score. The predicted victory margin is the predicted score for the winning team minus the predicted score for the losing team.

A sample was taken of ten college basketball games selected from USA Today, December 13, 1988, to investigate the accuracy of the predicted victory margin. The following table shows the data obtained. The strength ratings are in parentheses.

Visiting Team	Score	Home Team	Score
Eastern Michigan (77.06)	57	Michigan (101.07)	80
Jackson State (65.75)	71	Iowa (96.63)	86
Georgia Southern (80.01)	80	Eastern Kentucky (61.24)	69
Seton Hall (92.66)	96	Rutgers (72.22)	70
Niagara (70.32)	78	St. Bonaventure (70.57)	81
Fairfield (63.01)	48	Connecticut (85.15)	71
S. Carolina St. (73.29)	70	Clemson (80.13)	93
Monmouth (57.43)	70	Maryland (80.53)	74
Illinois-Chicago (73.95)	74	Michigan State (83.45)	96
Oral Roberts (71.37)	75	Georgetown (87.48)	91

a. Compute the correlation coefficient between the predicted victory margin and the actual victory margin.

b. Test for a significant relationship. Use $\alpha = .01$.

c. Develop an estimated regression equation using the predicted victory margin as the independent variable and the actual victory margin as the dependent variable.

d. What are the values of b_0 and b_1? Comment on what the values of β_0 and β_1 should be for an ideal system of predicting victory margins.

▬▬ COMPUTER EXERCISE

As part of a study of transportation safety, a department of transportation collected data on the number of fatal accidents and the percentage of licensed drivers under the age of 21. The values for the two variables for a 1-year period were obtained from a sample of 42 cities. In order to account for differences in the population sizes of the various cities, the measure for the number of fatal accidents was defined in terms of the number of fatal accidents per 1000 licenses. The data are given.

Percent Under 21	Number of Fatal Accidents per 1000 Licenses	Percent Under 21	Number of Fatal Accidents per 1000 Licenses
13	2.962	17	4.100
12	0.708	8	2.190
8	0.885	16	3.623
12	1.652	15	2.623
11	2.091	9	0.835
17	2.627	8	0.820
18	3.830	14	2.890
8	0.368	8	1.267
13	1.142	15	3.224
8	0.645	10	1.014
9	1.028	10	0.493
16	2.801	14	1.443
12	1.405	18	3.614
9	1.433	10	1.926
10	0.039	14	1.643
9	0.338	16	2.943
11	1.849	12	1.913
12	2.246	15	2.814
14	2.855	13	2.634
14	2.352	9	0.926
11	1.294	17	3.256

■ QUESTIONS

1. Develop numerical and graphical measures to summarize the data.

2. Does there appear to be a relationship between the number of fatal accidents per 1000 licenses and the percentage of drivers under the age of 21? Explain.

3. What conclusions and recommendations can you derive from your analysis?

APPENDIX

Calculus-Based Derivation of Least Squares Formulas

As mentioned in the chapter, the least squares method is a procedure for determining the values of b_0 and b_1 that minimize the sum of squared residuals. The sum of squared residuals is given by

$$\Sigma(y_i - \hat{y}_i)^2$$

Substituting $\hat{y}_i = b_0 + b_1 x_i$, we get

$$\Sigma(y_i - b_0 - b_1 x_i)^2 \tag{13A.1}$$

as the expression that must be minimized.

To minimize (13A.1) we must take the partial derivatives with respect to b_0 and b_1, set them equal to zero, and solve. Doing so we get

$$\frac{\partial \Sigma(y_i - b_0 - b_1 x_i)^2}{\partial b_0} = -2\Sigma(y_i - b_0 - b_1 x_i) = 0 \tag{13A.2}$$

$$\frac{\partial \Sigma(y_i - b_0 - b_1 x_i)^2}{\partial b_1} = -2\Sigma x_i(y_i - b_0 - b_1 x_i) = 0 \tag{13A.3}$$

Dividing (13A.2) by 2 and summing each term individually yields

$$-\Sigma y_i + \Sigma b_0 + \Sigma b_1 x_i = 0$$

Bringing Σy_i to the other side of the equal sign and noting that $\Sigma b_0 = nb_0$, we obtain

$$nb_0 + (\Sigma x_i)b_1 = \Sigma y_i \tag{13A.4}$$

Similar algebraic simplification applied to (13A.3) yields

$$(\Sigma x_i)b_0 + (\Sigma x_i^2)b_1 = \Sigma x_i y_i \qquad (13A.5)$$

(13A.4) and (13A.5) are known as the *normal equations*. Solving (13A.4) for b_0 yields

$$b_0 = \frac{\Sigma y_i}{n} - b_1 \frac{\Sigma x_i}{n} \qquad (13A.6)$$

Using (13A.6) to substitute for b_0 in (13A.5) provides

$$\frac{\Sigma x_i \Sigma y_i}{n} - \frac{(\Sigma x_i)^2}{n} b_1 + (\Sigma x_i^2)b_1 = \Sigma x_i y_i \qquad (13A.7)$$

Rearranging (13A.7), we obtain

$$b_1 = \frac{\Sigma x_i y_i - (\Sigma x_i \Sigma y_i)/n}{\Sigma x_i^2 - (\Sigma x_i)^2/n} \qquad (13A.8)$$

Since $\bar{y} = \Sigma y_i/n$ and $\bar{x} = \Sigma x_i/n$, we can rewrite (13A.6):

$$b_0 = \bar{y} - b_1\bar{x} \qquad (13A.9)$$

Equations (13A.8) and (13A.9) are the formulas we used in the chapter to compute the coefficients in the estimated regression equation.

Polaroid Corporation[*]

Cambridge, Massachusetts

Polaroid Corporation, a Fortune 500 company headquartered in Cambridge, Massachusetts, manufactures and markets worldwide a variety of products in instant image recording fields, primarily instant photographic cameras and films. The company's products are used in amateur and professional photography, industry, science, medicine, and education.

Polaroid's consumer photography business, which accounts for about half its revenues, began in 1947, when the company's founder, Dr. Edwin H. Land, announced a one-step dry process for producing a finished photograph within one minute after taking the picture. The first Polaroid Land camera and Polaroid Land film went on sale in 1948. Since then, Polaroid's continuous experimentation and development in chemistry, optics, and electronics have produced photographic systems of ever higher quality, reliability, and convenience. In April 1986, the company introduced its most advanced instant camera and film product—the Polaroid Spectra System.

Polaroid's other major business segment, technical and industrial photography, focuses on making Polaroid instant photography a key component of the growing number of imaging systems used in today's visual communications environment. To this end, Polaroid markets a wide variety of instant photographic systems, cameras, components, and films for professional, industrial, scientific, and medical uses.

Other businesses include magnetics, sunglasses, industrial polarizers, chemicals, custom coating, and holography. In its laboratories, Polaroid is exploring electronic photography and several new, non-photographic fields such as photovoltaics, fiber optics, and biomedicine.

▬▬ REGRESSION ANALYSIS AT POLAROID

Sensitometry is a key area in which Polaroid uses regression analysis. Sensitometry, the measurement of the sensitivity of photographic materials, provides information on many characteristics of film, such as its useful exposure range.

Within Polaroid's central sensitometry laboratory, scientists systematically sample and analyze instant films from the production floor that have been stored at temperature and humidity levels approximating those the films will be subjected to once in the hands of consumers. This is done to determine shelf life performance—that is, how well films maintain their sensitivity and provide optimal photographic results under actual environmental conditions and for how long.

Sensitometric testing is extensive and is done both throughout and beyond the life of the company's film products. Shelf life information when combined with data about average elapsed time between the purchase and use of film helps Polaroid to make

*The authors are indebted to Mr. Lawrence Friedman, Manager, Photographic Quality, Polaroid Corporation, for providing this application.

TABLE 13A.1   **Data from Polaroid Film Aging Study**

Sample No	Sample Size	Film Age (months)	Change in		
			Speed	Redbal	Bluebal
1	9	0	0	0	0
2	11	1	−20.5	3.4	7.9
3	9	2	−36.8	−0.5	20.7
4	11	3	−50.8	−6.6	37.6
5	10	4	−56.1	−19.3	37.7
6	10	5	−67.4	−22.1	44.6
7	10	6	−72.3	−21.5	50.7
8	10	7	−82.2	−20.7	63.4
9	10	8	−85.1	−22.8	67.6
10	10	9	−85.6	−36.1	66.3
11	9	10	−99.3	−35.6	73.2
12	7	11	−101.4	−39.4	83.4
13	12	12	−100.4	−32.2	85.3
14	8	13	−111.5	−34.8	91.1

manufacturing adjustments that ensure the greatest number of consumers will get the optimum performance from Polaroid film. In a word, sensitometric data serve to guide the company's film manufacturing philosophy.

One aspect of the testing is to examine the rates of change of film speed and color balance as a function of the age of the film. Film speed is a measure of sensitivity to light. Color balance is concerned with alterations in a film's ability to produce color—its color rendition—and is monitored closely as film ages. Color film is balanced by its manufacturer in order to produce certain colors under specific conditions.

In one such study of film aging (Table 13A.1), Polaroid's central sensitometry lab collected data on a Polaroid instant, extended range, color professional print film.

Fourteen separate film samples were taken, at monthly intervals, up to 13 months

Polaroid Corporation's new Spectra, an advanced photographic system designed to set new standards of quality in instant photography.

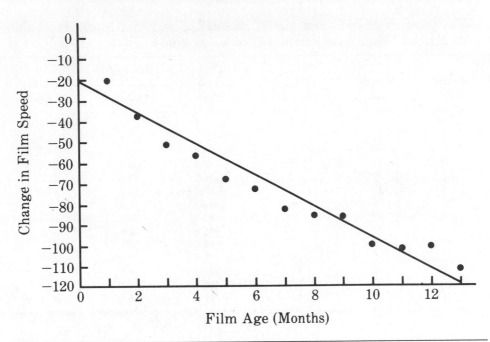

■■■ FIGURE 13A.1 **Regression Line Showing Change in Film Speed as a Function of Film Age**

after manufacture. A look at the data (which have been scaled for proprietary reasons) shows three general trends: (1) film speed decreases with age; (2) red balance becomes more negative with age, a technical way of saying that the film is becoming more cyan in color (red and cyan are complementary colors); (3) blue balance becomes more positive with age, that is, the film becomes more blue with age.

From the data collected in this experiment, three linear regression analyses were performed. In the regression line shown in Figure 13A.1, film age is the independent variable and change in film speed the dependent variable.* The estimated regression equation developed is:

$$\hat{y} = -19.817 - 7.604x$$

where

$$y = \text{change in film speed}$$

$$x = \text{film age}$$

A coefficient of determination of -0.931 indicates a very good fit. The equation shows that between the period from 0 to 13 months of age, the average decrease in film speed per month is 7.604 units.

The objectives of this segment of the research are to determine why film speed changes

---

*The regression line in Figure 13A.1 was selected for the purpose of illustration; given the shape of the data, curvilinear regression would have provided a better fit.

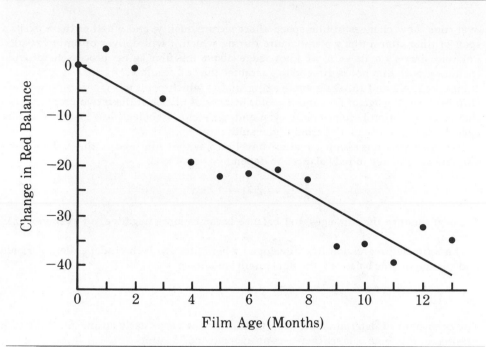

▬ FIGURE 13A.2   **Regression Line Showing Change in Red Balance as a Function of Film Age**

▬ FIGURE 13A.3   **Regression Line Showing Change in Blue Balance as a Function of Film Age**

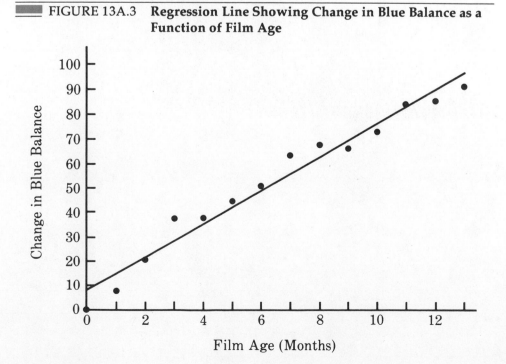

over time, how changes in film speed affect picture quality, and whether there exists a span of time after a film's manufacture during which it will deliver optimum results. Polaroid draws on its store of knowledge about the film aging process to provide customers with film precisely aged to guarantee the best results.

Figures 13A.2 and 13A.3 illustrate aging studies which point to a significant relationship between the age of film and its color balance. A glance at these two graphs shows that color rendition becomes more cyan and more blue ("cooler," less "warm") as film ages through the 13-month period under study.

The estimated regression equation developed where film age is the independent variable and change in red balance the dependent variable is:

$$\hat{y} = 0.591 - 3.258x$$

For every month the film ages, red balance becomes more negative, on average, 3.258 units.

The estimated regression line developed where film age is the independent variable and change in blue balance is the dependent variable is:

$$\hat{y} = 8.194 + 6.756x$$

The coefficient of determination here, 0.966, shows a particularly strong fit. With each passing month, blue balance increases an average of 6.756 units.

The information provided by all three linear regression analyses, when coupled with consumer purchase and use patterns, enables Polaroid to manufacture film that will have shifted those characteristics which determine picture quality to their optimum setting by the time the film is being used. In essence, Polaroid has the information to compensate in its manufacturing process for crucial alterations in film performance that happen as a result of the aging process.

CHAPTER 14

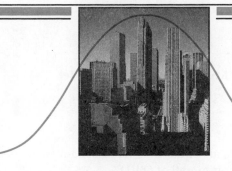

Multiple Regression

I n Chapter 13 we discussed how regression analysis can be used to develop a mathematical equation representing the relationship between two variables. Recall that the variable being predicted or explained by the mathematical equation is called the dependent variable; the variable being used to predict or explain the value of the dependent variable is called the independent variable. In this chapter we continue our study of regression analysis by considering situations which involve two or more independent variables. The study of regression models involving more than one independent variable is called multiple regression analysis.

14.1 ▦ THE MULTIPLE REGRESSION MODEL AND ITS ASSUMPTIONS

The probabilistic model for multiple regression analysis is a direct extension of the one introduced in the previous chapter for simple linear regression. To show this consider a situation involving the sale of a new product (y) in a certain region. Suppose that we believe sales are related to the population size (x_1) and the average disposable income (x_2) of people in the region by the following regression model:

$$y = \beta_0 + \beta_1 x_1 + \beta_2 x_2 + \epsilon \tag{14.1}$$

The relationship shown in (14.1) is a *multiple regression model* involving 2 independent variables. Note that if $\beta_2 = 0$, then x_2 is not related to y and hence the multiple regression model reduces to the 1-independent-variable model discussed in Chapter 13; that is, $y = \beta_0 + \beta_1 x_1 + \epsilon$.

The multiple regression model of (14.1) can be extended to the case of p

independent variables simply by adding more terms. Equation (14.2) shows the general case:

Multiple Regression Model

$$y = \beta_0 + \beta_1 x_1 + \beta_2 x_2 + \cdots + \beta_p x_p + \epsilon \qquad (14.2)$$

Note that if $\beta_3, \beta_4, \ldots, \beta_p$ all equal zero, (14.2) reduces to the 2-independent-variable multiple regression model of (14.1).

The assumptions made about the error term ϵ in Chapter 13 also apply in multiple regression analysis:

Assumptions About the Error Team ϵ in the Regression Model $y = \beta_0 + \beta_1 x_1 + \cdots + \beta_p x_p + \epsilon$

1. The error, ϵ, is a random variable with mean or expected value of 0; that is, $E(\epsilon) = 0$.

   Implication: For given values of $x_1, x_2, \ldots, x_p$, the expected, or average, value of y is given by

   $$E(y) = \beta_0 + \beta_1 x_1 + \beta_2 x_2 \cdots + \beta_p x_p \qquad (14.3)$$

   Equation (14.3) is referred to as the *multiple regression equation*. In this equation $E(y)$ represents the average of all possible values of y that could occur for the given values of $x_1, x_2, \ldots, x_p$.

2. The variance of ϵ is denoted by σ^2 and is the same for all values of the independent variables $x_1, x_2, \ldots, x_p$.

   Implication: The variance of y equals σ^2 and is the same for all values of $x_1, x_2, \ldots, x_p$.

3. The values of ϵ are independent.

   Implication: The size of the error for a particular set of values for the independent variables is not related to the size of the error for any other set of values.

4. The error ϵ is a normally distributed random variable reflecting the deviation between the y value and the expected value of y given by $\beta_0 + \beta_1 x_1 + \beta_2 x_2 + \cdots \beta_p x_p$.

   Implication: Since $\beta_0, \beta_1, \ldots, \beta_p$ are constants, for the given values of $x_1, x_2, \ldots, x_p$, the dependent variable y is also a normally distributed random variable.

To obtain more insight into the form of the relationship given by (14.3), consider for the moment the following 2-independent-variable multiple regression equation:

$$E(y) = \beta_0 + \beta_1 x_1 + \beta_2 x_2 \qquad (14.4)$$

The graph of this equation is a plane in 3 dimensional space. Figure 14.1 shows such a graph with x_1 and x_2 on the horizontal axis and y on the vertical axis. Note that ϵ is shown as the difference between the actual y value and the expected value of y, $E(y)$, when $x_1 = x_1^*$ and $x_2 = x_2^*$.

In regression analysis the term *response variable* is often used in place of the term *dependent variable*. Furthermore, since the multiple regression equation generates a plane or surface, its graph is referred to as a response surface.

In the previous chapter the least squares method was used to develop estimates of β_0 and β_1 for the simple linear regression model. In multiple regression analysis the least squares method is used in an analogous manner to develop estimates of the parameters β_0, β_1, β_2, ..., β_p. These estimates are denoted b_0, b_1, b_2, ..., b_p; the corresponding estimated regression equation is written as follows:

Estimated Regression Equation

$$\hat{y} = b_0 + b_1 x_1 + \cdots + b_p x_p \tag{14.5}$$

At this point we can begin to see the similarity between the concepts of multiple regression analysis and those of the previous chapter. The concepts of simple linear regression have simply been extended to the case involving more than one independent variable. In the next section we will begin to apply these concepts to the problem facing the Butler Trucking Company.

■ FIGURE 14.1 **Graph of the Regression Equation for Multiple Regression Analysis Involving Two Independent Variables**

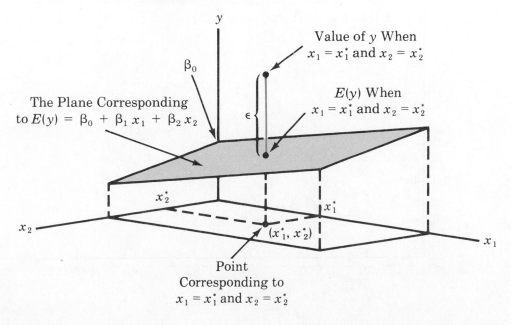

TABLE 14.1	Preliminary Data for the Butler Trucking Problem	
Day	Miles Traveled	Travel Time (hours)
1	100	9.3
2	50	4.8
3	100	8.9
4	100	5.8
5	50	4.2
6	80	6.8
7	75	6.6
8	80	5.9
9	90	7.6
10	90	6.1

14.2 ■ THE BUTLER TRUCKING PROBLEM

Butler Trucking is an independent trucking company located in southern California. A major portion of Butler's business involves deliveries throughout its local area.

To develop better work schedules management would like to use an estimated

■ FIGURE 14.2 **Scatter Diagram of Preliminary Data for the Butler Trucking Problem**

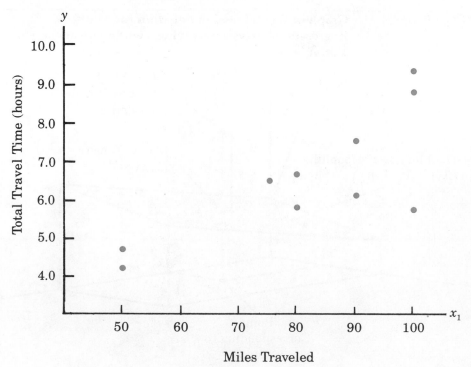

Miles Traveled

regression equation to help predict total daily travel time for its drivers. Initially management felt that travel time should be closely related to miles traveled. A random sample of 10 days of operation was taken; the data obtained are presented in Table 14.1, and the corresponding scatter diagram is shown in Figure 14.2.

The scatter diagram indicates that the number of miles traveled (x_1) and the travel time (y) appear to be positively related; as x_1 increases, y increases. After observing the scatter diagram, management hypothesized the following regression model:

$$y = \beta_0 + \beta_1 x_1 + \epsilon$$

Note that this is nothing more than the simple linear regression model with x_1 replacing x. As a result of this notational change we use x_{1i} to denote the ith observation for the independent variable x_1. Table 14.2 shows the application of the least squares formulas with these notational differences from Chapter 13 to compute b_0 and b_1. After rounding, the estimated regression equation relating travel time to miles traveled is given by $\hat{y} = 1.13 + .067x_1$.

Table 14.3 shows the computation of the residuals and SSE for the estimated regression equation $\hat{y} = 1.13 + .067x_1$; we see that SSE = 9.5669. Figure 14.3 shows a plot of the residuals versus $\hat{y}$. This residual plot appears to indicate that the assumption of constant variance (Assumption 2) is not satisfied; the variability of the y values about the estimated regression line is increasing as $\hat{y}$ gets larger. Although not shown, the value of SST = $\Sigma (y_i - \bar{y})^2$ = 24.0 and hence

TABLE 14.2　**Least Squares Calculations for the Model Involving One Independent Variable**

Day (i)	x_{1i} = Miles Traveled	y_i = Travel Time (hours)	$x_{1i}y_i$	x_{1i}^2
1	100	9.3	930	10,000
2	50	4.8	240	2500
3	100	8.9	890	10,000
4	100	5.8	580	10,000
5	50	4.2	210	2500
6	80	6.8	544	6400
7	75	6.6	495	5625
8	80	5.9	472	6400
9	90	7.6	684	8100
10	90	6.1	549	8100
	815	66.0	5594	69,625
	Σx_{1i}	Σy_i	$\Sigma x_{1i}y_i$	Σx_{1i}^2

$$\bar{x}_1 = \frac{815}{10} = 81.5 \qquad \bar{y} = \frac{66}{10} = 6.6$$

$$b_1 = \frac{\Sigma x_{1i}y_i - (\Sigma x_{1i}\Sigma y_i)/n}{\Sigma x_{1i}^2 - (\Sigma x_{1i})^2/n} = \frac{5594 - (815)(66)/10}{69,625 - (815)^2/10} = \frac{215}{3202.5} = .0671$$

$$b_0 = \bar{y} - b_1\bar{x}_1 = 6.6 - (.0671)(81.5) = 1.1314$$

TABLE 14.3 Computation of Residuals and SSE for Butler Trucking Using $\hat{y} = 1.13 + .067x_1$

Miles Traveled (x_{1i})	Travel Time (y_i)	Predicted Travel Time $(\hat{y}_i)$	Residual $(y_i - \hat{y}_i)$	$(y_i - \hat{y}_i)^2$
100	9.3	7.830	1.470	2.1609
50	4.8	4.480	.320	.1024
100	8.9	7.830	1.070	1.1449
100	5.8	7.830	−2.030	4.1209
50	4.2	4.480	− .280	.0784
80	6.8	6.490	.310	.0961
75	6.6	6.155	.445	.1980
80	5.9	6.490	− .590	.3481
90	7.6	7.160	.440	.1936
90	6.1	7.160	−1.060	1.1236
				9.5669

SSE

SSR = SST − SSE = 24.0 − 9.5669 = 14.4331. Thus the coefficient of determination is

$$r^2 = \frac{\text{SSR}}{\text{SST}} = \frac{14.4331}{24.0} = .60$$

FIGURE 14.3 Residual Plot for the Butler Trucking Problem Corresponding to $\hat{y} = 1.13 + .067x_1$

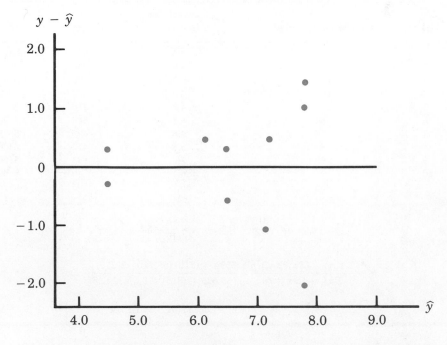

Since $r^2 = .60$, we conclude that 60% of the variability in travel time is explained by the relationship with miles traveled. However, note that 40% of the variance is still unexplained and, given the problem we observed with the residual plot, we should doubt the adequacy of our existing model.

Looking for an alternative, management suggested that perhaps the number of deliveries made could also be used to help predict travel time and hence improve the regression model. The data, with the addition of the number of deliveries, are shown in Table 14.4, where x_{2i} denotes the number of deliveries on day i.

With the number of deliveries included as a second independent variable, the following multiple regression model for Butler Trucking is obtained:

$$y = \beta_0 + \beta_1 x_1 + \beta_2 x_2 + \epsilon \tag{14.6}$$

where

$$x_1 = \text{miles traveled}$$

$$x_2 = \text{number of deliveries}$$

$$y = \text{travel time}$$

As we showed in Section 14.1, given the assumption that the expected value of the error term is zero, this model implies that the expected value of y is related to x_1 and x_2 as follows:

$$E(y) = \beta_0 + \beta_1 x_1 + \beta_2 x_2 \tag{14.7}$$

Of course, the values of the parameters β_0, β_1, and β_2 will not be known in practice; thus we will need to estimate them using the sample data in Table 14.4. The resulting estimated regression equation is

$$\hat{y} = b_0 + b_1 x_1 + b_2 x_2 \tag{14.8}$$

TABLE 14.4 Data for the Butler Trucking Problem with Miles Traveled (x_1) and Number of Deliveries (x_2) as the Independent Variables

Day	x_{1i} = Miles Traveled	x_{2i} = Number of Deliveries	y_i = Travel Time (hours)
1	100	4	9.3
2	50	3	4.8
3	100	4	8.9
4	100	2	5.8
5	50	2	4.2
6	80	1	6.8
7	75	3	6.6
8	80	2	5.9
9	90	3	7.6
10	90	2	6.1

In the next section we show how the least squares method is used to estimate b_0, b_1, and b_2.

14.3 ■■■ DEVELOPING THE ESTIMATED REGRESSION EQUATION

In Chapter 13 we presented formulas for estimating b_0 and b_1 for the regression model $y = \beta_0 + \beta_1 x + \epsilon$. In the general multiple regression case the usual presentation of formulas for computing the coefficients of the estimated regression equation involves the use of matrix algebra and is beyond the scope of this text. However, for the special case of 2 independent variables, we can show what is involved.

In this section we show how the least squares method can be used to compute b_0, b_1, and b_2 for the 2-independent-variable Butler Trucking problem. We then discuss how computer software packages such as Minitab can be used to obtain these solutions. In subsequent sections we will show how computer output can be used to determine the goodness of fit, to test for significance, and for prediction.

Including the effect of number of deliveries, we obtain the following estimated regression equation for Butler Trucking:

$$\hat{y} = b_0 + b_1 x_1 + b_2 x_2$$

where

$$x_1 = \text{miles traveled}$$
$$x_2 = \text{number of deliveries}$$

Using this notation, the predicted value for the ith observation is

$$\hat{y}_i = b_0 + b_1 x_{1i} + b_2 x_{2i}$$

where

$$x_{1i} = i\text{th value of } x_1$$
$$x_{2i} = i\text{th value of } x_2$$
$$\hat{y}_i = \text{predicted value of travel time (hours)}$$
$$\text{when } x_1 = x_{1i} \text{ and } x_2 = x_{2i}$$

For the case of 2 independent variables, the residuals are defined as follows:

$$\text{Residual for } i\text{th observation} = y_i - \hat{y}_i = y_i - (b_0 + b_1 x_{1i} + b_2 x_{2i}) \quad (14.9)$$

The least squares method determines the values of b_0, b_1, and b_2 that minimize the sum of squared residuals. Thus we must choose b_0, b_1, and b_2 to satisfy the following criterion:

$$\min \Sigma (y_i - b_0 - b_1 x_{1i} - b_2 x_{2i})^2 \quad (14.10)$$

TABLE 14.5 Calculation of Coefficients for Normal Equations

y_i	x_{1i}	x_{2i}	x_{1i}^2	x_{2i}^2	$x_{1i}x_{2i}$	$x_{1i}y_i$	$x_{2i}y_i$
9.3	100	4	10,000	16	400	930	37.2
4.8	50	3	2500	9	150	240	14.4
8.9	100	4	10,000	16	400	890	35.6
5.8	100	2	10,000	4	200	580	11.6
4.2	50	2	2500	4	100	210	8.4
6.8	80	1	6400	1	80	544	6.8
6.6	75	3	5625	9	225	495	19.8
5.9	80	2	6400	4	160	472	11.8
7.6	90	3	8100	9	270	684	22.8
6.1	90	2	8100	4	180	549	12.2
66.0	815	26	69,625	76	2165	5594	180.6

Using calculus it can be shown (see chapter appendix) that the values of b_0, b_1, and b_2 that minimize (14.10) must satisfy the following three equations, called the *normal equations*:

Normal Equations — Two Independent Variables

$$nb_0 + (\Sigma x_{1i})b_1 + (\Sigma x_{2i})b_2 = \Sigma y_i \qquad (14.11)$$

$$(\Sigma x_{1i})b_0 + (\Sigma x_{1i}^2)b_1 + (\Sigma x_{1i}x_{2i})b_2 = \Sigma x_{1i}y_i \qquad (14.12)$$

$$(\Sigma x_{2i})b_0 + (\Sigma x_{1i}x_{2i})b_1 + (\Sigma x_{2i}^2)b_2 = \Sigma x_{2i}y_i \qquad (14.13)$$

In order to apply the normal equations we must first use the data to find the values of the coefficients of b_0, b_1, b_2 and the values for the right hand sides of these equations. The necessary data for Butler Trucking are contained in Table 14.5.

Using the information in Table 14.5 we can substitute into the normal equations (14.11) to (14.13) to obtain the following normal equations for the Butler Trucking problem:

$$10b_0 + 815b_1 + 26b_2 = 66.0 \qquad (14.14)$$

$$815b_0 + 69,625b_1 + 2165b_2 = 5594.0 \qquad (14.15)$$

$$26b_0 + 2165b_1 + 76b_2 = 180.6 \qquad (14.16)$$

Since the least squares estimates b_0, b_1, and b_2 must satisfy these three equations simultaneously, in order to obtain values for b_0, b_1, and b_2 we will have to solve this system of three simultaneous linear equations in three variables. The solution* is given by $b_0 = .0367$, $b_1 = .0562$, and $b_2 = .7639$. Thus the estimated

*In the chapter appendix we show in detail how the solution is obtained.

regression equation for Butler Trucking is

$$\hat{y} = .0367 + .0562x_1 + .7639x_2 \tag{14.17}$$

Note on Interpretation of Coefficients

One observation can be made at this time concerning the relationship between the estimated regression equation with only miles traveled as an independent variable and the one which includes the number of deliveries as a second independent variable. The value of b_1 is not the same in both cases. In simple linear regression we interpret b_1 as the amount of change in y for a 1 unit change in the independent variable. In multiple regression analysis this interpretation must be modified somewhat. That is, in multiple regression analysis we interpret each regression coefficient as follows: b_i represents the change in y corresponding to a 1 unit change in x_i when all other independent variables are held constant. For example, in the Butler Trucking problem involving two independent variables $b_1 = .0562$. Thus .0562 hours is the expected increase in travel time corresponding to an increase of 1 mile in the distance traveled when the number of deliveries is held constant. Similarly, since $b_2 = .7639$, the expected increase in travel time corresponding to an increase of 1 delivery when the number of miles traveled is held constant is .7639 hours.

Computer Solution

It can be shown that for multiple regression problems involving p independent variables there are $p + 1$ normal equations that must be solved simultaneously for the estimated coefficients $b_0, b_1, b_2, \ldots, b_p$. The computational effort involved requires more sophisticated solution procedures than we have used in

---

▇▇▇ FIGURE 14.4   **MINITAB Output for the Butler Trucking Problem using Two Independent Variables**

```
The regression equation is
TIME = 0.04 + 0.0562 MILES + 0.764 DELIV

Predictor       Coef        Stdev     t-ratio        p
Constant       0.037        1.326        0.03    0.979
MILES        0.05616      0.01564        3.59    0.009
DELIV         0.7639       0.3053        2.50    0.041

s = 0.8494     R-sq = 79.0%     R-sq(adj) = 72.9%

Analysis of Variance

SOURCE        DF          SS           MS         F        p
Regression     2      18.9499       9.4749     13.13    0.004
Error          7       5.0501       0.7214
Total          9      24.0000

                       Y      PRED. Y    ST.DEV.
ROW    MILES        TIME       VALUE    PRED. Y    RESIDUAL    ST.RES.
  6       80       6.800       5.294      0.552       1.506       2.33R
```

$SSR + SSE = SST$ *(handwritten note)*

the solution of (14.14) to (14.16). Fortunately, computer software packages can be used to obtain these solutions with very little effort on the part of the user.

In Figure 14.4 we show the output from the Minitab computer package for the version of the Butler Trucking problem involving the two independent variables, miles traveled and number of deliveries. Note that in the column labeled "Coef" are the values for b_0, b_1, and b_2. Note also that the sum of squares due to regression (SSR = 18.9499) plus the sum of squares due to error (SSE = 5.0501) is equal to the total sum of squares (SST = 24.0). This relationship among SST, SSR, and SSE, which we introduced in Chapter 13 for simple linear regression analysis, also holds true for multiple regression analysis. We discuss the remainder of the output in the following sections.

EXERCISES

Note to student: Although some of the exercises involving data in this and subsequent sessions can be solved using "hand" calculations, they were designed to be solved using a computer software package.

1. A shoe store has developed the following estimated regression equation relating sales to inventory investment and advertising expenditures:

$$\hat{y} = 25 + 10x_1 + 8x_2$$

where

$$x_1 = \text{inventory investment (\$1000s)}$$

$$x_2 = \text{advertising expenditures (\$1000s)}$$

$$y = \text{sales (\$1000s)}$$

a. Estimate sales if there is a $15,000 investment in inventory and an advertising budget of $10,000.
b. Interpret the parameters (b_1 and b_2) in this estimated regression equation.

2. The owner of TAI Movie Theaters, Inc. would like to investigate the effect of television advertising on weekly gross revenue for special promotion films. The following historical data were developed:

Weekly Gross Revenue ($1000s)	Television Advertising ($1000s)
96	5.0
90	2.0
95	4.0
92	2.5
95	3.0
94	3.5
94	2.5
94	3.0

a. Using these data, develop an estimated regression equation relating weekly gross revenue to television advertising expenditure.
b. Estimate the weekly gross revenue in a week in which $3500 is spent on television advertising.

3. As an extension of Exercise 2 consider the possibility of incorporating the effect of

newspaper advertising as well as television advertising on weekly gross revenue. The following data were developed from historical records:

Weekly Gross Revenue ($1000s)	Television Advertising ($1000s)	Newspaper Advertising ($1000s)
96	5.0	1.5
90	2.0	2.0
95	4.0	1.5
92	2.5	2.5
95	3.0	3.3
94	3.5	2.3
94	2.5	4.2
94	3.0	2.5

a. Find an estimated regression equation relating weekly gross revenue to television and newspaper advertising.

b. Is the coefficient for television advertising expenditures the same in Exercise 2(a) and 3(a)? Interpret this coefficient in each case.

4. Heller Company manufactures lawn mowers and related lawn equipment. They believe that the quantity of lawnmowers sold depends on the price of their mower and the price of a competitor's mower. Let

$$y = \text{quantity sold (1000s)}$$

$$x_1 = \text{price of competitor's mower (dollars)}$$

$$x_2 = \text{price of Heller's mower (dollars)}$$

Management would like an estimated regression equation that relates quantity sold to the price of the Heller Mower and the competitor's mower. The following data are available concerning prices in ten different cities.

Competitor's Price (x_1)	Heller's Price (x_2)	Quantity Sold (y)
120	100	102
140	110	100
190	90	120
130	150	77
155	210	46
175	150	93
125	250	26
145	270	69
180	300	65
150	250	85

a. Determine the estimated regression equation that can be used to predict the quantity sold given the competitor's price and Heller's price.

b. Interpret b_1 and b_2.

c. Predict the quantity sold in a city where Heller prices its mower at $160 and the competitor prices its mower at $170.

5. The following data show the dollar value of prescriptions (in thousands) for 13 pharmacies in Iowa, the population of the city served by the given pharmacy, and the

average prescription inventory value. ("The use of categorical variables in Data Envelopment Analysis," R. Bonker and R. Morey, *Management Science*, December 1986).

Value ($1000s)	Population	Average Inventory Value ($)
61	1410	8,000
92	1523	9,000
93	1354	13,694
45	822	4,250
50	746	6,500
29	1281	7,000
56	1016	4,500
45	1070	5,000
183	1694	27,000
156	1910	21,560
120	1745	15,000
75	1353	8,500
122	1016	18,000

a. Determine the estimated regression equation that can be used to predict the dollar value of prescriptions (y) given the population size (x_1) and the average inventory value (x_2).

b. What other variables do you think might be useful in predicting y?

6. Two experts provided subjective lists of school districts that they think are among the best in the country. For each school district the average class size, the combined SAT score, and the percentage of schools that attended a 4-year college were provided (*Wall Street Journal*, March 31, 1989).

District	Average Class Size	Combined SAT Score	% Attend Four-Year College
Blue Springs, Mo.	25	1083	74
Garden City, N.Y.	18	997	77
Indianapolis, Ind.	30	716	40
Newport Beach, Calif.	26	977	51
Novi, Mich.	20	980	53
Piedmont, Calif.	28	1042	75
Pittsburg, Pa.	21	983	66
Scarsdale, N.Y.	20	1110	87
Wayne, Pa.	22	1040	85
Weston, Mass.	21	1031	89
Farmingdale, N.Y.	22	947	81
Mamaroneck, N.Y.	20	1000	69
Mayfield, Ohio	24	1003	48
Morristown, N.J.	22	972	64
New Rochelle, N.Y.	23	1039	55
Newtown Square, Pa.	17	963	79
Omaha, Neb.	23	1059	81
Shaker Heights, Ohio	23	940	82

a. Using these data, develop an estimated regression equation relating the percentage of students that attend a 4-year college to the average class size and the combined SAT score.

b. Estimate the percentage of students that attend a 4-year college if the average class size is 20 and the combined SAT score is 1000.

14.4 ▬ DETERMINING THE GOODNESS OF FIT

In Chapter 13 we used the coefficient of determination (r^2) to evaluate the strength of the regression relationship. Recall that r^2 was computed as

$$r^2 = \frac{SSR}{SST}$$

In multiple regression analysis we compute a similar quantity, called the multiple coefficient of determination:

Multiple Coefficient of Determination

$$R^2 = \frac{SSR}{SST} \qquad (14.18)$$

When multiplied by 100 the multiple coefficient of determination represents the percentage of variability in y that is explained by the estimated regression equation.

In the case of Butler Trucking (refer to Figure 14.4 for SSR and SST) we find

$$R^2 = \frac{18.9499}{24.0000} = .7896$$

Therefore 78.96% of the variability in y is explained by the relationship with miles traveled and number of deliveries. In Figure 14.4, this is rounded to show R$-$sq = 79.0%.

Refer now to Section 14.1. Note that the regression model with only miles traveled as the independent variable had $r^2 = .60$. Therefore, the percentage of variability explained has increased from 60% to 78.96%. In general it is always true that R^2 will increase as more independent variables are added to the regression model because adding variables to the model causes the prediction errors to be smaller, hence reducing SSE. Since SST = SSR + SSE, when SSE gets smaller SSR must get larger, causing R^2 = SSR/SST to increase.

Many analysts recommend adjusting R^2 for the number of independent variables in order to avoid overestimating the impact of adding an independent variable on the amount of explained variability. This *adjusted multiple coefficient of determination* is computed as follows:

Adjusted Multiple Coefficient of Determination

$$R_a^2 = 1 - (1 - R^2)\frac{n - 1}{n - p - 1} \qquad (14.19)$$

For the Butler Trucking problem we obtain

$$R_a^2 = 1 - (1 - .7896)\frac{10 - 1}{10 - 2 - 1}$$

$$= 1 - (.2104)(1.2857)$$

$$= .7295$$

Thus after adjusting for the number of independent varibles in the model, 72.95% of the variability in y has been accounted for. Note that both the value of R^2 and the value of R_a^2 are provided by the Minitab output shown in Figure 14.4.

■■■ EXERCISES

7. Shown is a partial computer printout from a multiple regression problem involving two independent variables.

```
s = 1.235

Analysis of Variance

SOURCE          DF     SS     MS
Regression      ___    ___    45.15
Error           12     ___    ___
Total           ___
```

a. Compute R^2.
b. Compute R_a^2.
c. Does the model appear to explain a large amount of the variability in the data?

8. In Exercise 1 the following estimated regression equation for relating sales to inventory investment and advertising expenditures was given:

$$\hat{y} = 25 + 10x_1 + 8x_2$$

The data used to develop the model came from a survey of 10 stores. In addition to the estimated regression equation, it was found as a result of a computer run that SST = 16,000 and SSR = 12,000.

a. For the estimated regression equation given, compute R^2.
b. Compute R_a^2.
c. Does the model appear to explain a large amount of variability in the data?

9. Refer to Exercise 6.

a. What are R^2 and R_a^2 for this problem?
b. Does the estimated regression equation provide a good fit to the data? Explain.

14.5 ■ TEST FOR A SIGNIFICANT RELATIONSHIP

The regression equation that we have assumed for the Butler Trucking problem involving two independent variables is

$$E(y) = \beta_0 + \beta_1 x_1 + \beta_2 x_2$$

Therefore the appropriate test for determining whether or not there is a significant relationship among x_1, x_2, and y is as follows:

H_0: $\beta_1 = \beta_2 = 0$

H_a: One or more of the coefficients is not equal to zero

If we reject H_0, we conclude that there is a significant relationship among x_1, x_2, and y.

The test used to determine if there is a significant relationship in the multiple regression case is an F test very similar to the one introduced in Chapter 13 for models involving one independent variable. First, we will show how it is used by applying it in the Butler Trucking problem. Then we will generalize the application of the test to models involving p independent variables.

The ANOVA Table for the Butler Trucking Problem

Recall that in Chapter 13 we discussed the use of the F test for determining whether or not the relationship between x and y is statistically significant. The test statistic used is $F = \text{MSR}/\text{MSE}$, where MSR denotes the mean square due to regression and MSE denotes the mean square due to error. Before we try to apply this test for the multiple regression case let us review the calculations needed to compute SST, SSR, and SSE.

In Chapter 13 we stated that the relationship among SST, SSR, and SSE is SST = SSR + SSE. As we indicated when discussing the computer output for the

TABLE 14.6   **Computation of Sum of Squares for Butler Trucking**
$(\hat{y} = .0367 + .0562x_1 + .7639x_2)$

y_i	$y_i - \bar{y}$	$(y_i - \bar{y})^2$	$\hat{y}_i$	$y_i - \hat{y}_i$	$(y_i - \hat{y}_i)^2$
9.3	2.7	7.29	8.7084	.5916	.3500
4.8	−1.8	3.24	5.1364	− .3364	.1132
8.9	2.3	5.29	8.7084	.1916	.0367
5.8	− .8	.64	7.1807	−1.3807	1.9063
4.2	−2.4	5.76	4.3725	− .1725	.0298
6.8	.2	.04	5.2936	1.5064	2.2692
6.6	.0	.00	6.5405	.0595	.0035
5.9	− .7	.49	6.0574	− .1574	.0248
7.6	1.0	1.00	7.3829	.2171	.0471
6.1	− .5	.25	6.6191	− .5191	.2695
66.0		24.00			5.0501

$\bar{y} = 66/10$ SST   SSE
$\quad = 6.6$

Butler Trucking problem, this equation also holds for multiple regression analysis. We can compute $SST = \Sigma(y_i - \bar{y})^2$ and $SSE = \Sigma(y_i - \hat{y}_i)^2$, then obtain SSR by subtraction; that is, $SSR = SST - SSE$. Table 14.6 shows the computation of SST and SSE for the Butler Trucking problem corresponding to the estimated regression equation $\hat{y} = .0367 + .0562x_1 + .7639x_2$. We see that SST = 24.00 and SSE = 5.0501. Using these two values we obtain SSR = 24.00 - 5.0501 = 18.9499. These are the same values shown in the computer output in Figure 14.4.

In the general multiple regression situation involving p independent variables, the numbers of degrees of freedom for SSR, SSE, and SST are as follows:

Sum of Squares	Degrees of Freedom (DF)
SSR	p
SSE	$n - p - 1$
SST	$n - 1$

Thus, for the Butler Trucking problem, SST has $10 - 1 = 9$ degrees of freedom. In addition, since we have $p = 2$ independent variables, SSR has 2 degrees of freedom. Finally, since $n - p - 1 = 10 - 2 - 1 = 7$, SSE has 7 degrees of freedom. Note that—just as before—the numbers of degrees of freedom for SSR and SSE add to $n - 1$, the degrees of freedom for SST.

Finally, recall that MSR and MSE are calculated by dividing the appropriate sum of squares by its degrees of freedom. That is,

$$MSR = \frac{SSR}{\text{Regression DF}} = \frac{SSR}{p}$$

$$MSE = \frac{SSE}{\text{Error DF}} = \frac{SSE}{n - p - 1}$$

For the Butler trucking problem then we see that

$$MSR = \frac{18.9499}{2} = 9.475$$

$$MSE = \frac{5.0501}{7} = .7214$$

TABLE 14.7   ANOVA Table for the Butler Trucking Problem with Two Independent Variables

Source of Variation	Sum of Squares	Degrees of Freedom	Mean Square
Regression	SSR = 18.9499	2	$MSR = \frac{18.9499}{2} = 9.475$
Error	SSE = 5.0501	7	$MSE = \frac{5.0501}{7} = .7214$
Total	SST = 24.00	9	

ᴛne sums of squares, degrees of freedom, and the corresponding mean squares are conveniently summarized in the ANOVA table shown in Table 14.7. Note that the same information was provided by the computer output shown in Figure 14.4.

The F Test for the Butler Trucking Problem

It can be shown that if the null hypothesis (H_0: $\beta_1 = \beta_2 = 0$) is true and the four underlying regression model assumptions are valid, then the sampling distribution of MSR/MSE follows an F distribution. The number of numerator degrees of freedom is equal to the degrees of freedom associated with the sum of squares due to regression, and the denominator degrees of freedom is equal to the degrees of freedom associated with the sum of squares due to error. Recall that the sum of squares due to regression (SSR) measures the amount of the variabilty in y explained by the regression model. Thus we would expect large values of $F = $ MSR/MSE to cast doubt on the null hypothesis (no relationship between the dependent and independent variables). Indeed, this is true; small values of F do not permit us to reject H_0, and large values lead to the rejection of H_0.

Let us return to the Butler Trucking problem and test the significance of the multiple regression model. First, we compute MSR/MSE:

$$\frac{\text{MSR}}{\text{MSE}} = \frac{18.9499/2}{5.0501/7}$$

$$= 13.133$$

To test the hypothesis that $\beta_1 = \beta_2 = 0$, we must determine whether or not 13.133 is a value that appears likely when random sampling from an F distribution with 2 numerator degrees of freedom and 7 denominator degrees of freedom.

Suppose that the level of significance is $\alpha = .05$. The critical value from the F distribution table (Table 4 of Appendix B) is 4.74. That is, if we are sampling randomly from an F distribution with numerator and denominator degrees of freedom equal to 2 and 7, respectively (as we would be if H_0 were true), then only 5% of the time would we get a value larger than 4.74. Figure 14.5 illustrates the determination of the critical region.

We can use the above analysis to formulate a rule for determining whether or not to reject H_0 at the .05 significance level as follows:

$$\text{Reject } H_0 \text{ if } \frac{\text{MSR}}{\text{MSE}} > 4.74$$

Since the value of MSR/MSE was 13.133, we can reject H_0. Hence we conclude that there is a significant relationship between total travel time and the two independent variables. Thus the estimated regression equation should be useful in predicting y for values of the independent variables within the range of those included in the sample.

The p-value provided by Minitab (see Figure 14.4) provides a very convenient way to perform this same test. Since the p-value provides the probability of obtaining a sample result more unlikely than what is observed, the p-value of .004 indicates that for a level of significance of $\alpha = .05$, the null hypothesis that

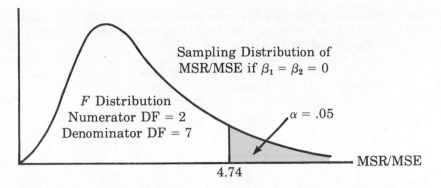

Sampling Distribution of
MSR/MSE if $\beta_1 = \beta_2 = 0$

F Distribution
Numerator DF = 2
Denominator DF = 7

$\alpha = .05$

MSR/MSE

4.74

FIGURE 14.5 **Determination of the Critical Value and Rejection Region Using $\alpha = .05$**

$\beta_1 = \beta_2 = 0$ should be rejected. In practice, the ease of use of the p-value approach to testing has made it the preferred method for performing hypothesis tests in regression analysis.

The General ANOVA Table and F Test

Now that we know how the F test can be applied for a multiple regression model with two independent variables, let us generalize our test to the case involving a model with p independent variables. The appropriate hypothesis test to determine if there is a significant relationship is as follows:

$$H_0: \quad \beta_1 = \beta_2 = \cdots = \beta_p = 0$$

$$H_a: \quad \text{One or more of the coefficients is not equal to zero}$$

Again, if we reject H_0 we can conclude that there is a significant relationship and that the estimated regression equation is useful for predicting or explaining the dependent variable y.

The general form of the ANOVA table for the multiple regression case involving p independent variables is shown in Table 14.8. The only change from the two-variable case is the degrees of freedom corresponding to SSR and SSE. Here the sum of squares due to regression has p degrees of freedom corresponding to the p independent variables; hence

$$\text{MSR} = \frac{\text{SSR}}{p} \tag{14.20}$$

In addition, the sum of squares due to error has $n - p - 1$ degrees of freedom; thus

$$\text{MSE} = \frac{\text{SSE}}{n - p - 1} \tag{14.21}$$

TABLE 14.8 ANOVA Table for a Multiple Regression Model Involving p Independent Variables

Source	Sum of Squares	Degrees of Freedom	Mean Square
Regression	SSR	p	$\text{MSR} = \dfrac{\text{SSR}}{p}$
Error	SSE	$n - p - 1$	$\text{MSE} = \dfrac{\text{SSE}}{n - p - 1}$
Total	SST	$n - 1$	

Hence the F statistic for the case of p independent variables is computed as follows:

$$F = \frac{\text{MSR}}{\text{MSE}} = \frac{\text{SSR}/p}{\text{SSE}/(n - p - 1)} \tag{14.22}$$

When looking up the critical value from the F distribution table the numerator degrees of freedom are p and the denominator degrees of freedom are $n - p - 1$. As we mentioned earlier, the decision to reject or not to reject H_0 can be made by comparing the computed F-statistic with the critical value.

t Test for Significance of Individual Parameters

If after using the F test we conclude that the multiple regression relationship is significant (that is, we conclude that at least one of the $\beta_i \neq 0$), it is often of interest to conduct further tests to see which individual parameters β_i are significant. The t test is a statistical method for testing the significance of the individual parameters.

The hypothesis test we wish to conduct is the same for the coefficient of each independent variable. It is stated as follows:

$$H_0: \quad \beta_i = 0$$

$$H_a: \quad \beta_i \neq 0$$

Recall that in Chapter 13 we learned how to conduct such a test for the case where there is only one independent variable. The hypotheses were:

$$H_0: \quad \beta_1 = 0$$

$$H_a: \quad \beta_1 \neq 0$$

To test these hypotheses we computed the sample statistic b_1/s_{b_1}, where b_1 was the least squares estimate of β_1 and s_{b_1} was an estimate of the standard deviation of the sampling distribution of b_1. We learned that the sampling distribution of b_1/s_{b_1} follows a t distribution with $n - 2$ degrees of freedom. Thus to conduct the hypothesis test we used the following rejection rule:

$$\text{Reject } H_0 \text{ if } \frac{b_1}{s_{b_1}} > t_{\alpha/2}$$

$$\text{or if } \frac{b_1}{s_{b_1}} < -t_{\alpha/2}$$

The procedure for testing individual parameters in the multiple regression case is essentially the same. The only differences are in the number of degrees of freedom for the appropriate t distribution and in the formula for computing s_{b_i}. The number of degrees of freedom is the same as for the sum of squares due to error. Thus we use $n - p - 1$ degrees of freedom, where p is the number of independent variables. (Note that for the case of one independent variable this reduces to the $n - 2$ degrees of freedom used in Chapter 13.) The formula for s_{b_i} is more involved, and we do not present it here; however, s_{b_i} is calculated and printed by most computer software packages for multiple regression analysis.

Let us return now to the Butler Trucking problem to test the significance of the parameters β_1 and β_2. Note that in the Minitab printout (Figure 14.4) the values of b_1, b_2, s_{b_1}, and s_{b_2} were given as

$$b_1 = .05616 \qquad s_{b_1} = .01564$$

$$b_2 = .7639 \qquad s_{b_2} = .3053$$

Therefore, for the parameters β_1 and β_2 we obtain

$$\frac{b_1}{s_{b_1}} = \frac{.05616}{.01564} = 3.59$$

$$\frac{b_2}{s_{b_2}} = \frac{.7639}{.3053} = 2.50$$

Note that both of these values were provided by the Minitab output of Figure 14.4 under the column labeled "t-ratio." Using $\alpha = .05$ and $10 - 2 - 1 = 7$ degrees of freedom we can find the appropriate t value for our hypothesis tests in Table 2 of Appendix B. We obtain

$$t_{.025} = 2.365$$

Now, since $b_1/s_{b_1} = 3.59 > 2.365$, we reject the hypothesis that $\beta_1 = 0$. Furthermore, since $b_2/s_{b_2} = 2.50 > 2.365$, we reject the hypothesis that $\beta_2 = 0$. Note also that the p-values provided by the Minitab outputs permit us to perform these tests very easily. For instance, the p-value of .009 for "MILES" indicates that the null hypothesis $H_0: \beta_1 = 0$ should be rejected for all values of α larger than .009.

Multicollinearity

We have used the term independent variable in regression analysis to refer to any variable being used to predict or explain the value of the dependent variable. The term does not mean, however, that the independent variables themselves are independent in any statistical sense. Quite the contrary, most independent variables in a multiple regression problem are correlated to some degree with one another. For example, in the Butler Trucking problem involving the two independent variables x_1 (miles traveled) and x_2 (number of deliveries), we could treat the miles traveled as the dependent variable and the number of deliveries as the independent variable in order to determine if these two variables are themselves related. We could then compute the sample correlation coefficient $r_{x_1x_2}$ to determine the extent to which these variables are related. We did so and obtained $r_{x_1x_2} = .28$. Thus there is some degree of linear association between the two independent variables. In multiple regression analysis we use the term *multicollinearity* to refer to the correlation among the independent variables.

To provide a better perspective of the potential problems of multicollinearity, let us consider a modification of the Butler Trucking problem. Instead of x_2 being the number of deliveries, let x_2 denote the number of gallons of gasoline consumed. Clearly, x_1 (the miles traveled) and x_2 are related; that is, we know that the number of gallons of gasoline used depends upon the number of miles traveled. Thus we would conclude logically that x_1 and x_2 are highly correlated independent variables.

Assume that we obtain the equation $\hat{y} = b_0 + b_1x_1 + b_2x_2$ and find that the F test shows that the regression is significant. Then suppose that we were to conduct a t test on β_1 to determine if $\beta_1 \neq 0$, and we cannot reject $H_0: \beta_1 = 0$. Does this mean that travel time is not related to miles traveled? Not necessarily. What it probably means is that with x_2 already in the model, x_1 does not contribute a significant addition toward determining the value of y. This would seem to make sense in our example, since if we know the amount of gasoline consumed we do not gain much additional information useful in predicting y by knowing the miles traveled. Similarly, a t test might lead us to conclude $\beta_2 = 0$ on the grounds that with x_1 in the model knowledge of the amount of gasoline consumed does not add much.

To summarize, the difficulty caused by multicollinearity in conducting t tests for the significance of individual parameters is that it is possible to conclude that none of the individual parameters are significantly different from zero when an F test on the overall multiple regression equation indicates a significant relationship. This problem is avoided when there is very little correlation among the independent variables.

Ordinarily multicollinearity does not affect the way in which we perform our regression analysis or interpret the output from a study. However, when multicollinearity is severe—that is, when two or more of the independent variables are highly correlated with one another—we can run into difficulties interpreting the results of t tests on the individual parameters. In addition to the type of problem illustrated above, severe cases of multicollinearity have been shown to result in least squares estimates that even have the wrong sign. That is, in simulated studies where researchers created the underlying regression model and then applied the least squares technique to develop estimates of β_0, β_1, β_2, and so on, it has been shown that under conditions of high multicollinearity the least squares estimates can even have a sign opposite to that of the parameter being estimated. For example, β_2 might actually be $+10$ and b_2, its estimate, might turn out to be -2. Thus little faith can be placed in the individual coefficients themselves if multicollinearity is present to a high degree.

Statisticians have developed several tests for determining whether or not multicollinearity is high enough to cause these types of problems. One simple test, referred to as the "rule of thumb" test, says that multicollinearity is a potential problem if the absolute value of the sample correlation coefficient exceeds .7 for any two of the independent variables. The other types of tests are more advanced and beyond the scope of this text.

If possible, every attempt should be made to avoid including independent variables that are highly correlated. In practice, however, it is rarely possible to adhere to this policy strictly. Thus the decision maker should be warned that when there is reason to believe that substantial multicollinearity is present it is difficult to separate out the effect of the individual independent variables on the dependent variable.

[handwritten margin note: Watch for multicollinearity above 0.7]

▬ EXERCISES

10. In Exercise 1 the following estimated regression equation for relating sales to inventory investment and advertising expenditures was given:

$$\hat{y} = 25 + 10x_1 + 8x_2$$

The data used to develop the model came from a survey of 10 stores. In addition to the estimated regression equation, it was found as a result of a computer run that SST = 16,000 and SSR = 12,000.
   a. Compute SSE, MSE, and MSR.
   b. Use an F test and an $\alpha = .05$ level of significance to determine if there is a relationship among the variables.

11. Refer to Exercise 3.
   a. Use a level of significance of $\alpha = .01$ to test the hypotheses

$$H_0: \quad \beta_1 = \beta_2 = 0$$

$$H_a: \quad \beta_1 \text{ or } \beta_2 \text{ is not equal to zero}$$

   for the model $y = \beta_0 + \beta_1 x_1 + \beta_2 x_2 + \epsilon$
where

$$x_1 = \text{television advertising (\$1000s)}$$

$$x_2 = \text{newspaper advertising (\$1000s)}$$

   b. Use a level of significance of $\alpha = .05$ to test the significance of β_1. Should x_1 be dropped from the model?
   c. Use a level of significance of $\alpha = .05$ to test the significance of β_2. Should x_2 be dropped from the model?

12. Refer to Exercise 4 involving the Heller Company. Test the significance of the overall model at $\alpha = .05$.

13. The following data show the price-earnings (P/E) ratio, the net profit margin, and the growth rate for 19 companies listed in "The *Forbes* 500s on Wall Street." (*Forbes*, May 1, 1989)

Firm	P/E Ratio	Profit Margin (%)	Growth Rate (%)
Exxon	11.3	6.5	10
Chevron	10.0	7.0	5
Texaco	9.9	3.9	5
Mobil	9.7	4.3	7
Amoco	10.0	9.8	8
Pfizer	11.9	14.7	12
Bristol Meyers	16.2	13.9	14
Merck	21.0	20.3	16
American Home Products	13.3	16.9	11
Abbott Laboratories	15.5	15.2	18
Eli Lilly	18.9	18.7	11
Upjohn	14.6	12.8	10
Warner-Lambert	16.0	8.7	7
Amdahl	8.4	11.9	4
Digital	10.4	9.8	19
Hewlett-Packard	14.8	8.1	18
NCR	10.1	7.3	6
Unisys	7.0	6.9	6
IBM	11.8	9.2	6

a. Determine the estimated regression equation that can be used to predict the price-earnings ratio given the net profit margin and the growth rate.

b. At the $\alpha = .05$ level of significance, determine if there is a relationship among the variables.

c. Does there appear to be any multicollinearity present in the data? Explain.

14. The following estimated regression equation was developed for a model involving two independent variables:

$$\hat{y} = 40.7 + 8.63x_1 + 2.71x_2$$

After dropping x_2 from the model the least squares method was used again to obtain an estimated regression equation involving only x_1 as an independent variable:

$$\hat{y} = 42.0 + 9.01x_1$$

a. Give an interpretation of the coefficient of x_1 in both models.

b. Could multicollinearity explain why the coefficient of x_1 differs in the two models? If so, how?

14.6 ■ ESTIMATION AND PREDICTION

Estimating the mean value of y and predicting an individual value of y in multiple regression is similar to that for the case of regression analysis involving one independent variable. First, recall that in Chapter 13 we showed that the point estimate of the expected value of y for a given value of x was the same as the point estimate of an individual value of y. In both cases we used $\hat{y} = b_0 + b_1x$ as the point estimate.

In multiple regression we use the same procedure. That is, we substitute the

TABLE 14.9 95% Confidence and Prediction Interval Estimates for the Butler Trucking Problem

Value of x_1	Value of x_2	Confidence Interval		Prediction Interval	
		Lower Limit	Upper Limit	Lower Limit	Upper Limit
50	2	3.082	5.663	1.985	6.760
50	3	3.711	6.562	2.673	7.600
80	1	3.987	6.600	2.897	7.690
80	2	5.295	6.820	3.909	8.206
100	2	6.074	8.288	4.887	9.475
100	4	7.481	9.936	6.354	11.063

given values of $x_1, x_2, \ldots, x_p$ into the estimated regression equation and use the corresponding value of $\hat{y}$ as the point estimate. For example, suppose that for the Butler Trucking problem we wanted to use the estimated regression equation involving x_1 (miles traveled) and x_2 (number of deliveries) to do the following:

1. Develop a *confidence interval estimate* of the mean travel time for all trucks that travel 50 miles and make two deliveries;

2. Develop a *prediction interval estimate* of the travel time for *one specific* truck that travels 50 miles and makes two deliveries.

Using the estimated regression equation $\hat{y} = .0367 + .0562x_1 + .7639x_2$ with $x_1 = 50$ and $x_2 = 2$, we obtain the following value of $\hat{y}$:

$$\hat{y} = .0367 + .0562(50) + .7639(2) = 4.3745$$

Hence our point estimate of travel time in both cases is approximately 4.4 hours.

To develop interval estimates for the mean value of y and for an individual value of y we use a procedure similar to that for the case of regression analysis involving one independent variable. The formulas required, however, are beyond the scope of the text. But, computer packages for multiple regression analysis will often provide confidence intervals once the values of $x_1, x_2 \ldots, x_p$ are specified by the user. In Table 14.9 we show 95% confidence and prediction interval estimates for the Butler Trucking problem for selected values of x_1 and x_2; these values were obtained using Minitab. Note that the interval estimate for an individual value of y is wider than the interval estimate for the expected value of y. This simply reflects the fact that for given values of x_1 and x_2 we can predict the mean travel time for all trucks with more precision than we can the travel time for one specific truck.

■ EXERCISES

15. The following estimated regression equation has been developed to predict annual sales for account executives:

$$\hat{y} = 160 + 8x_1 + 15x_2$$

where

$$\hat{y} = \text{sales (\$1000s)}$$

$$x_1 = \text{years of experience}$$

$$x_2 = \text{number of sales training programs attended}$$

a. Estimate expected annual sales for an employee with 3 years of experience who did not attend any sales training program.
b. Estimate annual sales for a given employee with 2 years of experience who attended one sales training program.
c. What is the expected increase in sales as a result of attending one sales training program?

16. Refer to Exercise 4.

a. Develop a 95% confidence interval estimate of the mean quantity sold if both the competitor's price and Heller's price are $175.
b. Would a prediction interval estimate have any meaning in this situation? Explain.

17. Refer to Exercise 6.

a. Develop a 95% confidence interval for the mean percentage of students that attend a 4-year college for a school district that has an average class size of 25 and whose students have a combined SAT score of 1000.
b. Suppose that a school district in Conway, South Carolina, has an average class size of 25 and a combined SAT score of 950. Develop a 95% prediction interval estimate of the percentage of students that attend a 4-year college.

14.7 ■ RESIDUAL ANALYSIS

In Chapter 13 we showed how residual plots could be used to determine whether or not the assumptions made regarding the error term and model form are valid. In the multiple regression case these same methods can be used to validate the assumptions made for the proposed model and provide additional information concerning the adequacy of the fitted least squares equation.

The statistical tests that we discussed in Section 14.5 are based on the assumptions presented in Section 14.1 for the error term, ϵ, and the assumption that the population regression equation has the linear form given by $E(y) = \beta_0 + \beta_1 x_1 + \cdots + \beta_p x_p$. Residual analysis will allow us to make a judgment about whether or not the model assumptions appear to be satisfied. In addition, residual analysis can often provide insight as to whether or not a different type of model—for example, one involving more variables or a different functional form—might better describe the observed relationship.

In simple linear regression we showed how a residual plot against the independent variable x and a residual plot against the predicted value $\hat{y}$ could be used to determine if the assumptions regarding the error term are appropriate. In multiple regression analysis these plots can be developed and interpreted similarly; the only difference in multiple regression is that plots against more than one independent variable can be developed. Statisticians usually look first at the residual plot against $\hat{y}$ and then review the plots against the independent variables if necessary.

Figure 14.6 shows three forms of residual plots versus $\hat{y}$, the value of y

A $\quad y - \hat{y}$

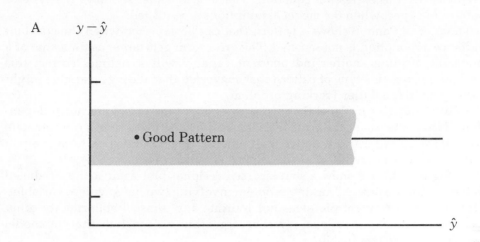

B $\quad y - \hat{y}$

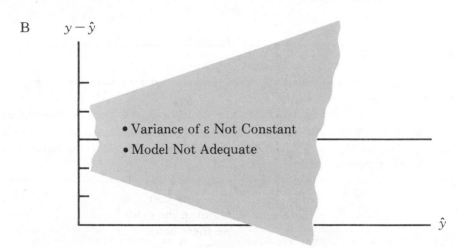

C $\quad y - \hat{y}$

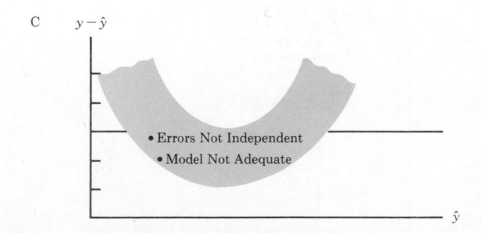

▬▬ FIGURE 14.6 **Possible Residual Patterns and Their Causes**

predicted by the regression equation. The plot in Panel A shows the type of pattern to expect when the model assumptions are satisfied.

The plot in Panel B shows a pattern that can be expected when the constant variance assumption is not satisfied. The error term gets larger as the value of $\hat{y}$ increases. Adding another independent variable will sometimes correct this problem. It was this type of pattern that suggested that adding a variable might be helpful in the Butler Trucking problem.

Finally, the plot in Panel C shows a case where the errors are not independent. When $\hat{y}$ is small, the error term is positive; when $\hat{y}$ assumes intermediate values, the error term is negative; and when $\hat{y}$ is large, the error term is again positive. Often a curvilinear model is needed in this case.

In Figure 14.7 we show a standardized residual plot against the predicted values, $\hat{y}$, for the Butler Trucking problem involving two independent variables. Note that this residual plot does not indicate any unusual abnormality other than the fact that one residual is rather large; thus we conclude that the model assumptions are reasonable.

▬ EXERCISES

18. Refer to Exercise 5. Plot the standardized residuals against $\hat{y}$. Does the residual plot support the assumptions regarding ϵ?

19. Refer to Exercise 6. Plot the standardized residuals against $\hat{y}$. Does the residual plot support the assumptions regarding ϵ?

20. Refer to Exercise 13. Plot the standardized residuals against $\hat{y}$. Does the residual plot support the assumptions regarding ϵ?

▬ FIGURE 14.7   **Standardized Residual Plot Against $\hat{y}$ for the Butler Trucking Problem with Two Independent Variables.**

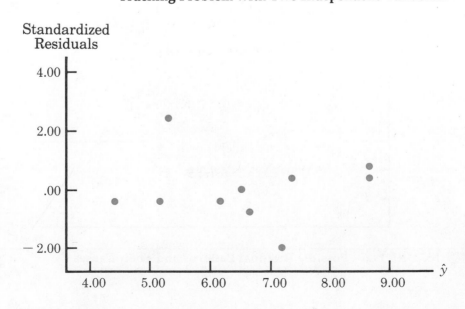

■ SUMMARY

In this chapter we have shown how extensions of the concepts of simple linear regression can be used to develop an estimated regression equation for predicting y that involves several independent variables. We noted that the interpretation of the coefficients had to be modified somewhat for this case. That is, we interpreted b_i as an estimate of the change in the dependent variable y that would result from a 1-unit change in independent variable x_i when the other independent variables do not change.

A key part of any multiple regression study is the use of a computer software package for carrying out the computational work. Many excellent packages exist and, after a short learning period, can be used to develop the estimated regression equation, conduct the appropriate significance tests, and prepare residual plots. We illustrated the use of the Minitab statistical package for the Butler Trucking problem.

■ GLOSSARY

Multiple regression model A regression model in which more than one independent variable is used to predict the dependent variable.

Multiple coefficient of determination (R^2) A measure of the goodness of fit for the estimated regression equation.

Adjusted multiple coefficient of determination (R_a^2) A measure of the goodness of fit for the estimated regression equation which accounts for the number of independent variables in the model.

Multicollinearity A term used to describe the case when the independent variables in a multiple regression model are correlated.

■ KEY FORMULAS

Multiple Regression Model

$$y = \beta_0 + \beta_1 x_1 + \beta_2 x_2 + \cdots + \beta_p x_p + \epsilon \tag{14.2}$$

Multiple Regression Equation

$$E(y) = \beta_0 + \beta_1 x_1 + \cdots + \beta_p x_p \tag{14.3}$$

Estimated Regression Equation

$$\hat{y} = b_0 + b_1 x_1 + \cdots + b_p x_p \tag{14.5}$$

Multiple Coefficient of Determination

$$R^2 = \frac{SSR}{SST} \tag{14.18}$$

Adjusted Multiple Coefficient of Determination

$$R_a^2 = 1 - (1 - R^2)\left(\frac{n-1}{n-p-1}\right) \tag{14.19}$$

Mean Square Due to Regression

$$MSR = \frac{SSR}{p} \tag{14.20}$$

Mean Square Due to Error

$$MSE = \frac{SSE}{n - p - 1} \qquad (14.21)$$

The F Statistic

$$F = \frac{MSR}{MSE} = \frac{SSR/p}{SSE/(n - p - 1)} \qquad (14.22)$$

■ SUPPLEMENTARY EXERCISES

21. The admissions' officer for Clearwater College developed the following estimated regression equation relating final college GPA to student's SAT mathematics scores and their high-school GPA.

$$\hat{y} = -1.41 + .0235x_1 + .00486x_2$$

where

$\quad x_1$ = high-school grade point average

$\quad x_2$ = SAT mathematics score

$\quad y$ = final college grade point average

a. Interpret the coefficients in this estimated regression equation.

b. Estimate the final college GPA for a student who has a high-school average of 84 and a score of 540 on the SAT mathematics test.

22. The personnel director for Electronics Associates developed the following estimated regression equation relating an employee's score on a job satisfaction test to his or her length of service and wage rate:

$$\hat{y} = 14.4 - 8.69x_1 + 13.5x_2$$

where

$\quad x_1$ = length of service (years)

$\quad x_2$ = wage rate (dollars)

$\quad y$ = job satisfaction test score (higher scores indicate more job satisfaction)

a. Interpret the coefficients in this estimated regression equation.

b. Develop an estimate of the job satisfaction test score for an employee that has 4 years of service and makes $6.50 per hour.

23. In a regression analysis involving 18 observations and four independent variables it was determined that SSR = 18,051.63 and SSE = 1014.3.

a. Determine R^2 and R_a^2.

b. Test for the significance of the relationship at the $\alpha = .01$ level of significance.

24. The following estimated regression equation involving three independent variables has been developed:

$$\hat{y} = 18.31 + 8.12x_1 + 17.9x_2 - 3.6x_3$$

Computer output indicates that $s_{b_1} = 2.1$, $s_{b_2} = 9.72$, and $s_{b_3} = .71$. There were 15 observations in the study.

a. Test $H_0: \beta_1 = 0$ at $\alpha = .05$.

b. Test $H_0: \beta_2 = 0$ at $\alpha = .05$.

c. Test $H_0: \beta_3 = 0$ at $\alpha = .05$.

d. Would you recommend dropping any of the independent variables from the model?

25. Shown is a partial computer output from a regression analysis:

```
The regression equation is
Y = 8.103 + 7.602 X1 + 3.111 X2

Predictor          Coef          Stdev        t-ratio
Constant           8.103         2.667
X1                 7.602         2.105
X2                 3.111         0.613

s = 3.35      R-sq = 92.3%      R-sq(adj) =

Analysis of Variance

SOURCE             DF            SS           MS
Regression         2            1612
Error              12
Total                          1746.48
```

a. Compute the appropriate t ratios.

b. Test for the significance of β_1 and β_2 at $\alpha = .05$.

c. Compute the entries in the DF, SS, and MS = SS/DF columns.

d. Compute R_a^2.

26. Recall that in Exercise 21, the admissions officer for Clearwater College developed the following estimated regression equation relating final college GPA to student's SAT mathematics scores and their high-school GPA.

$$\hat{y} = -1.41 + .0235x_1 + .00486x_2$$

where

$$x_1 = \text{high-school grade point average}$$

$$x_2 = \text{SAT mathematics score}$$

$$y = \text{final college grade point average}$$

A portion of the Minitab computer output is shown.

```
The regression equation is
Y = -1.41 + .0235 X1 + .00486 X2

Predictor              Coef        Stdev       t-ratio
Constant             -1.4053      0.4848       _____
X1                    0.023467    0.008666     _____
X2                    _____     0.001077     _____

s = 0.1298    R-sq = _____    R-sq(adj) = _____

Analysis of Variance

SOURCE           DF          SS          MS          F
Regression       _____       1.76209     _____       _____
Error            _____       _____       _____
Total            9           1.88000
```

a. Complete the missing entries in this output.

b. Compute F and test at the $\alpha = .05$ level whether or not a significant relationship exists.

c. Did the estimated regression equation provide a good fit to the data? Explain.

d. Use the t-test and $\alpha = .05$ to test $H_0: \beta_1 = 0$ and $H_0: \beta_2 = 0$.

27. Recall that in Exercise 22 the personnel director for Electronics Associates developed the following estimated regression equation relating an employee's score on a job satisfaction test to their length of service and wage rate:

$$\hat{y} = 14.4 - 8.69x_1 + 13.5x_2$$

where

$$x_1 = \text{length of service (years)}$$

$$x_2 = \text{wage rate (dollars)}$$

$$y = \text{job satisfaction test score (higher scores indicate more job satisfaction)}$$

A portion of the Minitab computer output is shown:

```
The regression equation is
Y = 14.4 - 8.69 X1 + 13.52 X2

Predictor              Coef        Stdev       t-ratio
Constant             14.448       8.191         1.76
X1                   -8.69        1.555        5.588
X2                   13.517       2.085        6.4829

s = 3.773    R-sq = 90.12%    R-sq(adj) = 98.02%
                          SSR
Analysis of Variance      ───
                          SST

SOURCE           DF          SS          MS          F
Regression       2           648.83      824.42      57.94
Error            5           71.17       14.23
Total            7           720.00
                                              F = MSR/MSE
```

a. Complete the missing entries in this output.

b. Compute F and test at the $\alpha = .05$ level whether a significant relationship exists or not.

c. Did the estimated regression equation provide a good fit to the data? Explain.

d. Use the t test and $\alpha = .05$ to $H_0: \beta_1 = 0$ and $H_0: \beta_2 = 0$.

28. Bauman Construction Company makes bids on a variety of projects. In an effort to estimate the bid to be made by one of its competitors Bauman has obtained data on 15 previous bids and developed the following estimated regression equation:

$$\hat{y} = 80 + 45x_1 - 3x_2$$

where

$$\hat{y} = \text{competitor's bid (\$1000s)}$$

$$x_1 = \text{square feet (1000s)}$$

$$x_2 = \text{local index of construction activity}$$

a. Estimate the competitor's bid on a project involving 50,000 square feet and an index of construction activity of 70.

b. If SSR = 19,780 and SST = 21,533, test at $\alpha = .01$ the significance of the relationship.

▬ COMPUTER EXERCISE

A government agency conducted a study to determine the relationship between income, household size, and the amount charged in the last 12 months on credit cards.

The objective of this study was to develop an equation that could be used to predict the amount charged on credit cards given the individual's income and household size. The following data was obtained.

Income ($1000s)	Household Size	Amount Charged	Income ($1000s)	Household Size	Amount Charged
33	3	2780	29	2	2491
32	2	2748	26	3	2300
35	1	3070	43	3	3690
34	2	2831	35	4	2935
35	2	3047	35	2	2946
37	1	3061	35	4	2670
38	2	3390	39	1	3485
35	2	3091	30	2	2722
41	2	3415	29	3	2620
47	1	4107	36	1	3126
42	1	3429	32	3	2726
33	5	2521	41	1	3486
38	1	3493	35	5	2720
38	3	3104	36	2	3172
34	3	2752	33	2	2769
30	4	2401	31	1	2804
27	2	2276	37	1	3170
34	2	2865	36	3	2923
29	3	2578	27	2	2270
37	1	3173	36	2	3053
38	5	3091	25	2	2266
33	3	2686	41	6	3187
38	5	2921	35	2	2947
28	4	2250	40	5	3224
35	5	2865	35	2	3105

■■■ QUESTIONS

1. Use numerical and graphical measures to summarize these data.

2. Does there appear to be any relationship between the amount charged and the individual's income? Between the amount charged and the household size?

3. Develop an estimated regression equation that can be used to predict the amount charged.

APPENDIX

Calculus-Based Derivation and Solution of Normal Equations for Regression with 2 Independent Variables

In order to show how the least squares method is applied in multiple regression we derive the normal equations for the 2-independent-variables case. The least squares criterion calls for the minimization of the following expression for the residual sum of squares:

$$\text{SSE} = \Sigma \, (y_i - b_0 - b_1 x_{1i} - b_2 x_{2i})^2 \tag{14A.1}$$

To minimize (14A.1) we must take the partial derivatives of SSE with respect to b_0, b_1, and b_2. We can then set the partial derivatives equal to zero and solve for the estimated regression coefficients b_0, b_1, and b_2. Taking the partial derivatives and setting them equal to zero provides

$$\frac{\partial \text{SSE}}{\partial b_0} = -2 \, \Sigma \, (y_i - b_0 - b_1 x_{1i} - b_2 x_{2i}) = 0 \tag{14A.2}$$

$$\frac{\partial \text{SSE}}{\partial b_1} = -2 \, \Sigma \, x_{1i} \, (y_i - b_0 - b_1 x_{1i} - b_2 x_{2i}) = 0 \tag{14A.3}$$

$$\frac{\partial \text{SSE}}{\partial b_2} = -2 \, \Sigma \, x_{2i} \, (y_i - b_0 - b_1 x_{1i} - b_2 x_{2i}) = 0 \tag{14A.4}$$

Dividing (14A.2) by 2 and summing the terms individually yields

$$-\Sigma \, y_i + \Sigma \, b_0 + \Sigma \, b_1 x_{1i} + \Sigma \, b_2 x_{2i} = 0$$

Bringing $\Sigma \, y_i$ to the right hand side of the equation and noting that $\Sigma \, b_0 = nb_0$, we obtain the normal equation given in this chapter as (14.11):

$$nb_0 + (\Sigma\, x_{1i})b_1 + (\Sigma x_{2i})b_2 = \Sigma\, y_i \tag{14.11}$$

Similar algebraic simplification applied to (14A.3) and (14A.4) leads to the normal equations given by (14.12) and (14.13):

$$(\Sigma\, x_{1i})b_0 + \quad (\Sigma\, x_{1i}^2)b_1 + (\Sigma\, x_{1i}x_{2i})b_2 = \Sigma\, x_{1i}y_i \tag{14.12}$$

$$(\Sigma\, x_{2i})b_0 + (\Sigma\, x_{1i}x_{2i})b_1 + \quad (\Sigma\, x_{2i}^2)b_2 = \Sigma\, x_{2i}y_i \tag{14.13}$$

Application of these procedures to a regression model involving p independent variables would lead to $p + 1$ normal equations of this type. However, matrix algebra is usually used for that type of derivation.

Solving the Normal Equations for the Butler Trucking Company

In the chapter we developed a regression model for Butler Trucking involving two independent variables, miles traveled and number of deliveries. Substituting the data for this problem (see Section 14.3) into (14.11) to (14.13) provided the following normal equations:

$$10b_0 + \quad 815b_1 + \quad 26b_2 = \quad 66.0 \tag{14.14}$$

$$815b_0 + 69{,}625b_1 + 2165b_2 = 5594.0 \tag{14.15}$$

$$26b_0 + \quad 2165b_1 + \quad 76b_2 = \quad 180.6 \tag{14.16}$$

By multiplying (14.14) by 81.5 and subtracting the result from (14.15), we can eliminate b_0 and obtain an equation involving only b_1 and b_2.

$$
\begin{aligned}
815b_0 + 69{,}625.0b_1 + 2165b_2 &= 5594.0 \\
-815b_0 - 66{,}422.5b_1 - 2119b_2 &= -5379.0 \\
\hline
3202.5b_1 + \quad 46b_2 &= 215.0
\end{aligned}
\tag{14A.5}
$$

Now multiply (14.14) by 2.6 and subtract the result from (14.16). This manipulation yields a second equation involving only b_1 and b_2.

$$
\begin{aligned}
26b_0 + 2165b_1 + 76.0b_2 &= 180.6 \\
-26b_0 - 2119b_1 - 67.6b_2 &= -171.6 \\
\hline
46b_1 + 8.4b_2 &= 9.0
\end{aligned}
\tag{14A.6}
$$

With equations (14A.5) and (14A.6) we can solve simultaneously for b_1 and b_2. Multiplying (14A.6) by 46/8.4 and subtracting the result from (14A.5) gives us an equation involving only b_1.

$$\begin{array}{rl} 3202.5000b_1 + 46b_2 = & 215.0000 \\ - \quad 251.9048b_1 - 46b_2 = & -49.2857 \\ \hline 2950.5952b_1 \qquad\qquad = & 165.7143 \end{array}$$

(14A.7)

Using (14A.7) to solve for b_1 we get

$$b_1 = \frac{165.7143}{2950.5952} = .056163$$

Using this value for b_1 we can substitute into (14A.6) to solve for b_2:

$$46(.056163) + 8.4b_2 = 9$$

$$2.583498 + 8.4b_2 = 9$$

$$8.4b_2 = 6.416502$$

$$b_2 = .7638693$$

Now we can substitute the values obtained for b_1 and b_2 into (14.14), thus obtaining b_0:

$$10b_0 + 815(.056163) + 26\,(.7638693) = 66.0$$

$$10b_0 \ + 45.77284 \ \quad + 19.860601 \quad\ = 66.0$$

$$10b_0 \qquad\qquad\qquad\qquad\qquad = .366554$$

$$b_0 \qquad\qquad\qquad\qquad\qquad = .0366554$$

Rounding to four significant digits, we obtain the following estimated regression equation for Butler Trucking:

$$\hat{y} = .0367 + .0562x_1 + .7639x_2$$

(14A.8)

Champion International Corporation*

Stamford, Connecticut

Champion International Corporation is one of the largest forest products companies in the world, employing over 41,000 people in the United States, Canada, and Brazil. Champion manages over 3 million acres of timberlands in the United States. Its objective is to maximize the return of this timber base by converting trees into three basic product groups: (1) building materials, such as lumber and plywood; (2) white paper products, including printing and writing grades of white paper; and (3) brown paper products, such as linerboard and corrugated containers. Given the highly competitive markets within the forest products industry, survival dictates that Champion must maintain its position as a low cost producer of quality products. This requires an ambitious capital program to improve the timber base and to build additional modern, cost effective timber conversion facilities.

THE MANAGEMENT SCIENCE FUNCTION

The Management Science function at Champion International Corporation is organizationally structured within the Management Information Services Department and operates as an internal consulting service within the company. Approximately 40% of the project activity is involved with facility and production planning, 30% with physical distribution, 20% with process improvement, and 10% with capital budgeting. The primary techniques used are mathematical programming (e.g., linear programming), simulation, and statistical analyses. Of the statistical analyses performed, multivariate (multiple) regression and forecasting techniques, such as exponential smoothing, are used most frequently.

A MULTIPLE REGRESSION APPLICATION

A pulp mill is a facility in which wood chips and chemicals are processed to produce wood pulp. The wood pulp is then used at a paper mill to produce paper products. Wood is made up primarily of cellulose fibers held together by a substance called lignin. These cellulose fibers will eventually become wood pulp. The pulping process used at a pulp mill chemically dissolves this lignin binding and separates the cellulose fibers. To start, the wood chips and some chemicals are cooked in a pressure cooker called a digester which softens the chips and dissolves the lignin binding; the softened chips are piped into a blow tower, where they are smashed against a steel target to separate the cellulose fibers, as shown in Figure 14A.1. The pulp is then chemically washed in a pulp washer,

*The authors are indebted to Marian Williams and Bill Griggs of Champion International Corporation for providing this application.

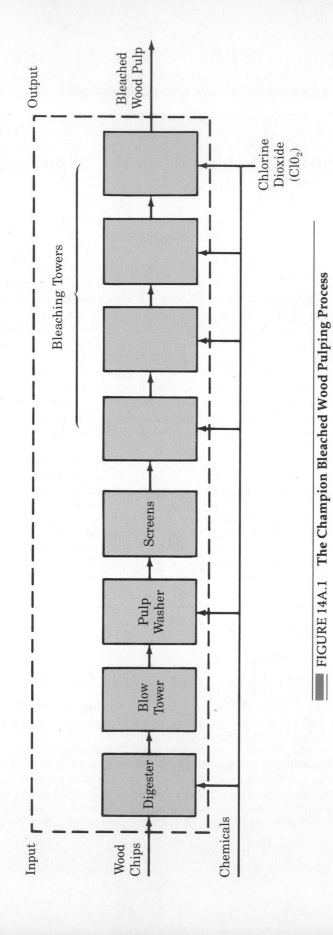

■ FIGURE 14A.1   The Champion Bleached Wood Pulping Process

from which it flows immediately into screens, which filter out any undesirable material. Although cellulose is composed of white fibers, wood pulp, no matter how well cooked and washed, contains a fair amount of lignin compounds, which gives the pulp a dark brownish color.* In order to produce white paper products, the pulp must be bleached to break down these lignin compounds. Therefore, the next step in the pulping process is bleaching. The pulp flows into a four stage bleaching process. In each tower in this process a different bleaching chemical is used to bleach the pulp gradually to a white color. The bleached pulp is now ready for further processing, such as the addition of dyes and other chemicals to prepare it for paper production.

A key bleaching agent used in the last tower of the bleaching process is chlorine dioxide (ClO_2). It can bleach pulp to a high white brightness, not obtainable with other bleaching agents, without significant loss of pulp fiber strength. Chlorine dioxide gas has a yellow green color with an irritating odor and is highly corrosive to most metals. Under certain conditions chlorine dioxide gas will decompose violently (i.e., explode). Owing to the combustible nature of chlorine dioxide gas, it is usually produced—at least at Champion pulp mill facilities—on site and is piped in solution form into the bleaching tower of the pulp mill.

One of Champion's pulp mill facilities presently has two ClO_2 generation processes—the R3 process and the sulfur dioxide (SO_2) process. The R3 process operates very efficiently, but the SO_2 process does not. A study was undertaken to look at process control and efficiency improvement of the SO_2 process. To begin the study, the acquisition of a basic understanding of the following areas was necessary:

▪▪▪ *SO_2 process.*   The SO_2 process requires the following four feed chemicals: (1) sulfuric acid (H_2SO_4), (2) sodium chlorate ($NaClO_3$), (3) sulfur dioxide gas (SO_2), and (4) air. These chemicals flow at metered rates into the ClO_2 generator, as shown in Figure 14A.2. ClO_2 gas is produced in the generator and flows into the ClO_2 absorber, where chilled water absorbs the ClO_2 gas to form a ClO_2 solution. This ClO_2 solution is piped into the pulp mill.

*John H. Ainsworth, *Paper—The Fifth Wonder*, Kaukaune, Wisconsin: Thomas Publishing Co., 1958.

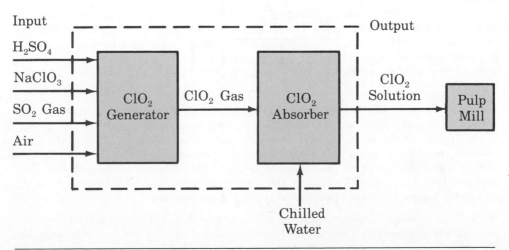

▪▪▪ FIGURE 14A.2   **A ClO_2 Generation Process at a Champion Pulp Mill Facility**

TABLE 14A.1   Sample ClO$_2$ Manufacturing Log Sheet

Time	Chemical Feed				Concentration Test			Production Rate
	Air (cfm)	SO$_2$ (cfm)	NaClO$_3$ (gpm)	H$_2$SO$_4$ (gpm)	NaClO$_3$ (gpl)	H$_2$SO$_4$ (gpl)	ClO$_2$ (gpl)	Chilled Water Flow (gpm)
Acceptable Range					40–45	600–620	15–16	
7 A.M.	400	51	6.1	3.6				300
8	400	51	6.1	3.6	42.0	610	14.9	300
9	400	56	6.5	3.7				350
10	400	56	6.5	3.7	45.0	615	15.5	350

■ *Process control.*   There is a control room for the generator which contains a dial for each feed chemical and also for chilled water flow. These dials regulate the amount of each chemical that is to be used to produce ClO$_2$ and are manually set and adjusted by a shift operator. Since the generator is run 24 hours a day, there are three operators—one each shift—who control the generator.

The demand for ClO$_2$ at the pulp mill dictates the amount produced. This causes the ClO$_2$ production rate to be variable. The chilled water flow rate to the absorber is measured in gallons per minute (gpm) and represents the production rate of ClO$_2$ solution. The pulp mill requires that the ClO$_2$ solution have a ClO$_2$ concentration between 15 and 16† grams per liter (gpl). If the chilled water flow is increased and the chemical feeds to the generator remain the same, then the concentration of the ClO$_2$ in the solution may fall out of the acceptable range. Consequently, any significant change in chilled water flow requires an adjustment of the chemical feeds that together produce ClO$_2$ gas.

It is the operator's task to maintain the appropriate ClO$_2$ production rate and concentration in the ClO$_2$ solution. Moreover, the operator must maintain concentrations of NaClO$_3$ and H$_2$SO$_4$ in the generator solution within acceptable ranges set by the chemical engineers. Controlling the concentration of these chemicals in the generator solution improves the efficiency and the stability of the chemical reactions that produce ClO$_2$ gas.

■ *Data recorded.*   Every hour, on the hour, the operator records on a log sheet the chemical feed rates, the chilled water flow rate, and some generator temperature data. Every 2 hours, on the hour, a sample of the generator solution and a sample of the ClO$_2$ solution are taken. Titrations are performed on the samples to determine concentrations of H$_2$SO$_4$ and NaClO$_3$ in the generator solution and the ClO$_2$ in the ClO$_2$ solution. The operator uses these results to adjust the feed rates if the concentrations are out of acceptable ranges and/or if a production rate change is necessary.

The basic problem in operating the generator revolves around how to set the chemical feed rates. The generator manufacturer provided suggested chemical feed rates only for operation near design capacity. Owing to the variable production rate environment in

†All numerical values have been modified to protect proprietary information.

which the generator must run, ClO_2 production is usually much less than design capacity. The chemical feed rates were being set by the operators based on experience and the concentration tests. As a result the process was becoming over-controlled by the operators. The chemical engineers at the mill requested that a set of control equations, one for each chemical feed, be developed from the operating log data to aid the operators in setting chemical feed rates.

The approach taken was to obtain a representative sample of the log data and apply multiple regression to develop an equation for each chemical feed. In order to do this, the following steps were taken:

1. Log sheets for 2½ months were obtained. A sample log sheet is presented in Table 14A.1. The data were entered into the computer, with each hour considered a separate observation.

2. The data were plotted so that engineers could look for unusual or nonrepresentative observations that should be eliminated.

3. A new data set was formed by only using those observations that contained concentration test data. Log variables were created for chemical feeds and the previous concentration tests. There were over 400 observations in this data set.

4. Taking one chemical feed at a time, an estimated regression equation was developed. For instance, an initial $NaClO_3$ multiple regression equation of the following form was developed:

$$\hat{y} = b_0 + b_1x_1 + b_2x_2$$

y = (dependent variable) difference between the amount of ClO_2 produced in time period t and the amount produced in time period $t - 2$

x_1 = average of the amount of $NaClO_3$ fed in time period $t - 1$ and time period $t - 2$

x_2 = concentration test for ClO_2 in time period $t - 2$

For this model, we found $R^2 = .50$

5. The residuals were plotted; they appeared random, and thus the model assumptions were judged reasonable.

▬▬ RESULTS

An SAS (Statistical Analysis System) computer program was developed to perform the above analysis. Four regression equations were developed and were programmed into a microcomputer at the mill. The operators key into the computer the concentration test data and the desired production rate. The computer then calculates the appropriate chemical feeds and displays them on a screen.

Since the operators have begun using the control equations, the generator efficiency has increased and the number of times that the concentrations fall within acceptable ranges has increased significantly.

CHAPTER 15

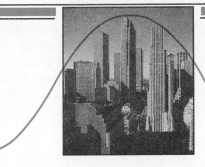

Regression Analysis: Model Building

M odel building in regression analysis is the process of developing the regression model that best describes the relationship among the independent and dependent variables. The major issues are determining the proper form (linear or curvilinear) of the relationship and variable selection. Variable selection involves determining which of a candidate set of independent variables should be included in the regression model.

In Chapters 13 and 14, we introduced and worked with two regression models:

Simple Linear Regression: $y = \beta_0 + \beta_1 x + \epsilon$

Multiple Regression: $y = \beta_0 + \beta_1 x_1 + \beta_2 x_2 + \cdots + \beta_p x_p + \epsilon$

Both of these models specify a linear relationship between the independent variable(s), the x's, and the dependent variable, y.

The primary method introduced in chapters 13 and 14 for determining whether or not a regression model was adequate was residual analysis. Essentially, if the residual plot looked like a horizontal band, we concluded that the assumptions concerning model form and the error term were satisfied. Otherwise, we concluded that the model was inadequate. When the regression model is judged inadequate, it may be because we have chosen the wrong functional form for the model (for example, the actual relationship is curvilinear) and/or the proper independent variables have not been included. In Chapter 14, we saw that when the residual plot for the Butler Trucking problem showed a nonconstant variance, a model that corrected this deficiency, was created by adding a second independent variable.

In this chapter, we focus on the model-building issues of identifying the proper model form and variable selection. Section 15.1, which establishes the framework for model building, introduces the concept of a *general linear model*. Surprisingly the general linear model makes it possible for us to accommodate curvilinear relationships between the independent variables and the dependent variable with no more computational difficulty than that involved in multiple regression. Section 15.2 extends our modeling capability by showing how qualitative independent variables may be incorporated into a regression model.

Section 15.3 provides the foundation for the more sophisticated computer-based procedures for variable selection. The issue of when one variable or a group of variables should be added to a regression model is examined. We show that the general approach to determining when to add or delete variables is based on the value of an F statistic. In Section 15.4, a large problem of the type often encountered in practice is introduced. It involves 25 observations on 8 independent variables. This larger problem provides an illustration for the computer-based variable selection procedures of Section 15.5. The stepwise, forward-selection, and backward-elimination procedures are explained.

In Section 15.6, we introduce some additional diagnostic procedures based on analysis of residuals. We show how the residuals can be used to identify potentially troublesome observations: outliers and observations that exert an unusually large amount of influence on the regression results. The Durban-Watson test is also introduced as a diagnostic for detecting serial or autocorrelation. The chapter concludes with a discussion of how regression analysis can be used for solving analysis of variance problems.

A caveat is in order before beginning this chapter. The "truly correct model" will never be found for real data. Several models may do an adequate job; the model-builder's goal is to find the best from a set of acceptable models.

15.1 ■ THE GENERAL LINEAR MODEL

Suppose we were faced with a situation in which the following regression model was appropriate:

$$y = \beta_0 + \beta_1 x^2 + \epsilon \tag{15.1}$$

Since x is squared, the relationship between y and x is said to be curvilinear. At first glance, it might not appear that the regression techniques developed in Chapters 13 and 14 are applicable to this model. But they are!

By making the substitution $z = x^2$, the preceding model can be converted into the following simple linear regression model with z as the independent variable:

$$y = \beta_0 + \beta_1 z + \epsilon \tag{15.2}$$

If there are n observations, each z_i is given by x_i^2. Using the data for y and the computed data for z, the least squares method can then be used to compute b_0 and b_1.

Let us see how the techniques we have already developed can be used with the type of curvilinear regression model in (15.1) by considering the problem faced by Nugent Industries. Management of Nugent Industries has been investigating the relationship between each customer's annual order size and

TABLE 15.1	Annual Order Size and Number of Sales Calls for Nugent Industries	
---	---	
Number of Sales Calls (x)	Annual Sales ($1000s) (y)
2	12
3	17
4	16
5	24
6	26
7	34
8	46

the number of sales calls made per year. A random sample of seven customer accounts was obtained; the data are presented in Table 15.1 and a scatter diagram is shown in Figure 15.1.

From the scatter diagram in Figure 15.1, it appears that a curvilinear relationship might be a better model of the relationship than a straight line. Note that

FIGURE 15.1   Scatter Diagram for Nugent Industries Problem Data

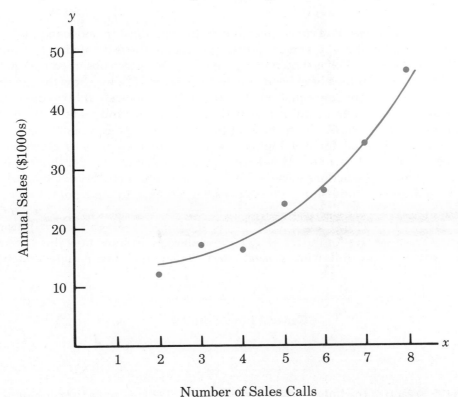

Number of Sales Calls

the curved line in the figure appears to provide a very good fit to the data. Suppose, after viewing the scatter diagram, we hypothesize that y is related to x by the following regression model:

$$y = \beta_0 + \beta_1 x^2 + \epsilon$$

Using the approach noted before we can make the substitution $z = x^2$ to obtain a simple linear regression model with the independent variable denoted by z:

$$y = \beta_0 + \beta_1 z + \epsilon$$

To develop estimates of β_0 and β_1 in this model, we simply substitute z_i for x_i^2 in the least squares formulas for b_0 and b_1. Doing so, we obtain

$$b_1 = \frac{\Sigma z_i y_i - (\Sigma z_i \Sigma y_i)/n}{\Sigma z_i^2 - (\Sigma z_i)^2/n}$$

$$b_0 = \overline{y} - b_1 \overline{z}.$$

Computing the values of b_1 and b_0 in these formulas requires that we substitute the value of x_i^2 everyplace z_i appears. The calculations of b_0 and b_1 for this problem are summarized in Table 15.2. Rounding to two decimal places and substituting x^2 for z leads to the estimated regression equation

$$\hat{y} = 9.64 + .53x^2$$

This equation represents a curvilinear (quadratic) relationship between x and y.

A residual plot of $y_i - \hat{y}_i$ versus $\hat{y}_i$ corresponding to the estimated regression equation $\hat{y} = 9.64 + .53x^2$ is shown in Figure 15.2. The pattern observed does not suggest any assumptions have been violated. In Figure 15.3, we show the graph of the estimated regression equation; it clearly fits the data well. It turns out that the relationship between y and x^2 is statistically significant and that the value of r^2 for this equation is .97. On the basis of this analysis, we recommend that $\hat{y} = 9.64 + .53x^2$ be used for developing predictions of annual sales given the number of sales calls per year. However, we caution that this model should not be used to make predictions outside the range of x values observed. Obviously one would expect diminishing returns per sales call beyond some point.

The method by which we developed a curvilinear relationship between x and y for the Nugent Industries problem can be generalized to handle curvilinear types of relationships for the multiple regression case. To show how this can be done, consider the following *general linear model* involving p independent variables.

General Linear Model

$$y = \beta_0 + \beta_1 z_1 + \beta_2 z_2 + \cdots + \beta_p z_p + \epsilon \qquad (15.3)$$

In (15.3) each of the independent variables z_j, $j = 1, 2, \ldots, p$ is a function of x_1, $x_2, \ldots, x_k$, the variables for which data have been collected. In some cases, each z_j

TABLE 15.2   Calculations for the Estimated Regression Equation
$\hat{y} = b_0 + b_1 x^2$

Customer (i)	x_i	$z_i = x_i^2$	y_i	$z_i y_i$	z_i^2
1	2	4	12	48	16
2	3	9	17	153	81
3	4	16	16	256	256
4	5	25	24	600	625
5	6	36	26	936	1296
6	7	49	34	1666	2401
7	8	64	46	2944	4096
Totals	35	203	175	6603	8771

$$\bar{z} = \frac{\Sigma z_i}{n} = \frac{203}{7} = 29$$

$$\bar{y} = \frac{\Sigma y_i}{n} = \frac{175}{7} = 25$$

$$b_1 = \frac{\Sigma z_i y_i - (\Sigma z_i \Sigma y_i)/n}{\Sigma z_i^2 - (\Sigma z_i)^2/n} = \frac{(6603) - (203)(175)/7}{(8771) - (203)^2/7} = .5298$$

$$b_0 = \bar{y} - b_1 \bar{z} = 25 - .5298(29) = 9.6358$$

$$\hat{y} = 9.6358 + .5298z$$

■ FIGURE 15.2   Residual Plot for the Estimated Regression Equation
$\hat{y} = 9.64 + .53x^2$

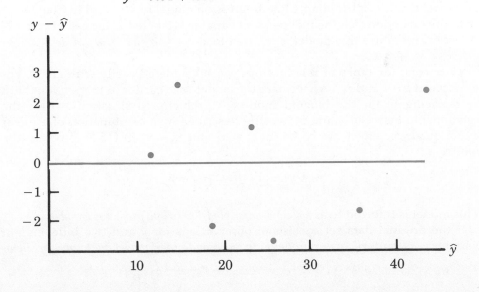

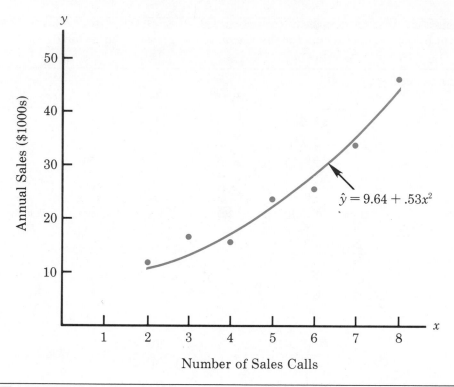

$$\hat{y} = 9.64 + .53x^2$$

FIGURE 15.3 **Graph of Estimated Regression Equation for Nugent Industries**

may be a function of only one x-variable. The simplest case occurs in which we have collected data for just one variable, x_1, and want to estimate y using a straight-line relationship. In this case $z_1 = x_1$ and (15.3) becomes

$$y = \beta_0 + \beta_1 x_1 + \epsilon$$

Note that this is just the simple liner regression model introduced in Chapter 13 with the exception that the independent variable is labeled x_1 instead of x. In the statistical literature this model is referred to as a *simple first-order model with one predictor variable.*

More complex types of relationships can be modeled easily with (15.3). For example, if $p = 1$ and $z_1 = x_1^2$, we have the model used earlier in this section (with x_1 replacing x) for the Nugent Industries problem. In situations where the relationship between y and x_1 is better described by a combination of a linear and a quadratic effect, we could set $z_1 = x_1$ and $z_2 = x_1^2$ in (15.3) to obtain the model

$$y = \beta_0 + \beta_1 x_1 + \beta_2 x_1^2 + \epsilon$$

This model is referred to as a *second-order model with one predictor variable.*

If the original data set consists of observations for y and two independent variables, x_1 and x_2, we can develop a *second-order model with two predictor variables* by setting $z_1 = x_1$, $z_2 = x_2$, $z_3 = x_1^2$, $z_4 = x_2^2$, and $z_5 = x_1 x_2$ in the general linear model of (15.3). The model obtained is

$$y = \beta_0 + \beta_1 x_1 + \beta_2 x_2 + \beta_3 x_1^2 + \beta_4 x_2^2 + \beta_5 x_1 x_2 + \epsilon$$

In this second-order model, the variable $z_5 = x_1 x_2$ is added to account for the potential effects of the two variables acting together. This type of effect is called *interaction*.

It should be apparent that many types of relationships can be modeled by (15.3). Each different function of the x variables generates a separate term, denoted by z_i. Thus the regression techniques with which we have been working are definitely not limited to linear, or straight-line, relationships.

In regression analysis (15.3) is referred to as the *general linear model*. The term linear in this context can be confusing, since it really refers only to the fact that $\beta_0, \beta_1, \ldots, \beta_p$ all have exponents of 1; it does not imply that the relationship between y and the x_is is linear. Indeed, in this section we have seen one example where (15.3) can be used to moded a curvilinear relationship. Let us now review an example involving interaction.

To understand better how interaction is treated with the general linear model, let us review the regression study conducted by Tyler Personal Care for one of its shampoo products. Two factors believed to have the most influence on sales were unit selling price and advertising expenditures. The following regression model was used.

$$y = \beta_0 + \beta_1 x_1 + \beta_2 x_2 + \beta_3 x_1 x_2 + \epsilon$$

where

$$y = \text{unit sales (1000s)}$$

$$x_1 = \text{selling price (\$)}$$

$$x_2 = \text{advertising expenditure (\$1000s)}$$

This model reflected Tyler's belief that sales depended linearly on selling price and advertising expenditures (the $\beta_1 x_1$ and $\beta_2 x_2$ terms) but that there was some interaction between the two (the $\beta_3 x_1 x_2$ term). Indeed, Tyler believed that at higher unit selling prices, the effect of increased advertising would diminish.

In order to develop an estimated regression equation, a general linear model involving three independent variables (z_1, z_2, and z_3) was used.

$$y = \beta_0 + \beta_1 z_1 + \beta_2 z_2 + \beta_3 z_3 + \epsilon$$

where

$$z_2 = x_1$$

$$z_2 = x_2$$

$$z_3 = x_1 x_2$$

The computer solution of the least squares equations for this model yielded $b_0 = 200$, $b_1 = -50$, $b_2 = 10$, and $b_3 = -3$. Substituting for z_1, z_2, and z_3 in the general linear model provided the following estimated regression equation for Tyler's shampoo product:

$$\hat{y} = 200 - 50x_1 + 10x_2 - 3x_1x_2$$

The term x_1x_2 is called a cross-product term; it accounts for the combined effect of x_1 and x_2 in predicting y. To understand better the interaction effect modeled by the cross-product term, consider the following values of $\hat{y}$ for this equation.

Selling Price x_1	Advertising x_2	Unit Sales $\hat{y}$
1.50	50	400
1.50	100	675
2.00	50	300
2.00	100	500

Holding price constant at $1.50 the difference in unit sales between an advertising expenditure of $100,000 and an advertising expenditure of $50,000 is $675,000 - 400,000 = 275,000$. Holding price constant at $2.00, the difference in unit sales between an advertising expenditure of $100,000 and an advertising expenditure of $50,000 is $500,000 - 300,000 = 200,000$. Thus, the regression results do show that the effect of advertising on unit sales depends on the level of the selling price.

The general linear model of (15.3) can be used to accommodate any situation in which z_1, z_2, . . . , z_p are functions of the x variables. However, it cannot be used in cases where the parameters, β_i, appear nonlinearly. Sometimes, however, a transformation can be applied to allow even these cases to be handled with a general linear model. One such transformation, the logarithmic transformation, is discussed briefly here.

The Logarithmic Transformation

Models in which the parameters $(\beta_0, \beta_1, \ldots, \beta_p)$ have exponents other than 1 are referred to as nonlinear models. However, for the case of the exponential model, it is possible to perform a transformation of variables that will permit us to perform regression analysis using (15.3), the general linear model. The exponential model involves the following regression equation:

$$E(y) = \beta_0 \beta_1^x \tag{15.4}$$

This model is appropriate in cases where the dependent variable y increases or decreases by a constant percentage, instead of by a fixed amount, as x increases.

As an example, suppose that sales for a product (y) were related to advertising expenditure x (in $1000s) according to the following exponential model:

$$E(y) = 500(1.2)^x$$

Thus, for $x = 1$, $E(y) = 500(1.2)^1 = 600$; for $x = 2$, $E(y) = 500(1.2)^2 = 720$; and for $x = 3$, $E(y) = 500(1.2)^3 = 864$. Note that $E(y)$ is not increasing by a constant amount in this case, but by a constant percentage; the percentage increase is 20%.

We can transform this nonlinear model to a linear model by taking the logarithm of both sides of (15.4):

$$\log E(y) = \log \beta_0 + x \log \beta_1 \qquad (15.5)$$

Now if we let $y' = \log E(y)$, $\beta_0' = \log \beta_0$, and $\beta_1' = \log \beta_1$, we can rewrite (15.5) as

$$y' = \beta_0' + \beta_1' x$$

It is clear that the formulas for simple linear regression can now be used to develop estimates of β_0' and β_1'. Denoting the estimates as b_0' and b_1' leads to the following estimated regression equation:

$$\hat{y}' = b_0' + b_1' x \qquad (15.6)$$

To obtain predictions of the original dependent variable y given a value of x, we would first substitute the value of x into (15.6) and compute $\hat{y}'$. The antilog of $\hat{y}'$ would be our prediction of y, or the expected value of y.

We should make it clear that there are many nonlinear models that cannot be transformed into an equivalent linear model. However, such models have had limited use in business and economic applications. Furthermore, the mathematical background needed for study of such models is beyond the scope of this text.

▬ EXERCISES

1. The highway department is doing a study on the relationship between traffic flow and speed. The following model has been hypothesized:

$$y = \beta_0 + \beta_1 x + \epsilon$$

where

$$y = \text{traffic flow in vehicles per hour}$$
$$x = \text{vehicle speed in miles per hour}$$

The following data have been collected during rush hour for six highways leading out of the city:

Traffic Flow (y)	Vehicle Speed (x)
1256	35
1329	40
1226	30
1335	45
1349	50
1124	25

a. Develop an estimated regression equation for these data.
b. Using $\alpha = .01$ test for a significant relationship.

2. In working further with the problem of Exercise 1 statisticians suggested the use of the following curvilinear estimated regression equation:

$$\hat{y} = b_0 + b_1 x + b_2 x^2$$

a. Use the data of Exercise 1 to estimate the parameters of this estimated regression equation.
b. Using $\alpha = .01$ test for a significant relationship.
c. Estimate the traffic flow in vehicles per hour at speeds of 38 miles per hour.

3. The following estimated regression equation has been developed for the relationship between y, sales ($1000s), and x, store size (square feet $\times$ 10,000):

$$\hat{y} = 150 + 100x - 10x^2$$

Ten stores were included in the sample. Values of SST = 168,000 and SSR = 140,000 were obtained.
a. Compute R^2 and R_a^2.
b. Using $\alpha = .05$ test for a significant relationship.

4. A study of emergency service facilities investigated the relationship between the number of facilities and the average distance travelled to provide the emergency service (*Management Science*, July 1988). The following data were collected:

Number of Facilities	Average Travel Distance (miles)
9	1.66
11	1.12
16	0.83
21	0.62
27	0.51
30	0.47

a. Develop a scatter diagram for these data treating average travel distance as the dependent variable.
b. Does a simple linear model appear to be appropriate? Explain.
c. Fit a model to the data that you believe will best explain the relationship between these two variables.

15.2 ▬ THE USE OF QUALITATIVE VARIABLES

So far the variables that we have used to build regression models have been quantitative variables; that is, variables that are measured in terms of how much or how many. Frequently, however, we will need to use variables that are not measured in these terms. We refer to such variables as *qualitative variables*. For instance, suppose that we were interested in predicting sales for a product which was available in either bottles or cans. Clearly the independent variable "container type" could influence the dependent variable "sales"—but container type is a qualitative, not a quantitative, variable. The distinguishing feature of qualitative variables is that there is no natural measure of "how much" or "how many"; these variables are used to refer to attributes that are either present or not present.

Let us see how qualitative variables might be used in the context of the Butler Trucking problem introduced in Chapter 14. For that problem we concluded that the travel time (y) could be predicted using the number of miles traveled (x_1) and the number of deliveries (x_2). The estimated regression equation we developed was $\hat{y} = .0367 + .0562x_1 + .7639x_2$. Suppose that management felt that the type of truck should also be considered in attempting to predict total travel time. Butler Trucking has only two types of trucks: pickups and vans. Thus truck type is an example of a qualitative variable. Table 15.3 shows the expanded data set for the Butler Trucking Company with the addition of truck type as a third independent variable.

To incorporate the effect of truck type into a model to predict total travel time, we define the following variable:

$$x_3 = \begin{cases} 0 & \text{if the truck is a pickup} \\ 1 & \text{if the truck is a van} \end{cases}$$

When preparing the data, whenever an observation involves a pickup truck we will set $x_3 = 0$; whenever an observation involves a van we will set $x_3 = 1$. In regression analysis this type of variable is commonly referred to as a *dummy*, or *indicator*, variable.

Adding this dummy variable to the previous regression equation for predicting travel time results in the following:

$$E(y) = \beta_0 + \beta_1 x_1 + \beta_2 x_2 + \beta_3 x_3 \tag{15.7}$$

We can see that when $x_3 = 0$, and hence the truck is a pickup, the regression equation reduces to

$$E(y) = \beta_0 + \beta_1 x_1 + \beta_2 x_2 + \beta_3(0)$$
$$= \beta_0 + \beta_1 x_1 + \beta_2 x_2 \tag{15.8}$$

However, if we want to predict y when a van is used, $x_3 = 1$ and the regression

TABLE 15.3 Data for the Butler Trucking Problem, Including Truck Type

Day (i)	Miles traveled (x_{1i})	Number of Deliveries (x_{2i})	Truck Type (x_{3i})		Travel Time (hours) y_i
1	100	4	Van	1	9.3
2	50	3	Pickup	0	4.8
3	100	4	Van	1	8.9
4	100	2	Pickup	0	5.8
5	50	2	Pickup	0	4.2
6	80	1	Van	1	6.8
7	75	3	Van	1	6.6
8	80	2	Pickup	0	5.9
9	90	3	Pickup	0	7.6
10	90	2	Van	1	6.1

equation becomes

$$E(y) = \beta_0 + \beta_1 x_1 + \beta_2 x_2 + \beta_3(1)$$

$$= \beta_0 + \beta_1 x_1 + \beta_2 x_2 + \beta_3 \quad (15.9)$$

If we subtract (15.8), the expected travel time for a pickup, from (15.9), the expected travel time for a van, we obtain

$$\underbrace{(\beta_0 + \beta_1 x_1 + \beta_2 x_2 + \beta_3)}_{\substack{\text{Expected travel} \\ \text{time for a van}}} - \underbrace{(\beta_0 + \beta_1 x_1 + \beta_2 x_2)}_{\substack{\text{Expected travel} \\ \text{time for a pickup}}} = \beta_3$$

Presence or absence of dummy variable

Thus β_3 can be interpreted as the difference in the expected travel time between a van and a pickup truck.

When we fit an estimated regression equation using the least squares method and incorporate the possible effect of truck type we obtain the following estimated regression equation:

$$\hat{y} = b_0 + b_1 x_1 + b_2 x_2 + b_3 x_3$$

As usual, b_3 turns out to be the least squares estimate of β_3 and hence our best estimate of the effect of truck type.

Computer Solution

In Figure 15.4 we show the computer output obtained using the Minitab statistical computer software system. The estimated regression equation is

$$\hat{y} = .522 + .0464 x_1 + .7102 x_2 + .9000 x_3$$

Thus we see that $b_3 = .9000$. Hence the best estimate of the difference in the

---

■■■■ FIGURE 15.4   **Minitab Output for the Butler Trucking Problem**

```
The regression equation is
TIME = 0.52 + 0.0464 MILES + 0.710 DELIV + 0.900 TYPE

Predictor       Coef        Stdev      t-ratio        p
Constant       0.522        1.210        0.43       0.681
MILES          0.04640      0.01500      3.09       0.021
DELIV          0.7102       0.2725       2.61       0.040
TYPE           0.9000       0.5281       1.70       0.139

s = 0.7531      R-sq = 85.8%      R-sq(adj) = 78.7%

Analysis of Variance

SOURCE        DF          SS          MS         F         p
Regression     3      20.5969      6.8656     12.10     0.006
Error          6       3.4031      0.5672
Total          9      24.0000
```

expected travel time when a van is used instead of a pickup truck is .9 hours, or 54 minutes.

To test for the significance of x_3 given that x_1 and x_2 are in the model, the appropriate hypotheses are

$$H_0: \quad \beta_3 = 0$$

$$H_a: \quad \beta_3 \neq 0$$

Using $\alpha = .05$ and the $p =$ value of .139, we cannot reject H_0 and must conclude that truck type is not a significant factor in predicting travel time once the effects of miles traveled and number of deliveries have been accounted for. Note that we concluded not that truck type is not significant, but that it is not significant once the effect of x_1 and x_2 have been accounted for.

In closing this section we should also point out that the concept of interaction and the method for dealing with it also applies when working with qualitative variables. All that needs to be done is to add cross-product terms of the form x_1x_2, where x_1, or x_2, or both may be qualitative variables.

▰▰▰ EXERCISES

5. The following regression model has been proposed to predict sales at a fast-food outlet:

$$y = \beta_0 + \beta_1x_1 + \beta_2x_2 + \beta_3x_3 + \epsilon$$

where

$x_1 =$ number of competitors within 1 mile

$x_2 =$ population within 1 mile (1000s)

$$x_3 = \begin{cases} 1 & \text{if drive-up window present} \\ 0 & \text{otherwise} \end{cases}$$

$y =$ sales ($1000s)

The following estimated regression equation was developed after 20 outlets were surveyed:

$$\hat{y} = 10.1 - 4.2x_1 + 6.8x_2 + 15.3x_3$$

a. What is the expected amount of sales attributable to the drive-up window?
b. Predict sales for a store with two competitors, a population of 8000 within 1 mile, and no drive-up window.
c. Predict sales for a store with one competitor, a population within 1 mile of 3000, and a drive-up window.

6. In order to investigate the relationship among the service time to repair a machine and (1) the number of months since the machine was serviced and (2) whether a mechanical

failure or an electrical failure had occurred, the following data were obtained:

Repair Time (hours)	Time Since Previous Service Call (months)	Type of Failure
2.9	2	Electrical
3.0	6	Mechanical
4.8	8	Electrical
1.8	3	Mechanical
2.9	2	Electrical
4.9	7	Electrical
4.2	9	Mechanical
4.8	8	Mechanical
4.4	4	Electrical
4.5	6	Electrical

a. Ignore for now the type of failure associated with the machine. Develop the estimated simple linear regression equation to predict the repair time given the number of months since the previous service call.

b. Does the equation that you developed in Exercise (a) provide a good fit for the observed data? Explain.

7. This problem is an extension of the situation described in Exercise 6. In an attempt to incorporate the possible effect of the type of failure the following dummy variable was added to the regression model:

$$x_2 = \begin{cases} 1 & \text{if the failure was electrical} \\ 0 & \text{if the failure was mechanical} \end{cases}$$

With the addition of this variable, the following regression equation was proposed:

$$E(y) = \beta_0 + \beta_1 x_1 + \beta_2 x_2$$

where

$$x_1 = \text{number of months since the previous service call}$$

$$y = \text{repair time (hours)}$$

a. What is the interpretation of β_2 in this regression equation?

b. Develop the estimated regression equation using both the number of months since the previous service call (x_1) and the type of failure associated with the machine (x_2).

c. At the $\alpha = .05$ level of significance, test whether or not the estimated regression equation developed in (b) represents a significant relationship between the independent variables and the dependent variable.

d. Does the estimated regression equation developed in (b) provide a better fit than the equation developed in Exercise 6? Explain.

e. Use the estimated regression equation developed in (b) to determine on the average how much longer it takes to service a machine involving an electrical failure than one with a mechanical failure.

8. The following data show the price-earnings ratio, the net profit margin, and the growth rate for 19 companies listed in "The *Forbes* 500s on Wall Street." (*Forbes*, May 1, 1989) The data in the column labeled "Industry" are simply codes used to define the industry for each company: 1 = energy–international oil; 2 = health-drugs, and 3 = electronics-computers.

Firm	P/E Ratio	Profit Margin	Growth Rate	Industry
Exxon	11.3	6.5	10	1
Chevron	10.0	7.0	5	1
Texaco	9.9	3.9	5	1
Mobil	9.7	4.3	7	1
Amoco	10.0	9.8	8	1
Pfizer	11.9	14.7	12	2
Bristol Meyers	16.2	13.9	14	2
Merck	21.0	20.3	16	2
American Home Products	13.3	16.9	11	2
Abbott Laboratories	15.5	15.2	18	2
Eli Lilly	18.9	18.7	11	2
Upjohn	14.6	12.8	10	2
Warner-Lambert	16.0	8.7	7	2
Amdahl	8.4	11.9	4	3
Digital	10.4	9.8	19	3
Hewlett-Packard	14.8	8.1	18	3
NCR	10.1	7.3	6	3
Unisys	7.0	6.9	6	3
IBM	11.8	9.2	6	3

a. Develop the estimated regression equation that can be used to predict the price-earnings ratio given the profit margin, growth rate, and type of industry.

b. Consider modifying the model developed in (a) to account for possible interaction involving profit margin and the type of industry. Is there significant interaction involving these variables? Explain.

15.3 ■■■ DETERMINING WHEN TO ADD OR DELETE VARIABLES

In this section we will show how an F test can be used to determine whether or not it is advantageous to add one variable—or a group of variables—to a multiple regression model. This test is based on a determination of the amount of reduction in the error sum of squares resulting from adding one or more independent variables to the model. We will first illustrate how the test might be used in the context of the Butler Trucking problem.

With miles traveled (x_1) as the only independent variable, the least squares procedure provided the following estimated regression equation:

$$\hat{y} = 1.13 + .067x_1$$

In Chapter 14 we showed that the error sum of squares for this model was SSE = 9.5669. When x_2, the number of deliveries, was added as a second independent variable we obtained the following estimated regression equation:

$$\hat{y} = .0367 + .0562x_1 + .7639x_2$$

The error sum of squares for this model was 5.0501. Clearly, adding x_2 resulted in a reduction of SSE. The question we want to answer is: Does adding the variable x_2 lead to a *significant* reduction in SSE?

We will use the notation SSE (x_1) to denote the error sum of squares when x_1 is the only independent variable, SSE(x_1, x_2) the error sum of squares when x_1 and x_2 are both independent variables, and so on. Hence the reduction in SSE resulting from adding x_2 to the model involving just x_1 is

$$\text{SSE}(x_1) - \text{SSE}(x_1, x_2) = 9.5669 - 5.0501 = 4.5168$$

An F test is conducted to determine whether or not this reduction is significant.

The numerator of the F statistic is the reduction in SSE divided by the number of variables added to the original model. Here only one variable, x_2, has been added; thus the numerator is

$$\frac{\text{SSE}(x_1) - \text{SSE}(x_1, x_2)}{1} = 4.5168$$

The numerator is a measure of the reduction in SSE per variable added to the model. The denominator of the F statistic is the mean square error for the model that includes all of the variables. For Butler Trucking this corresponds to the model containing both x_1 and x_2; thus $p = 2$ and hence

$$\text{MSE} = \frac{\text{SSE}(x_1, x_2)}{n - p - 1} = \frac{5.0501}{7} = .7214$$

The following F statistic provides the basis for testing whether or not the addition of x_2 is statistically significant:

$$F = \frac{\dfrac{\text{SSE}(x_1) - \text{SSE}(x_1, x_2)}{1}}{\dfrac{\text{SSE}(x_1, x_2)}{n - p - 1}} \tag{15.10}$$

The numerator degrees of freedom for this F test equal the number of variables added to the model, and the denominator degrees of freedom equal $n - p - 1$.

For the Butler Trucking problem, we obtain

$$F = \frac{\dfrac{4.5168}{1}}{\dfrac{5.0501}{7}} = \frac{4.5168}{.7214} = 6.26$$

Refer to Table 4 of Appendix B. We find that for a level of significance of $\alpha = .05$,

$$F_{.05} = 5.59$$

Since

$$F = 6.26 > F_{.05} = 5.59$$

we reject the null hypothesis that x_2 is not statistically significant; in other words, adding x_2 to the model involving only x_1 results in a significant reduction in the error sum of squares.

When we want to test for the significance of adding only one additional independent variable to an existing model, the result found with the F test just described could also be obtained by using the t test for the significance of an individual parameter (described in Section 14.5). Indeed, the F statistic we just computed is the square of the t statistic used to test the hypothesis that an individual parameter is zero.

Since the t test is equivalent to the F test when only one variable is being added to the model, we can now further clarify the proper use of the t test on the individual parameters. If an individual parameter is not significant, the corresponding variable can be dropped from the model. However, no more than one variable can ever be dropped from a model on the basis of a t test; if one variable is dropped, a second variable that was not significant initially might become significant.

We now turn briefly to a consideration of whether or not the addition of more than one variable—as a set—results in a significant reduction in the error sum of squares.

The General Case

Consider the following multiple regression model involving q independent variables, where $q < p$:

$$y = \beta_0 + \beta_1 x_1 + \beta_2 x_2 + \cdots + \beta_q x_q + \epsilon \tag{15.11}$$

If we add variables $x_{q+1}, x_{q+2}, \ldots, x_p$ to this model, we obtain a model involving p independent variables:

$$y = \beta_0 + \beta_1 x_1 + \beta_2 x_2 + \cdots + \beta_q x_q \tag{15.12}$$
$$+ \beta_{q+1} x_{q+1} + \beta_{q+2} x_{q+2} + \cdots + \beta_p x_p + \epsilon$$

To test whether or not the addition of $x_{q+1}, x_{q+2}, \ldots, x_p$ is statistically significant, the null and alternative hypotheses can be stated as follows:

H_0: $\beta_{q+1} = \beta_{q+2} = \cdots = \beta_p = 0$

H_a: One or more of the coefficients is not equal to zero.

The following F statistic provides the basis for testing whether or not the additional variables are statistically significant:

$$F = \frac{\dfrac{\text{SSE}(x_1, x_2, \ldots, x_q) - \text{SSE}(x_1, x_2, \ldots, x_q, x_{q+1}, \ldots, x_p)}{p - q}}{\dfrac{\text{SSE}(x_1, x_2, \ldots, x_q, x_{q+1}, \ldots, x_p)}{n - p - 1}} \tag{15.13}$$

This computed F value is then compared with F_α, the table value with $p - q$ numerator degrees of freedom and $n - p - 1$ denominator degrees of freedom. If $F > F_\alpha$, we reject H_0 and conclude that the set of additional variables is statistically significant. Note that for the special case where $q = 1$ and $p = 2$, (15.13) reduces to (15.10).

■ EXERCISES

9. In a regression analysis involving 27 observations, the following estimated regression equation was developed:

$$\hat{y} = 16.3 + 2.3x_1 + 12.1x_2 - 5.8x_3$$

Also, the following standard errors were obtained:

$$s_{b_1} = .53 \qquad s_{b_2} = 8.15 \qquad s_{b_3} = 1.30$$

At an $\alpha = .05$ level of significance conduct the following hypothesis tests:

a. $H_0: \beta_1 = 0$ versus $H_a: \beta_1 \neq 0$.
b. $H_0: \beta_2 = 0$ versus $H_a: \beta_2 \neq 0$.
c. $H_0: \beta_3 = 0$ versus $H_a: \beta_3 \neq 0$.
d. Can any of the variables be dropped from the model? Why or why not?

10. In a regression analysis involving 30 observations the following estimated regression equation was obtained:

$$\hat{y} = 17.6 + 3.8x_1 - 2.3x_2 + 7.6x_3 + 2.7x_4$$

For this model SST = 1805 and SSR = 1760.

a. Compute R^2.
b. Compute R_a^2.
c. At $\alpha = .05$ test the significance of the relationship among the variables.

11. Refer again to Exercise 10. Variables x_1 and x_4 were dropped from the model, and the following estimated regression equation was obtained:

$$\hat{y} = 11.1 - 3.6x_2 + 8.1x_3$$

For this model SST = 1805 and SSR = 1705.

a. Compute SSE(x_1, x_2, x_3, x_4).
b. Compute SSE(x_2, x_3)
c. Use an F test and an $\alpha = .05$ level of significance to determine if x_1 and x_4 contribute significantly to the model.

15.4 ■ FIRST STEPS IN ANALYSIS OF A LARGER PROBLEM: THE CRAVENS DATA

In introducing multiple regression analysis, we utilized the Butler Trucking problem extensively. Although the small size of this problem was an advantage when exploring introductory concepts, the limited size of the problem makes it difficult to illustrate some of the variable-selection issues involved in model building. To provide an illustration of the variable-selection procedures discussed in the next section, we now introduce a data set consisting of 25 observations on 8 independent variables. Permission to use these data was provided by Dr. David W. Cravens of the Department of Marketing at Texas Christian University. Consequently, we refer to the data set as the Cravens data.*

*For details see David W. Cravens, Robert B. Woodruff, and Joe C. Stamper, "An Analytical Approach for Evaluating Sales Territory Performance," *Journal of Marketing* 36 (January 1972): 31–37.

The Cravens data involve a company that sells products in a number of sales territories, each of which is assigned to a single sales represntative. It was desired to conduct a regression analysis to determine if sales in each territory could be explained using a variety of predictor (independent) variables. A random sample of 25 sales territories resulted in the data shown in Table 15.4; the variable definitions are shown in Table 15.5

As a preliminary step, let us consider the sample correlation coefficients between each pair of variables. Figure 15.5 shows the correlation matrix obtained using the Minitab correlation command. Note that the sample correlation coefficient between SALES and TIME is .623, between SALES and POTEN is .598, and so on.

Looking at the sample correlation coefficients between the independent variables, we see that the correlation between TIME and ACCTS is .758; thus, if ACCTS is used as an independent variable, TIME would not be able to provide much more explanatory power to the model. Recall from the discussion of multicollinearity in Section 14.5 that the rule-of-thumb test says that multicollinearity can cause problems if the absolute value of the sample correlation coefficient exceeds .7 for any two of the independent variables. If possible, then, we should avoid including both TIME and ACCTS in the same regression model. The sample correlation coefficient of .549 between CHANGE and RATING is also quite high and may warrant further consideration.

TABLE 15.4 The Cravens Data

Sales	Time	Poten	Adv	Share	Change	Accts	Work	Rating
3,669.88	43.10	74,065.1	4,582.9	2.51	0.34	74.86	15.05	4.9
3,473.95	108.13	58,117.3	5,539.8	5.51	0.15	107.32	19.97	5.1
2,295.10	13.82	21,118.5	2,950.4	10.91	−0.72	96.75	17.34	2.9
4,675.56	186.18	68,521.3	2,243.1	8.27	0.17	195.12	13.40	3.4
6,125.96	161.79	57,805.1	7,747.1	9.15	0.50	180.44	17.64	4.6
2,134.94	8.94	37,806.9	402.4	5.51	0.15	104.88	16.22	4.5
5,031.66	365.04	50,935.3	3,140.6	8.54	0.55	256.10	18.80	4.6
3,367.45	220.32	35,602.1	2,086.2	7.07	−0.49	126.83	19.86	2.3
6,519.45	127.64	46,176.8	8,846.2	12.54	1.24	203.25	17.42	4.9
4,876.37	105.69	42,053.2	5,673.1	8.85	0.31	119.51	21.41	2.8
2,468.27	57.72	36,829.7	2,761.8	5.38	0.37	116.26	16.32	3.1
2,533.31	23.58	33,612.7	1,991.8	5.43	−0.65	142.28	14.51	4.2
2,408.11	13.82	21,412.8	1,971.5	8.48	0.64	89.43	19.35	4.3
2,337.38	13.82	20,416.9	1,737.4	7.80	1.01	84.55	20.02	4.2
4,586.95	86.99	36,272.0	10,694.2	10.34	0.11	119.51	15.26	5.5
2,729.24	165.85	23,093.3	8,618.6	5.15	0.04	80.49	15.87	3.6
3,289.40	116.26	26,878.6	7,747.9	6.64	0.68	136.58	7.81	3.4
2,800.78	42.28	39,572.0	4,565.8	5.45	0.66	78.86	16.00	4.2
3,264.20	52.84	51,866.1	6,022.7	6.31	−0.10	136.58	17.44	3.6
3,453.62	165.04	58,749.8	3,721.1	6.35	−0.03	138.21	17.98	3.1
1,741.45	10.57	23,990.8	861.0	7.37	−1.63	75.61	20.99	1.6
2,035.75	13.82	25,694.9	3,571.5	8.39	−0.43	102.44	21.66	3.4
1,578.00	8.13	23,736.3	2,845.5	5.15	0.04	76.42	21.46	2.7
4,167.44	58.44	34,314.3	5,060.1	12.88	0.22	136.58	24.78	2.8
2,799.97	21.14	22,809.5	3,552.0	9.14	−0.74	88.62	24.96	3.9

TABLE 15.5 Minitab Variable Names for the Craven's Data

Variable	Definition
SALES	Total sales in units credited to the sales representative
TIME	Length in time employed in months
POTEN	Market potential; total industry sales in units for the sales territory[a]
ADV	Advertising expenditure in the sales territory
SHARE	Market share; weighted average for the past 4 years
CHANGE	Change in the market share over the previous 4 years
ACCTS	Number of accounts assigned to the sales representative[a]
WORK	Work load; a weighted index based on annual purchases and concentrations of accounts
RATING	Sales representative overall rating on eight performance dimensions; an aggregate rating on a 1–7 scale

[a]These data were coded to preserve confidentiality.

Looking at the sample correlation coefficients between SALES and each of the independent variables can provide us with a quick indication of which independent variables are, by themselves, good predictors. We see that the single best predictor of SALES is ACCTS, since it has the highest sample correlation coefficient (.754). Recall that for the case of one independent variable, the square of the sample correlation coefficient is the coefficient of determination. Thus, ACCTS can explain $(.754)^2(100)$, or 56.85%, of the variability in SALES. The next most important independent variables are TIME, POTEN, and ADV, each with a sample correlation coefficient of approximately .6.

FIGURE 15.5 Sample Correlation Coefficients for the Cravens Data (as Printed by Minitab)

	SALES	TIME	POTEN	ADV	SHARE	CHANGE	ACCTS	WORK
TIME	0.623							
POTEN	0.598	0.454						
ADV	0.596	0.249	0.174					
SHARE	0.484	0.106	-0.211	0.264				
CHANGE	0.489	0.251	0.268	0.377	0.085			
ACCTS	0.754	0.758	0.479	0.200	0.403	0.327		
WORK	-0.117	-0.179	-0.259	-0.272	0.349	-0.288	-0.199	
RATING	0.402	0.101	0.359	0.411	-0.024	0.549	0.229	-0.277

Although there are potential multicollinearity problems, let us for the moment consider developing an estimated regression equation using all eight independent variables. Using the Minitab computer package provided the results shown in Figure 15.6. The eight-variable multiple regression model has an adjusted coefficient of determination of 88.3%, which is very high for real data. Note, however, that the "p" column (the p-values for the t tests of individual parameters) shows that only POTEN, ADV, and SHARE are significant at the $\alpha = .05$ level, given the effect of all the others. Thus we might be inclined to investigate the results that would be obtained if we used just these three variables. Figure 15.7 shows the Minitab results obtained for the model that uses just these three variables. We see that the model using just three independent variables has an adjusted coefficient of determination of 82.7%, which, although not quite as good as for the eight-independent-variable model, is still very high.

How can we find a model that will do the best job given the data available? One approach sometimes advocated for determining the best model is to compute all possible regressions. That is, one could develop 8 one-variable models (each of which corresponds to one of the independent variables), 28 two-variable models (the number of combinations of 8 variables taken 2 at a time), and so on. In all, for the Cravens data, there are 255 different models involving one or more independent variables that would have to be fitted to the data.

With the excellent computer packages available today, it is possible to compute all possible regressions. But, doing so involves a great deal of computation and requires the model-builder to review a great deal of computer output, much of which is associated with obviously poor models. Statisticians usually prefer a more systematic approach to selecting the subset of independent

---

███ FIGURE 15.6   **Minitab Output for the Model Involving All Eight Independent Variables**

```
The regression equation is
SALES = - 1508 + 2.01 TIME + 0.0372 POTEN + 0.151 ADV + 199 SHARE + 291 CHANGE
          + 5.55 ACCTS + 19.8 WORK + 8 RATING

Predictor      Coef        Stdev     t-ratio       p
Constant      1507.8       778.6      -1.94     0.071
TIME           2.010       1.931       1.04     0.313
POTEN       0.037205    0.008202       4.54     0.000
ADV          0.15099     0.04711       3.21     0.006
SHARE         199.02       67.03       2.97     0.009
CHANGE        290.9        186.8       1.56     0.139
ACCTS          5.551       4.776       1.16     0.262
WORK          19.79        33.68       0.59     0.565
RATING          8.2        128.5       0.06     0.950

s = 449.0      R-sq = 92.2%      R-sq(adj) = 88.3%

Analysis of Variance

SOURCE        DF          SS           MS         F         p
Regression     8      38153568      4769196     23.65     0.000
Error         16       3225984       201624
Total         24      41379552
```

```
The regression equation is
SALES = - 1604 + 0.0543 POTEN + 0.167 ADV + 283 SHARE

Predictor        Coef        Stdev      t-ratio          p
Constant      -1603.6        505.6        -3.17      0.005
POTEN        0.054286     0.007474.        7.26      0.000
ADV           0.16748      0.04427         3.78      0.001
SHARE         282.75         48.76         5.80      0.000

s = 545.5       R-sq = 84.9%      R-sq(adj) = 82.7%

Analysis of Variance

SOURCE        DF          SS           MS          F          p
Regression     3      35130240     11710080      39.35     0.000
Error         21       6249310       297586
Total         24      41379552
```

■■■■ FIGURE 15.7   **Minitab Output for the Model Involving POTEN, ADV, and SHARE**

variables providing the best model. In the next section, we introduce some of the more popular approaches.

15.5 ■■■ VARIABLE-SELECTION PROCEDURES

In this section, we discuss three computer-based methods for selecting the independent variables in a regression model: stepwise regression, forward selection, and backward elimination. Given a data set involving several possible independent variables (the x_i's), these methods can be used to identify which independent variables provide the best model. All three methods are iterative; at each step a single variable is added or deleted and the new model is evaluated. The process continues until a stopping criterion indicates that the procedure cannot find a better model.

The criterion for selecting an independent variable to add or delete from the model at each step is based on the F statistic introduced in Section 15.3. Suppose, for instance, that we were considering adding x_3 to a model involving x_1 or deleting x_3 from a model involving x_1 and x_3. In Section 15.3, we showed that

$$F = \frac{\dfrac{\text{SSE}(x_1) - \text{SSE}(x_1, x_3)}{1}}{\dfrac{\text{SSE}(x_1, x_3)}{n - p - 1}}$$

can be used as a criterion for determining whether or not the presence of x_3 in the model causes a significant reduction in the error sum of squares. The value of this F statistic is the criterion used by all three methods to determine whether or not a variable should be added to or deleted from the regression model at each step. It is also used to indicate when the iterative procedure should stop. All three procedures stop when no more significant reduction in the error sum of squares can be obtained. As also noted in Section 15.3, when only one variable at a time is to be added or deleted, the t statistic (recall that $t^2 = F$) provides the same criterion.

With the stepwise regression procedure, a variable may be added or deleted at each step. The procedure stops when no more improvement can be obtained by adding or deleting a variable. With the forward-selection procedure, a variable is added at each step, but variables are never deleted. The procedure stops when no more improvement can be obtained by adding a variable. With the backward-elimination procedure, the procedure starts with a model involving all the possible independent variables. At each step a variable is eliminated. The procedure stops when no more improvement can be obtained by deleting a variable.

Stepwise Regression

We will illustrate the stepwise regression procedure using the Cravens data. To see how a step of the procedure is performed, suppose that after three steps the following three independent variables have been selected: ACCTS, ADV, and POTEN. At the next step, the procedure first determines if any of the variables *already in the model* should be deleted. It does so by first determining which of the three variables is the least significant addition in moving from a two- to three-independent-variable model. To determine this an F statistic is computed for each of the three variables. The F statistic for ACCTS enables us to test whether or not adding ACCTS to a model that already includes ADV and POTEN leads to a significant reduction is SSE. If not, the stepwise procedure will consider dropping ACCTS from the model. Before doing so, however, the same F statistic will be computed for ADV and POTEN. The variable with the smallest F statistic makes the least significant addition in moving from a two- to three-independent-variable regression model and becomes a candidate for deletion. If any variable is to be deleted, it will be the one chosen.

We will denote by FMIN the smallest of the F statistics for all variables in the regression model at the beginning of a new step. The variable with the smallest F statistic is the least significant addition to the model. If the value of FMIN is small enough not to be significant, the corresponding variable is deleted from the model. On the other hand, if FMIN is large enough to be significant, none of the variables are deleted from the model (none of the other variables can have smaller F statistics).

The user of a computer-based stepwise regression procedure must specify a cutoff value for the F statistic so that the method can determine when FMIN is large enough to be significant. With the Minitab package, the smallest significant F value is denoted by FREMOVE. If the user does not specify a value for FREMOVE, it is automatically set equal to 4 by Minitab. Anytime FMIN < FREMOVE, the stepwise procedure of Minitab will delete the corresponding variable from the model. If FMIN ≥ FREMOVE, no variable is deleted at that step of the procedure.

If no variable can be removed from the model, the stepwise procedure next checks to see if adding a variable can improve the model. For each variable *not in the model*, an F statistic is computed. The largest of these F statistics corresponds to the variable that will cause the largest reduction in SSE. This variable then becomes a candidate for inclusion in the model. We will denote the largest F

statistic for variables not currently in the model by FMAX. Again a cutoff value for the F statistic must be used to determine if FMAX is large enough for the corresponding variable to make a significant improvement in the model.

The cutoff value for determining when to add a variable is denoted by FENTER in the Minitab computer package. The user of the package may specify a cutoff value for FENTER; if the user does not, Minitab will automatically set FENTER = 4. If FMAX > FENTER, the corresponding variable is added to the model, and the stepwise regression procedure goes on to the next step. The procedure stops when no variables can be deleted and no variables can be added.

In summary, at each step of the stepwise regression procedure, the first consideration is to see if any variable can be removed. If none of the variables can be removed, the procedure then checks to see if any variables can be added. Because of the nature of the stepwise procedure, it is possible for a variable to enter the model at one step, be deleted at a subsequent step, and then reenter the model at a later step. The procedure stops when FMIN ≥ FREMOVE (no variables can be deleted) and FMAX ≤ FENTER (no variables can be added).

Figure 15.8 shows the result of using the Minitab stepwise regression procedure for the Cravens data. As we noted in Section 15.3, when only one variable is being added, the t statistic provides the same criterion as the F. One can show that $F = t^2$. The entries in the "T-RATIO" row are the t statistics. The values of FREMOVE and FENTER were both automatically set equal to 4. At step 1, there are no variables to consider for deletion. The variable providing the largest value for the F statistic is ACCTS, with $F = t^2 = (5.5)^2 = 30.25$. Since 30.25 > 4, ACCTS is added to the model. On the next three steps, ADV, POTEN, and SHARE are added to the model. After step 4, an F statistic was computed for each of the four variables in the model. The values of the F statistics were $t^2 = (3.22)^2 = 10.37$, $t^2 = 22.47$, $t^2 = 22.94$, and $t^2 = 14.59$ for ACCTS, ADV, POTEN,

---

▬ FIGURE 15.8 **Minitab Output Using Stepwise Regression for Cravens' Data**

STEPWISE REGRESSION OF SALES ON 8 PREDICTORS, WITH N = 25

STEP	1	2	3	4
CONSTANT	709.32	50.30	-327.23	-1441.93
ACCTS	21.7	19.0	15.6	9.2
T-RATIO	5.50	6.41	5.19	3.22
ADV		0.227	0.216	0.195
T-RATIO		4.50	4.77	4.74
POTEN			0.0219	0.0382
T-RATIO			2.53	4.79
SHARE				190
T-RATIO				3.82
S	881	650	583	454
R-SQ	56.85	77.51	82.77	90.04

and SHARE, respectively. Thus, FMIN = 10.37, and the corresponding variable is ACCTS. Since, 10.37 > 4, no variable is dropped from the model.

An F statistic was then computed for each of the other four variables not in the model. Since all of these F statistics were less than 4, no variables were added to the model. The stepwise procedure stopped at this point; no variables could be deleted and none could be added to improve the model. The results shown in Figure 15.8 were printed at this point. The estimated regression equation identified by the Minitab stepwise regression procedure is

$$\hat{y} = -1441.93 + 9.2 \text{ ACCTS} + 0.175 \text{ ADV} + .0382 \text{ POTEN} + 190 \text{ SHARE}$$

Note also in Figure 15.8 that, with the error sum of squares being reduced at each step, $s = \sqrt{\text{MSE}}$ has been reduced from 881 with the best one-variable model to 454 after four steps. The value of r^2 has been increased from 56.85 to 90.04.

Forward Selection

Forward selection is another computer-based procedure for variable selection. It is similar to the stepwise regression procedure except that it does not permit a variable to be deleted from the model once it has been added. The forward-selection procedure starts out with no independent variables. Then it adds variables, one at a time, as long as a significant reduction in the error sum of squares (SSE) can be achieved. When no variable can be added that will cause a further significant reduction in SSE, the procedure stops and prints out the results. For the Cravens data, the stepwise regression procedure added one variable at each step and did not delete any variables. Thus, for the Cravens' data, the forward-selection procedure leads to the same model as that provided by the stepwise procedure.

Backward Elimination

The backward-elimination procedure begins with a model including all the independent variables the model builder wants considered. (Figure 15.6 shows a regression model involving all eight independent variables for the Cravens data.) It then deletes one variable at a time using the same criterion as that for removing variables using the stepwise regression procedure. The variable with the smallest F statistic is deleted, provided F is less than the preestablished cutoff criterion (FREMOVE for Minitab). The major difference between the backward-elimination procedure and the stepwise procedure is that once a variable has been removed from the model, it cannot reenter at a later step.

Making the Final Choice

The analysis performed on the Cravens data to this point is good preparation for choosing a final model. But, more analysis should be conducted before making the final choice. As we have noted in Chapters 13 and 14, a careful analysis of the residuals should be made. We want the residual plot for the model chosen to resemble approximately a horizontal band. More will be said on analysis of residuals in the next section. For now let us assume that there is no difficulty

TABLE 15.6 Selected Models Involving ACCTS, ADV, POTEN, and SHARE

Model	Independent Variables	Adjusted R^2
1	ACCTS	55.0
2	ACCTS, ADV	75.5
3	POTEN, SHARE	72.3
4	ACCTS, ADV, POTEN	80.3
5	ADV, POTEN, SHARE	82.7
6	ACCTS, ADV, POTEN, SHARE	88.1

with the residuals and that we want to utilize the results of the stepwise regression procedure to help choose the model to be used.

The stepwise regression procedure has shown us that if we consider adding or deleting only one variable at a time, the best model utilizes four independent variables: ACCTS, ADV, POTEN, and SHARE. However, it may be that another model is better or that a simpler model using fewer variables will be preferred by the person using it. Table 15.6 is helpful in making the final choice. It shows several possible models consisting of some or all of these four independent variables.

From Table 15.6, we see that the model involving just ACCTS and ADV is pretty good. The adjusted R^2 is 75.5%, whereas the model using all four variables only provides a 12.6% increase. The simpler two-variable model might be preferred if, for instance, it is difficult to measure market potential (POTEN). On the other hand, if the data are readily available, the model builder would clearly prefer the model involving all four variables if highly accurate predictions of sales are needed.

NOTES AND COMMENTS

1. In the stepwise procedure, FENTER cannot be set smaller than FRE-MOVE. To see why, suppose this condition was not satisfied. For instance, suppose a model-builder set FENTER = 2 and FREMOVE = 4 and at some step of the procedure a variable with an F statistic of 3 was entered into the model. At the next step, any variable with an F statistic of 3 would be a candidate for removal from the model (since 3 < FREMOVE = 4). If the stepwise procedure deleted it, then at the very next step the variable would enter again, then it would be deleted, and so on. Thus, the procedure would cycle forever. To avoid this, the stepwise procedure requires that FENTER be greater than or equal to FREMOVE.

2. Functions of the independent variables may be used to create new independent variables for use with any of the procedures of this section.

continued on next page

For instance, if we desired x_1x_2 in the model to account for interaction, the data for x_1 and x_2 would be used to create the data for $z_i = x_{1i}x_{2i}$.

3. None of the procedures that add or delete variables one at a time can be guaranteed to identify the best regression model. But they are excellent approaches to finding good models—especially when there is not much multicollinearity present.

▬ EXERCISES

12. Two experts provided subjective lists of school districts that they think are among the best in the country. For each school district, the following data were obtained: average class size, instructional spending per student, average teacher salary, combined SAT score, percent of students taking the SAT, and the percentage of students that attended a 4-year college (*Wall Street Journal*, March 31, 1989).

	Average Class Size	Instructional Spending Per Student	Average Teacher Salary	Combined SAT Score (% Taking Test)	Attend Four-Year College
Blue Springs, Mo.	25	$3,060	$29,359	1083 (8%)	74%
Garden City, N.Y.	18	$9,700	$51,000	997 (99%)	77%
Indianapolis, Ind.	30	$3,222	$30,482	716 (42%)	40%
Newport Beach, Calif. (Newport-Mesa)	26	$4,028	$37,043	977 (46%)	51%
Novi, Mich.	20	$3,067	$39,797	980 (15%)	53%
Piedmont, Calif. (Piedmont City)	28	$4,208	$37,274	1042 (91%)	75%
Pittsburgh, Pa. (Fox Chapel area)	21	$4,884	$37,156	983 (80%)	66%
Scarsdale, N.Y. (Edgemont)	20	$9,853	$31,555	1110 (98%)	87%
Wayne, Pa. (Radnor Township)	22	$5,022	$40,406	1040 (95%)	85%
Weston, Mass.	21	$4,680	$39,800	1031 (99%)	89%
Farmingdale, N.Y.	22	$6,729	$45,846	947 (75%)	81%
Mamaroneck, N.Y.	20	$10,405	$49,625	1000 (90%)	69%
Mayfield, Ohio	24	$5,881	$36,228	1003 (25%)	48%
Morristown, N.J.	22	$6,300	$37,000	972 (80%)	64%
New Rochelle, N.Y.	23	$8,875	$41,650	1039 (80%)	55%
Newton Square, Pa. (Marple-Newton)	17	$5,313	$38,000	963 (75%)	79%
Omaha, Neb. (Westside)	23	$4,815	$32,500	1059 (31%)	81%
Shaker Heights, Ohio	23	$4,370	$38,639	940 (56%)	82%

Let the dependent variable be the percentage of students that attend a four-year college.
  a. Develop the best one-variable model.
  b. Use the stepwise procedure to develop the best model.
  c. Use the backward-elimination procedure to develop the best model.

13. Refer to Exercise 12. Let the dependent variable be the combined SAT score.

  a. Develop the best one-variable model.
  b. Use the stepwise procedure to develop the best model.
  c. Use the backward elimination procedure to develop the best model.

14. The following table shows some of the data available for 14 teams in the National Football League at the end of week 15 for the 1988 season.

Team	Won-Lost	Total Points	Rushing Yards	Passing Yards	Interceptions Made by Team	Interceptions Made by Opponent
Atlanta	5–10	305	1907	2473	19	23
Chicago	12–3	187	2134	2718	14	24
Dallas	3–12	358	1858	3386	24	10
Detroit	4–11	292	1184	1971	15	12
Green Bay	3–12	298	1274	3046	22	20
L.A. Rams	9–6	277	1882	3604	17	22
Minnesota	10–5	206	1744	3633	16	35
N. Orleans	9–6	274	1843	2963	15	17
N.Y. Giants	10–5	277	1492	3096	14	15
Philadelphia	9–6	312	1812	3247	17	29
Phoenix	7–8	372	1909	3633	19	14
San Fran.	10–5	256	2453	3131	14	21
Tampa Bay	4–11	340	1650	3169	33	18
Washington	7–8	367	1377	3930	24	14

Let the dependent variable be the number of wins
  a. Develop the best one-variable model.
  b. Use the stepwise procedure to develop the best model.
  c. Use the backward-elimination procedure to develop the best model.

15. Refer to Exercise 14. Let the dependent variable be the total points scored.

  a. Develop the best one-variable model.
  b. Use the stepwise procedure to develop the best model.
  c. Use the backward-elimination procedure to develop the best model.

16. Refer to Exercise 8.

  a. Use the stepwise procedure to develop a model that can be used to predict the price-earnings ratio.
  b. Use the backwards elimination procedure to develop a model that can be used to predict the price-earnings ratio.

15.6 ■ RESIDUAL ANALYSIS

In Chapters 13 and 14 we showed how residual plots could be used to determine when violations of assumptions concerning the regression model had occurred. We looked for violations of assumptions concerning the error term (ϵ) and the assumed functional form of the model. Some of the actions that can be taken when such violations are detected have been discussed in this chapter. When a different functional form is needed, curvilinear and interaction terms can be included through the use of the general linear model. When other (or more) variables need to be considered, some of the variable-selection procedures of the preceding section may be appropriate.

In this section, we discuss how residual analysis can be used to identify observations that can be classified as outliers or as being especially influential in determining the estimated regression equation. Some steps that should be taken when such observations have been found are noted. Also, in many regression studies involving economic data, a special type of correlation involving the error terms can cause problems; it is called *serial correlation*, or autocorrelation.

We show how residual analysis using the Durbin-Watson test can be used to determine when autocorrelation is a problem.

Detecting Outliers

In regression analysis, outliers are data points (observations) that do not fit the trend. Figure 15.9 shows an outlier for a data set involving a single independent variable. Outliers represent observations that are suspect and warrant careful examination. They may represent erroneous data; if so, the data should be corrected. They may signal a violation of model assumptions; if so, other models should be considered. And, finally, they may simply be unusual values that have occurred by chance. In this case they should be retained.

To illustrate the process of detecting outliers, consider the data set shown in Table 15.7; a scatter diagram is shown in Figure 15.10. Except for observation 4 ($x = 3$, $y = 75$), a pattern suggesting a negative linear relationship is apparent. Indeed, given the pattern of the rest of the data, we would have expected the y value for observation 4 to be much smaller and thus would identify the observation as an outlier. For the case of simple linear regression, one can usually detect outliers by simply examining the scatter diagram. In multiple regression models, the situation is a bit more difficult.

In the multiple regression case, we can use the standardized residuals to identify outliers. If the value of y for a particular x is unusually large or small (does not seem to follow the trend of the rest of the data) the corresponding standardized residual will be large in absolute value. Many computer packages automatically identify observations with standardized residuals that are large in absolute value. In Figure 15.11 we show the Minitab output from a regression analysis of the data in Table 15.7. The last line of the output (labeled "ROW 4" shows that the standardized residual for observation 4 is 2.67. Minitab considers a standardized residual less than -2 or greater than $+2$ to be an outlier; in such cases the observation is printed on a separate line with an R next to the standardized residual, as shown in Figure 15.11. Assuming normally distributed

---

■■■■ FIGURE 15.9   **A Data Set Involving an Outlier**

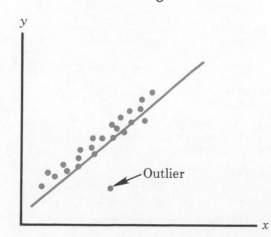

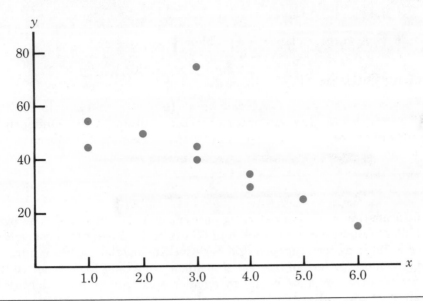

■ FIGURE 15.10   **Scatter Diagram for DataSet of Table 15.7**

errors, standardized residuals should fall outside these limits only approximately 5% of the time.

In deciding how to handle an outlier, we should first check to see if it is a valid observation. Perhaps an error has been made in initially recording the data or in entering the data into the computer system. For example, suppose that in checking the data for the outlier in Table 15.7, we find that an error has been made and that the correct value for observation 4 is $x = 3$, $y = 30$. Figure 15.12

■ FIGURE 15.11   **Minitab Output for Regression Analysis of Data Set with Outlier (Table 15.7)**

```
The regression equation is
Y = 65.0 - 7.33 X

Predictor       Coef      Stdev     t-ratio       p
Constant      64.958      9.258        7.02   0.000
X             -7.331      2.608       -2.81   0.023

s = 12.67      R-sq = 49.7%     R-sq(adj) = 43.4%

Analysis of Variance

SOURCE         DF         SS          MS         F       p
Regression      1     1268.2      1268.2      7.90   0.023
Error           8     1284.3       160.5
Total           9     2552.5

Unusual Observations
Obs.      X          Y      Fit Stdev.Fit   Residual   St.Resid
   4   3.00      75.00    42.97      4.04      32.03      2.67R

R denotes an obs. with a large st. resid.
```

TABLE 15.7	Data Set Illustrating The Effect Of An Outlier
x	y
1	45
1	55
2	50
3	75
3	40
3	45
4	30
4	35
5	25
6	15

shows the Minitab output obtained after correcting the value of y_4. We see that the effect of using an incorrect value for the dependent variable had a substantial effect on the goodness of fit. With the correct data, the value of r^2 has increased from 49.7% to 83.8% and the value of b_0 has decreased from 64.958 to 59.237. The slope of the line, however, has changed only from -7.331 to -6.949.

Detection of Influential Observations

In regression analysis, it sometimes happens that one or more observations have a strong influence on the results obtained. Figure 15.13 shows an example of an influential observation in simple linear regression. The estimated regression line has a negative slope. But, if the influential observation is dropped from the

■■■ FIGURE 15.12 Minitab Output for Revised Data Set in Table 15.7

```
The regression equation is
Y = 59.2 - 6.95 X

Predictor      Coef      Stdev     t-ratio       p
Constant     59.237      3.835      15.45     0.000
X            -6.949      1.080      -6.43     0.000

s = 5.248      R-sq = 83.8%      R-sq(adj) = 81.8%

Analysis of Variance

SOURCE      DF        SS          MS        F        p
Regression   1      1139.7      1139.7    41.38    0.000
Error        8       220.3        27.5
Total        9      1360.0
```

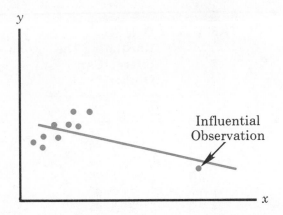

■■■ FIGURE 15.13   **A Data Set with an Influential Observation**

data set, the slope of the estimated regression line would change from negative to positive, and the y-intercept would be smaller. Clearly this one observation is much more influential in determining the estimated regression line than any of the others; dropping one of the other observations from the data set would have very little effect on the estimated regression equation.

Since influential observations can have such a dramatic effect on the estimated regression equation, it is important that they be examined carefully. We should first check to make sure that no error has been made in collecting or recording the data. If an error has occurred, it can be corrected and a new estimated regression equation can be developed. On the other hand, if the observation is valid, we might consider ourselves fortunate to have it. Such a point, if valid, can contribute to a better understanding of the appropriate model and can lead to a better estimated regression equation. The presence of the influential observation in Figure 15.13, if valid, would suggest trying to obtain data on intermediate values of x to understand better the relationship between x and y.

Influential observations can be identified from a scatter diagram when only one independent variable is present. An influential observation may be an outlier (an observation with a y value that deviates substantially from the trend); it may correspond to an x value far away from its mean (see, for example, Figure 15.13); or it may be caused by a combination of the two (a somewhat off-trend y value and a somewhat extreme x value). With multiple regression it is not as easy to identify influential observations. However, most computer packages offer diagnostics that help.

We have already seen how outliers are identified using Minitab. They are observations with large standardized residuals. Observations with extreme values for the independent variables are called high leverage points. The influential observation in Figure 15.13, caused by an extreme value of x, is a point with high leverage. The leverage of an observation is determined by how far the values of the independent variables are from their mean values. For the single-independent-variable case, the leverage of the ith observation, denoted h_i, can be computed using (15.14).

How far the values of indep. variables are from their mean values.

Leverage of Observation *i*

$$h_i = \frac{1}{n} + \frac{(x_i - \bar{x})^2}{\Sigma(x_i - \bar{x})^2} \tag{15.14}$$

From the formula, it is clear that the farther x_i is from its mean $\bar{x}$, the higher the leverage of observation *i*.

Many computer packages automatically identify observations with high leverage and large standardized residuals as part of the standard regression output. We have already seen how Minitab identifies observations with large standardized residuals. To provide an illustration of how computer packages identify points with high leverage, let us consider the data set presented in Table 15.8.

A scatter diagram for the data set in Table 15.8 is shown in Figure 15.14. From the scatter diagram, it is clear that observation 7 ($x = 70$, $y = 100$) is an observation with an extreme value of x. Thus we would expect it to be identified as a point with high leverage. For this observation, the leverage is computed using (15.14) as follows:

$$h_7 = \frac{1}{n} + \frac{(x_7 - \bar{x})^2}{\Sigma(x_i - \bar{x})^2} = \frac{1}{7} + \frac{(70 - 24.286)^2}{2621.43} = .94$$

The Minitab computer package identifies observations as having high leverage if $h_i > 2(p + 1)/n$, where p is the number of independent variables and n is the number of observations. For the data set in Table 15.8,

$$\frac{3(p + 1)}{n} = \frac{3(1 + 1)}{7} = \frac{6}{7} = .86$$

TABLE 15.8 Data Set Containing an Observation with High Leverage

x	y
10	125
10	130
15	120
20	115
20	120
25	110
70	100

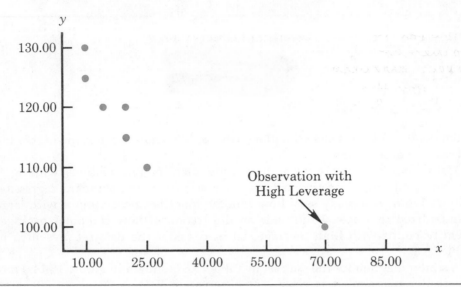

▬ FIGURE 15.14   **Scatter Diagram for the Data Set in Table 15.8**

Since $h_7 = .94 > .86$, Minitab will identify observation 7 as a high leverage point. Figure 15.15 shows the Minitab output for a regression analysis of this data set. Observation 7 ($x = 70$, $y = 100$) is identified as having high leverage; it is printed on a separate line at the bottom, with an "X" in the right-hand margin.

Influential observations are caused by an interaction of large residuals and high leverage. Diagnostic procedures are available that take both into account in determining when an observation is influential. One such measure is called Cook's D statistic. Cook's D statistic and other measures of influence are

▬ FIGURE 15.15   **Minitab Output for the Data Set in Table 15.8**

```
The regression equation is
Y = 127 -0.425 X

Predictor       Coef        Stdev      t-ratio        p
Constant      127.466       2.961       43.04      0.000
X            -0.42507      0.09537      -4.46      0.007

s = 4.883      R-sq = 79.9%     R-sq(adj) = 75.9%

Analysis of Variance

SOURCE          DF          SS           MS          F         p
Regression       1        473.65       473.65      19.87     0.007
Error            5        119.21        23.84
Total            6        592.86

Unusual Observations
Obs.      X           Y       Fit  Stdev.Fit   Residual    St.Resid
  7     70.0      100.00     97.71     4.73       2.29        1.91 X

X denotes an obs. whose X value gives it large influence.
```
high leverage pt.

discussed in texts on regression analysis (see Appendix A). For our purposes, we shall simply note that potentially influential observations can be identified as the observations with large standardized residuals and high leverage in the output of many computer regression packages. Such observations should be carefully reviewed to establish their validity and to evaluate their influence on the estimated regression equation.

Autocorrelation and the Durbin-Watson Test

The data used for regression studies in business and economics often have been collected over time. In such cases it is not uncommon for the value of y at time t, denoted by y_t, to be related to the value of y at previous time periods. When this occurs, we say autocorrelation (also called serial correlation) is present in the data. If the value of y in time period t is related to its value in time period $t - 1$, we say first-order autocorrelation is present. If the value of y in time period t is related to the value of y in time period $t - 2$, we say second-order autocorrelation is present, and so on.

When autocorrelation is present, one of the assumptions of the regression model is violated: The error terms are not independent. In the case of first-order autocorrelation, the error at time t, denoted by ϵ_t, will be related to the error at time period $t - 1$, denoted by ϵ_{t-1}. Two cases of first-order autocorrelation are shown in Figure 15.16. Panel A illustrates the case of positive autocorrelation; Panel B illustrates the case of negative autocorrelation. With positive autocorrelation we expect a positive residual in one period to be followed by a positive residual in the next period, a negative residual in one period to be followed by a negative residual in the next period, and so on. With negative autocorrelation, we expect a positive residual in one period to be followed by a negative residual in the next period, then a positive residual, and so on.

When autocorrelation is present, serious errors can be made in statistical inferences about the regression model. Thus it is important to be able to detect

---

■■■ FIGURE 15.16 **Two Data Sets with First-Order Autocorrelation**

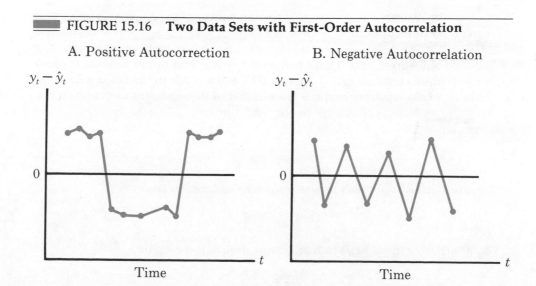

A. Positive Autocorrection

B. Negative Autocorrelation

autocorrelation and take corrective action. We will show how the Durbin-Watson statistic can be used to detect first-order autocorrelation.

Suppose that the values of ϵ are not independent but are related in the following manner:

$$\epsilon_t = \rho\epsilon_{t-1} + z_t \tag{15.15}$$

where ρ is a parameter with an absolute value less than 1 and z_t is a normally and independently distributed random variable with mean 0 and variance σ^2. From (15.15) we see that if $\rho = 0$, then the error terms are not related, and each has a mean of 0 and a variance of σ^2. In this case, there is no autocorrelation and the regression assumptions are satisfied. If $\rho > 0$, we have positive autocorrelation; if $\rho < 0$, we have negative autocorrelation. In either of these cases the regression assumptions concerning the error term are violated.

The Durbin-Watson test for autocorrelation uses the residuals to determine whether or not $\rho = 0$. To simplify the notation for the Durbin-Watson statistic, we shall denote the ith residual by $e_i = y_i - \hat{y}_i$. The Durbin-Watson statistic denoted d is given by

---

Durbin-Watson Statistic

$$d = \frac{\sum\limits_{t=2}^{n} (e_t - e_{t-1})^2}{\sum\limits_{t=1}^{n} e_t^2} \tag{15.16}$$

If successive values of the residuals are close together (positive autocorrelation), the Durbin-Watson statistic will be small. If successive values of the residuals are far apart (negative autocorrelation), the Durbin-Watson statistic will tend to be large.

The Durbin-Watson statistic ranges in value between 0 and 4, with a value of 2 indicating no autocorrelation is present. Durbin and Watson have developed tables that can be used to determine when their test statistic indicates the presence of autocorrelation. Table 15.9 shows lower and upper bounds (d_L and d_U) for hypothesis tests using $\alpha = .01$, $\alpha = .025$, and $\alpha = .05$; in the table, n denotes the number of observations and k is the number of independent variables in the model. The null hypothesis to be tested is always taken to be one of no autocorrelation:

$$H_0: \rho = 0$$

The alternative hypothesis to test for positive autocorrelation is

$$H_a: \rho > 0$$

The alternative hypothesis to test for negative autocorrelation is

$$H_a: \rho < 0$$

A two-sided test is also possible. In this case the alternative hypothesis is

$$H_a{:}\rho \neq 0$$

Figure 15.17 shows how the values of d_L and d_U in Table 15.9 are to be used to test for autocorrelation. Panel A illustrates the test for positive autocorrelation. If $d < d_L$, we conclude positive autocorrelation is present. If $d_L \leq d \leq d_U$, we say the test is inconclusive. If $d > d_U$, we conclude there is no evidence of positive autocorrelation.

Panel B illustrates the test for negative autocorrelation. If $d > 4 - d_L$, we conclude negative autocorrelation is present. If $4 - d_U \leq d \leq 4 - d_L$, we say the test is inconclusive. If $d < 4 - d_U$, we conclude there is no evidence of negative autocorrelation.

Panel C illustrates the two-sided test. If $d < d_L$ or $d > 4 - d_L$, we reject H_0 and conclude autocorrelation is present. If $d_L \leq d \leq d_U$ or $4 - d_U \leq d \leq 4 - d_L$, we say the test is inconclusive. If $d_U < d < 4 - d_U$, we conclude there is no evidence of autocorrelation.

If significant autocorrelation is identified, we should investigate whether we have omitted one or more key independent variables that have time-ordered

■■■■ FIGURE 15.17 **Hypothesis Test for Autocorrelation with Durbin-Watson Statistic**

A. Test for Positive Autocorrelation

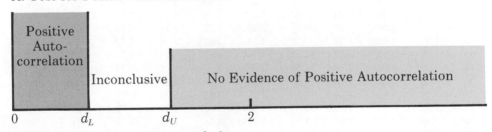

B. Test for Negative Autocorrelation

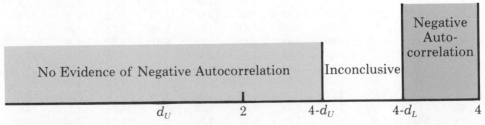

C. Two-sided Test for Autocorrelation

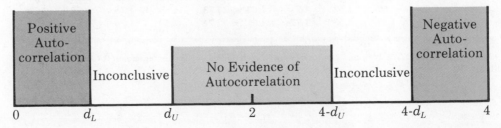

TABLE 15.9 Critical Values for the Durbin-Watson Test for Autocorrelation

Entries in the table give the critical values for a one-tailed Durbin-Watson test for autocorrelation. For a two-tailed test, the level of significance is doubled.

lower bound *upper bound*

		Significance Points of d_L and d_U: $\alpha = .05$								
					Number of Independent Variables					
	k	1		2		3		4		5	
n		d_L	d_U	d_L	d_U	d_L	d_U	d_L	d_U	d_L	d_U
15		1.08	1.36	0.95	1.54	0.82	1.75	0.69	1.97	0.56	2.21
16		1.10	1.37	0.98	1.54	0.86	1.73	0.74	1.93	0.62	2.15
17		1.13	1.38	1.02	1.54	0.90	1.71	0.78	1.90	0.67	2.10
18		1.16	1.39	1.05	1.53	0.93	1.69	0.82	1.87	0.71	2.06
19		1.18	1.40	1.08	1.53	0.97	1.68	0.86	1.85	0.75	2.02
20		1.20	1.41	1.10	1.54	1.00	1.68	0.90	1.83	0.79	1.99
21		1.22	1.42	1.13	1.54	1.03	1.67	0.93	1.81	0.83	1.96
22		1.24	1.43	1.15	1.54	1.05	1.66	0.96	1.80	0.86	1.94
23		1.26	1.44	1.17	1.54	1.08	1.66	0.99	1.79	0.90	1.92
24		1.27	1.45	1.19	1.55	1.10	1.66	1.01	1.78	0.93	1.90
25		1.29	1.45	1.21	1.55	1.12	1.66	1.04	1.77	0.95	1.89
26		1.30	1.46	1.22	1.55	1.14	1.65	1.06	1.76	0.98	1.88
27		1.32	1.47	1.24	1.56	1.16	1.65	1.08	1.76	1.01	1.86
28		1.33	1.48	1.26	1.56	1.18	1.65	1.10	1.75	1.03	1.85
29		1.34	1.48	1.27	1.56	1.20	1.65	1.12	1.74	1.05	1.84
30		1.35	1.49	1.28	1.57	1.21	1.65	1.14	1.74	1.07	1.83
31		1.36	1.50	1.30	1.57	1.23	1.65	1.16	1.74	1.09	1.83
32		1.37	1.50	1.31	1.57	1.24	1.65	1.18	1.73	1.11	1.82
33		1.38	1.51	1.32	1.58	1.26	1.65	1.19	1.73	1.13	1.81
34		1.39	1.51	1.33	1.58	1.27	1.65	1.21	1.73	1.15	1.81
35		1.40	1.52	1.34	1.58	1.28	1.65	1.22	1.73	1.16	1.80
36		1.41	1.52	1.35	1.59	1.29	1.65	1.24	1.73	1.18	1.80
37		1.42	1.53	1.36	1.59	1.31	1.66	1.25	1.72	1.19	1.80
38		1.43	1.54	1.37	1.59	1.32	1.66	1.26	1.72	1.21	1.79
39		1.43	1.54	1.38	1.60	1.33	1.66	1.27	1.72	1.22	1.79
40		1.44	1.54	1.39	1.60	1.34	1.66	1.29	1.72	1.23	1.79
45		1.48	1.57	1.43	1.62	1.38	1.67	1.34	1.72	1.29	1.78
50		1.50	1.59	1.46	1.63	1.42	1.67	1.38	1.72	1.34	1.77
55		1.53	1.60	1.49	1.64	1.45	1.68	1.41	1.72	1.38	1.77
60		1.55	1.62	1.51	1.65	1.48	1.69	1.44	1.73	1.41	1.77
65		1.57	1.63	1.54	1.66	1.50	1.70	1.47	1.73	1.44	1.77
70		1.58	1.64	1.55	1.67	1.52	1.70	1.49	1.74	1.46	1.77
75		1.60	1.65	1.57	1.68	1.54	1.71	1.51	1.74	1.49	1.77
80		1.61	1.66	1.59	1.69	1.56	1.72	1.53	1.74	1.51	1.77
85		1.62	1.67	1.60	1.70	1.57	1.72	1.55	1.75	1.52	1.77
90		1.63	1.68	1.61	1.70	1.59	1.73	1.57	1.75	1.54	1.78
95		1.64	1.69	1.62	1.71	1.60	1.73	1.58	1.75	1.56	1.78
100		1.65	1.69	1.63	1.72	1.61	1.74	1.59	1.76	1.57	1.78

Source: J. Durbin and G. S. Watson, "Testing for serial correlation in least square regression II," *Biometrika,* **38,** 1951, 159–178.

TABLE 15.9   (*Continued*)

Significance Points of d_L and d_U: $\alpha = .01$
Number of Independent Variables

n	k 1 d_L	d_U	2 d_L	d_U	3 d_L	d_U	4 d_L	d_U	5 d_L	d_U
15	0.81	1.07	0.70	1.25	0.59	1.46	0.49	1.70	0.39	1.96
16	0.84	1.09	0.74	1.25	0.63	1.44	0.53	1.66	0.44	1.90
17	0.87	1.10	0.77	1.25	0.67	1.43	0.57	1.63	0.48	1.85
18	0.90	1.12	0.80	1.26	0.71	1.42	0.61	1.60	0.52	1.80
19	0.93	1.13	0.83	1.26	0.74	1.41	0.65	1.58	0.56	1.77
20	0.95	1.15	0.86	1.27	0.77	1.41	0.68	1.57	0.60	1.74
21	0.97	1.16	0.89	1.27	0.80	1.41	0.72	1.55	0.63	1.71
22	1.00	1.17	0.91	1.28	0.83	1.40	0.75	1.54	0.66	1.69
23	1.02	1.19	0.94	1.29	0.86	1.40	0.77	1.53	0.70	1.67
24	1.04	1.20	0.96	1.30	0.88	1.41	0.80	1.53	0.72	1.66
25	1.05	1.21	0.98	1.30	0.90	1.41	0.83	1.52	0.75	1.65
26	1.07	1.22	1.00	1.31	0.93	1.41	0.85	1.52	0.78	1.64
27	1.09	1.23	1.02	1.32	0.95	1.41	0.88	1.51	0.81	1.63
28	1.10	1.24	1.04	1.32	0.97	1.41	0.90	1.51	0.83	1.62
29	1.12	1.25	1.05	1.33	0.99	1.42	0.92	1.51	0.85	1.61
30	1.13	1.26	1.07	1.34	1.01	1.42	0.94	1.51	0.88	1.61
31	1.15	1.27	1.08	1.34	1.02	1.42	0.96	1.51	0.90	1.60
32	1.16	1.28	1.10	1.35	1.04	1.43	0.98	1.51	0.92	1.60
33	1.17	1.29	1.11	1.36	1.05	1.43	1.00	1.51	0.94	1.59
34	1.18	1.30	1.13	1.36	1.07	1.43	1.01	1.51	0.95	1.59
35	1.19	1.31	1.14	1.37	1.08	1.44	1.03	1.51	0.97	1.59
36	1.21	1.32	1.15	1.38	1.10	1.44	1.04	1.51	0.99	1.59
37	1.22	1.32	1.16	1.38	1.11	1.45	1.06	1.51	1.00	1.59
38	1.23	1.33	1.18	1.39	1.12	1.45	1.07	1.52	1.02	1.58
39	1.24	1.34	1.19	1.39	1.14	1.45	1.09	1.52	1.03	1.58
40	1.25	1.34	1.20	1.40	1.15	1.46	1.10	1.52	1.05	1.58
45	1.29	1.38	1.24	1.42	1.20	1.48	1.16	1.53	1.11	1.58
50	1.32	1.40	1.28	1.45	1.24	1.49	1.20	1.54	1.16	1.59
55	1.36	1.43	1.32	1.47	1.28	1.51	1.25	1.55	1.21	1.59
60	1.38	1.45	1.35	1.48	1.32	1.52	1.28	1.56	1.25	1.60
65	1.41	1.47	1.38	1.50	1.35	1.53	1.31	1.57	1.28	1.61
70	1.43	1.49	1.40	1.52	1.37	1.55	1.34	1.58	1.31	1.61
75	1.45	1.50	1.42	1.53	1.39	1.56	1.37	1.59	1.34	1.62
80	1.47	1.52	1.44	1.54	1.42	1.57	1.39	1.60	1.36	1.62
85	1.48	1.53	1.46	1.55	1.43	1.58	1.41	1.60	1.39	1.63
90	1.50	1.54	1.47	1.56	1.45	1.59	1.43	1.61	1.41	1.64
95	1.51	1.55	1.49	1.57	1.47	1.60	1.45	1.62	1.42	1.64
100	1.52	1.56	1.50	1.58	1.48	1.60	1.46	1.63	1.44	1.65

Source: J. Durbin and G. S. Watson, "Testing for serial correlation in least squares regression II," *Biometrika*, **38,** 1951, 159–178.

TABLE 15.9 (*Continued*)

Significance Points of d_L and d_U: $\alpha = .025$
Number of Independent Variables

	k	1		2		3		4		5	
n		d_L	d_U	d_L	d_U	d_L	d_U	d_L	d_U	d_L	d_U
15		0.95	1.23	0.83	1.40	0.71	1.61	0.59	1.84	0.48	2.09
16		0.98	1.24	0.86	1.40	0.75	1.59	0.64	1.80	0.53	2.03
17		1.01	1.25	0.90	1.40	0.79	1.58	0.68	1.77	0.57	1.98
18		1.03	1.26	0.93	1.40	0.82	1.56	0.72	1.74	0.62	1.93
19		1.06	1.28	0.96	1.41	0.86	1.55	0.76	1.72	0.66	1.90
20		1.08	1.28	0.99	1.41	0.89	1.55	0.79	1.70	0.70	1.87
21		1.10	1.30	1.01	1.41	0.92	1.54	0.83	1.69	0.73	1.84
22		1.12	1.31	1.04	1.42	0.95	1.54	0.86	1.68	0.77	1.82
23		1.14	1.32	1.06	1.42	0.97	1.54	0.89	1.67	0.80	1.80
24		1.16	1.33	1.08	1.43	1.00	1.54	0.91	1.66	0.83	1.79
25		1.18	1.34	1.10	1.43	1.02	1.54	0.94	1.65	0.86	1.77
26		1.19	1.35	1.12	1.44	1.04	1.54	0.96	1.65	0.88	1.76
27		1.21	1.36	1.13	1.44	1.06	1.54	0.99	1.64	0.91	1.75
28		1.22	1.37	1.15	1.45	1.08	1.54	1.01	1.64	0.93	1.74
29		1.24	1.38	1.17	1.45	1.10	1.54	1.03	1.63	0.96	1.73
30		1.25	1.38	1.18	1.46	1.12	1.54	1.05	1.63	0.98	1.73
31		1.26	1.39	1.20	1.47	1.13	1.55	1.07	1.63	1.00	1.72
32		1.27	1.40	1.21	1.47	1.15	1.55	1.08	1.63	1.02	1.71
33		1.28	1.41	1.22	1.48	1.16	1.55	1.10	1.63	1.04	1.71
34		1.29	1.41	1.24	1.48	1.17	1.55	1.12	1.63	1.06	1.70
35		1.30	1.42	1.25	1.48	1.19	1.55	1.13	1.63	1.07	1.70
36		1.31	1.43	1.26	1.49	1.20	1.56	1.15	1.63	1.09	1.70
37		1.32	1.43	1.27	1.49	1.21	1.56	1.16	1.62	1.10	1.70
38		1.33	1.44	1.28	1.50	1.23	1.56	1.17	1.62	1.12	1.70
39		1.34	1.44	1.29	1.50	1.24	1.56	1.19	1.63	1.13	1.69
40		1.35	1.45	1.30	1.51	1.25	1.57	1.20	1.63	1.15	1.69
45		1.39	1.48	1.34	1.53	1.30	1.58	1.25	1.63	1.21	1.69
50		1.42	1.50	1.38	1.54	1.34	1.59	1.30	1.64	1.26	1.69
55		1.45	1.52	1.41	1.56	1.37	1.60	1.33	1.64	1.30	1.69
60		1.47	1.54	1.44	1.57	1.40	1.61	1.37	1.65	1.33	1.69
65		1.49	1.55	1.46	1.59	1.43	1.62	1.40	1.66	1.36	1.69
70		1.51	1.57	1.48	1.60	1.45	1.63	1.42	1.66	1.39	1.70
75		1.53	1.58	1.50	1.61	1.47	1.64	1.45	1.67	1.42	1.70
80		1.54	1.59	1.52	1.62	1.49	1.65	1.47	1.67	1.44	1.70
85		1.56	1.60	1.53	1.63	1.51	1.65	1.49	1.68	1.46	1.71
90		1.57	1.61	1.55	1.64	1.53	1.66	1.50	1.69	1.48	1.71
95		1.58	1.62	1.56	1.65	1.54	1.67	1.52	1.69	1.50	1.71
100		1.59	1.63	1.57	1.65	1.55	1.67	1.53	1.70	1.51	1.72

Source: J. Durbin and G. S. Watson, "Testing for serial correlation in least squares regression II," *Biometrika,* **38,** 1951, 159–178.

effects on the dependent variable. If no such variables can be identified, including an independent variable that measures the time of the observation (for instance, the value of this variable could be 1 for the first observation, 2 for the second observation, and so on) will sometimes eliminate or reduce the autocorrelation. When these attempts to reduce or remove autocorrelation do not work, transformations on the dependent or independent variables can prove helpful; a discussion of such transformations can be found in more advanced texts on regression analysis.

In closing this section we note that the Durbin-Watson tables list the smallest sample size as 15. The reason for this is that the test is generally inconclusive for small sample sizes. A rule of thumb suggested by statisticians is that the sample size should be at least 50 in order for the test to produce worthwhile results.

Suggested sample size of at least 50.

NOTES AND COMMENTS

Once an observation has been identified as potentially influential because of a large residual or high leverage, its impact on the estimated regression equation should be evaluated. More advanced texts discuss diagnostics for doing so. However, if one is not familiar with the more advanced material, a simple procedure is to run the regression analysis with and without the observation. Although more time consuming, this approach will reveal the influence of the observation on the results.

EXERCISES

17. Refer to the NFL data set presented in exercise 14. Let y, the dependent variable, be the number of wins.

   a. Develop a model that can be used to predict y given the number of interceptions made by the team.
   b. Develop a residual plot for the model developed in (a). Does the pattern of the residual plot appear acceptable? Explain.
   c. Are there any outliers?
   d. Are there any influential observations? If so, what effect do they have on the model?

18. Refer to the NFL data set presented in exercise 14. Let y, the dependent variable, be the number of wins.

   a. Develop a model that can be used to predict y given the number of interceptions made by the opponents.
   b. Develop a residual plot for the model developed in (a). Does the pattern of the residual plot appear acceptable? Explain.
   c. Are there any outliers?
   d. Are there any influential observations? If so, what effect do they have on the model?

19. Consider the data set presented in Exercise 12. Let y denote the percentage of students that attend a four-year college and x denote the combined SAT score.

   a. Develop a scatter diagram showing y given x.
   b. Develop a model that can be used to predict y given x.

c. Refer to the scatter diagram developed in (a). Do there appear to be any influential observations in the data set? If so, determine their effect on the model developed in (b).

20. Refer to the Cravens data set presented in Table 15.4.

   a. In Section 15.5 we indicated that the model developed using ACCTS and ADV did a good job in predicting SALES. For this model are there any influential observations? If so, what effect do they have on the fitted model?
   b. Consider a model using only ACCTS. For this model are there any influential observations? If so, what effect do they have on the fitted model?
   c. Consider a model using only ADV. For this model are there any influential observations? If so, what effect do they have on the fitted model?

21. Consider the data set presented in Exercise 8.

   a. Develop the estimated regression equation which can be used to predict the price-earnings ratio given the profit margin.
   b. Plot the residuals obtained from the model developed in (a) as a function of the order in which the data are presented. Does there appear to be any autocorrelation present in the data? Explain.
   c. At the 5% level of significance, test for any positive autocorrelation in the data.

22. Refer to the Cravens data set presented in Table 15.4. In Section 15.5 we showed that the model involving ACCTS, ADV, POTEN, and SHARE had an adjusted R^2 of 88.1%. At the 5% level of significance, use the Durbin-Watson test to determine if positive autocorrelation is present.

15.7 ▬ MULTIPLE REGRESSION APPROACH TO ANALYSIS OF VARIANCE

In Section 15.2 we discussed the use of dummy variables in multiple regression analysis. In this section we show how the use of dummy variables in a multiple regression equation can provide another approach to solving analysis of variance problems. We will demonstrate the multiple regression approach to analysis of variance by applying it to the GMAT experiment introduced in Chapter 12.

Recall that the objective of the GMAT study was to determine if the three preparation programs (3-hour review session, 1-day program, 10-week course) were different in terms of their effect on GMAT test scores. Sample observations were available from the population of students who took the 3-hour review session, the population of students who took the 1-day program, and the population of students who took the 10-week course.

We begin the regression approach to this problem by defining two dummy variables that will be used to indicate the population from which each sample observation was selected. Since there are three populations in the GMAT problem, we need two dummy variables. In general, if there are k populations we need to define $k - 1$ dummy variables. For the GMAT experiment we define x_1 and x_2 as shown in Table 15.10.

We can use the dummy variables x_1 and x_2 to relate the GMAT score of each student (y) to the type of GMAT preparation program:

$$E(y) = \text{expected value of the GMAT score}$$

$$= \beta_0 + \beta_1 x_1 + \beta_2 x_2$$

TABLE 15.10		GMAT Experiment Dummy Variables
x_1	x_2	These Values Are Used Whenever
0	0	Observation is associated with the 3 hour review session
1	0	Observation is associated with the 1 day program
0	1	Observation is associated with the 10 week course

Thus if we are interested in the expected value of the GMAT score for a student who completed the 3-hour review session, our procedure for assigning numerical values to the dummy variables x_1 and x_2 would result in setting $x_1 = x_2 = 0$. The multiple regression equation then reduces to

$$E(y) = \beta_0 + \beta_1(0) + \beta_2(0) = \beta_0$$

Thus we can interpret β_0 as the expected value of the GMAT score for students who complete the 3-hour review session.

Next let us consider the forms of the multiple regression equation for each of the other programs. For the 1-day program $x_1 = 1$ and $x_2 = 0$, and

$$E(y) = \beta_0 + \beta_1(1) + \beta_2(0) = \beta_0 + \beta_1$$

For the 10-week program $x_1 = 0$ and $x_2 = 1$, and

$$E(y) = \beta_0 + \beta_1(0) + \beta_2(1) = \beta_0 + \beta_2$$

We see that $\beta_0 + \beta_1$ represents the expected value of the GMAT score for students who complete the 1-day program, and $\beta_0 + \beta_2$ represents the expected value of the GMAT score for students who complete the 10-week course.

We now want to estimate the coefficients β_0, β_1, and β_2 and hence develop estimates of the expected value of the GMAT score for each program. The sample consisting of 15 observations of x_1, x_2, and y was entered into Minitab. The actual input data and the output from Minitab are shown in Table 15.11 and Figure 15.18, respectively.

Refer to Figure 15.18. We see that the estimates of β_0, β_1, and β_2 are $b_0 = 509$, $b_1 = 17$, and $b_2 = 43$. Thus our best estimate of the expected value of the GMAT score for each type of program is as follows:

Type of Program	Estimate of $E(y)$
3-hour review	$b_0 = 509$
1-day program	$b_0 + b_1 = 509 + 17 = 526$
10-week course	$b_0 + b_2 = 509 + 43 = 552$

TABLE 15.11　Input Data for the GMAT Problem

Observations Correspond to	x_1	x_2	y
	0	0	491
	0	0	579
3-hour review	0	0	451
	0	0	521
	0	0	503
	1	0	588
	1	0	502
1-day program	1	0	550
	1	0	520
	1	0	470
	0	1	533
	0	1	628
10-week course	0	1	502
	0	1	537
	0	1	561

Note that the best estimate of the expected value of the GMAT score for each program obtained from the regression analysis is the same as the sample means found earlier when applying the ANOVA procedure. That is, $\bar{x}_1 = 509$, $\bar{x}_2 = 526$, and $\bar{x}_3 = 552$.

Now let us see how we can use the output from the multiple regression package in order to perform the ANOVA test on the difference in the means for the three programs. First, we observe that if there is no difference in the means, then

$$E(y) \text{ for the 1 day program} - E(y) \text{ for the 3 hour review} = 0$$

$$E(y) \text{ for the 10 week course} - E(y) \text{ for the 3 hour review} = 0$$

FIGURE 15.18　Multiple Regression Output for the GMAT Problem

```
The regression equation is
Y = 509 + 17.0 X1 + 43.0 X2

Predictor      Coef      Stdev     t-ratio        p
Constant      509.00     20.81      24.46      0.000
X1             17.00     29.43       0.58      0.574
X2             43.00     29.43       1.46      0.170

s = 46.53      R-sq = 15.3%      R-sq(adj) = 1.2%

Analysis of Variance

SOURCE       DF        SS          MS         F        p
Regression    2       4690        2345      1.08     0.369
Error        12      25980        2165
Total        14      30670
```

Since β_0 equals $E(y)$ for the 3-hour review and $\beta_0 + \beta_1$ equals $E(y)$ for the 1-day program, the first difference above is equal to $(\beta_0 + \beta_1) - \beta_0 = \beta_1$. Moreover, since $\beta_0 + \beta_2$ equals $E(y)$ for the 10-week course, the second difference above is equal to $(\beta_0 + \beta_2) - \beta_0 = \beta_2$. Hence we would conclude that there is no difference in the three means if $\beta_1 = 0$ and $\beta_2 = 0$. Thus the null hypothesis for a test for difference of means can be stated as

$$H_0: \beta_1 = \beta_2 = 0$$

Recall that in order to test this type of null hypothesis about the significance of the regression relationship, we must compare the value of MSR/MSE to the critical value from an F distribution with numerator and denominator degrees of freedom equal to the degrees of freedom for the regression sum of squares and the error sum of squares, respectively. In the current problem the regression sum of squares has 2 degrees of freedom and the error sum of squares has 12 degrees of freedom. Thus the values for MSR and MSE are

$$MSR = \frac{SSR}{2} = \frac{4690}{2} = 2345$$

$$MSE = \frac{SSE}{12} = \frac{25,980}{12} = 2165$$

Hence the computed F value is

$$F = \frac{MSR}{MSE} = \frac{2345}{2165} = 1.0831$$

At the $\alpha = .05$ level of significance the critical value of F with 2 numerator degrees of freedom and 12 denominator degrees of freedom is 3.89. Since the observed value of F is less than the critical value of 3.89, we cannot reject the null hypothesis $H_0: \beta_1 = \beta_2 = 0$. Hence we cannot conclude that the means for the 3 preparation programs are different.

▬ EXERCISES

23. The Jacobs Chemical Company wants to estimate the mean time (minutes) required to mix a batch of material on machines produced by 3 different manufacturers. In order to limit the cost of testing, four batches of material were mixed on machines produced by each of the three manufacturers. The times needed to mix the material were recorded and are as follows:

Manufacturer 1	Manufacturer 2	Manufacturer 3
20	28	20
26	26	19
24	31	23
22	27	22

a. Write a multiple regression equation that can be used to analyze the data.

b. What are the best estimates of the coefficients in your regression equation?

c. In terms of the regression equation coefficients, what hypotheses do we have to test to see if the mean time to mix a batch of material is the same for each manufacturer?

d. For the α = .05 level of significance what conclusion should be drawn?

24. Four different paints are advertised as having the same drying time. In order to check the manufacturers' claims, five paint samples were tested for each brand of paint. The time in minutes until the paint was dry enough for a second coat to be applied was recorded for each sample. The following data were obtained:

Paint 1	Paint 2	Paint 3	Paint 4
128	144	133	150
137	133	143	142
135	142	137	135
124	146	136	140
141	130	131	153

a. For an α = .05 level of significance, test for a difference in mean drying times of the paints.

b. What is your estimate of mean drying time for paint 2? How is it obtained from the computer output?

■ SUMMARY

In this chapter we discussed several concepts used by model builders in identifying the best estimated regression equation. First, we introduced the concept of a general linear model in order to show how the methods discussed in Chapters 13 and 14 could be extended to handle curvilinear relationships and interaction effects. Then, we discussed how dummy variables could be used in model building to account for the effect of qualitative variables.

In applications of regression analysis to real problems, there are usually a large number of potential independent variables to consider. We presented a general approach, based on an F statistic for adding or deleting variables from a regression model. We then introduced a larger problem involving 25 observations and 8 potential independent variables. We saw that one issue encountered when solving larger problems is finding the best subset of the possible independent variables. To help in this regard, we discussed several variable-selection procedures, including stepwise regression, forward selection, and backward elimination.

In Section 15.6, we extended the applications of residual analysis to identify potentially troublesome observations (outliers and influential observations) and to show the Durbin-Watson test for autocorrelation. The chapter concluded with a discussion of how multiple regression models could be developed in order to provide another approach for solving analysis of variance problems.

■ GLOSSARY

General linear model. A model of the form $y = \beta_0 + \beta_1 z_1 + \beta_2 z_2 + \cdots + \beta_p z_p + \epsilon$, where each of the independent variables $z_j, j = 1, 2, \ldots, p$, is a function of x_1, $x_2, \ldots, x_k$, the variables for which data has been collected.

Interaction The joint effect of two variables acting together.

Qualitative variable A variable that is not measured in terms of how much or how many, but instead is assigned values to represent categories.

Dummy variable A variable that takes on the values 0 or 1 and is used to incorporate the effects of qualitative variables in a regression model.

Variable-selection procedures Computer-based methods for selecting a subset of the potential independent variables for a regression model

Outlier An observation with a residual that is far greater in magnitude than the rest of the residual values.

Influential observation An observation that has a great deal of influence in determining the estimated regression equation.

Leverage A measure designed to indicate how far an observation is from the others in terms of the values of the independent variables.

Autocorrelation Correlation in the errors that arises when the error terms at successive points in time are related. First-order autocorrelation is when ϵ_t and ϵ_{t-1} are related, second-order is when ϵ_t and ϵ_{t-2} are related, and so on.

Serial correlation Same as autocorrelation.

Durbin-Watson test A test to determine whether or not first-order autocorrelation is present.

▬ KEY FORMULAS

General Linear Model

$$y = \beta_0 + \beta_1 z_1 + \beta_2 z_2 + \cdots + \beta_p z_p + \epsilon \qquad (15.3)$$

General F Test for Adding or Deleting $p - q$ Variables

$$F = \frac{\dfrac{\text{SSE}(x_1, x_2, \cdots, x_q) - \text{SSE}(x_1, x_2, \cdots, x_q, x_{q+1}, \cdots, x_p)}{p - q}}{\dfrac{\text{SSE}(x_1, x_2, \cdots, x_q, x_{q+1}, \cdots, x_p)}{n - p - 1}} \qquad (15.13)$$

Autocorrelated Error Terms

$$\epsilon_t = \rho \epsilon_{t-1} + z_t \qquad (15.15)$$

Durbin-Watson Statistic

$$d = \frac{\sum\limits_{t=2}^{n} (e_t - e_{t-1})^2}{\sum\limits_{t=1}^{n} e_t^2} \qquad (15.16)$$

▬ SUPPLEMENTARY EXERCISES

25. Refer to the Cravens data set presented in Table 15.4.

a. Develop a scatter diagram showing SALES as a function of TIME.

b. Does a linear relationship between SALES and TIME appear to be appropriate? Explain.

c. Develop a model that can be used to predict SALES using just TIME or some appropriate function of TIME.

26. A study reported in the *Journal of Accounting Research* (Vol. 2, No. 2 Autumn 1987) investigated the relationship between audit delay (AUDELAY), the length of time from a company's fiscal year-end to the date of the auditor's report, and variables that describe

the client and the auditor. Some of the independent variables that were included in this study were

INDUS	A dummy variable which was coded as 1 if the firm was an industrial company or 0 if the firm was a bank, savings and loan, or insurance company.
PUBLIC	A dummy variable coded as 1 if the company was traded on an organized exchange or over the counter; otherwise coded 0.
ICQUAL	A measure of overall quality of internal controls, as judged by the auditor, using a five-point scale ranging from "virtually none" (1) to "excellent" (5).
INTFIN	A measure ranging from 1 to 4, as judged by the auditor, where 1 indicates "all work performed subsequent to year-end" and 4 indicates "most work performed prior to year-end."

Suppose that in a similar study a sample of 40 companies provided the following data:

AUDELAY	INDUS	PUBLIC	ICQUAL	INTFIN
62	0	0	3	1
45	0	1	3	3
54	0	0	2	2
71	0	1	1	2
91	0	0	1	1
62	0	0	4	4
61	0	0	3	2
69	0	1	5	2
80	0	0	1	1
52	0	0	5	3
47	0	0	3	2
65	0	1	2	3
60	0	0	1	3
81	1	0	1	2
73	1	0	2	2
89	1	0	2	1
71	1	0	5	4
76	1	0	2	2
68	1	0	1	2
68	1	0	5	2
86	1	0	2	2
76	1	1	3	1
67	1	0	2	3
57	1	0	4	2
55	1	1	3	2
54	1	0	5	2
69	1	0	3	3
82	1	0	5	1
94	1	0	1	1
74	1	1	5	2
75	1	1	4	3
69	1	0	2	2
71	1	0	4	4
79	1	0	5	2
80	1	0	1	4
91	1	0	4	1
92	1	0	1	4
46	1	1	4	3
72	1	0	5	2
85	1	0	5	1

a. Develop the estimated regression equation using all of the independent variables.

b. Did the model developed in (a) provide a good fit? Explain.

c. Develop a scatter diagram which shows AUDELAY as a function of INTFIN. What does this scatter diagram indicate about the relationship between AUDELAY and INTFIN?

d. Based upon your observations regarding the relationship between AUDELAY and INTFIN, develop an alternative model to the one developed in (a) in order to explain as much of the variability in AUDELAY as possible.

27. The following data set, reported in *Louis Rukeyser's Business Almanac* (1988 Simon and Schuster, p. 47), shows the percentage of management jobs held by women in various companies and the percentage of women in each company.

Industry/Company	Percentage of Management Jobs Held by Women	Percentage of Women
Industrial		
du Pont	7	22
Exxon	8	27
General Motors	6	19
Goodyear Tire and Rubber	25	39
Technology		
AT&T	32	48
General Electric	6	26
IBM	16	28
Xerox	23	38
Consumer products		
Johnson & Johnson	18	47
PepsiCo	28	46
Phillip Morris		
(excluding General Foods)	14	31
Proctor & Gamble	17	28
Retailing and trade		
Federated Department Stores	61	72
Kroger	16	47
Marriott	32	51
McDonald's	46	57
Sears, Roebuck	36	55
Media		
ABC (excluding Capital Cities)	36	43
Time	46	54
Times Mirror	27	37
Financial Services		
American Express	37	57
BankAmerica	64	72
Chemical Bank	34	57
Prudential Life Insurance	32	53
Wells Fargo Bank	58	71

a. Fit a simple linear regression model which can be used to predict the percentage of management jobs held by women given the percentage of women employed by the company.

b. Did the model developed in (a) provide a good fit to the data? Explain.

c. Use dummy variables to develop a model which relates the percentage of management jobs held by women to the type of industry (industrial, technology, and so on).

d. What conclusions can you reach based upon the model developed in (c)?

e. Develop a model which can be used to predict the percentage of management jobs held by women using the percentage of women employed by the company and the type of industry.

f. What final conclusions can be made regarding the percentage of management jobs held by women based upon your analyses?

28. Refer to the data set in Exercise 26.

a. Develop the best one-variable model which can be used to predict AUDELAY.

b. Use the stepwise procedure to develop the best model.

c. Use the backward-elimination procedure to develop the best model.

29. Refer to the data set in Exercise 27. In addition to the percentage of women employed in the company, create additional independent variables by using dummy variables to account for the type of industry.

a. Develop the best one-variable model which can be used to predict the percentage of management jobs held by women.

b. Use the stepwise procedure to develop the best model.

c. Use the backward-elimination procedure to develop the best model.

30. Refer to Exercise 26.

a. Develop a residual plot for the multiple regression model developed in (a) of Exercise 26. Does the pattern of the residual plot appear acceptable? Explain.

b. Are there any outliers?

c. Are there any influential observations? If so, what effect do they have on the model?

31. Refer to the NFL data set presented in exercise 14. Let y, the dependent variable, be the number of points scored.

a. Develop a model that can be used to predict y given the number of interceptions made by the team.

b. Develop a residual plot for the model developed in (a). Does the pattern of the residual plot appear acceptable? Explain.

c. Are there any outliers?

d. Are there any influential observations? If so, what effect do they have on the model?

32. Refer to the NFL data set presented in exercise 14. Let y, the dependent variable, be the number of points scored.

a. Develop a model that can be used to predict y given the number of interceptions made by the opponents.

b. Develop a residual plot for the model developed in (a). Does the pattern of the residual plot appear acceptable? Explain.

c. Are there any outliers?

d. Are there any influential observations? If so, what effect do they have on the model?

33. Refer to the data in Exercise 26. Consider a model in which only INDUS is used to predict AUDELAY. At the $\alpha = .01$ level of significance, test for any positive autocorrelation in the data.

34. Refer to the data in Exercise 26.

a. Develop an estimated regression equation which can be used to predict AUDELAY using INDUS and ICQUAL.

b. Plot the residuals obtained from the model developed in (a) as a function of the order in which the data are presented. Does there appear to be any autocorrelation present in the data? Explain.

c. At the 5% level of significance, test for any positive autocorrelation in the data.

35. Refer to the data in Exercise 27.

a. Develop an estimated regression equation which can be used to predict the percentage of management jobs held by women given the percentage of women employees in the company.

b. Plot the residuals obtained from the model developed in (a) as a function of the order in which the data are presented. Does there appear to be any autocorrelation present in the data? Explain.

c. At the 5% level of significance, test for any positive autocorrelation in the data.

36. A study was conducted to investigate the browsing activity by shoppers (*Journal of the Academy of Marketing Science*, Winter 1989). Shoppers were classified as non-browsers, light browsers and heavy browsers. For each shopper in the study a measure was obtained to determine how comfortable the shopper was in the store. Higher scores indicated greater comfort. Assume that the following data is from this study. Use a .05 level of significance to test for differences between comfort levels for the three types of browsers.

non-browser	light browser	heavy browser
4	5	5
5	6	7
6	5	5
3	4	7
3	7	4
4	4	6
5	6	5
4	5	7

███ COMPUTER EXERCISE

An article in the *Industrial and Labor Relations Review* (Vol. 41, No. 3, April 1988) reported the results of an investigation of factors related to the number of weeks a manufacturing worker was jobless since displacement (WEEKS). Some of the factors that were considered in this study were

AGE	The age of the worker
EDUC	The number of years of education
MARRIED	A dummy variable; 1 if married, 0 otherwise
HEAD	A dummy variable; 1 if the head-of-household, 0 otherwise
TENURE	The number of years on the old job
MGT	A dummy variable; 1 if management occupation, 0 otherwise
SALES	A dummy variable; 1 if sales occupation, 0 otherwise

Suppose that in a related study of 50 displaced workers that the following data were obtained.

WEEKS	AGE	EDUC	MARRIED	HEAD	TENURE	MGT	SALES
37	30	14	1	1	1	0	0
62	27	14	1	0	6	0	0
49	32	10	0	1	11	0	0
73	44	11	1	0	2	0	0
8	21	14	1	1	2	0	0
15	26	13	1	0	7	1	0
52	26	15	1	0	6	0	0
72	33	13	0	1	6	0	0
11	27	12	1	1	8	0	0
13	33	12	0	1	2	0	0
39	20	11	1	0	1	0	0
59	35	7	1	1	6	0	0
39	36	17	0	1	9	1	0
44	26	12	1	1	8	0	0
56	36	15	0	1	8	0	0
31	38	16	1	1	11	0	1
62	34	13	0	1	13	0	0
25	27	19	1	0	8	0	0
72	44	13	1	0	22	0	0
65	45	15	1	1	6	0	0
44	28	17	0	1	3	0	1
49	25	10	1	1	1	0	0
80	31	15	1	0	12	0	0
7	23	15	1	0	2	0	0
14	24	13	1	1	7	0	0
94	62	13	0	1	8	0	0
48	31	16	1	0	11	0	0
82	48	18	0	1	30	0	0
50	35	18	1	1	5	0	0
37	33	14	0	1	6	0	1
62	46	15	0	1	6	0	0
37	35	8	0	1	6	0	0
40	32	9	1	1	13	0	0
16	40	17	1	0	8	1	0
34	23	12	1	1	1	0	0
4	36	16	0	1	8	0	1
55	33	12	1	0	10	0	1
39	32	16	0	1	11	0	0
80	62	15	1	0	16	0	1
19	29	14	1	1	12	0	0
98	45	12	1	0	17	0	0
30	38	15	0	1	6	0	1
22	40	8	1	1	16	0	1
57	42	13	1	0	2	1	0
64	45	16	1	1	22	0	0
22	39	11	1	1	4	0	0
27	27	15	1	0	10	0	1
20	42	14	1	1	6	1	0
30	31	10	1	1	8	0	0
23	33	13	1	1	8	0	0

Use the methods presented in this chapter to analyze this data set. Present your findings, conclusions, and recommendations in the form of a brief report. Include in an appendix to your report any technical material, such as computer output, residual plots, etc., that you feel are appropriate.

APPLICATION

Monsanto Company*

St. Louis, Missouri

Monsanto Company traces its roots to an investment of $5,000 and a dusty warehouse on the Mississippi riverfront, where in 1901 John F. Queeny began manufacturing saccharin. From the maiden name of his wife, Olga Monsanto, came the name of Queeny's new company, and from his small investment and meager beginnings came what is today the nation's fourth largest chemical company.

Today more than a thousand products are made by Monsanto. Industrial chemical products include phenol, acetic acid, potassium hydroxide, maleic anhydride, and adipic acid. Monsanto also produces a long and widely varied list of petrochemicals, rubber chemicals, plasticizers, manmade fibers, agricultural products, nutrition chemicals, polymers, detergent intermediates, semiconductor materials, process control equipment, and instruments. AstroTurf® synthetic playing surface and Lasso® and Roundup® herbicides are some of the familiar products made by Monsanto.

World headquarters are located in St. Louis, Missouri. Monsanto is truly a worldwide corporation, with 146 manufacturing facilities, 20 laboratories or technical centers, and marketing operations in 65 countries. Approximately 52,800 people are employed worldwide, with about 38,000 employees in the United States.

▬ REGRESSION ANALYSIS AT MONSANTO

Monsanto utilizes regression analysis in many facets of its operations. Much of Monsanto's statistical analysis is done using prepared computer packages because of the complexity of many problems and the power and speed of computer analysis techniques. The following example, taken from product applications research in the Nutrition Chemicals Division, demonstrates the basic approach and utility of regression analysis.

Monsanto, through its Nutrition Chemicals Division, manufactures and markets a methionine supplement for use in poultry, swine, and cattle feeds. Poultry growers in particular work with very high volumes and low profit margins and have invested large amounts in the accurate definition of nutritional requirements for poultry. Optimal feed compositions result in more rapid growth and in a higher final body weight for a given feed intake. Feed efficiency, which relates gain in body weight to amount of feed consumed, is monitored closely over growth cycles. The success of poultry growers in improving feed efficiency is reflected in the low cost of poultry products relative to other meat products.

Monsanto conducts research at its own research farms and through several major universities in the United States to develop the growth response to methionine supplementation. The data shown in the scattergram (Figure 15A.1) give the results of one such experiment. Body weight after a fixed period of growth is shown as a function of the amount of methionine added to the feed, expressed as a total feed content of sulfur-containing amino acids.

With the data collected in this experiment a linear regression analysis was performed with supplemental methionine as the independent variable and body weight as the

*The authors are indebted to James R. Ryland and Robert M. Schisla, Senior Research Specialists, Monsanto Nutrition Chemical Division, St. Louis, Missouri, for providing this application.

dependent variable. The estimated regression equation developed is as follows:

$$y = 0.21 + 0.42x,$$

where

        y = body weight in kilograms,

        x = percentage of sulfur-containing amino acids in the feed.

The coefficient of determination, r^2, was 0.78, indicating a reasonably good fit. The F value of 81.5 pointed to a significant relationship between body weight and the percentage of supplemental methionine used in feed.

The use of linear models such as this is widespread in the poultry industry, but the dangers of using such a model to predict outside the range of the values of x are well recognized. An objective of the research was to find the relationship between methionine and body weight such that feed could be developed with the optimal levels of supplemental methionine to thereby provide the maximum possible body weight. Note that the regression model shows a positive relationship between x and y. Thus one might argue that increasing the supplemental methionine will continue to increase the body weight. However, this cannot be expected to continue without limit. In fact, further research has shown that as methionine supplementation is increased, body weight eventually levels off and may even decline as methionine is increased beyond nutritional requirements. The real point of interest—the level of supplemental methionine which provides peak growth performance—cannot be found with the above straight line relationship. Residual analysis here showed that a curvilinear relationship might provide a better model.

Experimentation was continued with the values of x (sulfur-containing amino acid percentage) increased up to 1.8%. Body weight began to level off as anticipated. Since the relationship was clearly nonlinear, Monsanto researchers and collaborating researchers

Chicks thrive on feed produced with Monsanto's Methionine supplement.

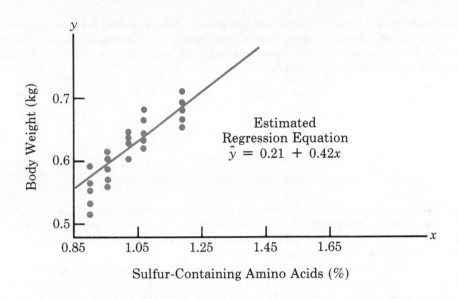

FIGURE 15A.1 **Body Weight as a Function of Sulfur-Containing Amino Acids (Sample of 25)**

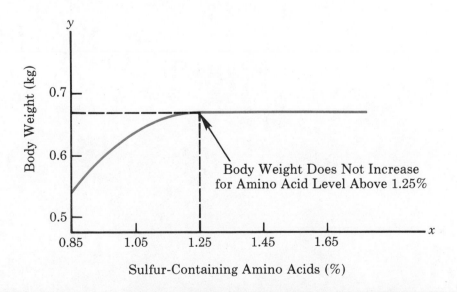

FIGURE 15A.2 **Nonlinear Regression Model for Body Weight as a Function of Sulfur-Containing Amino Acids**

in universities worked to obtain the best nonlinear regression model for the complete set of data. Using an advanced form of regression analysis and a computer package which uses the least squares method to develop a nonlinear regression model, the following estimated regression equation was developed:

$$\hat{y} = 0.55 + 0.12\,(1 - 3{,}410e^{-9.57x})$$

where

y = body weight in kilograms,

x = percentage sulfur-containing amino acids.

The F value again showed a significant relationship, and the coefficient of determination, r^2, of 0.88 showed a good fit for the complete set of data. A graph of the above equation is shown in Figure 15A.2. Of particular interest is the optimal body weight, which occurs at a sulfur-containing amino acid level of 1.25%.

Thus regression analysis was essential in enabling the poultry growers to obtain optimum weight gain per feed dollar. This application is typical of the important role regression analysis has played at Monsanto as the company works to develop quality products for its customers.

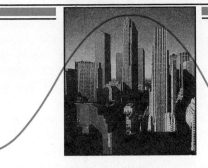

CHAPTER 16

Time Series Analysis and Forecasting

A critical aspect of managing any organization is planning for the future. Indeed, the long-run success of an organization is closely related to how well management is able to foresee the future and develop appropriate strategies. Good judgment, intuition, and an awareness of the state of the economy may give a manager a rough idea or "feeling" of what is likely to happen in the future. However, it is often difficult to convert this feeling into a number that can be used as next quarter's sales volume or next year's raw material cost per unit. The purpose of this chapter is to introduce several methods that can help predict many future aspects of a business operation.

Suppose that we have been asked to provide quarterly estimates of the sales volume for a particular product during the coming 1-year period. Production schedules, raw material purchasing plans, inventory policies, and sales quotas will all be affected by the quarterly estimates we provide. Consequently, poor estimates may result in poor planning and hence result in increased costs for the firm. How should we go about providing the quarterly sales volume estimates?

We will certainly want to review the actual sales data for the product in past periods. Suppose that we have actual sales data for each quarter over the past 3 years. Using these historical data we can identify the general level of sales and determine whether or not there is any trend, such as an increase or decrease in sales volume over time. A further review of the data might reveal a seasonal pattern, such as peak sales occurring in the third quarter of each year and sales volume bottoming out during the first quarter. By reviewing historical data over time we can often develop a better understanding of the pattern of past sales; often this can lead to better predictions of future sales for the product.

The historical sales data form what is called a *time series*. Specifically, a time series is a set of observations measured at successive points in time or over successive periods of time. In this chapter we will introduce several procedures that can be used to analyze time series data. The objective of this analysis will be to provide good *forecasts* or predictions of future values of the time series.

Forecasting methods can be classified as quantitative or qualitative. Quantitative forecasting methods are based on an analysis of historical data concerning a time series and possibly other related time series. If the historical data used are restricted to past values of the series that we are trying to forecast, the forecasting procedure is called a time series method. In this chapter we discuss three time series methods: smoothing (moving averages and exponential smoothing), trend projection, and trend projection adjusted for seasonal influence. If the historical data used in a quantitative forecasting method involve other time series that are believed to be related to the time series we are trying to forecast, we say that we are using a causal method. We discuss the use of multiple regression analysis as a causal forecasting method.

Qualitative forecasting methods generally utilize the judgment of experts to make forecasts. An advantage of these procedures is that they can be applied in situations where no historical data are available. We discuss some of these approaches in Section 16.6. Figure 16.1 provides an overview of the different types of forecasting methods.

16.1 ▰▰ THE COMPONENTS OF A TIME SERIES

In order to explain the pattern or behavior of the data in a time series it is often helpful to think of the time series as consisting of several components. The usual assumption is that four separate components—trend, cyclical, seasonal, and irregular—combine to make the time series take on specific values. Let us look more closely at each of these components of a time series.

▰▰ FIGURE 16.1 **An Overview of Forecasting Methods**

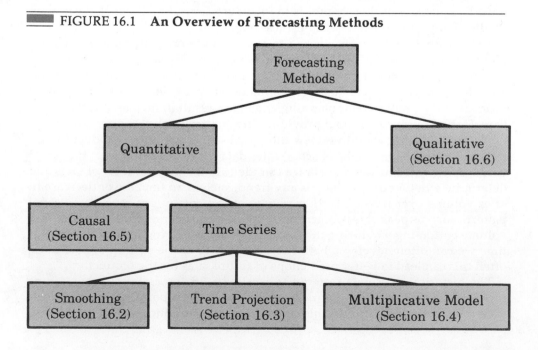

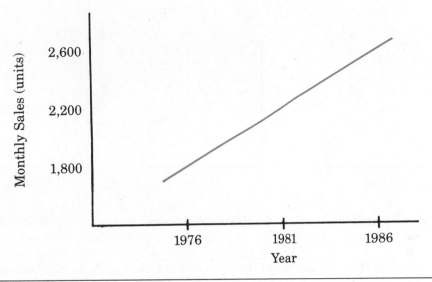

▬▬▬ FIGURE 16.2   **Linear Trend of Camera Sales**

Trend Component

In time series analysis the measurements may be taken every hour, day, week, month, or year or at any other regular interval.* Although time series data generally exhibit random fluctuations, the time series may still show gradual shifts or movements to relatively higher or lower values over a longer period of time. The gradual shifting of the time series, which is usually due to long-term factors such as changes in the population, changes in demographic characteristics of the population, changes in technology, and changes in consumer preferences, is referred to as the *trend* in the time series.

For example, a manufacturer of photographic equipment may see substantial month-to-month variability in the number of cameras sold. However, in reviewing the sales over the past 10 to 15 years this manufacturer may find a gradual increase in the annual sales volume. Suppose that the sales volume was approximately 1800 cameras per month in 1979, 2200 cameras per month in 1984, and 2600 cameras per month in 1989. While actual month-to-month sales volumes may vary substantially, this gradual growth in sales over time shows an upward trend for the time series. Figure 16.2 shows a straight line that may be a good approximation of the trend in the sales data. While the trend for camera sales appears to be linear and increasing over time, sometimes the trend in a time series is better described by other patterns.

Figure 16.3 shows some other possible time series trend patterns. In Panel A of this figure we see a nonlinear trend. The curve shown describes a time series showing very little growth initially, followed by a period of rapid growth, and then a leveling off. This might be a good approximation to sales for a product from introduction through a growth period and into a period of market saturation. The linear decreasing trend in Panel B of Figure 16.3 is useful for time series displaying a steady decrease over time. The horizontal line in Panel

---

*We restrict our attention here to time series where the values of the series are recorded at equal intervals. Treatment of cases where the observations are not made at equal intervals is beyond the scope of this text.

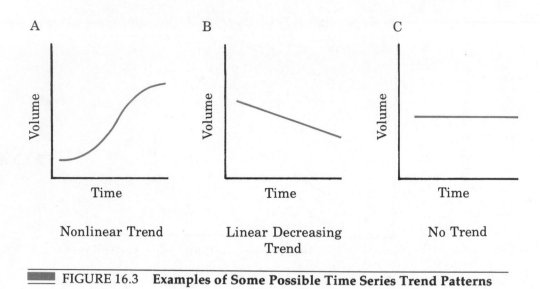

A   Nonlinear Trend

B   Linear Decreasing Trend

C   No Trend

FIGURE 16.3   Examples of Some Possible Time Series Trend Patterns

C of Figure 16.3 is used for a time series that does not show any consistent increase or decrease over time. It is actually the case of no trend.

Cyclical Component

Although a time series may exhibit a gradual shifting or trend pattern over long periods of time, we cannot expect all future values of the time series to be exactly on the trend line. In fact, time series often show alternating sequences of points below and above the trend line. Any regular pattern of sequences of points above and below the trend line lasting more than one year is attributable to the *cyclical component* of the time series. Figure 16.4 shows the graph of a time series with an obvious cyclical component. The observations are taken at intervals 1 year apart.

FIGURE 16.4   Trend and Cyclical Components of a Time Series. Data Points Are 1 Year Apart.

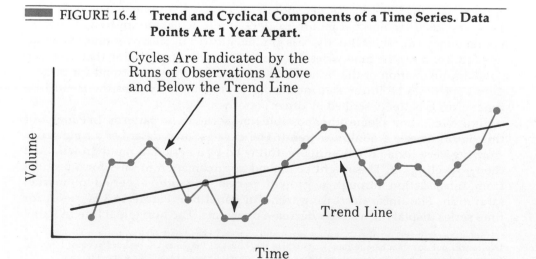

Many time series exhibit cyclical behavior with regular runs of observations below and above the trend line. The general belief is that this component of the time series represents multiyear cyclical movements in the economy. For example, periods of moderate inflation followed by periods of rapid inflation can lead to many time series that alternate below and above a generally increasing trend line (e.g., housing costs). Many time series in the late 1970s and early 1980s displayed this type of behavior.

Seasonal Component

While the trend and cyclical components of a time series are identified by analyzing multiyear movements in historical data, many time series show a regular pattern of variability within 1-year periods. For example, a manufacturer of swimming pools expects low sales activity in the fall and winter months, with peak sales occurring in the spring and summer months. Manufacturers of snow removal equipment and heavy clothing, however, expect just the opposite yearly pattern. It should not be surprising that the component of the time series that represents the variability in the data due to seasonal influences is called the *seasonal component*. Although we generally think of seasonal movement in a time series as occurring within 1 year, the seasonal component can also be used to represent any regularly repeating pattern that is less than 1 year in duration. For example, daily traffic volume data show within-the-day "seasonal" behavior, with peak levels occurring during rush hours, moderate flow during the rest of the day and early evening, and light flow from midnight to early morning.

Irregular Component

The *irregular component* of the time series is the residual, or "catchall," factor that accounts for the deviation of the actual time series value from what we would expect if the trend, cyclical, and seasonal components completely explained the time series. It accounts for the random variability in the time series. The irregular component is caused by the short-term, unanticipated, and nonrecurring factors that affect the time series. Since this component accounts for the random variability in the time series, it is unpredictable. We cannot attempt to predict its impact on the time series in advance.

16.2 ■■■ FORECASTING USING SMOOTHING METHODS

In this section we discuss forecasting techniques that are appropriate for a fairly stable time series; that is, one that exhibits no significant trend, cyclical, or seasonal effects. In such situations the objective of the forecasting method is to "smooth out" the irregular component of the time series through an averaging process. We begin with a consideration of the method known as moving averages.

Moving Averages

The *moving averages* method uses the average of the *most recent n* data values in the time series as the forecast for the next period. Mathematically, the moving average calculation is made as follows:

Moving Average

$$\text{Moving average} = \frac{\Sigma\,(\text{most recent } n \text{ data values})}{n} \qquad (16.1)$$

The term *moving* average is based on the fact that as a new observation becomes available for the time series, it replaces the oldest observation in (16.1), and a new average is computed. As a result the average will change, or move, as new observations become available.

To illustrate the moving averages method, consider the 12 weeks of data presented in Table 16.1 and Figure 16.5. These data show the number of gallons of gasoline sold by a gasoline distributor in Bennington, Vermont, over the past 12 weeks.

In order to use moving averages to forecast gasoline sales, we must first select the number of data values to be included in the moving average. As an example, let us compute forecasts using a 3-week moving average. The moving average calculation for the first 3 weeks of the gasoline sales time series is as follows:

$$\text{Moving Average (Weeks 1–3)} = \frac{17 + 21 + 19}{3} = 19$$

This moving average value is then used as the forecast for week 4. Since the actual value observed in week 4 is 23, we see that the forecast error in week 4 is $23 - 19 = 4$. In general, the error associated with any forecast is the difference between the observed value of the time series and the forecast value.

TABLE 16.1   Gasoline Sales Time Series

Week	Sales (1000s of gallons)
1	17
2	21
3	19
4	23
5	18
6	16
7	20
8	18
9	22
10	20
11	15
12	22

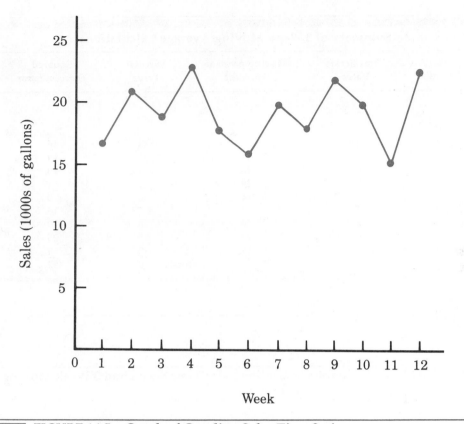

· The calculation for the second 3 week moving average is shown below:

$$\text{Moving Average (Weeks 2–4)} = \frac{21 + 19 + 23}{3} = 21$$

Hence the forecast for week 5 is 21. The error associated with this forecast is 18 − 21 = −3. Thus we see that the forecast error can be positive or negative depending upon whether the forecast is too low or too high. A complete summary of the 3-week moving average calculations for the gasoline sales time series is shown in Table 16.2 and Figure 16.6.

An important consideration in using any forecasting method is the accuracy of the forecast. Clearly, we would like the forecast errors to be small. The last two columns of Table 16.2, which contain the forecast errors and the forecast errors squared, can be used to develop measures of accuracy.

One measure of forecast accuracy you might think of using would be to simply sum the forecast errors over time. The problem with this measure is that if the errors are random (as they should be if the forecasting method selected is appropriate), some errors will be positive and some errors will be negative, resulting in a sum near zero regardless of the size of the individual errors. Indeed, we see from Table 16.2 that the sum of forecast errors for the gasoline sales time series is zero. This difficulty can be avoided by squaring each of the individual forecast errors.

TABLE 16.2   **Summary of 3 Week Moving Average Calculations**

Week	Time Series Value	Moving Average Forecast	Forecast Error	Squared Forecast Error
1	17			
2	21			
3	19			
4	23	19	4	16
5	18	21	−3	9
6	16	20	−4	16
7	20	19	1	1
8	18	18	0	0
9	22	18	4	16
10	20	20	0	0
11	15	20	−5	25
12	22	19	3	9
		Totals	0	92

FIGURE 16.6   **Graph of Gasoline Sales Time Series and 3 Week Moving Average Forecasts**

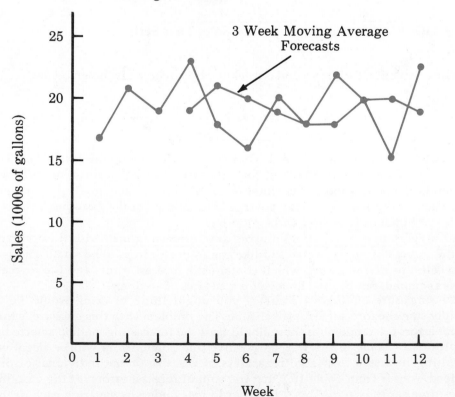

For the gasoline sales time series we can use the last column of Table 16.2 to compute the average of the sum of the squared errors. Doing so we obtain

$$\text{Average of the Sum of Squared Errors} = \frac{92}{9} = 10.22$$

This average of the sum of squared errors is commonly referred to as the *mean squared error* (MSE). The mean squared error is an often used measure of the accuracy of a forecasting method and is the one we use in this chapter.

As we indicated previously, in order to use the moving averages method we must first select the number of data values to be included in the moving average. It should not be too surprising that for a particular time series different length moving averages will differ in their ability to accurately forecast the time series. One possible approach to choosing the number of values to be included is to use trial and error to identify the length that minimizes the MSE. Then, if we are willing to assume that the length which is best for the past will also be best for the future, we would forecast the next value in the time series using the number of data values that minimized the MSE for the historical time series. Exercise 2 at the end of the section will ask you to consider 4 week and 5 week moving averages for the gasoline sales data. A comparison of the mean square error for each will indicate the number of weeks of data you may want to include in the moving average calculation.

Weighted Moving Averages

In the moving averages method each observation in the moving average calculation receives the same weight. One possible variation, known as *weighted moving averages*, involves selecting different weights for each data value and then computing a weighted mean as the forecast. In most cases the most recent observation receives the most weight, and the weight decreases for older data values. For example, using the gasoline sales time series, let us illustrate the computation of a weighted 3-week moving average, where the most recent observation receives a weight 3 times as great as that given the oldest observation, and the next oldest observation receives a weight twice as great as the oldest. The weighted moving average forecast for week 4 would be computed as follows:

Weighted Moving Average Forecast For Week 4

$$= \tfrac{3}{6}(19) + \tfrac{2}{6}(21) + \tfrac{1}{6}(17) = 19.33$$

Note that for the weighted moving average the sum of the weights is equal to 1. This was also true for the simple moving average, where each weight was $\tfrac{1}{3}$. However, recall that the simple or unweighted moving average provided a forecast of 19. Exercise 3 at the end of the section asks you to calculate the remaining values for the 3-week weighted moving average and compare the forecast accuracy with what we have obtained for the unweighted moving average.

Exponential Smoothing

Exponential smoothing is a forecasting technique that uses a weighted average of past time series values to forecast the value of the time series in the next period. The basic exponential smoothing model is as follows:

Exponential Smoothing Model

$$F_{t+1} = \alpha Y_t + (1 - \alpha)F_t \qquad (16.2)$$

where

F_{t+1} = the forecast of the time series for period $t + 1$

Y_t = actual value of the time series in period t

F_t = forecast of the time series for period t

α = *smoothing constant* $(0 \leq \alpha \leq 1)$

To see that the forecast for any period is a weighted average of *all the previous actual values* for the time series, suppose that we have a time series consisting of three periods of data, Y_1, Y_2, and Y_3. To get the exponential smoothing calculations started, we let F_1 equal the actual value of the time series in period 1; that is, $F_1 = Y_1$. Hence the forecast for period 2 is written as follows:

$$F_2 = \alpha Y_1 + (1 - \alpha)F_1$$
$$= \alpha Y_1 + (1 - \alpha)Y_1$$
$$= Y_1$$

In general, then, the exponential smoothing forecast for period 2 is equal to the actual value of the time series in period 1.

To obtain the forecast for period 3, we substitute $F_2 = Y_1$ in the expression for F_3; the result is

$$F_3 = \alpha Y_2 + (1 - \alpha)Y_1$$

Finally, substituting this expression for F_3 in the expression for F_4, we obtain

$$F_4 = \alpha Y_3 + (1 - \alpha)[\alpha Y_2 + (1 - \alpha)Y_1]$$
$$= \alpha Y_3 + \alpha(1 - \alpha)Y_2 + (1 - \alpha)^2 Y_1$$

Hence we see that F_4 is a weighted average of the first three time series values. The sum of the coefficients or weights for Y_1, Y_2, and Y_3 equals 1. A similar argument can be made to show that any forecast F_{t+1} is a weighted average of the previous t time series values.

An advantage of exponential smoothing is that it is a simple procedure and requires very little historical data for its use. Once the smoothing constant α has been selected, only two pieces of information are required in order to compute the forecast for the next period. Referring to (16.2), we see that with a given α we can compute the forecast for period $t + 1$ simply by knowing the actual and forecast time series values for period t, that is, Y_t and F_t.

To illustrate the exponential smoothing approach to forecasting, consider the gasoline sales time series presented previously in Table 16.1 and Figure 16.5. As we indicated in the discussion above, the exponential smoothing forecast for period 2 is equal to the actual value of the time series in period 1. Thus, with $Y_1 = 17$, we will set $F_2 = 17$ to get the exponential smoothing computations started. Referring to the time series data in Table 16.1, we find an actual time series value in period 2 of $Y_2 = 21$. Thus period 2 has a forecast error of $21 - 17 = 4$.

Continuing with the exponential smoothing computations provides the following forecast for period 3:

$$F_3 = 0.2Y_2 + 0.8F_2 = 0.2(21) + 0.8(17) = 17.8$$

Once the actual time series value in period 3, $Y_3 = 19$, is known, we can generate a forecast for period 4 as follows:

$$F_4 = 0.2Y_3 + 0.8F_3 = 0.2(19) + 0.8(17.8) = 18.04$$

By continuing the exponential smoothing calculations we are able to determine the weekly forecast values and the corresponding weekly forecast errors, as shown in Table 16.3. Note that we have not shown an exponential smoothing forecast or the forecast error for period 1, because F_1 was set equal to Y_1 in order to begin the smoothing computations. For week 12, we have $Y_{12} = 22$ and $F_{12} = 18.48$. Can you use this information to generate a forecast for week 13 before the actual value of week 13 becomes known? Using the exponential smoothing

TABLE 16.3 Summary of the Exponential Smoothing Forecasts and Forecast Errors For Gasoline Sales With Smoothing Constant $\alpha = 0.2$

Week (t)	Time Series Value (Y_t)	Exponential Smoothing Forecast (F_t)	Forecast Error ($Y_t - F_t$)
1	17		
2	21	17.00	4.00
3	19	17.80	1.20
4	23	18.04	4.96
5	18	19.03	−1.03
6	16	18.83	−2.83
7	20	18.26	1.74
8	18	18.61	−0.61
9	22	18.49	3.51
10	20	19.19	0.81
11	15	19.35	−4.35
12	22	18.48	3.52

model, we have

$$F_{13} = 0.2Y_{12} + 0.8F_{12} = 0.2(22) + 0.8(18.48) = 19.18$$

Thus the exponential smoothing forecast of the amount sold in week 13 is 19.18, or 19,180 gallons of gasoline. With this forecast the firm can make plans and decisions accordingly. The accuracy of the forecast will not be known until the firm conducts its business through week 13.

Figure 16.7 shows the plot of the actual and the forecast time series values. Note in particular how the forecasts "smooth out" the irregular fluctuations in the time series.

In the preceding exponential smoothing calculations we used a smoothing constant of $\alpha = 0.2$, although any value of α between 0 and 1 is acceptable. However, some values will yield better forecasts than others. Some insight into choosing a good value for α can be obtained by rewriting the basic exponential smoothing model as follows:

$$F_{t+1} = \alpha Y_t + (1 - \alpha)F_t$$

$$F_{t+1} = \alpha Y_t + F_t - \alpha F_t$$

$$F_{t+1} = \underbrace{F_t}_{\substack{\text{Forecast} \\ \text{in Period } t}} + \alpha \underbrace{(Y_t - F_t)}_{\substack{\text{Forecast Error} \\ \text{in Period } t}} \tag{16.3}$$

■■■ FIGURE 16.7　**Graph of Actual and Forecast Gasoline Sales Time Series with Smoothing Constant $\alpha = .2$**

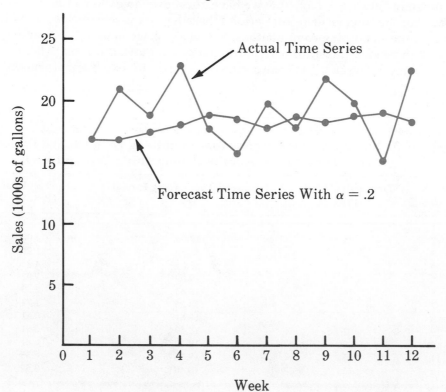

Thus we see that the new forecast F_{t+1} is equal to the previous forecast F_t plus an adjustment, which is α times the most recent forecast error, $Y_t - F_t$. That is, the forecast in period $t + 1$ is obtained by adjusting the forecast in period t by a fraction of the forecast error. If the time series contains substantial random variability, a small value of the smoothing constant is preferred. The reason for this choice is that since much of the forecast error is due to random variability, we do not want to overreact and adjust the forecasts too quickly. For a time series with relatively little random variability, larger values of the smoothing constant have the advantage of quickly adjusting the forecasts when forecasting errors occur and therefore allowing the forecast to react faster to changing conditions.

The criterion we will use to determine a desirable value for the smoothing constant α is the same as the criterion we proposed earlier for determining the number of periods of data to include in the moving averages calculation. That is, we choose the value of α that minimizes the mean squared error (MSE). A summary of the MSE calculations for the exponential smoothing forecast of gasoline sales with $\alpha = 0.2$ is shown in Table 16.4. Note that there is one less squared error term than the number of time periods, because we had no past values with which to make a forecast for period 1. Would a different value of α have provided better results in terms of a lower MSE value? Perhaps the most straightforward way to answer this question is simply to try another value for α. We will then compare its mean squared error with the MSE value of 8.98 obtained using a smoothing constant of 0.2.

The exponential smoothing results with $\alpha = 0.3$ are shown in Table 16.5. With MSE = 9.35, we see that for the current data set a smoothing constant of $\alpha = 0.3$ results in less forecast accuracy than a smoothing constant of $\alpha = 0.2$. Thus we would be inclined to prefer the original smoothing constant of 0.2. With a trial-and-error calculation with other values of α, a "good" value for the

TABLE 16.4 Mean Squared Error Computations for Forecasting Gasoline Sales with $\alpha = 0.2$

Week (t)	Time Series Value (Y_t)	Forecast (F_t)	Forecast Error ($Y_t - F_t$)	Squared Forecast Error ($Y_t - F_t)^2$
1	17			
2	21	17.00	4.00	16.00
3	19	17.80	1.20	1.44
4	23	18.04	4.96	24.60
5	18	19.03	−1.03	1.06
6	16	18.83	−2.83	8.01
7	20	18.26	1.74	3.03
8	18	18.61	−0.61	0.37
9	22	18.49	3.51	12.32
10	20	19.19	0.81	0.66
11	15	19.35	−4.35	18.92
12	22	18.48	3.52	12.39
			Total	98.80

$$\text{Mean squared error (MSE)} = \frac{98.80}{11} = 8.98$$

TABLE 16.5 **Mean Squared Error Computations for Forecasting Gasoline Sales with $\alpha = 0.3$**

Week (t)	Time Series Value (Y_t)	Forecast (F_t)	Forecast Error ($Y_t - F_t$)	Squared Forecast Error ($Y_t - .F_t)^2$
1	17			
2	21	17.00	4.00	16.00
3	19	18.20	0.80	0.64
4	23	18.44	4.56	20.79
5	18	19.81	−1.81	3.28
6	16	19.27	−3.27	10.69
7	20	18.29	1.71	2.92
8	18	18.80	−0.80	0.64
9	22	18.56	3.44	11.83
10	20	19.59	0.41	0.17
11	15	19.71	−4.71	22.18
12	22	18.30	3.70	13.69
			Total	102.83

$$\text{Mean Squared Error (MSE)} = \frac{102.83}{11} = 9.35$$

smoothing constant can be found. This value can be used in the exponential smoothing model to provide forecasts for the future. At a later date, after a number of new time series observations have been obtained, it is good practice to analyze the newly collected time series data to see if the smoothing constant should be revised to provide better forecasting results.

NOTES AND COMMENTS

Another commonly used measure of forecast accuracy is the *mean absolute deviation* (MAD). This measure is simply the average of the sum of the absolute values of all the forecast errors. Using the errors given in Table 16.2, we obtain

$$\text{Mean Absolute Deviation (MAD)} = \frac{4 + 3 + 4 + 1 + 0 + 4 + 0 + 5 + 3}{9}$$

$$= 2.67$$

One major difference between MSE and MAD is that the MSE measure is influenced much more by large forecast errors than by small errors (since for the MSE measure the errors are squared). The selection of the best measure of forecasting accuracy is not a simple matter. Indeed, forecasting experts often disagree as to which measure should be used. In this chapter we will use the MSE measure.

■■ EXERCISES

1. Corporate Triple A Bond interest rates for the 12 months of 1988 are shown below (*The Media General Financial Weekly*, April 17, 1989).

   9.5   9.3   9.4   9.6   9.8   9.7   9.8   10.5   9.9   9.7   9.6   9.6

   a. Develop three-month and four-month moving averages for this time series. Does the three-month or four-month moving average provide the better forecasts? Explain.

   b. What is the moving averages forecast for January 1989?

2. Refer to the gasoline sales time series data in Table 16.1.

   a. Compute 4-week and 5-week moving averages for the time series.

   b. Compute MSE for the 4-week and 5-week moving average forecasts.

   c. What appears to be the best number of weeks of past data to use in the moving average computation? Remember that the MSE for the 3-week moving average is 10.22.

3. Refer again to the gasoline sales time series data in Table 16.1.

   a. Using a weight of $\frac{1}{2}$ for the most recent observation, $\frac{1}{3}$ for the second most recent, and $\frac{1}{6}$ for third most recent, compute a 3-week weighted moving average for the time series.

   b. Compute MSE for the weighted moving average in part (a). Do you prefer this weighted moving average to the unweighted moving average? Remember that MSE for the unweighted moving average is 10.22.

   c. Suppose you are allowed to choose any weight as long as they sum to one. Could you always find a set of weights that would make MSE smaller for a weighted moving average than an unweighted moving average? Why or why not?

4. Use the gasoline time series data from Table 16.1 to show the exponential smoothing forecasts using $\alpha = .1$. Using the mean squared error criterion, would you prefer a smoothing constant of $\alpha = .1$ or $\alpha = .2$ for the gasoline sales time series?

5. Using a smoothing constant of $\alpha = .2$, equation (16.2) shows that the forecast for the 13th week of the gasoline sales data from Table 16.1 is given by $F_{13} = .2Y_{12} + .8F_{12}$. However, the forecast for week 12 is given by $F_{12} = .2Y_{11} + .8F_{11}$. Thus we could combine these two results to show that the forecast for the 13th week can be written

$$F_{13} = .2Y_{12} + .8(.2Y_{11} + .8F_{11}) = .2Y_{12} + .16Y_{11} + .64F_{11}$$

   a. Making use of the fact that $F_{11} = .2Y_{10} + .8F_{10}$ (and similarly for F_{10} and F_9), continue to expand the expression for F_{13} until it is written in terms of the past data values $Y_{12}, Y_{11}, Y_{10}, Y_9, Y_8$, and the forecast for period 8.

   b. Refer to the coefficients or weights for the past data $Y_{12}, Y_{11}, Y_{10}, Y_9$, and Y_8; what observation do you make about how exponential smoothing weights past data values in arriving at new forecasts? Compare this weighting pattern with the weighting pattern of the moving averages method.

6. Alabama building contracts by month for 1988 are shown below (*Alabama Business*, June 1989). Data are in millions of dollars.

   240   350   230   260   280   320   220   310   240   310   240   230

   a. Compare a three-month moving averages forecast with an exponential smoothing forecast using $\alpha = .2$. Which provides the better forecasts?

   b. What is the forecast for January 1989?

7. The following time series shows the sales of a particular product over the past 12 months:

Month	Sales
1	105
2	135
3	120
4	105
5	90
6	120
7	145
8	140
9	100
10	80
11	100
12	110

a. Use $\alpha = .3$ to compute the exponential-smoothing values for the time series.

b. Use a smoothing constant of .5 to compute the exponential smoothing values. Does a smoothing constant of .3 or .5 appear to provide the better forecasts?

8. The Dow Jones Industrial Average is based on common stock prices of 30 industrial stocks. This average is used to describe what is happening in the stock market. The weekly closing levels of the Dow Jones average for twelve weeks during the period May to July of 1989 are shown below (*Wall Street Journal*, July 31, 1989).

Week	Dow Jones	Week	Dow Jones
1	2480	7	2520
2	2470	8	2470
3	2475	9	2440
4	2510	10	2480
5	2500	11	2530
6	2480	12	2550

a. Compute the exponential smoothing forecasts using $\alpha = .2$.

b. Compute the exponential smoothing forecasts using $\alpha = .3$.

c. Which exponential smoothing model provides the better forecasts? What is the forecast of the Dow Jones Industrial Average for week 13?

16.3 ■ FORECASTING TIME SERIES USING TREND PROJECTION

In this section we will see how to forecast the values of a time series that exhibits a long-term linear trend. Specifically, let us consider the time series data for bicycle sales of a particular manufacturer over the past 10 years, as shown in Table 16.6 and Figure 16.8. Note that 21,600 bicycles were sold in year 1, 22,900 were sold in year 2, and so on; in year 10, the most recent year, 31,400 bicycles

TABLE 16.6	Bicycle Sales Data
Year (*t*)	Sales (1000s) (*Y~t~*)
1	21.6
2	22.9
3	25.5
4	21.9
5	23.9
6	27.5
7	31.5
8	29.7
9	28.6
10	31.4

were sold. Although the graph in Figure 16.8 shows some up-and-down movement over the past 10 years, the time series seems to have an overall increasing or upward trend in the number of bicycles sold.

We do not want the trend component of a time series to follow each and every "up" and "down" movement. Rather, the trend component should reflect the gradual shifting—in our case, growth—of the time series values. After we view

■■■■ FIGURE 16.8   Graph of the Bicycle Sales Time Series

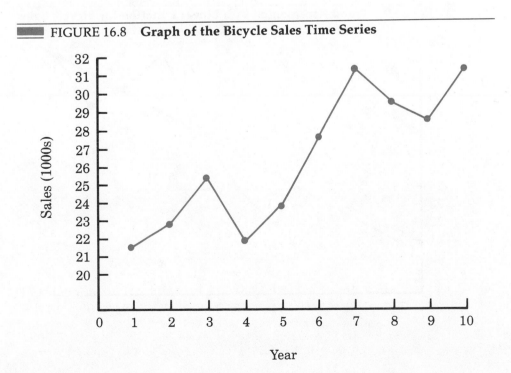

the time series data in Table 16.6 and the graph in Figure 16.8 we might agree that a linear trend as shown in Figure 16.9 has the potential of providing a reasonable description of the long-run movement in the series. Thus we can now concentrate on finding the linear function that best approximates the trend.

Using the bicycle sales data to illustrate the calculations involved, we will now describe how regression analysis can be used to identify a linear trend for a time series. Recall in the discussion of simple linear regression in Chapter 13 we described how the least squares method was used to find the best straight line relationship between two variables. This is the methodology we will use to develop the trend line for the bicycle sales time series. Specifically, we will be using regression analysis to estimate the relationship between time and sales volume.

In Chapter 13 the estimated regression equation describing a straight-line relationship between an independent variable x and a dependent variable y was written

$$\hat{y} = b_0 + b_1 x \qquad (16.4)$$

In forecasting, in order to better focus on the fact that the independent variable is time, we will use t in (16.4) instead of x; in addition, we will use T_t in place of $\hat{y}$. Thus for a linear trend the estimated sales volume expressed as a function of

▬ FIGURE 16.9 **Trend Represented by a Linear Function for Bicycle Sales**

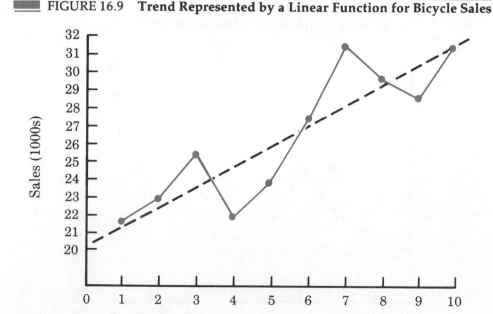

time can be written as follows:

Equation for Linear Trend

$$T_t = b_0 + b_1 t \qquad (16.5)$$

where

T_t = forecast value (based upon trend) of the time series in period t

b_0 = intercept of the trend line

b_1 = slope of the trend line

t = point in time

In (16.5), we will let $t = 1$ for the time of the first observation on the time series data, $t = 2$ for the time of the second observation, and so on. Note that for the time series on bicycle sales $t = 1$ corresponds to the oldest time series value and $t = 10$ corresponds to the most recent year's data. Formulas for computing the estimated regression coefficients (b_1 and b_0) in (16.4) were presented in Chapter 13; they are shown again below, with t replacing x and Y_t replacing y_i.

Computing the Slope (b_1) and Intercept (b_0)

$$b_1 = \frac{\Sigma t Y_t - (\Sigma t \, \Sigma Y_t)/n}{\Sigma t^2 - (\Sigma t)^2/n} \qquad (16.6)$$

$$b_0 = \overline{Y} - b_1 \bar{t} \qquad (16.7)$$

where

Y_t = actual value of the time series in period t

n = number of periods

$\overline{Y}$ = average value of the time series; that is, $\overline{Y} = \Sigma Y_t / n$

$\bar{t}$ = average value of t; that is, $\bar{t} = \Sigma t / n$

Using these relationships for b_0 and b_1 and the bicycle sales data of Table 16.6, we have the following calculations:

t	Y_t	tY_t	t^2
1	21.6	21.6	1
2	22.9	45.8	4
3	25.5	76.5	9
4	21.9	87.6	16
5	23.9	119.5	25
6	27.5	165.0	36
7	31.5	220.5	49
8	29.7	237.6	64
9	28.6	257.4	81
10	31.4	314.0	100
Totals 55	264.5	1545.5	385

$$\bar{t} = \frac{55}{10} = 5.5 \text{ years}$$

$$\overline{Y} = \frac{264.5}{10} = 26.45 \text{ thousands}$$

$$b_1 = \frac{1545.5 - (55)(264.5)/10}{385 - (55)^2/10} = \frac{90.75}{82.50} = 1.10$$

$$b_0 = 26.45 - 1.10 (5.5) = 20.4$$

Therefore,

$$T_t = 20.4 + 1.1t \tag{16.8}$$

is the expression for the linear trend component for the bicycle sales time series.

Trend Projections

The slope of 1.1 indicates that over the past 10 years the firm has experienced an average growth in sales of around 1100 units per year. If we assume that the past 10-year trend in sales is a good indicator of the future, then (16.8) can be used to project the trend component of the time series. For example, substituting $t = 11$ into (16.8) yields next year's trend projection, T_{11}:

$$T_{11} = 20.4 + 1.1(11) = 32.5$$

Thus using the trend component only we would forecast sales of 32,500 bicycles next year.

The use of a linear function to model the trend is common. However, as we discussed earlier, sometimes time series exhibit a curvilinear, or nonlinear, trend similar to those shown in Figure 16.10. In Chapter 15 we discussed how regression analysis can be used to model curvilinear relationships of the type shown in Figure 16.10(a). More advanced texts discuss in detail how to develop

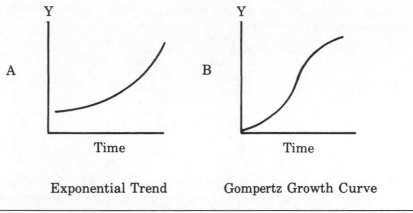

Exponential Trend   Gompertz Growth Curve

■■■ FIGURE 16.10   **Some Possible Functional Forms for Nonlinear Trend Patterns**

regression models for more complex relationships such as the one shown in Figure 16.10(b). For our purposes it is sufficient to note that the analyst should choose the function that provides the best fit to the data.

■■■ EXERCISES

9. The enrollment data for a state college for the past 6 years are shown:

Year	Enrollment
1	20,500
2	20,200
3	19,500
4	19,000
5	19,100
6	18,800

Develop the equation for the linear trend component for this time series. Comment on what is happening to enrollment at this institution.

10. The time series for the retail price index of consumer goods and services is shown below (*Accountancy*, July 1989).

Year	Price Index
1980	66.9
1981	74.8
1982	81.2
1983	85.0
1984	89.2
1985	94.6
1986	97.8
1987	101.9
1988	106.9

Use trend projection to forecast the retail price index for 1989 and 1990.

11. Average attendance figures at home football games for a major university show the following pattern for the past 7 years:

Year	Attendance
1	28,000
2	30,000
3	31,500
4	30,400
5	30,500
6	32,200
7	30,800

Develop the equation for the linear trend component (16.5) for this time series.

12. Automobile sales at B. J. Scott Motors, Inc., provided the following 10-year time series:

Year	Sales
1	400
2	390
3	320
4	340
5	270
6	260
7	300
8	320
9	340
10	370

Plot the time series and comment on the appropriateness of a linear trend. What type of functional form do you believe would be most appropriate for the trend pattern of this time series?

13. The president of a small manufacturing firm has been concerned about the continual growth in manufacturing costs over the past several years. Shown below is a time series of the cost per unit for the firm's leading product over the past 8 years:

Year	Cost/Unit ($)
1	20.00
2	24.50
3	28.20
4	27.50
5	26.60
6	30.00
7	31.00
8	36.00

a. Show a graph of this time series. Does a linear trend appear to exist?
b. Develop the equation for the linear trend component for the above time series. What is the average cost increase that the firm has been realizing per year?

14. Earnings per share for the Walgreen Company for the most recent 10-year period are as follows (*The Value Line*, April 21, 1989).

.84 .73 .94 1.14 1.33 1.53 1.67 1.68 2.10 2.50

a. Use a linear trend projection to forecast this time series for the coming year.
b. What does this time series analysis tell you about the Walgreen Company? Do the historical data indicate the Walgreen Company is a good investment?

15. The gross revenue data for Delta Airlines for the ten-year period from 1979 to 1988 are shown below (*Moody's Transportation News Report*, May 26, 1989). Data are in millions of dollars.

Year	Revenue
1979	2428
1980	2951
1981	3533
1982	3618
1983	3616
1984	4264
1985	4738
1986	4460
1987	5318
1988	6915

a. Develop a linear trend expression for this time series. Comment on what the expression tells about the gross revenue for Delta Airlines for the ten-year period.
b. Provide the 1989 and 1990 forecasts for gross revenue.

16.4 ■■■ FORECASTING A TIME SERIES WITH TREND AND SEASONAL COMPONENTS

In the previous section we showed how to forecast a time series that had a trend component. In this section we expand the discussion by showing how to forecast a time series that has both trend and seasonal components. The approach we will take is first to remove the seasonal effect or seasonal component from the time series. This step is referred to as *deseasonalizing* the time series. After deseasonalizing, the time series will have only a trend component. As a result we can use the method described in the previous section to identify the trend component of the time series. Then, using a trend projection calculation, we will be able to forecast the trend component of the time series in future periods. The final step in developing the forecast will be to incorporate the seasonal component by using a seasonal index to adjust the trend projection. In this manner, we will be able to identify the trend and seasonal components and consider both in forecasting the time series.

In addition to a trend component (T) and a seasonal component (S), we will assume that the time series also has an irregular component (I). The irregular component accounts for any random effects in the time series that cannot be explained by the trend and seasonal components. Using T_t, S_t, and I_t to identify the trend, seasonal, and irregular components at time t, we will assume that the actual time series value, denoted by Y_t, can be described by the following *multiplicative time series model*:

$$Y_t = T_t \times S_t \times I_t \tag{16.9}$$

In this model T_t is the trend measured in units of the item being forecast. However, the S_t and I_t components are measured in relative terms, with values above 1.00 indicating effects above the normal or average level. Values below 1.00 indicate below-average levels for each component. In order to illustrate the use of (16.9) to model a time series, suppose that we have a trend projection of 540 units. In addition, suppose that $S_t = 1.10$ shows a seasonal effect 10% above average and $I_t = 0.98$ shows an irregular effect 2% below average. Using these values in (16.9), the time series value would be $Y_t = 540(1.10)(0.98) = 582$.

In this section we will illustrate the use of the multiplicative model with trend, seasonal, and irregular components by working with the quarterly data presented in Table 16.7 and Figure 16.11. These data show the television set sales (in thousands of units) for a particular manufacturer over the past 4 years. We begin by showing how to identify the seasonal component of the time series.

Calculating the Seasonal Indexes

Looking at Figure 16.11 we observe that sales are lowest in the second quarter of each year, followed by higher sales levels in quarters 3 and 4. Thus, we conclude that a seasonal pattern exists for the television-set sales. The computational procedure used to identify each quarter's seasonal influence begins by computing a moving average to isolate the combined seasonal and irregular components, S_t and I_t.

In using moving averages to do this, we use 1 year of data in each calculation. Since we are working with a quarterly series, we will use 4 data values in each moving average. The moving average calculation for the first 4 quarters of the television set sales data is as follows:

$$\text{First Moving Average} = \frac{4.8 + 4.1 + 6.0 + 6.5}{4} = \frac{21.4}{4} = 5.35$$

TABLE 16.7 Quarterly Data for Television Set Sales

Year	Quarter	Sales (1000s)
1	1	4.8
	2	4.1
	3	6.0
	4	6.5
2	1	5.8
	2	5.2
	3	6.8
	4	7.4
3	1	6.0
	2	5.6
	3	7.5
	4	7.8
4	1	6.3
	2	5.9
	3	8.0
	4	8.4

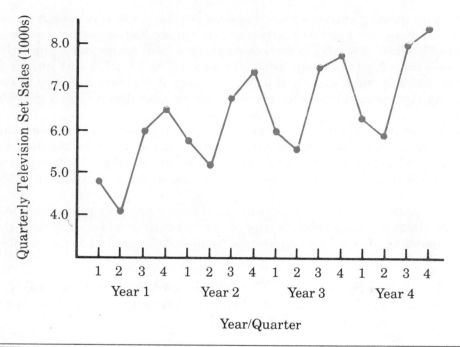

━━ FIGURE 16.11   **Graph of Quarterly Television-Set Sales Time Series**

Note that the moving average calculation for the first 4 quarters yields the average quarterly sales over the first year of the time series. Continuing the moving average calculation, we next add the 5.8 value for the first quarter of year 2 and drop the 4.8 for the first quarter of year 1. Thus the second moving average is

$$\text{Second Moving Average} = \frac{4.1 + 6.0 + 6.5 + 5.8}{4} = \frac{22.4}{4} = 5.6$$

Similarly, the third moving average calculation is $(6.0 + 6.5 + 5.8 + 5.2)/4 = 5.875$.

Before we proceed with the moving average calculations for the entire time series, let us return to the first moving average calculation, which resulted in a value of 5.35. The 5.35 value represents an average quarterly sales volume (across all seasons) for year 1. As we look back at the calculation of the 5.35 value, perhaps it makes sense to associate 5.35 with the "middle" quarter of the moving average group. However, note that some difficulty in identifying the middle quarter is encountered; with 4 quarters in the moving average, there is no middle quarter. The 5.35 value corresponds to the last half of quarter 2 and the first half of quarter 3. Similarly, if we go to the next moving average value of 5.60, the middle corresponds to the last half of quarter 3 and the first half of quarter 4.

Recall that the reason we are computing moving averages is to isolate the combined seasonal and irregular components. However, the moving average values we have computed do not correspond directly to the original quarters of the time series. We can resolve this difficulty by using the midpoints between successive moving average values. For example, since 5.35 corresponds to the first half of quarter 3 and 5.60 corresponds to the last half of quarter 3, we will use $(5.35 + 5.60)/2 = 5.475$ as the moving average value for quarter 3. Similarly, we associate a moving average value of $(5.60 + 5.875)/2 = 5.738$ with quarter 4. What results is called a centered moving average. A complete summary of the moving average calculations for the television set sales data is shown in Table 16.8.

Note that if the number of data points in a moving average calculation is an odd number, the middle point will correspond to one of the periods in the time series. In such cases, we would not have to center the moving average values to correspond to a particular time period, as we did in the calculations in Table 16.8.

Let us pause for a moment to consider what the moving averages in Table 16.8 tell us about this time series. A plot of the actual time series values and the corresponding centered moving average is shown in Figure 16.12. Note particu-

TABLE 16.8 Moving Average Calculations for the Television Set Sales Time Series

Year	Quarter	Sales (1000s)	Four-Quarter Moving Average	Centered Moving Average
1	1	4.8		
	2	4.1		
			5.350	
	3	6.0		5.475
			5.600	
	4	6.5		5.738
			5.875	
2	1	5.8		5.975
			6.075	
	2	5.2		6.188
			6.300	
	3	6.8		6.325
			6.350	
	4	7.4		6.400
			6.450	
3	1	6.0		6.538
			6.625	
	2	5.6		6.675
			6.725	
	3	7.5		6.763
			6.800	
	4	7.8		6.838
			6.875	
4	1	6.3		6.938
			7.000	
	2	5.9		7.075
			7.150	
	3	8.0		
	4	8.4		

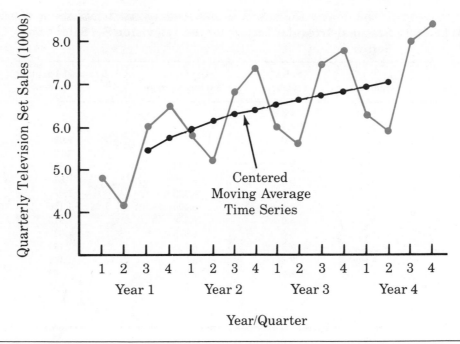

Graph of Quarterly Television-Set Sales Time Series and Centered Moving Average

larly how the centered moving average values tend to "smooth out" the fluctuations in the time series. Since the moving average values were computed for four quarters of data, they do not include the fluctuations due to seasonal influences. Each point in the centered moving average represents what the value of the time series would be if there were no seasonal or irregular influence.

By dividing each time series observation by the corresponding centered moving average value we can identify the seasonal-irregular effect in the time series. For example, the third-quarter of year 1 shows 6.0/5.475 = 1.096 as the combined seasonal-irregular component. The resulting seasonal-irregular values for the entire time series values are summarized in Table 16.9.

Consider the third quarter. The results from years 1, 2, and 3 show third-quarter values of 1.096, 1.075, and 1.109, respectively. Thus in all cases the seasonal-irregular component appears to have an above average influence in the third quarter. Since the year-to-year fluctuations in the seasonal-irregular component can be attributed primarily to the irregular component, we can average the computed values to eliminate the irregular influence and obtain an estimate of the third-quarter seasonal influence:

$$\text{Seasonal Effect of Third Quarter} = \frac{1.096 + 1.075 + 1.109}{3} = 1.09$$

We refer to 1.09 as the *seasonal index* for the third quarter. In Table 16.10 we summarize the calculations involved in computing the seasonal indexes for the television set sales time series. Thus, we see that the seasonal indexes for all four

TABLE 16.9   Seasonal-Irregular Factors for the Television Set Sales Time
Series

Year	Quarter	Sales (1000s)	Centered Moving Average	Seasonal-Irregular Component
1	1	4.8		
	2	4.1		
	3	6.0	5.475	1.096
	4	6.5	5.738	1.133
2	1	5.8	5.975	.971
	2	5.2	6.188	.840
	3	6.8	6.325	1.075
	4	7.4	6.400	1.156
3	1	6.0	6.538	.918
	2	5.6	6.675	.839
	3	7.5	6.763	1.109
	4	7.8	6.838	1.141
4	1	6.3	6.938	.908
	2	5.9	7.075	.834
	3	8.0		
	4	8.4		

quarters are as follows: quarter 1, .93; quarter 2, .84; quarter 3, 1.09; and quarter 4, 1.14.

Interpretation of the values in Table 16.10 provides some observations about the "seasonal" component in television-set sales. The best sales quarter is the fourth quarter, with sales averaging 14% above the average quarterly value. The worst, or slowest, sales quarter is the second quarter, with its seasonal index at .84, showing the sales average 16% below the average quarterly sales. The seasonal component corresponds nicely to the intuitive expectation that television viewing interest and thus television purchase patterns tend to peak in the fourth quarter, with its coming winter season and fewer outdoor activities. The low second-quarter sales reflect the reduced television interest resulting from the spring and presummer activities of the potential customers.

TABLE 16.10   Seasonal Index Calculations for the Television-Set Sales Time Series

Quarter	Seasonal-Irregular Component Values $(S_t I_t)$	Seasonal Index (S_t)
1	.971, .918, .908	.93
2	.840, .839, .834	.84
3	1.096, 1.075, 1.109	1.09
4	1.133, 1.156, 1.141	1.14

One final adjustment is sometimes necessary in obtaining the seasonal indexes. The multiplicative model requires that the average seasonal index equal 1.00; that is, the sum of the four seasonal indexes in Table 16.10 must equal 4.00. This is necessary if the seasonal effects are to even out over the year, as they must. The average of the seasonal indexes in our example is equal to 1.00, and hence this type of adjustment is not necessary. In other cases a slight adjustment may be necessary. The adjustment can be made by simply multiplying each seasonal index by the number of seasons divided by the sum of the unadjusted seasonal indexes. For example, for quarterly data we would multiply each seasonal index by 4/(sum of the unadjusted seasonal indexes). Some of the exercises will require this adjustment in order to obtain the appropriate seasonal indexes.

Deseasonalizing the Time Series

Often the purpose of finding seasonal indexes is to remove the seasonal effects from a time series. This process is referred to as *deseasonalizing* the time series. Economic time series adjusted for seasonal variations (deseasonalized time series) are often reported in publications such as the *Survey of Current Business* and the *Wall Street Journal*. Using the notation of the multiplicative model, we have

$$Y_t = T_t \times S_t \times I_t$$

By dividing each time series observation by the corresponding seasonal index, we have removed the effect of season from the time series. The deseasonalized time series for television set sales is summarized in Table 16.11. A graph of the deseasonalized television set sales time series is shown in Figure 16.13.

TABLE 16.11 Deseasonalized Values for the Television-Set Sales Time Series

Year	Quarter	Sales (1000s) (Y_t)	Seasonal Index (S_t)	Deseasonalized Sales $(Y_t/S_t = T_t I_t)$
1	1	4.8	.93	5.16
	2	4.1	.84	4.88
	3	6.0	1.09	5.50
	4	6.5	1.14	5.70
2	1	5.8	.93	6.24
	2	5.2	.84	6.19
	3	6.8	1.09	6.24
	4	7.4	1.14	6.49
3	1	6.0	.93	6.45
	2	5.6	.84	6.67
	3	7.5	1.09	6.88
	4	7.8	1.14	6.84
4	1	6.3	.93	6.77
	2	5.9	.84	7.02
	3	8.0	1.09	7.34
	4	8.4	1.14	7.37

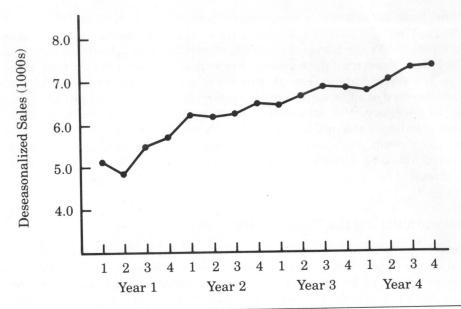

■ FIGURE 16.13 **Deseasonalized Television-Set Sales Time Series**

Using the Deseasonalized Time Series to Identify Trend

Looking at Figure 16.13, we see that while the graph shows some up-and-down movement over the past 16 quarters, the time series seems to have an upward linear trend. To identify this trend, we will use the same procedure we introduced for identifying trend when forecasting with annual data; in this case, since we have deseasonalized the data, quarterly sales values can be used. Thus for a linear trend the estimated sales volume expressed as a function of time can be written

$$T_t = b_0 + b_1 t$$

where

T_t = trend value for television-set sales in period t

b_0 = intercept of the trend line

b_1 = slope of the trend line

As we did before, we will let $t = 1$ for the time of the first observation on the time series data, $t = 2$ for the time of the second observation, and so on. Thus for the deseasonalized television-set sales time series $t = 1$ corresponds to the first deseasonalized quarterly sales value and $t = 16$ corresponds to the most recent deseasonalized quarterly sales value. The formulas for computing the value of b_0

and the value of b_1 are shown again

$$b_1 = \frac{\Sigma\, tY_t - (\Sigma\, t\, \Sigma\, Y_t)/n}{\Sigma\, t^2 - (\Sigma\, t)^2/n}$$

$$b_0 = \overline{Y} - b_1 \overline{t}$$

Note, however, that Y_t now refers to the deseasonalized time series value at time t and not the actual value of the time series. Using the given relationships for b_0 and b_1 and the deseasonalized sales data of Table 16.11, we have the following calculations:

t	Y_t (Deseasonalized)	tY_t	t^2
1	5.16	5.16	1
2	4.88	9.76	4
3	5.50	16.50	9
4	5.70	22.80	16
5	6.24	31.20	25
6	6.19	37.14	36
7	6.24	43.68	49
8	6.49	51.92	64
9	6.45	58.05	81
10	6.67	66.70	100
11	6.88	75.68	121
12	6.84	82.08	144
13	6.77	88.01	169
14	7.02	98.28	196
15	7.34	110.10	225
16	7.37	117.92	256
Totals 136	101.74	914.98	1496

$$\overline{t} = \frac{136}{16} = 8.5$$

$$\overline{Y} = \frac{101.74}{16} = 6.359$$

$$b_1 = \frac{914.98 - (136)(101.74)/16}{1496 - (136)^2/16} = \frac{50.19}{340} = .148$$

$$b_0 = 6.359 - .148\,(8.5) = 5.101$$

Therefore,

$$T_t = 5.101 + .148t$$

is the expression for the linear trend component of the time series.

The slope of .148 indicates that over the past 16 quarters, the firm has experienced an average deseasonalized growth in sales of around 148 sets per quarter. If we assume that the past 16-quarter trend in sales data is a reasonably good indicator of the future, then this equation can be used to project the trend component of the time series for future quarters. For example, substituting $t = 17$ into the equation yields next quarter's trend projection, T_{17}:

$$T_{17} = 5.101 + .148(17) = 7.617$$

Using the trend component only, we would forecast sales of 7617 television sets for the next quarter. In a similar fashion, if we use the trend component only, we would forecast sales of 7765, 7913, and 8061 television sets in quarters 18, 19, and 20, respectively.

Seasonal Adjustments

Now that we have a forecast of sales for each of the next four quarters based upon trend, we must adjust these forecasts to account for the effect of season. For example, since the seasonal index for the first quarter of year 5($t = 17$) is .93, the quarterly forecast can be obtained by multiplying the forecast based upon trend ($T_{17} = 7617$) times the seasonal index (.93). Thus the forecast for the next quarter is 7617(.93) = 7084. Table 16.12 shows the quarterly forecast for quarters 17, 18, 19, and 20. The quarterly forecasts show the high volume fourth quarter with a 9190 unit forecast, while the low volume second quarter has a 6523-unit forecast.

Models Based on Monthly Data

The television-set sales example provided in this section used quarterly data to illustrate the computation of seasonal indexes with relatively few computations. Many businesses use monthly rather than quarterly forecasts. In such cases the procedures introduced in this section can be applied with minor modifications. First, a 12-month moving average replaces the 4-quarter moving average; second, 12 monthly seasonal indexes, rather than 4 quarterly seasonal indexes, will need to be computed. Other than these changes, the computational and forecasting procedures are identical. Exercise 17 at the end of this section asks you to develop monthly seasonal indexes for a situation requiring monthly forecasts.

TABLE 16.12   Quarterly Forecasts for the Television Set Sales Time Series

Year	Quarter	Trend Forecast	Seasonal Index (see Table 16.10)	Quarterly Forecast
5	1	7617	.93	(7617)(.93) = 7084
	2	7765	.84	(7765)(.84) = 6523
	3	7913	1.09	(7913)(1.09) = 8625
	4	8061	1.14	(8061)(1.14) = 9190

Cyclical Component

Mathematically the multiplicative model of (16.9) can be expanded to include a cyclical component as follows:

$$Y_t = T_t \times C_t \times S_t \times I_t \tag{16.10}$$

Just as with the seasonal component, the cyclical component is expressed as a percent of trend. As mentioned in Section 16.1, this component is attributable to multiyear cycles in the time series. It is analogous to the seasonal component, but over a longer period of time. However, because of the length of time involved, it is often difficult to obtain enough relevant data to estimate the cyclical component. Another difficulty is that the length of cycles usually varies. We leave further discussion of the cyclical component to texts on forecasting methods.

▬ EXERCISES

16. The quarterly sales data (number of copies sold) for a college textbook over the past 3 years are as follows:

Quarter	Year 1	Year 2	Year 3
1	1690	1800	1850
2	940	900	1100
3	2625	2900	2930
4	2500	2360	2615

a. Show the four-quarter moving average values for this time series. Plot both the original time series and the moving averages on the same graph.
b. Compute seasonal indexes for the four quarters.
c. When does the textbook publisher experience the largest seasonal index? Does this appear reasonable? Explain.

17. Identify the monthly seasonal indexes for the following 3 years of expenses for a 6-unit apartment house in southern Florida. Use a 12-month moving average calculation.

Month	Year 1	Year 2	Year 3
January	170	180	195
February	180	205	210
March	205	215	230
April	230	245	280
May	240	265	290
June	315	330	390
July	360	400	420
August	290	335	330
September	240	260	290
October	240	270	295
November	230	255	280
December	195	220	250

18. Air pollution control specialists in southern California monitor the amount of ozone, carbon dioxide and nitrogen dioxide in the air on an hourly basis (*Los Angeles Times*, July

21, 1989). The hourly time series data exhibit seasonality with the levels of pollutants showing similar patterns over the hours in the day. On July 15, 16, and 17 the observed levels of nitrogen dioxide in the downtown area for the twelve hours from 6:00 a.m. to 6:00 p.m. were as follows:

July 15:	25	28	35	50	60	60	40	35	30	25	25	20
July 16:	28	30	35	48	60	65	50	40	35	25	20	20
July 17:	35	42	45	70	72	75	60	45	40	25	25	25

a. Identify the hourly seasonal factors for the 12-hour daily readings.

b. Using the seasonal factors from part (a), the data was deseasonalized; the trend equation developed for the deseasonalized data was $T_t = 32.983 + 0.3922\,t$. Using the trend component only, develop forecasts for the twelve hours for July 18.

c. Use the seasonal factors from part (a) to adjust the trend forecasts developed in part (b).

16.5 ■ FORECASTING TIME SERIES USING REGRESSION MODELS

In our discussion of regression analysis in Chapters 13, 14, and 15 we showed how one or more independent variables could be used to predict the value of a single dependent variable. Looking at regression analysis as a forecasting tool, the time series value that we would like to forecast can be viewed as the dependent variable. Thus if we can identify a good set of related independent, or predictor, variables we may be able to develop an estimated regression equation for predicting or forecasting the time series.

The approach we used in Section 16.3 to fit a linear trend line to the bicycle sales time series is a special case of regression analysis. In that example two variables—bicycle sales and time—were shown to be linearly related.* The inherent complexity of most real-world problems necessitates the consideration of more than one variable to predict the variable of interest. The statistical technique known as multiple regression analysis can be used in such situations.

Recall that in order to develop an estimated regression equation we need a sample of observations for the dependent variable and all independent variables. In time series analysis the n periods of time series data provide a sample of n observations on each variable that can be used in the analysis. For a function involving k independent variables we use the following notation:

$$Y_t = \text{actual value of the time series in period } t$$
$$x_{1t} = \text{value of independent variable 1 in period } t$$

---

*In a purely technical sense the number of bicycles sold is not thought of as being related to time; instead, time is used as a surrogate for variables that the number of bicycles sold is actually related to but that are either unknown or too difficult or too costly to measure.

$$x_{2t} = \text{value of independent variable 2 in period } t$$

.

.

.

$$x_{kt} = \text{value of independent variable } k \text{ in period } t$$

The n periods of data necessary to develop the estimated regression equation would appear as follows:

Period	Time Series Value (Y_t)	x_{1t}	x_{2t}	x_{3t}	.	.	.	x_{kt}
				Value of Independent Variables				
1	Y_1	x_{11}	x_{21}	x_{31}	.	.	.	x_{k1}
2	Y_2	x_{12}	x_{22}	x_{32}	.	.	.	k_{k2}
.	.	.	.	.	.	.	.	.
.	.	.	.	.	.	.	.	.
n	Y_n	x_{1n}	x_{2n}	x_{3n}	.	.	.	x_{kn}

As you might imagine, there are a number of possible choices for the independent variables in a forecasting model. One possible choice for an independent variable is simply time. This is the choice we made in Section 16.3 when we estimated the trend of the time series using a linear function of the independent variable time. Letting

$$x_{1t} = t$$

we obtain an estimated regression equation of the form

$$\hat{Y}_t = b_0 + b_1 t$$

where $\hat{Y}_t$ is the estimate of the time series value Y_t and where b_0 and b_1 are the estimated regression coefficients. In a more complex model additional terms could be added corresponding to time raised to other powers. For example, if

$$x_{2t} = t^2$$

and

$$x_{3t} = t^3$$

the estimated regression equation would then become

$$\hat{Y}_t = b_0 + b_1 x_{1t} + b_2 x_{2t} + b_3 x_{3t}$$
$$= b_0 + b_1 t + b_2 t^2 + b_3 t^3$$

Note that this model provides a forecast of a time series with curvilinear characteristics over time.

Other regression-based forecasting models employ a mixture of economic and demographic independent variables. For example, in forecasting the sale of refrigerators we might select independent variables such as the following:

x_{1t} = price in period t

x_{2t} = total industry sales in period $t - 1$

x_{3t} = number of building permits for new houses in period $t - 1$

x_{4t} = population forecast for period t

x_{5t} = advertising budget for period t

According to the usual multiple regression procedure, an estimated regression equation with five independent variables would be used to develop forecasts.

Whether or not a regression approach provides a good forecast depends largely on how well we are able to identify and obtain data for independent variables that are closely related to the time series. Generally, during the development of an estimated regression equation we will want to consider many possible sets of independent variables. Thus part of the regression analysis procedure should focus on the selection of the set of independent variables that provides the best forecasting model.

In the chapter introduction we stated that *causal forecasting models* utilized time series related to the one being forecast in an effort to better explain the cause of a time series behavior. Regression analysis is the tool most often used in developing these causal models. The related time series become the independent variables, and the time series being forecast is the dependent variable.

Another type of regression based forecasting model occurs whenever the independent variables are all previous values of the same time series. For example, if the time series values are denoted by $Y_1, Y_2, \ldots, Y_n$, then with a dependent variable Y_t we might try to find an estimated regression equation relating Y_t to the most recent time series values Y_{t-1}, Y_{t-2}, and so on. With the three most recent periods as independent variables, the estimated regression equation would be

$$\hat{Y}_t = b_0 + b_1 Y_{t-1} + b_2 Y_{t-2} + b_3 Y_{t-3}$$

Regression models where the independent variables are previous values of the time series are referred to as *autoregressive models*.

Finally, another regression based forecasting approach is one that incorporates a mixture of the independent variables previously discussed. For example, we might select a combination of time variables, some economic/demographic variables, and some previous values of the time series variable itself.

16.6 ■ QUALITATIVE APPROACHES TO FORECASTING

In the previous sections we have discussed several types of quantitative forecasting methods. Since each of these techniques requires historical data on the variable of interest, in situations where no historical data are available these techniques cannot be applied. Furthermore, even when historical data are available a significant change in environmental conditions affecting the time series may make the use of past data questionable in predicting future values of the time series. For example, a government imposed gasoline rationing program would cause one to question the validity of a gasoline sales forecast based on past data. Qualitative forecasting techniques offer an alternative in these, and other, cases.

One of the most commonly used qualitative forecasting methods is the *Delphi approach*. This technique, originally developed by a research group at the Rand Corporation, attempts to obtain forecasts through "group consensus." In the usual application of this technique the members of a panel of experts—all of whom are physically separated from and unknown to each other—are asked to respond to a series of questionnaires. The responses from the first questionnaire are tabulated and used to prepare a second questionnaire which contains information and opinions of the whole group. Each respondent is then asked to reconsider and possibly revise his or her previous response in light of the group information that has been provided. This basic process continues until the coordinator feels that some degree of consensus has been reached. Note that the goal of the Delphi approach is not to produce a single answer as output but to produce instead a relatively narrow spread of opinions within which the "majority" of experts concur.

The qualitative procedure referred to as *scenario writing* consists of developing a conceptual scenario of the future based upon a well defined set of assumptions. Thus by starting with a different set of assumptions many different future scenarios can be presented. The job of the decision maker is to decide which scenario is most likely to occur in the future and then to make decisions accordingly.

Subjective or *intuitive qualitative approaches* are based upon the ability of the human mind to process a variety of information that is, in most cases, difficult to quantify. These techniques are often used in group work, wherein a committee or panel seeks to develop new ideas or solve complex problems through a series of "brainstorming sessions." In such sessions individuals are freed from the usual group restrictions of peer pressure and criticism, since any idea or opinion can be presented without regard to its relevancy and, even more importantly, without fear of criticism.

▬ SUMMARY

The purpose of this chapter has been to provide an introduction to the basic methods of time series analysis and forecasting. First, we showed that in order to explain the behavior of a time series, it is often helpful to think of the time series as consisting of four separate components: trend, cyclical, seasonal, and irregular. By isolating these components and measuring their apparent effect, it is possible to forecast future values of the time series.

We discussed how smoothing methods can be used to forecast a time series that exhibits no significant trend, seasonal, or cyclical effect. The moving averages approach consists of computing an average of past data values and then using this average as the forecast for the next period. The exponential smoothing method is a more preferred technique which uses a weighted average of past time series values to compute a forecast.

When the time series exhibits only a long-term trend, we showed how regression analysis could be used to make trend projections. When both trend and seasonal influences are significant, we showed how to isolate the effects of the two factors and prepare better forecasts. Finally, regression analysis was described as a procedure for developing so-called causal forecasting models. A causal forecasting model is one that relates the time series value (dependent variable) to other independent variables that are believed to explain (cause) the time series behavior.

Qualitative forecasting methods were discussed as approaches that could be used when little or no historical data were available. These methods are also considered most appropriate when the past pattern of the time series is not expected to continue into the future.

It is important to realize that time series analysis and forecasting is a major field in its own right. In this chapter we have just scratched the surface of the field of time series and forecasting methodology.

■ GLOSSARY

Time series   A set of observations measured at successive points in time or over successive periods of time.

Forecast   A projection or prediction of future values of a time series.

Trend   The long-run shift or movement in the time series observable over several periods of data.

Cyclical component   The component of the time series model that results in periodic above-trend and below-trend behavior of the time series lasting more than 1 year.

Seasonal component   The component of the time series model that shows a periodic pattern over 1 year or less.

Irregular component   The component of the time series model that reflects the random variation of the actual time series values beyond what can be explained by the trend, cyclical, and seasonal components.

Moving averages   A method of forecasting or smoothing a time series by averaging each successive group of data points. The moving averages method can be used to isolate the seasonal component of the time series.

Mean squared error (MSE)   One approach to measuring the accuracy of a forecasting model. This measure is the average of the sum of the squared difference between the forecast values and the actual time series values.

Weighted moving averages   A method of forecasting or smoothing a time series by computing a weighted average of past data values. The sum of the weights must equal one.

Exponential smoothing   A forecasting technique that uses a weighted average of past time series values in order to arrive at smoothed time series values which can be used as forecasts.

Smoothing constant   A parameter of the exponential-smoothing model which provides the weight given to the most recent time series value in the calculation of the forecast value.

Multiplicative time series model   A model that assumes that the separate components of the time series can be multiplied together to identify the actual time series value. When the 4 components of trend, cyclical, seasonal, and irregular are assumed present we obtain: $Y_t = T_t \times C_t \times S_t \times I_t$. When cyclical is not modeled we obtain: $Y_t = T_t \times S_t \times I_t$.

Deseasonalized time series   A time series that has had the effect of season removed by dividing each original time series observation by the corresponding seasonal index.

Causal forecasting methods   Forecasting methods that relate a time series to other variables that are believed to explain or cause its behavior.

Autoregressive model   A time series model that uses a regression relationship based on past time series values to predict the future time series values.

Delphi approach   A qualitative forecasting method that obtains forecasts through "group consensus."

Scenario writing A qualitative forecasting method which consists of developing a conceptual scenario of the future based upon a well defined set of assumptions.

 ## KEY FORMULAS

Moving Average

$$\text{Moving average} = \frac{\Sigma (\text{Most recent } n \text{ data values})}{n} \tag{16.1}$$

Exponential Smoothing Model

$$F_{t+1} = \alpha Y_t + (1 - \alpha)F_t \tag{16.2}$$

or

$$F_{t+1} = F_t + \alpha(Y_t - F_t) \tag{16.3}$$

Linear Trend Relationship

$$T_t = b_0 + b_1 t \tag{16.5}$$

Multiplicative Model with Seasonal Component

$$Y_t = T_t \times S_t \times I_t \tag{16.9}$$

Multiplicative Model with Seasonal and Cyclical Components

$$Y_t = T_t \times C_t \times S_t \times I_t \tag{16.10}$$

▬ SUPPLEMENTARY EXERCISES

19. Data below show the monthly percentage of all shipments that are received on time during 1988 (*Purchasing*, February 23, 1989).

80   82   84   83   83   84   85   84   82   83   84   83

a. Compare a three-month moving averages forecast with an exponential smoothing forecast using $\alpha = .2$. Which provides the better forecasts?
b. What is the forecast for January 1989?

20. The number of component parts used in a production process the last 10 weeks are shown below:

Week	Parts	Week	Parts
1	200	6	210
2	350	7	280
3	250	8	350
4	360	9	290
5	250	10	320

Using a smoothing constant of .25, develop the exponential smoothing values for this time series. Indicate your forecast for next week.

21. A chain of grocery stores experienced the following weekly demand (cases) for a particular brand of automatic-dishwasher detergent:

Week	Demand	Week	Demand
1	22	6	24
2	18	7	20
3	23	8	19
4	21	9	18
5	17	10	21

Use exponential smoothing with $\alpha = .2$ in order to develop a forecast for week 11.

22. United Dairies, Inc., supplies milk to several independent grocers throughout Dade County, Florida. Management of United Dairies would like to develop a forecast of the number of half-gallons of milk sold per week. Sales data for the past 12 weeks are as follows:

Week	Sales (Units)	Week	Sales (Units)
1	2750	7	3300
2	3100	8	3100
3	3250	9	2950
4	2800	10	3000
5	2900	11	3200
6	3050	12	3150

Using exponential smoothing with $\alpha = .4$, develop a forecast of demand for the 13th week.

23. Ten weeks of data on the Commodity Futures Index are shown below (*Security Traders Handbook*, April 21, 1989):

    7.35   7.40   7.55   7.56   7.60   7.52   7.52   7.70   7.62   7.55

a. Compute the exponential smoothing forecasts using $\alpha = .2$.
b. Compute the exponential smoothing forecasts using $\alpha = .3$.
c. Which exponential smoothing model provides the better forecasts? What is the forecast for the next week?

24. The vacancy rate for office rentals is reported in terms of the percentage of available offices that are not rented. Office vacancy rates for downtown Philadelphia from 1980 to 1987 are shown below (*Business Review*, May–June 1989).

Year	Vacancy Rate
1980	5.9
1981	4.6
1982	6.4
1983	9.5
1984	9.2
1985	9.5
1986	10.8
1987	11.0

a. Develop a linear trend for this time series.
b. Provide forecasts of the vacancy rate for 1988, 1989 and 1990.
c. Should city planners be concerned with the forecasts of office vacancy? What conclusion should be reached and what possible actions should the city planners consider?

25. Canton Supplies, Inc., is a service firm that employs approximately 100 individuals. Because of the necessity of meeting monthly cash obligations, management of Canton Supplies would like to develop a forecast of monthly cash requirements. Because of a recent change in operating policy, only the past 7 months of data were considered to be relevant. Use the historical data shown below to develop a forecast of cash requirements for each of the next 2 months using trend projection.

Month	1	2	3	4	5	6	7
Cash Required ($1000s)	205	212	218	224	230	240	246

26. Data below show the time series of the most recent quarterly capital expenditures in billions of dollars for the 1000 largest manufacturing firms (*Manufacturing Investment Outlook,* Spring 1989).

$$24 \quad 25 \quad 23 \quad 24 \quad 22 \quad 26 \quad 28 \quad 31 \quad 29 \quad 32 \quad 37 \quad 42$$

a. Develop a linear trend expression for the above time series.
b. Show a graph of the time series and the linear trend expression.
c. Using the time series, what appears to be happening to the capital expenditures? What is the forecast one year or four quarters into the future?

27. The Costello Music Company has been in business for 5 years. During this time the sale of electric organs has grown from 12 units in the first year to 76 units in the most recent year. Fred Costello, the firm's owner, would like to develop a forecast of organ sales for the coming year. The historical data are shown below:

Year	1	2	3	4	5
Sales	12	28	34	50	76

a. Show a graph of this time series. Does a linear trend appear to exist?
b. Develop the equation for the linear trend component for the above time series. What is the average increase in sales that the firm has been realizing per year?

28. Hudson Marine has been an authorized dealer for C&D marine radios for the past 7 years. The number of radios sold each year is shown below:

Year	1	2	3	4	5	6	7
Number Sold	35	50	75	90	105	110	130

a. Show a graph of this time series. Does a linear trend appear to exist?
b. Develop the equation for the linear trend component for the above time series.
c. Use the linear trend developed in part (b) and prepare a forecast for annual sales in year 8.

29. Aggregate personal income data by month for 1988 are as follows (*Wall Street Journal,* July 31, 1989).

Jan	3.92	July	4.05
Feb	3.96	Aug	4.08
Mar	4.00	Sept	4.10
Apr	4.01	Oct	4.12
May	4.02	Nov	4.18
June	4.04	Dec	4.22

a. Develop a linear trend expression for the above time series. What was happening to aggregate personal income during 1988?

b. Provide estimates of aggregate personal income for the first six months of 1989.

c. In June 1989, the actual level of aggregate personal income turned out to be 4.41. Comment on the forecasting error based on your 1988 trend projection.

30. Refer to Exercise 28. Suppose that the quarterly sales values for the 7 years of historical data are as follows:

Year	Quarter 1	Quarter 2	Quarter 3	Quarter 4	Total Sales
1	6	15	10	4	35
2	10	18	15	7	50
3	14	26	23	12	75
4	19	28	25	18	90
5	22	34	28	21	105
6	24	36	30	20	110
7	28	40	35	27	130

a. Show the 4-quarter moving average values for this time series. Plot both the original time series and the moving average series on the same graph.

b. Compute the seasonal indexes for the four quarters.

c. When does Hudson Marine experience the largest seasonal effect? Does this seem reasonable? Explain.

31. Consider the Costello Music Company problem presented in Exercise 27. The quarterly sales data are shown below:

Year	Quarter 1	Quarter 2	Quarter 3	Quarter 4	Total Yearly Sales
1	4	2	1	5	12
2	6	4	4	14	28
3	10	3	5	16	34
4	12	9	7	22	50
5	18	10	13	35	76

a. Compute the seasonal indexes for the four quarters.

b. When does Costello Music experience the largest seasonal effect? Does this appear reasonable? Explain.

32. Refer to the Hudson Marine data presented in Exercise 30.

a. Deseasonalize the data and use the deseasonalized time series to identify the trend.

b. Use the results of part (a) to develop a quarterly forecast for next year based upon trend.

c. Use the seasonal indexes developed in Exercise 19 to adjust the forecasts developed in part (b) to account for the effect of season.

33. Consider the Costello Music Company time series presented in Exercise 31.

a. Deseasonalize the data and use the deseasonalized time series to identify the trend.

b. Use the results of part (a) to develop a quarterly forecast for next year based upon trend.

c. Use the seasonal indexes developed in Exercise 31 to adjust the forecasts developed in part (b) to account for the effect of season.

The Cincinnati Gas & Electric Company*

Cincinnati, Ohio

The Cincinnati Gas Light and Coke Company was chartered by the State of Ohio on April 3, 1837. Under this charter the company manufactured gas by distillation of coal and sold it for lighting purposes. During the last quarter of the 19th century the company successfully marketed gas for lighting, heating, and cooking and as fuel for gas engines.

In 1901 the Cincinnati Gas Light and Coke Company and the Cincinnati Electric Light Company merged to form the Cincinnati Gas & Electric Company (CG&E). This new company was able to shift from manufactured gas to natural gas and adopt the rapidly emerging technologies in generating and distributing electricity. CG&E operated as a subsidiary of the Columbia Gas Electric Company from 1909 until 1944.

Today CG&E is a privately owned public utility serving approximately 370,000 gas customers and 600,000 electric customers. The company's service area covers approximately 3,000 square miles in and around the Greater Cincinnati area. In 1981 the Company's revenues exceeded 1 billion dollars and its assets totaled approximately 2.5 billion dollars.

FORECASTING AT CG&E

As in any modern company, forecasting at CG&E is an integral part of operating and managing the business. Depending upon the decision to be made, the forecasting techniques used range from judgment and graphical trend projections to sophisticated multiple regression models.

Forecasting in the utility industry offers some unique perspectives as compared to other industries. Since there are no finished-goods or in-process inventories of electricity, this product must be generated to meet the instantaneous requirements of the customers. Electrical shortages are not just lost sales, but "brownouts" or "blackouts." This situation places an unusual burden on the utility forecaster. On the positive side, the demand for energy and the sale of energy is more predictable than for many other products. Also, unlike the situation in a multiproduct firm, a great amount of forecasting effort and expertise can be concentrated on the two products: gas and electricity.

FORECASTING ELECTRIC ENERGY AND PEAK LOADS

The two types of forecasts discussed in this section are the long range forecasts of electric peak load and electric energy. The largest observed electric demand for any given period,

*The authors are indebted to Dr. Richard Evans, The Cincinnati Gas & Electric Company, Cincinnati, Ohio for providing this application.

such as an hour, a day, a month, or a year, is defined as the peak load. The cumulative amount of energy generated and used over the period of an hour is referred to as electric energy.

Until the mid 1970s the seasonal pattern of both electric energy and electric peak load were very regular; the time series for both of these exhibited a fairly steady exponential growth. Business cycles had little noticeable effect on either. Perhaps the most serious shift in the behavior of these time series came from the increasing installation of air conditioning units in the Greater Cincinnati area. This fact caused an accelerated growth in the trend component and also in the relative magnitude of the summer peaks. Nevertheless, the two time series were very regular and generally quite predictable.

Trend projection was the most popular method used to forecast electric energy and electric peak load. The forecast accuracy was quite acceptable and even enviable when compared to forecast errors experienced in other industries.

■ A NEW ERA IN FORECASTING

In the mid-1970s a variety of actions by the government, the off-and-on energy shortages, and price signals to the consumer began to affect the consumption of electric energy. As a result the behavior of the peak load and electric energy time series became more and more unpredictable. Hence a simple trend projection forecasting model was no longer adequate. As a result a special forecasting model—referred to as an econometric model—was developed by CG&E to better account for the behavior of these time series.

The purpose of the econometric model is to forecast the annual energy consumption by residential, commercial, and industrial classes of service. These forecasts are then used to develop forecasts of summer and winter peak loads. First, energy consumption in the industrial and commercial classes is forecast. For an assumed level of economic activity, the projection of electric energy is made along with a forecast of employment in the area. The employment forecast is converted to a forecast of adult population through the use of unemployment rates and labor force participation rates. Household forecasts are then developed through the use of demographic statistics on the average number of persons per household. The resulting forecast of households is used as an indicator of residential customers.

At this point a comparison is made with the demographic projections for the area

Cincinnati Gas & Electric Company linesman works on electric high-voltage transmission tower

population. The differences between the residential customers forecast and the population forecast are reconciled to produce the final forecast of residential customers. This forecast becomes the principal independent variable in forecasting residential electric energy.

Summer and winter peak loads are then forecast by applying class peak contribution factors to the energy forecasts. The contributions that each class makes toward the peak are summed to establish the peak forecast.

A number of economic and demographic time series are used in the construction of the above econometric model. Simply speaking, the entire forecasting system is a compilation of several statistically verified multiple regression equations.

■■■ IMPACT AND VALUE OF THE FORECASTS

The forecast of the annual electric peak load guides the timing decisions for constructing future generating units. The financial impact of these decisions is great. For example, the last generating unit built by the company cost nearly 600 million dollars, and the interest rate on a recent first mortgage bond was 16%. At this rate, annual interest costs would be nearly 100 million dollars. Obviously, a timing decision that leads to having the unit available no sooner than necessary is crucial.

The energy forecasts are important in other ways also. For example, purchases of coal and nuclear fuel for the generating units are based on the forecast levels of energy needed. The revenue from the electric operations of the Company is determined from forecasted sales, which in turn enters into the planning of rate changes and external financing. These planning and decision-making processes are among the most important management activities in the company. It is imperative that the decision makers have the best forecast information available to assist them in arriving at these decisions.

CHAPTER 17

Tests of Goodness of Fit and Independence

I n Chapter 11 we introduced the chi-square distribution and illustrated how it could be used in estimation and hypothesis tests about a population variance. In this chapter we introduce two additional hypothesis-testing procedures, both based on the use of the chi-square distribution. As with other hypothesis-testing procedures, these tests compare sample results with those that are expected when the null hypothesis is true. The hypothesis test is based upon how "close" the sample results are to the expected results.

In the following section we introduce a goodness of fit test involving a multinomial population. Later we discuss the test for independence using contingency tables and then show goodness of fit tests for Poisson and normal probability distributions.

17.1 ▆▆▆ GOODNESS OF FIT TEST – A MULTINOMIAL POPULATION

In this section we consider the case where each element of a population is assigned to one and only one of several classes or categories. Such a population is a *multinomial population*. For example, consider the market analysis being conducted by the J. Scott and Associates market research firm. The study involves a market-share evaluation. Over the past year market shares have stabilized, with 30% for company A, 50% for company B, and 20% for company C. Recently company C has developed a "new and improved" product that will replace its current entry in the market. Management of company C has asked J. Scott and Associates to determine if the new product will cause a shift in the market shares of the three competitors.

In this case the population of interest is a multinomial population, since each customer is classified as buying from company A, company B, or company C. Thus we have a multinomial population with three classifications or categories. Let us define the following notation:

$$p_A = \text{market share for company A}$$

$$p_B = \text{market share for company B}$$

$$p_C = \text{market share for company C}$$

Based upon the assumption that company C's new product will not alter the market shares, the null and alternative hypotheses would be stated as follows:

H_0: $p_A = .30$, $p_B = .50$, and $p_C = .20$

H_a: The population proportions are not

        $p_A = .30$, $p_B = .50$, and $p_C = .20$

If the sample results lead to the rejection of H_0, the firm will have evidence that the introduction of the new product has had an impact on the market shares.

Let us assume that the market research firm will use a consumer panel of 200 customers for the study. Each individual will be asked to specify a purchase preference among the three alternatives: company A's product, company B's product, and company C's new product. The 200 purchase preference responses are summarized below:

Company A's Product	Company B's Product	Company C's New Product
48	98	54

We now want to demonstrate a *goodness of fit test* that will determine if the sample of 200 customer purchase preferences is consistent with the null hypothesis. The goodness of fit test is based on a comparison of the sample of ·*observed* results such as those shown above with the *expected* results under the assumption that the null hypothesis is true. Thus the next step is to compute expected purchase preferences for the 200 customers under the assumption that $p_A = .30$, $p_B = .50$, and $p_C = .20$. Doing this provides the expected results:

Company A's Product	Company B's Product	Company C's New Product
200(.30) = 60	200(.50) = 100	200(.20) = 40

Thus we see that the expected frequency for each category is found by multiplying the sample size of 200 by the hypothesized proportion for the category.

The goodness of fit test now focuses on the differences between the observed frequencies and the expected frequencies. Large differences between observed and expected frequencies cast doubt on the assumption that the hypothesized proportions or market shares are correct. Whether the differences between the observed and expected frequencies are "large" or "small" is a question answered with the aid of the following test statistic:

Test Statistic for Goodness of Fit

$$\chi^2 = \sum_{i=1}^{k} \frac{(f_i - e_i)^2}{e_i} \qquad (17.1)$$

where

f_i = observed frequency for category i

e_i = expected frequency for category i based on the assumption that H_0 is true

k = the number of categories

Note: The test statistic has a chi-square distribution with $k - 1$ degrees of freedom provided the expected frequencies are 5 *or more* for all categories.

Let us return to the market share data for the three companies. Since the expected frequencies are all 5 or more, we can proceed with the computation of the chi-square test statistic as follows:

$$\chi^2 = \frac{(48 - 60)^2}{60} + \frac{(98 - 100)^2}{100} + \frac{(54 - 40)^2}{40} = 2.40 + .04 + 4.90 = 7.34$$

Suppose that we test the null hypothesis that the multinomial population has the proportions of $p_A = .30$, $p_B = .50$, and $p_C = .20$ at the $\alpha = .05$ level of significance. Since we will reject the null hypothesis if the differences between the observed and expected frequencies are *large,* we will place a rejection area of .05 in the upper tail of chi-square distribution. Checking the chi-square distribution table (Table 3 of Appendix B), we find that with $k - 1 = 3 - 1 = 2$ degrees of freedom $\chi^2_{.05} = 5.99$. Since $7.34 > 5.99$, we reject H_0. In rejecting H_0 we are concluding that the introduction of the new product by company C will alter the current market share structure. While the goodness of fit test itself permits no further conclusions, we can informally compare the observed and expected frequencies to obtain an idea of how the market share structure has changed.

Considering company C, we find that the observed frequency of 54 is larger than the expected frequency of 40. Since the expected frequency was based on current market shares, the larger observed frequency suggests that the new product will have a positive effect on company C's market share. Comparisons of the observed and expected frequencies for the other two companies indicate that company C's gain in market share will hurt company A more than company B.

As illustrated in the example, the goodness of fit test uses the chi-square distribution to determine if a hypothesized probability distribution for a population provides a good fit. The hypothesis test is based upon differences between observed frequencies in a sample and the expected frequencies based on the assumed population distribution. Let us outline the general steps that can be used to conduct a goodness of fit test for any hypothesized multinomial population distribution:

1. Formulate a null hypothesis indicating a hypothesized multinomial distribution for the population.

2. Use a simple random sample of n items and record the observed frequencies for each of k classes or categories.

3. Based upon the assumption that the null hypothesis is true, determine the probability or proportion associated with each of the classes.

4. Multiply the category proportions in step 3 by the sample size to determine the expected class frequencies.

5. Use the observed and expected frequencies in (17.1) to compute a χ^2 value for the test.

6. Complete the test by using the following rejection rule:

$$\text{Reject } H_0 \text{ if } \chi^2 > \chi_\alpha^2$$

where α is the level of significance for the test.

■ EXERCISES

1. During the first 13 weeks of the television season, the Saturday evening 8:00 P.M. to 9:00 P.M. audience proportions were recorded as ABC, 29%, CBS, 28%, NBC, 25%, and independents, 18%. A sample of 300 homes 2 weeks after a Saturday night schedule revision showed the following viewing audience data: ABC, 95 homes, CBS, 70 homes, NBC, 89 homes, and independents, 46 homes. Test with $\alpha = .05$ to determine if the viewing audience proportions have changed.

2. Where do America's millionaires live? Assume a sample of 300 millionaires showed 64 living in the Northeast, 62 living in the Midwest, 100 living in the South, and 75 living in the West. The sample results are based on data from *Louis Rukeyser's Business Almanac*, 1988. Conduct a hypothesis test for H_0: $p_1 = p_2 = p_3 = p_4 = .25$ with $\alpha = .05$. What is your conclusion about where America's millionaires live?

3. A new container design has been adopted by a manufacturer. Color preferences indicated in a sample of 150 individuals are as follows:

Red	Blue	Green
40	64	46

Test using $\alpha = .10$ to see if the color preferences are different. *Hint:* Formulate the null hypothesis as H_0: $p_1 = p_2 = p_3 = \frac{1}{3}$.

4. Consumer panel preferences for three proposed store displays are as follows:

Display A	Display B	Display C
43	53	39

Use $\alpha = .05$ and test to see if there is a difference in preference among the three display designs.

5. Grade-distribution guidelines for a statistics course at a major university are as follows: 10% A, 30% B, 40% C, 15% D, and 5% F. A sample of 120 statistics grades at the end of a semester showed 18 A's, 30 B's, 40 C's, 22 D's, and 10 F's. Use $\alpha = .05$ and test to see if the actual grades deviate significantly from the grade-distribution guidelines.

17.2 ■■■ TEST OF INDEPENDENCE – CONTINGENCY TABLES

Another important application of the chi-square distribution involves using sample data to test for the independence of two variables. Let us illustrate the test of independence by considering the study conducted by the Alber's Brewery of Tucson, Arizona. Alber's manufactures and distributes three types of beers: a light beer, a regular beer, and a dark beer. In an analysis of the market segments for the three beers, the firm's market research group has raised the question of whether or not preferences for the three beers differ among male and female beer drinkers. If beer preference is independent of the sex of the beer drinker, one advertising campaign will be initiated for all of Alber's beers. However, if beer preference depends upon the sex of the beer drinker, the firm will tailor its promotions toward different target markets.

A test of independence addresses the question of whether or not the beer preference (light, regular, or dark) is independent of the sex of the beer drinker (male, female). The hypotheses for this test of independence are as follows:

H_0: Beer preference is independent of the sex of the beer drinker

H_a: Beer preference is not independent of the sex of the beer drinker

Table 17.1 can be used to describe the situation being studied. By identifying the population to be all male and female beer drinkers, a sample can be selected and each individual asked to state his or her preference for the three Alber's beers. Every individual in the sample will be classified in one of the 6 cells in the table. For example, an individual may be a male preferring regular beer (cell 2), a female preferring light beer (cell 4), a female preferring dark beer (cell 6), and so on. Since we have listed all possible combinations of beer preference and sex, or, in other words, listed all possible contingencies, Table 17.1 is called a *contingency table*. The test of independence makes use of the contingency table format and for this reason is sometimes referred to as a *contingency table test*.

TABLE 17.1 Contingency Table – Beer Preference and Sex of Beer Drinker

	Beer Preference		
Sex	*Light*	*Regular*	*Dark*
Male	(cell 1)	(cell 2)	(cell 3)
Female	(cell 4)	(cell 5)	(cell 6)

TABLE 17.2 **Sample Results of Beer Preferences for Male and Female Beer Drinkers (Observed Frequencies)**

Sex	Beer Preference			Total
	Light	Regular	Dark	
Male	20	40	20	80
Female	30	30	10	70
Total	50	70	30	150

Let us assume that a simple random sample of 150 beer drinkers has been selected. After they have taste tested each beer the individuals in the sample are asked to state their preference or first choice. The contingency table in Table 17.2 summarizes the responses for the study. As we see in the contingency table, the data for the test of independence are collected in terms of counts or frequencies for each cell or category. Thus of the 150 individuals in the sample 20 were men who favored light beer, 40 were men who favored regular beer, 20 were men who favored dark beer, and so on.

Note that the data in Table 17.2 contain the sample or observed frequencies for each of 6 classes or categories. If we can determine the expected frequencies under the assumption of independence between beer preference and sex of the beer drinker, we can use the chi-square distribution, just as we did in the previous section, to determine whether or not there is a significant difference between observed and expected frequencies.

Expected frequencies for the cells of the contingency table are based on the following rationale: First, we assume that the null hypothesis of independence between beer preference and sex of the beer drinker is true. Then, we note that the sample of 150 beer drinkers showed a total of 50 preferring light beer, 70 preferring regular beer, and 30 preferring dark beer. In terms of fractions we conclude that $50/150 = 1/3$ of the beer drinkers prefer light beer, $70/150 = 7/15$ prefer regular beer, and $30/150 = 1/5$ prefer dark beer. If the *independence* assumption is valid, we argue that these same fractions must be applicable to both male and female beer drinkers. Thus under the assumption of independence we would expect the sample of 80 male beer drinkers to show that $(1/3)80 = 26.67$ prefer light beer, $(7/15)80 = 37.33$ prefer regular beer, and $(1/5)80 = 16$ prefer dark beer. Application of these same fractions to the 70 female beer drinkers provides the expected frequencies as shown in Table 17.3.

Let e_{ij} denote the expected frequency for the contingency table category in row i and column j. With this notation let us reconsider the expected frequency calculation for males (row $i = 1$) who prefer regular beer (column $j = 2$)—that is, expected frequency e_{12}. Following our previous argument for the computation of expected frequencies, we showed that

$$e_{12} = (7/15)\ 80 = 37.33$$

TABLE 17.3	Expected Frequencies if Beer Preference is Independent of the Sex of the Beer Drinker

Sex	Beer Preference			Total
	Light	*Regular*	*Dark*	**Total**
Male	26.67	37.33	16.00	80
Female	23.33	32.67	14.00	70
Total	50	70	30	150

Writing this slightly differently, we find

$$e_{12} = (^7/_{15})\,80 = (^{70}/_{150})\,80 = \frac{(80)\,(70)}{150} = 37.33$$

Note that the 80 in the above expression is the total number of males (row 1 total), the 70 is the total number of males and females preferring regular beer (column 2 total), and the 150 is the total sample size. Thus we see that

$$e_{12} = \frac{(\text{Row 1 total})\,(\text{Column 2 total})}{\text{Sample size}}$$

Generalization of the above expression shows that the following formula provides the expected frequencies for a contingency table in the test of independence.

Expected Frequencies for Contingency Tables Under the Assumption of Independence

$$e_{ij} = \frac{(\text{Row } i \text{ Total})\,(\text{Column } j \text{ Total})}{\text{Sample Size}} \tag{17.2}$$

Using the above formula for male beer drinkers who prefer dark beer we find an expected frequency of $e_{13} = (80)(30)/150 = 16.00$, as shown previously in Table 17.3. Use (17.2) to verify that it provides the other expected frequencies shown in Table 17.3.

The test procedure for comparing the observed frequencies of Table 17.2 with the expected frequencies of Table 17.3 is similar to the goodness of fit calculations made in the previous section. Specifically, the χ^2 value based on the observed and expected frequencies is computed as follows:

Test Statistic for Independence

$$\chi^2 = \sum_i \sum_j \frac{(f_{ij} - e_{ij})^2}{e_{ij}} \qquad (17.3)$$

where

f_{ij} = observed frequency for contingency table category in row i and column j

e_{ij} = expected frequency for contingency table category in row i and column j based on the assumption of independence

Note: With n rows and m columns in the contingency table, the test statistic has a chi-square distribution with $(n - 1)(m - 1)$ degrees of freedom provided the expected frequencies are 5 *or more* for all categories.

The double summation in (17.3) is used to indicate that the calculation must be made for all the cells in the contingency table.

By reviewing the expected frequencies in Table 17.3, we see that the expected frequencies are 5 or more for each category. Thus we proceed with the computation of the chi-square test statistic. The resulting value is as follows:

$$\chi^2 = \frac{(20 - 26.67)^2}{26.67} + \frac{(40 - 37.33)^2}{37.33} + \cdot \cdot \cdot + \frac{(10 - 14.00)^2}{14.00}$$

$$= 1.67 + .19 + \cdot \cdot \cdot + 1.14 = 6.13$$

The number of degrees of freedom for the appropriate chi-square distribution is computed by multiplying the *number of rows minus* 1 times the *number of columns minus 1.* With 2 rows and 3 columns, we have $(2 - 1)(3 - 1) = (1)(2) = 2$ degrees of freedom for the test of independence of beer preference and the sex of the beer drinker. With $\alpha = .05$ for the level of significance of the test, Table 3 of Appendix B shows an upper-tail χ^2 value of $\chi^2_{.05} = 5.99$. Note here that again we are using the upper-tail value because we will reject the null hypothesis only if the differences between observed and expected frequencies provide a large χ^2 value. In our example $\chi^2 = 6.13$ is greater than the critical value of $\chi^2_{.05} = 5.99$. Thus we reject the null hypothesis of independence and conclude that the preference for the beers is not independent of the sex of the beer drinkers.

Although the test for independence allows only the above conclusion, again we can informally compare the observed and expected frequencies in order to obtain an idea of how the dependence between the beer preference and sex of the beer drinker comes about. Refer to Tables 17.2 and 17.3. We see that male beer drinkers have higher observed than expected frequencies for both regular and dark beers, while female beer drinkers have a higher observed than expected frequency for only the light beer. These observations give us an insight into the beer preference differences between the male and female beer drinkers.

■ EXERCISES

6. The number of units sold by three salespersons over a 3-month period are as follows:

	Product		
Salesperson	A	B	C
Troutman	14	12	4
Kempton	21	16	8
McChristian	15	5	10

Use $\alpha = .05$ and test for the independence of salesperson and type of product. What is your conclusion?

7. Starting positions for business and engineering gradutes are classified by industry as shown below:

	Industry			
Degree Major	Oil	Chemical	Electrical	Computer
Business	30	15	15	40
Engineering	30	30	20	20

Use $\alpha = .01$ and test for independence of degree major and industry type.

8. The *GMAT Occasional Papers* (March 1988) provided data on the primary reason for application to an MBA program by full-time and part-time students. Do the data suggest full-time and part-time students differ in the reason for application to MBA programs? Explain. Use $\alpha = .01$.

	Primary Reason for Application		
Student Status	Program Quality	Convenience/Cost	Other
Full-time	421	393	76
Part-time	400	593	46

9. A sport preference poll shows the following data for men and women:

	Favorite Sport		
Sex	Baseball	Basketball	Football
Men	19	15	24
Women	16	18	16

Use $\alpha = .05$ and test for similar sport preferences by men and women. What is your conclusion?

10. Three suppliers provide the following data on defective parts:

	Part Quality		
Supplier	Good	Minor Defect	Major Defect
A	90	3	7
B	170	18	7
C	135	6	9

Use $\alpha = .05$ and test for independence between supplier and part quality. What does the result of your analysis tell the purchasing department?

11. A study of educational levels of voters and their political party affiliations showed the following results:

| | Party Affiliation | | |
Educational Level	Democratic	Republican	Independent
Did not complete high school	40	20	10
High school degree	30	35	15
College degree	30	45	25

Use $\alpha = .01$ and test if party affiliation is independent of the educational level of the voters.

12. *Personnel Administrator* (January 1984) provided the following data as an example of selection among 40 male and 40 female applicants for 12 open positions.

Applicant	Selected	Not Selected	Total
Male	7	33	40
Female	5	35	40
Total	12	68	80

a. The chi-square test of independence was suggested as a way of determining if the decision to hire 7 males and 5 females should be interpreted as having a selection bias in favor of males. Conduct the test of independence using $\alpha = .10$. What is your conclusion?

b. Using the same test, would the decision to hire 8 males and 4 females suggest concern for a selection bias?

c. How many males could be hired for the 12 open positions before the procedure would suggest concern for a selection bias?

17.3 ■ GOODNESS OF FIT TEST – POISSON AND NORMAL DISTRIBUTIONS

In Section 17.1 we introduced the goodness of fit test involving a multinomial population. In general the goodness of fit test can be used with any hypothesized probability distribution. In this section we illustrate the goodness of fit test procedure for cases where the population is hypothesized to have a Poisson or a normal distribution. As we shall see, the goodness of fit test and the use of the chi-square distribution for these tests follow the same general procedure used for the goodness of fit test in Section 17.1.

A Poisson Distribution

Let us illustrate the goodness of fit test for the case where the hypothesized population distribution is a Poisson distribution. As an example consider the arrivals of customers at Dubek's Food Market in Tallahassee, Florida. Dubek's management makes staffing decisions involving the number of clerks, the number of checkout lanes, and so on based on the anticipated arrivals of customers at the store. Because of some recent staffing problems Dubek's management has asked a local consulting firm to assist with the scheduling of clerks for the checkout lanes. The general objective of the consulting firm's work is to provide enough clerks to achieve a good level of service while maintaining a reasonable total payroll cost.

After reviewing the checkout lane operation the consulting firm makes a recommendation for a clerk-scheduling procedure. The procedure, based on a mathematical analysis of waiting lines, is applicable only in situations where the number of customers arriving during a specified time period follows the Poisson probability distribution. Thus, before the scheduling process is implemented, data on customer arrivals must be collected and a statistical test conducted to see if an assumption of a Poisson distribution for arrivals appears reasonable.

We define the arrivals at the store in terms of the *number of customers* entering the store during 5-minute intervals. Thus the following null and alternative hypotheses are appropriate for the Dubek study:

H_0: The number of customers entering the store during 5-minute intervals has a Poisson probability distribution

H_a: The number of customers entering the store during 5-minute intervals does not have a Poisson distribution

If a sample of customer arrivals indicates H_0 cannot be rejected, Dubek will proceed with the implementation of the consulting firm's scheduling procedure. However, if the sample leads to the rejection of H_0, the assumption of the Poisson distribution for the arrivals cannot be made and other scheduling procedures will have to be considered.

In order to test the assumption of a Poisson distribution for the number of arrivals during weekday morning hours, a store employee randomly selects a sample of 128 5-minute intervals during weekday mornings over a 3-week period. For each 5-minute interval in the sample, the store employee records the number of customer arrivals. In summarizing the data, the employee determines the number of 5-minute intervals having no arrivals, the number of 5-minute intervals having one arrival, the number of 5-minute intervals having two arrivals, and so on. These data are summarized in Table 17.4. The number of customers arriving are the category descriptions, with the observed frequencies recorded in the column showing the number of 5-minute intervals having the corresponding number of customer arrivals.

Table 17.4 provides the observed frequencies for the ten categories. We now want to use a goodness of fit test to determine whether or not the sample of 128 time periods supports the hypothesized Poisson probability distribution. In order to conduct the goodness of fit test, we need to consider the expected frequency for each of the 10 categories under the assumption that the Poisson distribution of arrivals is true. That is, we need to compute the expected number of time periods that no customers, 1 customer, 2 customers, and so on would arrive if, in fact, the customer arrivals have a Poisson distribution.

TABLE 17.4   Observed Frequencies of Dubek Customer Arrivals for a Sample of 128 5-Minute Time Periods

Category Description: Number of Customers Arriving	Observed Frequency
0	2
1	8
2	10
3	12
4	18
5	22
6	22
7	16
8	12
9	6
Total	128

The Poisson probability function, which was first introduced in Chapter 5, is as follows:

$$f(x) = \frac{\mu^x e^{-\mu}}{x!} \tag{17.4}$$

In this function, μ represents the mean or expected number of customers arriving per 5-minute period, and x is the random variable indicating the number of customers arriving during a 5-minute period. In this case x may be equal to 0, 1, 2, and so on. Finally, $f(x)$ is the probability that x customers will arrive in a 5-minute interval.

Before we use (17.4) to compute Poisson probabilities, we must obtain an estimate of μ, the mean number of customer arrivals during a 5-minute time period. The sample mean for the data in Table 17.4 provides this estimate. With no customers arriving in two 5-minute time periods, one customer arriving in eight 5-minute time periods, and so on, the total number of customers that arrived during the sample of 128 5-minute time periods is given by $0(2) + 1(8) + 2(10) + \cdots + 9(6) = 640$. The 640 customer arrivals over the sample of 128 periods provides a mean arrival rate of $\mu = 640/128 = 5$ customers per 5-minute period. With this value for the mean of the Poisson probability distribution an estimate of the Poisson probability function for Dubek's Food Market is

$$f(x) = \frac{5^x e^{-5}}{x!} \tag{17.5}$$

Assume that the Poisson probability distribution is appropriate for the Dubek customer arrivals. The above probability function then can be evaluated for different values of x in order to determine the probability associated with each

TABLE 17.5 **Expected Frequencies of Dubek Customer Arrivals, Assuming a Poisson Probability Distribution with $\mu = 5$**

Category Description: Number of Customers Arriving (x)	Poisson Probability, f (x)	Expected Number of 5-Minute Time Periods with x Arrivals, 128 f(x)
0	.0067	.8576
1	.0337	4.3136
2	.0842	10.7776
3	.1404	17.9712
4	.1755	22.4640
5	.1755	22.4640
6	.1462	18.7136
7	.1044	13.3632
8	.0653	8.3584
9	.0363	4.6464
10 or more	.0318	4.0704
	Total	128.0000

category of arrivals. These probabilities, which can also be found in Table 7 of Appendix B, are shown in Table 17.5. For example, the probability of 0 customers arriving during a 5-minute interval is $f(0) = .0067$, the probability of 1 customer arriving during a 5-minute interval is $f(1) = .0337$, and so on. As we saw in Section 17.1, the expected frequencies for the categories are found by multiplying the probabilities by the sample size. For example, the expected number of periods with 0 arrivals is given by $(.0067)(128) = .8576$, the expected number of periods with 1 arrival is given by $(.0337)(128) = 4.3136$, and so on.

Before we make the usual chi-square calculations to compare the observed and expected frequencies, note that in Table 17.5, four of the categories have an expected frequency less than 5. This condition violates the requirements necessary for use of the chi-square distribution. However, expected category frequencies less than 5 cause no difficulty, since categories can be combined to satisfy the "at least 5" expected frequency requirement. In particular, we will combine 0 and 1 into a single category and then combine 9 with "10 or more" into another single category. Thus the rule of a minimum expected frequency of 5 in each category is satisfied.

Let us combine the categories for the observed results in Table 17.4 accordingly. Now, we can compare the observed frequencies with the expected frequencies. This comparison is summarized in Table 17.6.

As in Section 17.1 the goodness of fit test focuses on the differences between observed and expected frequencies, $f_i - e_i$. Obviously, large differences cast doubt on the assumption that the customer arrivals have a Poisson distribution.

Using the observed and expected frequencies shown in Table 17.6, we again compute the chi-square test statistic:

$$\chi^2 = \sum_{i=1}^{k} \frac{(f_i - e_i)^2}{e_i}$$

TABLE 17.6 Comparison of the Observed and Expected Frequencies for the Dubek Customer Arrivals

Category Description: Number of Customers Arriving	Observed Frequency (f_i)	Expected Frequency (e_i)	Difference ($f_i - e_i$)
0 or 1	10	5.1712	4.8288
2	10	10.7776	−.7776
3	12	17.9712	−5.9712
4	18	22.4640	−4.4640
5	22	22.4640	−.4640
6	22	18.7136	3.2864
7	16	13.3632	2.6368
8	12	8.3584	3.6416
9 or more	6	8.7168	−2.7168
Total	128	128.0000	

Doing so, we have

$$\chi^2 = \frac{(4.8288)^2}{5.1712} + \frac{(-.7776)^2}{10.7776} + \cdots + \frac{(-2.7168)^2}{8.7168}$$

$$= 4.5091 + .0561 + \cdots + .8468 = 10.98$$

We need to determine the appropriate degrees of freedom associated with this goodness of fit test. In general, the chi-square distribution for a goodness of fit test has $k - p - 1$ degrees of freedom, where k is the number of categories and p is the number of population parameters estimated from the sample data. For the Poisson distribution goodness of fit test we are considering, Table 17.6 shows $k = 9$ categories. Since the sample data were used to estimate the mean of the Poisson distribution, $p = 1$. Thus $k - p - 1 = 9 - 1 - 1 = 7$ degrees of freedom exist for the chi-square distribution of interest in the Dubek study.

Suppose that we test the hypothesis that the probability distribution for the customer arrivals is Poisson with a .05 level of significance ($\alpha = .05$). This is an upper-tail test, since we will reject the null hypothesis only if the difference between the observed and expected frequencies—and thus the χ^2 value—becomes large. We check the χ^2 values in Table 3 of Appendix B and find that with 7 degrees of freedom $\chi^2_{.05} = 14.07$. As with similar one-tailed tests, we will reject the null hypothesis only if the computed value of χ^2 exceeds the value of χ^2_α.

Checking the previous calculations for the Dubek study, we find a computed $\chi^2 = 10.98$. Since this value is less than the critical value of 14.07, we cannot reject the null hypothesis. Thus for the Dubek study the assumption of a Poisson probability distribution for weekday morning customer arrivals cannot be rejected. With this statistical finding, Dubek will proceed with the consulting firm's scheduling procedure for weekday mornings.

A Normal Distribution

The goodness of fit test for a normal probability distribution is also based on the use of the chi-square distribution. It is very similar to the procedure we have just discussed for the Poisson distribution. In particular, observed frequencies for several categories of sample data will be compared to expected frequencies under the assumption that the population has a normal distribution. Since the normal probability distribution is continuous, we must modify the way that the categories are defined and how the expected frequencies are computed. Let us demonstrate the goodness of fit test for a normal probability distribution by considering the job applicant test data for Chemline, Inc., shown in Table 17.7.

Chemline hires approximately 400 new employees annually for its four plants located throughout the United States. Standardized tests are given by the personnel department, with performance on the test being a major factor in the employee-hiring decision. With numerous tests being given annually, the personnel director has asked if a normal distribution could be applied to the population of test scores. If such a distribution can be applied, use of the distribution would be most helpful in evaluating specific test scores. That is, scores in the upper 20%, lower 40%, and so on, could quickly be identified. Thus we would like to test the null hypothesis that the population of aptitude test scores follows a normal probability distribution.

Let us first use the data in Table 17.7 to develop estimates of the mean and standard deviation of the normal distribution that will be considered in the null hypothesis. We use the sample mean $\bar{x}$ and the sample standard deviation s as point estimators of the mean and standard deviation of the normal distribution. Calculations are as follows:

$$\bar{x} = \frac{\Sigma\, x_i}{n} = \frac{3421}{50} = 68.42$$

$$s = \sqrt{\frac{\Sigma\, (x_i - \bar{x})^2}{49}} = 10.41$$

TABLE 17.7   **Chemline Employee Aptitude Test Scores for 50 Randomly Chosen Job Applicants**

71	66	61	65	54
93	60	86	70	70
73	73	55	63	56
62	76	54	82	79
76	68	53	58	85
80	56	61	61	64
65	62	90	69	76
79	77	54	64	74
65	65	61	56	63
80	56	71	79	84

Using these values, we state the following hypotheses about the distribution of the job applicant test scores:

H_0: The population of test scores has a normal distribution with mean 68.42 and standard deviation 10.41

H_a: The population of test scores does not have a normal distribution with mean 68.42 and standard deviation 10.41

The hypothesized normal distribution is shown in Figure 17.1.

Now let us consider a way of defining the categories for a goodness of fit test involving a normal distribution. For the discrete probability distribution in the Poisson distribution test, the categories were readily defined in terms of the number of customers arriving, such as 0, 1, 2, and so on. However, with a continuous probability distribution, such as the normal, we will have to come up with a different procedure for defining the categories. Actually, we will need to define the categories in terms of *intervals* of test scores.

Recall the rule of thumb for an expected frequency of at least 5 in each interval or category. We will have to define the categories of test scores such that the expected frequencies will be at least 5 for each category. With a sample size of 50, one way of doing this is to divide the normal distribution into ten equal-probability intervals (see Figure 17.2). With a sample size of 50, we would expect five outcomes in each interval or category, and the rule of thumb for expected frequencies would be satisfied. This is the procedure we will follow for determining the number of categories for the goodness of fit test whenever a continuous probability distribution is being considered. Namely, we will break the assumed population distribution into equal-probability intervals such that at least five observations are expected in each.

Let us look more closely at the procedure used to calculate the category boundaries. Since the normal probability distribution is being assumed, the standard normal probability tables can be used to determine these boundaries.

■ FIGURE 17.1 **Hypothesized Normal Distribution of Test Scores for the Chemline Job Applicants**

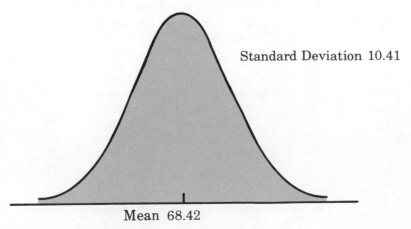

Standard Deviation 10.41

Mean 68.42

First consider the test score cutting off the lowest 10% of the test scores. From Table 1 of Appendix B we find that the z value for this test score is approximately -1.28. Therefore, the test score of $x = 68.42 - 1.28(10.41) = 55.10$ provides this cutoff value for the lowest 10% of the scores. For the lowest 20%, we find $z = -.84$, and thus $x = 68.42 - .84(10.41) = 59.68$. Working through the normal distribution in a similar manner provides the following test score values:

Lower 10%:	$68.42 - 1.28(10.41)$	$= 55.10$
Lower 20%:	$68.42 - .84(10.41)$	$= 59.68$
Lower 30%:	$68.42 - .52(10.41)$	$= 63.01$
Lower 40%:	$68.42 - .25(10.41)$	$= 65.82$
Mid-score:	$68.42 + 0(10.41)$	$= 68.42$
Upper 40%:	$68.42 + .25(10.41)$	$= 71.02$
Upper 30%:	$68.42 + .52(10.41)$	$= 73.83$
Upper 20%:	$68.42 + .84(10.41)$	$= 77.16$
Upper 10%:	$68.42 + 1.28(10.41)$	$= 81.74$

These cutoff or interval boundary points have been identified on the graph in Figure 17.2.

With the categories or intervals of test scores now defined and with the known expected frequencies of 5 per category, we can return to the sample data of Table 17.7 and determine the observed frequencies for the categories. Doing so provides the results in Table 17.8. Also note that Table 17.8 contains a column of differences between the observed and expected frequencies.

With the results in Table 17.8, the goodness of fit calculations proceed exactly

---

■ FIGURE 17.2 **Normal Probability Distribution for the Chemline Example with 10 Equal-Probability Intervals**

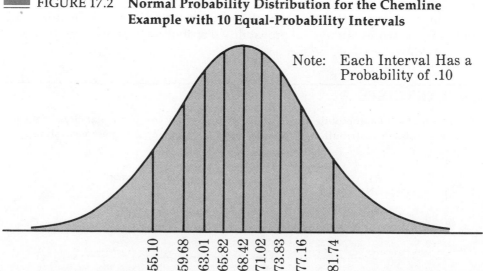

Note: Each Interval Has a Probability of .10

TABLE 17.8   **Observed and Expected Frequencies for Chemline Job-Applicant Test Scores**

Test Score Interval	Observed Frequency (f_i)	Expected Frequency (e_i)	Difference $(f_i - e_i)$
Less than 55.10	5	5	0
55.10 to 59.68	5	5	0
59.68 to 63.01	9	5	4
63.01 to 65.82	6	5	1
65.82 to 68.42	2	5	-3
68.42 to 71.02	5	5	0
71.02 to 73.83	2	5	-3
73.83 to 77.16	5	5	0
77.16 to 81.74	5	5	0
81.74 and Over	6	5	1
Total	50	50	

as before. Namely, we compare the observed and expected results by computing a χ^2 value as follows:

$$\chi^2 = \sum_{i=1}^{k} \frac{(f_i - e_i)^2}{e_i} = \frac{0^2}{5} + \frac{0^2}{5} + \frac{4^2}{5} + \cdots + \frac{1^2}{5} = 7.2$$

To determine whether or not the computed χ^2 value of 7.20 is large enough to reject H_0, we need to refer to the appropriate chi-square probability distribution tables. Using the rule for computing the number of degrees of freedom for the goodness of fit test, we have $k - p - 1 = 10 - 2 - 1 = 7$ degrees of freedom, where there are $k = 10$ categories and $p = 2$ parameters (mean and standard deviation) estimated from the sample data. Using a .10 level of significance for this hypothesis test, we have $\chi^2_{.10} = 12.017$ for the upper-tail rejection region. With $7.20 < 12.017$ we conclude that the null hypothesis cannot be rejected. Thus the hypothesis that the probability distribution for the Chemline job applicant test scores is a normal probability distribution cannot be rejected.

▬ EXERCISES

13. The number of automobile accidents occurring per day in a particular city is believed to have a Poisson distribution. A sample of 80 days during the past year shows the following data:

Number of Accidents	Observed Frequency (days)
0	34
1	25
2	11
3	7
4	3

Do these data support the belief that the number of accidents per day has a Poisson distribution? Use $\alpha = .05$.

14. The number of incoming phone calls occurring at a company switchboard during 1 minute intervals is believed to have a Poisson distribution. Use $\alpha = .10$ and the following data to test the assumption that the incoming phone calls have a Poisson distribution:

Number of Incoming Phone Calls During a 1-Minute Interval	Observed Frequency
0	15
1	31
2	20
3	15
4	13
5	4
6	2
Total	100

15. The weekly demand for a product is believed to be normally distributed. Use a goodness of fit test and the data in the following sample to test this assumption:

18	25	26	27	26	25	20	22	23	25	25	28	22	27	20
19	31	26	27	25	24	21	29	28	22	24	26	25	25	24

Use $\alpha = .10$. The sample mean is 24.5, and the sample standard deviation is 3.

16. Use $\alpha = .01$ and conduct a goodness of fit test to see if the following sample appears to have been selected from a normal distribution:

55	86	94	58	55
95	55	52	69	95
90	65	87	50	56
55	57	98	58	79
92	62	59	88	65

After you complete the goodness of fit calculations, construct a histogram of the data. Does the histogram representation support the conclusion reached with the goodness of fit test? (*Note:* $\bar{x} = 71$ and $s = 17$)

■ SUMMARY

In this chapter we introduced the goodness of fit test and the test of independence, both of which are based on the use of the chi-square distribution. The purpose of the goodness of fit test is to determine whether or not a hypothesized probability distribution can be rejected as a distribution for a particular population of interest. The computations for conducting the goodness of fit test involve comparing observed frequencies from a sample with expected frequencies when the hypothesized probability distribution is assumed true. A chi-square distribution is used to determine if the differences in observed and expected frequencies are sufficient to reject the hypothesized probability distribution. We illustrated the goodness of fit test for assumed multinomial, Poisson, and normal probability distributions.

A test of independence for two variables is a straightforward extension of the methodology employed in the goodness of fit test for a multinomial population. A contingency table is used to determine the observed and expected frequencies. Then a chi-square value is computed. Large chi-square values, caused by large differences between observed and expected frequencies, lead to the rejection of the null hypothesis of independence.

■ GLOSSARY

Goodness of fit test   A statistical test conducted to determine whether or not to reject a hypothesized probability distribution for a population.

Contingency table   A table used to summarize observed and expected frequencies for a test of independence of population characteristics.

■ KEY FORMULAS

Test Statistic for Goodness of Fit

$$\chi^2 = \sum_{i=1}^{k} \frac{(f_i - e_i)^2}{e_i} \tag{17.1}$$

Expected Frequencies for Contingency Tables Under the Assumption of Independence

$$e_{ij} = \frac{(\text{Row } i \text{ Total})(\text{Column } j \text{ Total})}{\text{Sample Size}} \tag{17.2}$$

Test Statistic for Independence

$$\chi^2 = \sum_i \sum_j \frac{(f_{ij} - e_{ij})^2}{e_{ij}} \tag{17.3}$$

■ SUPPLEMENTARY EXERCISES

17. In setting sales quotas, the marketing manager makes the assumption that order potentials are the same in each of four sales territories. A sample of 200 sales shows the following number of orders from each region:

Sales Territories

I	II	III	IV
60	45	59	36

Should the manager's assumption be rejected? Use $\alpha = .05$.

18. An October 1988 poll sponsored by the *Cincinnati Post* used a random sample of 616 registered voters throughout the state of Ohio to determine how Ohioans rate their local schools. The following rating categories and results were reported:

School Rating Category	Frequency
Excellent	135
Good	234
Fair	129
Poor	98

Assume Ohio school administrators had hypothesized rating percentages of 20% excellent, 40% good, 25% fair, and 15% poor for the population of Ohio registered voters. Use the goodness fit test and the survey data to determine if the administrator's hypothesis should be rejected. Use $a = .05$.

19. At Ontario University entering freshmen have historically selected the following colleges:

College	Percentage
Business	15%
Education	20%
Engineering	30%
Liberal Arts	25%
Science	10%

Data obtained for the most recent class show that 68 students selected business, 110 selected education, 140 selected engineering, 122 selected liberal arts, and 48 selected science. Use $\alpha = .10$ to see if the historical percentages have changed.

20. A regional transit authority was concerned about the number of riders on one of its bus routes. In setting up the route, the assumption was that the number of riders was uniformly distributed from Monday through Friday. The following historical data were obtained:

Day	Number of Riders
Monday	13
Tuesday	16
Wednesday	28
Thursday	17
Friday	16

Test using $\alpha = .05$ to determine if the transit authority's assumption appears to be incorrect.

21. A sample of parts provided the following contingency table data concerning part quality and production shift:

Shift	Number Good	Number Defective
First	368	32
Second	285	15
Third	176	24

Use $\alpha = .05$ and test the hypothesis that part quality is independent of the production shift. What is your conclusion?

22. The Graduate Management Admission Council (GMAC) sponsored a survey of MBA students to learn about characteristics of the population of students interested in graduate education in business administration. The following table was published in the *GMAT Occasional Papers* (March 1988). Do the data suggest males and females differ in the reason for application to MBA programs? Explain. Use $\alpha = .05$.

	Primary Reason for Application		
	Program Quality	*Convenience/Cost*	*Other*
Male	519	599	86
Female	298	390	36

23. A lending institution shows the following data regarding loan approvals by four different loan officers:

Loan Officer	Loan Approval Decision	
	Approved	Rejected
Miller	24	16
McMahon	17	13
Games	35	15
Runk	11	9

Use $\alpha = .05$ and test to determine if the loan approval decision is independent of the loan officer reviewing the loan application.

24. An analysis of attendance records and performance on the final examination was made for a first-year mathematics course. The following results were obtained.

Number of Classed Missed	Grade on Final		
	80 or Above	70s	Below 70
None	18	11	6
1–5	14	12	6
More than 5	3	9	20

Use $\alpha = .05$ and test for independence between number of classes missed and the grade on the final examination. What is your conclusion?

25. As part of the standard course evaluation, students are asked to rate the course as either poor, good, or excellent. The course evaluation form also asks students to indicate whether or not the course taken was a required part of their academic program or was taken as an elective. The dean of the college is interested in determining if the rating of the course is independent of the reason for taking the course. The following results were obtained.

Reason for Taking the Course	Rating		
	Poor	Good	Excellent
Required	16	38	16
Elective	4	10	16

How would you respond to the dean? Use $\alpha = .01$.

26. The following data were collected on the number of emergency ambulance calls for an urban county and a rural county in Virginia (*Journal of The Operational Research Society*, November 1986):

County	Day of Week							Total
	Sun	Mon	Tue	Wed	Thur	Fri	Sat	
Urban	61	48	50	55	63	73	43	393
Rural	7	9	16	13	9	14	10	78
Total	68	57	66	68	72	87	53	471

Conduct a test for independence using $\alpha = .05$. What is your conclusion?

27. A random sample of final examination grades for a college course is shown below:

55 85 72 99 48 71 88 70 59 98 80 74 93 85 74 82 90 71 83 60
95 77 84 73 63 72 95 79 51 85 76 81 78 65 75 87 86 70 80 64

Use $\alpha = .05$ and test to determine if a normal distribution should be rejected as being population distribution of grades.

28. A salesperson makes four calls per day. A sample of 100 days shows the following frequencies of sales volumes:

Number of Sales	Observed Frequency (days)
0	30
1	32
2	25
3	10
4	3
	100

Assume the population is a binomial distribution with a probability of purchase equal to $p = .30$. Recall that in Chapter 5 the binomial probabilities were given by

$$f(x) = \frac{n!}{x!(n-x)!}\, p^x(1-p)^{n-x}$$

For this exercise $n = 4$, $p = .30$, and $x = 0, 1, 2, 3,$ and 4.

a. Compute the expected frequencies for $x = 0, 1, 2, 3$ and 4. Combine categories if necessary to satisfy the requirement that the expected frequencies are 5 or more for all categories.

b. Should the assumption of a binomial distribution be rejected? Use $\alpha = .05$.

■■■ COMPUTER EXERCISE

The computer exercise in Chapter 2 described a study conducted by Consolidated Foods, Inc. The company had taken a sample of 100 customers in order to learn about how customers were paying for their food purchases. The data collected for each customer included how much was spent on the purchase and how the customer paid for the purchase. The alternative payment methods included cash, an approved check, or a credit card. In addition, data were also collected on the sex of the customer. The data obtained for the 100 customers are shown below.

Amount Spent ($)	Sex	Method of Payment	Amount Spent ($)	Sex	Method of Payment
84.12	Male	Check	86.34	Female	Check
34.66	Male	Credit card	20.23	Female	Credit card
37.27	Female	Credit card	108.70	Female	Check
38.82	Female	Credit card	45.36	Female	Credit card
46.50	Female	Credit card	83.31	Male	Check
99.67	Female	Check	64.45	Male	Credit card
70.18	Female	Check	54.33	Female	Credit card
99.21	Male	Check	16.78	Female	Cash
138.42	Female	Check	115.96	Male	Check
93.68	Female	Check	95.83	Female	Check
120.89	Female	Check	19.76	Female	Cash
10.14	Female	Cash	35.37	Male	Cash
74.51	Male	Check	111.98	Female	Check
17.91	Male	Check	103.95	Female	Check
49.59	Male	Check	90.40	Male	Credit card

Amount Spent ($)	Sex	Method of Payment	Amount Spent ($)	Sex	Method of Payment
4.74	Male	Cash	6.68	Male	Cash
48.14	Male	Cash	32.09	Male	Credit card
65.67	Male	Credit card	79.70	Male	Credit card
89.66	Female	Check	96.08	Male	Credit card
96.40	Female	Check	20.60	Male	Cash
54.16	Female	Credit card	78.81	Female	Check
79.55	Female	Check	123.62	Female	Check
67.95	Female	Check	125.01	Female	Check
30.69	Male	Cash	41.58	Male	Credit card
151.89	Female	Check	36.73	Male	Credit card
130.41	Female	Check	52.07	Female	Credit card
98.80	Female	Check	19.78	Male	Cash
23.59	Female	Cash	66.44	Female	Check
104.67	Female	Check	5.08	Male	Cash
90.04	Female	Check	50.15	Male	Credit card
77.62	Female	Check	114.42	Female	Check
36.01	Male	Cash	97.26	Male	Credit card
88.17	Female	Check	22.75	Male	Cash
66.76	Female	Credit card	53.63	Female	Credit card
23.50	Male	Cash	132.31	Female	Check
127.34	Female	Check	105.54	Male	Check
26.02	Male	Cash	66.09	Male	Check
79.77	Male	Check	62.24	Female	Check
29.35	Male	Check	97.93	Female	Check
71.31	Female	Credit card	10.57	Female	Cash
43.57	Female	Credit card	51.21	Male	Credit card
76.18	Female	Credit card	90.17	Female	Check
59.38	Male	Credit card	24.08	Male	Credit card
72.99	Male	Credit card	42.72	Male	Cash
19.24	Male	Cash	97.72	Female	Check
80.20	Female	Check	112.67	Female	Check
55.79	Female	Cash	14.30	Female	Cash
134.27	Female	Check	28.76	Male	Credit card
64.68	Female	Credit card	81.85	Female	Check
75.54	Female	Check	56.84	Female	Credit card

■ QUESTIONS

1. Develop a contingency table showing the sex of the customer and the method of payment.

2. Is method of payment independent of the sex of the customer? Explain.

APPLICATION

United Way*
Rochester, New York

The United Way of Greater Rochester is a nonprofit fund-raising and social planning organization dedicated to improving the quality of life of residents in the 6 counties it serves. The annual United Way/Red Cross campaign, conducted each spring, helps support more than 140 human service agencies. These agencies meet a wide variety of human needs—physical, mental, and social—and serve people of all ages, backgrounds, and economic means.

The United Way relies on thousands of dedicated volunteers, many of whom serve year-round. Because of this volunteer involvement, the United Way is able to hold its operating costs to less than nine cents of every dollar raised—a remarkably low level.

MARKET RESEARCH STUDY

The United Way of Greater Rochester was interested in determining Rochester community perceptions of charities. Although a national survey of attitudes had been conducted by the United Way of America, it was believed that some of the general conclusions drawn from the national survey might not be applicable to the local United Way organizations. It was suggested that the United Way of Rochester conduct its own market research study to evaluate community perceptions for the purpose of recommending appropriate adjustments.

The initial research began with focus group interviews; subject groups included professional, service, and general worker categories. Conclusions drawn from these interviews included the following:

1. The more people are informed, the better they feel about giving to charities;

2. There exists a misconception by many individuals regarding how much charities allocate to administrative funds;

3. There is a great mistrust of large charities, including the feeling that many of their activities are dishonest.

After the completion of the focus group interviews, questionnaire development was started. The bases for selecting questions were the focus group interviews and the specific needs and interests that were set as guidelines by United Way personnel. These guidelines were stated as goals, which included increasing giving among groups now contributing, improving the effectiveness of public relations, and increasing the amount of money raised by identifying new target markets.

*The authors are indebted to Dr. Philip R. Tyler, Marketing Consultant to the United Way of Greater Rochester, New York for providing this application.

A draft of the questionnaire was discussed with United Way personnel. Minor adjustments were made in wording, and redundant questions were eliminated. The instrument was then tested for effectiveness on a small test group. The final questionnaire was then distributed to 440 individuals at 18 different organizations. From this group 323 completed questionnaires were obtained.

■■■ STATISTICAL ANALYSIS

SPSS (Statistical Package for the Social Sciences) is an integrated system of computer programs for statistical analysis. SPSS was selected to analyze the data because it is easy to use and provides a wide range of options for both analysis and output. The specific analysis of the data utilized the following SPSS procedures:

1. *Procedure FREQUENCIES.* This procedure produces one-way frequency distribution tables for specified variables. Using this procedure, the data were summarized in both graphical and tabular form.

2. *Procedure CROSSTABS.* Essentially this is a joint frequency distribution of two or more variables, with appropriate statistical analysis. For example, the chi-square test for contingency tables presented in this chapter was used to test for the independence of two variables.

As an illustration of the use of the chi-square test for independence in this study, consider the following question asked of respondents:

Of the funds collected, what percentage do you feel goes to United Way Administrative Expenses:

( )   Up to 10%
( )   11%–20%
( )   21% and over*

*This category was originally 21%–30%, 31%–50%, and 51% and over; these three classes were combined during the statistical analysis in order to obtain expected frequencies appropriate for the chi-square test.

Poster Child for the annual United Way campaign.

TABLE 17A.1 **Perception of United Way Administrative Expenses by Occupation**

			UP TO 10% 1	11–20% 2	21% AND OVER 3	ROW TOTAL
PROD LINE	COUNT ROW PCT COL PCT TOT PCT	1.	3 17.65 2.86 1.03	3 17.65 3.61 1.03	11 64.71 10.78 3.79	17 5.86
MAINT-WARE HSE	COUNT ROW PCT COL PCT TOT PCT	2.	3 20.00 2.86 1.03	7 46.67 8.43 2.41	5 33.33 4.90 1.72	15 5.17
CRAFTS-FOREMEN	COUNT ROW PCT COL PCT TOT PCT	3.	7 46.67 6.67 2.41	5 33.33 6.02 1.72	3 20.00 2.94 1.03	15 5.17
CLERICAL	COUNT ROW PCT COL PCT TOT PCT	4.	15 19.48 14.29 5.17	27 35.06 32.53 9.31	35 45.45 34.31 12.07	77 26.55
SALES	COUNT ROW PCT COL PCT TOT PCT	5.	3 15.79 2.86 1.03	8 42.11 9.64 2.76	8 42.11 7.84 2.76	19 6.55
MANAGERS	COUNT ROW PCT COL PCT TOT PCT	6.	37 64.91 35.24 12.76	13 22.81 15.66 4.48	7 12.28 6.86 2.41	57 19.66
PRO-TECH	COUNT ROW PCT COL PCT TOT PCT	7.	23 43.40 21.90 7.93	9 16.98 10.84 3.10	21 39.62 20.59 7.24	53 18.28
OTHER	COUNT ROW PCT COL PCT TOT PCT	8.	14 37.84 13.33 4.83	11 29.73 13.25 3.79	12 32.43 11.76 4.14	37 12.76
	COLUMN TOTAL		105 36.21	83 28.62	102 35.17	290 100.00

CHI SQUARE = 49.76840 WITH 14 DEG OF FREEDOM SIGNIFICANCE = .00001

Each respondent was also asked to indicate their occupation according to the following classification:

( ) Production line/assemblers
( ) Maintenance/warehouse
( ) Craftsmen and foremen
( ) Clerical worker
( ) Sales worker
( ) Managers and administrators
( ) Professional and technical workers
( ) Other _____

In this case a test of independence addresses the question of whether the perception of United Way administrative expenses is independent of the occupation of the respondent. The hypotheses for the test of independence are as follows:

H_0:   Perception of United Way administrative expenses is independent of the occupation of the respondent,

H_a: Perception of United Way administrative expenses is not independent of the occupation of the respondent.

The contingency table in Table 17A.1 summarizes the responses to the study and provides the necessary output statistics. We see that 290 useful responses were obtained. Note that there are $(8 - 1)(3 - 1) = 14$ degrees of freedom. With $\alpha = .05$ for the level of significance, Table 3 of Appendix B shows an upper-tail χ^2 value of $\chi^2_{.05} = 23.6848$. Since $\chi^2 = 49.7684$ is greater than the critical value of $\chi^2_{.05} = 23.6848$, we reject the null hypothesis of independence and conclude that perception of United Way administrative expenses is not independent of the respondent's occupation.

We can informally compare the observed and expected frequencies in order to obtain an idea of how the dependence between perception of administrative expenses and occupation comes about. As was pointed out earlier, actual administrative costs are less than 9%. Thus 36% of the respondents have an accurate perception of United Way administrative expenses. Note that 35% have a *very* inaccurate perception (21% and above); In this group, production-line employees, clerical, sales, and professional-technical employees have more inaccurate perceptions than other groups. Certainly, other types of general observations such as this can be made.

▬ MAIN FINDINGS

The study described and the resulting statistical analysis enabled conclusions to be drawn regarding the perceptions of Rochester people on several important issues. Some of the major findings include the following:

1. Perceptions of the United Way vary quite dramatically from company to company.

2. The United Way has the highest percentage of contributors of the charities studied. Percentage contribution varies significantly by occupation and education.

3. The perception of the funds used for administrative expenses varies significantly by occupation and education.

4. The most important considerations in deciding the size of a gift are financial constraints.

5. In-work solicitation is by far the most preferred method of collection.

6. Only 11.4% of the respondents do not feel well-enough informed to make a proper United Way giving decision. This varies significantly by occupation.

7. Perceptions about the United Way in general vary significantly by occupation.

The general conclusions developed in this study were instrumental in defining adjustments to the United Way of Greater Rochester. The result has been improved communications and campaign efforts.

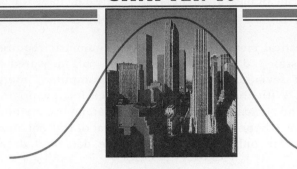

CHAPTER 18

Nonparametric Methods

T he statistical methods presented thus far in the text are generally referred to as *parametric methods.* In this chapter we introduce several statistical methods that are referred to as *nonparametric methods.* These nonparametric methods are often applicable in situations where the parametric methods of the preceding chapters are not. Nonparametric methods typically require less restrictive assumptions concerning the level of data measurement, and/or fewer assumptions concerning the form of the probability distributions generating the sample data.

One consideration used to determine whether a parametric or a nonparametric method should be used is the scale of measurement used to generate the data. As discussed in Chapter 1, there are four scales of measurement: nominal, ordinal, interval, and ratio. All data are generated by one of these four scales of measurement; thus all statistical analyses are conducted with either nominal, ordinal, interval, or ratio data.

Let us again define the four scales of measurement. Examples of each were provided in Chapter 1.

1. *Nominal scale.* The scale of measurement is nominal if the data are simply labels or categories used to define an attribute of the element.

2. *Ordinal scale.* The scale of measurement is ordinal if the data have the properties of nominal data and the data can be used to rank, or order, the observations.

3. *Interval scale.* The scale of measurement is interval if the data have the properties of ordinal data and the interval between observations is expressed in terms of a fixed unit of measure.

4. *Ratio scale.* The scale of measurement is ratio if the data have the properties of interval data and the ratio of observations is meaningful.

Most of the statistical methods referred to as parametric require the use of interval or ratio scaled data. With these levels of measurement, means, variances, standard deviations, and so on, can be computed, interpreted, and used in the analysis. With nominal or ordinal data, it is inappropriate to compute means, variances, and standard deviations; hence parametric methods normally cannot be used. Thus whenever the data are nominal or ordinal, nonparametric methods are often the only way to analyze the data and draw statistical conclusions.

Another consideration used to determine whether a parametric method or a nonparametric method should be employed is the assumption concerning the population from which the data were obtained. For example, a parametric procedure for testing a hypothesis about the difference between the means of two populations was presented in Chapter 10. In the small-sample case, the *t* distribution can be used for this test provided we are willing to assume that the populations are normally distributed with equal variances. If this assumption about the populations is not appropriate, the parametric method based on the use of the *t* distribution should not be used. However, nonparametric methods, which require no assumptions about the population probability distributions, are available for testing for differences between two populations. Because of this and other cases in which no population assumptions are required, nonparametric methods are often referred to as *distribution-free* methods.

In general, for a statistical method to be classified as nonparametric it must satisfy at least one of the following conditions:*

1. The method may be used with nominal data.

2. The method may be used with ordinal data.

3. The method may be used with interval or ratio data when no assumption can be made about the population probability distribution.

If the level of data measurement is interval or ratio and if the necessary probability distribution assumptions for the population are appropriate, parametric methods provide a more powerful or more discerning statistical procedure. In many cases where a nonparametric method as well as a parametric method can be applied, the nonparametric method is almost as good or almost as powerful as the parametric method. In cases where the data are nominal or ordinal or in cases where the assumptions required by parametric methods are inappropriate, only nonparametric methods are available. Because of the less restrictive data-measurement requirements and the fewer assumptions needed concerning the population distribution, nonparametric methods are regarded as more generally applicable than parametric methods. The sign test, the Wilcoxon signed-rank test, the Mann-Whitney-Wilcoxon test, the Kruskal-Wallis test, and

---

*See W. J. Conover, *Practical Nonparametric Statistics*, 2d ed. (New York: John Wiley and Sons, 1980).

Spearman rank correlation are the nonparametric methods that are presented in this chapter.

18.1 ▰▰▰ SIGN TEST

A common market research application of the *sign test* involves using a sample of n potential customers to identify a preference for one of two brands of a product such as coffee, soft drinks, detergents, and so on. The n preferences are nominal data because the consumer simply names, or labels, a preference. Given these data our objective is to determine whether or not a difference in preference exists for the two items being compared. As we will see, the sign test is a nonparametric statistical procedure for answering this question.

Small-Sample Case

The small-sample case for the sign test is appropriate whenever $n \leq 20$. Let us illustrate the use of the sign test for the small-sample case by considering a study conducted for Sun Coast Farms. Sun Coast Farms produces an orange juice product marketed under the name "Citrus Valley." A competitor of Sun Coast Farms has begun producing a new orange juice product known as "Tropical Orange." In a study of consumer preferences for the two brands, 12 individuals were given unmarked samples of the two brands of orange juice. The brand each individual tasted first was randomly selected. After tasting the two products, the individuals were asked to state a preference for one of the two brands. The purpose of the study was to determine whether or not preferences for the two products differ. Letting p indicate the proportion of the population of consumers favoring Citrus Valley, we want to test the following hypotheses:

$$H_0: \quad p = .50$$

$$H_a: \quad p \neq .50$$

If H_0 cannot be rejected, there will be no evidence indicating that a difference in preference exists for the two brands of orange juice. However, if H_0 can be rejected, the conclusion can be made that the consumer preferences are different for the two brands. In this case, the brand selected by the greater number of consumers can be concluded the "most preferred" brand.

In the following discussion we will show how the small-sample version of the sign test can be used to test these hypotheses and draw a conclusion about consumer preferences. In recording the preference data for the 12 individuals participating in the study, a + sign will be recorded if the individual expresses a preference for Citrus Valley and a − sign will be recorded if the individual expresses a preference for Tropical Orange. Using this procedure the data will be recorded in terms of the + or − signs; thus the nonparametric test is referred to as the sign test.

Under the assumption that H_0 is true ($p = .50$), the number of + values follows a binomial probability distribution with $p = .50$. With a sample size of $n = 12$, Table 5 in Appendix B shows the following probabilities for the binomial probability distribution with $p = .50$:

Number of + Signs	Binomial Probability
0	.0002
1	.0029
2	.0161
3	.0537
4	.1208
5	.1934
6	.2256
7	.1934
8	.1208
9	.0537
10	.0161
11	.0029
12	.0002

A graphical representation of this binomial probability distribution is shown in Figure 18.1. This probability distribution shows the probability of the number of + signs under the assumption H_0 is true and is thus the appropriate sampling distribution for the hypothesis test. We use this sampling distribution to determine a rule for rejecting H_0; our approach will be similar to the method we used to develop rejection rules for hypothesis testing in Chapter 9. For example,

▬▬ FIGURE 18.1   **Binomial Probabilities for the Number of + Signs when $n = 12$ and $p = .50$**

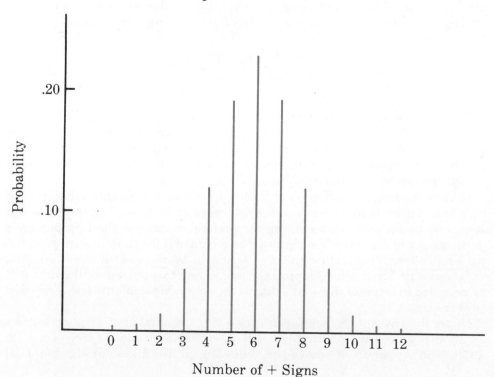

using $\alpha = .05$, we would place a rejection region or area of approximately .025 in each tail of the distribution shown in Figure 18.1. Starting at the lower end of the distribution, we see that the probability of obtaining 0, 1, or 2 + signs is $.0002 + .0029 + .0161 = .0192$. Note that we stop at 2 + signs because adding the probability of 3 + signs would make the area in the lower tail equal to $.0192 + .0537 = .0729$, which substantially exceeds the desired area of .025. At the upper end of the distribution, we find the same probably of .0192 corresponding to 10, 11, or 12 + signs. Thus, the closest we can come to $\alpha = .05$ is $.0192 + .0192 = .0384$. As a result, we adopt the following rejection rule:

Reject H_0 if the number of + signs is less than 3
or greater than 9

The preference data that were obtained for the Sun Coast Farms example are shown in Table 18.1. Since only 2 + signs were observed, the null hypothesis is rejected. There is evidence from this study that consumer preference differs for the two brands of orange juice. We would advise Sun Coast Farms that consumers indicate a preference for the competitor's Tropical Orange brand.

In the Sun Coast Farms example, all 12 individuals in the study were able to state a preference. In many situations, one or more individuals in the sample may not be able to state a definite preference. In such cases the individual's response of no preference can be removed from the study and the analysis conducted with a smaller sample size.

The binomial probability distribution as shown in Table 5 of Appendix B can be used to provide the decision rule for any sign test up to a sample size of $n = 20$. Using the null hypothesis $p = .50$ and the sample size n, the decision rule can be established for any level of significance. In addition, by considering the probabilities in only the lower or upper tail of the binomial probability distribution, rejection rules can also be developed for one-tailed tests. Appendix B does not provide binomial probability distribution tables for sample sizes greater than 20. In these cases we can use the large-sample normal approxima-

TABLE 18.1. **Preference Data for the Sun Coast Farms' Taste Test**

Individual	Brand Preference	Recorded Data
1	Tropical Orange	−
2	Tropical Orange	−
3	Citrus Valley	+
4	Tropical Orange	−
5	Tropical Orange	−
6	Tropical Orange	−
7	Tropical Orange	−
8	Tropical Orange	−
9	Citrus Valley	+
10	Tropical Orange	−
11	Tropical Orange	−
12	Tropical Orange	−

tion of binomial probabilities to determine the appropriate rejection rule for the sign test.

Large-Sample Case

Using the null hypothesis H_0: $p = .50$ and a sample size of $n > 20$, the normal approximation of the sampling distribution for the number of + signs is as follows:

Normal Approximation of the Sampling Distribution of the Number of Plus Signs when No Preference Exists

$$\text{Mean:} \quad \mu = .50n \quad\quad\quad (18.1)$$

$$\text{Standard Deviation:} \quad \sigma = \sqrt{.25n} \quad\quad\quad (18.2)$$

Distribution Form: Approximately normal provided $n > 20$

Let us consider an application of the sign test to political polling. A poll taken during a recent presidential election campaign asked 200 registered voters to rate the Democratic and Republican candidates in terms of best overall foreign policy. Results of the poll showed 72 rated the Democratic candidate higher, 103 rated the Republican candidate higher, and 25 indicated no difference between the candidates. Does the poll indicate that there is a significant difference between public opinion of the foreign policies of the two candidates?

Using the sign test, we see that $n = 200 - 25 = 175$ individuals were able to indicate the candidate they believed had the best overall foreign policy. Using (18.1) and (18.2), we find the sampling distribution of the number of plus signs has the following properties.

$$\mu = .50n = .50(175) = 87.5$$

$$\sigma = .25n = \sqrt{.25(175)} = 6.6$$

In addition, with $n = 175$ we can assume that the sampling distribution is approximately normal. This distribution is shown in Figure 18.2. Since the distribution is approximately normal, we can use the table of areas for the standard normal probability distribution to develop the rejection rule for the test. With $\alpha = .05$, the rejection rule for this two-tailed test can be written as follows:

$$\text{Reject } H_0 \text{ if } z < -1.96 \text{ or if } z > +1.96$$

Using the number of times the Democratic candidate received the higher foreign policy rating as the number of plus signs $x = 72$, we have the following value of the test statistic:

$$z = \frac{x - \mu}{\sigma} = \frac{72 - 87.5}{6.6} = -2.35$$

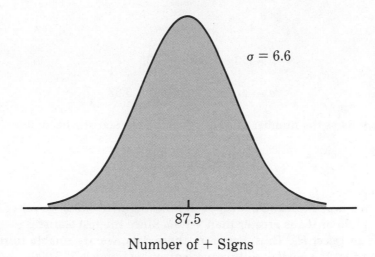

$\sigma = 6.6$

87.5

Number of + Signs

■■■ FIGURE 18.2 **Probability Distribution of the Number of + Signs for a Sign Test with $n = 175$**

With $z = -2.35$ less than -1.96, the hypothesis of no difference in foreign policy for the two candidates should be rejected at the .05 level of significance. Based on this study the Republican candidate is perceived to have the higher-rated foreign policy.

Hypothesis Tests About a Median

In Chapter 9 we described how hypothesis tests can be used to make an inference about a population mean. We now show how the sign test can be used to conduct hypothesis tests about a population median. Recall that the median splits a population such that 50% of the values fall at the median or above and 50% fall at the median or below. We can apply the sign test to conduct a hypothesis test about the value of a median by using a plus sign whenever the data in the sample is above the median and a minus sign whenever the data in the sample is below the median. Any data exactly equal to the hypothesized value of the median should be discarded. The computations for the sign test are done in exactly the same manner as before.

For example, the following hypothesis test is being conducted about the median price of new homes in St. Louis, Missouri.

$$H_0: \quad \text{Median} = \$75,000$$

$$H_a: \quad \text{Median} \neq \$75,000$$

From a sample of 62 new homes, 34 had prices above $75,000, 26 had prices below $75,000, and 2 had prices of exactly $75,000.

Using (18.1) and (18.2) for the $n = 60$ homes with prices different than $75,000, we have

$$\mu = .50n = .50(60) = 30$$

$$\sigma = \sqrt{.25n} = \sqrt{.25(60)} = 3.87$$

Using $x = 34$ as the number of plus signs, the test statistic becomes

$$z = \frac{x - \mu}{\sigma} = \frac{34 - 30}{3.87} = 1.03$$

With a two-tailed test and a level of significance of $\alpha = .05$, we reject H_0 if z is less than -1.96 or if z is greater than $+1.96$. Since the test statistic $z = 1.03$, we are unable to reject H_0. Thus, based on these data, we are unable to reject the assumption that the median selling price of a new home is $75,000.

■■■ EXERCISES

1. The following data show the preferences indicated by 10 individuals in taste tests involving 2 brands of coffee.

Individual	Brand A Versus Brand B	Individual	Brand A Versus Brand B
1	+	6	+
2	+	7	−
3	+	8	+
4	−	9	−
5	+	10	+

With $\alpha = .05$, test for a signficant difference in the preferences for the 2 brands. A plus indicates a preference for brand A over brand B.

2. Researchers studied physical contact between parents and children in the same family (*Journal of Marriage and the Family*, August 1984). Based on this study, assume that a sample of interactions for 20 different families with two children showed 14 cases where the mother was touched more than the father, 4 cases where the father was touched more than the mother, and 2 cases where the touching was judged equal.

    a. What are the null and alternative hypotheses for the sign test?
    b. Using $\sigma = .05$, what is the rejection rule?
    c. What is your conclusion?

3. In a television preference poll a sample of 180 individuals was asked to state a preference for one of the two shows aired at the same time on Friday evenings. "Big Town Detective" was favored by 100, 65 favored "The Friday Variety Special," and 15 were unable to state a preference for one over the other. Is there evidence of a significant difference in the preferences for the two shows? Use $\alpha = .05$ for the test.

4. Menu planning at the Hampshire House Restaurant involves the question of customer preferences for steak and seafood. Two hundred fifty customers were asked to state a preference for the two menu items. A preference for steak was stated by 140, and 110 stated a preference for seafood. Use $\alpha = .05$ and test for a difference in the preference for the two menu items.

5. The nationwide median hourly wage for a particular labor group is $14.50 per hour. A sample of 200 individuals in this labor group was taken in a city. 134 individuals had a wage rate less than $14.50 per hour; 54 individuals had a wage rate greater than $14.50 per hour; 12 individuals had a wage rate of $14.50. Test the null hypothesis that the median hourly wage in this city is the same as the nationwide median hourly wage. Use a .02 level of significance.

6. In a sample of 150 college basketball games, it was found that the home team won 98 games. Test to see if this data supports the claim that there is a home-team advantage in college basketball. Use a .05 level of significance. What is your conclusion?

7. The median number of part-time employees at fast-food restaurants in a particular city was known to be 15 last year. City officials think the use of part-time employees may be increasing. A sample of 9 fast-food restaurants showed that there were more than 15 part-time employees at 7 of the restaurants, 1 restaurant had exactly 15 part-time employees, and 1 had fewer than 15 part-time employees. Test at $\alpha = .05$ whether or not it can be concluded that there has been an increase in the median number of part-time employees.

8. The *Wall Street Journal*, October 22, 1988, reported that the median age at first marriage for men is 25.9 years and for women is 23.6 years. Suppose a sample of 225 first marriages in a certain Ohio county showed 122 cases where men were less than 25.9 years of age and 103 cases where men were more than 25.9 years of age. Test the hypothesis that the median age at first marriage for men in the sampled county is the same as the reported 25.9 years. Use $\alpha = .05$. What is your conclusion?

18.2 ■■■ WILCOXON SIGNED-RANK TEST

The Wilcoxon signed-rank test is the nonparametric alternative to the parametric matched-sample test presented in Chapter 10. In the matched-sample situation, each experimental unit generates two paired or matched observations, one from population 1 and one from population 2. The differences between the matched observations provide insight concerning the differences between the two populations.

The methodology of the parametric matched-sample analysis (the *t* test on paired differences) requires interval data and the assumption that the population of differences between the pairs of observations is *normally distributed*. With this assumption, the *t* distribution can be used to test the hypothesis of no difference between population means. If some question exists concerning the appropriateness of the assumption of normally distributed differences, the nonparametric Wilcoxon signed-rank test can be used. We illustrate this nonparametric test for data used to compare the effectiveness of two production methods.

A manufacturing firm is attempting to determine if a difference in task-completion times exists for two production methods. A sample of 11 workers was selected, and each worker completed a production task using each of the two production methods. The production method that each worker used first was selected randomly. Thus each worker in the sample provided a pair of observations, as shown in Table 18.2. A positive difference in task completion times indicates that method 1 required more time and a negative difference in times indicates that method 2 required more time. Do the data indicate that the methods are significantly different in terms of task-completion times?

TABLE 18.2   **Production Task Completion Times (Minutes)**

Worker	Method 1	Method 2	Difference
1	10.2	9.5	.7
2	9.6	9.8	−.2
3	9.2	8.8	.4
4	10.6	10.1	.5
5	9.9	10.3	−.4
6	10.2	9.3	.9
7	10.6	10.5	.1
8	10.0	10.0	.0
9	11.2	10.6	.6
10	10.7	10.2	.5
11	10.6	9.8	.8

The question raised is whether or not the two methods provide differences in task-completion times. In effect, we have two populations of task-completion times, one population associated with each method. The hypotheses that will be tested are

H_0:   The populations are identical

H_a: The populations are not identical

If H_0 cannot be rejected, there will be insufficient evidence to conclude that the task-completion times differ for the two methods. However, if H_0 can be rejected, we will conclude that the two methods differ in terms of task-completion times.

The first step of the Wilcoxon signed-rank test requires a ranking of the *absolute value* of the differences between the two methods. To do this we first discard any differences of zero and then rank the remaining absolute differences from lowest to highest. Tied differences are assigned average rank values. The ranking of the absolute values of differences is shown in the fourth column of Table 18.3. Note that the difference of 0 for worker 8 is discarded from the rankings; then the smallest absolute difference of .1 is assigned the rank of 1. This ranking of absolute differences continues with the largest absolute difference of .9 assigned the rank of 10. The absolute differences of .4 for workers 3 and 5 are assigned the average rank of 3.5, while the absolute differences of .5 for workers 4 and 10 are assigned the average rank of 5.5

Once the ranks of the absolute differences have been determined, the ranks are given the sign of the original difference in the data. For example, the .1 difference for worker 7, which was assigned the rank of 1, is given the value of +1 because the observed difference between the two methods was positive. The .2 difference, which was assigned the rank of 2, is given the value of −2 because the observed difference between the two methods was negative for this individual. The complete list of signed ranks, together with their sum, is shown in the last column of Table 18.3.

TABLE 18.3   **Ranking of Absolute Differences for the Production Task-Completion Time Example**

Worker	Difference	Absolute Value of Difference	Rank	Signed Rank
1	.7	.7	8	+ 8
2	−.2	.2	2	− 2
3	.4	.4	3.5	+ 3.5
4	.5	.5	5.5	+ 5.5
5	−.4	.4	3.5	− 3.5
6	.9	.9	10	+10
7	.1	.1	1	+ 1
8	0	0	—	—
9	.6	.6	7	+ 7
10	.5	.5	5.5	+ 5.5
11	.8	.8	9	+ 9
			Sum of Signed Ranks	+44.0

Let us return to the original hypothesis of identical population task-completion times for the two methods. If the populations representing task-completion times for each of the two methods are identical, we would expect the positive ranks and the negative ranks to cancel each other, so that the sum of the signed rank values would be approximately 0. Thus the test for significance under the Wilcoxon signed-rank test involves determining whether or not the computed sum of signed ranks (+44 in our example) is significantly different than 0.

Let T denote the sum of the signed-rank values in a Wilcoxon signed-rank test. It can be shown that if the two populations are identical and the number of matched pairs of data is 10 or more, the sampling distribution of T can be approximated as follows.

Sampling Distribution of T for Identical Populations

$$\text{Mean:} \quad \mu_T = 0 \tag{18.3}$$

$$\text{Standard Deviation:} \quad \sigma_T = \sqrt{\frac{n(n + 1)(2n + 1)}{6}} \tag{18.4}$$

Distribution Form: Approximately normal provided $n \geq 10$

For the example, we have $n = 10$, since we discarded the observation with the difference of 0 (worker 8). Thus using (18.4), we have

$$\sigma_T = \sqrt{\frac{10(11)(21)}{6}} = 19.62$$

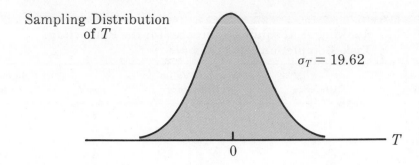

Sampling Distribution of T

$\sigma_T = 19.62$

0

T

■■■ FIGURE 18.3   **Sampling Distribution of the Wilcoxon T for the Production Task Completion Time Example**

The sampling distribution of T under the assumption of identical populations is shown in Figure 18.3.

The value of the test statistic z is as follows.

$$z = \frac{T - \mu_T}{\sigma_T} = \frac{44 - 0}{19.62} = 2.24$$

Testing the null hypothesis of no difference using a level of significance of $\alpha = .05$, we reject H_0 if $z < -1.96$ or if $z > 1.96$. With the value of $z = 2.24$, we reject H_0 and conclude that the two populations are not identical and that the methods differ in terms of task-completion times. The fact that method 2 showed the shorter completion times for 8 of the 11 workers would lead us to conclude that differences between the two populations indicate method 2 to be the better production method.

■■■ EXERCISES

9. A sample of ten individuals was used in a study to test the effects of a relaxant on the time required to fall asleep for male adults. Data for ten subjects showing the number of minutes required to fall asleep with and without the relaxant are given. Use a .05 level of significance to determine if the relaxant reduces the time required to fall asleep. What is your conclusion?

Subject	Without Relaxant	With Relaxant
1	15	10
2	12	10
3	22	12
4	8	11
5	10	9
6	7	5
7	8	10
8	10	7
9	14	11
10	9	6

10. Shown below are the number of baggage-related complaints per 1000 passengers for ten airlines during the months of December 1988 and January 1989 (*U.S. Department of Transportation*, March 1989). Use $\alpha = .05$ and the Wilcoxon signed-rank test to determine if the data indicate the number of baggage related complaints for the airline industry has *decreased* over the two months studied. What is your conclusion?

Airline	December Complaints	January Complaints
American	8.9	8.0
Delta	8.2	7.9
Continental	7.9	8.2
Eastern	7.5	7.8
Northwest	9.6	6.5
Pan American	5.0	5.1
Piedmont	12.3	11.0
TWA	11.2	10.9
United	7.7	7.4
USAir	8.6	7.7

11. A test is conducted of two overnight mail-delivery services. Two samples of identical deliveries are set up such that both delivery services are notified of the need for a delivery at the same time. The number of hours required to make the delivery is recorded for each service. Do the data shown suggest a difference in the delivery times for the two services? Use a .05 level of significance for the test.

Delivery	Service 1	Service 2
1	24.5	28.0
2	26.0	25.5
3	28.0	32.0
4	21.0	20.0
5	18.0	19.5
6	36.0	28.0
7	25.0	29.0
8	21.0	22.0
9	24.0	23.5
10	26.0	29.5
11	31.0	30.0

12. Harding Investors, Inc. provides a 6-week training program for newly hired management trainees. As part of the program evaluation procedure, the firm gives each trainee a pretest and post-test. Use a one-tailed test with $\alpha = .05$ and analyze the following data as part of the evaluation of the firm's management training program. What is your conclusion?

Trainee	Pretest Score	Post-test Score
1	45	65
2	60	70
3	65	63
4	60	67
5	52	60
6	62	58
7	57	70
8	70	65
9	72	80
10	66	88
11	78	74

13. Ten test-market cities were selected as part of a market research study designed to evaluate the effectiveness of a particular advertising campaign. The sales dollars for each city were recorded for the week prior to the promotional program. Then the campaign was conducted for 2 weeks, with new sales data collected for the week immediately following the campaign. The resulting data with sales in thousands of dollars are shown.

City	Precampaign Sales	Postcampaign Sales
Kansas City	130	160
Dayton	100	105
Cincinnati	120	140
Columbus	95	90
Cleveland	140	130
Indianapolis	80	82
Louisville	65	55
St. Louis	90	105
Pittsburgh	140	152
Peoria	125	140

Use $\alpha = .05$. What conclusion would you draw concerning the value of the advertising program?

18.3 ■ MANN-WHITNEY-WILCOXON TEST

In this section we present another nonparametric method that can be used to determine if there is a difference between two populations. This test, unlike the signed-rank test, is not based on a matched sample. Two independent samples, one from each population, are used. This test, developed jointly by Mann, Whitney and Wilcoxon, is sometimes referred to as the *Mann-Whitney test* and is sometimes referred to as the *Wilcoxon rank-sum test*. Both the Mann-Whitney and Wilcoxon versions of this test are equivalent; thus we refer to it as the *Mann-Whitney-Wilcoxon (MWW) test*.

The MWW test is based upon independent random samples from each population. Recall that in Chapter 10 we conducted a parametric test for the difference between the means of two populations. The hypotheses tested were as follows:

$$H_0: \mu_1 - \mu_2 = 0$$

$$H_a: \mu_1 - \mu_2 \neq 0$$

In the small-sample case, the parametric method used was based on two assumptions:

1. Both populations are normally distributed.
2. The variances of the two populations are equal.

The nonparametric MWW test does not require either of the above assumptions. The only requirement of the MWW test is that the measurement scale for the data generated by the two independent random samples is at least ordinal.

Instead of testing for the difference between the means of the two populations, the MWW test is to determine whether or not the two populations are identical. The hypotheses for the Mann-Whitney-Wilcoxon test are as follows:

H_0: The two populations are identical

H_a: The two populations are not identical

We first demonstrate how the MWW test can be applied by showing an application for the small-sample size case.

Small-Sample Case

The small-sample size case for the MWW test is appropriate whenever the sample sizes for both populations are less than or equal to 10. We illustrate the use of the MWW test for the small-sample case by considering the academic achievement of students attending Johnston High School. The majority of students attending Johnston High School previously attended either Garfield Junior High School or Mulberry Junior High School. The question raised by school administrators was whether or not the population of students that attended Garfield were identical to the population of students that attended Mulberry in terms of academic potential. The hypotheses under consideration were expressed as follows:

H_0: The two populations are identical in terms
      of academic potential

H_a: The two populations are not identical in terms
      of academic potential

Using high school records, Johnston High School administrators selected a random sample of four high-school students who had attended Garfield Junior High and another random sample of five students who had attended Mulberry Junior High. The current high-school class standing was recorded for each of the 9 students used in the study. The ordinal class standing for the nine students is shown in Table 18.4.

The first step in the MWW procedure is to rank the *combined* data from the two

TABLE 18.4. **High-School Class-Standing Data**

Garfield Students		Mulberry Students	
Student	*Class Standing*	*Student*	*Class Standing*
Fields	8	Hart	70
Clark	52	Phipps	202
Jones	112	Kirkwood	144
Tibbs	21	Abbott	175
		Guest	146

samples from low to high. The lowest value (class standing 8) receives a rank of 1 and the highest value (classing standing 202) receives a rank of 9. The complete ranking of the 9 students is as follows:

Student	Class Standing	Combined Sample Rank
Fields	8	1
Tibbs	21	2
Clark	52	3
Hart	70	4
Jones	112	5
Kirkwood	144	6
Guest	146	7
Abbott	175	8
Phipps	202	9

The next step is to sum the ranks for each sample separately. This calculation is shown in Table 18.5. The MWW procedure may utilize the sum of the ranks for either sample. In the following discussion we use the sum of the ranks for the sample of 4 students from Garfield. We denote this sum by the symbol T. Thus for our example, $T = 11$.

Let us consider for a moment the properties of the sum of the ranks for the Garfield sample. Since there are 4 students in the sample, Garfield could have the top four ranking students in the study. If this were the case, $T = 1 + 2 + 3 + 4 = 10$ would be the smallest value possible for the rank sum T. On the other hand, Garfield could have the bottom four ranking students, in which case $T = 6 + 7 + 8 + 9 = 30$ would be the largest value possible for T. Thus, T for the Garfield sample must take on a value between 10 and 30.

Note that values of T near 10 imply Garfield has the significantly better, or higher-ranking, students, whereas values of T near 30 imply Garfield has the significantly weaker, or lower-ranking, students. Thus, if the two populations of students were identical in terms of academic potential, we would expect the value of T to be near the average of the above two values, or $(10 + 30)/2 = 20$.

Critical values of the MWW T statistic are provided in Table 10 of Appendix B for cases where both sample sizes are less than or equal to 10.* In these tables, n_1 refers to the sample size corresponding to the sample whose rank sum is being used in the test. The value of T_L is read directly from the tables and the value of T_U is computed from (18.5).

$$T_U = n_1(n_1 + n_2 + 1) - T_L \qquad (18.5)$$

Neither the value of T_L nor the value of T_U are in the rejection region. The null hypothesis of identical populations should be rejected only if T is strictly less than T_L or strictly greater than T_U.

For example, using Table 10 of Appendix B with a .05 level of significance we see that the lower-tail critical value for the MWW statistic with $n_1 = 4$ (Garfield) and $n_2 = 5$ (Mulberry) is $T_L = 12$. The upper-tail critical value for the MWW

*A more complete table of critical values for the Mann-Whitney-Wilcoxon test can be found in *Practical Nonparametric Statistics*, by W. J. Conover.

TABLE 18.5. **Rank Sums for High School Students From Each Junior High School**

Garfield Students			Mulberry Students		
Student	*Class Standing*	*Sample Rank*	*Student*	*Class Standing*	*Sample Rank*
Fields	8	1	Hart	70	4
Clark	52	3	Phipps	202	9
Jones	112	5	Kirkwood	144	6
Tibbs	21	2	Abbott	175	8
			Guest	146	7
Sum of Ranks		11			34

statistic is computed using (18.5) as follows:

$$T_U = 4(4 + 5 + 1) - 12 = 28$$

Thus the MWW decision rule indicates that the null hypothesis of identical populations can be rejected if the sum of the ranks for the first sample (Garfield) is less than 12 or greater than 28. The rejection rule can be written as follows:

$$\text{Reject } H_0 \text{ if } T < 12 \text{ or if } T > 28$$

Referring to Table 18.5, we see that $T = 11$. Thus, the null hypothesis H_0 is rejected and we can conclude that the population of students at Garfield differs from the population of students at Mulberry in terms of academic potential. The higher class ranking obtained by the sample of Garfield students indicates that

TABLE 18.6 **Account Balances for Two Branches of the Third National Bank**

Branch 1		Branch 2	
Sampled Account	*Account Balance ($)*	*Sampled Account*	*Account Balance ($)*
1	1095	1	885
2	955	2	850
3	1200	3	915
4	1195	4	950
5	925	5	800
6	950	6	750
7	805	7	865
8	945	8	1000
9	875	9	1050
10	1055	10	935
11	1025		
12	975		

Garfield students appear to be better prepared for high school than the Mulberry students.

Large-Sample Case

In the case were both sample sizes are greater than or equal to 10, a normal approximation of the distribution of T can be used to conduct the analysis for the MWW test. We show an example of the large-sample case by considering an application at the Third National Bank.

The Third National Bank has two branch offices. Data collected from two independent simple random samples, one from each branch, are shown in Table 18.6. Do the data indicate that the populations of checking account balances at the two branch banks are or are not identical?

The first step in the MWW test is to rank the *combined* data from the lowest to the highest values. Using the combined set of 22 observations shown in Table 18.6, we find the lowest data value of $750 (sixth item of sample 2) and assign to it a rank of 1. Continuing the ranking, we have the following.

Account Balance	Item	Assigned Rank
$ 750	6th of sample 2	1
$ 800	5th of sample 2	2
$ 805	7th of sample 1	3
$ 850	2nd of sample 2	4
:	:	:
:	:	:
$1195	4th of sample 1	21
$1200	3rd of sample 1	22

In ranking the combined data, we may find that two or more data values are the same. In this case, these same values are given the *average* ranking of their

---

TABLE 18.7 **Combined Ranking of the Data in the Two Samples From the Third National Bank**

	Branch 1			Branch 2	
Sampled Account	*Account Balance ($)*	*Rank*	*Sampled Account*	*Account Balance ($)*	*Rank*
1	1095	20	1	885	7
2	955	14	2	850	4
3	1200	22	3	915	8
4	1195	21	4	950	12.5
5	925	9	5	800	2
6	950	12.5	6	750	1
7	805	3	7	865	5
8	945	11	8	1000	16
9	875	6	9	1050	18
10	1055	19	10	935	10
11	1025	17		Sum of Ranks	83.5
12	975	15			
	Sum of Ranks	169.5			

positions in the combined data set. This situations of *ties* occurs with the ranking of the 22 account balances from the two branch banks. For example, the balance of $945 (eighth item of sample 1) will be assigned the rank of 11. However, the next 2 values in the data set are tied with values of $950 (see the sixth item of sample 1 and the fourth item of sample 2). Since these two values will be considered for assigned ranks of 12 and 13, they are both given the assigned rank of 12.5. At the next highest data value of $955, we continue the ranking process by assigning $955 the rank of 14. Table 18.7 shows the entire data set with the assigned rank of each observation.

The next step in the MWW test is to sum the ranks for each sample. These sums are shown in Table 18.7. The test procedure can be based upon the sum of the ranks for either sample. We use the sum of the ranks for the sample from Branch 1. Thus for this example, $T = 169.5$.

Given that the sample sizes are $n_1 = 12$ and $n_2 = 10$, we can use the normal approximation to the sampling distribution of the rank sum, T. The appropriate sampling distribution is given by the following:

Sampling Distribution of T for Identical Populations

$$\text{Mean:} \quad \mu_T = \tfrac{1}{2}\, n_1(n_1 + n_2 + 1) \tag{18.6}$$

$$\text{Standard Deviation:} \quad \sigma_T = \sqrt{\tfrac{1}{12}\, n_1 n_2(n_1 + n_2 + 1)} \tag{18.7}$$

Distribution Form: Approximately normal provided $n_1 \geq 10$ and $n_2 \geq 10$

For Branch 1, we have

$$\mu_T = \tfrac{1}{2}\, 12(12 + 10 + 1) = 138$$

$$\sigma_T = \sqrt{\tfrac{1}{12}\, 12(10)(12 + 10 + 1)} = 15.17$$

The sampling distribution of T is shown in Figure 18.4. Following the usual hypothesis-testing procedure, we compute the test statistic z to determine if the observed value of T appears to be from the sampling distribution of Figure 18.4. If T does not appear to be from this distribution, we will reject the null hypothesis and conclude that the populations are not identical. Computing the test statistic we have

$$z = \frac{T - \mu_T}{\sigma_T} = \frac{169.5 - 138}{15.17} = 2.08$$

At an $\alpha = .05$ level of significance, we know that, in order to reject H_0, z must be less than -1.96 or greater than $+1.96$. Since $z = 2.08$, we reject H_0. Thus we conclude that the two populations are not identical. That is, the populations of account balances at the two branches are not the same.

In summary, the Mann-Whitney-Wilcoxon rank-sum test follows the steps outlined below in order to determine if two independent random samples are selected from identical populations.

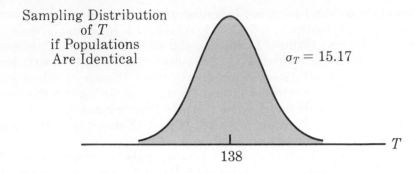

Sampling Distribution
of T
if Populations
Are Identical

$\sigma_T = 15.17$

138

T

■■■ FIGURE 18.4 **Sampling Distribution of T for the Third National Bank Example**

1. Rank the combined sample observations from lowest to highest, with tied values being assigned the average of the tied rankings.

2. Compute T, the sum of the ranks for the first sample.

3. In the large-sample case, make the test for significant differences between the two populations by using the observed value of T and comparing it to the sampling distribution of T for identical populations (see equations (18.6) and (18.7)). The value of the standardized test statistic z will provide the basis for deciding whether or not to reject H_0. In the small-sample case, use Table 9 in Appendix B to find the critical values for the test.

■■■   NOTES AND COMMENTS   ■■■

The nonparametric test discussed in this section is used to determine whether or not two populations are identical. Parametric statistical tests, such as the t test described in Chapter 10, test the equality of two population means. When we reject the hypothesis that the means are equal, we conclude that the populations differ only in their means. When we reject the hypothesis that the populations are identical using the MWW test, we cannot state how they differ. The populations could have different means, different variances, and/or different forms. Nonetheless, if we believe that the populations are the same in every way except for the means, a rejection of H_0 using the nonparametric method implies that the means differ. The major advantages of the MWW test, compared to the parametric t test, are that it does not require any assumptions about the form of the probability distribution from which the measurements come and the test may be used with ordinal data.

═══ **EXERCISES**

14. Anderson Company has sent two groups of employees to a privately run program providing word-processing training. One group was from the data-processing depart-

ment; the other was from the typing pool. At the completion of the program, Anderson Company received a report showing the class rank for each of its employees. Of the 70 persons finishing the program, the class ranks of the 13 employees of Anderson Company are given (by group).

Data-Processing Group	Typists
1	17
12	26
15	29
23	33
30	45
33	51
	62

Use $\alpha = .10$ and test to see whether or not there is a performance difference between the two groups in the word-processing program.

15. Mileage performance tests were conducted for two models of automobiles. Twelve automobiles of each model were randomly selected and a miles-per-gallon rating for each model was developed based upon 1000 miles of highway driving. The data are shown.

Model 1		Model 2	
Automobile	Miles per Gallon	Automobile	Miles per Gallon
1	20.6	1	21.3
2	19.9	2	17.6
3	18.6	3	17.4
4	18.9	4	18.5
5	18.8	5	19.7
6	20.2	6	21.1
7	21.0	7	17.3
8	20.5	8	18.8
9	19.8	9	17.8
10	19.8	10	16.9
11	19.2	11	18.0
12	20.5	12	20.1

Use $\alpha = .10$ and test for a significant difference in the populations of miles-per-gallon ratings for the two models.

16. Insurance costs for men and women were reported in *Newsweek*, May 8, 1989. Assume that the following data show the annual cost of $100,000, 5-year term insurance policies for nonsmoking men and nonsmoking women. Use the MWW to test for a significant difference between the costs for men and women. Use $\alpha = .05$.

Men	Women
167	146
175	162
160	164
165	148
172	166
180	158
185	150
170	150
163	140
184	142

17. The following data from police records show the number of daily crime reports from a sample of days during the winter months and a sample of days during the summer months. Using a .05 level of significance, determine if there is a significant difference between the number of crime reports in winter and summer months.

Winter	Summer
18	28
20	18
15	24
16	32
21	18
20	29
12	23
16	38
19	28
20	18

18. A certain brand of microwave oven was priced at 10 stores in Dallas and 13 stores in San Antonio. The data are as follows.

Dallas	San Antonio
445	460
489	451
405	435
485	479
439	475
449	445
436	429
420	434
430	410
405	422
	425
	459
	430

Use a .05 level of significance and test whether or not prices for the microwave oven are the same in the two cities.

18.4 ■■■ THE KRUSKAL-WALLIS TEST

The MWW test in Section 18.3 can be used to test whether or not two populations are identical. This test has been extended to the case of three or more populations by Kruskal and Wallis. The hypotheses for the *Kruskal-Wallis test* with $k \geq 3$ populations can be written as follows:

$$H_0: \quad \text{All } k \text{ populations are identical}$$

$$H_a: \quad \text{Not all populations are identical}$$

The Kruskal-Wallis test is based on the analysis of independent random samples from each of the k populations.

TABLE 18.8 **Performance Evaluation Ratings for 20 Employees**

College A	College B	College C
25	60	50
70	20	70
60	30	60
85	15	80
95	40	90
90	35	70
80		75

In Chapter 12 we introduced the completely randomized experimental design as a procedure that could be used to test for the equality of means among three or more populations. The parametric procedure known as analysis of variance (ANOVA) was used to analyze the data and conduct the test. The ANOVA model requires interval or ratio data, the populations to all be normally distributed, and the variances of the populations to be equal.

The nonparametric Kruskal-Wallis test can be used with ordinal data as well as with interval or ratio data. In addition, the Kruskal-Wallis test does not require the assumptions of normality and equal variances that are required by the parametric analysis of variance procedure. Thus, whenever the data from $k \geq 3$ independent random samples is ordinal or whenever the assumptions of normality and equal variances are questionable, the Kruskal-Wallis test provides an alternate statistical procedure for testing whether or not the populations are identical. Let us demonstrate the Kruskal-Wallis test by showing how it can be used in an employee-selection application.

Williams Manufacturing Company hires employees for its management staff from three area colleges. Recently the company's personnel department has been collecting and reviewing annual performance ratings in an attempt to determine if there are differences in performance among the managers hired from the three area colleges. Performance-rating data are available from independent samples of seven employees from college A, six employees from college B, and seven employees from college C. These data are summarized in Table

TABLE 18.9 **Rankings for the 20 Employees**

College A	Rank	College B	Rank	College C	Rank
25	3	60	9	50	7
70	12	20	2	70	12
60	9	30	4	60	9
85	17	15	1	80	15.5
95	20	40	6	90	18.5
90	18.5	35	5	70	12
80	15.5			75	14
Sum of Ranks	95		27		88

18.8; the overall performance rating of each manager is given on a 0 to 100 scale, with 100 being the highest possible performance evaluation.

Suppose that we are interested in testing whether or not the three populations are identical with respect to performance evaluations. The Kruskal-Wallis test statistic, which is based on the sum of ranks for each of the samples, can be computed as follows:

Kruskal-Wallis Test Statistic

$$W = \frac{12}{n_T(n_T + 1)} \sum_{i=1}^{k} \frac{R_i^2}{n_i} - 3(n_T + 1) \qquad (18.8)$$

where

$$k = \text{the number of populations}$$

$$n_i = \text{the number of items in sample } i$$

$$n_T = \Sigma\, n_i = \text{total number of items in all samples}$$

$$R_i = \text{sum of the ranks for sample } i$$

Kruskal and Wallis were able to show that under the null hypothesis that the populations are identical, the sampling distribution of W can be approximated by a chi-square distribution with $k - 1$ degrees of freedom. This approximation is generally acceptable if each of the sample sizes is greater than or equal to 5.

In order to compute the W statistic for our example, we must first rank-order all 20 data items. The lowest data value of 15 from the college B sample receives a rank of 1, whereas the highest data value of 95 from the college A sample receives a rank of 20. The data values, their associated ranks and the sum of the ranks for the three samples are shown in Table 18.9. Note that we assign the average rank to tied items;* for example the data values of 60, 70, 80, and 90 had ties.

The sample sizes are:

$$n_1 = 7 \qquad n_2 = 6 \qquad n_3 = 7$$

and

$$n_T = \Sigma\, n_i = 7 + 6 + 7 = 20$$

Using (18.8) the W statistic is computed as follows:

$$W = \frac{12}{20(21)} \left[\frac{(95)^2}{7} + \frac{(27)^2}{6} + \frac{(88)^2}{7} \right] - 3(20 + 1) = 8.92$$

---

*If numerous tied ranks are observed, (18.8) must be modified; the modified formula can be found in *Practical Nonparametric Statistics*, by W.J. Conover.

The chi-square distribution table (Table 3 of Appendix B) shows that with $k - 1 = 2$ degrees of freedom and $\alpha = .05$ in the upper tail of the distribution, the critical chi-square value is $\chi^2 = 5.99147$. Since the test statistic $W = 8.92$ is greater than 5.99147, we reject the null hypothesis that the 3 populations are identical. As a result we conclude that manager performance differs significantly depending on the college attended. Furthermore, since the performance ratings were the lowest for college B, it would appear reasonable for the company to either cut back on its recruiting from college B or at least do a more thorough evaluation of graduates from this college.

NOTES AND COMMENTS

The Kruskal-Wallis procedure illustrated in this example began with the collection of interval-scaled data showing employee-performance evaluation ratings. However, the procedure would have also worked had the data been the ordinal rankings of the 20 employees. In this case the Kruskal-Wallis test could have been applied directly to the original data; the step of constructing the rank orderings from the performance evaluation ratings would have been omitted.

▬ EXERCISES

19. Three products received the following performance ratings by a panel of 15 consumers.

Performance Ratings

Product A	50	62	75	48	65
Product B	80	95	98	87	90
Product C	60	45	30	58	57

Use the Kruskal-Wallis test and $\alpha = .05$ to determine if there is a significant difference in the performance evaluations for the products.

20. Three different admission-test-preparation programs are being evaluated. The scores obtained by a sample of 20 people utilizing the test-preparation programs yielded these results.

Program A	Program B	Program C
540	450	600
400	540	630
490	400	580
530	410	490
490	480	590
610	370	620
	550	570

Use the Kruskal-Wallis test to determine if there is a significant difference in the three test preparation programs. Use $\alpha = .01$.

21. In Chapter 12, the ANOVA procedure was used to test for significant differences in gas mileage for three types of automobiles. The data obtained from tests on five automobiles of each type are shown again.

	Miles per Gallon				
Automobile A	19	21	20	19	21
Automobile B	19	20	22	21	23
Automobile C	24	26	23	25	27

a. Use the Kruskal-Wallis test with $\alpha = .05$ to determine if there is a significant difference in the gasoline mileage for the three automobiles.

b. What information available in the data is used by the ANOVA procedure and not the Kruskal-Wallis test.

22. A large corporation has been sending many of its first-level managers to an off-site supervisory skills course. Four different management-development centers offer this course and the corporation is interested in determining if there are differences in the quality of training provided. A sample of 20 employees who have attended these programs has been obtained and the employees ranked with respect to supervisory skills. The results are shown:

	Supervisory Skills Rank				
Supervisory skills course 1	3	14	10	12	13
Supervisory skills course 2	2	7	1	5	11
Supervisory skills course 3	19	16	9	18	17
Supervisory skills course 4	20	4	15	6	8

Note that the top-ranked supervisor attended supervisory skills course 2 and the lowest-ranked supervisor attended course 4. Use $\alpha = .05$ and test to see if there is a significant difference in the training provided by the four programs.

18.5 ■ RANK CORRELATION

Correlation was introduced in Chapter 13 as a measure of the linear association between two variables for which interval or ratio data are available. In this section we consider measures of association between two variables when only ordinal data are available. The *Spearman rank-correlation coefficient, r_s*, has been developed for this purpose.

The formula for the Spearman rank-correlation coefficient is as follows.

Spearman Rank-Correlation Coefficient

$$r_s = 1 - \frac{6 \Sigma d_i^2}{n(n^2 - 1)}$$

(18.9)

where

n = the number of items or individuals being ranked

x_i = the rank of item i with respect to one variable

y_i = the rank of item i with respect to a second variable

$d_i = x_i - y_i$

Let us illustrate the use of the Spearman rank-correlation coefficient in the following example. A company wants to determine if individuals who were expected at the time of employment to be better salespersons actually turn out to have better sales records. To investigate this question, the vice president in charge of personnel has carefully reviewed the original job interview summaries, academic records, and letters of recommendations for ten current members of the firm's sales force. Based on the review of this information, the vice president ranked the ten individuals in terms of their potential for success, basing the assessment solely upon the information available at the time of employment. Then a list was obtained of the number of units sold by each salesperson over the first 2 years. Based on actual sales performance, a second ranking of the ten salespersons was carried out. Table 18.10 shows the relevant data and the two rankings. The statistical question involves determining whether or not there is agreement between the ranking of potential at the time of employment and the ranking based upon the actual sales performance over the first 2 years.

Let us compute the Spearman rank-correlation coefficient for the data in Table 18.10. The computations for the rank-correlation coefficient are summarized in Table 18.11. Here we see that the rank-correlation coefficient is a positive .73. The Spearman rank-correlation coefficient ranges from -1.0 to $+1.0$, with an interpretation similar to the sample correlation coefficient in that positive values near 1.0 indicate a strong association between the rankings; as one rank increases, the other rank increases. On the other hand, rank correlations near -1.0 indicate a strong negative association in the ranks (as one rank increases, the other rank decreases). The value $r_s = .73$ indicates a positive correlation between potential and actual performance. Individuals ranked high on potential tend to rank high on performance.

A Test for Significant Rank Correlation

At this point, we have seen how sample results can be used to compute the sample rank-correlation coefficient. As with many other statistical procedures,

---

TABLE 18.10 **Sales Potential and Actual 2-Year Sales Data for 10 Salespersons**

Salesperson	Ranking of Potential	2-Year Sales (units)	Ranking According to 2-Year sales
A	2	400	1
B	4	360	3
C	7	300	5
D	1	295	6
E	6	280	7
F	3	350	4
G	10	200	10
H	9	260	8
I	8	220	9
J	5	385	2

TABLE 18.11 **Computation of the Spearman Rank-Correlation Coefficient for Sales Potential and Sales Performance**

Salesperson	x_i = Ranking of Potential	y_i = Ranking of Sales Performance	$d_i = x_i - y_i$	d_i^2
A	2	1	1	1
B	4	3	1	1
C	7	5	2	4
D	1	6	−5	25
E	6	7	−1	1
F	3	4	−1	1
G	10	10	0	0
H	9	8	1	1
I	8	9	−1	1
J	5	2	3	9
				$\Sigma\, d_i^2 = 44$

$$r_s = 1 - \frac{6\Sigma\, d_i^2}{n(n^2 - 1)} = 1 - \frac{6(44)}{10(100 - 1)} = .73$$

we may wish to use the sample results to make an inference about the population rank correlation, ρ_s, between two variables. In our example, the population rank-correlation coefficient could be obtained by making the rank-correlation coefficient computations for all members of the sales force. However, we would like to avoid all this data collection and make an inference about the population rank-correlation based on the sample rank-correlation coefficient, r_s. To make this inference, we must test the following hypotheses:

$$H_0: \ \rho_s = 0$$

$$H_a: \ \rho_s \neq 0$$

Under the null hypothesis of no rank correlation ($\rho_s = 0$), the rankings are independent and the sampling distribution of r_s is as follows.

Sampling Distribution of r_s

Mean: $\quad \mu_{r_s} = 0$ (18.10)

Standard deviation: $\quad \sigma_{r_s} = \sqrt{\dfrac{1}{(n-1)}}$ (18.11)

Form: Approximately normal provided $n \geq 10$

The sample rank-correlation coefficient for sales potential and sales performance in our example was $r_s = .73$. Use this value to test for a significant rank correlation.

From (18.10) we have $\mu_{r_s} = 0$, and from (18.11) we have $\sigma_{r_s} = \sqrt{1/(10 - 1)} = .33$. Using the test statistic, we have

$$z = \frac{r_s - \mu_{r_s}}{\sigma_{r_s}} = \frac{.73 - 0}{.33} = 2.21$$

Using a level of significance of $\alpha = .05$, we see that the null hypothesis of no correlation will be rejected if $z < -1.96$ or if $z > 1.96$. Since $z = 2.21 > 1.96$, we reject the hypothesis of no rank correlation. Thus we can conclude that a significant rank correlation exists between sales potential and sales performance.

▬ **EXERCISES**

23. Consider the following two sets of rankings for 6 items.

	Case One			**Case Two**	
Item	*First Ranking*	*Second Ranking*	*Item*	*First Ranking*	*Second Ranking*
A	1	1	A	1	6
B	2	2	B	2	5
C	3	3	C	3	4
D	4	4	D	4	3
E	5	5	E	5	2
F	6	6	F	6	1

Note that in the first case the rankings are identical, whereas in the second case the rankings are exactly opposite. What value should you expect for the Spearman rank-correlation coefficient for each of these cases? Explain. Calculate the rank-correlation coefficient for each case.

24. Airline passengers file complaints about lost, stolen, damaged, and delayed baggage. How do the airlines compare? Using the U.S. Department of Transportation (March 1989) data in Exercise 10, ten airlines can be ranked from fewest to most complaints per 1000 passengers. Shown below are the rankings of the airlines for December 1988 and January 1989.

December 1988	**January 1989**
Pan American	Pan American
Eastern	Northwest
United	United
Continental	USAir
Delta	Eastern
USAir	Delta
American	American
Northwest	Continental
TWA	TWA
Piedmont	Piedmont

a. Compute the rank correlation for the airlines for the two months of data.
b. Test for significant rank correlation using $\alpha = .05$. What is your conclusion?

25. In the baseball draft eight players are ranked by a scout in terms of speed and then in terms of power hitting.

Player	Speed Ranking	Power-Hitting Ranking
A	1	8
B	2	5
C	3	6
D	4	7
E	5	2
F	6	3
G	7	4
H	8	1

Use the Sperman rank-correlation coefficient to measure the association between speed and power. Use $\alpha = .05$ and test for the significance of this correlation coefficient.

26. In a poll of men and women television viewers, preferences for the top 10 shows led to the following rankings. Is there a relationship between the rankings provided for the two groups? Use $\alpha = .10$.

Television Show	Ranking by Men	Ranking by Women
1	1	5
2	5	10
3	8	6
4	7	4
5	2	7
6	3	2
7	10	9
8	4	8
9	6	1
10	9	3

27. A student organization surveyed both recent graduates and current students in an attempt to obtain information on the quality of teaching at a particular university. An analysis of the responses provided the following rankings for 10 professors on the basis of teaching ability.

Professor	Ranking by Current Students	Ranking by Recent Graduates
1	4	6
2	6	8
3	8	5
4	3	1
5	1	2
6	2	3
7	5	7
8	10	9
9	7	4
10	9	10

Do the rankings given by the current students agree with the rankings given by the recent graduates? Use $\alpha = .10$ and test for a significant rank correlation.

■ SUMMARY

In this chapter we have presented several statistical procedures that are classified as nonparametric methods. The parametric methods of the earlier chapters generally required interval or ratio data and were often based on assumptions concerning the population; e.g.., the probability distribution was normal. Since nonparametric methods can be applied to ordinal data as well as interval and ratio data and since nonparametric methods do not require population-distribution assumptions, nonparametric methods expand the class of problems that can be subjected to statistical analysis.

The sign test provides a nonparametric procedure for identifying differences between two populations when the only data available are nominal data. In the small-sample case, the binominal probability distribution can be used to determine the critical values for the sign test; in the large-sample case, a normal approximation may be used. The Wilcoxon signed-rank test provides a procedure for analyzing matched sample data whenever interval or ratio scaled data are available for each matched pair. The Wilcoxon procedure tests the hypothesis that the two populations being considered are identical.

The Mann-Whitney-Wilcoxon test provides a nonparametric method for testing for a difference between 2 populations based on 2 independent random samples. Tables were presented for the small-sample case and a normal approximation was provided for the large-sample case. The Kruskal-Wallis test extended the Mann-Whitney-Wilcoxon test to the case of 3 or more populations. The Kruskal-Wallis test is the nonparametric version of the parametric ANOVA test for differences among population means.

In the last section of this chapter we introduced the Spearman rank-correlation coefficient as a measure of association for two ordinal or rank-ordered sets of items.

■ GLOSSARY

Nonparametric methods   A collection of statistical methods that generally require very few, if any, assumptions about the population distributions and the level of measurement. These methods can be applied when nominal or ordinal data are available.

Distribution-free methods   Another name for nonparametric statistical methods suggested by the lack of assumptions required concerning the population distribution.

Sign test   A nonparametric statistical test for identifying differences between two populations based on the analysis of nominal data.

Wilcoxon signed-rank test   A nonparametric statistical test for identifying differences between two populations based on the analysis of two matched or paired samples.

Mann-Whitney-Wilcoxon (MWW) test   A nonparametric statistical test for identifying differences between two populations based on the analysis of two independent samples.

Kruskal-Wallis test   A nonparametric test for identifying differences among 3 or more populations.

Spearman rank-correlation coefficient   A correlation measure based on rank-order data for two variables.

■ KEY FORMULAS

Sign Test (Large Sample Case-Normal Approximation)

$$\text{Mean:} \quad \mu = .50n \tag{18.1}$$

$$\text{Standard deviation:} \quad \sigma = \sqrt{.25n} \tag{18.2}$$

Wilcoxon Signed-Rank Test

$$\mu_T = 0 \tag{18.3}$$

$$\sigma_T = \sqrt{\frac{n(n + 1)(2n + 1)}{6}} \qquad\qquad (18.4)$$

Mann-Whitney-Wilcoxon (Normal Approximation)

$$\mu_T = \tfrac{1}{2}\, n_1(n_1 + n_2 + 1) \qquad\qquad (18.6)$$

$$\sigma_T = \sqrt{\tfrac{1}{12}\, n_1 n_2(n_1 + n_2 + 1)} \qquad\qquad (18.7)$$

Kruskal-Wallis Test Statistic

$$W = \frac{12}{n_T\,(n_T + 1)} \sum_{i=1}^{k} \frac{R_i^2}{n_i} - 3(n_T + 1) \qquad\qquad (18.8)$$

Spearman Rank Correlation Coefficient

$$r_s = 1 - \frac{6\Sigma\, d_i^2}{n(n^2 - 1)} \qquad\qquad (18.9)$$

■ SUPPLEMENTARY EXERCISES

28. Mueller Beverage Products of Milwaukee, Wisconsin, has conducted a market research study designed to determine if there is a consumer preference for Mueller's Old Brew Beer over the individual consumer's usual beer. Each individual participating in the test was provided with a glass of his or her usual beer and a glass of Mueller's Old Brew. The two glasses were not labeled, and thus the individuals had no way of knowing beforehand which of the two glasses was Mueller's Old Brew and which was the individual's usual brand. The glass that each individual tasted first was randomly selected. After tasting the beer in each glass, the individuals were asked to indicate their *preferred* beer. The test results from a sample of 24 individuals are shown:

Individual	Brand Preferred	Value Recorded
1	Old Brew	+
2	Old Brew	+
3	Usual Brand	−
4	Old Brew	+
5	Usual Brand	−
6	Old Brew	+
7	Usual Brand	−
8	Old Brew	+
9	Old Brew	+
10	Usual Brand	−
11	Old Brew	+
12	Usual Brand	−
13	Usual Brand	−
14	Usual Brand	−
15	Old Brew	+
16	Usual Brand	−
17	Old Brew	+
18	Old Brew	+
19	Old Brew	+
20	Usual Brand	−
21	Old Brew	+
22	Old Brew	+
23	Usual Brand	−
24	Old Brew	+

If an individual selected Mueller's Old Brew as the preferred beer, a plus sign was recorded. On the other hand, if the individual stated a preference for his or her usual brand, a minus sign was recorded. Do the data for the 24 individuals indicate a significant difference in the preferences for the beers? Use $\alpha = .05$.

29. Two pilots for a prime-time television show (a western and a mystery show) are being tested. Both have been shown to a group of 12 viewers. The viewer preferences are shown.

Viewer	Preference	Viewer	Preference
1	Mystery	7	Mystery
2	Mystery	8	Western
3	Mystery	9	Mystery
4	Western	10	Mystery
5	Mystery	11	Mystery
6	Western	12	Mystery

Using $\alpha = .05$, test to see if there is a significant difference in preferences.

30. In a soft-drink taste test, 48 individuals stated a preference for one of two well-known brands. Results showed 28 favoring brand A, 16 favoring brand B, and 4 undecided. Use the sign test with $\alpha = .10$ and determine whether or not there is a significant difference in the preferences for the two brands of soft-drinks.

31. Use the sign test and perform the statistical analysis that will help us determine whether or not the task-completion times for the two production methods differ. The data are as follows. Use $\alpha = .05$.

Worker	Method 1 (minutes)	Method 2 (minutes)
1	10.2	9.5
2	9.6	9.8
3	9.2	8.8
4	10.6	10.1
5	9.9	10.3
6	10.2	9.3
7	10.6	10.5
8	10.0	10.0
9	11.2	10.6
10	10.7	10.2
11	10.6	9.8

32. The national median price of new homes for 1988 was $123,500 (*U.S. News & World Report*, September 19, 1988). Assume that data on the prices of new homes were obtained from samples of loans recorded in Chicago and Dallas–Fort Worth. Use the data to test the hypothesis that the median price of homes in each of the two cities is the same as the national median price. Use $\alpha = .05$. State your conclusion for each city.

	Greater Than $123,500	Equal to $123,500	Less Than $123,500
Chicago	55	6	28
Dallas–Forth Worth	42	3	36

33. Mayfield Products, Inc. has collected data on preferences of 12 individuals concerning cleaning power of 2 brands of detergent. The individuals and their preferences are

shown below. A plus indicates a preference for brand A.

Individual	Brand A Versus Brand B	Individual	Brand A Versus Brand B
1	−	7	−
2	+	8	+
3	+	9	+
4	+	10	−
5	−	11	+
6	+	12	+

With $\alpha = .10$, test for a significant difference in the preference for the 2 brands.

34. Twelve homemakers were asked to estimate the retail selling price of two models of refrigerators. The estimates of selling price provided by the homemakers are shown.

Homemaker	Model 1	Model 2
1	$650	$ 900
2	760	720
3	740	690
4	700	850
5	590	920
6	620	800
7	700	890
8	690	920
9	900	1000
10	500	690
11	610	700
12	720	700

Use these data and test at the .05 level of significance to determine if there is a difference in the homemaker's perception of selling price for the two models.

35. A study was designed to evaluate the weight-gain potential of a new poultry feed. A sample of 12 chickens was used in a 6-week study. The weight of each chicken was recorded before and after the 6-week test period. The difference between the before and after weights of each chicken are as follows: 1.5, 1.2, −.2, .0, .5, .7, .8, 1.0, .0, .6, .2, −.01. A negative value indicates a weight loss during the test period, whereas .0 indicates no weight change over the period. Use a .05 level of significance to determine if the new feed appears to provide a weight gain for the chickens.

36. The following data show product weights for items produced on two production lines. Test for a difference between the product weights for the two lines. Use $\alpha = .10$.

Production Line 1	Production Line 2
13.6	13.7
13.8	14.1
14.0	14.2
13.9	14.0
13.4	14.6
13.2	13.5
13.3	14.4
13.6	14.8
12.9	14.5
14.4	14.3
	15.0
	14.9

37. It is desired to determine if there is a significant difference in the time required to complete a program evaluation with the 3 different methods that are in common use. The

times (in hours) required for each of 18 evaluators to conduct a program evaluation are given below.

Method 1	Method 2	Method 3
68	62	58
74	73	67
65	75	69
76	68	57
77	72	59
72	70	62

Use $\alpha = .05$ and test to see if there is a significant difference in the time required by the 3 methods.

38. A sample of 20 engineers, who have been with a company for 3 years, has been taken and they have been rank-ordered with respect to managerial potential. Some of the engineers have attended the company's management-development course, others have attended an off-site management-development program at a local university, and the remainder have not attended any program. Use the rankings given and $\alpha = .025$ to test for a significant difference in the managerial potential of the three groups.

No Program	Company Program	Off-Site Program
16	12	7
9	20	1
10	17	4
15	19	2
11	6	3
13	18	8
	14	5

39. Shown below are course evaluation ratings for 4 instructors. Use $\alpha = .05$ and the Kruskal-Wallis procedure to test for a significant difference in teaching abilities.

Instructor	Course-Evaluation Rating								
Black	88	80	79	68	96	69			
Jennings	87	78	82	85	99	99	85	94	
Swanson	88	76	68	82	85	82	84	83	81
Wilson	80	85	56	71	89	87			

40. Wisman investment analysts ranked 12 companies, first with respect to book value and then with respect to growth potential.

Company	Ranking of Book Value	Ranking of Growth Potential
1	12	2
2	2	9
3	8	6
4	1	11
5	9	4
6	7	5
7	3	12
8	11	1
9	4	7
10	5	10
11	6	8
12	10	3

For these data does a relationship exist between the companies' book values and growth potentials? Use $\alpha = .05$.

41. Two individuals provided the following preference rankings of seven soft drinks. Compute the rank correlation for the two individuals.

Soft Drink	Ranking by Individual 1	Ranking by Individual 2
A	1	3
B	3	2
C	5	5
D	6	7
E	7	6
F	4	1
G	2	4

42. A sample of 15 students obtained the following rankings on midterm and final examinations in a statistics course.

Midterm Rank	Final Rank
1	4
2	7
3	1
4	3
5	8
6	2
7	5
8	12
9	6
10	9
11	14
12	15
13	11
14	10
15	13

Compute the Spearman rank-correlation coefficient for the data and test for a significant correlation with $\alpha = .10$.

APPLICATION

West Shell Realtors*

Cincinnati, Ohio

West Shell Realtors was founded in 1958 with one office and a sales staff of three people. The company's first-year sales were $900,000. In 1964 the Company began a long-term expansion program, with new offices being added almost yearly. Since that time West Shell has grown to its current size, with 17 offices located in southwest Ohio, southeast Indiana, and northern Kentucky.

As part of the company's expansion program, a 1968 merger with the Robson-Middendorf Company added a commercial and industrial division to the previously residentially focused company. Expansion in 1971 added property management to West Shell's range of services. Currently, West Shell is a recognized leader in the industry, with a motto that describes a full line of realty services for its customers.

West Shell is heavily involved with employee transfers and relocations for a wide variety of corporations. Currently, West Shell's relocation department accounts for approximately 47% of the Company's residential business. Selling houses for employees transferred to other locations and finding houses for employees transferred into the greater Cincinnati area are part of West Shell's total service for its customers.

STATISTICS IN REAL ESTATE

As you might expect, a real estate firm such as West Shell must monitor sales performance closely if it is to remain competitive. Monthly reports are generated for each of West Shell's 17 offices as well as for the total company. Statistical summaries of total sales dollars, number of units sold, mean selling price per unit, and so on are essential in keeping both office managers and the company's top management informed of progress and/or trouble spots in the organization. Monthly progress reports also show statistical summaries of the percentage over or under budget for each office in the Company.

In addition to monthly statistical summaries for ongoing operations, the company uses statistical considerations to guide corporate plans and strategies. Managers must determine where to focus sales efforts from among the total area served by the company. Statistical considerations, such as the ratio of total units sold to total units in a particular area provides information which guides the sales effort. Phone calls and canvassing efforts by the sales force are targeted in the high potential areas.

STATISTICAL ANALYSIS FOR OFFICE LOCATION

West Shell has implemented a strategy of planned expansion over the past 20 years. Each time an expansion plan calls for the establishment of a new sales office, the Company

*The authors are indebted to Rodney Fightmaster, Vice President, West Shell Realtors, Cincinnati, Ohio, for providing this application.

must address the question of the best place to locate the new office. Selling prices of homes, turnover rates, forecast sales volumes, and so on are the types of data that assist in evaluating and comparing alternative office location sites.

In one such instance the company had identified two areas as prime candidates for a new office location: Clifton and Roselawn. The statistical issues involved determining in what ways the two areas were alike and in what ways they differed. There were a variety of factors to be considered in comparing the two areas, but let us focus on one such factor: the selling price of homes in the two areas. Were the selling prices in the two areas similar or different? The actual sales price of units sold over a period of time could be viewed as a sample of sales for the area. In cases where the number of units sold were relatively small and where a normal distribution assumption for selling prices was inappropriate, nonparametric statistical methods could be employed to help answer the question concerning differences between the two areas.

For example, if a sample of 25 sales in the Clifton area showed a mean selling price of $75,250 and a sample of 18 sales in the Roselawn area showed a mean selling price of $70,375, the Mann-Whitney-Wilcoxon rank-sum test could be used to determine statistically whether or not the population of sales in the two areas appeared identical or not. Using the test with the rank-sum for the Clifton area we find $\mu_T = 550$ and $\sigma_T = 40.6$. The rank-sum for the Clifton area is $T = 595.2$ which yields $z = 1.11$. At the .05 level of significance, the Mann-Whitney-Wilcoxon test did not reject the hypothesis that the two populations are identical. The selection basis for the location of the new office should now focus on criteria other than unit selling price, since the areas do not differ significantly on this factor.

The real estate business has been and continues to be extremely competitive. At West Shell statistical data and statistical considerations play a meaningful role in helping the company maintain a leadership role in the industry.

Residential sales are a large portion of West Shell's business

CHAPTER 19

Decision Analysis

Decision analysis can be used to determine optimal strategies when a decision maker is faced with several decision alternatives and an uncertain or risk-filled pattern of future events. For example, a manufacturer of a new style or line of seasonal clothing would like to manufacture large quantities of the product if consumer acceptance and consequently demand for the product are going to be high. However, the manufacturer would like to produce much smaller quantities if consumer acceptance and demand for the product are going to be low. Unfortunately, seasonal clothing items require the manufacturer to make a production quantity decision before the demand is known. Actual consumer acceptance of the new product will not be determined until the items have been placed in the stores and buyers have had the opportunity to purchase them. Selection of the best production volume decision from among several production volume alternatives when the decision maker is faced with the uncertainty of future demand is a problem suited for decision analysis.

We begin the study of decision analysis by considering problems in which there are reasonably few decision alternatives and possible future events. The concepts of a payoff table and a decision tree are introduced to provide a structure for this type of decision situation and to illustrate the fundamentals involved in decision analysis. The discussion is then extended to show how additional information obtained through experimentation can be combined with the decision maker's preliminary information in order to develop an optimal decision strategy.

19.1 ▬ STRUCTURING THE DECISION PROBLEM

In order to illustrate the decision analysis approach, let us consider the case of Political Systems, Inc. (PSI), a newly formed computer service firm specializing in information services such as surveys and data analysis for individuals running for political office. PSI is in the final stages of selecting a computer system for its Midwest branch, located in Chicago. While the firm has decided on a computer manufacturer, it is currently attempting to determine the size of the computer system that would be the most economical to lease. We will use decision analysis to help PSI make its computer leasing decision.

The first step in the decision analysis approach is to identify the alternatives considered by the decision maker. For PSI, the final decision will be to lease one of three computer systems, which differ in size and capacity. The three *decision alternatives*, denoted by d_1, d_2, and d_3, are as follows:

$$d_1 = \text{lease the large computer system}$$

$$d_2 = \text{lease the medium-sized computer system}$$

$$d_3 = \text{lease the small computer system}$$

Obviously, the selection of the *best* decision alternative will depend on what PSI management foresees as the possible market acceptance of the service and consequently the possible demand or load on the PSI computer system. Often the future events associated with a decision situation are uncertain. That is, while a decision maker may have an idea of the variety of possible future events, the decision maker will often be unsure as to which particular event will occur. Thus the second step in a decision analysis approach is to identify the future events that might occur. These future events, which are not under the control of the decision maker, are referred to as the *states of nature*. It is assumed that the list of possible states of nature includes everything that can happen and that the individual states of nature do not overlap; that is, the states of nature are defined so that one and only one of the listed states of nature will occur.

When asked about the states of nature for the PSI decision problem, management viewed the possible acceptance of the PSI service as an either-or situation. That is, management believed that the firm's overall level of acceptance in the market place would be one of two possibilities: high acceptance or low acceptance. Thus the PSI states of nature, denoted s_1 and s_2, are as follows:

$$s_1 = \text{high customer acceptance of PSI services}$$

$$s_2 = \text{low customer acceptance of PSI services}$$

Given the three decision alternatives and the two states of nature, which computer system should PSI lease? In order to answer this question, we will need information on the profit associated with each combination of a decision alternative and a state of nature. For example, what profit would PSI experience if the firm decided to lease the large computer system (d_1) and market acceptance was high (s_1)? What profit would PSI experience if the firm decided to lease the large computer system (d_1) and market acceptance was low (s_2)?

Payoff Tables

In decision analysis terminology, we refer to the outcome resulting from making a certain decision and the occurrence of a particular state of nature as the *payoff*. Using the best information available, management has estimated the

TABLE 19.1   Payoff Table for the PSI Computer Leasing Problem

| Decision Alternatives | | States of Nature | |
		High Acceptance s_1	*Low Acceptance* s_2
Lease a large system	d_1	200,000	−20,000
Lease a medium-sized system	d_2	150,000	20,000
Lease a small system	d_3	100,000	60,000

Profit or payoff in $

payoffs or profits for the PSI computer leasing problem. These estimates are presented in Table 19.1. A table of this form is referred to as a *payoff table*. In general, entries in a payoff table can be stated in terms of profits, costs, or any other measure of output that may be appropriate for the particular situation being analyzed. The notation we will use for the entries in the payoff table is $V(d_i, s_j)$, which denotes the payoff associated with decision alternative d_i and state of nature s_j. Using this notation we see that $V(d_3, s_1) = \$100,000$.

Decision Trees

A *decision tree* provides a graphical representation of the decision-making process. Figure 19.1 shows a decision tree for the PSI computer leasing problem. Note that the tree shows the natural or logical progression that will occur over time. First the firm must make its decision (d_1, d_2, or d_3); then, once the decision

FIGURE 19.1   Decision Tree for the PSI Problem

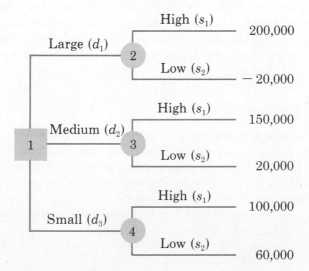

is implemented, the state of nature (s_1 or s_2) will occur. The number at each end point of the tree represents the payoff associated with a particular chain of events. For example, the topmost payoff of 200,000 arises whenever management makes the decision to purchase a large system (d_1) and customer acceptance turns out to be high (s_1). The next-lower terminal point of $-20,000$ is reached when management has made the decision to lease the large system (d_1) and the state of nature turns out to be a low degree of customer acceptance (s_2). Thus we see that each possible sequence of events for the PSI problem is represented in the decision tree.

Using the general terminology associated with decision trees, we will refer to the intersection or junction points of the tree as *nodes* and the arcs or connectors between the nodes as *branches*. Figure 19.1 shows the PSI decision tree with the nodes numbered 1 to 4. When the branches *leaving* a given node are decision branches, we refer to the node as a *decision node*. Decision nodes are denoted by squares. Similarly, when the branches leaving a given node are state-of-nature branches, we refer to the node as a *state-of-nature node*. State-of-nature nodes are denoted by circles. Using this node-labeling procedure, node 1 is a decision node, whereas nodes 2, 3, and 4 are state-of-nature nodes.

The identification of the decision alternatives, the states of nature, and the determination of the payoff associated with each decision alternative and state of nature combination are the first three steps in the decision analysis process. The question we now turn to is the following: How can the decision maker best utilize the information presented in the payoff table or the decision tree to arrive at a decision? As we will see, there are several approaches that may be used.

19.2 ■■■ DECISION MAKING WITHOUT PROBABILITIES

In this section we consider approaches to decision making that do not require knowledge of the probabilities of the states of nature. These approaches are appropriate in situations where the decision maker has very little confidence in his or her ability to assess the probabilities of the various states of nature, or where it is desirable to consider best- and worst-case analyses that are independent of state-of-nature probabilities. Because different approaches sometimes lead to different decision recommendations, it is important for the decision maker to understand the approaches available and then select the specific approach that, according to the decision maker's judgment, is the most appropriate.

Optimistic Approach

The *optimistic approach* evaluates each decision alternative in terms of the *best* payoff that can occur. The decision alternative that is recommended is the one that provides the best possible payoff. For a problem in which it is desired to maximize profit, as it is in the PSI leasing problem, the optimistic approach would lead the decision maker to choose the alternative corresponding to the largest profit. For problems involving minimization, this approach leads to choosing the alternative with the smallest payoff.

To illustrate the use of the optimistic approach we will show how it can be

TABLE 19.2 PSI Maximum Payoff ($) for Each Decision Alternative

Decision Alternatives		Maximum Payoff	
Large system	d_1	200,000 ◄————————Maximum of the
Medium system	d_2	150,000	maximum payoff values
Small system	d_3	100,000	

used to develop a recommendation for the PSI leasing problem. First we determine the maximum payoff possible for each of the decision alternatives; then we select the decision alternative that provides the overall maximum profit. This is just a systematic way of identifying the decision alternative that provides the largest possible profit. Table 19.2 illustrates this calculation for the PSI problem.

Since $200,000, corresponding to d_1, yields the maximum of the maximum payoffs, the decision to lease a large system is the recommended decision alternative using the optimistic approach. It is easy to see why this is called an optimistic approach. It simply recommends the decision alternative that provides the best of all payoffs, $200,000.

Conservative Approach

The *conservative approach* evaluates each decision alternative in terms of the *worst* payoff that can occur. The decision alternative recommended is the one that provides the best of the worst possible payoffs. For a problem in which the output measure is profit, as it is in the PSI leasing problem, the conservative approach would lead the decision maker to choose the alternative that maximizes the minimum possible profit that could be obtained. For problems involving minimization, this approach identifies the alternative that will minimize the maximum payoff.

To illustrate the use of the conservative approach, we will show how it can be used to develop a recommendation for the PSI leasing problem. First the decision maker would identify the minimum payoff for each of the decision alternatives; then, the decision maker would select the alternative that maximizes the minimum payoff. Table 19.3 illustrates this approach for the PSI problem.

TABLE 19.3 PSI Minimum Payoff ($) for Each Decision Alternative

Decision Alternatives		Minimum Payoff	
Large system	d_1	−20,000	Maximum of the
Medium system	d_2	20,000	minimum payoff values
Small system	d_3	60,000 ◄	

Since $60,000, corresponding to d_3, yields the maximum of the minimum payoffs, the decision alternative to lease a small system is recommended. This decision approach is considered conservative because it concentrates on the worst possible payoffs and then recommends the decision alternative that avoids the possibility of extremely "bad" payoffs. In using the conservative approach, PSI is guaranteed a profit of at least $60,000. While PSI may still make more, it *cannot* make less than $60,000.

Minimax Regret Approach

Minimax regret is another approach to decision making without probabilities. This approach is neither purely optimistic or purely conservative. Let us illustrate the minimax regret approach by showing how it can be used to select a decision alternative for the PSI leasing problem.

Suppose that we make the decision to lease the small system (d_3) and afterwards learn that customer acceptance of the PSI service is high (s_1). Table 19.1 shows the resulting profit to be $100,000. However, now that we know that state of nature s_1 has occurred, we see that the large system decision (d_1), yielding a profit of $200,000, would have been the optimal decision. The difference between the optimal payoff ($200,000) and the payoff experienced ($100,000) is referred to as the *opportunity loss* or *regret* associated with the d_3 decision when state of nature s_1 occurs ($200,000 − $100,000 = $100,000). If we had made decision d_2 and state of nature s_1 had occurred, the opportunity loss or regret would have been $200,000 − $150,000 = $50,000.

In maximization problems the general expression for opportunity loss or regret is given by

---

Opportunity Loss or Regret

$$R(d_i, s_j) = V^*(s_j) - V(d_i, s_j) \qquad (19.1)$$

where

$R(d_i, s_j)$ = regret associated with decision alternative d_i and state of nature s_j

$V^*(s_j)$ = best payoff value* under state of nature s_j

$V(d_i, s_j)$ = payoff associated with decision alternative d_i and state of nature s_j

---

Using equation (19.1) and the payoffs in Table 19.1, we can compute the regret associated with all combinations of decision alternatives d_i and states of nature s_j. We simply replace each entry in the payoff table with the value found by subtracting the entry from the largest entry in its column. Table 19.4 shows the regret, or opportunity loss, table for the PSI problem.

The next step in applying the minimax regret approach requires the decision maker to identify the maximum regret for each decision alternative. These data

---

*In cost minimization problems $V^*(s_j)$ will be the smallest entry in column j. Thus for minimization problems, equation (19.1) must be changed to $R(d_i, s_j) = V(d_i, s_j) - V^*(s_j)$.

TABLE 19.4   Regret or Opportunity Loss ($) for the PSI Problem

| | | States of Nature | |
| | | High Acceptance s_1 | Low Acceptance s_2 |
Decision Alternatives			
Large system	d_1	0	80,000
Medium system	d_2	50,000	40,000
Small system	d_3	100,000	0

are shown in Table 19.5. The choice of a best decision is made by selecting the alternative corresponding to the *mini*mum of the *maxi*mum *regret* values; hence the name *minimax regret*. For the PSI problem the decision to lease a medium-sized computer system, with a corresponding regret of $50,000, is the recommended minimax regret decision.

Note that the three approaches discussed in this section have provided different recommendations. This is not in itself bad. It simply reflects the difference in decision-making philosophies that underlie the various approaches. Ultimately, the decision maker will have to choose the most appropriate approach and then make the final decision accordingly. The major criticism of the approaches discussed in this section is that they do not consider any information about the probabilities of the various states of nature. In the next section we discuss an approach that utilizes probability information in selecting a decision alternative.

TABLE 19.5   PSI Maximum Regret or Opportunity Loss ($) for Each Decision Alternative

Decision Alternatives		Maximum Regret or Opportunity Loss	
Large system	d_1	80,000	
Medium system	d_2	50,000 ◄———Minimum of the
Small system	d_3	100,000	maximum regret

■ EXERCISES

1. Suppose that a decision maker faced with four decision alternatives and four states of nature develops the following profit payoff table:

| | | States of Nature | | | |
		s_1	s_2	s_3	s_4
	d_1	14	9	10	5
Decision	d_2	11	10	8	7
Alternatives	d_3	9	10	10	11
	d_4	8	10	11	13

a. If the decision maker knows nothing about the chances or probability of occurrence of the four states of nature, what is the recommended decision using the optimistic, conservative, and minimax regret approaches?

b. Which approach do you prefer? Explain. Is it important for the decision maker to establish the most appropriate approach before analyzing the problem? Explain.

c. Assume that the payoff table provides *cost* rather than profit payoffs. What is the recommended decision using the optimistic, conservative, and minimax regret approaches?

2. Southland Corporation's decision to produce a new line of recreational products has resulted in the need to construct either a small plant or a large plant. The decision as to which plant size to select depends on how the marketplace reacts to the new product line. In order to conduct an analysis, marketing management has decided to view the possible long-run demand as either low, medium, or high. The following payoff table shows the projected profit in millions of dollars.

		Low	Medium	High
		Long-Run Demand		
Decision	Small plant	150	200	200
Alternatives	Large plant	50	200	500

a. Construct a decision tree for this problem.

b. Determine the recommended decision using the optimistic, conservative, and minimax regret approaches.

3. McHuffter Condominiums, Inc. of Pensacola, Florida, recently purchased land near the Gulf of Mexico and is attempting to determine the size of the condominium development it should build. Three sizes of developments are being considered: small, d_1; medium, d_2; and large, d_3. At the same time an uncertain economy makes it difficult to ascertain the demand for the new condominiums. McHuffter's management realizes that a large development followed by a low demand could be very costly to the company. However, if McHuffter makes a conservative small development decision and then finds a high demand, the firm's profits will be lower than they might have been. With the three levels of demand—low, medium, and high—McHuffter's management has prepared the following payoff table:

		Low	Medium	High
		Demand		
	Small	400	400	400
Decision	Medium	100	600	600
Alternatives	Large	−300	300	900

Profit in $ × 10³

a. Construct a decision tree for this problem.

b. If nothing is known about the demand probabilities, what are the decision recommendations using the optimistic, conservative, and minimax regret approaches?

19.3 ■■■ DECISION MAKING WITH PROBABILITIES

In many decision-making situations it is possible to obtain probability estimates for each of the possible states of nature. When such probabilities are available, the *expected value approach* can be used to identify the best decision alternative.

The expected value approach evaluates each decision alternative in terms of its expected value. The decision alternative that is recommended is the one that provides the best expected value. Let us first define the expected value of a decision alternative and then show how it can be used for the PSI decision problem.

Let

$$N = \text{the number of possible states of nature}$$

$$P(s_j) = \text{the probability of state of nature } s_j$$

Since one and only one of the N states of nature can occur, the associated probabilities must satisfy the following two conditions:

$$P(s_j) \geq 0 \qquad \text{for all states of nature} \tag{19.2}$$

$$\sum_{j=1}^{N} P(s_j) = P(s_1) + P(s_2) + \cdots + P(s_N) = 1 \tag{19.3}$$

The expected value (EV) of decision alternative d_i is defined as follows:

Expected Value of Decision Alternative d_i

$$EV(d_i) = \sum_{j=1}^{N} P(s_j) V(d_i, s_j) \tag{19.4}$$

In words, the expected value of a decision alternative is the sum of weighted payoffs for the alternative. The weight for a payoff is the probability of the associated state of nature and therefore the probability that the payoff occurs. Let us now return to the PSI problem to see how the expected value approach can be applied.

Suppose that PSI management believes that s_1, the high-acceptance state of nature, has a .3 probability of occurrence and that s_2, the low-acceptance state of nature, has a .7 probability. Thus $P(s_1) = .3$ and $P(s_2) = .7$. Using the payoff values $V(d_i, s_j)$ shown in Table 19.1 and equation (19.4), expected values for the three decision alternatives can be calculated:

$$EV(d_1) = .3(200,000) + .7(-20,000) = \$46,000$$

$$EV(d_2) = .3(150,000) + .7(20,000) = \$59,000$$

$$EV(d_3) = .3(100,000) + .7(60,000) = \$72,000$$

Thus, according to the expected value approach, since d_3 has the highest expected value (\$72,000), d_3 is the recommended decision.

The calculations required to identify the decision alternative with the best expected value can be conveniently carried out on a decision tree. Figure 19.2 shows the decision tree for the PSI problem with state-of-nature branch

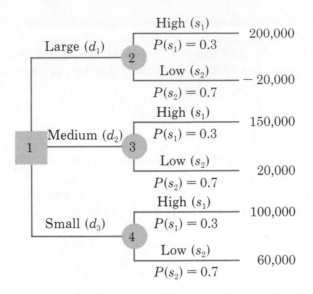

Large (d_1)

High (s_1)
$P(s_1) = 0.3$ 200,000

2

Low (s_2)
$P(s_2) = 0.7$ – 20,000

Medium (d_2)

High (s_1)
$P(s_1) = 0.3$ 150,000

3

Low (s_2)
$P(s_2) = 0.7$ 20,000

1

Small (d_3)

High (s_1)
$P(s_1) = 0.3$ 100,000

4

Low (s_2)
$P(s_2) = 0.7$ 60,000

■■■ FIGURE 19.2   **PSI Decision Tree with State-of-Nature Branch Probabilities**

probabilities. We will now use the branch probabilities and the expected value approach to arrive at the optimal decision for PSI.

Working backward through the decision tree, we first compute the expected value at each state-of-nature node. That is, at each state-of-nature node we weight each possible payoff by its chance of occurrence. By doing this we obtain the expected values for nodes 2, 3, and 4 as shown in Figure 19.3.

Since the decision maker controls the branch leaving decision node 1 and since we are trying to maximize expected profits, the best decision branch at node 1 is d_3. Thus the decision tree analysis leads us to recommend d_3 with an

■■■ FIGURE 19.3   **Applying the Expected Value Approach Using Decision Trees**

Large (d_1)

2   EV = 0.3 (200,000) + 0.7 (– 20,000) = $46,000

Medium (d_2)

1

3   EV = 0.3 (150,000) + 0.7 (20,000) = $59,000

Small (d_3)

4   EV = 0.3 (100,000) + 0.7 (60,000) = $72,000

expected value of $72,000. Note that this is the same recommendation that was obtained using the expected value approach in conjunction with the payoff table.

We have seen how decision trees can be used to analyze decisions with state-of-nature probabilities. While other decision problems may be substantially more complex than the PSI problem, if there are a reasonable number of decision alternatives and states of nature, the decision tree approach outlined in this section can be used. First the analyst must draw a decision tree consisting of decision and state-of-nature nodes and branches that describe the sequential nature of the problem. Assuming that the expected value approach is to be used, the next step is to determine the probabilities for each of the state-of-nature branches and compute the expected value at each state-of-nature node. The decision branch leading to the state-of-nature node with the best expected value is then selected. The decision alternative associated with this branch is the decision recommended.

■ EXERCISES

4. The payoff table presented in Exercise 1 is repeated here:

| | | States of Nature | | | |
		s_1	s_2	s_3	s_4
	d_1	14	9	10	5
Decision	d_2	11	10	8	7
Alternatives	d_3	9	10	10	11
	d_4	8	10	11	13

Suppose that the decision maker obtains information that enables the following probability estimates to be made: $P(s_1) = .5, P(s_2) = .2, P(s_3) = .2, P(s_4) = .1$.

a. Use the expected value approach to determine the optimal decision.

b. Now assume that the entries in the payoff table are costs; use the expected value approach to determine the optimal decision.

5. Hale's TV Productions is considering producing a pilot for a comedy series for a major television network. While the network may reject the pilot and the series, it may also purchase the program for 1 or 2 years. Hale may decide to produce the pilot or transfer the rights for the series to a competitor for $100,000. Hale's profits are summarized in the following payoff table:

| | | States of Nature | | |
		Reject	1 Year	2 Years
Produce pilot	d_1	−100	50	150 ◀——— Profit in
Sell to competitor	d_2	100	100	100	$ × 10^3

If the probability estimates for the states of nature are $P(\text{reject}) = .2, P(1 \text{ year}) = .3$, and $P(2 \text{ years}) = .5$, what should the company do?

6. Consider the McHuffter Condominium problem presented in Exercise 3. If $P(\text{low}) = .20, P(\text{medium}) = .35$, and $P(\text{high}) = .45$, what is the decision recommended using the expected value approach?

7. Martin's Service Station is considering investing in a heavy-duty snowplow this fall. Martin has analyzed the situation carefully and feels that this would be a very profitable

investment if the snowfall is heavy. A small profit could still be made if the snowfall is moderate, but Martin would lose money if snowfall is light. Specifically, Martin forecasts a profit of $7000 if snowfall is heavy and $2000 if it is moderate, and a $9000 loss if it is light. Based on the weather bureau's long-range forecast, Martin estimates that P(heavy snowfall) = .4, P(moderate snowfall) = .3, and P(light snowfall) = .3.

  a. Prepare a decision tree for Martin's problem.
  b. What is the expected value at each state-of-nature node?
  c. Using the expected value approach, would you recommend that Martin invest in the snowplow?

19.4 ■ SENSITIVITY ANALYSIS

For the PSI problem, management provided a .3 probability for s_1, the high-acceptance state of nature, and a .7 probability for s_2, the low-acceptance state of nature. Using these probabilities we found that decision alternative d_3 had the highest expected value and was the recommended decision. In this section we consider how changes in the probability estimates for the states of nature affect or alter the recommended decision. The study of the effect of such changes is referred to as *sensitivity analysis*.

One approach to sensitivity analysis is to consider different probabilities for the states of nature and then recompute the expected value for each of the decision alternatives. Repeating this several times, we can begin to learn how changes in the probabilities for the states of nature affect the recommended decision. For example, suppose that we consider a change in the probabilities for the states of nature such that $P(s_1) = .6$ and $P(s_2) = .4$. Using these probabilities and repeating the expected value computations, we find the following:

$$EV(d_1) = .6(200,000) + .4(-20,000) = \$112,000$$

$$EV(d_2) = .6(150,000) + .4(20,000) \quad = \$\ 98,000$$

$$EV(d_3) = .6(100,000) + .4(60,000) \quad = \$\ 84,000$$

Thus, with these probabilities, the recommended decision alternative is d_1, with an expected value of $112,000.

Obviously, we could continue to modify the probabilities of the states of nature and begin to learn more about how such changes affect the recommended decision. The only drawback to this approach is that there will be numerous calculations required to evaluate the effect of several possible changes in the state of nature probabilities.

For the special case of decision analysis with two states of nature, the sensitivity analysis computations can be eased substantially through the use of a graphical procedure. Let us demonstrate this procedure by further analyzing the PSI problem. We begin by denoting the probability of state of nature s_1 by p. That is,

$$P(s_1) = p$$

and thus

$$P(s_2) = 1 - P(s_1) = 1 - p$$

The expected value for decision alternative d_1 can then be written as a function of p.

$$
\begin{aligned}
\text{EV}(d_1) &= P(s_1)(200{,}000) + P(s_2)(-20{,}000) \\
&= p(200{,}000) + (1 - p)(-20{,}000) \\
&= 220{,}000p - 20{,}000
\end{aligned}
\tag{19.5}
$$

Repeating the expected value computation for decision alternatives d_2 and d_3, we obtain the following expressions for expected value as a function of p:

$$\text{EV}(d_2) = 130{,}000p + 20{,}000 \tag{19.6}$$

$$\text{EV}(d_3) = 40{,}000p + 60{,}000 \tag{19.7}$$

Thus we have developed three linear equations that express the expected value of the three decision alternatives as a function of the probability of state of nature s_1.

Let us continue by developing a graph with values of p on the horizontal axis and the associated expected values on the vertical axis. Since expressions (19.5), (19.6), and (19.7) are all linear equations, we can graph each equation, or line, by finding any two points on the line and drawing the line through the points. Using $\text{EV}(d_1)$ in (19.5) as an example, we first let $p = 0$ and find that $\text{EV}(d_1) = -20{,}000$. Then, letting $p = 1$, we find that $\text{EV}(d_1) = 200{,}000$. Connecting these two points, $(0, -20{,}000)$ and $(1, 200{,}000)$, provides the line labeled $\text{EV}(d_1)$ in Figure 19.4. This figure also shows a line labeled $\text{EV}(d_2)$ and a line labeled $\text{EV}(d_3)$; these are the graphs of (19.6) and (19.7), respectively.

Figure 19.4 can now be used for sensitivity analysis. Recall that PSI is seeking to maximize profit. Note that for small values of p, decision alternative d_3 provides the largest expected value and is thus the recommended decision. Similarly, for large values of p, we see that decision alternative d_1 provides the largest expected value and is thus the recommended decision. Further, note that there is no section of the graph for which decision alternative d_2 provides the largest expected value. Thus, with the exception of the point where the three expected value lines intersect and all three expected values are equal, decision alternative d_2 can never be the decision alternative recommended by the expected value approach.

Referring to Figure 19.4 again, we see that the three lines intersect at a value of p between .4 and .5. In other words, at some point between .4 and .5, all three decision alternatives provide the same expected value.

Whenever two or more lines intersect on a sensitivity analysis graph, we can set the equations for two of the intersecting lines equal to each other and solve for the value of p. Using the equations for decision alternative d_1 and decision alternative d_3, we obtain

$$220{,}000p - 20{,}000 = 40{,}000p + 60{,}000$$

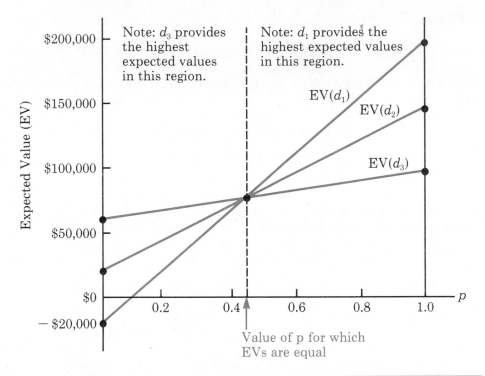

Note: d_3 provides the highest expected values in this region.

Note: d_1 provides the highest expected values in this region.

EV(d_1)

EV(d_2)

EV(d_3)

Value of p for which EVs are equal

FIGURE 19.4　Expected Value as a Function of p

Thus

$$180{,}000p = 80{,}000$$

$$p = \frac{80{,}000}{180{,}000} = .44$$

Hence, whenever $p = .44$, each decision alternative will provide the same expected value. Using this value of p in Figure 19.4, we can now conclude that for $p < .44$, decision alternative d_3 provides the largest expected value; and that for $p > .44$, decision alternative d_1 provides the largest expected value. Since p is simply the probability of state of nature s_1 and $(1 - p)$ is the probability of state of nature s_2, we now have the sensitivity analysis information that tells us how changes in the state-of-nature probabilities affect the recommended decision alternative.

The benefit of performing sensitivity analysis is that it can provide a better perspective on management's original judgment regarding the state-of-nature probabilities. Management originally estimated the probability of high customer acceptance as $P(s_1) = .3$. As a result, decision alternative d_3 was recommended. After carrying out the sensitivity analysis, we can now tell management that the original estimate of $P(s_1)$ is not extremely critical in order for d_3 to be the recommended decision. In fact, as long as $P(s_1) < .44$, the d_3 decision alternative remains optimal.

Note that in the PSI problem the three lines for the three decision alternatives graphed in Figure 19.4 intersect at the same point ($p = .44$). Similar sensitivity analysis computations for other decision analysis problems with two states of nature and three decision alternatives should not be expected to result in the same type of graph. With a different sensitivity analysis graph, d_1 could be the best decision alternative for certain values of p, d_2 the best decision alternative of other values of p, and d_3 the best decision alternative for the remaining values of p. This situation is demonstrated in the sensitivity analysis computation for Exercise 8 at the end of the section.

The graphical sensitivity analysis procedure we have described for the PSI problem applies only to decision analysis problems with two states of nature. However, sensitivity analysis is important in problems with more than two states of nature. In these cases a computer software package can be used to assist with the computations. Basically, we return to the approach of testing a variety of likely changes for the state-of-nature probabilities. The software package is helpful in making the necessary expected value computations and providing the decision alternative recommendations with a minimum of time and effort on the part of the analyst.

▬ EXERCISES

8. The payoff table showing profit for a decision problem with two states of nature and three decision alternatives is presented below:

		States of Nature	
		s_1	s_2
Decision	d_1	80	50
Alternatives	d_2	65	85
	d_3	30	100

Use graphical sensitivity analysis to determine the values of the probability of state of nature s_1 for which each of the decision alternatives has the largest expected value.

9. Milford Trucking, located in Chicago, has requests to haul two shipments, one to St. Louis and one to Detroit. Because of a scheduling problem, Milford will be able to accept only one of these assignments. The St. Louis customer has guaranteed a return shipment, but the Detroit customer has not. Thus if Milford accepts the Detroit shipment and cannot find a Detroit-to-Chicago return shipment, the truck will return to Chicago empty. The payoff table showing profit is as follows:

Shipment		Return Shipment from Detroit s_1	No Return Shipment from Detroit s_2
St. Louis	d_1	2000	2000
Detroit	d_2	2500	1000

a. If the probability of a Detroit return shipment is .4, what should Milford do?
b. Use graphical sensitivity analysis to determine the values of the probability of state of nature s_1 for which d_1 has the largest expected value.

19.5 ▬ EXPECTED VALUE OF PERFECT INFORMATION

Suppose that PSI had the opportunity to conduct a market research study that would evaluate consumer needs for the PSI service. Such a study could help by improving the current probability assessments for the states of nature. However, if the cost of obtaining the market research information exceeds its value, PSI should not conduct the market research study.

To determine the maximum possible value that PSI should pay for additional information, let us suppose that PSI could obtain perfect information regarding the states of nature; that is, we will assume that PSI could determine with certainty which state of nature will occur. To make use of perfect information we need to develop a decision strategy for PSI to follow. As we will show, a decision strategy is simply a policy or decision rule that is to be followed by the decision maker. In computing the *expected value of perfect information* (*EVPI*), the decision strategy is a rule that specifies which decision alternative should be selected given each state of nature.

To help determine the optimal decision strategy for PSI we have reproduced PSI's payoff table as Table 19.6. We see that if state of nature s_1 occurs, then the best decision alternative is d_1 with a profit of $200,000. Similarly, if state of nature s_2 occurs, then the best decision alternative is d_3 with a profit of $60,000. Thus the optimal decision strategy PSI should follow if perfect information were available can be stated as follows:

If s_1 occurs, then select d_1.

If s_2 occurs, then select d_3.

What is the expected value for this decision strategy? Since $P(s_1) = .3$ and $P(s_2) = .7$, we see that there is a .3 probability that PSI will make $200,000 and a .7 probability PSI will make $60,000. Thus the expected value of the decision strategy that uses perfect information is

$$(.3)(\$200,000) + (.7)(\$60,000) = \$102,000$$

Recall that when perfect information was not available, the expected value approach resulted in recommending decision alternative d_3 with an expected value of $72,000. Since $72,000 is the expected value without perfect information and $102,000 is the expected value with perfect information, $102,000 —

TABLE 19.6 **Payoff Table for the PSI Problem**

| | | States of Nature | |
| | | High Acceptance | Low Acceptance |
Decision Alternatives		s_1	s_2
Large system	d_1	200,000	−20,000
Medium system	d_2	150,000	20,000
Small system	d_3	100,000	60,000

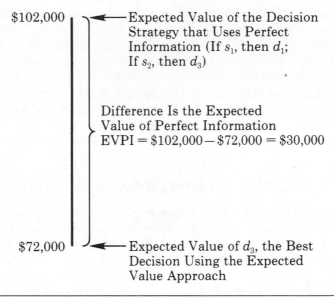

$102,000 ◀—— Expected Value of the Decision
Strategy that Uses Perfect
Information (If s_1, then d_1;
If s_2, then d_3)

Difference Is the Expected
Value of Perfect Information
EVPI = $102,000 − $72,000 = $30,000

$72,000 ◀—— Expected Value of d_3, the Best
Decision Using the Expected
Value Approach

FIGURE 19.5   The Expected Value of Perfect Information

$72,000 = $30,000 represents the expected value of perfect information (EVPI); that is,

$$EVPI = \$102,000 - \$72,000 = \$30,000$$

In other words, $30,000 represents the additional expected value that can be obtained if perfect information were available about the states of nature. Figure 19.5 provides a summary of the computation of the EVPI for the PSI problem.

Generally speaking, a market research study will not provide "perfect" information; however, the information provided might be worth a good portion of the $30,000. In any case, PSI's management knows it should never pay more than $30,000 for any information, no matter how good. Provided the market survey cost is reasonably small—say, $5000 to $10,000—it appears economically desirable for PSI to consider the market research study.

■ EXERCISES

10. Consider the Hale's TV Productions problem (Exercise 5). What is the maximum that Hale should be willing to pay for inside information on what the network will do?

11. Consider the McHuffter Condominium problem (Exercise 6). What is the expected value of perfect information?

12. Refer again to the investment problem faced by Martin's Service Station (Exercise 7). Martin can purchase a blade to attach to his service truck that can also be used to plow driveways and parking lots. Since this truck must also be available to start cars, etc., Martin will not be able to generate as much revenue plowing snow if he elects this alternative. But he will keep his loss smaller if there is light snowfall. Under this alternative Martin forecasts a profit of $3500 if snowfall is heavy and $1000 if it is moderate and a $1500 loss if snowfall is light.

    a. Prepare a new decision tree showing all three alternatives.
    b. Using the expected value approach, what is the optimal decision?
    c. What is the expected value of perfect information?

13. Consider the Milford Trucking problem (Exercise 9). What is the expected value of perfect information that would tell Milford Trucking whether or not Detroit has a return shipment?

19.6 ■■■ DECISION ANALYSIS WITH SAMPLE INFORMATION

In applying the expected value approach, we have seen how probability information about the states of nature affects the expected value calculations and thus the decision recommendation. Frequently decision makers have preliminary or prior probability estimates for the states of nature that are initially the best probability values available. However, in order to make the best possible decision, the decision maker may want to seek additional information about the states of nature. This new information can be used to revise or update the prior probabilities so that the final decision is based on more accurate probability estimates for the states of nature.

The seeking of additional information is most often accomplished through experiments designed to provide sample information or more current data about the states of nature. Raw material sampling, product testing, and test market research are examples of experiments that may enable a revision or updating of the state-of-nature probabilities. In the following discussion we will reconsider the PSI computer leasing problem and show how sample information can be used to revise the state-of-nature probabilities. We will then show how the revised probabilities can be used to develop an optimal decision strategy for PSI.

Recall that management had assigned a probability of $P(s_1) = .3$ to state of nature s_1 and a probability of $P(s_2) = .7$ to state of nature s_2. At this point we will refer to these initial probability estimates, $P(s_1)$ and $P(s_2)$, as the *prior probabilities* for the states of nature. Using these prior probabilities we found that d_3, the decision to lease the small system was optimal, yielding an expected value of $72,000. Recall also that we showed that the expected value of new information about the states of nature could potentially be worth as much as EVPI = $30,000.

Suppose that PSI decides to consider hiring a market research firm to study the potential acceptance of the PSI service. The market research study will provide new information that can be combined with the prior probabilities through a Bayesian procedure to obtain updated or revised probability estimates for the states of nature. These *revised* probabilities are called *posterior probabilities*. The process of revising probabilities is depicted in Figure 19.6.

We will refer to the new information obtained through research or experimentation as an *indicator*. Since in many cases the experiment conducted to obtain the additional information will consist of taking a statistical sample, the new information is also often referred to as *sample information*.

Using the indicator terminology, we can denote the outcomes of the PSI marketing research study as follows:

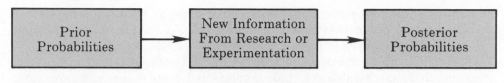

I_1 = favorable market research report (i.e., in the market research study the individuals contacted generally express interest in PSI's services)

I_2 = unfavorable market research report (i.e., in the market research study the individuals contacted generally express little interest in PSI's services)

Given one of these possible indicators, our objective is to provide improved estimates of the probabilities of the two states of nature. The end result of the *Bayesian revision* process depicted in Figure 19.6 is a set of posterior probabilities of the form $P(s_j|I_k)$, where $P(s_j|I_k)$ represents the conditional probability that state of nature s_j will occur given that the outcome of the market research study was indicator I_k.

To make effective use of this indicator information, we must know something about the probability relationships between the indicators and the states of nature. For example, in the PSI problem, given that the state of nature ultimately turns out to be high customer acceptance, what is the probability that the market research study will result in a favorable report? In this case we are asking about the conditional probability of indicator I_1 given state of nature s_1, written $P(I_1|s_1)$. In order to carry out the analysis, we will need conditional probabilities for all indicators given all states of nature, that is, $P(I_1|s_1)$, $P(I_1|s_2)$, $P(I_2|s_1)$, and $P(I_2|s_2)$.

In the PSI example the past record of the marketing research company on similar studies has led to the following estimates of the relevant conditional probabilities:

States of Nature	Market Research Report	
	Favorable I_1	*Unfavorable I_2*
High acceptance s_1	$P(I_1\|s_1) = .8$	$P(I_2\|s_1) = .2$
Low acceptance s_2	$P(I_1\|s_2) = .1$	$P(I_2\|s_2) = .9$

Note that these probability estimates indicate that a good degree of confidence can be placed in the market research report. When the true state of nature is s_1, the market research report will be favorable 80% of the time and unfavorable only 20%. When the true state is s_2, the report will make the correct indication 90% of the time. Now let us see how this additional information can be incorporated into the decision-making process.

19.7 ▬▬▬ DEVELOPING A DECISION STRATEGY

A decision strategy is a policy or decision rule that is to be followed by the decision maker. In the PSI case, with the market research study, a decision strategy is a rule that recommends a particular decision based on whether the market research report is favorable or unfavorable. We will employ a decision tree analysis to find the optimal decision strategy for PSI.

Figure 19.7 shows the decision tree for the PSI computer leasing problem provided that a market research study is conducted. Note that as you move from left to right, the tree shows the natural or logical order that will occur in the decision-making process. First the firm will obtain the market research report indicator (I_1 or I_2); then a decision (d_1, d_2, or d_3) will be made; finally, the state of nature (s_1 or s_2) will occur. The decision and the state of nature combine to provide the final profit or payoff.

▬ FIGURE 19.7   **The PSI Decision Tree Incorporating the Results of the Market Research Study**

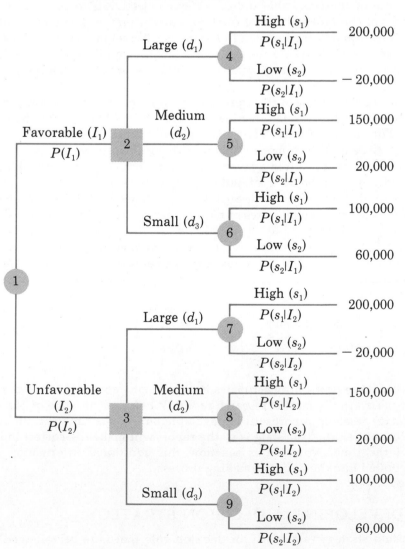

Using decision tree terminology, we have now introduced an *indicator node*, node 1, and *indicator branches*, I_1 and I_2. Since the branches emanating from indicator nodes are not under the control of the decision maker but are determined by chance, these nodes are depicted by a circle just like the state-of-nature nodes. We see that nodes 2 and 3 are decision nodes, while nodes 4, 5, 6, 7, 8, and 9 are state-of-nature nodes. For decision nodes the decision maker must select the specific branch d_1, d_2, or d_3 that will be taken. Selecting the best decision branch is equivalent to making the best decision. However, since the indicator and state-of-nature branches are not controlled by the decision maker, the specific branch leaving an indicator or a state-of-nature node will depend on the probability associated with the branch. Thus, before we can carry out an analysis of the decision tree and develop a decision strategy, we must compute the probability of each indicator branch and the probability of each state-of-nature branch. Note from the decision tree that the state-of-nature branches occur *after* the indicator branches. Thus when we attempt to compute state-of-nature branch probabilities, we will need to consider which indicator was previously observed. That is, we will express state-of-nature probabilities in terms of the probability of state of nature s_j *given* indicator I_k was observed. Thus all state-of-nature probabilities will be expressed in a $P(s_j|I_k)$ form.

Computing Branch Probabilities

The prior probabilities for the states of nature in the PSI problem were given as $P(s_1) = .3$ and $P(s_2) = .7$. In Section 19.6 we identified the relationships between the market research indicators and states of nature with the conditional probabilities

$$P(I_1|s_1) = .8 \qquad P(I_2|s_1) = .2$$
$$P(I_1|s_2) = .1 \qquad P(I_2|s_2) = .9$$

In order to develop a decision strategy utilizing the decision tree in Figure 19.7, we need indicator branch probabilities $P(I_k)$ and state-of-nature branch probabilities $P(s_j|I_k)$. The problem now facing us is determining how to use the given prior probability estimates $P(s_j)$ and conditional probability estimates $P(I_k|s_j)$ to calculate the branch probabilities $P(I_k)$ and $P(s_j|I_k)$. In this section we will show how the Bayesian revision process discussed in Chapter 4 and referred to in Figure 19.6 can be used to calculate the branch probabilities $P(I_k)$ and $P(s_j|I_k)$.

In order to see how this Bayesian procedure is applied and at the same time understand how the procedure works, let us look closely at the calculation of the indicator branch probability, $P(I_1)$, for the PSI market research study. First, note that there are only two ways in which the outcome I_1 can occur:

1. The market research report is favorable (I_1) *and* the state of nature turns out to be high acceptance (s_1), written ($I_1 \cap s_1$).

2. The market research report is favorable (I_1) *and* the state of nature turns out to be low acceptance (s_2), written ($I_1 \cap s_2$).

The probabilities of these two outcomes are written $P(I_1 \cap s_1)$ and $P(I_1 \cap s_2)$, respectively. We can now add these two probabilities to obtain the following branch probability:

$$P(I_1) = P(I_1 \cap s_1) + P(I_1 \cap s_2) \qquad (19.8)$$

The multiplication law of probability provides the following formulas for $P(I_1 \cap s_1)$ and $P(I_1 \cap s_2)$:

$$P(I_1 \cap s_1) = P(I_1|s_1)P(s_1) \qquad (19.9)$$

$$P(I_1 \cap s_2) = P(I_1|s_2)P(s_2) \qquad (19.10)$$

Finally, substituting the above expressions for $P(I_1 \cap s_1)$ and $P(I_1 \cap s_2)$ in equation (19.8), we obtain

$$P(I_1) = P(I_1|s_1)P(s_1) + P(I_1|s_2)P(s_2) \qquad (19.11)$$

Generalizing the above expression for any indicator branch probability, $P(I_k)$, and N states of nature, $s_1, s_2, \ldots, s_N$, we have

$$P(I_k) = P(I_k|s_1)P(s_1) + P(I_k|s_2)P(s_2) + \cdots + P(I_k|s_N)P(s_N) \qquad (19.12)$$

or

$$P(I_k) = \sum_{j=1}^{N} P(I_k|s_j)P(s_j) \qquad (19.13)$$

Returning to the PSI problem with the two prior probabilities $P(s_1) = .3$ and $P(s_2) = .7$ and the conditional probabilities $P(I_1|s_1) = .8, P(I_1|s_2) = .1, P(I_2|s_1) = .2$, and $P(I_2|s_2) = .9$, we can use equation (19.13) to compute the two indicator branch probabilities. These calculations are as follows:

$$P(I_1) = P(I_1|s_1)P(s_1) + P(I_1|s_2)P(s_2)$$
$$= (.8)(.3) + (.1)(.7) = .31$$

and

$$P(I_2) = P(I_2|s_1)P(s_1) + P(I_2|s_2)P(s_2)$$
$$= (.2)(.3) + (.9)(.7) = .69$$

The above probabilities indicate that the probability of I_1, a favorable market research report, is .31 and the probability of I_2, an unfavorable market research report, is .69.

Now that we know the indicator branch probabilities, let us show how the Bayesian process enables us to compute the revised, or posterior, state-of-nature branch probabilities $P(s_j|I_k)$. We will illustrate this procedure by considering the state-of-nature branch probability $P(s_1|I_1)$, the probability the market acceptance is high (s_1) given that the market research report is favorable (I_1). The fundamental conditional probability relationship as presented in Chapter 4 can be written

$$P(s_1|I_1) = \frac{P(I_1 \cap s_1)}{P(I_1)} \tag{19.14}$$

Using equation (19.9) for $P(I_1 \cap s_1)$, we have

$$P(s_1|I_1) = \frac{P(I_1|s_1)P(s_1)}{P(I_1)} \tag{19.15}$$

With known probabilities $P(I_1|s_1) = .8$, $P(s_1) = .3$, and $P(I_1) = .31$, the revised state-of-nature probability, $P(s_1|I_1)$, becomes

$$P(s_1|I_1) = \frac{(.8)(.3)}{0.31} = \frac{.24}{.31} = .7742$$

Recall that the prior probability of a high market acceptance was $P(s_1) = .3$. The preceding probability information now tells us that if the market research indicator is favorable, the probability of a high market acceptance should be revised to $P(s_1|I_1) = .7742$.

Generalizing (19.15) for any state of nature s_j and any indicator I_k provides

$$P(s_j|I_k) = \frac{P(I_k|s_j)P(s_j)}{P(I_k)} \tag{19.16}$$

Thus, we can use (19.16) to compute the revised or posterior state-of-nature branch probabilities. For example, the revised probability of low market acceptance, s_2, given the market research indicator is favorable, I_1, becomes

$$P(s_2|I_1) = \frac{P(I_1|s_2)P(s_2)}{P(I_1)} = \frac{(.1)(.7)}{.31} = \frac{.07}{.31} = .2258$$

Similar calculations for an unfavorable market research indicator, I_2, will provide the revised state-of-nature branch probabilities $P(s_1|I_2) = .0870$ and $P(s_2|I_2) = .9130$. Figure 19.8 shows the PSI decision tree after all indicator and all revised state-of-nature branch probabilities have been computed.

Although the above procedure can be used to compute branch probabilities, the calculations can become quite cumbersome as the problem size grows larger. Thus in order to assist in applying Bayes' theorem to compute branch probabilities, we present the following tabular procedure that will make it easier to carry out the computations, especially for large decision analysis problems.

Computing Branch Probabilities: A Tabular Procedure

The procedure used for computing the probabilities of the indicator and state-of-nature branches can be carried out by utilizing the tabular approach for Bayes' theorem discussed in Chapter 4. First, for each indicator I_k we form a table

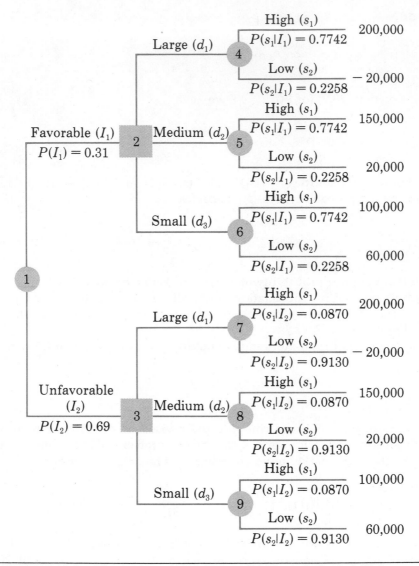

■■■ FIGURE 19.8   **The PSI Decision Tree with Branch Probabilities**

consisting of the following five column headings:

■ Column 1   States of nature s_j
■ Column 2   Prior probabilities $P(s_j)$
■ Column 3   Conditional probabilities $P(I_k|s_j)$
■ Column 4   Joint probabilities $P(I_k \cap s_j)$
■ Column 5.   Posterior probabilities $P(s_j|I_k)$

Then, given any indicator I_k, the following procedure can be used to calculate $P(I_k)$ and the $P(s_j|I_k)$ values.

Step 1. In column 1 list the states of nature appropriate to the problem being analyzed.

Step 2. In column 2 enter the prior probability corresponding to each state of nature listed in column 1.

Step 3. In column 3 enter the appropriate value of $P(I_k|s_j)$ for each state of nature specified in column 1.

Step 4. To compute each entry in column 4, multiply each entry in column 2 by the corresponding entry in column 3.

Step 5. Add the entries in column 4. The sum is the value of $P(I_k)$. For convenience, write the sum below column 4.

Step 6. To compute each entry in column 5 divide the corresponding entry in column 4 by $P(I_k)$.

We will now use this procedure to compute $P(I_1)$ and the revised state-of-nature probabilities $P(s_j|I_1)$ for the PSI problem.

Steps 1, 2, and 3.

s_j	$P(s_j)$	$P(I_1|s_j)$	$P(I_1 \cap s_j)$	$P(s_j|I_1)$
s_1	.3	.8		
s_2	.7	.1		

Steps 4 and 5.

s_j	$P(s_j)$	$P(I_1|s_j)$	$P(I_1 \cap s_j)$	$P(s_j|I_1)$
s_1	.3	.8	.24	
s_2	.7	.1	.07	
			$P(I_1) = .31$	

Step 6.

s_j	$P(s_j)$	$P(I_1|s_j)$	$P(I_1 \cap s_j)$	$P(s_j|I_1)$
s_1	.3	.8	.24	.24/.31 = 0.7742
s_2	.7	.1	.07	.07/.31 = 0.2258
			$P(I_1) = \overline{.31}$	

Note that $P(I_1)$, $P(s_1|I_1)$, and $P(s_2|I_1)$ are exactly the same as we calculated by applying equations (19.13) and (19.16) directly. The preceding tabular computations could be repeated in order to compute $P(I_2)$ and the revised state-of-nature probabilities $P(s_j|I_2)$.

An Optimal Decision Strategy

Regardless of the method used to compute the branch probabilities, we can now use the branch probabilities and the expected value approach to arrive at the optimal decision for PSI. Working *backward* through the decision tree, we first compute the expected value at each state-of-nature node. That is, at each state-of-nature node the possible payoffs are weighted by their chance of occurrence. Thus the expected values for nodes 4 through 9 are computed as follows:

$$EV(\text{node } 4) = (.7742)(200{,}000) + (.2258)(-20{,}000) = 150{,}324$$

$$EV(\text{node } 5) = (.7742)(150{,}000) + (.2258)(20{,}000) = 120{,}646$$

$$EV(\text{node } 6) = (.7742)(100{,}000) + (.2258)(60{,}000) = 90{,}968$$

$$EV(\text{node } 7) = (.0870)(200{,}000) + (.9130)(-20{,}000) = -860$$

$$EV(\text{node } 8) = (.0870)(150{,}000) + (.9130)(20{,}000) = 31{,}310$$

$$EV(\text{node } 9) = (.0870)(100{,}000) + (.9130)(60{,}000) = 63{,}480$$

Figure 19.9 shows the above calculations directly on the decision tree. Since the decision maker controls the branch leaving a decision node and since we are trying to maximize expected profits, the optimal decision at node 2 is d_1. Thus since d_1 leads to an expected value of \$150,324, we say that EV(node 2) = \$150,324.

A similar analysis of decision node 3 shows that the optimal decision branch at this node is d_3. Thus EV(node 3) becomes \$63,480, provided that the optimal decision of d_3 is made.

▬▬ FIGURE 19.9 **Developing a Decision Strategy for the PSI Problem**

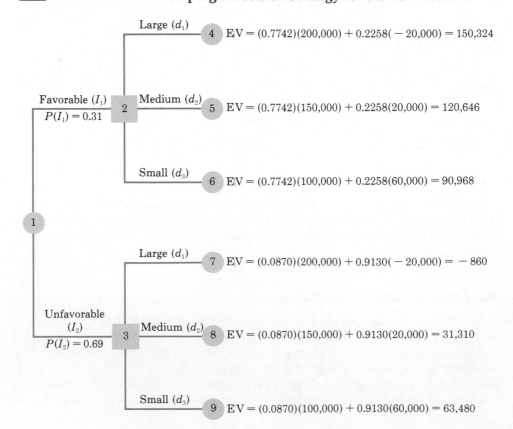

As a final step, we can continue working backward to the indicator node and establish its expected value. We see that since node 1 has probability branches, we cannot select the best branch. Rather we must compute the expected value over all possible branches. Thus we have

$$\text{EV (node 1)} = (.31)\,\text{EV (node 2)} + (.69)\,\text{EV (node 3)}$$

$$= (.31)(\$150{,}324) + (.69)(\$63{,}480) = \$90{,}402$$

The value of $90,402 is viewed as the expected value of the optimal decision strategy when the market research study is used. In other words, it is the expected value using the sample information provided by the market research report.

Note that the final decision has not yet been determined. We will need to know the results of the market research study before deciding to lease a large system (d_1) or a small system (d_3). The results of the decision theory analysis at this point, however, have provided us with the following optimal *decision strategy* if the market research study is conducted:

If	*Then*
Report favorable (I_1)	Lease Large system (d_1)
Report unfavorable (I_2)	Lease small system (d_3)

Thus we have seen how the decision tree approach can be used to develop optimal decision strategies when sample information is available. While other decision analysis problems may not be as simple as the PSI problem, the approach we have outlined is still applicable. First, draw a decision tree consisting of indicator, decision, and state-of-nature nodes and branches such that the tree describes the specific sequence of decisions and chance outcomes. Posterior probability calculations must be made in order to establish indicator and state-of-nature branch probabilities. Then, by working backward through the tree, computing expected values at state-of-nature and indicator nodes, and selecting the best decision branch at decision nodes, the analyst can determine an optimal decision strategy and its associated expected value.

■ EXERCISES

14. Suppose that you are given a decision situation with three possible states of nature: s_1, s_2, and s_3. The prior probabilities are $P(s_1) = .2$, $P(s_2) = .5$, and $P(s_3) = .3$. Indicator information I is obtained and it is known that $P(I|s_1) = .1$, $P(I|s_2) = .05$, and $P(I|s_3) = .2$. Compute the revised or posterior probabilities: $P(s_1|I)$, $P(s_2|I)$, and $P(s_3|I)$.

15. The payoff table showing profit for a decision problem with two states of nature and three decision alternatives is presented below:

	s_1	s_2
d_1	15	10
d_2	10	12
d_3	8	20

The prior probabilities for s_1 and s_2 are $P(s_1) = .8$ and $P(s_2) = .2$.

a. Using only the prior probabilities and the expected value approach, find the optimal decision.

b. Use graphical sensitivity analysis to determine the values of the probability of state of nature s_1 for which each of the decision alternatives has the largest expected value.

c. Find the EVPI.

d. Suppose that some indicator information I is obtained with $P(I|s_1) = .2$ and $P(I|s_2) = .75$. Find the posterior probabilities $P(s_1|I)$ and $P(s_2|I)$. Recommend a decision alternative based on these probabilities.

16. Consider the following decision tree representation of a decision analysis problem with two indicators, two decision alternatives, and two states of nature:

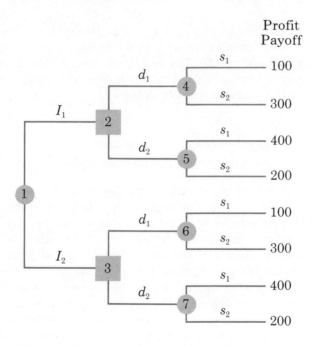

Assume that the following probability information is given:

$$P(s_1) = .4 \qquad P(I_1|s_1) = .8 \qquad P(I_2|s_1) = .2$$

$$P(s_2) = .6 \qquad P(I_1|s_2) = .4 \qquad P(I_2|s_2) = .6$$

a. What are the values for $P(I_1)$ and $P(I_2)$?

b. What are the values of $P(s_1|I_1)$, $P(s_2|I_1)$, $P(s_1|I_2)$, and $P(s_2|I_2)$?

c. Use the decision tree approach and determine the optimal decision strategy. What is the expected value of your solution?

19.8 ■ EXPECTED VALUE OF SAMPLE INFORMATION

In the PSI problem, management now has a decision strategy of leasing the large computer system if the market research report is favorable and leasing the small computer system if the market research report is unfavorable. Since the additional information provided by the market research firm will result in an added cost for PSI in terms of the fee paid to the research firm, PSI management may question the value of this market research information.

The value of sample information is often measured by calculating what is referred to as the *expected value of sample information (EVSI)*. For maximization problems,*

Expected Value of Sample Information

$$\text{EVSI} = \begin{bmatrix} \text{Expected Value of the} \\ \text{Optimal Decision } with \\ \text{Sample Information} \end{bmatrix} - \begin{bmatrix} \text{Expected Value of the} \\ \text{Optimal Decision } without \\ \text{Sample Information} \end{bmatrix} \quad (19.17)$$

For PSI the market research information is considered the "sample" information. The decision tree calculations indicated that the expected value of the optimal decision with the market research information was $90,402, while the expected value of the optimal decision without the market research information was $72,000. Using equation (19.17), the expected value of the market research report is

$$\text{EVSI} = \$90{,}402 - \$72{,}000 = \$18{,}402$$

Thus PSI should be willing to pay up to $18,402 for the market research information.

Efficiency of Sample Information

In Section 19.5 we saw that the expected value of perfect information (EVPI) for the PSI problem was $30,000. While we never expected the market research report to obtain perfect information, we can use an *efficiency* measure to express the value of the report. With perfect information having an efficiency rating of 100%, the efficiency rating E for sample information is computed as follows:

Efficiency of Sample Information

$$E = \frac{\text{EVSI}}{\text{EVPI}} \times 100 \quad (19.18)$$

---

*In minimization problems the expected value of the optimal decision with sample information will be less than or equal to the expected value of the optimal decision without sample information. Thus in minimization problems,

$$\text{EVSI} = \begin{bmatrix} \text{Expected Value of the} \\ \text{Optimal Decision } Without \\ \text{Sample Information} \end{bmatrix} - \begin{bmatrix} \text{Expected Value of the} \\ \text{Optimal Decision } with \\ \text{Sample Information} \end{bmatrix}$$

For the PSI example

$$E = \frac{18,402}{30,000} \times 100 = 61.3\%$$

In other words, the information from the market research firm is 61.3% as "efficient" as perfect information.

Low efficiency ratings for sample information might lead the decision maker to look for other types of information. On the other hand, high efficiency ratings indicate that the sample information is almost as good as perfect information, and additional sources of information should not be worthwhile.

■ EXERCISES

17. The payoff table for Exercise 16 is as follows:

	s_1	s_2
d_1	100	300
d_2	400	200

a. What is your decision without the indicator information?
b. What is the expected value of the indicator or sample information, EVSI?
c. What is the expected value of perfect information, EVPI?
d. What is the efficiency of the indicator information?

18. The payoff table for Hale's TV Productions (Exercise 5) is as follows:

Decision Alternatives		States of Nature		
		s_1	s_2	s_3
Produce pilot	d_1	−100	50	150
Sell to competitor	d_2	100	100	100
Probability of states of nature		.2	.3	.5

For a consulting fee of $2500, an agency will review the plans for the comedy series and indicate the overall chances of a favorable network reaction to the series. If the special agency review results in a favorable (I_1) or an unfavorable (I_2) evaluation, what should Hale's decision strategy be? Assume that Hale believes that the following conditional probabilities are realistic appraisals of the agency's evaluation accuracy:

$$P(I_1|s_1) = .3 \quad P(I_2|s_1) = .7$$

$$P(I_1|s_2) = .6 \quad P(I_2|s_2) = .4$$

$$P(I_1|s_3) = .9 \quad P(I_2|s_3) = .1$$

a. Show the decision tree for this problem.
b. What is the recommended decision strategy and the expected value, assuming that the agency information is obtained?
c. What is the EVSI? Is the $2500 consulting fee worth the information? What is the maximum that Hale should be willing to pay for the consulting information?

19. McHuffter Condominiums (Exercise 3) is conducting a survey that will help evaluate

the demand for the new condominium development. McHuffter's payoff table (profit) is as follows:

			States of Nature		
			Low (s_1)	Medium (s_2)	High (s_3)
Decision Alternatives	Small	d_1	400	400	400
	Medium	d_2	100	600	600
	Large	d_3	−300	300	900
Probability of states of nature			.20	.35	.45

The survey will result in three indicators of demand [weak (I_1), average (I_2), or strong (I_3)], where the conditional probabilities are as follows:

	$P(I_k|s_k)$		
	I_1	I_2	I_3
s_1	.6	.3	.1
s_2	.4	.4	.2
s_3	.1	.4	.5

a. What is McHuffter's optimal strategy?
b. What is the value of the survey information?
c. What are the EVPI and the efficiency of the survey information?

20. The payoff table for Martin's Service Station (Exercises 7 and 12) is as follows:

		Snowfall		
Decision Alternatives		Heavy s_1	Moderate s_2	Light s_3
Purchase snowplow	d_1	7000	2000	−9000
Do not invest	d_2	0	0	0
Purchase snowplow with blade	d_3	3500	1000	−1500
Probabilities of states of nature		.4	.3	.3

Suppose that Martin decides to wait to check the September temperature pattern before making a final decision. Estimates of the probabilities associated with an unseasonably cold September (I_1) are as follows: $P(I_1|s_1) = .30$, $P(I_1|s_2) = .20$, $P(I_1|s_3) = .05$. If Martin observes an unseasonably cold September, what is the recommended decision? If Martin does not observe an unseasonably cold September (I_2), what is the recommended decision?

19.9 ■ OTHER TOPICS IN DECISION ANALYSIS

In discussing decision analysis we have considered only situations where there are a finite number of states of nature. The next step would be to consider situations where the states of nature were so numerous that it would be impractical, if not impossible, to treat the states of nature as a discrete random variable consisting of a finite number of values. For example, let us suppose that we are attempting to price a new product and are concerned with the potential sales volume we might experience at different prices. We might think of the states of nature as being all possible sales volumes from 0 to 200,000 units. Although there are a finite number of states of nature, no units sold, 1 unit sold, and so on, we recognize that attempting to deal with this large number of possible states of nature is extremely impractical. The solution procedure that is used in such circumstances is to treat the state of nature as a continuous random variable. For example, perhaps a reasonable approximation of the state of nature (that is, sales volume) is that sales are normally distributed with a mean of 100,000 units and a standard deviation of 25,000 units. Although decision analysis techniques have been developed to handle such situations, we shall not attempt to present these procedures in this chapter.

Another area of decision analysis is concerned with alternative measures of the payoffs. In the PSI example we used profit in dollars as the measure of the payoff. Then the expected value approach was used to select the best decision. While decision analysis applications are often based on payoffs measured in monetary values, perhaps there are other measures of payoff that should be used.

For example, let us consider a situation in which we have two alternative investments. Investment A yields a certain profit of $50,000. Investment B yields a 50% chance of making $100,002 but also a 50% chance of making nothing. Thus the expected value for B is

$$E(B) = .5(100,002) + .5(0)$$

$$= \$50,001$$

Using the expected value approach with profits in dollars as the measure of the payoff, we would select decision alternative B. However, many decision makers, if not most, would select alternative A; that is, some decision makers prefer the no-risk $50,000 profit over the higher expected value but risky 50–50 chance of a $100,002 profit. In decision theory terminology, if alternative A is preferred, we say that alternative A has a higher *utility*, where utility is a measure of the decision maker's preference considering monetary value as well as the risk involved. Ideally we would like to measure payoffs in terms of the decision maker's utility and select optimal decision strategies based on expected utility rather than expected monetary value. More advanced books on decision analysis present the details associated with this approach to decision making.

■ SUMMARY

In this chapter we have emphasized how decision analysis can be used to solve problems with a limited number of decision alternatives and a limited number of possible states of nature. The goal of decision analysis is to identify the best decision alternative given an uncertain or risk-filled pattern of future events (that is, states of nature).

We presented three approaches to decision making without probabilities and discussed the use of the expected value approach for solving problems with probabilities. Then we showed how additional information about the states of nature can be used to revise or update the probability estimates and develop an optimal decision strategy for the problem. The notions of expected value of sample information, expected value of perfect information, and efficiency of information were used to evaluate the contribution of the sample information.

▬ GLOSSARY

States of nature The uncontrollable future events that can affect the payoff associated with a decision.

Payoff The outcome measure, such as profit, cost, time, and so on. Each combination of a decision alternative and a state of nature has an associated payoff.

Payoff table A tabular representation of the payoffs for a decision problem.

Optimistic approach An approach to choosing a decision alternative without using probabilities. For a maximization problem it leads to choosing the alternative corresponding to the largest payoff; for a minimization problem it leads to choosing the alternative corresponding to the smallest payoff.

Conservative approach An approach to choosing a decision alternative without using probabilities. For a maximization problem it leads to choosing the alternative that maximizes the minimum payoff; for a minimization problem it leads to choosing the alternative that minimizes the maximum payoff.

Minimax regret An approach to choosing a decision alternative without using probabilities. For each alternative, the maximum regret is computed. This approach leads to choosing the alternative that minimizes the maximum regret.

Opportunity loss or regret The amount of loss (lower profit or higher cost) due to not making the best decision for each state of nature.

Decision tree A graphical representation of the decision-making situation from decision to state-of-nature to payoff.

Nodes The intersection or junction points of the decision tree.

Branches Lines or arcs connecting nodes of the decision tree.

Expected value For a decision alternative, it is the weighted average of the payoffs. The weights are the state-of-nature probabilities.

Expected value of perfect information (EVPI) The expected value of information that would tell the decision maker exactly which state of nature was going to occur (that is, perfect information).

Prior probabilities The probabilities of the states of nature prior to obtaining sample information.

Posterior (revised) probabilities The probabilities of the states of nature after using Bayes' theorem to adjust the prior probabilities based on given indicator information.

Indicators Information about the states of nature. An indicator may be the result of a sample.

Bayesian revision The process of revising prior probabilities to create the posterior probabilities based on sample information.

Expected value of sample information (EVSI) The difference between the expected value of an optimal strategy based on new information and the "best" expected value without any new information. It is a measure of the value of new information.

Efficiency The ratio of EVSI to EVPI; perfect information is 100% efficient.

▬ KEY FORMULAS

Opportunity Loss or Regret

$$R(d_i,s_j) = V^*(s_j) - V(d_i,s_j) \tag{19.1}$$

Expected Value of Decision Alternative d_i

$$EV(d_i) = \sum_{j=1}^{N} P(s_j)V(d_i,s_j). \tag{19.4}$$

Calculation of Indicator Branch Probability

$$P(I_k) = \sum_{j=1}^{N} P(I_k \mid s_j) P(s_j) \qquad (19.13)$$

Calculation of State-of-Nature Branch Probability

$$P(s_j \mid I_k) = \frac{P(I_k \mid s_j) P(s_j)}{P(I_k)} \qquad (19.16)$$

Efficiency of Sample Information

$$E = \frac{\text{EVSI}}{\text{EVPI}} (100) \qquad (19.18)$$

■ SUPPLEMENTARY EXERCISES

21. In order to save on gasoline expenses, Rona and Jerry agreed to form a carpool for traveling to and from work. After limiting the travel routes to two alternatives, Rona and Jerry could not agree on the best way to travel to work. Jerry preferred the expressway, since it was usually the fastest; however, Rona pointed out that traffic jams on the expressway sometimes led to long delays. Rona preferred the somewhat longer but more consistent Queen City Avenue. While Jerry still preferred the expressway, he agreed with Rona that they should take Queen City Avenue if the expressway had a traffic jam. Unfortunately, they do not know the state of the expressway ahead of time. The following payoff table provides the one-way time estimates for traveling to or from work:

		States of Nature	
Route		Expressway Open s_1	Expressway Jammed s_2
Expressway	d_1	25	45
Queen City Avenue	d_2	30	30
			Travel time in minutes

a. After driving to work on the expressway for 1 month (20 days), they found the expressway jammed three times. Assuming that these days are representative of future days, should they continue to use the expressway for traveling to work? Explain.

b. Use graphical sensitivity analysis to determine the values of the probability of state of nature s_1 for which d_1 has the best expected value.

c. Would it make sense not to adopt the expected value approach for this particular problem? Explain.

22. The Gorman Manufacturing Company must decide whether it should purchase a component part from a supplier or manufacture the component at its Milan, Michigan, plant. If demand is high, it would be to Gorman's advantage to manufacture the component. However, if demand is low, Gorman's unit manufacturing cost will be high

due to underutilization of equipment. The projected profit in thousands of dollars for Gorman's make or buy decision is shown below:

Decision Alternatives	Demand		
	Low	*Medium*	*High*
Manufacture component	−20	40	100
Purchase component	10	45	70

The states of nature have the following probabilities: P(low demand) = .35, P(medium demand) = .35, and P(high demand) = .30.

a. Use a decision tree to recommend a decision.

b. Use EVPI to determine whether Gorman should attempt to obtain a better estimate of demand

23. In Exercise 21, suppose that Rona and Jerry wished to determine the best way to return home in the evenings. In 20 days of traveling home on the expressway, they found the expressway jammed six times.

a. Using the travel time table shown in Exercise 21, what route would you recommend they take on their way home in the evening?

b. If they had perfect information about the traffic condition of the expressway, what would be their savings in terms of expected travel time?

24. A firm produces a perishable food product at a cost of $10 per case. The product sells for $15 per case. For planning purposes the company is considering possible demands of 100, 200, or 300 cases. If the demand is less than production, the excess production is lost. If demand is more than production, the firm, in an attempt to maintain a good service image, will satisfy the excess demand with a special production run at a cost of $18 per case. The product, however, always sells at $15 per case.

a. Set up the payoff table for this problem.

b. If $P(100)$ = .2, $P(200)$ = .2, and $P(300)$ = .6, should the company produce 100, 200, or 300 cases?

c. What is the EVPI?

25. Sealcoat, Inc. has a contract with one of its customers to supply a unique liquid chemical product that will be used by the customer in the manufacture of a lubricant for airplane engines. Because of the chemical process used by Sealcoat, batch sizes for the liquid chemical product must be 1000 pounds. The customer has agreed to adjust manufacturing to the full batch quantities and will order either one, two, or three batches every 3 months. Since an aging process of 1 month exists for the product, Sealcoat will have to make its production (how much to make) decision before the customer places an order. Thus Sealcoat can list the product demand alternatives of 1000, 2000, or 3000 pounds, but the exact demand is unknown.

Sealcoat's manufacturing costs are $150 per pound, and the product sells at the fixed contract price of $200 per pound. If the customer orders more than Sealcoat has produced, Sealcoat has agreed to absorb the added cost of filling the order by purchasing a higher-quality substitute product from another chemical firm. The substitute product, including transportation expenses, will cost Sealcoat $240 per pound. Since the product cannot be stored more than 2 months without spoilage, Sealcoat cannot inventory excess production until the customer's next 3-month order. Therefore, if the customer's current order is less than Sealcoat has produced, the excess production will be reprocessed and is valued at $50 per pound.

The inventory decision in this problem is how much should Sealcoat produce given the above costs and the possible demands of 1000, 2000, or 3000 pounds? Based on historical data and an analysis of the customer's future demands, Sealcoat has assessed the following probability distribution for demand:

Demand	Probability
1000	0.3
2000	0.5
3000	0.2
Total	1.0

a. Develop a payoff table for the Sealcoat problem.
b. How many batches should Sealcoat produce every 3 months?
c. How much of a discount should Sealcoat be willing to allow the customer for specifying in advance exactly how many batches will be purchased?

26. A quality-control procedure involves 100% inspection of parts received from a supplier. Historical records show that the following defective rates have been observed:

Percent Defective	Probability
0	.15
1	.25
2	.40
3	.20

The cost to inspect 100% of the parts received is $250 for each shipment of 500 parts. If the shipment is not 100% inspected, defective parts will cause rework problems later in the production process. The rework cost is $25 for each defective part.

a. Complete the following payoff table, where the entries represent the total cost of inspection and reworking:

Inspection	Percent Defective			
	0	*1*	*2*	*3*
100% inspection	$250	$250	$250	$250
No inspection				

b. The plant manager is considering eliminating the inspection process in order to save the $250 inspection cost per shipment. Do you support this action? Use expected value to justify your answer.
c. Show the decision tree for this problem.

27. A food processor considers daily production runs of 100, 200, or 300 cases. Possible demands for the product are 100, 200, or 300 cases. The payoff table is as follows:

			Demand		
			100	*200*	*300*
			s_1	s_2	s_3
	d_1	100	500	200	− 100
Production	d_2	200	−400	800	700
	d_3	300	−1000	−200	1600

a. If $P(s_1) = 0.20$, $P(s_2) = 0.20$, and $P(s_3) = 0.60$, what is your recommended production quantity?
b. On some days the firm receives phone calls for advance orders and on some days it does not. Let I_1 = advance orders are received and I_2 = no advance orders are received.

If $P(I_2|s_1) = .80$, $P(I_2|s_2) = .40$, and $P(I_2|s_3) = .10$, what is your recommended production quantity for days the company does not receive any advance orders?

28. The Gorman Manufacturing Company (Exercise 22) has the following payoff table for a make-or-buy decision:

		Demand		
		Low	Medium	High
Decision Alternatives		s_1	s_2	s_3
Manufacture component	d_1	-20	40	100
Purchase component	d_2	10	45	70
Probabilities of states of nature		.35	.35	.30

A test market study of the potential demand for the product is expected to report either a favorable (I_1) or unfavorable (I_2) condition. The relevant conditional probabilities are as follows:

$$P(I_1 \mid s_1) = .10 \qquad P(I_2 \mid s_1) = .90$$

$$P(I_1 \mid s_2) = .40 \qquad P(I_2 \mid s_2) = .60$$

$$P(I_1 \mid s_3) = .60 \qquad P(I_2 \mid s_3) = .40$$

a. What is the probability that the market research report will be favorable?
b. What is Gorman's optimal decision strategy?
c. What is the expected value of the market research information?
d. What is the efficiency of the information?

29. The traveling time to work for Rona and Jerry has the following time payoff table (see Exercise 21):

		States of Nature for Expressway	
		Open	Jammed
Route		s_1	s_2
Expressway	d_1	25	45
Queen City Avenue	d_2	30	30
Probabilities of states of nature		.85	.15

After a period of time Rona and Jerry noted that the weather seemed to affect the traffic conditions on the expressway. They identified three weather conditions (indicators) with the following conditional probabilities:

$$I_1 = \text{clear}$$

$$I_2 = \text{overcast}$$

$$I_3 = \text{rain}$$

$$P(I_1 \mid s_1) = .8 \qquad P(I_2 \mid s_1) = .2 \qquad P(I_3 \mid s_1) = 0$$

$$P(I_1 \mid s_2) = .1 \qquad P(I_2 \mid s_2) = .3 \qquad P(I_3 \mid s_2) = .6$$

a. Show the decision tree for the problem of traveling to work.
b. What is the optimal decision strategy and the expected travel time?
c. What is the efficiency of the weather information?

30. The research and development manager for Beck Company is trying to decide whether or not to fund a project to develop a new lubricant. It is assumed that the project will be either a major technical success, a minor technical success, or a failure. The company has estimated that the value of a major technical success is $150,000, since the lubricant can be used in a number of products the company is making. If the project is a minor technical success, its value is $10,000, since Beck feels that the knowledge gained will benefit some other ongoing projects. If the project is a failure, it will cost the company $100,000.

Based on the opinion of the scientists involved and the manager's own subjective assessment, the assigned prior probabilities are as follows:

$$P(\text{major success}) = .15$$

$$P(\text{minor success}) = .45$$

$$P(\text{failure}) = .40$$

a. Using the expected value approach, should the project be funded?
b. Suppose that a group of expert scientists from a research institute could be hired as consultants to study the project and make a recommendation. If this study will cost $30,000, should the Beck Company consider hiring the consultants?

31. Consider again the problem faced by the R&D manager of Beck Company (Exercise 30). Suppose that an experiment can be conducted to shed some light on the technical feasibility of the project. There are three possible outcomes for the experiment:

I_1 = prototype lubricant works well at all temperatures

I_2 = prototype lubricant works well only at temperatures above 10°F

I_3 = prototype lubricant does not work well at any temperature

Suppose that we can determine the following conditional probabilities:

$$P(I_1 \mid \text{major success}) = .70$$

$$P(I_1 \mid \text{minor success}) = .10$$

$$P(I_1 \mid \text{failure}) = .10$$

$$P(I_2 \mid \text{major success}) = .25$$

$$P(I_2 \mid \text{minor success}) = .70$$

$$P(I_2 \mid \text{failure}) = .30$$

$$P(I_3 \mid \text{major success}) = .05$$

$$P(I_3 \mid \text{minor success}) = .20$$

$$P(I_3 \mid \text{failure}) = .60$$

a. Assuming that the experiment is conducted and the prototype lubricant works well at all temperatures, should the project be funded?
b. Assuming that the experiment is conducted and the prototype lubricant works well only at temperatures above 10°F, should the project be funded?
c. Develop a decision strategy that Beck's R&D manager can use to recommend a funding decision based on the outcome of the experiment.
d. Find the EVSI for the experiment. How efficient is the information in the experiment?

32. The payoff table for the Sealcoat, Inc. (Exercise 25) is as follows:

Production Quantity		Demand 1,000 s_1	2,000 s_2	3,000 s_3
1,000	d_1	50,000	10,000	−30,000
2,000	d_2	−50,000	100,000	60,000
3,000	d_3	−150,000	0	150,000
Probabilities		.30	.50	.20

Sealcoat has identified a pattern in the demand for the product based on the customer's previous order quantity. Let

$$I_1 = \text{customer's last order was 1000 pounds}$$

$$I_2 = \text{customer's last order was 2000 pounds}$$

$$I_3 = \text{customer's last order was 3000 pounds}$$

The conditional probabilities are as follows:

$$P(I_1 \mid s_1) = .10 \qquad P(I_2 \mid s_1) = .40 \qquad P(I_3 \mid s_1) = .50$$

$$P(I_1 \mid s_2) = .22 \qquad P(I_2 \mid s_2) = .68 \qquad P(I_3 \mid s_2) = .10$$

$$P(I_1 \mid s_3) = .80 \qquad P(I_2 \mid s_3) = .20 \qquad P(I_3 \mid s_3) = .00$$

a. Develop an optimal decision strategy for Sealcoat.
b. What is the EVSI?
c. What is the efficiency of the information for the most recent order?

33. Milford Trucking Co. (Exercise 9) has the following payoff table:

Shipment		Return Shipment from Detroit s_1	No Return Shipment from Detroit s_2
St. Louis	d_1	2000	2000
Detroit	d_2	2500	1000
Probabilities		.40	.60

a. Milford can phone a Detroit truck dispatch center and determine if the general Detroit shipping activity is busy (I_1) or slow (I_2). If the report is busy, the chances of obtaining a return shipment will increase. Suppose that the following conditional probabilities are given:

$$P(I_1 \mid s_1) = .6 \qquad P(I_2 \mid s_1) = .4$$

$$P(I_1 \mid s_2) = .3 \qquad P(I_2 \mid s_2) = .7$$

What should Milford do?
b. If the Detroit report is busy (I_1), what is the probability that Milford will obtain a return shipment if it makes the trip to Detroit?
c. What is the efficiency of the phone information?

34. The quality control inspection process (Exercise 26) has the following payoff table:

		Percentage Defective			
		0	1	2	3
Inspection		s_1	s_2	s_3	s_4
100% inspection	d_1	250	250	250	250
No inspection	d_2	0	125	250	375
Probabilities		.15	.25	.40	.20

Suppose that a sample of five parts is selected from the shipment and one defect is found.

a. Let I = one defect in a sample of five. Use the binomial probability distribution to compute $P(I \mid s_1)$, $P(I \mid s_2)$, $P(I \mid s_3)$, and $P(I \mid s_4)$, where the state of nature identifies the value for p.

b. If I occurs, what are the revised probabilities for the states of nature?

c. Should the entire shipment be 100% inspected whenever one defect is found in a sample of size 5?

d. What is the cost saving associated with the sample information?

Ohio Edison Company*

Akron, Ohio

Ohio Edison Company is an investor-owned electric utility headquartered in northeastern Ohio. Ohio Edison and a Pennsylvania subsidiary provide electrical service to over 2 million people. Most of this electricity is generated by coal-fired power plants. In order to meet evolving air–quality standards, Ohio Edison has embarked on a program to replace existing pollution control equipment on most of its generating plants with more efficient equipment. The combination of this program to upgrade air-quality control equipment with the continuing need to construct new generating plants to meet future power requirements has resulted in a large capital investment program.

▋ A DECISION ANALYSIS APPLICATION

The flue gas emitted by coal-fired power plants contains small ash particles and sulfur dioxide (SO_2). Federal and state regulatory agencies have established emission limits for both particulates and sulfur dioxide. Recently, Ohio Edison developed a plan to comply with new air-quality standards at one of its largest power plants. This plant consists of seven coal-fired units and constitutes about one-third of the generating capacity of Ohio Edison and the subsidiary company. Most of these units had been constructed in the 1960s. Although all of the units had initially been constructed with equipment to control particulate emissions, that equipment was not capable of meeting new particulate emission requirements.

A decision had already been made to burn low-sulfur coal in four of the smaller units (units 1 to 4) at the plant in order to meet SO_2 emission standards. Fabric filters were to be installed on these units to control particulate emissions. Fabric filters, also known as baghouses, use thousands of fabric bags to filter out the particulates; they function in much the same way as a household vacuum cleaner.

It was considered likely, although not certain, that the three larger units (units 5 to 7) at this plant would burn medium- to high-sulfur coal. A method of controlling particulate emissions at these units had not yet been selected. Preliminary studies had narrowed the particulate control equipment choice to a decision between fabric filters and electrostatic precipitators (which remove particulates suspended in the flue gas as charged particles by passing the flue gas through a strong electric field). This decision was affected by a number of uncertainties, including the following:

▋ Uncertainty in the way some air-quality laws and regulations might be interpreted. Certain interpretations could require that either low-sulfur coal or high-sulfur Ohio coal (or neither) be burned in units 5 to 7.

*The authors are indebted to Thomas J. Madden and M. S. Hyrnick of Ohio Edison Company, Akron, Ohio for providing this application.

■ Potential future changes in air quality laws and regulations.

■ An overall plant reliability improvement program was underway at this plant. The outcome of this program would affect the operating costs of whichever pollution control technology was installed in these units.

■ Construction costs of the equipment were uncertain, particularly since limited space at the plant site made it necessary to install the equipment on a massive bridge deck over a four-lane highway immediately adjacent to the power plant.

■ The costs associated with replacing the electrical power required to operate the particulate control equipment were uncertain.

■ Various uncertain factors, including potential accidents and chronic operating problems that could increase the costs of operating the generating units, were identified. The degree to which each of these factors affected operating costs varied with the choice of technology and with the sulfur content of the coal.

DECISION ANALYSIS

The decision to be made involved a choice between two types of particulate control equipment (fabric filters or electrostatic precipitators) for units 5 to 7. Because of the complexity of the problem, the high degree of uncertainty associated with factors affecting the decision, and the importance (because of potential reliability and cost impact on Ohio Edison) of the choice, decision analysis was used in the selection process.

The decision measure used to evaluate the outcomes of the particulate technology decision analysis was the annual revenue requirements for the three large units over their remaining lifetime. Revenue requirements are the monies that would have to be

Decision analysis aided in the selection of particulate control equipment being installed at Ohio Edison's W. H. Sammis Electric Generating Plant.

collected from the utility customers in order to recover costs resulting from the decision. They include not only direct costs but also the cost of capital and return on investment.

A decision tree was constructed to represent the particulate control decision, its uncertainties and costs. A simplified version of this decision tree is shown in Figure 19A.1. The decision and state-of-nature nodes are indicated. Note that to conserve space a type of shorthand notation is used. The coal sulfur content state-of-nature node should actually be located at the end of each branch of the capital cost state-of-nature node, as the dotted lines indicate. Each of the indicated state-of-nature nodes actually represents several probabilistic cost models or submodels. The total revenue requirements calculated are the sum of the revenue requirements for capital and operating costs. Costs associated with these models were obtained from engineering calculations or estimates. Probabilities were obtained from existing data or the subjective assessments of knowledgeable persons.

■ RESULTS

A decision tree similar to that shown in Figure 19A.1 was used to generate cumulative probability distributions for the annual revenue requirements outcomes calculated for

■ FIGURE 19A.1 **Simplified Particulate Control Equipment Decision Tree**

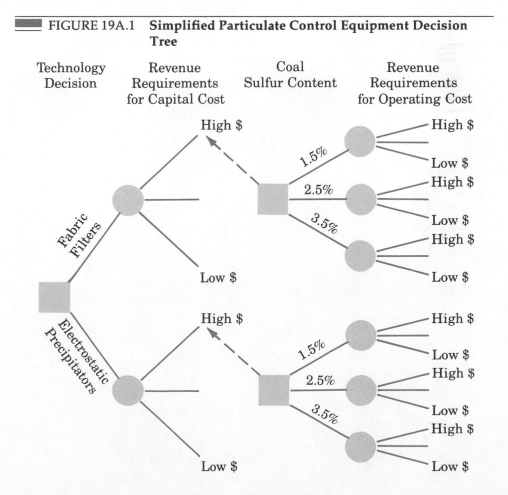

each of the two particulate control alternatives. Careful study of these results led to the following conclusions:

■ The expected value of annual revenue requirements for the electrostatic precipitator technology was approximately $1 million lower than that for the fabric filters.

■ The fabric filter alternative had a higher "upside risk"—that is, a higher probability of high revenue requirements—than did the precipitator alternative.

■ The precipitator technology had nearly an 80% probability of having lower annual revenue requirements than the fabric filters.

■ Although the capital cost of the fabric filter equipment (the cost of installing the equipment) was lower than for the precipitator, this was more than offset by the higher operating costs associated with the fabric filter.

These results led Ohio Edison to select the electrostatic precipitator technology for the generating units in question. Had the decision analysis not been performed, the particulate control decision might have been based chiefly on capital cost, a decision measure that would have favored the fabric filter equipment. Decision analysis offers a means for effectively analyzing the uncertainties involved in a decision. Because of this, it is felt that the use of decision analysis methodology in this application resulted in a decision that yielded both lower expected revenue requirements and lower risk.

CHAPTER 20

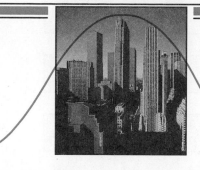

Statistical Methods for Quality Control

I n recent years the Japanese commitment to producing high-quality products has caused many companies to reevaluate their views regarding quality and how it can be achieved. Organizations recognize that to be competitive in today's marketplace, they must strive to reach high levels of quality. As a result there has been an increased emphasis on methods for monitoring and maintaining quality.

Quality control consists of making a series of inspections and measurements in order to determine if quality standards are being met. In this chapter we present two statistical methods that are used in quality control. The first method, referred to as *acceptance sampling,* is used in situations where a decision has to be made to accept or reject a group of items based upon the quality found in a sample. The second method, referred to as *statistical process control,* uses graphical displays known as *control charts* to monitor a production process; the goal is to determine whether the process should be continued or whether the process should be adjusted in order to achieve the desired quality level.

20.1 ▥ ACCEPTANCE SAMPLING

In acceptance sampling the items of interest can be incoming shipments of raw materials or purchased parts as well as finished goods from final assembly. Suppose that we want to decide whether to accept or reject the group of items based upon specified quality characteristics. In quality-control terminology, the group of items is referred to as a *lot,* and *acceptance sampling* is a statistical method that enables us to make the accept-reject decision based upon inspecting a sample of the items from the lot.

The general steps of acceptance sampling are shown in Figure 20.1. After a lot is received, a sample of items is selected for inspection. The results of the

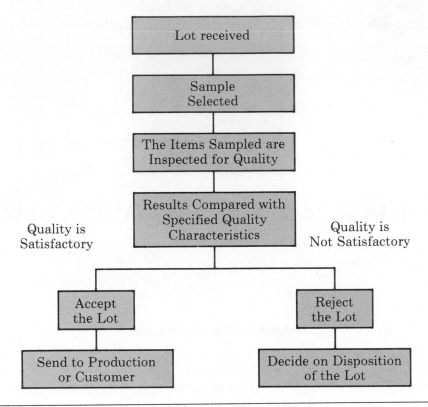

■■■ FIGURE 20.1   **Acceptance Sampling Procedure**

inspection are compared to specified quality characteristics. If the quality characteristics are satisfied, the lot is accepted and sent to production or shipped to customers. If the lot is rejected, management must decide on the disposition of the lot. In some cases the decision may be to keep the lot and remove the unacceptable or nonconforming items during production. In other cases the lot may be returned to the supplier at the supplier's expense; the extra work and cost placed on the supplier often provides good motivation for increasing the supplier's commitment to providing high-quality lots. Finally, in those instances where the rejected lot contains finished goods, the goods must be scrapped or reworked to bring them up to acceptable quality standards.

The statistical procedure of acceptance sampling is based on the hypothesis-testing methodology presented in Chapter 9. The null and alternative hypotheses are stated as follows:

$$H_0: \quad \text{Good-quality lot}$$

$$H_a: \quad \text{Poor-quality lot}$$

The states of nature (good-quality lot, poor-quality lot) and the decisions to accept the lot or reject the lot are shown in Table 20.1. This table, just like the one presented in Chapter 9, shows the results of a hypothesis-testing procedure. Note that correct decisions correspond to accepting a good-quality lot and rejecting a poor-quality lot. However, as with other hypothesis-testing proce-

TABLE 20.1   **The Outcomes of Acceptance Sampling**

Decision	States of Nature	
	H_0 True Good-Quality Lot	H_0 False Poor-Quality Lot
Accept the Lot	Correct Decision	Type II Error (accepting a poor-quality lot)
Reject the Lot	Type I Error (rejecting a good-quality lot)	Correct Decision

dures, we need to be aware of the possibilities of making a Type I error (rejecting a good-quality lot) or a Type II error (accepting a poor-quality lot).

Since the probability of a Type I error creates a risk for the producer of the lot, it is referred to as the *producer's risk*. For instance, a producer's risk of .05 indicates that there is a 5% chance that a good-quality lot will be erroneously rejected and returned to the producer. On the other hand, since the probability of a Type II error creates a risk for the consumer of the lot, it is referred to as the *consumer's risk*. For example, a consumer's risk of .10 means that there is a 10% chance that a poor-quality lot will be erroneously accepted and thus used in production or shipped to the customer. Specific values for the producer's risk and the consumer's risk can be controlled by the person designing the acceptance sampling procedure. To illustrate how this is done, let us consider the problem faced by KALI, Inc.

KALI Inc. — An Example of Acceptance Sampling

KALI, Inc. manufactures home appliances that are marketed under a variety of trade names. However, KALI does not manufacture every component used in the products. Several components are purchased directly from suppliers. For example, one of the components that KALI purchases for use in home air conditioners is an overload protector, a device that turns off the compressor if it overheats. Since the compressor can be seriously damaged if the overload protector does not function properly, KALI is very concerned about the quality of the overload protectors. In order to assure quality, one alternative would be to test every component received; this approach is referred to as 100% inspection. However, in order to determine if an overload protector functions properly, it must be subjected to tests that ensure that it will shut off a compressor properly. Since the tests are time-consuming and expensive to conduct, KALI cannot justify testing every overload protector purchased.

Instead, KALI uses an acceptance sampling plan to monitor the quality of the overload protectors. The acceptance sampling plan requires that KALI's quality-control inspectors select and test a sample of overload protectors from each shipment. If very few defective units are found in the sample, the lot is probably of good quality and should be accepted. However, if a large number of defective units are found in the sample, the lot is probably of poor quality and should be rejected.

An *acceptance sampling plan* consists of a sample size (n) and an acceptance criterion (c). The *acceptance criterion* is the maximum number of defective items that can be found in the sample and still indicate the lot can be accepted. For instance, for the KALI problem let us assume for the moment that a sample of 15 items will be selected from each incoming shipment or lot. Furthermore, assume that the manager of quality control states that the lot can be accepted only if no defective items are found. Thus, in this case the acceptance sampling plan established by the quality-control manager is $n = 15$ and $c = 0$.

This acceptance sampling plan is easy for the quality-control inspector to implement. The inspector simply selects a sample of 15 items, performs the tests, and reaches a conclusion based on the following decision rule:

■ *Accept the lot* if 0 defective items are found.
■ *Reject the lot* if 1 or more defective items are found.

Before implementing this acceptance sampling plan, the quality-control manager will be interested in evaluating the risks or errors possible under the plan. The plan will be implemented only if both the producer's risk (Type I error) and the consumer's risk (Type II error) are controlled at reasonable levels.

Computing the Probability of Accepting a Lot

The key to analyzing both the producer's risk and the consumer's risk is based on a what-if type of analysis. That is, we will assume that a lot has some known percentage of defective items and compute the probability of accepting the lot for a given sampling plan. By varying the assumed percentage of defective items, we can examine the effect the sampling plan has on both types of risks.

Let us begin by assuming that a large shipment of overload protectors has been received and that 5% of the overload protectors in the shipment are defective. For a shipment or lot with 5% defective, what is the probability that the $n = 15$, $c = 0$ sampling plan will lead us to accept the lot? Since each overload protector tested will be either defective or nondefective and since the lot size is large, the number of defective items in a sample of 15 has a *binomial probability distribution*. The binomial probability function, which was presented in Chapter 5, is as follows:

The Binomial Probability Function for Acceptance Sampling

$$f(x) = \frac{n!}{x!(n - x)!} p^x (1 - p)^{(n-x)} \qquad (20.1)$$

where

> n = the sample size
>
> p = the proportion of defective items in the entire lot
>
> x = the number of defective items in the sample
>
> $f(x)$ = the probability of finding x defective items in a sample of n items

---

For the KALI acceptance sampling plan, $n = 15$; thus, for a lot with 5% defective ($p = .05$) we have

$$f(x) = \frac{15!}{x!(15 - x)!} (.05)^x (1 - .05)^{(15-x)} \tag{20.2}$$

Using (20.2), $f(0)$ will provide the probability that 0 overload protectors will be defective and thus the lot will be accepted. In using (20.2) recall that $0! = 1$. Thus the probability computation for $f(0)$ is as follows:

$$f(0) = \frac{15!}{0!(15 - 0)!} (.05)^0 (1 - .05)^{(15-0)}$$

$$= \frac{15!}{0!(15)!} (.05)^0 (.95)^{15} = (.95)^{15} = .4633$$

We now know that the $n = 15$, $c = 0$ sampling plan has a .4633 probability of accepting a lot with 5% defective. Thus there must be a corresponding $1 - .4633 = .5367$ probability of rejecting a lot with 5% defective.

In Table 20.2 we show the probability that the $n = 15$, $c = 0$ sampling plan will lead to the acceptance of lots with 1%, 2%, 3%, . . . defective. The probabilities in the table were computed by using $p = .01$, $p = .02$, $p = .03$, . . . in the binomial probability function (20.1).

Tables of binomial probabilities (See Table 5, Appendix B) can help reduce the computational effort in determining the probabilities of accepting lots. Selected binomial probabilities for $n = 15$ and $n = 20$ are shown in Table 20.3. Use Table 20.3 to verify that if the lot contained 10% defective, there would be a .2059 probability that the $n = 15$, $c = 0$ sampling plan would indicate the lot should be accepted.

Using the data in Table 20.2, a graph of the probability of accepting the lot versus the percent defective in the lot can be drawn as shown in Figure 20.2. This graph, or curve, is called the *operating characteristic (OC) curve* for the $n = 15$, $c = 0$ acceptance sampling plan.

TABLE 20.2 **Probability of Accepting the Lot for the KALI Inc. Example with $n = 15$ and $c = 0$.**

Percent Defective in the Lot	Probability of Accepting the Lot
1%	.8601
2%	.7386
3%	.6333
4%	.5421
5%	.4633
10%	.2059
15%	.0874
20%	.0352
25%	.0134

TABLE 20.3   Selected Binomial Probabilities for Samples of Sizes 15 and 20

n	x	.05	.10	.15	.20	p .25	.30	.35	.40	.45	.50
15	0	.4633	.2059	.0874	.0352	.0134	.0047	.0016	.0005	.0001	.0000
	1	.3658	.3432	.2312	.1319	.0668	.0305	.0126	.0047	.0016	.0005
	2	.1348	.2669	.2856	.2309	.1559	.0916	.0476	.0219	.0090	.0032
	3	.0307	.1285	.2184	.2501	.2252	.1700	.1110	.0634	.0318	.0139
	4	.0049	.0428	.1156	.1876	.2252	.2186	.1792	.1268	.0780	.0417
	5	.0006	.0105	.0449	.1032	.1651	.2061	.2123	.1859	.1404	.0916
	6	.0000	.0019	.0132	.0430	.0917	.1472	.1906	.2066	.1914	.1527
	7	.0000	.0003	.0030	.0138	.0393	.0811	.1319	.1771	.2013	.1964
	8	.0000	.0000	.0005	.0035	.0131	.0348	.0710	.1181	.1647	.1964
	9	.0000	.0000	.0001	.0007	.0034	.0116	.0298	.0612	.1048	.1527
	10	.0000	.0000	.0000	.0001	.0007	.0030	.0096	.0245	.0515	.0916
	11	.0000	.0000	.0000	.0000	.0001	.0006	.0024	.0074	.0191	.0417
	12	.0000	.0000	.0000	.0000	.0000	.0001	.0004	.0016	.0052	.0139
	13	.0000	.0000	.0000	.0000	.0000	.0000	.0001	.0003	.0010	.0032
	14	.0000	.0000	.0000	.0000	.0000	.0000	.0000	.0000	.0001	.0005
	15	.0000	.0000	.0000	.0000	.0000	.0000	.0000	.0000	.0000	.0000
20	0	.3585	.1216	.0388	.0115	.0032	.0008	.0002	.0000	.0000	.0000
	1	.3774	.2702	.1368	.0576	.0211	.0068	.0020	.0005	.0001	.0000
	2	.1887	.2852	.2293	.1369	.0669	.0100	.0100	.0031	.0008	.0002
	3	.0596	.1901	.2428	.2054	.1339	.0716	.0323	.0123	.0040	.0011
	4	.0133	.0898	.1821	.2182	.1897	.1304	.0738	.0350	.0139	.0046
	5	.0022	.0319	.1028	.1746	.2023	.1789	.1272	.0746	.0365	.0148
	6	.0003	.0089	.0454	.1091	.1686	.1916	.1712	.1244	.0746	.0370
	7	.0000	.0020	.0160	.0545	.1124	.1643	.1844	.1659	.1221	.0739
	8	.0000	.0004	.0046	.0222	.0609	.1144	.1614	.1797	.1623	.1201
	9	.0000	.0001	.0011	.0074	.0271	.0654	.1158	.1597	.1771	.1602
	10	.0000	.0000	.0002	.0020	.0099	.0308	.0686	.1171	.1593	.1762
	11	.0000	.0000	.0000	.0005	.0030	.0120	.0336	.0710	.1185	.1602
	12	.0000	.0000	.0000	.0001	.0008	.0039	.0136	.0355	.0727	.1201
	13	.0000	.0000	.0000	.0000	.0002	.0010	.0045	.0146	.0366	.0739
	14	.0000	.0000	.0000	.0000	.0000	.0002	.0012	.0049	.0150	.0370
	15	.0000	.0000	.0000	.0000	.0000	.0000	.0003	.0013	.0049	.0148
	16	.0000	.0000	.0000	.0000	.0000	.0000	.0000	.0003	.0013	.0046
	17	.0000	.0000	.0000	.0000	.0000	.0000	.0000	.0000	.0002	.0011
	18	.0000	.0000	.0000	.0000	.0000	.0000	.0000	.0000	.0000	.0002
	19	.0000	.0000	.0000	.0000	.0000	.0000	.0000	.0000	.0000	.0000
	20	.0000	.0000	.0000	.0000	.0000	.0000	.0000	.0000	.0000	.0000

Perhaps we should consider other sampling plans, ones with different sample sizes n and/or different acceptance criteria c. First consider the case where the sample size remains $n = 15$ but the acceptance criterion increases from $c = 0$ to $c = 1$. That is, we will now accept the lot if 0 or 1 defective components are found in the sample. For a lot with 5% defective ($p = .05$), the binomial probability function in (20.2) can be used to compute $f(0)$ and $f(1)$. Summing these two probabilities provides the probability that the $n = 15$, $c = 1$ sampling plan will

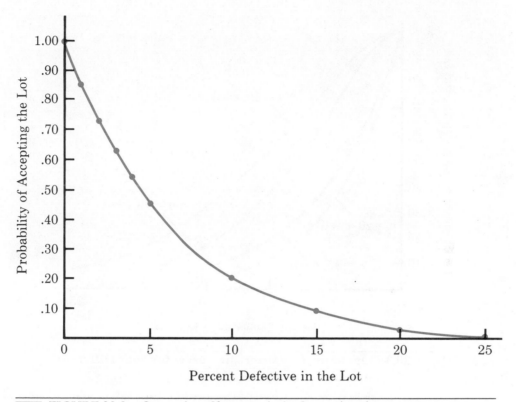

FIGURE 20.2   Operating Characteristic Curve for the $n = 15, c = 0$ Acceptance Sampling Plan

accept the lot. Alternatively, using Table 20.3, we find that with $n = 15$ and $p = .05$, $f(0) = .4633$ and $f(1) = .3658$. Thus there is a $.4633 + .3658 = .8291$ probability that the $n = 15$, $c = 1$ plan will lead to the acceptance of a lot with 5% defective.

Figure 20.3 shows the operating characteristic curves for four alternative acceptance sampling plans for the KALI, Inc. example. Both samples of size 15 and 20 are considered. Note that regardless of the percent defective in the lot, the $n = 15$, $c = 1$ sampling plan provides the highest probabilities of accepting the lot. On the other hand, the $n = 20$, $c = 0$ sampling plan provides the lowest probabilities of accepting the lot; however, this plan also provides the highest probabilities of rejecting the lot.

Selecting an Acceptance Sampling Plan

Now that we know how to use the binomial probability distribution to compute the probability of accepting a lot with a given percent defective, we are ready to select the values of n and c that determine the desired acceptance sampling plan for the application being studied. In order to do this, management must specify two values for the fraction defective in the lot. One value, denoted p_0, will be used to control for the producer's risk, and the other value, denoted p_1, will be used to control for the consumer's risk.

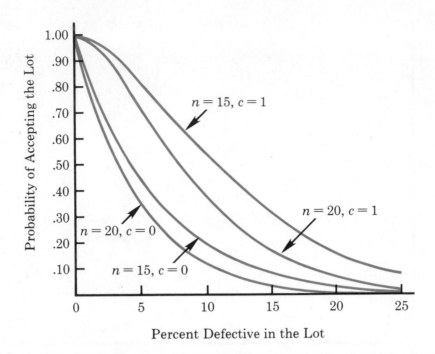

FIGURE 20.3 **Operating Characteristic Curves for Four Different Acceptance Sampling Plans**

In showing how this can be done, we will use the following notation:

α = the producer's risk; the probability that a lot with p_0 defective will be rejected

β = the consumer's risk; the probability that a lot with p_1 defective will be accepted

For instance, suppose that for the KALI, Inc. example management specifies that p_0 = .03 and p_1 = .15. From the OC curve in Figure 20.4, we see that p_0 = .03 provides a producer's risk of approximately $1 - .63$ = .37. Using p_1 = .15 shows a consumer's risk of approximately .09. Thus, if management is willing to tolerate both a .37 probability of rejecting a lot with 3% defective and a .09 probability of accepting a lot with 15% defective, the n = 15, c = 0 acceptance sampling plan would be acceptable.

Suppose, however, that management requests a producer's risk of α = .10 and a consumer's risk of β = .20. Thus we see that n = 15, c = 0 sampling plan has a better-than-desired consumer's risk but an unacceptably large producer's risk. The fact that α = .37 indicates that 37% of the lots will have the error of being rejected when only 3% of the items in the lot are defective. The producer's risk is too high, and a different acceptance sampling plan should be considered.

Using p_0 = .03, α = .10, p_1 = .15, and β = .20 in Figure 20.3 shows that the acceptance sampling plan with n = 20 and c = 1 comes the closest to meeting both the producer's risk and the consumer's risk requirements. Exercise 1 at the end of this section will ask you to compute the producer's risk and the consumer's risk for the n = 20, c = 1 sampling plan.

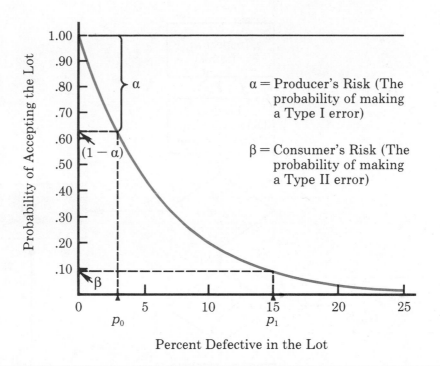

FIGURE 20.4 **Operating Characteristic Curve for $n = 15$, $c = 0$ with $p_0 = .03$ and $p_1 = .15$**

As shown in this section, several computations and several operating characteristic curves may need to be considered in order to determine the sampling plan with the desired producer's and consumer's risk. Fortunately, in order to assist in the application of acceptance sampling, tables of sampling plans have been published. For instance, the American Military Standard Table, *MIL-STD-105D*, provides information helpful in designing acceptance sampling plans. More advanced texts on quality control, such as those described in the bibliography at the end of the text, describe the use of such tables. In addition, these more advanced texts also discuss the role of sampling costs in determining the optimal sampling plan.

Multiple Sampling Plans

The acceptance sampling procedure that we have presented for the KALI, Inc. example is a *single-sample* plan. It is called a single-sample plan because only one sample or sampling stage is used. After determining the number of defective components in the sample, a decision must be made to accept or to reject the lot. An alternative to the single-sample plan is a multiple sampling plan, in which two or more stages of sampling are used. At each stage a decision is made among three possibilities: stop sampling and accept the lot, stop sampling and reject the lot, or continue sampling. Although more complex in analysis, multiple sampling plans often result in a smaller total sample size than single-sample plans with the same α and β probabilities.

The logic of a double-, or two-stage, sampling plan is shown in Figure 20.5.

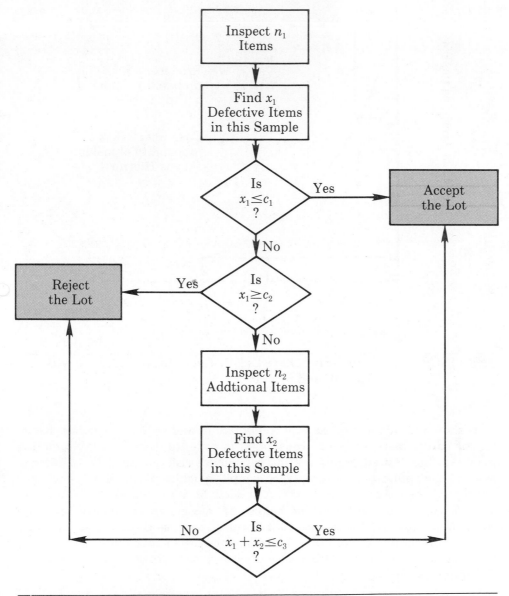

■■■■■ FIGURE 20.5   **A Two-Stage Acceptance Sampling Plan**

Initially a sample of n_1 items is selected. If the number of defective components, x_1, is less than or equal to c_1, accept the lot. If x_1 is greater than or equal to c_2, reject the lot. If x_1 is between c_1 and c_2 ($c_1 < x_1 < c_2$), select a second sample of n_2 items. Determine the combined, or total, number of defective components from the first sample (x_1) and from the second sample (x_2). If $x_1 + x_2 \le c_3$, accept the lot; otherwise reject the lot. The development of the double-sampling plan is more difficult due to the fact that the sample sizes n_1 and n_2 and the acceptance numbers c_1, c_2, and c_3 must be determined in order to meet both the desired producer's and the desired consumer's risk.

| ■■■ | NOTES AND COMMENTS | ■■■ |

1. The use of the binomial probability distribution for acceptance sampling is based on the assumption of large lots. In situations where the lot size is small, the hypergeometric probability distribution is the appropriate distribution. Experts in the field of quality control indicate that the Poisson distribution provides a good approximation for acceptance sampling when the sample size is at least 16, the lot size is at least ten times the sample size, and p is less than 0.1.* For larger sample sizes, the normal approximation to the binomial probability distribution can be used.

2. In the MIL-ST-105D sampling tables, p_0 is referred to as the acceptable quality level (AQL). In some sampling tables, p_1 is called the lot tolerance percent defective (LTPD) or the rejectable quality level (RQL). Many of the published sampling plans also use quality indices such as the indifference quality level (IQL) and the average outgoing quality limit (AOQL). The more advanced texts listed in the bibliography provide a complete discussion of these other indicies.

3. In this section we provided an introduction to *attributes sampling plans*. In these plans each item sampled is classified as nondefective or defective. In *variables sampling plans*, a sample is taken and a measurement of the quality characteristic is taken. For instance, in gold jewelry the measurement of quality may be the amount of gold in the jewelry. A simple statistic such as the average amount of gold for the jewelry in the sample is computed and compared with an allowable value in order to determine whether to accept or reject the lot.

■■■ EXERCISES

1. Refer to the KALI, Inc. example presented in this section. The quality-control manager requested a producer's risk of .10 when p_0 was .03 and a consumer's risk of .20 when p_1 was .15. Consider the acceptance sampling plan based on a sample size of 20 and an acceptance number of 1. Answer the following questions:

   a. What is the producer's risk for the $n = 20$, $c = 1$ sampling plan?
   b. What is the consumer's risk for the $n = 20$, $c = 1$ sampling plan?
   c. Does the $n = 20$, $c = 1$ sampling plan satisfy the risk's requested by the quality control manager? Discuss.

2. Consider an acceptance sampling plan with $n = 20$ and $c = 0$. Compute the producer's risk for each of the following cases.

   a. The lot has a defective rate of 2%.
   b. The lot has a defective rate of 6%.

3. Repeat Exercise 2 for the acceptance sampling plan with $n = 20$ and $c = 1$. What happens to the producer's risk as the acceptance number c is increased? Explain.

4. To inspect incoming shipments of raw materials, a manufacturer is considering samples of size 10, 15, and 20. Use the binomial probabilities from Table 5 of Appendix B

---

*Juran, J. M. and Frank M. Gryna, Jr., *Quality Planning and Analysis*, McGraw-Hill, New York, 1980, p. 412.

to select a sampling plan that provides a producer's risk of $\alpha = .03$ when p_0 is .05 and a consumer's risk of $\beta = .12$ when p_1 is .30.

5. A domestic manufacturer of watches purchases quartz crystals from a Swiss firm. The crystals are shipped in lots of 1000. The acceptance sampling procedure uses 20 randomly selected crystals.

 a. Construct operating characteristic curves for acceptance numbers of 0, 1, and 2.

 b. If p_0 is .01 and $p_1 = .08$, what are the producer's and consumer's risks for each sampling plan in part a?

20.2 ■ STATISTICAL PROCESS CONTROL

In this section we consider quality-control procedures for a production process where goods are being manufactured continuously. Based upon sampling and inspection of production output, a decision will be made either to continue the production process or to adjust the production process in order to bring the items or goods being produced up to acceptable quality standards.

Despite the efforts spent in designing quality into manufacturing and production operations, machine tools will invariably wear out, vibrations will cause machine settings to fall out of adjustment, purchased materials will be defective, and human operators will make mistakes. Any or all of these factors can result in poor-quality output. Fortunately, quality-control procedures are available for monitoring the quality of production output so that poor quality can be detected early and the production process adjusted or corrected.

If the variation in the quality of the production output is due to *assignable causes* such as tools wearing out, incorrect machine settings, poor-quality raw materials, or operator error, the process should be adjusted or corrected as soon as possible. Alternatively, if the variation in quality is due to randomly occurring variations in materials, temperature, humidity, and so on, which the manufacturer cannot possibly account for, these causes are referred to as *common causes*. The main objective of statistical process control is to determine if variations in output are due to assignable causes or common causes.

Whenever assignable causes are detected, we will conclude that the process is *out of control*. In this case, corrective action will be taken to bring the process back to an acceptable level of quality. However, if the variation in the output of a production process is due only to common causes, we will conclude that the process is in *statistical control*, or simply in *control*; in such cases no changes or adjustments are necessary.

The statistical procedures for process control are based on the hypothesis-testing methodology presented in Chapter 9. The null hypothesis, H_0, is formulated in terms of the production process being in control. The alternative hypothesis, H_a, is formulated in terms of the production process being out of control. The states of nature are that the production process is in control and that the production process is out of control; the decision alternatives are to continue the production process or to adjust the production process. Table 20.4 shows that correct decisions of continuing an in-control process and adjusting an out-of-control process are possible. However, as with other hypothesis-testing procedures, there also exists the possibility of making a Type I error (adjusting an in-control process) as well as the possibility of making a Type II error (allowing an out-of-control process to continue).

TABLE 20.4   **The Outcomes of Statistical Process Control**

Decision	States of Nature	
	H_0 True *Process in Control*	H_0 False *Process Out of Control*
Continue Process	Correct Decision	Type II Error (Allowing an out-of-control process to continue)
Adjust Process	Type I Error (Adjusting an in-control process)	Correct Decision

Control Charts

A *control chart* is a graphical tool used to help determine if a process is in control or out of control. As such, a control chart provides a basis for deciding whether the variation in the output is due to common causes (in control) or is due to assignable causes (out of control). Whenever an out-of-control situation is detected, adjustments and/or other corrective action will be taken to bring the process back into control.

Control charts can be classified by the type of data they contain. For instance, an $\bar{x}$ chart is used for cases where the quality of the output is measured in terms of a variable such as length, weight, temperature, and so on. In this case, the decision to continue or to adjust the production process will be based on the mean value found in a sample of the output. Since $\bar{x}$ is used to denote the sample mean, the control chart used is called an $\bar{x}$ chart. To introduce some of the concepts common to all control charts, let us consider some of the specific features of an $\bar{x}$ chart.

Figure 20.6 shows the general structure of an $\bar{x}$ chart. The center line of the chart corresponds to the mean of the process when the process is *in control.* The vertical line identifies the scale of measurement for the variable of interest. Each time a sample is taken from the production process, a value of $\bar{x}$ is computed and

■■■■ FIGURE 20.6 $\bar{x}$ **Chart Structure**

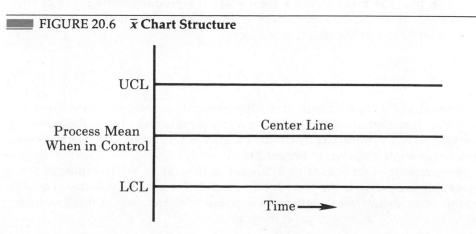

a data point showing the value of $\bar{x}$ for the sample is plotted on the control chart.

The two lines labeled UCL and LCL play an important role in determining if the process is in control or out of control. The lines are referred to as the *upper control limit* and the *lower control limit*, respectively. These limits are chosen so that when the process is in control, there will be a high probability that the value of $\bar{x}$ will fall between the two control limits. Values outside the control limits provide strong statistical evidence that the process is out of control and corrective action should be taken.

Over time, more and more data points will be added to the control chart. The order of the data points will appear from left to right as the process is sampled. In essence, every time a point is plotted on the control chart, we are carrying out a hypothesis test to see if the process is in control.

In addition to the $\bar{x}$ chart, there are also control charts that can be used to monitor the range of the measurements in the sample (R chart), the percent defective in the sample (p chart), and the number of defects in the sample (c chart). In each case, the general structure of the control chart follows the format of the $\bar{x}$ chart shown in Figure 20.6. The major difference in each control chart is the measurement scale used; for instance, in a p chart the measurement scale denotes the percentage of defective items in the sample instead of the sample mean. In the following discussion we will illustrate the construction and use of the $\bar{x}$ chart and the p chart.

$\bar{x}$ Chart

To illustrate the construction of an $\bar{x}$ chart, let us consider the situation involving KJW Packaging. KJW operates a production line where cartons of cereal are filled. Suppose that KJW knows that when the process is operating correctly—and hence the system is in control—the mean filling weight is $\mu = 16.05$ ounces and the process standard deviation is $\sigma = .10$ ounces. In addition, suppose that the distribution of filling weights is normally distributed. This distribution is shown in Figure 20.7.

The sampling distribution of $\bar{x}$, as presented in Chapter 7, can be used to determine the variation that can be expected in $\bar{x}$ values for a process that is in control. To show how this can be done, let us first briefly review the properties of the sampling distribution of $\bar{x}$. First, recall that the expected value or mean of $\bar{x}$ is equal to μ, the mean filling weight when the production line is in control. For samples of size n, the formula for the standard deviation of $\bar{x}$, referred to as the standard error of the mean, is as follows:

$$\sigma_{\bar{x}} = \frac{\sigma}{\sqrt{n}} \tag{20.3}$$

In addition, since the distribution of filling weights is normally distributed, the sampling distribution of $\bar{x}$ is normal for any sample size. Thus, the sampling distribution of $\bar{x}$ is normally distributed with mean μ and standard deviation $\sigma_{\bar{x}}$. This distribution is shown in Figure 20.8.

The sampling distribution of $\bar{x}$ is used to determine what values of $\bar{x}$ are reasonable to observe if the process is in control. The general practice in quality control is to define reasonable as any value of $\bar{x}$ that is within 3 standard

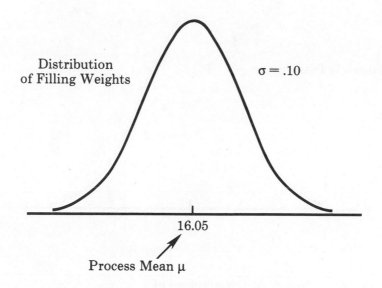

Distribution
of Filling Weights

$\sigma = .10$

16.05

Process Mean μ

■■■ FIGURE 20.7   **Distribution of Cereal-Carton-Filling Weights**

deviations above or below the mean value. Recall from the study of the normal probability distribution that approximately 99.7% of the values of a normally distributed random variable lie within ±3 standard deviations of its mean value. Thus, if a value of $\bar{x}$ falls within the interval $\mu - 3\sigma_{\bar{x}}$ to $\mu + 3\sigma_{\bar{x}}$, we will assume that the process is in control. In summary, then, the control limits for an $\bar{x}$ chart are computed as follows:

Control Limits for an $\bar{x}$ Chart

$$\text{UCL} = \mu + 3\sigma_{\bar{x}} \qquad\qquad (20.4)$$

$$\text{LCL} = \mu - 3\sigma_{\bar{x}} \qquad\qquad (20.5)$$

Reconsider the KJW Packaging example with the process distribution of filling weights shown in Figure 20.7 and the sampling distribution of $\bar{x}$ shown in Figure 20.8. Assume that a quality-control inspector periodically samples six cartons and uses the sample mean filling weight to determine if the process is in control or out of control. Using (20.3), the standard error of the mean is $\sigma_{\bar{x}} = \sigma/\sqrt{n} = .10/\sqrt{6} = .04$. Thus with the process mean at 16.05, the control limits are UCL = 16.05 + 3(.04) = 16.17 and LCL = 16.05 − 3(.04) = 15.93. The control chart, along with the results of ten samples taken over a 10-hour period, are shown in Figure 20.9. For convenience in reading the chart, the sample numbers from 1 to 10 are shown below the chart.

Refer to Figure 20.9; note that the mean for the fifth sample shows that the process was out of control. In other words, the sample mean $\bar{x} = 15.89$ provides an indication that assignable causes in output variation were detected and that underfilling was occurring. As a result, corrective action was taken at this point

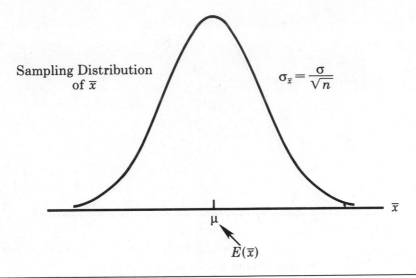

Sampling Distribution of $\bar{x}$

$$\sigma_{\bar{x}} = \frac{\sigma}{\sqrt{n}}$$

μ

$E(\bar{x})$

$\bar{x}$

■ FIGURE 20.8   **Sampling Distribution of $\bar{x}$**

to bring the process back into control. The fact that the remaining points on the $\bar{x}$ chart fall within the upper and lower control limits indicates that the corrective action was successful.

In the KJW Packaging example we assumed that the mean and standard deviation of filling weights were known. In most situations these values are not known but are estimated using sample data. To discuss how this is done, assume that KJW did not know the values of μ and σ when the process is in control. To estimate these values, inspectors would select samples from the process when it is believed to be operating in control. For instance, they might select a random sample of five boxes each morning and five boxes each afternoon for 10 days of in-control operation. For each subgroup, or sample, the mean of the sample and

■ FIGURE 20.9   **The $\bar{x}$ Chart for the Cereal-Carton-Filling Process**

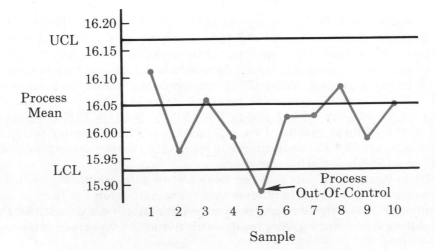

UCL

Process Mean

LCL

Process Out-Of-Control

16.20
16.15
16.10
16.05
16.00
15.95
15.90

1   2   3   4   5   6   7   8   9   10

Sample

the range of the sample is computed. The overall average of the sample means is used as the estimate of μ, the process mean when the system is in control. Using control-chart tables, the upper and lower control limits can be computed based upon the overall mean and the average value of the range for the samples. Although the details are beyond the scope of this chapter, we note that the use of such tables provides a convenient way of indirectly estimating σ using the ranges observed in the samples. The advantage of this approach is that upper and lower control limits can be established with little computational effort.

A more accurate approach to establishing the control chart when μ and σ are unknown is to treat all the sample data collected as one large sample. The sample mean and sample standard deviation for this combined data set can be used to estimate μ and σ. With the increasing use of computer software packages for quality control, these calculations can be done with little effort on the part of the user.

p Chart

Let us consider the case where the output quality is measured in terms of the items being either nondefective or defective. The decision to continue the production process or to adjust the production process will be based on $\bar{p}$, the proportion of defective items found in a sample of the output. The control chart used for proportion defective data is called a *p chart*.

To illustrate the construction of a p chart, consider the situation involving the use of automated mail-sorting machines in a post office. These automated machines scan the ZIP code on letters and divert each letter to its proper carrier route. Even when the machine is operating properly, some letters are diverted to incorrect routes. Assume that when the machine is operating correctly, or in a state of control, 3% of the letters are incorrectly diverted. Thus p, the fraction defective when the process is in control, is .03.

The sampling distribution of $\bar{p}$, as presented in Chapter 7, can be used to determine the variation that can be expected in $\bar{p}$ values for a process that is in control. Recall that the expected value or mean of $\bar{p}$ is p, the fraction defective when the process is in control. With samples of size n, the formula for the standard deviation of $\bar{p}$, referred to as the standard error of the proportion, is as follows:

$$\sigma_{\bar{p}} = \sqrt{\frac{p(1-p)}{n}} \qquad (20.6)$$

We also learned in Chapter 7 that the sampling distribution of $\bar{p}$ can be approximated by a normal probability distribution whenever the sample size is large. With $\bar{p}$, the sample size can be considered large whenever the following two conditions are satisfied:

$$np \geq 5$$

$$n(1-p) \geq 5$$

In summary, then, whenever the sample size is large, the sampling distribution of $\bar{p}$ can be approximated by a normal distribution with mean p and standard deviation $\sigma_{\bar{p}}$. This distribution is shown in Figure 20.10.

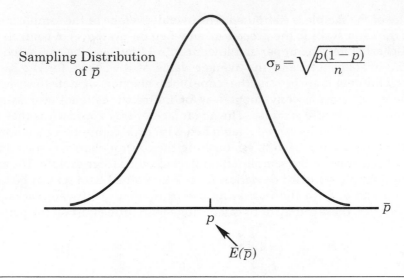

Sampling Distribution of $\bar{p}$

$\sigma_{\bar{p}} = \sqrt{\dfrac{p(1-p)}{n}}$

p

$\bar{p}$

$E(\bar{p})$

■■ FIGURE 20.10   **Sampling Distribution of $\bar{p}$**

To establish control limits for a p chart, we follow the same procedure that we used to establish control limits for an $\bar{x}$ chart. That is, the limits for the control chart are set at three standard deviations, or errors, above and below the percent defective when the process is in control. Thus we have the following control limits:

Control Limits for a p Chart

$$\text{UCL} = p + 3\sigma_{\bar{p}} \qquad\qquad (20.7)$$

$$\text{LCL} = p - 3\sigma_{\bar{p}} \qquad\qquad (20.8)$$

Using $p = .03$ and samples of size $n = 200$, (20.6) shows that the standard error is

$$\sigma_{\bar{p}} = \sqrt{\frac{.03(1 - .03)}{200}} = .0121$$

Thus the control limits are UCL $= .03 + 3(.0121) = .0662$ and LCL $= .03 - 3(.0121) = -.0063$. Since LCL is negative, LCL is set equal to 0 in the control chart.

The control chart for the automated sorting process is shown in Figure 20.11. The points plotted show a series of proportion defective found in samples of 200 letters taken from the process. Since all points are within the control limits, there is no evidence to conclude that the sorting process is out of control. In fact, the control chart indicates that the process should continue operation.

In cases where the percentage of defective items for a process that is in control is not known, this value is first estimated using sample data. Suppose, for

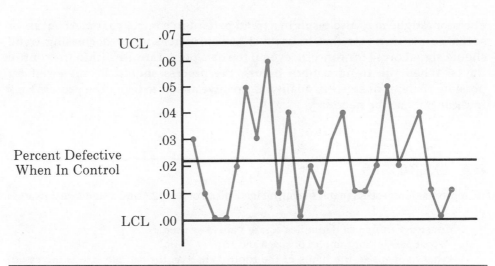

■■■■■ FIGURE 20.11   **Control Chart for the Proportion Defective in a Mail-Sorting Process**

example, that M different samples, each of size n, are selected from a process that is in control. The fraction or proportion of defective items in each sample is then determined. Treating all the data collected as one large sample, we can determine the average number of defective items for all the data; this value can then be used to provide an estimate of p, the percentage of defective items observed when the process is in control. Note that this estimate of p also enables us to estimate the standard error of the proportion; then, upper and lower control limits can be established.

Interpretation of Control Charts

The location of points and the pattern of points in a control chart enable us to determine, with a small probability of error, whether or not a process is in statistical control. A primary indication that a process may be out of control occurs whenever a data point is outside the control limits. This situation was observed for point 5 in Figure 20.9. When such a point is found, there is strong statistical evidence that the process it out of control; in such cases corrective action should be taken as soon as possible.

In addition to points outside the control limits, the pattern of the points within the control limits can signal concern for maintaining quality. For example, assume that all the data points are within the control limits but that a large number of points are on one side of the center line. This may indicate that an equipment problem, a change in materials, or some other assignable cause of a shift in quality has occurred. When this pattern occurs, careful investigation of the production process should be undertaken to determine if indeed a change in quality has occurred.

Another pattern to watch for in control charts is a gradual shift, or trend, over time. For example, as tools wear out, the dimensions of machined parts will gradually deviate more and more from their designed levels. Gradual changes in temperature or humidity, general equipment deterioration, dirt buildup, and/or operator fatigue may also result in a trend pattern in control charts. About six or seven points in a row that indicate either an increasing or decreasing trend should signal cause for concern, even if the data points are all within the control limits. When the trend pattern occurs, the process should be reviewed for possible changes or shifts in quality. Corrective action to bring the process back into control may be necessary.

▬ EXERCISES

6. A production process that is in control has a mean of $\mu = 12.5$ and a standard deviation of $\sigma = .8$.

   a. Construct an $\bar{x}$ chart if samples of size 4 are to be used.

   b. Repeat part (a) for samples of size 8 and 16.

   c. What happens to the limits of the control chart as the sample size is increased? Discuss why this is reasonable.

7. Temperature is used to measure the output of a production process. When the process is in control, the mean of the process is $\mu = 128.5$ and the standard deviation is $\sigma = .4$.

   a. Construct an $\bar{x}$ chart if samples of size 6 are to be used.

   b. Is the process in control for a sample providing the following data: 128.8, 128.2, 129.1, 128.7, 128.4, and 129.2?

   c. Is the process in control for a sample providing the following data: 129.3, 128.7, 128.6, 129.2, 129.5 and 129.0?

8. A soap manufacturer uses a weight to measure the output of a laundry detergent powder. The control limits are set at UCL = 20.12 and LCL = 19.90. Sample of size 5 are used for the sampling and inspection process. What are the process mean and process standard deviation for the manufacturing operation?

9. Twenty-five samples, each of size five, were selected from a process that was in control. The sum of all the data collected was 677.5 pounds.

   a. What is an estimate of the process mean (in terms of pounds per unit) when the process is in control?

   b. Develop the control chart for this process if samples of size five will be used. Assume that the process standard deviation is .5 when the process is in control.

10. Twenty-five samples of 100 items each were inspected when a manufacturing process was considered to be operating satisfactorily. In the 25 samples, a total of 135 items were found to be defective.

    a. What is an estimate the process proportion defective when the process is in control?

    b. What is the standard error of the proportion if samples of size 100 will be used for statistical process control?

    c. Compute the upper and lower control limits for the control chart.

11. Over several weeks of normal, or in-control, operation, 20 samples of 150 packages each of synthetic-gut tennis strings were tested for breaking strength. A total of 141 packages failed to conform to the manufacturers specifications.

    a. What is an estimate of the process proportion defective when the system is in control?

    b. Compute the upper and lower control limits for a p chart.

    c. What conclusion should be made about the process if tests for a new sample of 150 packages found 12 defective? Do there appear to be assignable causes present in this situation?

12. An automotive industry supplier produces pistons for several models of automobiles. Twenty samples, each consisting of 200 pistons, were selected when the process was known to be operating correctly. The number of defective pistons found in each sample is as follows:

8	10	6	4	5	7	8	12	8	15
14	10	10	7	5	8	6	10	4	8

a. What is an estimate of the process proportion defective for the piston-manufacturing process when it is in control?
b. Construct a p chart for the manufacturing process, assuming each sample will have 200 pistons.
c. What conclusion should be made if a sample of 200 has 20 defective pistons?

■ SUMMARY

In this chapter we discussed how statistical methods can be used to assist in the control of quality. We first considered the technique referred to as acceptance sampling. With this quality-control procedure, a sample of the lot is selected and inspected. The number of defective items in the sample provides the basis for accepting or rejecting the lot. The sample size and the acceptance criterion can be adjusted to control both the producer's risk (Type I error) and the consumer's risk (Type II error) associated with the acceptance sampling procedure.

Control charts were presented as a graphical aid to monitoring the quality of a production process. For an $\bar{x}$ chart and a p chart, the center line of the chart corresponds to the process mean or percent defective under normal, or in-control, operation. Control limits are established three standard errors above and below the process mean or the process percent defective. Samples are selected periodically and the data points plotted on the control chart. Data points outside the control limits indicate that the process is out of control and corrective action should be taken. The pattern of the data points within the control limits can also indicate concern for quality and suggest that corrective action may be warranted.

■ GLOSSARY

Quality control   Consists of making a series of inspections and measurements in order to determine if quality standards are being met.

Lot   A group of items such as incoming shipments of raw materials or purchased parts as well as finished goods from final assembly.

Acceptance sampling   A statistical quality-control procedure in which the number of defective items found in a sample is used to determine whether a lot should be accepted or rejected.

Producer's risk   In acceptance sampling, the risk of rejecting a good-quality lot. This is the Type I error.

Consumer's risk   In acceptance sampling, the risk of accepting a poor-quality lot. This is the Type II error.

Acceptance criterion   The maximum number of defective items that can be found in the sample and still enable acceptance of the lot.

Operating characteristic curve   A graph showing the probability of accepting the lot as a function of the percent defective in the lot. This curve can be used to help determine if a particular acceptance sampling plan meets both the producer's risk and consumer's risk requirements.

Multiple sampling plan A form of acceptance sampling where more than one sample or stage is used. Based on the number of defective items found in a sample, a decision will be made to accept the lot, reject the lot, or continue sampling.

Common causes Normal or natural variations in process outputs that are due purely to chance. No corrective action is necessary when output variations are due to common causes.

Assignable causes Variations in process outputs that are due to factors such as machine tools wearing out, incorrect machine settings, poor-quality raw materials, operator error, and so on. Corrective action should be taken when assignable causes of output variation are detected.

Control chart A graphical tool used to help determine if a process is in control or out of control.

$\bar{x}$ chart A control chart used when the output of a process is measured in terms of the mean value of a variable such as a length, weight, temperature, and so on.

p chart A control chart used when the output of a process is measured in terms of the percent defective.

■ KEY FORMULAS

Binomial Probability Function for Acceptance Sampling

$$f(x) = \frac{n!}{x!(n-x)!} \, p^x(1-p)^{n-x} \tag{20.1}$$

Standard Error of the Mean

$$\sigma_{\bar{x}} = \frac{\sigma}{\sqrt{n}} \tag{20.3}$$

Control Limits for an $\bar{x}$ Chart

$$\text{UCL} = \mu + 3\sigma_{\bar{x}} \tag{20.4}$$

$$\text{LCL} = \mu - 3\sigma_{\bar{x}} \tag{20.5}$$

Standard Error of the Proportion

$$\sigma_{\bar{p}} = \sqrt{\frac{p(1-p)}{n}} \tag{20.6}$$

Control Limits for a p chart

$$\text{UCL} = p + 3\sigma_{\bar{p}} \tag{20.7}$$

$$\text{LCL} = p - 3\sigma_{\bar{p}} \tag{20.8}$$

■ SUPPLEMENTARY EXERCISES

13. An $n = 10$, $c = 2$ acceptance sampling plan is being considered; assume that $p_0 = .05$ and $p_1 = .20$.

a. Compute the producer's risk and consumer's risk for this acceptance sampling plan.

b. Would either the producer or the consumer, or both, be unhappy with the proposed sampling plan?

c. What change in the sampling plan, if any, would you recommend?

14. An acceptance sampling plan with $n = 15$ and $c = 1$ has been designed with a producer's risk of .075.

a. Was the value of p_0 .01, .02, .03, .04 or .05? What does this value mean?
b. What is the consumer's risk associated with this plan if p_1 is .25?

15. A manufacturer produces lots of a canned food product. Let p denote the proportion of the lot that do not meet the product quality specifications. An $n = 25, c = 0$ acceptance sampling plan will be used.

a. Compute points on the operating characteristic curve when $p = .01, .03, .10$ and .20.
b. Plot the operating characteristic curve.
c. What is the probability the acceptance sampling plan will reject a lot that has .01 defective?

16. Sometimes an acceptance sampling plan will be based on a large sample. In this case the normal approximation to the binomial probability distribution can be used to compute the producer's risk and consumer's risk associated with the plan. Referring to Chapter 6, the normal distribution used to approximate binomial probabilities has a mean of np and a standard deviation of $\sqrt{np(1 - p)}$. Assume that an acceptance sampling plan is $n = 250, c = 10$.

a. What is the producer's risk if p_0 is .02? As discussed in Chapter 6, a continuity correction factor should be used in this case. Thus the probability of acceptance is based on the normal probability x being less than or equal to 10.5.
b. What is the consumer's risk if p_1 is .08?
c. What is an advantage of a large sample size for acceptance sampling? What is a disadvantage?

17. Samples of size 5 provided the following 20 sample means for a production process that is known to be in control.

$$
\begin{array}{llllllll}
95.72 & 95.24 & 95.18 & 95.44 & 95.46 & 95.32 & 95.40 & 95.44 \\
95.08 & 95.50 & 95.80 & 95.22 & 95.56 & 95.22 & 95.04 & 95.72 \\
94.82 & 95.46 & 95.60 & 95.78 &&&&
\end{array}
$$

a. Based on these data, what is an estimate of the mean when the process is in control?
b. Assuming that the process standard deviation is $\sigma = .50$, develop a control chart for this production process.
c. Do any of the 20 sample mean given above indicate that the process was out of control?

18. Product filling weights are normally distributed with a mean of 350 grams and a standard deviation of 15 grams.

a. Develop the control limits for samples of size 10, 20, and 30.
b. What happens to the control limits as the sample size is increased?
c. What happens when a Type I error is made?
d. What happens when a Type II error is made?
e. What is the probability of a Type I error for samples of size 10, 20, and 30?
f. What is the advantage of increasing the sample size for control-chart purposes? What error probability is reduced as the sample size is increased?

19. Consider the two situations described next. For each, comment on whether or not there is reason for concern about the quality of the process.

a. A p chart has LCL = 0 and UCL = .068. When the process is in control, the proportion defective is .033. Plot the following seven sample results: .035, .062, .055, .049, .058, .066, and .055. Discuss.
b. An $\bar{x}$ chart has LCL = 22.2 and UCL = 24.5. The mean is $\mu = 23.35$ when the process

is in control. Plot the following seven sample results: 22.4, 22.6, 22.65, 23.2, 23.4, 23.85 and 24.1. Discuss.

20. Managers of 1200 different retail outlets make twice-a-month restocking orders from a central warehouse. Past experience has shown that 4% of the orders contain one or more errors such as wrong item shipped, wrong quantity shipped, item requested but not shipped and so on. Random samples of 200 orders are selected monthly and checked for accuracy.

    a. Construct a control chart for this situation.

    b. Six months of data show the following number of orders with one or more errors: 10, 15, 6, 13, 8, and 17. Plot the data on the control chart. What does your plot indicate about the order-filling process?

APPLICATION

Dow Chemical U.S.A.*

Freeport, Texas

Dow Chemical U.S.A., Texas Operations, began in 1940 when The Dow Chemical Company purchased 800 acres of Texas land on the Gulf Coast to build a magnesium production facility. That original site has expanded to cover more than 5000 acres and is now one of the largest petrochemical complexes in the world.

Texas Operations products include magnesium, styrene, plastics, adhesives, solvents, glycerine, glycol, chlorine, caustic, and many others. While some products are made solely for use in other processes, many end up as essential ingredients in such products as pharmaceuticals, toothpastes, dog foods, water hoses, ice chests, milk cartons, garbage bags, shampoos, and furniture.

Dow's Texas Operations are located principally in the Brazosport area but also include sites in La Porte, Texas; Russellville, Arkansas; Denver, Colorado; and Bayonne, New Jersey.

Annually, Texas Operations manufactures approximately 50 percent of Dow products sold in the United States and about 25 percent of those sold on a global basis.

STATISTICAL QUALITY CONTROL FOR PROCESS IMPROVEMENT

Dow's Texas Operations produces over 30 percent of the world's magnesium, a very lightweight metal first produced to meet the increased demand for aircraft production during World War II. Magnesium is found in products ranging from tennis racquets to suitcases to "mag" wheels. The Magnesium Department was the first group in Texas Operations to train its technical people and managers in the use of statistical quality control (SQC). Some of the earliest successful applications of SQC were quality improvements in chemical processing.

Figure 20A.1 shows the results of an $\bar{x}$ chart for a drier operation before and after statistical quality control. The lower values of $\bar{x}$ and the lower variation in the $\bar{x}$ values for different shifts were both significant improvements in the process control. In addition, the $\bar{x}$ chart revealed the fact that differences existed between operators. The blackened circles in Figure 20A.1 represent observations from one operator in question; the open circles represent observations from other operators. Upon examination of the control chart information, the operator in question was retrained. The second chart in Figure 20A.1 shows the improvement in quality that was obtained.

*The authors are indebted to Clifford B. Wilson, Magnesium Technical Manager, The Dow Chemical Company, for providing this application.

**Chemical processing facilities at Dow Chemical
U.S.A., Texas Operations**

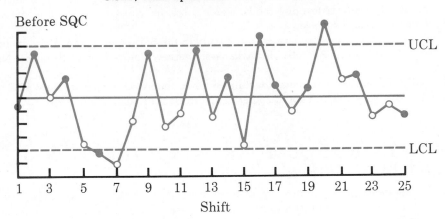

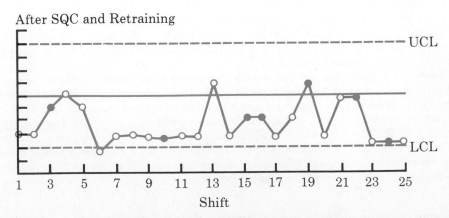

■■■ FIGURE 20A.1 $\bar{x}$ **Charts for Drier Analysis Before and After Statistical
Quality Control**

Before SQC

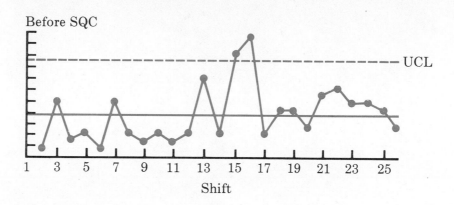

After SQC

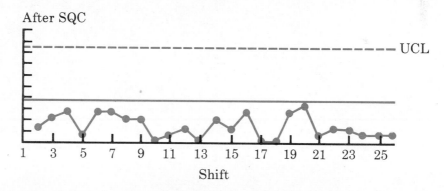

▬▬▬ FIGURE 20A.2   **Range Charts for Drier Analysis Before and After Statistical Quality Control**

Figure 20A.2 shows range charts for the same drier operation before and after statistical quality control. The use of these charts made the operators realize that their attempts to fine-tune the process affected the variation in the drier output. A comparison of the before and after range charts shows the reduction in variation that was obtained through the use of SQC.

Dow Chemical Company has experienced quality improvements everywhere statistical quality control has been used in the magnesium processing. Documented savings of several hundred thousand dollars per year have been realized, and new applications are continually being discovered.

Appendices

APPENDIX

A

References and Bibliography

▬ GENERAL

Freedman, D., R. Pisani, and R. Purves, *Statistics*, New York, W. W. Norton, 1978.

Freund, J. E., and R. E. Walpole, *Mathematical Statistics*, 3d ed., Englewood Cliffs, N.J., Prentice-Hall, 1980.

Hoaglin, D. C., F. Mosteller, and J. W. Tukey, *Understanding Robust and Exploratory Data Analysis*, New York, John Wiley & Sons, 1983.

Hogg, R. V., and A. T. Craig, *Introduction to Mathematical Statistics*, 4th ed., New York, Macmillian, 1978.

Mood, A. M., F. A. Graybill, and D. C. Boes, *Introduction to the Theory of Statistics*, 3d ed., New York, McGraw-Hill, 1974.

Neter, J., W. Wasserman, and G. A. Whitmore, *Applied Statistics*, 3rd. ed., Boston, Allyn & Bacon, 1987.

Ryan, T. A., B. L. Joiner, and B. F. Ryan, *Minitab Handbook*, 2d ed., Boston, PWS-Kent, 1985.

Winkler, R. L., and W. L. Hays, *Statistics: Probability, Inference, and Decision*, 2d ed., New York, Holt, Rinehart & Winston, 1975.

▬ PROBABILITY

Barr, D. R., and P. W. Zehna, *Probability: Modeling Uncertainty*, Reading, Mass., Addison-Wesley, 1983.

Feller, W., *An Introduction to Probability Theory and Its Applications*, Vol. I, 3d ed., New York, John Wiley & Sons, 1968.

Feller, W., *An Introduction to Probability Theory and Its Applications*, Vol. II, 2d ed., New York, John Wiley & Sons, 1971.

Hoel, P. G., S. C. Port, and C. J. Stone, *Introduction to Probability Theory*, Boston, Houghton Mifflin, 1971.

Mendenhall, W., R. L. Scheaffer, and D. Wackerly, *Mathematical Statistics with Applications*, 3rd ed., Boston, PWS-Kent, 1986.

Parzen, E., *Modern Probability Theory and Its Applications*, New York, John Wiley & Sons, Inc., 1960.

Wadsworth, G. P., and J. G. Bryan, *Applications of Probability and Random Variables*, 2d ed., New York, McGraw-Hill, 1974.

Zehna, P. W., *Probability Distributions and Statistics*, Boston, Allyn & Bacon, 1970.

■ SAMPLING METHODS

Cochran, W. G., *Sampling Techniques*, 3d ed., New York, John Wiley & Sons, 1977.

Kish, L., *Survey Sampling*, New York, John Wiley & Sons, 1965.

Scheaffer, R. L., W. Mendenhall, and L. Ott, *Elementary Survey Sampling*, 2d ed., North Scituate, Mass., Duxbury Press, 1979.

Williams, B., *A Sampler on Sampling*, New York, John Wiley & Sons, 1978.

■ EXPERIMENTAL DESIGN

Anderson, V. L., and R. A. McLean, *Design of Experiments: A Realistic Approach*, New York, Marcel Dekker, 1974.

Box, G. E. P., W. G. Hunter, and J. S. Hunter, *Statistics for Experimenters*, New York, John Wiley & Sons, 1978.

Cochran, W. G., and G. M. Cox, *Experimental Designs*, 2d ed., New York, John Wiley & Sons, 1957.

Hicks, C. R., *Fundamental Concepts in the Design of Experiments*, 2d ed., New York, Holt, Rinehart and Winston 1973.

Mendenhall, W., *Introduction to Linear Models and the Design and Analysis of Experiments*, Belmont, Calif., Duxbury Press, 1968.

Montgomery, D. C., *Design and Analysis of Experiments*, New York, John Wiley & Sons, 1976.

Winer, B. J., *Statistical Principles in Experimental Design*, 2d ed., New York, McGraw-Hill, 1971.

■ REGRESSION ANALYSIS

Belsley, D. A., E. Kuh, and R. Welsch, *Regression Diagnostics: Identifying Influential Data and Sources of Collinearity*, New York, John Wiley & Sons, Inc., 1980.

Chatterjee, S., and B. Price, *Regression Analysis by Example*, New York, John Wiley & Sons, Inc., 1978.

Cook, R. D., and S. Weisberg, *Residuals and Influence in Regression*, New York, Chapman and Hall, 1982.

Daniel, C., and F. Wood, *Fitting Equations to Data*, 2d ed., New York, John Wiley & Sons, 1980.

Draper, N. R., and H. Smith, *Applied Regression Analysis*, 2d ed., New York, John Wiley & Sons, 1981.

Gunst, R. F., and R. L. Mason, *Regression Analysis and Its Application: A Data-Oriented Approach*, New York, Marcel Dekker, 1980.

Kleinbaum, D. G., and L. L. Kupper, *Applied Regression Analysis and Other Multivariable Methods*, North Scituate, Mass., Duxbury Press, 1978.

Mendenhall, W., *Introduction to Linear Models and the Design and Analysis of Experiments*, Belmont, Calif., Wadsworth Publishing, 1968.

Mosteller, F., and J. W. Tukey, *Data Analysis and Regression: A Second Course in Statistics*, Reading, Mass., Addison-Wesley, 1977.

Neter, J., W. Wasserman and M. H. Kutner, *Applied Linear Statistical Models*, 2d ed., Homewood, Ill., Richard D. Irwin, 1985.

Weidberg, S., *Applied Linear Regression*, 2d ed., New York, John Wiley & Sons, 1985.

Wesolowsky, G. O., *Multiple Regression and Analysis of Variance*, New York, John Wiley & Sons, 1976.

Wonnacott, T. H., and R. J. Wonnacott, *Regression: A Second Course in Statistics*, New York, John Wiley & Sons, 1981.

■ NONPARAMETRIC METHODS

Conover, W. J., *Practical Nonparametric Statistics*, 2d ed., New York, John Wiley & Sons, 1980.

Gibbons, J. D., *Nonparametric Statistical Inference*, New York, McGraw-Hill, 1971.

Gibbons, J. D., I. Olkin, and M. Sobel, *Selecting and Ordering Populations: A New Statistical Methodology*, New York, John Wiley & Sons, 1977.

Hollander, M., and D. A. Wolfe, *Nonparametric Statistical Methods*, New York, John Wiley & Sons, 1973.

Lehmann, E. L., *Nonparametrics: Statistical Methods Based on Ranks*, San Francisco, Holden-Day, 1975.

Mosteller, F., and R. E. K. Rourke, *Sturdy Statistics*, Reading, Mass., Addison-Wesley, 1973.

Siegel, S., *Nonparametric Statistics for the Behavioral Sciences*, New York, McGraw-Hill, 1956.

■ FORECASTING

Bowerman, B. L., and R. T. O'Connell, *Time Series and Forecasting: An Applied Approach*, North Scituate, Mass., Duxbury Press, 1979.

Box, G. E. P., and G. M. Jenkins, *Time Series Analysis, Forecasting and Control*, rev. ed., San Francisco, Holden-Day, 1976.

Brown, R. G., *Smoothing, Forecasting, and Prediction*, Englewood Cliffs, N.J., Prentice-Hall, 1963.

Gilchrist, W. G., *Statistical Forecasting*, New York, John Wiley & Sons, 1976.

Makridakis, S., and S. C. Wheelwright, *Forecasting: Methods and Applications*, New York, John Wiley & Sons, 1978.

Nelson, C. R., *Applied Time Series Analysis for Managerial Forecasting*, San Francisco, Holden-Day, 1973.

Thomopoulos, N. T., *Applied Forecasting Methods*, Englewood Cliffs, N.J., Prentice-Hall, 1980.

Wheelwright, S. C., and S. Makridakis, *Forecasting Methods for Management*, 3rd ed., New York, John Wiley & Sons, 1980.

■ DECISION THEORY

Behn, R. D., and J. W. Vaupel, *Quick Analysis for Busy Decision Makers*, New York, Basic Books, Inc., 1982.

Brown, R. V., A. S. Kahn, and C. Peterson, *Decision Analysis for the Manager*, New York, Holt, Rinehart & Winston, 1974.

Hadley, G., *Introduction to Probability and Statistical Decision Theory*, San Francisco, Holden-Day, 1967.

Luce, R. D., and H. Raiffa, *Games and Decisions: Introduction and Critical Survey*, New York, John Wiley & Sons, 1957.

Raiffa, H., *Decision Analysis: Introductory Lectures on Choices under Uncertainty*, Reading, Mass., Addison-Wesley, 1968.

Winkler, R. L., *An Introduction to Bayesian Inference and Decision*, New York, Holt, Rinehart & Winston, 1972.

■■■ QUALITY CONTROL

Deming, W. E., *Quality, Productivity, and Competitive Position*, Cambridge, Mass., MIT Center for Advanced Engineering Study, 1982.

Duncan, A. J., *Quality Control and Industrial Statistics*, 5th ed. Homewood, IL, Irwin, 1986.

Evans, J. R., and W. M. Lindsay, *The Management and Control of Quality*, St. Paul, West, 1989.

Ishikawa, Kaoru, *Guide to Quality Control*, 2d rev. ed., New York, Quality Resources, 1986.

Juran, J. M., and F. M. Gryna, Jr., *Quality Planning and Analysis*, 2d ed., New York, McGraw-Hill, 1980.

B

Tables

TABLE 1 Standard Normal Distribution

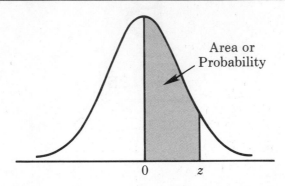

Area or Probability

0   z

Entries in the table give the area under the curve between the mean and z standard deviations above the mean. For example, for z = 1.25 the area under the curve between the mean and z is .3944.

z	.00	.01	.02	.03	.04	.05	.06	.07	.08	.09
.0	.0000	.0040	.0080	.0120	.0160	.0199	.0239	.0279	.0319	.0359
.1	.0398	.0438	.0478	.0517	.0557	.0596	.0636	.0675	.0714	.0753
.2	.0793	.0832	.0871	.0910	.0948	.0987	.1026	.1064	.1103	.1141
.3	.1179	.1217	.1255	.1293	.1331	.1368	.1406	.1443	.1480	.1517
.4	.1554	.1591	.1628	.1664	.1700	.1736	.1772	.1808	.1844	.1879
.5	.1915	.1950	.1985	.2019	.2054	.2088	.2123	.2157	.2190	.2224
.6	.2257	.2291	.2324	.2357	.2389	.2422	.2454	.2486	.2518	.2549
.7	.2580	.2612	.2642	.2673	.2704	.2734	.2764	.2794	.2823	.2852
.8	.2881	.2910	.2939	.2967	.2995	.3023	.3051	.3078	.3106	.3133
.9	.3159	.3186	.3212	.3238	.3264	.3289	.3315	.3340	.3365	.3389
1.0	.3413	.3438	.3461	.3485	.3508	.3531	.3554	.3577	.3599	.3621
1.1	.3643	.3665	.3686	.3708	.3729	.3749	.3770	.3790	.3810	.3830
1.2	.3849	.3869	.3888	.3907	.3925	.3944	.3962	.3980	.3997	.4015
1.3	.4032	.4049	.4066	.4082	.4099	.4115	.4131	.4147	.4162	.4177
1.4	.4192	.4207	.4222	.4236	.4251	.4265	.4279	.4292	.4306	.4319
1.5	.4332	.4345	.4357	.4370	.4382	.4394	.4406	.4418	.4429	.4441
1.6	.4452	.4463	.4474	.4484	.4495	.4505	.4515	.4525	.4535	.4545
1.7	.4554	.4564	.4573	.4582	.4591	.4599	.4608	.4616	.4625	.4633
1.8	.4641	.4649	.4656	.4664	.4671	.4678	.4686	.4693	.4699	.4706
1.9	.4713	.4719	.4726	.4732	.4738	.4744	.4750	.4756	.4761	.4767
2.0	.4772	.4778	.4783	.4788	.4793	.4798	.4803	.4808	.4812	.4817
2.1	.4821	.4826	.4830	.4834	.4838	.4842	.4846	.4850	.4854	.4857
2.2	.4861	.4864	.4868	.4871	.4875	.4878	.4881	.4884	.4887	.4890
2.3	.4893	.4896	.4898	.4901	.4904	.4906	.4909	.4911	.4913	.4916
2.4	.4918	.4920	.4922	.4925	.4927	.4929	.4931	.4932	.4934	.4936
2.5	.4938	.4940	.4941	.4943	.4945	.4946	.4948	.4949	.4951	.4952
2.6	.4953	.4955	.4956	.4957	.4959	.4960	.4961	.4962	.4963	.4964
2.7	.4965	.4966	.4967	.4968	.4969	.4970	.4971	.4972	.4973	.4974
2.8	.4974	.4975	.4976	.4977	.4977	.4978	.4979	.4979	.4980	.4981
2.9	.4981	.4982	.4982	.4983	.4984	.4984	.4985	.4985	.4986	.4986
3.0	.4986	.4987	.4987	.4988	.4988	.4989	.4989	.4989	.4990	.4990

TABLE 2 *t* Distribution

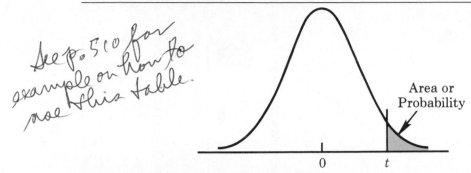

Area or Probability

0   *t*

Entries in the table give *t* values for an area or probability in the upper tail of the *t* distribution. For example, with 10 degrees of freedom and a .05 area in the upper tail, $t_{.05} = 1.812$.

See p. 500 for example on how to use this table.

Degrees of Freedom	Area in Upper Tail				
	.10	.05	.025	.01	.005
1	3.078	6.314	12.706	31.821	63.657
2	1.886	2.920	4.303	6.965	9.925
3	1.638	2.353	3.182	4.541	5.841
4	1.533	2.132	2.776	3.747	4.604
5	1.476	2.015	2.571	3.365	4.032
6	1.440	1.943	2.447	3.143	3.707
7	1.415	1.895	2.365	2.998	3.499
8	1.397	1.860	2.306	2.896	3.355
9	1.383	1.833	2.262	2.821	3.250
10	1.372	1.812	2.228	2.764	3.169
11	1.363	1.796	2.201	2.718	3.106
12	1.356	1.782	2.179	2.681	3.055
13	1.350	1.771	2.160	2.650	3.012
14	1.345	1.761	2.145	2.624	2.977
15	1.341	1.753	2.131	2.602	2.947
16	1.337	1.746	2.120	2.583	2.921
17	1.333	1.740	2.110	2.567	2.898
18	1.330	1.734	2.101	2.552	2.878
19	1.328	1.729	2.093	2.539	2.861
20	1.325	1.725	2.086	2.528	2.845
21	1.323	1.721	2.080	2.518	2.831
22	1.321	1.717	2.074	2.508	2.819
23	1.319	1.714	2.069	2.500	2.807
24	1.318	1.711	2.064	2.492	2.797
25	1.316	1.708	2.060	2.485	2.787
26	1.315	1.706	2.056	2.479	2.779
27	1.314	1.703	2.052	2.473	2.771
28	1.313	1.701	2.048	2.467	2.763
29	1.311	1.699	2.045	2.462	2.756
30	1.310	1.697	2.042	2.457	2.750
40	1.303	1.684	2.021	2.423	2.704
60	1.296	1.671	2.000	2.390	2.660
120	1.289	1.658	1.980	2.358	2.617
∞	1.282	1.645	1.960	2.326	2.576

TABLE 3   Chi-Square Distribution

Entries in the table give χ_α^2 values, where α is the area or probability in the upper tail of the chi-square distribution. For example, with 10 degrees of freedom and a .01 area in the upper tail, $\chi_{.01}^2 = 23.2093$.

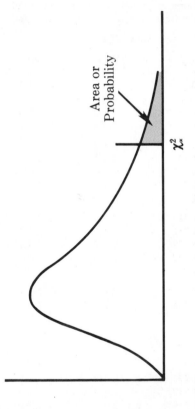

Area or Probability

χ_α^2

Degrees of Freedom	.995	.99	.975	.95	Area in Upper Tail .90	.10	.05	.025	.01	.005
1	392.704×10^{-10}	157.088×10^{-9}	982.069×10^{-9}	393.214×10^{-8}	.0157908	2.70554	3.84146	5.02389	6.63490	7.87944
2	.0100251	.0201007	.0506356	.102587	.210720	4.60517	5.99147	7.37776	9.21034	10.5966
3	.0717212	.114832	.215795	.351846	.584375	6.25139	7.81473	9.34840	11.3449	12.8381
4	.206990	.297110	.484419	.710721	1.063623	7.77944	9.48773	11.1433	13.2767	14.8602
5	.411740	.554300	.831211	1.145476	1.61031	9.23635	11.0705	12.8325	15.0863	16.7496
6	.675727	.872085	1.237347	1.63539	2.20413	10.6446	12.5916	14.4494	16.8119	18.5476
7	.989265	1.239043	1.68987	2.16735	2.83311	12.0170	14.0671	16.0128	18.4753	20.2777
8	1.344419	1.646482	2.17973	2.73264	3.48954	13.3616	15.5073	17.5346	20.0902	21.9550
9	1.734926	2.087912	2.70039	3.32511	4.16816	14.6837	16.9190	19.0228	21.6660	23.5893

10	2.15585	2.55821	3.24697	3.94030	4.86518	15.9871	18.3070	20.4831	23.2093	25.1882
11	2.60321	3.05347	3.81575	4.57481	5.57779	17.2750	19.6751	21.9200	24.7250	26.7569
12	3.07382	3.57056	4.40379	5.22603	6.30380	18.5494	21.0261	23.3367	26.2170	28.2995
13	3.56503	4.10691	5.00874	5.89186	7.04150	19.8119	22.3621	24.7356	27.6883	29.8194
14	4.07468	4.66043	5.62872	6.57063	7.78953	21.0642	23.6848	26.1190	29.1413	31.3193
15	4.60094	5.22935	6.26214	7.26094	8.54675	22.3072	24.9958	27.4884	30.5779	32.8013
16	5.14224	5.81221	6.90766	7.96164	9.31223	23.5418	26.2962	28.8454	31.9999	34.2672
17	5.69724	6.40776	7.56418	8.67176	10.0852	24.7690	27.5871	30.1910	33.4087	35.7185
18	6.26481	7.01491	8.23075	9.39046	10.8649	25.9894	28.8693	31.5264	34.8053	37.1564
19	6.84398	7.63273	8.90655	10.1170	11.6509	27.2036	30.1435	32.8523	36.1908	38.5822
20	7.43386	8.26040	9.59083	10.8508	12.4426	28.4120	31.4104	34.1696	37.5662	39.9968
21	8.03366	8.89720	10.28293	11.5913	13.2396	29.6151	32.6705	35.4789	38.9321	41.4010
22	8.64272	9.54249	10.9823	12.3380	14.0415	30.8133	33.9244	36.7807	40.2894	42.7958
23	9.26042	10.19567	11.6885	13.0905	14.8479	32.0069	35.1725	38.0757	41.6384	44.1813
24	9.88623	10.8564	12.4011	13.8484	15.6587	33.1963	36.4151	39.3641	42.9798	45.5585
25	10.5197	11.5240	13.1197	14.6114	16.4734	34.3816	37.6525	40.6465	44.3141	46.9278
26	11.1603	12.1981	13.8439	15.3791	17.2919	35.5631	38.8852	41.9232	45.6417	48.2899
27	11.8076	12.8786	14.5733	16.1513	18.1138	36.7412	40.1133	43.1944	46.9630	49.6449
28	12.4613	13.5648	15.3079	16.9279	18.9392	37.9159	41.3372	44.4607	48.2782	50.9933
29	13.1211	14.2565	16.0471	17.7083	19.7677	39.0875	42.5569	45.7222	49.5879	52.3356
30	13.7867	14.9535	16.7908	18.4926	20.5992	40.2560	43.7729	46.9792	50.8922	53.6720
40	20.7065	22.1643	24.4331	26.5093	29.0505	51.8050	55.7585	59.3417	63.6907	66.7659
50	27.9907	29.7067	32.3574	34.7642	37.6886	63.1671	67.5048	71.4202	76.1539	79.4900
60	35.5346	37.4848	40.4817	43.1879	46.4589	74.3970	79.0819	83.2976	88.3794	91.9517
70	43.2752	45.4418	48.7576	51.7393	55.3290	85.5271	90.5312	95.0231	100.425	104.215
80	51.1720	53.5400	57.1532	60.3915	64.2778	96.5782	101.879	106.629	112.329	116.321
90	59.1963	61.7541	65.6466	69.1260	73.2912	107.565	113.145	118.136	124.116	128.299
100	67.3276	70.0648	74.2219	77.9295	82.3581	118.498	124.342	129.561	135.807	140.169

Reprinted by permission of Biometrika Trustees from Table 8, Percentage Points of the χ^2 Distribution, by E. S. Pearson and H. O. Hartley, *Biometrika Tables for Statisticians*, Vol. I, 3rd Edition, 1966.

TABLE 4 *F* Distribution

See p. 504 for example of how to use this table.

P. 570 gives an example of mult. regression

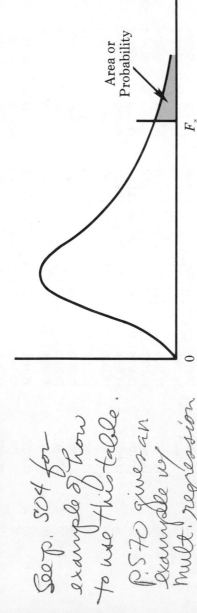

Entries in the table give F_α values, where α is the area or probability in the upper tail of the *F* distribution. For example, with 12 numerator degrees of freedom, 15 denominator degrees of freedom, and a .05 area in the upper tail, $F_{.05} = 2.48$.

Table of $F_{.05}$ Values

Denominator Degrees of Freedom	Numerator Degrees of Freedom																		
	1	2	3	4	5	6	7	8	9	10	12	15	20	24	30	40	60	120	∞
1	161.4	199.5	215.7	224.6	230.2	234.0	236.8	238.9	240.5	241.9	243.9	245.9	248.0	249.1	250.1	251.1	252.2	253.3	254.3
2	18.51	19.00	19.16	19.25	19.30	19.33	19.35	19.37	19.38	19.40	19.41	19.43	19.45	19.45	19.46	19.47	19.48	19.49	19.50
3	10.13	9.55	9.28	9.12	9.01	8.94	8.89	8.85	8.81	8.79	8.74	8.70	8.66	8.64	8.62	8.59	8.57	8.55	8.53
4	7.71	6.94	6.59	6.39	6.26	6.16	6.09	6.04	6.00	5.96	5.91	5.86	5.80	5.77	5.75	5.72	5.69	5.66	5.63
5	6.61	5.79	5.41	5.19	5.05	4.95	4.88	4.82	4.77	4.74	4.68	4.62	4.56	4.53	4.50	4.46	4.43	4.40	4.36

6	5.99	5.14	4.76	4.53	4.39	4.28	4.21	4.15	4.10	4.06	4.00	3.94	3.87	3.84	3.81	3.77	3.74	3.70	3.67
7	5.59	4.74	4.35	4.12	3.97	3.87	3.79	3.73	3.68	3.64	3.57	3.51	3.44	3.41	3.38	3.34	3.30	3.27	3.23
8	5.32	4.46	4.07	3.84	3.69	3.58	3.50	3.44	3.39	3.35	3.28	3.22	3.15	3.12	3.08	3.04	3.01	2.97	2.93
9	5.12	4.26	3.86	3.63	3.48	3.37	3.29	3.23	3.18	3.14	3.07	3.01	2.94	2.90	2.86	2.83	2.79	2.75	2.71
10	4.96	4.10	3.71	3.48	3.33	3.22	3.14	3.07	3.02	2.98	2.91	2.85	2.77	2.74	2.70	2.66	2.62	2.58	2.54
11	4.84	3.98	3.59	3.36	3.20	3.09	3.01	2.95	2.90	2.85	2.79	2.72	2.65	2.61	2.57	2.53	2.49	2.45	2.40
12	4.75	3.89	3.49	3.26	3.11	3.00	2.91	2.85	2.80	2.75	2.69	2.62	2.54	2.51	2.47	2.43	2.38	2.34	2.30
13	4.67	3.81	3.41	3.18	3.03	2.92	2.83	2.77	2.71	2.67	2.60	2.53	2.46	2.42	2.38	2.34	2.30	2.25	2.21
14	4.60	3.74	3.34	3.11	2.96	2.85	2.76	2.70	2.65	2.60	2.53	2.46	2.39	2.35	2.31	2.27	2.22	2.18	2.13
15	4.54	3.68	3.29	3.06	2.90	2.79	2.71	2.64	2.59	2.54	2.48	2.40	2.33	2.29	2.25	2.20	2.16	2.11	2.07
16	4.49	3.63	3.24	3.01	2.85	2.74	2.66	2.59	2.54	2.49	2.42	2.35	2.28	2.24	2.19	2.15	2.11	2.06	2.01
17	4.45	3.59	3.20	2.96	2.81	2.70	2.61	2.55	2.49	2.45	2.38	2.31	2.23	2.19	2.15	2.10	2.06	2.01	1.96
18	4.41	3.55	3.16	2.93	2.77	2.66	2.58	2.51	2.46	2.41	2.34	2.27	2.19	2.15	2.11	2.06	2.02	1.97	1.92
19	4.38	3.52	3.13	2.90	2.74	2.63	2.54	2.48	2.42	2.38	2.31	2.23	2.16	2.11	2.07	2.03	1.98	1.93	1.88
20	4.35	3.49	3.10	2.87	2.71	2.60	2.51	2.45	2.39	2.35	2.28	2.20	2.12	2.08	2.04	1.99	1.95	1.90	1.84
21	4.32	3.47	3.07	2.84	2.68	2.57	2.49	2.42	2.37	2.32	2.25	2.18	2.10	2.05	2.01	1.96	1.92	1.87	1.81
22	4.30	3.44	3.05	2.82	2.66	2.55	2.46	2.40	2.34	2.30	2.23	2.15	2.07	2.03	1.98	1.94	1.89	1.84	1.78
23	4.28	3.42	3.03	2.80	2.64	2.53	2.44	2.37	2.32	2.27	2.20	2.13	2.05	2.01	1.96	1.91	1.86	1.81	1.76
24	4.26	3.40	3.01	2.78	2.62	2.51	2.42	2.36	2.30	2.25	2.18	2.11	2.03	1.98	1.94	1.89	1.84	1.79	1.73
25	4.24	3.39	2.99	2.76	2.60	2.49	2.40	2.34	2.28	2.24	2.16	2.09	2.01	1.96	1.92	1.87	1.82	1.77	1.71
26	4.23	3.37	2.98	2.74	2.59	2.47	2.39	2.32	2.27	2.22	2.15	2.07	1.99	1.95	1.90	1.85	1.80	1.75	1.69
27	4.21	3.35	2.96	2.73	2.57	2.46	2.37	2.31	2.25	2.20	2.13	2.06	1.97	1.93	1.88	1.84	1.79	1.73	1.67
28	4.20	3.34	2.95	2.71	2.56	2.45	2.36	2.29	2.24	2.19	2.12	2.04	1.96	1.91	1.87	1.82	1.77	1.71	1.65
29	4.18	3.33	2.93	2.70	2.55	2.43	2.35	2.28	2.22	2.18	2.10	2.03	1.94	1.90	1.85	1.81	1.75	1.70	1.64
30	4.17	3.32	2.92	2.69	2.53	2.42	2.33	2.27	2.21	2.16	2.09	2.01	1.93	1.89	1.84	1.79	1.74	1.68	1.62
40	4.08	3.23	2.84	2.61	2.45	2.34	2.25	2.18	2.12	2.08	2.00	1.92	1.84	1.79	1.74	1.69	1.64	1.58	1.51
60	4.00	3.15	2.76	2.53	2.37	2.25	2.17	2.10	2.04	1.99	1.92	1.84	1.75	1.70	1.65	1.59	1.53	1.47	1.39
120	3.92	3.07	2.68	2.45	2.29	2.17	2.09	2.02	1.96	1.91	1.83	1.75	1.66	1.61	1.55	1.50	1.43	1.35	1.25
∞	3.84	3.00	2.60	2.37	2.21	2.10	2.01	1.94	1.88	1.83	1.75	1.67	1.57	1.52	1.46	1.39	1.32	1.22	1.00

(continues)

Reprinted by permission of the Biometrika Trustees from Table 18, Percentage Points of the F Distribution, by E. S. Pearson and H. O. Hartley, *Biometrika Tables for Statisticians*, Vol. I, 3rd Edition, 1966.

TABLE 4 (Continued)

Table of F$_{.025}$ Values

Denominator Degrees of Freedom	Numerator Degrees of Freedom																		
	1	2	3	4	5	6	7	8	9	10	12	15	20	24	30	40	60	120	∞
1	647.8	799.5	864.2	899.6	921.8	937.1	948.2	956.7	963.3	968.6	976.7	984.9	993.1	997.2	1,001	1,006	1,010	1,014	1,018
2	38.51	39.00	39.17	39.25	39.30	39.33	39.36	39.37	39.39	39.40	39.41	39.43	39.45	39.46	39.46	39.47	39.48	39.49	39.50
3	17.44	16.04	15.44	15.10	14.88	14.73	14.62	14.54	14.47	14.42	14.34	14.25	14.17	14.12	14.08	14.04	13.99	13.95	13.90
4	12.22	10.65	9.98	9.60	9.36	9.20	9.07	8.98	8.90	8.84	8.75	8.66	8.56	8.51	8.46	8.41	8.36	8.31	8.26
5	10.01	8.43	7.76	7.39	7.15	6.98	6.85	6.76	6.68	6.62	6.52	6.43	6.33	6.28	6.23	6.18	6.12	6.07	6.02
6	8.81	7.26	6.60	6.23	5.99	5.82	5.70	5.60	5.52	5.46	5.37	5.27	5.17	5.12	5.07	5.01	4.96	4.90	4.85
7	8.07	6.54	5.89	5.52	5.29	5.12	4.99	4.90	4.82	4.76	4.67	4.57	4.47	4.42	4.36	4.31	4.25	4.20	4.14
8	7.57	6.06	5.42	5.05	4.82	4.65	4.53	4.43	4.36	4.30	4.20	4.10	4.00	3.95	3.89	3.84	3.78	3.73	3.67
9	7.21	5.71	5.08	4.72	4.48	4.32	4.20	4.10	4.03	3.96	3.87	3.77	3.67	3.61	3.56	3.51	3.45	3.39	3.33
10	6.94	5.46	4.83	4.47	4.24	4.07	3.95	3.85	3.78	3.72	3.62	3.52	3.42	3.37	3.31	3.26	3.20	3.14	3.08
11	6.72	5.26	4.63	4.28	4.04	3.88	3.76	3.66	3.59	3.53	3.43	3.33	3.23	3.17	3.12	3.06	3.00	2.94	2.88
12	6.55	5.10	4.47	4.12	3.89	3.73	3.61	3.51	3.44	3.37	3.28	3.18	3.07	3.02	2.96	2.91	2.85	2.79	2.72
13	6.41	4.97	4.35	4.00	3.77	3.60	3.48	3.39	3.31	3.25	3.15	3.05	2.95	2.89	2.84	2.78	2.72	2.66	2.60
14	6.30	4.86	4.24	3.89	3.66	3.50	3.38	3.29	3.21	3.15	3.05	2.95	2.84	2.79	2.73	2.67	2.61	2.55	2.49
15	6.20	4.77	4.15	3.80	3.58	3.41	3.29	3.20	3.12	3.06	2.96	2.86	2.76	2.70	2.64	2.59	2.52	2.46	2.40
16	6.12	4.69	4.08	3.73	3.50	3.34	3.22	3.12	3.05	2.99	2.89	2.79	2.68	2.63	2.57	2.51	2.45	2.38	2.32
17	6.04	4.62	4.01	3.66	3.44	3.28	3.16	3.06	2.98	2.92	2.82	2.72	2.62	2.56	2.50	2.44	2.38	2.32	2.25
18	5.98	4.56	3.95	3.61	3.38	3.22	3.10	3.01	2.93	2.87	2.77	2.67	2.56	2.50	2.44	2.38	2.32	2.26	2.19
19	5.92	4.51	3.90	3.56	3.33	3.17	3.05	2.96	2.88	2.82	2.72	2.62	2.51	2.45	2.39	2.33	2.27	2.20	2.13
20	5.87	4.46	3.86	3.51	3.29	3.13	3.01	2.91	2.84	2.77	2.68	2.57	2.46	2.41	2.35	2.29	2.22	2.16	2.09
21	5.83	4.42	3.82	3.48	3.25	3.09	2.97	2.87	2.80	2.73	2.64	2.53	2.42	2.37	2.31	2.25	2.18	2.11	2.04
22	5.79	4.38	3.78	3.44	3.22	3.05	2.93	2.84	2.76	2.70	2.60	2.50	2.39	2.33	2.27	2.21	2.14	2.08	2.00
23	5.75	4.35	3.75	3.41	3.18	3.02	2.90	2.81	2.73	2.67	2.57	2.47	2.36	2.30	2.24	2.18	2.11	2.04	1.97
24	5.72	4.32	3.72	3.38	3.15	2.99	2.87	2.78	2.70	2.64	2.54	2.44	2.33	2.27	2.21	2.15	2.08	2.01	1.94
25	5.69	4.29	3.69	3.35	3.13	2.97	2.85	2.75	2.68	2.61	2.51	2.41	2.30	2.24	2.18	2.12	2.05	1.98	1.91
26	5.66	4.27	3.67	3.33	3.10	2.94	2.82	2.73	2.65	2.59	2.49	2.39	2.28	2.22	2.16	2.09	2.03	1.95	1.88
27	5.63	4.24	3.65	3.31	3.08	2.92	2.80	2.71	2.63	2.57	2.47	2.36	2.25	2.19	2.13	2.07	2.00	1.93	1.85
28	5.61	4.22	3.63	3.29	3.06	2.90	2.78	2.69	2.61	2.55	2.45	2.34	2.23	2.17	2.11	2.05	1.98	1.91	1.83
29	5.59	4.20	3.61	3.27	3.04	2.88	2.76	2.67	2.59	2.53	2.43	2.32	2.21	2.15	2.09	2.03	1.96	1.89	1.81
30	5.57	4.18	3.59	3.25	3.03	2.87	2.75	2.65	2.57	2.51	2.41	2.31	2.20	2.14	2.07	2.01	1.94	1.87	1.79
40	5.42	4.05	3.46	3.13	2.90	2.74	2.62	2.53	2.45	2.39	2.29	2.18	2.07	2.01	1.94	1.88	1.80	1.72	1.64
60	5.29	3.93	3.34	3.01	2.79	2.63	2.51	2.41	2.33	2.27	2.17	2.06	1.94	1.88	1.82	1.74	1.67	1.58	1.48
120	5.15	3.80	3.23	2.89	2.67	2.52	2.39	2.30	2.22	2.16	2.05	1.94	1.82	1.76	1.69	1.61	1.53	1.43	1.31
∞	5.02	3.69	3.12	2.79	2.57	2.41	2.29	2.19	2.11	2.05	1.94	1.83	1.71	1.64	1.57	1.48	1.39	1.27	1.00

(continues)

Table of F.01 Values

Denominator Degrees of Freedom	Numerator Degrees of Freedom																		
	1	2	3	4	5	6	7	8	9	10	12	15	20	24	30	40	60	120	∞
1	4,052	4,999.5	5,403	5,625	5,764	5,859	5,928	5,982	6,022	6,056	6,106	6,157	6,209	6,235	6,261	6,287	6,313	6,339	6,366
2	98.50	99.00	99.17	99.25	99.30	99.33	99.36	99.37	99.39	99.40	99.42	99.43	99.45	99.46	99.47	99.47	99.48	99.49	99.50
3	34.12	30.82	29.46	28.71	28.24	27.91	27.67	27.49	27.35	27.23	27.05	26.87	26.69	26.60	26.50	26.41	26.32	26.22	26.13
4	21.20	18.00	16.69	15.98	15.52	15.21	14.98	14.80	14.66	14.55	14.37	14.20	14.02	13.93	13.84	13.75	13.65	13.56	13.46
5	16.26	13.27	12.06	11.39	10.97	10.67	10.46	10.29	10.16	10.05	9.89	9.72	9.55	9.47	9.38	9.29	9.20	9.11	9.06
6	13.75	10.92	9.78	9.15	8.75	8.47	8.26	8.10	7.98	7.87	7.72	7.56	7.40	7.31	7.23	7.14	7.06	6.97	6.88
7	12.25	9.55	8.45	7.85	7.46	7.19	6.99	6.84	6.72	6.62	6.47	6.31	6.16	6.07	5.99	5.91	5.82	5.74	5.65
8	11.26	8.65	7.59	7.01	6.63	6.37	6.18	6.03	5.91	5.81	5.67	5.52	5.36	5.28	5.20	5.12	5.03	4.95	4.86
9	10.56	8.02	6.99	6.42	6.06	5.80	5.61	5.47	5.35	5.26	5.11	4.96	4.81	4.73	4.65	4.57	4.48	4.40	4.31
10	10.04	7.56	6.55	5.99	5.64	5.39	5.20	5.06	4.94	4.85	4.71	4.56	4.41	4.33	4.25	4.17	4.08	4.00	3.91
11	9.65	7.21	6.22	5.67	5.32	5.07	4.89	4.74	4.63	4.54	4.40	4.25	4.10	4.02	3.94	3.86	3.78	3.69	3.60
12	9.33	6.93	5.95	5.41	5.06	4.82	4.64	4.50	4.39	4.30	4.16	4.01	3.86	3.78	3.70	3.62	3.54	3.45	3.36
13	9.07	6.70	5.74	5.21	4.86	4.62	4.44	4.30	4.19	4.10	3.96	3.82	3.66	3.59	3.51	3.43	3.34	3.25	3.17
14	8.86	6.51	5.56	5.04	4.69	4.46	4.28	4.14	4.03	3.94	3.80	3.66	3.51	3.43	3.35	3.27	3.18	3.09	3.00
15	8.68	6.36	5.42	4.89	4.56	4.32	4.14	4.00	3.89	3.80	3.67	3.52	3.37	3.29	3.21	3.13	3.05	2.96	2.87
16	8.53	6.23	5.29	4.77	4.44	4.20	4.03	3.89	3.78	3.69	3.55	3.41	3.26	3.18	3.10	3.02	2.93	2.84	2.75
17	8.40	6.11	5.18	4.67	4.34	4.10	3.93	3.79	3.68	3.59	3.46	3.31	3.16	3.08	3.00	2.92	2.83	2.75	2.65
18	8.29	6.01	5.09	4.58	4.25	4.01	3.84	3.71	3.60	3.51	3.37	3.23	3.08	3.00	2.92	2.84	2.75	2.66	2.57
19	8.18	5.93	5.01	4.50	4.17	3.94	3.77	3.63	3.52	3.43	3.30	3.15	3.00	2.92	2.84	2.76	2.67	2.58	2.49
20	8.10	5.85	4.94	4.43	4.10	3.87	3.70	3.56	3.46	3.37	3.23	3.09	2.94	2.86	2.78	2.69	2.61	2.52	2.42
21	8.02	5.78	4.87	4.37	4.04	3.81	3.64	3.51	3.40	3.31	3.17	3.03	2.88	2.80	2.72	2.64	2.55	2.46	2.36
22	7.95	5.72	4.82	4.31	3.99	3.76	3.59	3.45	3.35	3.26	3.12	2.98	2.83	2.75	2.67	2.58	2.50	2.40	2.31
23	7.88	5.66	4.76	4.26	3.94	3.71	3.54	3.41	3.30	3.21	3.07	2.93	2.78	2.70	2.62	2.54	2.45	2.35	2.26
24	7.82	5.61	4.72	4.22	3.90	3.67	3.50	3.36	3.26	3.17	3.03	2.89	2.74	2.66	2.58	2.49	2.40	2.31	2.21
25	7.77	5.57	4.68	4.18	3.85	3.63	3.46	3.32	3.22	3.13	2.99	2.85	2.70	2.62	2.54	2.45	2.36	2.27	2.17
26	7.72	5.53	4.64	4.14	3.82	3.59	3.42	3.29	3.18	3.09	2.96	2.81	2.66	2.58	2.50	2.42	2.33	2.23	2.13
27	7.68	5.49	4.60	4.11	3.78	3.56	3.39	3.26	3.15	3.06	2.93	2.78	2.63	2.55	2.47	2.38	2.29	2.20	2.10
28	7.64	5.45	4.57	4.07	3.75	3.53	3.36	3.23	3.12	3.03	2.90	2.75	2.60	2.52	2.44	2.35	2.26	2.17	2.06
29	7.60	5.42	4.54	4.04	3.73	3.50	3.33	3.20	3.09	3.00	2.87	2.73	2.57	2.49	2.41	2.33	2.23	2.14	2.03
30	7.56	5.39	4.51	4.02	3.70	3.47	3.30	3.17	3.07	2.98	2.84	2.70	2.55	2.47	2.39	2.30	2.21	2.11	2.01
40	7.31	5.18	4.31	3.83	3.51	3.29	3.12	2.99	2.89	2.80	2.66	2.52	2.37	2.29	2.20	2.11	2.02	1.92	1.80
60	7.08	4.98	4.13	3.65	3.34	3.12	2.95	2.82	2.72	2.63	2.50	2.35	2.20	2.12	2.03	1.94	1.84	1.73	1.60
120	6.85	4.79	3.95	3.48	3.17	2.96	2.79	2.66	2.56	2.47	2.34	2.19	2.03	1.95	1.86	1.76	1.66	1.53	1.38
∞	6.63	4.61	3.78	3.32	3.02	2.80	2.64	2.51	2.41	2.32	2.18	2.04	1.88	1.79	1.70	1.59	1.47	1.32	1.00

TABLE 5 Binomial Probabilities

Entries in the table give the probability of x successes in n trials of a binomial experiment, where p is the probability of a success on one trial. For example, with six trials and $p = .05$, the probability of two successes is .0305.

						p					
n	x	.01	.02	.03	.04	.05	.06	.07	.08	.09
2	0	.9801	.9604	.9409	.9216	.9025	.8836	.8649	.8464	.8281
	1	.0198	.0392	.0582	.0768	.0950	.1128	.1302	.1472	.1638
	2	.0001	.0004	.0009	.0016	.0025	.0036	.0049	.0064	.0081
3	0	.9703	.9412	.9127	.8847	.8574	.8306	.8044	.7787	.7536
	1	.0294	.0576	.0847	.1106	.1354	.1590	.1816	.2031	.2236
	2	.0003	.0012	.0026	.0046	.0071	.0102	.0137	.0177	.0221
	3	.0000	.0000	.0000	.0001	.0001	.0002	.0003	.0005	.0007
4	0	.9606	.9224	.8853	.8493	.8145	.7807	.7481	.7164	.6857
	1	.0388	.0753	.1095	.1416	.1715	.1993	.2252	.2492	.2713
	2	.0006	.0023	.0051	.0088	.0135	.0191	.0254	.0325	.0402
	3	.0000	.0000	.0001	.0002	.0005	.0008	.0013	.0019	.0027
	4	.0000	.0000	.0000	.0000	.0000	.0000	.0000	.0000	.0001
5	0	.9510	.9039	.8587	.8154	.7738	.7339	.6957	.6591	.6240
	1	.0480	.0922	.1328	.1699	.2036	.2342	.2618	.2866	.3086
	2	.0010	.0038	.0082	.0142	.0214	.0299	.0394	.0498	.0610
	3	.0000	.0001	.0003	.0006	.0011	.0019	.0030	.0043	.0060
	4	.0000	.0000	.0000	.0000	.0000	.0001	.0001	.0002	.0003
	5	.0000	.0000	.0000	.0000	.0000	.0000	.0000	.0000	.0000
6	0	.9415	.8858	.8330	.7828	.7351	.6899	.6470	.6064	.5679
	1	.0571	.1085	.1546	.1957	.2321	.2642	.2922	.3164	.3370
	2	.0014	.0055	.0120	.0204	.0305	.0422	.0550	.0688	.0833
	3	.0000	.0002	.0005	.0011	.0021	.0036	.0055	.0080	.0110
	4	.0000	.0000	.0000	.0000	.0001	.0002	.0003	.0005	.0008
	5	.0000	.0000	.0000	.0000	.0000	.0000	.0000	.0000	.0000
	6	.0000	.0000	.0000	.0000	.0000	.0000	.0000	.0000	.0000
7	0	.9321	.8681	.8080	.7514	.6983	.6485	.6017	.5578	.5168
	1	.0659	.1240	.1749	.2192	.2573	.2897	.3170	.3396	.3578
	2	.0020	.0076	.0162	.0274	.0406	.0555	.0716	.0886	.1061
	3	.0000	.0003	.0008	.0019	.0036	.0059	.0090	.0128	.0175
	4	.0000	.0000	.0000	.0001	.0002	.0004	.0007	.0011	.0017
	5	.0000	.0000	.0000	.0000	.0000	.0000	.0000	.0001	.0001
	6	.0000	.0000	.0000	.0000	.0000	.0000	.0000	.0000	.0000
	7	.0000	.0000	.0000	.0000	.0000	.0000	.0000	.0000	.0000
8	0	.9227	.8508	.7837	.7214	.6634	.6096	.5596	.5132	.4703
	1	.0746	.1389	.1939	.2405	.2793	.3113	.3370	.3570	.3721
	2	.0026	.0099	.0210	.0351	.0515	.0695	.0888	.1087	.1288
	3	.0001	.0004	.0013	.0029	.0054	.0089	.0134	.0189	.0255
	4	.0000	.0000	.0001	.0002	.0004	.0007	.0013	.0021	.0031
	5	.0000	.0000	.0000	.0000	.0000	.0000	.0001	.0001	.0002
	6	.0000	.0000	.0000	.0000	.0000	.0000	.0000	.0000	.0000
	7	.0000	.0000	.0000	.0000	.0000	.0000	.0000	.0000	.0000
	8	.0000	.0000	.0000	.0000	.0000	.0000	.0000	.0000	.0000

(continues)

TABLE 5 (*Continued*)

n	x	.01	.02	.03	.04	.05	.06	.07	.08	.09
					p					
9	0	.9135	.8337	.7602	.6925	.6302	.5730	.5204	.4722	.4279
	1	.0830	.1531	.2116	.2597	.2985	.3292	.3525	.3695	.3809
	2	.0034	.0125	.0262	.0433	.0629	.0840	.1061	.1285	.1507
	3	.0001	.0006	.0019	.0042	.0077	.0125	.0186	.0261	.0348
	4	.0000	.0000	.0001	.0003	.0006	.0012	.0021	.0034	.0052
	5	.0000	.0000	.0000	.0000	.0000	.0001	.0002	.0003	.0005
	6	.0000	.0000	.0000	.0000	.0000	.0000	.0000	.0000	.0000
	7	.0000	.0000	.0000	.0000	.0000	.0000	.0000	.0000	.0000
	8	.0000	.0000	.0000	.0000	.0000	.0000	.0000	.0000	.0000
	9	.0000	.0000	.0000	.0000	.0000	.0000	.0000	.0000	.0000
10	0	.9044	.8171	.7374	.6648	.5987	.5386	.4840	.4344	.3894
	1	.0914	.1667	.2281	.2770	.3151	.3438	.3643	.3777	.3851
	2	.0042	.0153	.0317	.0519	.0746	.0988	.1234	.1478	.1714
	3	.0001	.0008	.0026	.0058	.0105	.0168	.0248	.0343	.0452
	4	.0000	.0000	.0001	.0004	.0010	.0019	.0033	.0052	.0078
	5	.0000	.0000	.0000	.0000	.0001	.0001	.0003	.0005	.0009
	6	.0000	.0000	.0000	.0000	.0000	.0000	.0000	.0000	.0001
	7	.0000	.0000	.0000	.0000	.0000	.0000	.0000	.0000	.0000
	8	.0000	.0000	.0000	.0000	.0000	.0000	.0000	.0000	.0001
	9	.0000	.0000	.0000	.0000	.0000	.0000	.0000	.0000	.0001
	10	.0000	.0000	.0000	.0000	.0000	.0000	.0000	.0000	.0001
12	0	.8864	.7847	.6938	.6127	.5404	.4759	.4186	.3677	.3225
	1	.1074	.1922	.2575	.3064	.3413	.3645	.3781	.3837	.3827
	2	.0060	.0216	.0438	.0702	.0988	.1280	.1565	.1835	.2082
	3	.0002	.0015	.0045	.0098	.0173	.0272	.0393	.0532	.0686
	4	.0000	.0001	.0003	.0009	.0021	.0039	.0067	.0104	.0153
	5	.0000	.0000	.0000	.0001	.0002	.0004	.0008	.0014	.0024
	6	.0000	.0000	.0000	.0000	.0000	.0000	.0001	.0001	.0003
	7	.0000	.0000	.0000	.0000	.0000	.0000	.0000	.0000	.0000
	8	.0000	.0000	.0000	.0000	.0000	.0000	.0000	.0000	.0000
	9	.0000	.0000	.0000	.0000	.0000	.0000	.0000	.0000	.0000
	10	.0000	.0000	.0000	.0000	.0000	.0000	.0000	.0000	.0000
	11	.0000	.0000	.0000	.0000	.0000	.0000	.0000	.0000	.0000
	12	.0000	.0000	.0000	.0000	.0000	.0000	.0000	.0000	.0000
15	0	.8601	.7386	.6333	.5421	.4633	.3953	.3367	.2863	.2430
	1	.1303	.2261	.2938	.3388	.3658	.3785	.3801	.3734	.3605
	2	.0092	.0323	.0636	.0988	.1348	.1691	.2003	.2273	.2496
	3	.0004	.0029	.0085	.0178	.0307	.0468	.0653	.0857	.1070
	4	.0000	.0002	.0008	.0022	.0049	.0090	.0148	.0223	.0317
	5	.0000	.0000	.0001	.0002	.0006	.0013	.0024	.0043	.0069
	6	.0000	.0000	.0000	.0000	.0000	.0001	.0003	.0006	.0011
	7	.0000	.0000	.0000	.0000	.0000	.0000	.0000	.0001	.0001
	8	.0000	.0000	.0000	.0000	.0000	.0000	.0000	.0000	.0000
	9	.0000	.0000	.0000	.0000	.0000	.0000	.0000	.0000	.0000
	10	.0000	.0000	.0000	.0000	.0000	.0000	.0000	.0000	.0000
	11	.0000	.0000	.0000	.0000	.0000	.0000	.0000	.0000	.0000
	12	.0000	.0000	.0000	.0000	.0000	.0000	.0000	.0000	.0000
	13	.0000	.0000	.0000	.0000	.0000	.0000	.0000	.0000	.0000
	14	.0000	.0000	.0000	.0000	.0000	.0000	.0000	.0000	.0000
	15	.0000	.0000	.0000	.0000	.0000	.0000	.0000	.0000	.0000

(continues)

TABLE 5 (*Continued*)

n	x	.01	.02	.03	.04	.05	.06	.07	.08	.09
					P					
18	0	.8345	.6951	.5780	.4796	.3972	.3283	.2708	.2229	.1831
	1	.1517	.2554	.3217	.3597	.3763	.3772	.3669	.3489	.3260
	2	.0130	.0443	.0846	.1274	.1683	.2047	.2348	.2579	.2741
	3	.0007	.0048	.0140	.0283	.0473	.0697	.0942	.1196	.1446
	4	.0000	.0004	.0016	.0044	.0093	.0167	.0266	.0390	.0536
	5	.0000	.0000	.0001	.0005	.0014	.0030	.0056	.0095	.0148
	6	.0000	.0000	.0000	.0000	.0002	.0004	.0009	.0018	.0032
	7	.0000	.0000	.0000	.0000	.0000	.0000	.0001	.0003	.0005
	8	.0000	.0000	.0000	.0000	.0000	.0000	.0000	.0000	.0001
	9	.0000	.0000	.0000	.0000	.0000	.0000	.0000	.0000	.0000
	10	.0000	.0000	.0000	.0000	.0000	.0000	.0000	.0000	.0000
	11	.0000	.0000	.0000	.0000	.0000	.0000	.0000	.0000	.0000
	12	.0000	.0000	.0000	.0000	.0000	.0000	.0000	.0000	.0000
	13	.0000	.0000	.0000	.0000	.0000	.0000	.0000	.0000	.0000
	14	.0000	.0000	.0000	.0000	.0000	.0000	.0000	.0000	.0000
	15	.0000	.0000	.0000	.0000	.0000	.0000	.0000	.0000	.0000
	16	.0000	.0000	.0000	.0000	.0000	.0000	.0000	.0000	.0000
	17	.0000	.0000	.0000	.0000	.0000	.0000	.0000	.0000	.0000
	18	.0000	.0000	.0000	.0000	.0000	.0000	.0000	.0000	.0000
20	0	.8179	.6676	.5438	.4420	.3585	.2901	.2342	.1887	.1516
	1	.1652	.2725	.3364	.3683	.3774	.3703	.3526	.3282	.3000
	2	.0159	.0528	.0988	.1458	.1887	.2246	.2521	.2711	.2818
	3	.0010	.0065	.0183	.0364	.0596	.0860	.1139	.1414	.1672
	4	.0000	.0006	.0024	.0065	.0133	.0233	.0364	.0523	.0703
	5	.0000	.0000	.0002	.0009	.0022	.0048	.0088	.0145	.0222
	6	.0000	.0000	.0000	.0001	.0003	.0008	.0017	.0032	.0055
	7	.0000	.0000	.0000	.0000	.0000	.0001	.0002	.0005	.0011
	8	.0000	.0000	.0000	.0000	.0000	.0000	.0000	.0001	.0002
	9	.0000	.0000	.0000	.0000	.0000	.0000	.0000	.0000	.0000
	10	.0000	.0000	.0000	.0000	.0000	.0000	.0000	.0000	.0000
	11	.0000	.0000	.0000	.0000	.0000	.0000	.0000	.0000	.0000
	12	.0000	.0000	.0000	.0000	.0000	.0000	.0000	.0000	.0000
	13	.0000	.0000	.0000	.0000	.0000	.0000	.0000	.0000	.0000
	14	.0000	.0000	.0000	.0000	.0000	.0000	.0000	.0000	.0000
	15	.0000	.0000	.0000	.0000	.0000	.0000	.0000	.0000	.0000
	16	.0000	.0000	.0000	.0000	.0000	.0000	.0000	.0000	.0000
	17	.0000	.0000	.0000	.0000	.0000	.0000	.0000	.0000	.0000
	18	.0000	.0000	.0000	.0000	.0000	.0000	.0000	.0000	.0000
	19	.0000	.0000	.0000	.0000	.0000	.0000	.0000	.0000	.0000
	20	.0000	.0000	.0000	.0000	.0000	.0000	.0000	.0000	.0000

(continues)

TABLE 5 (*Continued*)

n	x	.10	.15	.20	.25	.30	.35	.40	.45	.50
2	0	.8100	.7225	.6400	.5625	.4900	.4225	.3600	.3025	.2500
	1	.1800	.2550	.3200	.3750	.4200	.4550	.4800	.4950	.5000
	2	.0100	.0225	.0400	.0625	.0900	.1225	.1600	.2025	.2500
3	0	.7290	.6141	.5120	.4219	.3430	.2746	.2160	.1664	.1250
	1	.2430	.3251	.3840	.4219	.4410	.4436	.4320	.4084	.3750
	2	.0270	.0574	.0960	.1406	.1890	.2389	.2880	.3341	.3750
	3	.0010	.0034	.0080	.0156	.0270	.0429	.0640	.0911	.1250
4	0	.6561	.5220	.4096	.3164	.2401	.1785	.1296	.0915	.0625
	1	.2916	.3685	.4096	.4219	.4116	.3845	.3456	.2995	.2500
	2	.0486	.0975	.1536	.2109	.2646	.3105	.3456	.3675	.3750
	3	.0036	.0115	.0256	.0469	.0756	.1115	.1536	.2005	.2500
	4	.0001	.0005	.0016	.0039	.0081	.0150	.0256	.0410	.0625
5	0	.5905	.4437	.3277	.2373	.1681	.1160	.0778	.0503	.0312
	1	.3280	.3915	.4096	.3955	.3602	.3124	.2592	.2059	.1562
	2	.0729	.1382	.2048	.2637	.3087	.3364	.3456	.3369	.3125
	3	.0081	.0244	.0512	.0879	.1323	.1811	.2304	.2757	.3125
	4	.0004	.0022	.0064	.0146	.0284	.0488	.0768	.1128	.1562
	5	.0000	.0001	.0003	.0010	.0024	.0053	.0102	.0185	.0312
6	0	.5314	.3771	.2621	.1780	.1176	.0754	.0467	.0277	.0156
	1	.3543	.3993	.3932	.3560	.3025	.2437	.1866	.1359	.0938
	2	.0984	.1762	.2458	.2966	.3241	.3280	.3110	.2780	.2344
	3	.0146	.0415	.0819	.1318	.1852	.2355	.2765	.3032	.3125
	4	.0012	.0055	.0154	.0330	.0595	.0951	.1382	.1861	.2344
	5	.0001	.0004	.0015	.0044	.0102	.0205	.0369	.0609	.0938
	6	.0000	.0000	.0001	.0002	.0007	.0018	.0041	.0083	.0156
7	0	.4783	.3206	.2097	.1335	.0824	.0490	.0280	.0152	.0078
	1	.3720	.3960	.3670	.3115	.2471	.1848	.1306	.0872	.0547
	2	.1240	.2097	.2753	.3115	.3177	.2985	.2613	.2140	.1641
	3	.0230	.0617	.1147	.1730	.2269	.2679	.2903	.2918	.2734
	4	.0026	.0109	.0287	.0577	.0972	.1442	.1935	.2388	.2734
	5	.0002	.0012	.0043	.0115	.0250	.0466	.0774	.1172	.1641
	6	.0000	.0001	.0004	.0013	.0036	.0084	.0172	.0320	.0547
	7	.0000	.0000	.0000	.0001	.0002	.0006	.0016	.0037	.0078
8	0	.4305	.2725	.1678	.1001	.0576	.0319	.0168	.0084	.0039
	1	.3826	.3847	.3355	.2670	.1977	.1373	.0896	.0548	.0312
	2	.1488	.2376	.2936	.3115	.2965	.2587	.2090	.1569	.1094
	3	.0331	.0839	.1468	.2076	.2541	.2786	.2787	.2568	.2188
	4	.0046	.0185	.0459	.0865	.1361	.1875	.2322	.2627	.2734
	5	.0004	.0026	.0092	.0231	.0467	.0808	.1239	.1719	.2188
	6	.0000	.0002	.0011	.0038	.0100	.0217	.0413	.0703	.1094
	7	.0000	.0000	.0001	.0004	.0012	.0033	.0079	.0164	.0312
	8	.0000	.0000	.0000	.0000	.0001	.0002	.0007	.0017	.0039
9	0	.3874	.2316	.1342	.0751	.0404	.0207	.0101	.0046	.0020
	1	.3874	.3679	.3020	.2253	.1556	.1004	.0605	.0339	.0176
	2	.1722	.2597	.3020	.3003	.2668	.2162	.1612	.1110	.0703
	3	.0446	.1069	.1762	.2336	.2668	.2716	.2508	.2119	.1641
	4	.0074	.0283	.0661	.1168	.1715	.2194	.2508	.2600	.2461

(continues)

TABLE 5 (*Continued*)

n	x	.10	.15	.20	.25	.30	.35	.40	.45	.50
9	5	.0008	.0050	.0165	.0389	.0735	.1181	.1672	.2128	.2461
	6	.0001	.0006	.0028	.0087	.0210	.0424	.0743	.1160	.1641
	7	.0000	.0000	.0003	.0012	.0039	.0098	.0212	.0407	.0703
	8	.0000	.0000	.0000	.0001	.0004	.0013	.0035	.0083	.0176
	9	.0000	.0000	.0000	.0000	.0000	.0001	.0003	.0008	.0020
10	0	.3487	.1969	.1074	.0563	.0282	.0135	.0060	.0025	.0010
	1	.3874	.3474	.2684	.1877	.1211	.0725	.0403	.0207	.0098
	2	.1937	.2759	.3020	.2816	.2335	.1757	.1209	.0763	.0439
	3	.0574	.1298	.2013	.2503	.2668	.2522	.2150	.1665	.1172
	4	.0112	.0401	.0881	.1460	.2001	.2377	.2508	.2384	.2051
	5	.0015	.0085	.0264	.0584	.1029	.1536	.2007	.2340	.2461
	6	.0001	.0012	.0055	.0162	.0368	.0689	.1115	.1596	.2051
	7	.0000	.0001	.0008	.0031	.0090	.0212	.0425	.0746	.1172
	8	.0000	.0000	.0001	.0004	.0014	.0043	.0106	.0229	.0439
	9	.0000	.0000	.0000	.0000	.0001	.0005	.0016	.0042	.0098
	10	.0000	.0000	.0000	.0000	.0000	.0000	.0001	.0003	.0010
12	0	.2824	.1422	.0687	.0317	.0138	.0057	.0022	.0008	.0002
	1	.3766	.3012	.2062	.1267	.0712	.0368	.0174	.0075	.0029
	2	.2301	.2924	.2835	.2323	.1678	.1088	.0639	.0339	.0161
	3	.0853	.1720	.2362	.2581	.2397	.1954	.1419	.0923	.0537
	4	.0213	.0683	.1329	.1936	.2311	.2367	.2128	.1700	.1208
	5	.0038	.0193	.0532	.1032	.1585	.2039	.2270	.2225	.1934
	6	.0005	.0040	.0155	.0401	.0792	.1281	.1766	.2124	.2256
	7	.0000	.0006	.0033	.0115	.0291	.0591	.1009	.1489	.1934
	8	.0000	.0001	.0005	.0024	.0078	.0199	.0420	.0762	.1208
	9	.0000	.0000	.0001	.0004	.0015	.0048	.0125	.0277	.0537
	10	.0000	.0000	.0000	.0000	.0002	.0008	.0025	.0068	.0161
	11	.0000	.0000	.0000	.0000	.0000	.0001	.0003	.0010	.0029
	12	.0000	.0000	.0000	.0000	.0000	.0000	.0000	.0001	.0002
15	0	.2059	.0874	.0352	.0134	.0047	.0016	.0005	.0001	.0000
	1	.3432	.2312	.1319	.0668	.0305	.0126	.0047	.0016	.0005
	2	.2669	.2856	.2309	.1559	.0916	.0476	.0219	.0090	.0032
	3	.1285	.2184	.2501	.2252	.1700	.1110	.0634	.0318	.0139
	4	.0428	.1156	.1876	.2252	.2186	.1792	.1268	.0780	.0417
	5	.0105	.0449	.1032	.1651	.2061	.2123	.1859	.1404	.0916
	6	.0019	.0132	.0430	.0917	.1472	.1906	.2066	.1914	.1527
	7	.0003	.0030	.0138	.0393	.0811	.1319	.1771	.2013	.1964
	8	.0000	.0005	.0035	.0131	.0348	.0710	.1181	.1647	.1964
	9	.0000	.0001	.0007	.0034	.0116	.0298	.0612	.1048	.1527
	10	.0000	.0000	.0001	.0007	.0030	.0096	.0245	.0515	.0916
	11	.0000	.0000	.0000	.0001	.0006	.0024	.0074	.0191	.0417
	12	.0000	.0000	.0000	.0000	.0001	.0004	.0016	.0052	.0139
	13	.0000	.0000	.0000	.0000	.0000	.0001	.0003	.0010	.0032
	14	.0000	.0000	.0000	.0000	.0000	.0000	.0000	.0001	.0005
	15	.0000	.0000	.0000	.0000	.0000	.0000	.0000	.0000	.0000

(continues)

TABLE 5 (*Continued*)

n	x	.10	.15	.20	.25	.30	.35	.40	.45	.50
18	0	.1501	.0536	.0180	.0056	.0016	.0004	.0001	.0000	.0000
	1	.3002	.1704	.0811	.0338	.0126	.0042	.0012	.0003	.0001
	2	.2835	.2556	.1723	.0958	.0458	.0190	.0069	.0022	.0006
	3	.1680	.2406	.2297	.1704	.1046	.0547	.0246	.0095	.0031
	4	.0700	.1592	.2153	.2130	.1681	.1104	.0614	.0291	.0117
	5	.0218	.0787	.1507	.1988	.2017	.1664	.1146	.0666	.0327
	6	.0052	.0301	.0816	.1436	.1873	.1941	.1655	.1181	.0708
	7	.0010	.0091	.0350	.0820	.1376	.1792	.1892	.1657	.1214
	8	.0002	.0022	.0120	.0376	.0811	.1327	.1734	.1864	.1669
	9	.0000	.0004	.0033	.0139	.0386	.0794	.1284	.1694	.1855
	10	.0000	.0001	.0008	.0042	.0149	.0385	.0771	.1248	.1669
	11	.0000	.0000	.0001	.0010	.0046	.0151	.0374	.0742	.1214
	12	.0000	.0000	.0000	.0002	.0012	.0047	.0145	.0354	.0708
	13	.0000	.0000	.0000	.0000	.0002	.0012	.0045	.0134	.0327
	14	.0000	.0000	.0000	.0000	.0000	.0002	.0011	.0039	.0117
	15	.0000	.0000	.0000	.0000	.0000	.0000	.0002	.0009	.0031
	16	.0000	.0000	.0000	.0000	.0000	.0000	.0000	.0001	.0006
	17	.0000	.0000	.0000	.0000	.0000	.0000	.0000	.0000	.0001
	18	.0000	.0000	.0000	.0000	.0000	.0000	.0000	.0000	.0000
20	0	.1216	.0388	.0115	.0032	.0008	.0002	.0000	.0000	.0000
	1	.2702	.1368	.0576	.0211	.0068	.0020	.0005	.0001	.0000
	2	.2852	.2293	.1369	.0669	.0278	.0100	.0031	.0008	.0002
	3	.1901	.2428	.2054	.1339	.0716	.0323	.0123	.0040	.0011
	4	.0898	.1821	.2182	.1897	.1304	.0738	.0350	.0139	.0046
	5	.0319	.1028	.1746	.2023	.1789	.1272	.0746	.0365	.0148
	6	.0089	.0454	.1091	.1686	.1916	.1712	.1244	.0746	.0370
	7	.0020	.0160	.0545	.1124	.1643	.1844	.1659	.1221	.0739
	8	.0004	.0046	.0222	.0609	.1144	.1614	.1797	.1623	.1201
	9	.0001	.0011	.0074	.0271	.0654	.1158	.1597	.1771	.1602
	10	.0000	.0002	.0020	.0099	.0308	.0686	.1171	.1593	.1762
	11	.0000	.0000	.0005	.0030	.0120	.0336	.0710	.1185	.1602
	12	.0000	.0000	.0001	.0008	.0039	.0136	.0355	.0727	.1201
	13	.0000	.0000	.0000	.0002	.0010	.0045	.0146	.0366	.0739
	14	.0000	.0000	.0000	.0000	.0002	.0012	.0049	.0150	.0370
	15	.0000	.0000	.0000	.0000	.0000	.0003	.0013	.0049	.0148
	16	.0000	.0000	.0000	.0000	.0000	.0000	.0003	.0013	.0046
	17	.0000	.0000	.0000	.0000	.0000	.0000	.0000	.0002	.0011
	18	.0000	.0000	.0000	.0000	.0000	.0000	.0000	.0000	.0002
	19	.0000	.0000	.0000	.0000	.0000	.0000	.0000	.0000	.0000
	20	.0000	.0000	.0000	.0000	.0000	.0000	.0000	.0000	.0000

TABLE 6   Values of $e^{-\mu}$

μ	$e^{-\mu}$	μ	$e^{-\mu}$	μ	$e^{-\mu}$
.0	1.0000	3.1	.0450	8.0	.000335
.1	.9048	3.2	.0408	9.0	.000123
.2	.8187	3.3	.0369	10.0	.000045
.3	.7408	3.4	.0334		
.4	.6703	3.5	.0302		
.5	.6065	3.6	.0273		
.6	.5488	3.7	.0247		
.7	.4966	3.8	.0224		
.8	.4493	3.9	.0202		
.9	.4066	4.0	.0183		
1.0	.3679	4.1	.0166		
1.1	.3329	4.2	.0150		
1.2	.3012	4.3	.0136		
1.3	.2725	4.4	.0123		
1.4	.2466	4.5	.0111		
1.5	.2231	4.6	.0101		
1.6	.2019	4.7	.0091		
1.7	.1827	4.8	.0082		
1.8	.1653	4.9	.0074		
1.9	.1496	5.0	.0067		
2.0	.1353	5.1	.0061		
2.1	.1225	5.2	.0055		
2.2	.1108	5.3	.0050		
2.3	.1003	5.4	.0045		
2.4	.0907	5.5	.0041		
2.5	.0821	5.6	.0037		
2.6	.0743	5.7	.0033		
2.7	.0672	5.8	.0030		
2.8	.0608	5.9	.0027		
2.9	.0550	6.0	.0025		
3.0	.0498	7.0	.0009		

TABLE 7 Poisson Probabilities

Entries in the table give the probability of x occurrences for a Poisson process with a mean μ. For example, when $\mu = 2.5$, the probability of four occurrences is .1336.

x	0.1	0.2	0.3	0.4	μ 0.5	0.6	0.7	0.8	0.9	1.0
0	.9048	.8187	.7408	.6703	.6065	.5488	.4966	.4493	.4066	.3679
1	.0905	.1637	.2222	.2681	.3033	.3293	.3476	.3595	.3659	.3679
2	.0045	.0164	.0333	.0536	.0758	.0988	.1217	.1438	.1647	.1839
3	.0002	.0011	.0033	.0072	.0126	.0198	.0284	.0383	.0494	.0613
4	.0000	.0001	.0002	.0007	.0016	.0030	.0050	.0077	.0111	.0153
5	.0000	.0000	.0000	.0001	.0002	.0004	.0007	.0012	.0020	.0031
6	.0000	.0000	.0000	.0000	.0000	.0000	.0001	.0002	.0003	.0005
7	.0000	.0000	.0000	.0000	.0000	.0000	.0000	.0000	.0000	.0001

x	1.1	1.2	1.3	1.4	μ 1.5	1.6	1.7	1.8	1.9	2.0
0	.3329	.3012	.2725	.2466	.2231	.2019	.1827	.1653	.1496	.1353
1	.3662	.3614	.3543	.3452	.3347	.3230	.3106	.2975	.2842	.2707
2	.2014	.2169	.2303	.2417	.2510	.2584	.2640	.2678	.2700	.2707
3	.0738	.0867	.0998	.1128	.1255	.1378	.1496	.1607	.1710	.1804
4	.0203	.0260	.0324	.0395	.0471	.0551	.0636	.0723	.0812	.0902
5	.0045	.0062	.0084	.0111	.0141	.0176	.0216	.0260	.0309	.0361
6	.0008	.0012	.0018	.0026	.0035	.0047	.0061	.0078	.0098	.0120
7	.0001	.0002	.0003	.0005	.0008	.0011	.0015	.0020	.0027	.0034
8	.0000	.0000	.0001	.0001	.0001	.0002	.0003	.0005	.0006	.0009
9	.0000	.0000	.0000	.0000	.0000	.0000	.0001	.0001	.0001	.0002

x	2.1	2.2	2.3	2.4	μ 2.5	2.6	2.7	2.8	2.9	3.0
0	.1225	.1108	.1003	.0907	.0821	.0743	.0672	.0608	.0550	.0498
1	.2572	.2438	.2306	.2177	.2052	.1931	.1815	.1703	.1596	.1494
2	.2700	.2681	.2652	.2613	.2565	.2510	.2450	.2384	.2314	.2240
3	.1890	.1966	.2033	.2090	.2138	.2176	.2205	.2225	.2237	.2240
4	.0992	.1082	.1169	.1254	.1336	.1414	.1488	.1557	.1622	.1680

(continues)

TABLE 7 *(Continued)*

x	2.1	2.2	2.3	2.4	μ 2.5	2.6	2.7	2.8	2.9	3.0
5	.0417	.0476	.0538	.0602	.0668	.0735	.0804	.0872	.0940	.1008
6	.0146	.0174	.0206	.0241	.0278	.0319	.0362	.0407	.0455	.0540
7	.0044	.0055	.0068	.0083	.0099	.0118	.0139	.0163	.0188	.0216
8	.0011	.0015	.0019	.0025	.0031	.0038	.0047	.0057	.0068	.0081
9	.0003	.0004	.0005	.0007	.0009	.0011	.0014	.0018	.0022	.0027
10	.0001	.0001	.0001	.0002	.0002	.0003	.0004	.0005	.0006	.0008
11	.0000	.0000	.0000	.0000	.0000	.0001	.0001	.0001	.0002	.0002
12	.0000	.0000	.0000	.0000	.0000	.0000	.0000	.0000	.0000	.0001

x	3.1	3.2	3.3	3.4	μ 3.5	3.6	3.7	3.8	3.9	4.0
0	.0450	.0408	.0369	.0344	.0302	.0273	.0247	.0224	.0202	.0183
1	.1397	.1304	.1217	.1135	.1057	.0984	.0915	.0850	.0789	.0733
2	.2165	.2087	.2008	.1929	.1850	.1771	.1692	.1615	.1539	.1465
3	.2237	.2226	.2209	.2186	.2158	.2125	.2087	.2046	.2001	.1954
4	.1734	.1781	.1823	.1858	.1888	.1912	.1931	.1944	.1951	.1954
5	.1075	.1140	.1203	.1264	.1322	.1377	.1429	.1477	.1522	.1563
6	.0555	.0608	.0662	.0716	.0771	.0826	.0881	.0936	.0989	.1042
7	.0246	.0278	.0312	.0348	.0385	.0425	.0466	.0508	.0551	.0595
8	.0095	.0111	.0129	.0148	.0169	.0191	.0215	.0241	.0269	.0298
9	.0033	.0040	.0047	.0056	.0066	.0076	.0089	.0102	.0116	.0132
10	.0010	.0013	.0016	.0019	.0023	.0028	.0033	.0039	.0045	.0053
11	.0003	.0004	.0005	.0006	.0007	.0009	.0011	.0013	.0016	.0019
12	.0001	.0001	.0001	.0002	.0002	.0003	.0003	.0004	.0005	.0006
13	.0000	.0000	.0000	.0000	.0001	.0001	.0001	.0001	.0002	.0002
14	.0000	.0000	.0000	.0000	.0000	.0000	.0000	.0000	.0000	.0001

x	4.1	4.2	4.3	4.4	μ 4.5	4.6	4.7	4.8	4.9	5.0
0	.0166	.0150	.0136	.0123	.0111	.0101	.0091	.0082	.0074	.0067
1	.0679	.0630	.0583	.0540	.0500	.0462	.0427	.0395	.0365	.0337
2	.1393	.1323	.1254	.1188	.1125	.1063	.1005	.0948	.0894	.0842
3	.1904	.1852	.1798	.1743	.1687	.1631	.1574	.1517	.1460	.1404
4	.1951	.1944	.1933	.1917	.1898	.1875	.1849	.1820	.1789	.1755
5	.1600	.1633	.1662	.1687	.1708	.1725	.1738	.1747	.1753	.1755
6	.1093	.1143	.1191	.1237	.1281	.1323	.1362	.1398	.1432	.1462
7	.0640	.0686	.0732	.0778	.0824	.0869	.0914	.0959	.1002	.1044
8	.0328	.0360	.0393	.0428	.0463	.0500	.0537	.0575	.0614	.0653
9	.0150	.0168	.0188	.0209	.0232	.0255	.0280	.0307	.0334	.0363

(continues)

TABLE 7 (*Continued*)

x	4.1	4.2	4.3	4.4	μ 4.5	4.6	4.7	4.8	4.9	5.0
10	.0061	.0071	.0081	.0092	.0104	.0118	.0132	.0147	.0164	.0181
11	.0023	.0027	.0032	.0037	.0043	.0049	.0056	.0064	.0073	.0082
12	.0008	.0009	.0011	.0014	.0016	.0019	.0022	.0026	.0030	.0034
13	.0002	.0003	.0004	.0005	.0006	.0007	.0008	.0009	.0011	.0013
14	.0001	.0001	.0001	.0001	.0002	.0002	.0003	.0003	.0004	.0005
15	.0000	.0000	.0000	.0000	.0001	.0001	.0001	.0001	.0001	.0002

x	5.1	5.2	5.3	5.4	μ 5.5	5.6	5.7	5.8	5.9	6.0
0	.0061	.0055	.0050	.0045	.0041	.0037	.0033	.0030	.0027	.0025
1	.0311	.0287	.0265	.0244	.0225	.0207	.0191	.0176	.0162	.0149
2	.0793	.0746	.0701	.0659	.0618	.0580	.0544	.0509	.0477	.0446
3	.1348	.1293	.1239	.1185	.1133	.1082	.1033	.0985	.0938	.0892
4	.1719	.1681	.1641	.1600	.1558	.1515	.1472	.1428	.1383	.1339
5	.1753	.1748	.1740	.1728	.1714	.1697	.1678	.1656	.1632	.1606
6	.1490	.1515	.1537	.1555	.1571	.1584	.1594	.1601	.1605	.1606
7	.1086	.1125	.1163	.1200	.1234	.1267	.1298	.1326	.1353	.1377
8	.0692	.0731	.0771	.0810	.0849	.0887	.0925	.0962	.0998	.1033
9	.0392	.0423	.0454	.0486	.0519	.0552	.0586	.0620	.0654	.0688
10	.0200	.0220	.0241	.0262	.0285	.0309	.0334	.0359	.0386	.0413
11	.0093	.0104	.0116	.0129	.0143	.0157	.0173	.0190	.0207	.0225
12	.0039	.0045	.0051	.0058	.0065	.0073	.0082	.0092	.0102	.0113
13	.0015	.0018	.0021	.0024	.0028	.0032	.0036	.0041	.0046	.0052
14	.0006	.0007	.0008	.0009	.0011	.0013	.0015	.0017	.0019	.0022
15	.0002	.0002	.0003	.0003	.0004	.0005	.0006	.0007	.0008	.0009
16	.0001	.0001	.0001	.0001	.0001	.0002	.0002	.0002	.0003	.0003
17	.0000	.0000	.0000	.0000	.0000	.0001	.0001	.0001	.0001	.0001

x	6.1	6.2	6.3	6.4	μ 6.5	6.6	6.7	6.8	6.9	7.0
0	.0022	.0020	.0018	.0017	.0015	.0014	.0012	.0011	.0010	.0009
1	.0137	.0126	.0116	.0106	.0098	.0090	.0082	.0076	.0070	.0064
2	.0417	.0390	.0364	.0340	.0318	.0296	.0276	.0258	.0240	.0223
3	.0848	.0806	.0765	.0726	.0688	.0652	.0617	.0584	.0552	.0521
4	.1294	.1249	.1205	.1162	.1118	.1076	.1034	.0992	.0952	.0912
5	.1579	.1549	.1519	.1487	.1454	.1420	.1385	.1349	.1314	.1277
6	.1605	.1601	.1595	.1586	.1575	.1562	.1546	.1529	.1511	.1490
7	.1399	.1418	.1435	.1450	.1462	.1472	.1480	.1486	.1489	.1490
8	.1066	.1099	.1130	.1160	.1188	.1215	.1240	.1263	.1284	.1304
9	.0723	.0757	.0791	.0825	.0858	.0891	.0923	.0954	.0985	.1014

(continues)

TABLE 7　(*Continued*)

x	6.1	6.2	6.3	6.4	μ 6.5	6.6	6.7	6.8	6.9	7.0
10	.0441	.0469	.0498	.0528	.0558	.0588	.0618	.0649	.0679	.0710
11	.0245	.0265	.0285	.0307	.0330	.0353	.0377	.0401	.0426	.0452
12	.0124	.0137	.0150	.0164	.0179	.0194	.0210	.0227	.0245	.0264
13	.0058	.0065	.0073	.0081	.0089	.0098	.0108	.0119	.0130	.0142
14	.0025	.0029	.0033	.0037	.0041	.0046	.0052	.0058	.0064	.0071
15	.0010	.0012	.0014	.0016	.0018	.0020	.0023	.0026	.0029	.0033
16	.0004	.0005	.0005	.0006	.0007	.0008	.0010	.0011	.0013	.0014
17	.0001	.0002	.0002	.0002	.0003	.0003	.0004	.0004	.0005	.0006
18	.0000	.0001	.0001	.0001	.0001	.0001	.0001	.0002	.0002	.0002
19	.0000	.0000	.0000	.0000	.0000	.0000	.0000	.0001	.0001	.0001

x	7.1	7.2	7.3	7.4	μ 7.5	7.6	7.7	7.8	7.9	8.0
0	.0008	.0007	.0007	.0006	.0006	.0005	.0005	.0004	.0004	.0003
1	.0059	.0054	.0049	.0045	.0041	.0038	.0035	.0032	.0029	.0027
2	.0208	.0194	.0180	.0167	.0156	.0145	.0134	.0125	.0116	.0107
3	.0492	.0464	.0438	.0413	.0389	.0366	.0345	.0324	.0305	.0286
4	.0874	.0836	.0799	.0764	.0729	.0696	.0663	.0632	.0602	.0573
5	.1241	.1204	.1167	.1130	.1094	.1057	.1021	.0986	.0951	.0916
6	.1468	.1445	.1420	.1394	.1367	.1339	.1311	.1282	.1252	.1221
7	.1489	.1486	.1481	.1474	.1465	.1454	.1442	.1428	.1413	.1396
8	.1321	.1337	.1351	.1363	.1373	.1382	.1388	.1392	.1395	.1396
9	.1042	.1070	.1096	.1121	.1144	.1167	.1187	.1207	.1224	.1241
10	.0740	.0770	.0800	.0829	.0858	.0887	.0914	.0941	.0967	.0993
11	.0478	.0504	.0531	.0558	.0585	.0613	.0640	.0667	.0695	.0722
12	.0283	.0303	.0323	.0344	.0366	.0388	.0411	.0434	.0457	.0481
13	.0154	.0168	.0181	.0196	.0211	.0227	.0243	.0260	.0278	.0296
14	.0078	.0086	.0095	.0104	.0113	.0123	.0134	.0145	.0157	.0169
15	.0037	.0041	.0046	.0051	.0057	.0062	.0069	.0075	.0083	.0090
16	.0016	.0019	.0021	.0024	.0026	.0030	.0033	.0037	.0041	.0045
17	.0007	.0008	.0009	.0010	.0012	.0013	.0015	.0017	.0019	.0021
18	.0003	.0003	.0004	.0004	.0005	.0006	.0006	.0007	.0008	.0009
19	.0001	.0001	.0001	.0002	.0002	.0002	.0003	.0003	.0003	.0004
20	.0000	.0000	.0001	.0001	.0001	.0001	.0001	.0001	.0001	.0002
21	.0000	.0000	.0000	.0000	.0000	.0000	.0000	.0000	.0001	.0001

(continues)

TABLE 7 (*Continued*)

x	8.1	8.2	8.3	8.4	μ 8.5	8.6	8.7	8.8	8.9	9.0
0	.0003	.0003	.0002	.0002	.0002	.0002	.0002	.0002	.0001	.0001
1	.0025	.0023	.0021	.0019	.0017	.0016	.0014	.0013	.0012	.0011
2	.0100	.0092	.0086	.0079	.0074	.0068	.0063	.0058	.0054	.0050
3	.0269	.0252	.0237	.0222	.0208	.0195	.0183	.0171	.0160	.0150
4	.0544	.0517	.0491	.0466	.0443	.0420	.0398	.0377	.0357	.0337
5	.0882	.0849	.0816	.0784	.0752	.0722	.0692	.0663	.0635	.0607
6	.1191	.1160	.1128	.1097	.1066	.1034	.1003	.0972	.0941	.0911
7	.1378	.1358	.1338	.1317	.1294	.1271	.1247	.1222	.1197	.1171
8	.1395	.1392	.1388	.1382	.1375	.1366	.1356	.1344	.1332	.1318
9	.1256	.1269	.1280	.1290	.1299	.1306	.1311	.1315	.1317	.1318
10	.1017	.1040	.1063	.1084	.1104	.1123	.1140	.1157	.1172	.1186
11	.0749	.0776	.0802	.0828	.0853	.0878	.0902	.0925	.0948	.0970
12	.0505	.0530	.0555	.0579	.0604	.0629	.0654	.0679	.0703	.0728
13	.0315	.0334	.0354	.0374	.0395	.0416	.0438	.0459	.0481	.0504
14	.0182	.0196	.0210	.0225	.0240	.0256	.0272	.0289	.0306	.0324
15	.0098	.0107	.0116	.0126	.0136	.0147	.0158	.0169	.0182	.0194
16	.0050	.0055	.0060	.0066	.0072	.0079	.0086	.0093	.0101	.0109
17	.0024	.0026	.0029	.0033	.0036	.0040	.0044	.0048	.0053	.0058
18	.0011	.0012	.0014	.0015	.0017	.0019	.0021	.0024	.0026	.0029
19	.0005	.0005	.0006	.0007	.0008	.0009	.0010	.0011	.0012	.0014
20	.0002	.0002	.0002	.0003	.0003	.0004	.0004	.0005	.0005	.0006
21	.0001	.0001	.0001	.0001	.0001	.0002	.0002	.0002	.0002	.0003
22	.0000	.0000	.0000	.0000	.0001	.0001	.0001	.0001	.0001	.0001

x	9.1	9.2	9.3	9.4	μ 9.5	9.6	9.7	9.8	9.9	10
0	.0001	.0001	.0001	.0001	.0001	.0001	.0001	.0001	.0001	.0000
1	.0010	.0009	.0009	.0008	.0007	.0007	.0006	.0005	.0005	.0005
2	.0046	.0043	.0040	.0037	.0034	.0031	.0029	.0027	.0025	.0023
3	.0140	.0131	.0123	.0115	.0107	.0100	.0093	.0087	.0081	.0076
4	.0319	.0302	.0285	.0269	.0254	.0240	.0226	.0213	.0201	.0189
5	.0581	.0555	.0530	.0506	.0483	.0460	.0439	.0418	.0398	.0378
6	.0881	.0851	.0822	.0793	.0764	.0736	.0709	.0682	.0656	.0631
7	.1145	.1118	.1091	.1064	.1037	.1010	.0982	.0955	.0928	.0901
8	.1302	.1286	.1269	.1251	.1232	.1212	.1191	.1170	.1148	.1126
9	.1317	.1315	.1311	.1306	.1300	.1293	.1284	.1274	.1263	.1251

(continues)

TABLE 7 (*Continued*)

x	9.1	9.2	9.3	9.4	μ 9.5	9.6	9.7	9.8	9.9	10.0
10	.1198	.1210	.1219	.1228	.1235	.1241	.1245	.1249	.1250	.1251
11	.0991	.1012	.1031	.1049	.1067	.1083	.1098	.1112	.1125	.1137
12	.0752	.0776	.0799	.0822	.0844	.0866	.0888	.0908	.0928	.0948
13	.0526	.0549	.0572	.0594	.0617	.0640	.0662	.0685	.0707	.0729
14	.0342	.0361	.0380	.0399	.0419	.0439	.0459	.0479	.0500	.0521
15	.0208	.0221	.0235	.0250	.0265	.0281	.0297	.0313	.0330	.0347
16	.0118	.0127	.0137	.0147	.0157	.0168	.0180	.0192	.0204	.0217
17	.0063	.0069	.0075	.0081	.0088	.0095	.0103	.0111	.0119	.0128
18	.0032	.0035	.0039	.0042	.0046	.0051	.0055	.0060	.0065	.0071
19	.0015	.0017	.0019	.0021	.0023	.0026	.0028	.0031	.0034	.0037
20	.0007	.0008	.0009	.0010	.0011	.0012	.0014	.0015	.0017	.0019
21	.0003	.0003	.0004	.0004	.0005	.0006	.0006	.0007	.0008	.0009
22	.0001	.0001	.0002	.0002	.0002	.0002	.0003	.0003	.0004	.0004
23	.0000	.0001	.0001	.0001	.0001	.0001	.0001	.0001	.0002	.0002
24	.0000	.0000	.0000	.0000	.0000	.0000	.0000	.0001	.0001	.0001

x	11	12	13	14	μ 15	16	17	18	19	20
0	.0000	.0000	.0000	.0000	.0000	.0000	.0000	.0000	.0000	.0000
1	.0002	.0001	.0000	.0000	.0000	.0000	.0000	.0000	.0000	.0000
2	.0010	.0004	.0002	.0001	.0000	.0000	.0000	.0000	.0000	.0000
3	.0037	.0018	.0008	.0004	.0002	.0001	.0000	.0000	.0000	.0000
4	.0102	.0053	.0027	.0013	.0006	.0003	.0001	.0001	.0000	.0000
5	.0224	.0127	.0070	.0037	.0019	.0010	.0005	.0002	.0001	.0001
6	.0411	.0255	.0152	.0087	.0048	.0026	.0014	.0007	.0004	.0002
7	.0646	.0437	.0281	.0174	.0104	.0060	.0034	.0018	.0010	.0005
8	.0888	.0655	.0457	.0304	.0194	.0120	.0072	.0042	.0024	.0013
9	.1085	.0874	.0661	.0473	.0324	.0213	.0135	.0083	.0050	.0029
10	.1194	.1048	.0859	.0663	.0486	.0341	.0230	.0150	.0095	.0058
11	.1194	.1144	.1015	.0844	.0663	.0496	.0355	.0245	.0164	.0106
12	.1094	.1144	.1099	.0984	.0829	.0661	.0504	.0368	.0259	.0176
13	.0926	.1056	.1099	.1060	.0956	.0814	.0658	.0509	.0378	.0271
14	.0728	.0905	.1021	.1060	.1024	.0930	.0800	.0655	.0514	.0387
15	.0534	.0724	.0885	.0989	.1024	.0992	.0906	.0786	.0650	.0516
16	.0367	.0543	.0719	.0866	.0960	.0992	.0963	.0884	.0772	.0646
17	.0237	.0383	.0550	.0713	.0847	.0934	.0963	.0936	.0863	.0760
18	.0145	.0256	.0397	.0554	.0706	.0830	.0909	.0936	.0911	.0844
19	.0084	.0161	.0272	.0409	.0557	.0699	.0814	.0887	.0911	.0888

(*continues*)

TABLE 7 (*Continued*)

x	11	12	13	14	μ 15	16	17	18	19	20
20	.0046	.0097	.0177	.0286	.0418	.0559	.0692	.0798	.0866	.0888
21	.0024	.0055	.0109	.0191	.0299	.0426	.0560	.0684	.0783	.0846
22	.0012	.0030	.0065	.0121	.0204	.0310	.0433	.0560	.0676	.0769
23	.0006	.0016	.0037	.0074	.0133	.0216	.0320	.0438	.0559	.0669
24	.0003	.0008	.0020	.0043	.0083	.0144	.0226	.0328	.0442	.0557
25	.0001	.0004	.0010	.0024	.0050	.0092	.0154	.0237	.0336	.0446
26	.0000	.0002	.0005	.0013	.0029	.0057	.0101	.0164	.0246	.0343
27	.0000	.0001	.0002	.0007	.0016	.0034	.0063	.0109	.0173	.0254
28	.0000	.0000	.0001	.0003	.0009	.0019	.0038	.0070	.0117	.0181
29	.0000	.0000	.0001	.0002	.0004	.0011	.0023	.0044	.0077	.0125
30	.0000	.0000	.0000	.0001	.0002	.0006	.0013	.0026	.0049	.0083
31	.0000	.0000	.0000	.0000	.0001	.0003	.0007	.0015	.0030	.0054
32	.0000	.0000	.0000	.0000	.0001	.0001	.0004	.0009	.0018	.0034
33	.0000	.0000	.0000	.0000	.0000	.0001	.0002	.0005	.0010	.0020
34	.0000	.0000	.0000	.0000	.0000	.0000	.0001	.0002	.0006	.0012
35	.0000	.0000	.0000	.0000	.0000	.0000	.0000	.0001	.0003	.0007
36	.0000	.0000	.0000	.0000	.0000	.0000	.0000	.0001	.0002	.0004
37	.0000	.0000	.0000	.0000	.0000	.0000	.0000	.0000	.0001	.0002
38	.0000	.0000	.0000	.0000	.0000	.0000	.0000	.0000	.0000	.0001
39	.0000	.0000	.0000	.0000	.0000	.0000	.0000	.0000	.0000	.0001

TABLE 8 Random Digits

63271	59986	71744	51102	15141	80714	58683	93108	13554	79945
88547	09896	95436	79115	08303	01041	20030	63754	08459	28364
55957	57243	83865	09911	19761	66535	40102	26646	60147	15702
46276	87453	44790	67122	45573	84358	21625	16999	13385	22782
55363	07449	34835	15290	76616	67191	12777	21861	68689	03263
69393	92785	49902	58447	42048	30378	87618	26933	40640	16281
13186	29431	88190	04588	38733	81290	89541	70290	40113	08243
17726	28652	56836	78351	47327	18518	92222	55201	27340	10493
36520	64465	05550	30157	82242	29520	69753	72602	23756	54935
81628	36100	39254	56835	37636	02421	98063	89641	64953	99337
84649	48968	75215	75498	49539	74240	03466	49292	36401	45525
63291	11618	12613	75055	43915	26488	41116	64531	56827	30825
70502	53225	03655	05915	37140	57051	48393	91322	25653	06543
06426	24771	59935	49801	11082	66762	94477	02494	88215	27191
20711	55609	29430	70165	45406	78484	31639	52009	18873	96927
41990	70538	77191	25860	55204	73417	83920	69468	74972	38712
72452	36618	76298	26678	89334	33938	95567	29380	75906	91807
37042	40318	57099	10528	09925	89773	41335	96244	29002	46453
53766	52875	15987	46962	67342	77592	57651	95508	80033	69828
90585	58955	53122	16025	84299	53310	67380	84249	25348	04332
32001	96293	37203	64516	51530	37069	40261	61374	05815	06714
62606	64324	46354	72157	67248	20135	49804	09226	64419	29457
10078	28073	85389	50324	14500	15562	64165	06125	71353	77669
91561	46145	24177	15294	10061	98124	75732	00815	83452	97355
13091	98112	53959	79607	52244	63303	10413	63839	74762	50289
73864	83014	72457	22682	03033	61714	88173	90835	00634	85169
66668	25467	48894	51043	02365	91726	09365	63167	95264	45643
84745	41042	29493	01836	09044	51926	43630	63470	76508	14194
48068	26805	94595	47907	13357	38412	33318	26098	82782	42851
54310	96175	97594	88616	42035	38093	36745	56702	40644	83514
14877	33095	10924	58013	61439	21882	42059	24177	58739	60170
78295	23179	02771	43464	59061	71411	05697	67194	30495	21157
67524	02865	39593	54278	04237	92441	26602	63835	38032	94770
58268	57219	68124	73455	83236	08710	04284	55005	84171	42596
97158	28672	50685	01181	24262	19427	52106	34308	73685	74246
04230	16831	69085	30802	65559	09205	71829	06489	85650	38707
94879	56606	30401	02602	57658	70091	54986	41394	60437	03195
71446	15232	66715	26385	91518	70566	02888	79941	39684	54315
32886	05644	79316	09819	00813	88407	17461	73925	53037	91904
62048	33711	25290	21526	02223	75947	66466	06232	10913	75336
84534	42351	21628	53669	81352	95152	08107	98814	72743	12849
84707	15885	84710	35866	06446	86311	32648	88141	73902	69981
19409	40868	64220	80861	13860	68493	52908	26374	63297	45052
57978	48015	25973	66777	45924	56144	24742	96702	88200	66162
57295	98298	11199	96510	75228	41600	47192	43267	35973	23152
94044	83785	93388	07833	38216	31413	70555	03023	54147	06647
30014	25879	71763	96679	90603	99396	74557	74224	18211	91637
07265	69563	64268	88802	72264	66540	01782	08396	19251	83613
84404	88642	30263	80310	11522	57810	27627	78376	36240	48952
21778	02085	27762	46097	43324	34354	09369	14966	10158	76089

TABLE 9   Critical Values for the Durbin-Watson Test for Autocorrelation

Entries in the table give the critical values for a one-tailed Durbin-Watson test for autocorrelation.
For a two-tailed test, the level of significance is doubled.

Significance Points of d_L and d_U: $\alpha = .05$
Number of Independent Variables

	k	1		2		3		4		5	
n		d_L	d_U	d_L	d_U	d_L	d_U	d_L	d_U	d_L	d_U
15		1.08	1.36	0.95	1.54	0.82	1.75	0.69	1.97	0.56	2.21
16		1.10	1.37	0.98	1.54	0.86	1.73	0.74	1.93	0.62	2.15
17		1.13	1.38	1.02	1.54	0.90	1.71	0.78	1.90	0.67	2.10
18		1.16	1.39	1.05	1.53	0.93	1.69	0.82	1.87	0.71	2.06
19		1.18	1.40	1.08	1.53	0.97	1.68	0.86	1.85	0.75	2.02
20		1.20	1.41	1.10	1.54	1.00	1.68	0.90	1.83	0.79	1.99
21		1.22	1.42	1.13	1.54	1.03	1.67	0.93	1.81	0.83	1.96
22		1.24	1.43	1.15	1.54	1.05	1.66	0.96	1.80	0.86	1.94
23		1.26	1.44	1.17	1.54	1.08	1.66	0.99	1.79	0.90	1.92
24		1.27	1.45	1.19	1.55	1.10	1.66	1.01	1.78	0.93	1.90
25		1.29	1.45	1.21	1.55	1.12	1.66	1.04	1.77	0.95	1.89
26		1.30	1.46	1.22	1.55	1.14	1.65	1.06	1.76	0.98	1.88
27		1.32	1.47	1.24	1.56	1.16	1.65	1.08	1.76	1.01	1.86
28		1.33	1.48	1.26	1.56	1.18	1.65	1.10	1.75	1.03	1.85
29		1.34	1.48	1.27	1.56	1.20	1.65	1.12	1.74	1.05	1.84
30		1.35	1.49	1.28	1.57	1.21	1.65	1.14	1.74	1.07	1.83
31		1.36	1.50	1.30	1.57	1.23	1.65	1.16	1.74	1.09	1.83
32		1.37	1.50	1.31	1.57	1.24	1.65	1.18	1.73	1.11	1.82
33		1.38	1.51	1.32	1.58	1.26	1.65	1.19	1.73	1.13	1.81
34		1.39	1.51	1.33	1.58	1.27	1.65	1.21	1.73	1.15	1.81
35		1.40	1.52	1.34	1.58	1.28	1.65	1.22	1.73	1.16	1.80
36		1.41	1.52	1.35	1.59	1.29	1.65	1.24	1.73	1.18	1.80
37		1.42	1.53	1.36	1.59	1.31	1.66	1.25	1.72	1.19	1.80
38		1.43	1.54	1.37	1.59	1.32	1.66	1.26	1.72	1.21	1.79
39		1.43	1.54	1.38	1.60	1.33	1.66	1.27	1.72	1.22	1.79
40		1.44	1.54	1.39	1.60	1.34	1.66	1.29	1.72	1.23	1.79
45		1.48	1.57	1.43	1.62	1.38	1.67	1.34	1.72	1.29	1.78
50		1.50	1.59	1.46	1.63	1.42	1.67	1.38	1.72	1.34	1.77
55		1.53	1.60	1.49	1.64	1.45	1.68	1.41	1.72	1.38	1.77
60		1.55	1.62	1.51	1.65	1.48	1.69	1.44	1.73	1.41	1.77
65		1.57	1.63	1.54	1.66	1.50	1.70	1.47	1.73	1.44	1.77
70		1.58	1.64	1.55	1.67	1.52	1.70	1.49	1.74	1.46	1.77
75		1.60	1.65	1.57	1.68	1.54	1.71	1.51	1.74	1.49	1.77
80		1.61	1.66	1.59	1.69	1.56	1.72	1.53	1.74	1.51	1.77
85		1.62	1.67	1.60	1.70	1.57	1.72	1.55	1.75	1.52	1.77
90		1.63	1.68	1.61	1.70	1.59	1.73	1.57	1.75	1.54	1.78
95		1.64	1.69	1.62	1.71	1.60	1.73	1.58	1.75	1.56	1.78
100		1.65	1.69	1.63	1.72	1.61	1.74	1.59	1.76	1.57	1.78

(continues)

Source: J. Durbin and G. S. Watson, "Testing for serial correlation in least square regression II," *Biometrika,* **38,** 1951, 159–178.

TABLE 9   (*Continued*)

Significance Points of d_L and d_U: $\alpha = .025$
Number of Independent Variables

n	k 1 d_L	d_U	2 d_L	d_U	3 d_L	d_U	4 d_L	d_U	5 d_L	d_U
15	0.95	1.23	0.83	1.40	0.71	1.61	0.59	1.84	0.48	2.09
16	0.98	1.24	0.86	1.40	0.75	1.59	0.64	1.80	0.53	2.03
17	1.01	1.25	0.90	1.40	0.79	1.58	0.68	1.77	0.57	1.98
18	1.03	1.26	0.93	1.40	0.82	1.56	0.72	1.74	0.62	1.93
19	1.06	1.28	0.96	1.41	0.86	1.55	0.76	1.72	0.66	1.90
20	1.08	1.28	0.99	1.41	0.89	1.55	0.79	1.70	0.70	1.87
21	1.10	1.30	1.01	1.41	0.92	1.54	0.83	1.69	0.73	1.84
22	1.12	1.31	1.04	1.42	0.95	1.54	0.86	1.68	0.77	1.82
23	1.14	1.32	1.06	1.42	0.97	1.54	0.89	1.67	0.80	1.80
24	1.16	1.33	1.08	1.43	1.00	1.54	0.91	1.66	0.83	1.79
25	1.18	1.34	1.10	1.43	1.02	1.54	0.94	1.65	0.86	1.77
26	1.19	1.35	1.12	1.44	1.04	1.54	0.96	1.65	0.88	1.76
27	1.21	1.36	1.13	1.44	1.06	1.54	0.99	1.64	0.91	1.75
28	1.22	1.37	1.15	1.45	1.08	1.54	1.01	1.64	0.93	1.74
29	1.24	1.38	1.17	1.45	1.10	1.54	1.03	1.63	0.96	1.73
30	1.25	1.38	1.18	1.46	1.12	1.54	1.05	1.63	0.98	1.73
31	1.26	1.39	1.20	1.47	1.13	1.55	1.07	1.63	1.00	1.72
32	1.27	1.40	1.21	1.47	1.15	1.55	1.08	1.63	1.02	1.71
33	1.28	1.41	1.22	1.48	1.16	1.55	1.10	1.63	1.04	1.71
34	1.29	1.41	1.24	1.48	1.17	1.55	1.12	1.63	1.06	1.70
35	1.30	1.42	1.25	1.48	1.19	1.55	1.13	1.63	1.07	1.70
36	1.31	1.43	1.26	1.49	1.20	1.56	1.15	1.63	1.09	1.70
37	1.32	1.43	1.27	1.49	1.21	1.56	1.16	1.62	1.10	1.70
38	1.33	1.44	1.28	1.50	1.23	1.56	1.17	1.62	1.12	1.70
39	1.34	1.44	1.29	1.50	1.24	1.56	1.19	1.63	1.13	1.69
40	1.35	1.45	1.30	1.51	1.25	1.57	1.20	1.63	1.15	1.69
45	1.39	1.48	1.34	1.53	1.30	1.58	1.25	1.63	1.21	1.69
50	1.42	1.50	1.38	1.54	1.34	1.59	1.30	1.64	1.26	1.69
55	1.45	1.52	1.41	1.56	1.37	1.60	1.33	1.64	1.30	1.69
60	1.47	1.54	1.44	1.57	1.40	1.61	1.37	1.65	1.33	1.69
65	1.49	1.55	1.46	1.59	1.43	1.62	1.40	1.66	1.36	1.69
70	1.51	1.57	1.48	1.60	1.45	1.63	1.42	1.66	1.39	1.70
75	1.53	1.58	1.50	1.61	1.47	1.64	1.45	1.67	1.42	1.70
80	1.54	1.59	1.52	1.62	1.49	1.65	1.47	1.67	1.44	1.70
85	1.56	1.60	1.53	1.63	1.51	1.65	1.49	1.68	1.46	1.71
90	1.57	1.61	1.55	1.64	1.53	1.66	1.50	1.69	1.48	1.71
95	1.58	1.62	1.56	1.65	1.54	1.67	1.52	1.69	1.50	1.71
100	1.59	1.63	1.57	1.65	1.55	1.67	1.53	1.70	1.51	1.72

(continues)

TABLE 9 (*Continued*)

Significance Points of d_L and d_U: $\alpha = .01$
Number of Independent Variables

	k	1		2		3		4		5	
n		d_L	d_U	d_L	d_U	d_L	d_U	d_L	d_U	d_L	d_U
15		0.81	1.07	0.70	1.25	0.59	1.46	0.49	1.70	0.39	1.96
16		0.84	1.09	0.74	1.25	0.63	1.44	0.53	1.66	0.44	1.90
17		0.87	1.10	0.77	1.25	0.67	1.43	0.57	1.63	0.48	1.85
18		0.90	1.12	0.80	1.26	0.71	1.42	0.61	1.60	0.52	1.80
19		0.93	1.13	0.83	1.26	0.74	1.41	0.65	1.58	0.56	1.77
20		0.95	1.15	0.86	1.27	0.77	1.41	0.68	1.57	0.60	1.74
21		0.97	1.16	0.89	1.27	0.80	1.41	0.72	1.55	0.63	1.71
22		1.00	1.17	0.91	1.28	0.83	1.40	0.75	1.54	0.66	1.69
23		1.02	1.19	0.94	1.29	0.86	1.40	0.77	1.53	0.70	1.67
24		1.04	1.20	0.96	1.30	0.88	1.41	0.80	1.53	0.72	1.66
25		1.05	1.21	0.98	1.30	0.90	1.41	0.83	1.52	0.75	1.65
26		1.07	1.22	1.00	1.31	0.93	1.41	0.85	1.52	0.78	1.64
27		1.09	1.23	1.02	1.32	0.95	1.41	0.88	1.51	0.81	1.63
28		1.10	1.24	1.04	1.32	0.97	1.41	0.90	1.51	0.83	1.62
29		1.12	1.25	1.05	1.33	0.99	1.42	0.92	1.51	0.85	1.61
30		1.13	1.26	1.07	1.34	1.01	1.42	0.94	1.51	0.88	1.61
31		1.15	1.27	1.08	1.34	1.02	1.42	0.96	1.51	0.90	1.60
32		1.16	1.28	1.10	1.35	1.04	1.43	0.98	1.51	0.92	1.60
33		1.17	1.29	1.11	1.36	1.05	1.43	1.00	1.51	0.94	1.59
34		1.18	1.30	1.13	1.36	1.07	1.43	1.01	1.51	0.95	1.59
35		1.19	1.31	1.14	1.37	1.08	1.44	1.03	1.51	0.97	1.59
36		1.21	1.32	1.15	1.38	1.10	1.44	1.04	1.51	0.99	1.59
37		1.22	1.32	1.16	1.38	1.11	1.45	1.06	1.51	1.00	1.59
38		1.23	1.33	1.18	1.39	1.12	1.45	1.07	1.52	1.02	1.58
39		1.24	1.34	1.19	1.39	1.14	1.45	1.09	1.52	1.03	1.58
40		1.25	1.34	1.20	1.40	1.15	1.46	1.10	1.52	1.05	1.58
45		1.29	1.38	1.24	1.42	1.20	1.48	1.16	1.53	1.11	1.58
50		1.32	1.40	1.28	1.45	1.24	1.49	1.20	1.54	1.16	1.59
55		1.36	1.43	1.32	1.47	1.28	1.51	1.25	1.55	1.21	1.59
60		1.38	1.45	1.35	1.48	1.32	1.52	1.28	1.56	1.25	1.60
65		1.41	1.47	1.38	1.50	1.35	1.53	1.31	1.57	1.28	1.61
70		1.43	1.49	1.40	1.52	1.37	1.55	1.34	1.58	1.31	1.61
75		1.45	1.50	1.42	1.53	1.39	1.56	1.37	1.59	1.34	1.62
80		1.47	1.52	1.44	1.54	1.42	1.57	1.39	1.60	1.36	1.62
85		1.48	1.53	1.46	1.55	1.43	1.58	1.41	1.60	1.39	1.63
90		1.50	1.54	1.47	1.56	1.45	1.59	1.43	1.61	1.41	1.64
95		1.51	1.55	1.49	1.57	1.47	1.60	1.45	1.62	1.42	1.64
100		1.52	1.56	1.50	1.58	1.48	1.60	1.46	1.63	1.44	1.65

TABLE 10 T_L Values for the Mann-Whitney-Wilcoxon Test

Reject the hypothesis of identical populations if the sum of the ranks for the n_1 items is *less than* the value T_L shown in the following table or if the sum of the ranks for the n_1 items is *greater than* the value T_U where

$$T_U = n_1(n_1 + n_2 + 1) - T_L$$

$\alpha = .05$		n_2								
		2	3	4	5	6	7	8	9	10
n_1	2	3	3	3	3	3	3	4	4	4
	3	6	6	6	7	8	8	9	9	10
	4	10	10	11	12	13	14	15	15	16
	5	15	16	17	18	19	21	22	23	24
	6	21	23	24	25	27	28	30	32	33
	7	28	30	32	34	35	37	39	41	43
	8	37	39	41	43	45	47	50	52	54
	9	46	48	50	53	56	58	61	63	66
	10	56	59	61	64	67	70	73	76	79

$\alpha = .10$		n_2								
		2	3	4	5	6	7	8	9	10
n_1	2	3	3	3	4	4	4	5	5	5
	3	6	7	7	8	9	9	10	11	11
	4	10	11	12	13	14	15	16	17	18
	5	16	17	18	20	21	22	24	25	27
	6	22	24	25	27	29	30	32	34	36
	7	29	31	33	35	37	40	42	44	46
	8	38	40	42	45	47	50	52	55	57
	9	47	50	52	55	58	61	64	67	70
	10	57	60	63	67	70	73	76	80	83

C

Summation Notation

·

Summations

Definition

$$\sum_{i=1}^{n} x_i = x_1 + x_2 + \cdots + x_n \qquad \text{(C.1)}$$

Example: $x_1 = 5$, $x_2 = 8$, $x_3 = 14$:

$$\sum_{i=1}^{3} x_i = x_1 + x_2 + x_3$$
$$= 5 + 8 + 14$$
$$= 27$$

Result 1

For a constant c:

$$\sum_{i=1}^{n} c = \underbrace{(c + c + \cdots + c)}_{n \text{ times}} = nc \qquad \text{(C.2)}$$

Example: $c = 5$, $n = 10$:

$$\sum_{i=1}^{10} 5 = 10\,(5) = 50$$

Example: $c = \bar{x}$:

$$\sum_{i=1}^{n} \bar{x} = n\bar{x}$$

Result 2

$$\sum_{i=1}^{n} cx_i = cx_1 + cx_2 + \cdots + cx_n$$
$$= c(x_1 + x_2 + \cdots + x_n) = c \sum_{i=1}^{n} x_i \qquad \text{(C.3)}$$

Example: $x_1 = 5, x_2 = 8, x_3 = 14, c = 2$:

$$\sum_{i=1}^{3} 2x_i = 2 \sum_{i=1}^{3} x_i = 2(27) = 54$$

Result 3

$$\sum_{i=1}^{n} (ax_i + by_i) = a \sum_{i=1}^{n} x_i + b \sum_{i=1}^{n} y_i \qquad \text{(C.4)}$$

Example: $x_1 = 5, x_2 = 8, x_3 = 14, a = 2, y_1 = 7, y_2 = 3, y_3 = 8, b = 4$:

$$\sum_{i=1}^{3} (2x_i + 4y_i) = 2 \sum_{i=1}^{3} x_i + 4 \sum_{i=1}^{3} y_i$$
$$= 2(27) + 4(18)$$
$$= 54 + 72$$
$$= 126$$

Double Summations

Consider the following data involving the variable x_{ij}, where i is the subscript denoting the row position and j is the subscript denoting the column position:

		Column		
		1	2	3
Row	1	$x_{11} = 10$	$x_{12} = 8$	$x_{13} = 6$
	2	$x_{21} = 7$	$x_{22} = 4$	$x_{23} = 12$

Definition

$$\sum_{i=1}^{n} \sum_{j=1}^{m} x_{ij} = (x_{11} + x_{12} + \cdots + x_{1m}) + (x_{21} + x_{22} + \cdots + x_{2m})$$
$$+ (x_{31} + x_{32} + \cdots + x_{3m}) + \cdots + (x_{n1} + x_{n2} + \cdots + x_{nm}) \qquad \text{(C.5)}$$

Example:

$$\sum_{i=1}^{2} \sum_{j=1}^{3} x_{ij} = x_{11} + x_{12} + x_{13} + x_{21} + x_{22} + x_{23}$$
$$= 10 + 8 + 6 + 7 + 4 + 12$$
$$= 47$$

Definition

$$\sum_{i=1}^{n} x_{ij} = x_{1j} + x_{2j} + \cdots + x_{nj} \qquad \text{(C.6)}$$

Example:

$$\sum_{i=1}^{2} x_{i2} = x_{12} + x_{22}$$
$$= 8 + 4$$
$$= 12$$

Shorthand Notation

Sometimes when a summation is for all values of the subscript, we use the following shorthand notations:

$$\sum_{i-1}^{n} x_i = \sum x_i \tag{C.7}$$

$$\sum_{i-1}^{n} \sum_{j-1}^{m} x_{ij} = \sum \sum x_{ij} \tag{C.8}$$

$$\sum_{i-1}^{n} x_{ij} = \sum_{i} x_{ij} \tag{C.9}$$

Answers to Even-Numbered Exercises

Chapter 1

2. a. 7
  b. 4
  c. 7
4. a. $923,714
  b. No, the firms are not representative.
6. a. 10
  b. The Fortune 500
  c. $2,093,740,000
8. a. Qualitative
  b. Nominal
10. a. Quantitative, ratio
  b. Qualitative, nominal
  c. Qualitative, ordinal
  d. Qualitative, nominal
  e. Quantitative, ratio
14. a. Product test data, test market data
18. a. The 250 households
  b. 250
  c. Number of pounds of garbage produced per day
  d. All households in Cedar Bluff
20. a. 40% in the sample died of heart disease.
  b. Quantitative
24. a. All viewers reached by the television station
  b. The viewers contacted in the telephone survey
26. a. correct
  b. challenge
  c. correct
  d. challenge
  e. challenge

Chapter 2

2. a.

Major	Relative Frequency
Management	.25
Accounting	.24
Finance	.13
Marketing	.38
Total	1.00

4. a.

Movie	Frequency	Relative Frequency
Coming to America	13	.22
Big	11	.18
Crocodile Dundee II	5	.08
Who Framed Roger Rabbit	18	.30
Good Morning Viet Nam	10	.17
Other	3	.05
	60	1.00

c. Who Framed Roger Rabbit was most successful.

6. a. Ordinal
  b.

Rating	Frequency	Relative Frequency
Poor	2	.03
Fair	4	.07
Good	12	.20
Very Good	24	.40
Excellent	18	.30
Totals	60	1.00

8. a, b & d

Score	Frequency	Relative Frequency	Cumulative Frequency
25–34	3	.12	3
35–44	1	.04	4
45–54	2	.08	6
55–64	6	.24	12
65–74	4	.16	16
75–84	6	.24	22
85–94	2	.08	24
95–104	1	.04	25
Totals	25	1.00	

10. a & b

Computer Usage (hours)	Frequency	Relative Frequency
0.0–2.9	14	.28
3.0–5.9	6	.12
6.0–8.9	6	.12
9.0–11.9	7	.14
12.0–14.9	14	.28
15.0–17.9	3	.06
	50	1.00

12. a, b & d

Salary ($1000s)	Frequency	Relative Frequency	Cumulative Frequency
0–499	1	.04	1
500–999	9	.36	10
1000–1499	11	.44	21
1500–1999	2	.08	23
2000–2499	2	.08	23
Totals	25	1.00	

14. a & b

Miles per Gallon	Frequency	Relative Frequency
24.0–25.9	2	.07
26.0–27.9	5	.17
28.0–29.9	10	.33
30.0–31.9	9	.30
32.0–33.9	3	.10
34.0–35.9	1	.03
Totals	30	1.00

d. Civic

16. Data is in ten-cent increments.

```
−0 | 5
 0 | 8
 1 | 0  1  3  4  5  7  8
 2 | 1  6  7
 3 | 0  5
 4 |
 5 | 7
 6 | 0  3
 7 |
 8 | 2
 9 | 9
10 | 0
```

18. a

```
1 | 5  6  8  9
2 | 1  2
3 | 4
4 | 1  6  8
5 |
6 | 5  6  8
7 | 1
8 | 9  9
9 | 2  8
```

20. a & b

Industry	Frequency	Relative Frequency
Beverage	2	.10
Chemicals	3	.15
Electronics	6	.30
Food	7	.35
Aerospace	2	.10
Totals	20	1.00

22. a

Party Affiliation	Frequency	Relative Frequency
Democrat	17	.425
Republican	17	.425
Independent	6	.150
Totals	40	1.000

24. a–d

Sales	Frequency	Relative Frequency	Cumulative Frequency	Cumulative Relative Frequency
0–4999	7	.175	7	.175
5000–9999	11	.275	18	.450
10000–14999	4	.100	22	.550
15000–19999	4	.100	26	.650
20000–24999	4	.100	30	.750
25000–29999	1	.025	31	.775
30000–34999	1	.025	32	.800
35000–39999	2	.050	34	.850
40000–44999	5	.125	39	.975
45000–49999	1	.025	40	1.000
Totals	40	1.000		

26. a, b

Closing Price	Frequency	Relative Frequency	Cumulative Frequency	Cumulative Relative Frequency
0–9⁷⁄₈	9	.225	9	.225
10–19⁷⁄₈	10	.250	19	.475
20–29⁷⁄₈	5	.125	24	.600
30–39⁷⁄₈	11	.275	35	.875
40–49⁷⁄₈	2	.050	37	.925
50–59⁷⁄₈	2	.050	39	.975
60–69⁷⁄₈	0	.000	39	.975
70–79⁷⁄₈	1	.025	40	1.000
Totals	40	1.000		

28. a & b

Room and Board Cost ($)	Frequency	Relative Frequency
1500–1999	3	.06
2000–2499	8	.16
2500–2999	14	.28
3000–3499	14	.28
3500–3999	7	.14
4000–4999	3	.06
4500–4999	1	.02
Totals	50	1.00

30. a.

```
0 | 8 9
1 | 0 2 2 2 3 4 4 4
1 | 5 5 6 6 6 6 7 7 8 8 8 8 9 9 9
2 | 0 1 2 2 2 3 4 4 4
2 | 5 6 8
3 | 0 1 3
```

32. a.

```
4 | 1 2 5 5 6 7 7 9 9
5 | 1 2 4 4 6 7 8 9
6 | 2 5
7 | 1 2 8
8 | 4 7
```

b.

```
2 | 3 4 6 7 7 9
3 | 0 1 1 2 3 3 4 5 6 7 8
4 | 1 2 8
5 | 9
6 | 1 2 6
```

Chapter 3

2. a. 38.75 29
b. 38.611 38.375
c. median = 38.5
d. 29.5 47.5
e. 31
4. a. 178
b. 178
c. Do not report a mode.
d. 184
6. a. 48.33 49 Do not report a mode.
b. 45 55
c. 45 55
8.   City: mean = 15.58, median = 15.9,
                    mode = 15.3
        Country: mean = 18.92, median = 18.7,
                    mode = 18.6 and 19.4

10. a. 13.125 14.5 15
b. 5 17
c. $Q_1 = 8.5$ $Q_3 = 16.5$
L. Hinge = 8.5 U. Hinge = 16.5
12. a. Range = 32 IQR = 10
b. 92.75 9.63
14. Dawson: Range = 2 s = .6749
        Clark: Range = 8 s = 2.5841

16. a. 142,288
b. 6,305
c. 43,727.1
18. Quarter milers: s = .056 Coef. of Var. = 5.8
        Milers: s = .130 Coef. of Var. = 2.9
20. a. $\bar{x} = 77.5$ s = 9.86
b. z = 3.30 An outlier
c. 16% 2.5%

22. a. 75%
b. 94%
c. 75% Almost all
24. 100 145 175 202.5 220
26. a. 1.9 10.0 14.5 20.4 55.9
b. Inner fences: −5.6 36.0
Outer fences: −21.2 51.6
c. 46.1 and 49.9 are mild outliers.
55.9 is an extreme outlier.
28. a. 484 1061 2472 4514 32,249
b. Inner fences: −4118.50 9693.5
Outer fences: −9298 14,873
c. Yes, Dow Chemical is a mild outlier;
DuPont is an extreme outlier.
30. a. 17.48
b. $s^2 = 11.71$ s = 3.42
32. a. 34.35
b. $s^2 = 268.76$ s = 16.39
34. a. 64.7
b. 367.67
c. 19.17
36. a. $\bar{x} = 1028.18$ median = 1000 no mode
b. Range = 510 IQR = 220
c. $s^2 = 24,256.36$ s = 155.74
d. No outliers
38. a. 12.33
b. 12
c. 18
d. 10
e. 17
f. 30.75
g. 5.545
40. $\bar{x} = 1583.33$, median = 1550, mode = 1500,
Range = 800, $s^2 = 69,666.67$, s = 263.944
42. a. 68%
b. 95%
c. yes
44. a. Public: 32 Auto: 32
b. Public: 4.64 Auto: 1.83
c. Auto has less variability.
46. a. 94 104 120 135.5 144
48. $\bar{x} = 12.3$, $s^2 = 28.8$, s = 5.37
50. a. 55.72
b. $s^2 = 32.42$ s = 5.69

Chapter 4

2. b. eight: (p, p, p), (p, p, np), (p, np, p),
(p, np, np), (np, p, p), (np, p, np),
(np, np, p), (np, np, np)
c. 16
4. a. 9
6. 20
8. a. 9
c. 5
d. 1
10. a. 1,000,000
b. 5,760,000
c. More

12. a. Relative frequency approach
   b. 1106/2038, 826/2038, 106/2038
14. a. No, probabilities do not sum to one.
   b. Revise probabilities so they sum to one.
16. No, there are 4 equally likely outcomes:
   (H, T) (H, H) (T, T) and (T, H)
18. No some designs are more often selected.
20. d. $P(\text{Ace}) = .08$, $P(\text{Club}) = .25$, $P(\text{face card}) = .23$
22. a. 36
   c. 1/6
   d. 5/18
   e. No, $P(\text{odd}) = P(\text{even}) = 1/2$
   f. Classical
24. a. .31
   b. .69
26. .72
28. .35, .26
30. a. .40; .50; .60
   b. $\{E_1, E_2, E_4, E_6, E_7\}$; $P(A \cup B) = .65$
   c. $\{E_4\}$ $P(A \cap B) = .25$
   d. Yes
   e. $\{E_1, E_3, E_5, E_6\}$ $P(B^c) = .50$
32. a. A = first car starts, B = second car starts
   $P(A) = .80$ $P(B) = .40$ $P(A \cap B) = .30$
   b. .90
   c. .10

34. a.

	Single	Married	Total
Under 30	.55	.10	.65
30 or over	.20	.15	.35
Total	.75	.25	1.00

   b. higher probability of under 30
   c. higher probability of single
   d. .55
   e. .8462
   f. No

36. a.

Own U.S. Car

		Yes	No	
Own Foreign Car	Yes	.15	.05	.20
	No	.75	.05	.80
		.90	.10	1.00

   b. 20% own a foreign car; 90% own a U.S. car
   c. .15
   d. .95
   e. .17
   f. .75
   g. No
38. c. .72
   d. .40
   e. No

40. a. 0
   b. 0
   c. No
   d. Mutually exclusive events are dependent.
42. a. .10; .20; .09
   b. .51
   c. .26; .51; .23
44. a. .21
   b. Yes
46. .6007
48. a. Three
   b. .10, .50, .40
50. a. .02
   b. .64
   c. .93
   d. yes
52. a. .76
   b. .24
54. a. $\{E_1, E_2, E_3, E_4\}$
   b. $\{E_1, E_5, E_6, E_7, E_8\}$
   c. $\{E_2\}$
   d. $\{E_7, E_8\}$
   e. empty
   f. $\{E_4, E_5, E_6, E_7, E_8\}$
   g. $\{E_1, E_2, E_3, E_4\}$
   h. $\{E_1, E_2, E_3, E_4\}$
   i. $\{E_1, E_2, E_3\}$
   j. No
   k. Yes
56. b. .2022
   c. .4618
   d. .4005
58. a. .35; .80; .67
   b. No
60. a. .90; 0
   b. No

62. a.

	Smoker	Nonsmoker	Total
Record of Heart Disease	.10	.08	.18
No Record of Heart Disease	.20	.62	.82
Total	.30	.70	1.00

   b. .10
   c. 30%-smokers; 70%-nonsmokers; 18%-record of heart disease; 82%-no record of heart disease
   d. .333
   e. .114
   f. No
   g. Smokers have a higher probability of having a record of heart disease.
64. a. .25
   b. Yes
   c. No

66. a. .25; .40; .10
  b. .25
  c. B and S are independent; program appears to have no effect.
68. a. .20
  b. .35
  c. 65%
70. Call back since P(sale|call back) = .21.
72. 3.44%
74. a. .12
  b. .625
  c. .305
76. a. .0625
  b. .0132
  c. Three

Chapter 5

2. 0, 1, 2, 3, 4, 5
4. a. $(H, H), (H, T), (T, H), (T, T)$
  b. x = number of heads

c. Outcome	(H, H)	(H, T)	(T, H)	(T, T)
Value of x	2	1	1	0

6. a. x	1	2	3	4
$f(x)$	.15	.25	.40	.20

  c. $f(x) \geq 0, \Sigma f(x) = 1$
8. a. It is a proper probability distribution.
  b. .60
10. a. Yes
  b. .65
12. a. .05
  b. .70
  c. .40
14. a. 13.474%
  b. $\sigma^2 = 71.79$ $\sigma = 8.47$
  c. 6.964, .9864
16. a. 83
  b. -47; concern is to protect against the big accident.
18. a. 445
  b. 1,250 loss
20. a. Medium: 145; Large: 140
  b. Medium: 2725; Large: 12,400
22. a. 17
  b. 3.33; 1.82
24. a. Probability of a defective part must be .03 for each trial; trials must be independent.
  c. 2

d. Number of defects	0	1	2
Probability	.9409	.0582	.0009

26. $f(1) = .4116; f(2) = .2646; f(0) = .2401$
28. a. .90
  b. .99
  c. .999
  d. Yes
30. a. .5905
  b. .1937
  c. .6082
34. 212.5; 31.88
36. a. .0821
  b. .065
  c. .384
38. a. .000045
  b. .010245
  c. .0821
  d. .9179
40. a. .50
  b. .067
  c. .4667
  d. .30
42. .00582
44. a. .01
  b. .07
  c. .92
  d. .07
46. a. 1.6
  b. $120
48. a. 2.2
  b. 1.16
50. a. $f(x) \geq 0$ and $\Sigma f(x) = 1$
  b. $17.25
  c. $1.25, 7.8%
  d. 1.3875
52. a. 3 hours
  b. 1
54. a. 22.80, 5.625, 2.37
  b. 228, 56.25, 7.5
56. a. .1712
  b. .2528
58. a. .2785
  b. .3417
60. a. .9510
  b. .0480
  c. .0490
62. .1912
64. a. .2241
  b. .5767

Chapter 6

2. b. .50
  c. .60
  d. 15
  e. 8.33
4. b. .50
  c. .30
  d. .40

6. a. .40
   b. .64
   c. .68
8. b. .6826
   c. .9544
10. a. .3413
    b. .4332
    c. .4772
    d. .4938
    e. .4986
12. a. .6640
    b. .1903
    c. .1091
14. a. −.80
    b. 1.66
    c. .26
    d. 2.56
    e. −.50
16. a. .3830
    b. .1056
    c. .0062
    d. .1603
18. a. .7745
    b. 36.32
    c. 19%
20. a. 2.865 years
    b. .6247
22. a. 11
    b. .5780
    c. .1170
    d. .6950
24. a. .0377
    b. .0823
    c. .9108
26. a. .8336
    b. .0049
28. a. .9412
    b. .5200
30. a. 50 hours
    b. .3935
    c. .1353
32. a. .341
    b. .372
    c. .189
34. b. .20
    c. 37 minutes
    d. 8.33, 2.89
36. a. .2642
    b. .2736
    c. .2207
    d. .1539
38. a. 1.645
    b. .84
    c. −.53
    d. −.25
    e. .72
40. .0062
42. a. .5899
    b. 30 or more

44. a. 47.06%
    b. .0475
    c. 42,480
46. a. .0228
    b. Do not use the die.
    c. 3.36733
48. a. 90
    b. .0668
    c. .3085
    d. .0014
50. a. 2 minutes
    b. .221
    c. .3935
    d. .0821

Chapter 7

2. a. L.A. Gear, New Balance, Adidas, Avia, Reebok
   b. 0
   c. 252
4. 2782, 493, 825, 1807, 289
6. 447, 348, 499, 568, 055, 392
   126, 036, 599, 294, 570, 159
8. a. Randomly select a page (1 to 853) and then randomly select a line (1 to 400) on the sampled page.
   b. Skip or ignore inappropriate lines and repeat the sampling procedure of part (a).
10. finite, infinite, infinite, infinite, finite
12. a. 8.875
    b. 1.31
14. .0288
16. a. 32
    b. .91
18. a. .6680
    b. .8294
20. a. Normal with $E(\bar{x}) = 12.55$ and $\sigma_{\bar{x}} = .6325$
    b. .8858
    c. .5704
22. 170, 4.43
24. a. .6528
    b. .3616
26. a. 1.41
    b. 1.41
    c. 1.41
    d. 1.34
28. Normal with $E(\bar{x}) = 18$ and $\sigma_{\bar{x}} = .57$.
30. a. Normal with $E(\bar{x}) = 14.25$ and $\sigma_{\bar{x}} = .2828$
    b. .9441
    c. .6212
    d. .9878, .7888
32. a. No, since $n/N = .01$.
    b. 1.29, 1.30; little difference.
34. a. Normal with $E(\bar{x}) = 300$ and $\sigma_{\bar{x}} = 13.69$
    b. 13.69
    c. .8558
    d. .3557

36. a. .9756
  b. .7372
38. a. Normal with $E(\bar{p}) = .80$ and $\sigma_{\bar{p}} = .02$
  b. .8664
  c. .9596
40. a. Normal with $E(\bar{p}) = .15$ and $\sigma_{\bar{p}} = .0505$
  b. .4448
  c. .8389
42. 4324, 2875, 318, 538, 4771
44. a. .4714
  b. .7198
  c. 693
46. a. Normal with $E(\bar{x}) = 12$ and $\sigma_{\bar{x}} = .0566$
  b. .50
48. a. 67
  b. 1.5
  c. Normal with $E(\bar{x}) = 67$ and $\sigma_{\bar{x}} = 1.5$
  d. .9082
  e. .4972
50. a. No, since $n/N = .01$.
  b. Use $\sigma_{\bar{x}} = .0566$
  c. .9232
52. 246
54. a. 625
  b. .7888
56. a. Assume population has a normal distribution.
  b. .9266
  c. Increase n to at least 30.
58. a. .5990
  b. .7660
  c. 385
60. .9525
62. .4714
64. All are convenience samples.

Chapter 8

2. a. 23,480 to 24,520
  b. 23,380 to 24,620
  c. 23,186 to 24,814
4. a. 46.15 to 47.85
  b. 42.53 to 45.47
  c. men
6. a. 6
  b. 96
8. a. 24.31 to 26.15
  b. 24.14 to 26.32
10. a. .95
  b. .90
  c. .01
  d. .05
  e. .95
  f. .85
12. 7.26 to 9.74
14. a. 13.2
  b. 7.8
  c. 7.62 to 18.78
  d. Wide interval; larger sample desirable

16. a. 21.15 to 23.65
  b. 21.12 to 23.68
  c. Intervals are essentially the same.
18. 4.51 to 6.59
20. a. 50, 89, 200
  b. Only if $E = 1$ is essential.
22. 59
24. a. 49
  b. 30 to 34
26. .3233 to .3767
28. .6007 to .7393
30. a. .2271 to .3529
  b. .2537 to .3263
  c. .2619 to .3181
  d. The interval width decreases indicating a better precision.
32. a. .4081 to .5319
  b. 383
34. a. 72
  b. 129
  c. 289
  d. 801
  e. n becomes larger
36. 55.15 to 56.05
38. a. 1483.1
  b. 374.9
  c. 1136.4 to 1829.8
40. 9.19 to 14.81
42. 710
44. 37
46. 54
48. .0438 to .2362
50. a. 1267
  b. 1508
52. .5134 to .5866
54. a. 406
  b. .7356 to .8044
  c. .6008 to .6792
  d. Yes

Chapter 9

2. a. $H_0: \mu \le 14$
    $H_a: \mu > 14$
4. a. $H_0: \mu \ge 220$
    $H_a: \mu < 220$
6. a. $H_0: \mu \le 1$
    $H_a: \mu > 1$
  b. Claiming $\mu > 1$ when it is not true
  c. Claiming $\mu \le 1$ when it is not true
8. a. $H_0: \mu \ge 220$
    $H_a: \mu < 220$
  b. Claiming $\mu < 220$ when it is not true
  c. Claiming $\mu \ge 220$ when it is not true
10. a. $z = 1.77$; do not reject H_0.
12. $z = -2.74$; reject H_0.
    p-value = .0031
14. a. Reject H_0 if $z < -1.645$.
  b. $z = -1.98$; reject H_0.
  c. .0239

16. a. $z = -1.06$; do not reject H_0.
   b. .2892
18. $z = 6.37$; reject H_0.
20. a. \$50,367 to \$53,633
   b. Reject H_0.
22. a. \$8.66 or less
   b. Reject H_0.
24. $t = 1.55$; do not reject H_0.
26. a. $t = -3.33$; reject H_0.
   b. p-value is less than .005
28. $\bar{x} = 2.4$, $s = .52$, $t = 2.43$
   Reject H_0.
30. $z = -1.25$; do not reject H_0.
   p-value = .1056.
32. a. $z = 1.38$; reject H_0.
   b. .0838
34. $z = -3.64$; reject H_0.
36. $z = 1.63$; do not reject H_0.
   p-value = .0516
38. a. Concluding $\mu \le 15$ when it is not true
   b. .2676
   c. .0179
40. a. Concluding $\mu = 28$ when it is not true.
   b. .0853, .6179, .6179, .0853
   c. .9147
42. .1151, .0015
   Increasing n reduces β.
44. 109
46. 324
48. a. $z = 2.68$; reject H_0.
   b. .0037
   c. 27.16 to 28.04
50. $z = 2.26$; reject H_0.
52. a. $z = 1.64$; do not reject H_0.
   b. .0505
   c. 349 to 375
54. a. .0143
   b. Reject H_0.
56. $z = 2.52$; reject H_0.
   p-value = .0059
58. $z = 1.07$; do not reject H_0.
   p-value = .1423
60. $z = -1.66$; do not reject H_0.
   p-value = .0485
62. 219

Chapter 10

2. 2548.73 to 2951.27
4. a. 1200
   b. 438 to 1962
   c. Populations normal with equal variances
6. a. 1.63
   b. 1.48 to 3.52
8. $z = 4.99$; reject H_0.
10. $z = 2.66$; reject H_0.
12. a. Populations normal with equal variances
   b. 1,514,286
   c. $t = 1.11$; do not reject H_0.

14. a. $t = 2.38$; reject H_0.
   b. .03 to 1.39
16. a. $t = 5.88$; reject H_0.
   b. 1.4 to 3.0
18. −.0032 to .0866
20. a. $z = 2.42$; reject H_0.
   p-value = .0078
   b. Be aware of potential for less spending.
   c. .0156 to .1444
22. $z = 5.67$; reject H_0.
24. a. $z = 2.33$; reject H_0.
   b. .02 to .22
26. 17.55 to 22.15
28. $t = -1.69$; do not reject H_0.
30. $t = 2.30$; reject H_0.
32. a. H_0: $p_1 - p_2 \le 0$
     H_a: $p_1 - p_2 > 0$
   b. $z = 1.80$; reject H_0.
     p-value = .0359
34. $z = -2.19$; reject H_0.

Chapter 11

2. a. .22 to .71
   b. .47 to .84
4. a. 7.56 to 12.77
   b. 7.95 to 11.81
   c. 8.35 to 11.03
   d. The estimate is more precise for larger n.
6. $\chi^2 = 31.5$; reject H_0.
8. $\chi^2 = 10.16$; do not reject H_0.
10. $F = 1.13$; do not reject H_0.
12. $F = 2.20$; reject H_0.
14. a. $F = 4$; reject H_0.
   b. Drive carefully on wet pavement.
16. 10.72 to 24.68
18. a. $\chi^2 = 27.44$; reject H_0.
   b. .00012 to .00042
20. a. 15
   b. 6.25 to 11.13
22. $F = 1.39$; do not reject H_0.
24. $F = 2.12$; do not reject H_0.
   $s^2 = 63.67$

Chapter 12

2. No significant difference; $F = 2.54 < F_{.05} = 3.24$.
4. Significant difference; $F = 17.5 > F_{.05} = 3.89$.

6.

Source	SS	DF	MS	F
Treatments	300	4	75	12.5
Error	180	30	6	
Total	480	34		

8. No significant difference; $F = 0.27 < F_{.05} = 3.47$.

10.

Source	SS	DF	MS	F
Treatments	4560	2	2280	9.87
Error	6240	27	231.11	
Total	10800	29		

There is a significant difference; $F = 9.87 > F_{.05} = 3.35$.

12. Significant difference; $F = 19.99 > F_{.05} = 3.10$.

14.

Source	SS	DF	MS	F
Treatments	1200	3	400	35.34
Error	600	53	11.32	
Total	1800	56		

There is a significant difference; $F = 35.34 > F_{.05}$.

16. Significant difference; $F = 20.57 > F_{.05} = 18.51$.

18.

Source	SS	DF	MS	F
Treatments	310	4	77.5	17.71
Blocks	85	2	42.5	
Error	35	8	4.375	
Total	430	14		

There is a significant difference; $F = 17.71 > F_{.05} = 3.84$.

20.

Source	SS	DF	MS	F
Treatments	900	3	300	12.60
Blocks	400	7	57.14	
Error	500	21	23.81	
Total	1800	31		

There is a significant difference; $F = 12.60 > F_{.05} = 3.07$.

22. Interval: -1.54 to 5.54. Since this interval includes 0 we cannot reject the hypothesis that the means are equal.

24. Interval: -2.95 to 0.95. Since this interval includes 0 we cannot reject the hypothesis that the means are equal.

26. Factor A is significant; $F = 10.75 > F_{.05} = 5.14$.
Factor B is not significant; $F = 3.00 < F_{.05} = 5.99$.
Interaction is not significant; $F = 1.75 < F_{.05} = 5.14$.

28. Factor A, Factor B, and Interaction are significant; F values are 3.72, 4.94, and 12.52, respectively.

30. The means of the k populations have to be equal.

32. MSTR is based upon the variation between sample means whereas MSE is computed based upon the variation within each sample.

34. **a.** No significant difference.
b. Since $F = .18 < F_{.05}$ there is no significant difference.

36. Significant difference; $F = 10.59 > F_{.05}$.

38. No significant difference; $F = 1.48 < F_{.05} = 3.35$.

40. No significant difference; $F = 1.42 < F_{.05} = 4.26$.

42. **a.** Significant difference; $F = 5.19 > F_{.05} = 4.26$.
b. Intervals: -31.66 to -2.34. Since this interval does not include 0 we can conclude that the means are not equal.

44. No, since there was no significant difference using the ANOVA procedure.

46. No significant difference without blocking.

48. No significant difference; $F = 1.67 < F_{.01}$ 10.92.

50. Factor A is significant; $F = 13 > F_{.05} = 7.71$. Factor B and Interaction are not significant.

Chapter 13

2. **b.** $\hat{y} = 30.33 - 1.88x$

4. Relationship looks curvilinear.

6. **c.** There does appear to be a linear relationship.
d. $\hat{y} = 6.38 + 45.68x$
e. \$129,716

8. **b.** $\hat{y} = 47.61 + .44x$
c. 85.01

10. **a.** SSE = 51, SST = 656, SSR = 605
c. $r^2 = .92$

12. **a.** SSE = 54,054.0543 SST = 340,000 SSR = 285,944.4543
c. $r^2 = .84$

14. **a.** $\hat{y} = -54.84 + 98.80x$
b. $r^2 = .992$
c. 38.13

16. **a.** $\bar{x} = 3.8, \bar{y} = 23.2$
c. Yes

18. **a.** $\hat{y} = .75 + .51x$
b. Cannot conclude there is a significant relationship.
c. $F = 3.33 < F_{.05} = 10.13$; not significant.

20. Significant relationship; $F = 48.17 > F_{.01} = 34.12$.

22. Significant relationship; $F = 106.92 > F_{.05} = 5.32$.

24. 35.14% to 53.58%

26. \$97.810 to \$115,950

28. \$112,830 to \$155,730

30. a. $\hat{y} = 39.016 - .2861x$
b. Significant relationship; $F = 38.72 > F_{.05} = 5.32$
c. $r^2 = .8288$
d. 23.10 to 26.32
e. Larger interval = 20.10 to 29.32
32. a. $\hat{y} = 6.1092 + .8951x$
b. Significant relationship; $t = 6.01 > t_{.025} = 2.306$.
c. $28.49
34. b. Yes
c. $\hat{y} = -33.88 + .09253x$
d. Reject H_0; $B_1 = 0$ since $F = 10.85 > F_{.05} = 4.84$.
e. 104.915; 24.9 to 184.93
36. a. $\hat{y} = 2.32 + .64x$
b. Variance appears to increase for larger values of x
38. a. $\hat{y} = 343 + 0.051x$
c. The assumptions concerning ϵ should be questioned.
40. b. Appears to be a negative linear relationship.
c. $s_{xy} = -60$
d. $r = -0.97$
42. $r = 0.69$
44. a. $r = -.94$
b. Significant, since $-6.75 < -3.707$
46. a. $r = -.91$
b. Significant, since $-6.22 < -3.355$
48. In regression analysis we are interested in developing a model of the relationship between a dependent variable and one or more independent variables to help in estimating values of the dependent variable. In correlation analysis we are only concerned with determining if variables are related and not in specifying an equation relating the variables.
50. The estimate of a mean value is an estimate of the average of all y values associated with the same x. The estimate of an individual y value is an estimate of only one of the y values associated with a particular x.
52. To determine whether or not there is a significant relationship between x and y.
54. a. $r^2 = .78$; SR = .78 (SST) = 35.1; SSE = 9.9.
b. Significant; $F = 99.27 > F_{.01} = 7.64$.
c. $r_{xy} = +.88$
56. a. There seems to be an inverse linear relationship between distance to work and days absent.
b. $\hat{y} = 8.10 - .34x$
c. Significant relationship; $F = 19.70 > F_{.05} = 5.32$.
d. $r^2 = .71$
e. 5.22 to 7.58 days
58. $r = .934$
60. $r = .9369$

62. a. $r = .127$
b. Not significant ($F = 0.13$)
d. $b_0 = 14.4, b_1 = .112$
For an ideal system $\beta_0 = 0$ and $\beta_1 = 1$.

Chapter 14

2. a. $\hat{y} = 88.6 + 1.60x$
b. $94,200
4. a. $\hat{y} = 66.5 + .414x_1 - .270x_2$
b. $b_1 = .414$ is the marginal increase in sales of the Heller mower with respect to a change in price of the competitor's mower with the price of the Heller mower held constant.
c. 93,680
6. a. $\hat{y} = 26.7 - 1.43x_1 + .0757x_2$ where x_1 = size and x_2 = SAT score.
b. 73.8%
8. a. .75
b. .68
10. a. SSE = 4000; $s^2 = 571.43$; MSR = 6000
b. Significant relationship; $F = 10.50 > F_{.05} = 4.74$
12. Significant relationship; $F = 6.58 > F_{.05} = 4.74$
14. a. In the two independent variables case the coefficient of x_1 represents the expected change in y corresponding to a one unit increase in x_1 when x_2 is held constant.
b. Yes
16. a. 71,130 to 112,350
b. 42,870 to 140,610
18. No unusual patterns
20. Residual plot supports ϵ assumptions.
22. b. $\hat{y} = 14.4 - 8.69(4) + 13.5(6.5) = 67.39$
24. a. Reject H_0: $\beta_1 = 0$ since $t = 3.87 > t_{.025} = 2.201$
b. Cannot reject H_0: $\beta_2 = 0$ since $t = 1.84 < t_{.025} = 2.201$
c. Reject H_0: $\beta_3 = 0$ since $t = -5.07 < -2.201$
d. Yes
26. b. Significant relationship; $F = 52.44 > F_{.05} = 4.74$.
c. $R^2 = .937$; $R_a^2 = .919$
d. Reject H_0: $\beta_1 = 0$; $t = 2.71 > t_{.025} = 2.365$. Reject H_0: $\beta_2 = 0$; $t = 4.51 > t_{.025} = 2.365$.
28. a. $2,120,000
b. Significant relationship; $F = 67.70 > F_{.01} = 6.93$.

Chapter 15

2. a. $y = 433 + 37.4x - 0.383x^2$
b. Significant relationship since the relationship between x and y was significant (Exercise 1).

c. Confidence interval: 1270.41 to 1333.61
Prediction interval: 1242.55 to 1361.47

4. b. No; curvilinear.
c. Several possible models; e.g. $\hat{y} = 2.90 - 0.185x + .00351x^2$ ($R_a^2 = .91$) or $\ln y = 2.60 - 0.998 \ln x$ ($R_a^2 = .98$)

6. a. $\hat{y} = 2.15 + .304x_1$ where $x_1 =$ months since previous service call
b. No; $R^2 = .53$

8. a.

D1	D2	Industry
0	0	1
1	0	2
0	1	3

$\hat{y} = $ P/E $= 7.54 + 0.183$ %PROFIT $+ .213$ %GROWTH $+ 2.98$ D1 $- .84$ D2
b. Consider the interaction term %PROFXD1 = %PROF times D1
$\hat{y} = $ P/E $= 10.3 + 0.373$ %PROFXD1 ($R_a^2 = .636$)

10. a. $R^2 = .975$
b. $R_a^2 = .971$
c. Significant relationship; $F = 244.44 > F_{.05} = 2.76$.

12. a. $\hat{y} = $ %COLLEGE $= -26.6 + 0.0970$SATSCORE
c. $\hat{y} = $ %COLLEGE $= -26.93 + 0.084$SATSCORE $+ 0.204$ %TAKETEST
d. Same as part (c).

14. a. $\hat{y} = $ WINS $= 14.3 - 0.373$TEAMINT
b. $\hat{y} = 11.199 - 0.28$TEAMINT $+ 0.00288$PASSING $- 0.026$POINTS
c. Same as part (b).

16. a. $\hat{y} = 10.30 + 0.373$ %PROFXD1
b. Same as part (a).

18. a. $\hat{y} = $ WINS $= 2.95 + 0.222$OPPONINT
b. No unusual patterns

20. a. Observation 7
b. Observation 7
c. Observation 15

22. $d = 1.60$; test is inconclusive

24. a.

D1	D2	D3	Paint
0	0	0	1
1	0	0	2
0	1	0	3
0	0	1	4

$\hat{y} = $ TIME $= 133 + 6$D1 $+ 3$D2 $+ 11$D3
Not significant
b. 139

26. a. $\hat{y} = $ AUDELAY $= 80.4 + 11.9$INDUS $- 4.82$PUBLIC $- 2.62$ICQUAL $- 4.07$INTFIN
b. $R_a^2 = .312$; not evidence of a good fit.
c. Possible curvilinear relationship between AUDELAY and INTFIN

d. Let INTFINSQ = (INTFIN)^2
$\hat{y} = 112.79 + 11.6$INDUS $- 2.49$ICQUAL $- 36.6$INTFIN $+ 6.6$INTFINSQ
$R^2 = 59.05$

28. a. $\hat{y} = $ AUDELAY $= 63.0 + 11.1$INDUS
b. $\hat{y} = 112.79 + 11.6$INDUS $- 2.49$ICQUAL $- 36.6$INTFIN $+ 6.6$INTFINSQ
c. Same as part (b).

30. b. No outliers
c. No influential observations

32. a. POINTS $= 394 - 5.10$OPPONINT
b. Two unusual points
c. Outlier: observation 2
d. Large Influence: Observation 7

34. a. AUDELAY $= 70.6 + 12.7$INDUS $- 2.92$ICQUAL
b. No obvious pattern indicative of positive autocorrelation
c. $d = 1.43$; test is inconclusive.

36.

D1	D2	Type of Browser
0	0	non browser
1	0	light browser
0	1	heavy browser

SCORE $= 4.25 + 1$D1 $+ 1.5$D2
Significant relationship

Chapter 16

2. a.

Week	4-Week	5-Week
10	19.00	18.80
11	20.00	19.20
12	18.75	19.00

b. 9.65, 7.41
c. 5-week

4. Weeks 10, 11 and 12: 18.48, 18.63, 18.27
MSE $= 9.25$; $\alpha = .2$ is better.

6. a.

Month	3-Month	$\alpha = .2$
10	256.67	265.51
11	286.67	274.41
12	263.33	267.53

Using months 4 to 12 for both, $\alpha = .2$ is better.
MSE (3-month) $= 1998.72$
MSE ($\alpha = .2$) $= 1632.72$
b. 260

8. a. Months 10, 11 and 12: 2478.44, 2478.75, 2489.00
b. Months 10, 11 and 12: 2474.31, 2476.02, 2492.21
c. MSE ($\alpha = .2$) $= 1075$; MSE ($\alpha = .3$) $= 1092$; Use $\alpha = .2$; $F_{13} = 2501$

10. 112.4; 117.1

12. Consider a nonlinear trend

14. a. $T_t = .365 + .193t$; $2.49

  b. EPS increasing by an average of $.193 per year.

16. a. 1938.75, 1966.25, 1956.25, 2025.00, 1990.00, 2002.50, 2052.50, 2060.00, 2123.75

  b. .900, .486, 1.396, 1.217

  c. 3rd Quarter; Yes, fall back-to-school demand.

18. a. 0.771, 0.864, 0.954, 1.392, 1.571, 1.667, 1.207, 0.994, 0.850, 0.647, 0.579, 0.504

  b. 47.49, 47.89, 48.28, 48.67, 49.06, 49.46, 49.85, 50.24, 50.63, 51.02, 51.42, 51.81

  c. 37, 41, 46, 68, 77, 82, 60, 50, 43, 33, 30, 26

20. Forecast for weeks 8, 9, 10 and 11: 258.64, 281.48, 283.61 and 292.71.

22. 3117

24. a. $T_t = 4.3071 + .9012t$

  b. 12.4, 13.3, 14.2

  c. Yes; vacancy rate is increasing.

26. a. $T_t = 18.97 + 1.479t$

  c. Capital expenditures are increasing; 42.63

28. a. Yes

  b. $T_t = 22.857 + 15.536t$

  c. 147.15

30. a. Selected centered moving averages for $t = 5, 10, 15$ and 20 are 11.125, 18.125, 22.875 and 27.000.

  b. 0.899, 1.362, 1.118, 0.621

  c. Quarter 2, prior to summer boating season.

32. a. $T_t = 6.329 + 1.055t$

  b. 36.92, 37.98, 39.03, 40.09

  c. 33.23, 51.65, 43.71, 24.86

Chapter 17

2. $\chi^2 = 12.2$; reject H_0.

4. $\chi^2 = 2.31$; do not reject H_0.

6. $\chi^2 = 6.31$; do not reject H_0.

8. $\chi^2 = 37.17$; reject H_0.

10. $\chi^2 = 7.96$; do not reject H_0.

12. a. $\chi^2 = 0.39$; do not reject H_0.

  b. $\chi^2 = 1.59$; do not reject H_0.

  c. 9 males and 3 females
  $\chi^2 = 3.53$; reject H_0.

14. $\chi^2 = 4.98$; do not reject H_0.

16. $\chi^2 = 11.2$; reject H_0.

18. $\chi^2 = 6.15$; do not reject H_0.

20. $\chi^2 = 7.44$; do not reject H_0.

22. $\chi^2 = 5.26$; do not reject H_0.

24. $\chi^2 = 22.87$; reject H_0.

26. $\chi^2 = 6.20$; do not reject H_0.

28. a. 24.01, 41.16, 26.46, and 8.37. *Note:* $x = 3$ and $x = 4$ are combined to make $e_i > 5$.

  b. $\chi^2 = 6.17$; do not reject H_0.

Chapter 18

2. a. H_0: $p = .50$
  H_a: $p \neq .50$

  b. If less than 5 or greater than 13

  c. Reject H_0

4. $z = 1.90$; do not reject H_0.

6. $z = 3.76$; reject H_0.

8. $z = 1.27$; do not reject H_0.

10. $z = 1.89$; reject H_0.

12. $z = -2.05$; reject H_0.

14. $T = 28.5$; reject H_0.

16. $z = 3.33$; reject H_0.

18. $z = -.25$; do not reject H_0.

20. $W = 9.06$; do not reject H_0.

22. $W = 8.03$; reject H_0

24. a. $+.60$

  b. $z = 1.82$; do not reject H_0.

26. $r_s = .04$; $z = .12$; do not reject H_0.

28. $z = .82$; do not reject H_0.

30. $z = 1.81$; reject H_0.

32. $z = 2.96$; reject H_0.
  $z = 0.68$; do not reject H_0.

34. $z = -2.59$; reject H_0.

36. $z = -2.97$; reject H_0.

38. $W = 12.61$; reject H_0.

40. $r_s = -.92$; $z = -3.05$; reject H_0.

42. $r_s = .76$; $z = 2.84$; reject H_0.

Chapter 19

2. b. Optimistic: d_2, conservative: d_1, minimax: d_2

4. a. d_1

  b. d_4

6. d_2 (medium)

8.

Value(s) of p	Best Decision(s)
$0 \leq p < 0.3$	d_3
0.3	d_2 or d_3
$0.3 < p < 0.7$	d_2
0.7	d_2 or d_1
$0.7 < p \leq 1.0$	d_1

10. $25,000

12. b. d_3

  c. $2150

14. $P(s_1|I) = .1905$, $P(s_2|I) = .2381$, $P(s_3|I) = .5714$

16. a. $P(I_1) = .56$ $P(I_2) = .44$

  b. .57143, .42857, .18182, .81818

18. b. If I_1, then d_1. If I_2, then d_2.
  EV (node 1) = $101.50

  c. EVSI = $1500

20. If unseasonably cold, then purchase snowplow.
If not unseasonably cold, then purchase blade.

22. a. Purchase the component.
b. EVPI = $9000

24. a.

		Demand		
		100	200	300
	100	500	200	−100
Prod.	200	−500	1000	700
	300	−1500	0	1500

b. Produce 300 cases
c. EVPI = $600

26. a. 0, 125, 250, 375
b. Yes, EV(no inspection) = 206.25

28. a. .355
b. If I_1, then d_1. If I_2, then d_2.
c. EVSI = $3650
d. 40.6%

30. a. No.
b. No. EVPI = $27,000

32. a. If I_1, then d_3.
If I_2, then d_2.
If I_3, then d_1.
b. EVSI = $10,900
c. 22.7%

34. a. 0, .048, .092, .133
b. 0, .159, .488, .353
c. Yes
d. $24.25

Chapter 20

2. a. .3324
b. .7099
4. $n = 15, c = 2$ is close with $\alpha = .0361, \beta = .1268$
$n = 20, c = 3$ with $\alpha = .0158$ and $\beta = .1070$ is recommended.
6. a. 11.3, 13.7
b. 11.65, 13.35
11.90, 13.10
c. Closer together
8. 20.01, .082
10. a. .0540
b. .0226
c. 0, .1218
12. a. .0413
b. 0, .0836
c. .10, out of control
14. a. $p = .03$ with $\alpha = .0729$
b. .0802
16. a. .0064
b. .0136
c. Low α and β, but more expensive
18. a. 335.77, 364.23
339.94, 360.06
343.78, 358.22
b. Control limits are closer to process mean.
c. Adjust an in-control process.
d. An out-of-control process continues.
e. .0028
f. Type II error
20. a. 0, .0817
b. Only last month out of control
$\bar{p} = .085$.

INDEX

t Distribution

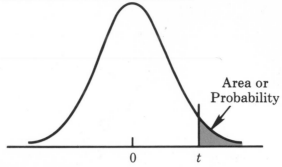

Area or Probability

0   *t*

Entries in the table give *t* values for an area or probability in the upper tail of the *t* distribution. For example, with 10 degrees of freedom and a .05 area in the upper tail, $t_{.05} = 1.812$.

Degrees of Freedom	Area in Upper Tail				
	.10	.05	.025	.01	.005
1	3.078	6.314	12.706	31.821	63.657
2	1.886	2.920	4.303	6.965	9.925
3	1.638	2.353	3.182	4.541	5.841
4	1.533	2.132	2.776	3.747	4.604
5	1.476	2.015	2.571	3.365	4.032
6	1.440	1.943	2.447	3.143	3.707
7	1.415	1.895	2.365	2.998	3.499
8	1.397	1.860	2.306	2.896	3.355
9	1.383	1.833	2.262	2.821	3.250
10	1.372	1.812	2.228	2.764	3.169
11	1.363	1.796	2.201	2.718	3.106
12	1.356	1.782	2.179	2.681	3.055
13	1.350	1.771	2.160	2.650	3.012
14	1.345	1.761	2.145	2.624	2.977
15	1.341	1.753	2.131	2.602	2.947
16	1.337	1.746	2.120	2.583	2.921
17	1.333	1.740	2.110	2.567	2.898
18	1.330	1.734	2.101	2.552	2.878
19	1.328	1.729	2.093	2.539	2.861
20	1.325	1.725	2.086	2.528	2.845
21	1.323	1.721	2.080	2.518	2.831
22	1.321	1.717	2.074	2.508	2.819
23	1.319	1.714	2.069	2.500	2.807
24	1.318	1.711	2.064	2.492	2.797
25	1.316	1.708	2.060	2.485	2.787
26	1.315	1.706	2.056	2.479	2.779
27	1.314	1.703	2.052	2.473	2.771
28	1.313	1.701	2.048	2.467	2.763
29	1.311	1.699	2.045	2.462	2.756
30	1.310	1.697	2.042	2.457	2.750
40	1.303	1.684	2.021	2.423	2.704
60	1.296	1.671	2.000	2.390	2.660
120	1.289	1.658	1.980	2.358	2.617
∞	1.282	1.645	1.960	2.326	2.576

Reprinted by permission of Biometrika Trustees from Table 12, Percentage Points of the *t* Distribution, 3rd. Edition, 1966. E. S. Pearson and H. O. Hartley, *Biometrika Tables for Statisticians*, Vol. I.